ADVANCED ENGINEERING MATHEMATICS

ADVANCED ENGINEERING MATHEMATICS

Dean G. Duffy

CRC Press
Boca Raton Boston New York Washington London

Library of Congress Cataloging-in-Publication Data

Duffy Dean G.
Advanced engineering mathematics / Dean G. Duffy.
p. cm.
Includes bibliographical references and index.
ISBN 0-8493-7854-0 (alk. paper)
1. Mathematics. I. Title.
QA37.2.D78 1997
510—dc21
for Library of Congress 97-33991
CIP

No claim to original U.S. Government works
International Standard Book Number 0-8493-7854-0
Library of Congress Card Number 97-33991
Printed in the United States of America 1 2 3 4 5 6 7 8 9 0
Printed on acid-free paper

Introduction

This book grew out of a two-semester course given to sophomore and junior engineering majors at the U.S. Naval Academy. These students had just completed three semesters of traditional calculus and a fourth semester of ordinary differential equations. Consequently, it was assumed that they understood single and multivariable calculus, the calculus of single-variable, vector-valued functions, and how to solve a constant coefficient, ordinary differential equation.

The first five chapters were taught to system and electrical engineers because they needed transform methods to solve ordinary differential and difference equations. The last six chapters served mechanical, aeronautical, and other engineering majors. These students focused on the general topics of boundary-value problems, linear algebra, and vector calculus.

The book has been designed so that the instructor may inject his own personality into the course. For example, the instructor who enjoys the more theoretical aspects may dwell on them during his lecture with the confidence that the mechanics of how to solve the problems are completely treated in the text. Those who enjoy working problems may choose from a wealth of problems and topics. References are given to original sources and classic expositions so that the theoretically inclined may deepen their understanding of a given subject.

Overall this book consists of two parts. The first half involves advanced topics in single variable calculus, either with real or complex variables, while the second portion involves advanced topics in multivariable calculus. Unlike most engineering mathematics books, we begin with complex variables because they provide powerful techniques in understanding and computing Fourier, Laplace, and z-transforms.

Chapter 1 starts by reviewing complex numbers; in particular, we find all of the roots of a complex number, $z^{1/n}$, where n is an integer and z is a complex number. This naturally leads to complex algebra and complex functions. Finally, we define the derivative of a complex function.

The remaining portion of Chapter 1 is devoted to contour integration on the complex plane. First, we compute contour integrals by straightforward line integration. Focusing on closed contours, we introduce the Cauchy-Goursat theorem, Cauchy's integral theorem, and Cauchy's residue theorem to greatly facilitate the evaluation of these integrals. This analysis includes the classification of singularities. Although Chapter 1 is not necessary for most of this book, some sections or portions of some sections (2.5, 2.6, 3.1–3.6, 4.5, 4.10, 5.1, 5.3–5.5, 6.1, 6.5, 7.5–7.6, 8.4, 8.7, 9.4, 9.6, 11.6) require this material and must therefore be excluded when encountered. If the students have had elementary complex arithmetic (Section 1.1), the affected sections drop to 3.4, 3.6, 4.10, 5.3, 5.5, 7.5, 8.4, and 9.6.

Chapter 2 lays the foundation for transform methods and the solution of partial differential equations. We begin by deriving the classic Fourier series and working out some interesting problems. Next we investigate the properties of Fourier series, including Gibbs phenomena, and whether we can differentiate or integrate a Fourier series. Then we reexpress the classic Fourier series in alternative forms. Finally we use Fourier series to solve ordinary differential equations with periodic forcing. As a postscript we apply Fourier series to situations where there is a finite number of data values.

In Chapter 3 we introduce the Fourier transform. We compute some Fourier transforms and find their inverse by partial fractions and contour integration. Furthermore, we explore various properties of this transform, including convolution. Finally, we find the particular solution of an ordinary differential equation using Fourier transforms.

Chapter 4 presents Laplace transforms. This chapter includes finding a Laplace transform from its definition and using various theorems. We find the inverse by partial fractions, convolution, and contour integration. With these tools, the student can then solve an ordinary differential equation with initial conditions and a piece-wise continuous forcing. We also include systems of ordinary differential equations. Finally, we examine the importance of the transfer function, impulse response, and step response.

With the rise of digital technology and its associated difference equations, a version of the Laplace transform, the z-transform, was developed. In Chapter 5 we find a z-transform from its definition or by using various theorems. We also illustrate how to compute the inverse by long division, partial fractions, and contour integration. Finally, we use z-transforms to solve difference equations, especially with respect to the stability of the system.

Chapter 6 is a transitional chapter. We expand the concept of Fourier series so that it includes solutions to the Sturm-Liouville problem and show how any piece-wise continuous function can be reexpressed in terms of an expansion of these solutions. In particular, we focus on expansions that involve Bessel functions and Legendre polynomials.

Chapter 7, 8, and 9 deal with solutions to the wave, heat, and Laplace's equations, respectively. They serve as prototypes of much wider classes of partial differential equations. Of course, considerable attention is given to the technique of separation of variables. However, additional methods such as Laplace and Fourier transforms and integral representations are also included. Finally, we include a section on the numerical solution of each of these equations.

Chapter 10 is devoted to vector calculus. In this book we focus on the use of the del operator. This includes such topics as line integrals, surface integrals, the divergence theorem, and Stokes' theorem.

Finally, in Chapter 11 we present some topics from linear algebra. From this vast field of mathematics we study the solution of systems of linear equations because this subject is of greatest interest to engineers. Consequently, we shall cover such topics as matrices, determinants, and Cramer's rule. For the solution of systems of ordinary differential equations we discuss the classic eigenvalue problem.

This book contains a wealth of examples. Furthermore, in addition to the standard rote problems, I have sought to include many problems from the scientific and engineering literature. I have formulated many of the more complicated problems or computations as multistep projects. These problems may be given outside of class to deepen the students' understanding of a particular topic.

The answers to the odd problems are given in the back of the book while the worked solutions to all of the problems are available from the publisher. It is hoped that by including problems from the open literature some of the academic staleness that often pervades college texts will be removed.

Acknowledgments

I would like to thank the many midshipmen and cadets who have taken engineering mathematics from me. They have been willing or unwilling guinea pigs in testing out many of the ideas and problems in this book.

Special thanks goes to Dr. Mike Marcozzi for his many useful and often humorous suggestions for improving this book. The three-dimensional plots were done on MATLAB. Finally, I would like to express my appreciation to all those authors and publishers who allowed me the use of their material from the scientific and engineering literature.

Dedicated to the Brigade of Midshipmen
and the Corps of Cadets

Contents

Chapter 1

Complex Variables

The theory of complex variables was originally developed by mathematicians as an aid in understanding functions. Functions of a complex variable enjoy many powerful properties that their real counterparts do not. That is *not* why we will study them. For us they provide the keys for the complete mastery of transform methods and differential equations.

In this chapter all of our work points to one objective: integration on the complex plane by the method of residues. For this reason we will minimize discussions of limits and continuity which play such an important role in conventional complex variables in favor of the computational aspects. We begin by introducing some simple facts about complex variables. Then we progress to differential and integral calculus on the complex plane.

1.1 COMPLEX NUMBERS

A *complex number* is any number of the form $a + bi$, where a and b are real and $i = \sqrt{-1}$. We denote any member of a *set* of complex numbers by the *complex variable* $z = x + iy$. The real part of z, usually denoted by $\mathrm{Re}(z)$, is x while the imaginary part of z, $\mathrm{Im}(z)$, is y. The *complex conjugate*, $\overline{z}$ or z^*, of the complex number $a + bi$ is $a - bi$.

Complex numbers obey the fundamental rules of algebra. Thus, two complex numbers $a + bi$ and $c + di$ are equal if and only if $a = c$ and $b = d$. Just as real numbers have the fundamental operations of addition, subtraction, multiplication, and division, so too do complex numbers. These operations are defined:

Addition

$$(a + bi) + (c + di) = (a + c) + (b + d)i \tag{1.1.1}$$

Subtraction

$$(a + bi) - (c + di) = (a - c) + (b - d)i \tag{1.1.2}$$

Multiplication

$$(a + bi)(c + di) = ac + bci + adi + i^2bd = (ac - bd) + (ad + bc)i \tag{1.1.3}$$

Division

$$\frac{a + bi}{c + di} = \frac{a + bi}{c + di}\frac{c - di}{c - di} = \frac{ac - adi + bci - bdi^2}{c^2 + d^2} = \frac{ac + bd + (bc - ad)i}{c^2 + d^2}. \tag{1.1.4}$$

The *absolute value* or *modulus* of a complex number $a + bi$, written $|a + bi|$, equals $\sqrt{a^2 + b^2}$. Additional properties include:

$$|z_1 z_2 z_3 \cdots z_n| = |z_1||z_2||z_3| \cdots |z_n| \tag{1.1.5}$$

$$|z_1/z_2| = |z_1|/|z_2| \quad \text{if} \quad z_2 \neq 0 \tag{1.1.6}$$

$$|z_1 + z_2 + z_3 + \cdots + z_n| \leq |z_1| + |z_2| + |z_3| + \cdots + |z_n| \tag{1.1.7}$$

and

$$|z_1 + z_2| \geq |z_1| - |z_2|. \tag{1.1.8}$$

The use of inequalities with complex variables has meaning only when they involve absolute values.

It is often useful to plot the complex number $x + iy$ as a point (x, y) in the xy plane, now called the *complex plane*. Figure 1.1.1 illustrates this representation.

This geometrical interpretation of a complex number suggests an alternative method of expressing a complex number: the polar form. From the polar representation of x and y,

$$x = r\cos(\theta) \quad \text{and} \quad y = r\sin(\theta), \tag{1.1.9}$$

where $r = \sqrt{x^2 + y^2}$ is the *modulus*, *amplitude*, or *absolute value* of z and θ is the *argument* or *phase*, we have that

$$z = x + iy = r[\cos(\theta) + i\sin(\theta)]. \tag{1.1.10}$$

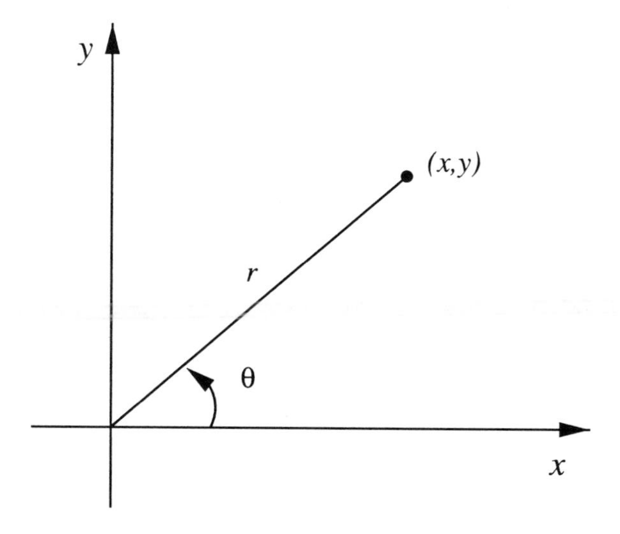

Figure 1.1.1: The complex plane.

However, from the Taylor expansion of the exponential in the real case,

$$e^{i\theta} = \sum_{k=0}^{\infty} \frac{(\theta i)^k}{k!}. \tag{1.1.11}$$

Expanding (1.1.11),

$$e^{i\theta} = 1 - \frac{\theta^2}{2!} + \frac{\theta^4}{4!} - \frac{\theta^6}{6!} + \cdots + i\left(\theta - \frac{\theta^3}{3!} + \frac{\theta^5}{5!} - \frac{\theta^7}{7!} + \cdots\right) \tag{1.1.12}$$

$$= \cos(\theta) + i\sin(\theta). \tag{1.1.13}$$

Equation (1.1.13) is *Euler's formula*. Consequently, we may express (1.1.10) as

$$z = re^{i\theta}, \tag{1.1.14}$$

which is the *polar form* of a complex number. Furthermore, because

$$z^n = r^n e^{in\theta} \tag{1.1.15}$$

by the law of exponents,

$$z^n = r^n[\cos(n\theta) + i\sin(n\theta)]. \tag{1.1.16}$$

Equation (1.1.16) is *De Moivre's theorem*.

• **Example 1.1.1**

Let us simplify the following complex number:

$$\frac{3-2i}{-1+i} = \frac{3-2i}{-1+i} \times \frac{-1-i}{-1-i} = \frac{-3-3i+2i+2i^2}{1+1} = \frac{-5-i}{2} = -\frac{5}{2} - \frac{i}{2}. \tag{1.1.17}$$

• **Example 1.1.2**

Let us reexpress the complex number $-\sqrt{6} - i\sqrt{2}$ in polar form. From (1.1.9) $r = \sqrt{6+2}$ and $\theta = \tan^{-1}(b/a) = \tan^{-1}(1/\sqrt{3}) = \pi/6$ or $7\pi/6$. Because $-\sqrt{6} - i\sqrt{2}$ lies in the third quadrant of the complex plane, $\theta = 7\pi/6$ and

$$-\sqrt{6} - i\sqrt{2} = 2\sqrt{2}e^{7\pi i/6}. \tag{1.1.18}$$

Note that (1.1.18) is not a unique representation because $\pm 2n\pi$ may be added to $7\pi/6$ and we still have the same complex number since

$$e^{i(\theta \pm 2n\pi)} = \cos(\theta \pm 2n\pi) + i\sin(\theta \pm 2n\pi) = \cos(\theta) + i\sin(\theta) = e^{i\theta}. \tag{1.1.19}$$

For uniqueness we will often choose $n = 0$ and define this choice as the *principal branch*. Other branches correspond to different values of n.

• **Example 1.1.3**

Find the curve described by the equation $|z - z_0| = a$.
From the definition of the absolute value,

$$\sqrt{(x-x_0)^2 + (y-y_0)^2} = a \tag{1.1.20}$$

or

$$(x-x_0)^2 + (y-y_0)^2 = a^2. \tag{1.1.21}$$

Equation (1.1.21), and hence $|z - z_0| = a$, describes a circle of radius a with its center located at (x_0, y_0). Later on, we shall use equations such as this to describe curves in the complex plane.

• **Example 1.1.4**

As an example in manipulating complex numbers, let us show that

$$\left|\frac{a+bi}{b+ai}\right| = 1. \tag{1.1.22}$$

We begin by simplifying

$$\frac{a+bi}{b+ai} = \frac{a+bi}{b+ai} \times \frac{b-ai}{b-ai} = \frac{2ab}{a^2+b^2} + \frac{b^2-a^2}{a^2+b^2}i. \tag{1.1.23}$$

Therefore,

$$\left|\frac{a+bi}{b+ai}\right| = \sqrt{\frac{4a^2b^2}{(a^2+b^2)^2} + \frac{b^4-2a^2b^2+a^4}{(a^2+b^2)^2}} = \sqrt{\frac{a^4+2a^2b^2+b^4}{(a^2+b^2)^2}} = 1. \tag{1.1.24}$$

Problems

Simplify the following complex numbers. Represent the solution in the Cartesian form $a + bi$:

1. $\dfrac{5i}{2+i}$

2. $\dfrac{5+5i}{3-4i} + \dfrac{20}{4+3i}$

3. $\dfrac{1+2i}{3-4i} + \dfrac{2-i}{5i}$

4. $(1-i)^4$

5. $i(1-i\sqrt{3})(\sqrt{3}+i)$

Represent the following complex numbers in polar form:

6. $-i$

7. -4

8. $2+2\sqrt{3}\,i$

9. $-5+5i$

10. $2-2i$

11. $-1+\sqrt{3}\,i$

12. By the law of exponents, $e^{i(\alpha+\beta)} = e^{i\alpha}e^{i\beta}$. Use Euler's formula to obtain expressions for $\cos(\alpha+\beta)$ and $\sin(\alpha+\beta)$ in terms of sines and cosines of α and β.

13. Using the property that $\sum_{n=0}^{N} q^n = (1-q^{N+1})/(1-q)$ and the geometric series $\sum_{n=0}^{N} e^{int}$, obtain the following sums of trigonometric functions:

$$\sum_{n=0}^{N} \cos(nt) = \cos\left(\frac{Nt}{2}\right) \frac{\sin[(N+1)t/2]}{\sin(t/2)}$$

and

$$\sum_{n=1}^{N} \sin(nt) = \sin\left(\frac{Nt}{2}\right) \frac{\sin[(N+1)t/2]}{\sin(t/2)}.$$

These results are often called *Lagrange's trigonometric identities.*

14. (a) Using the property that $\sum_{n=0}^{\infty} q^n = 1/(1-q)$, if $|q| < 1$, and the geometric series $\sum_{n=0}^{\infty} \epsilon^n e^{int}$, $|\epsilon| < 1$, show that

$$\sum_{n=0}^{\infty} \epsilon^n \cos(nt) = \frac{1 - \epsilon \cos(t)}{1 + \epsilon^2 - 2\epsilon \cos(t)}$$

and

$$\sum_{n=1}^{\infty} \epsilon^n \sin(nt) = \frac{\epsilon \sin(t)}{1 + \epsilon^2 - 2\epsilon \cos(t)}.$$

(b) Let $\epsilon = e^{-a}$, where $a > 0$. Show that

$$2\sum_{n=1}^{\infty} e^{-na} \sin(nt) = \frac{\sin(t)}{\cosh(a) - \cos(t)}.$$

1.2 FINDING ROOTS

The concept of finding roots of a number, which is rather straightforward in the case of real numbers, becomes more difficult in the case of complex numbers. By finding the *roots* of a complex number, we wish to find all the solutions w of the equation $w^n = z$, where n is a positive integer for a given z.

We begin by writing z in the polar form:

$$z = re^{i\varphi} \tag{1.2.1}$$

while we write

$$w = Re^{i\Phi} \tag{1.2.2}$$

for the unknown. Consequently,

$$w^n = R^n e^{in\Phi} = re^{i\varphi} = z. \tag{1.2.3}$$

We satisfy (1.2.3) if

$$R^n = r \quad \text{and} \quad n\Phi = \varphi + 2k\pi, \quad k = 0, \pm1, \pm2, \ldots, \tag{1.2.4}$$

because the addition of any multiple of 2π to the argument is also a solution. Thus, $R = r^{1/n}$, where R is the uniquely determined real positive root, and

$$\Phi_k = \frac{\varphi}{n} + \frac{2\pi k}{n}, \qquad k = 0, \pm1, \pm2, \ldots \tag{1.2.5}$$

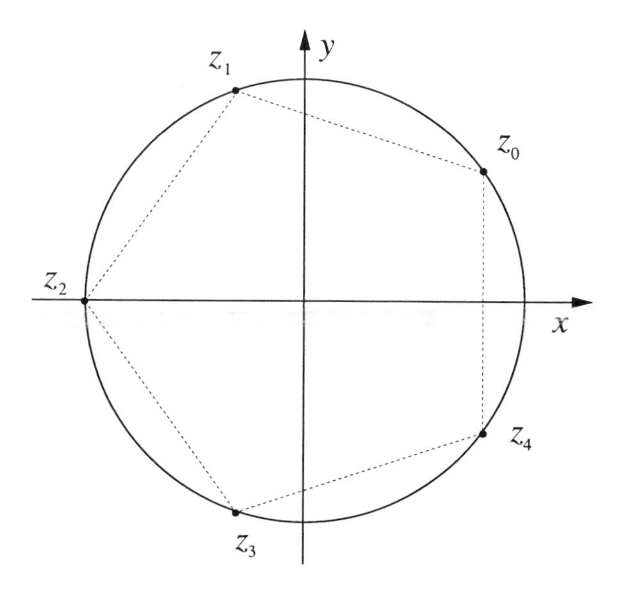

Figure 1.2.1: The zeros of $z^5 = -32$.

Because $w_k = w_{k \pm n}$, it is sufficient to take $k = 0, 1, 2, \ldots, n-1$. Therefore, there are exactly n solutions:

$$w_k = Re^{\Phi_k i} = r^{1/n} \exp\left[i\left(\frac{\varphi}{n} + \frac{2\pi k}{n}\right)\right] \tag{1.2.6}$$

with $k = 0, 1, 2, \ldots, n-1$. They are the n roots of z. Geometrically we can locate these w_k's on a circle, centered at the point (0,0), with radius R and separated from each other by $2\pi/n$ radians. These roots also form the vertices of a regular polygon of n sides inscribed inside of a circle of radius R. (See Example 1.2.1.)

In summary, the method for finding the n roots of a complex number z_0 is as follows. First, write z_0 in its polar form: $z_0 = re^{i\varphi}$. Then multiply the polar form by $e^{2i\pi k}$. Using the law of exponents, take the $1/n$ power of both sides of the equation. Finally, using Euler's formula, evaluate the roots for $k = 0, 1, \ldots, n-1$.

• **Example 1.2.1**

Let us find all of the values of z for which $z^5 = -32$ and locate these values on the complex plane.

Because

$$-32 = 32e^{\pi i} = 2^5 e^{\pi i}, \tag{1.2.7}$$

$$z_k = 2\exp\left(\frac{\pi i}{5} + \frac{2\pi i k}{5}\right), \qquad k = 0, 1, 2, 3, 4, \tag{1.2.8}$$

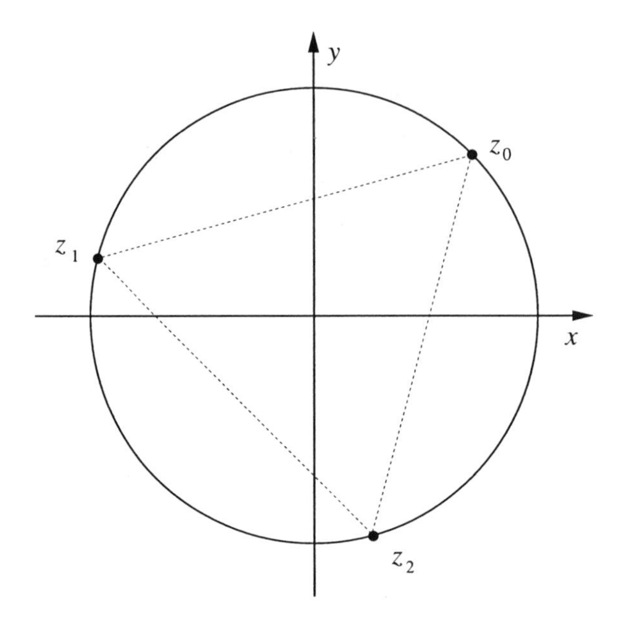

Figure 1.2.2: The zeros of $z^3 = -1 + i$.

or

$$z_0 = 2\exp\left(\frac{\pi i}{5}\right) = 2\left[\cos\left(\frac{\pi}{5}\right) + i\sin\left(\frac{\pi}{5}\right)\right], \tag{1.2.9}$$

$$z_1 = 2\exp\left(\frac{3\pi i}{5}\right) = 2\left[\cos\left(\frac{3\pi}{5}\right) + i\sin\left(\frac{3\pi}{5}\right)\right], \tag{1.2.10}$$

$$z_2 = 2\exp(\pi i) = -2, \tag{1.2.11}$$

$$z_3 = 2\exp\left(\frac{7\pi i}{5}\right) = 2\left[\cos\left(\frac{7\pi}{5}\right) + i\sin\left(\frac{7\pi}{5}\right)\right] \tag{1.2.12}$$

and

$$z_4 = 2\exp\left(\frac{9\pi i}{5}\right) = 2\left[\cos\left(\frac{9\pi}{5}\right) + i\sin\left(\frac{9\pi}{5}\right)\right]. \tag{1.2.13}$$

Figure 1.2.1 shows the location of these roots in the complex plane.

• **Example 1.2.2**

Let us find the cube roots of $-1 + i$ and locate them graphically.

Because $-1+i = \sqrt{2}\,\exp(3\pi i/4)$,

$$z_k = 2^{1/6}\exp\left(\frac{\pi i}{4} + \frac{2i\pi k}{3}\right), \qquad k = 0, 1, 2, \tag{1.2.14}$$

or

$$z_0 = 2^{1/6}\exp\left(\frac{\pi i}{4}\right) = 2^{1/6}\left[\cos\left(\frac{\pi}{4}\right) + i\sin\left(\frac{\pi}{4}\right)\right], \tag{1.2.15}$$

$$z_1 = 2^{1/6}\exp\left(\frac{11\pi i}{12}\right) = 2^{1/6}\left[\cos\left(\frac{11\pi}{12}\right) + i\sin\left(\frac{11\pi}{12}\right)\right] \tag{1.2.16}$$

and

$$z_2 = 2^{1/6}\exp\left(\frac{19\pi i}{12}\right) = 2^{1/6}\left[\cos\left(\frac{19\pi}{12}\right) + i\sin\left(\frac{19\pi}{12}\right)\right]. \tag{1.2.17}$$

Figure 1.2.2 gives the location of these zeros on the complex plane.

Problems

Extract all of the possible roots of the following complex numbers:

1. $8^{1/6}$
2. $(-1)^{1/3}$
3. $(-i)^{1/3}$
4. $(-27i)^{1/6}$

5. Find algebraic expressions for the square roots of $a - bi$, where $a > 0$ and $b > 0$.

6. Find all of the roots for the algebraic equation $z^4 - 3iz^2 - 2 = 0$.

7. Find all of the roots for the algebraic equation $z^4 + 6iz^2 + 16 = 0$.

1.3 THE DERIVATIVE IN THE COMPLEX PLANE: THE CAUCHY-RIEMANN EQUATIONS

In the previous two sections, we have done complex arithmetic. We are now ready to introduce the concept of function as it applies to complex variables.

We have already introduced the complex variable $z = x + iy$, where x and y are variable. We now define another complex variable $w = u + iv$ so that for each value of z there corresponds a value of $w = f(z)$. From all of the possible complex functions that we might invent, we will focus on those functions where for each z there is one, and only one, value of w. These functions are *single-valued*. They differ from functions such

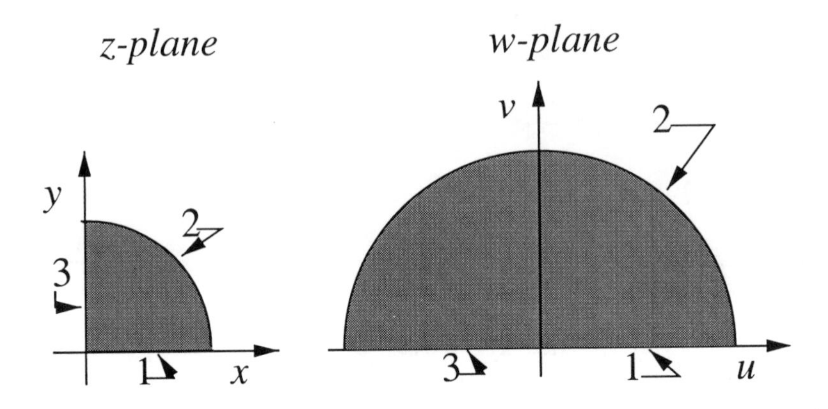

Figure 1.3.1: The complex function $w = z^2$.

as the square root, logarithm, and inverse sine and cosine, where there are multiple answers for each z. These *multivalued functions* do arise in various problems. However, they are beyond the scope of this book and we shall always assume that we are dealing with single-valued functions.

A popular method for representing a complex function involves drawing some closed domain in the z-plane and then showing the corresponding domain in the w-plane. This procedure is called *mapping* and the z-plane illustrates the *domain* of the function while the w-plane illustrates its *image* or *range*. Figure 1.3.1 shows the z-plane and w-plane for $w = z^2$; a pie-shaped wedge in the z-plane maps into a semicircle on the w-plane.

• Example 1.3.1

Given the complex function $w = e^{-z^2}$, let us find the corresponding $u(x,y)$ and $v(x,y)$.

From Euler's formula,

$$w = e^{-z^2} = e^{-(x+iy)^2} = e^{y^2-x^2}e^{-2ixy} = e^{y^2-x^2}[\cos(2xy) - i\sin(2xy)]. \tag{1.3.1}$$

Therefore, by inspection,

$$u(x,y) = e^{y^2-x^2}\cos(2xy) \qquad \text{and} \qquad v(x,y) = -e^{y^2-x^2}\sin(2xy). \tag{1.3.2}$$

Note that there is no i in the expression for $v(x,y)$. The function $w = f(z)$ is single-valued because for each distinct value of z, there is an unique value of $u(x,y)$ and $v(x,y)$.

• Example 1.3.2

As counterpoint, let us show that $w = \sqrt{z}$ is a multivalued function.

We begin by writing $z = re^{i\theta+2\pi ik}$, where $r = \sqrt{x^2+y^2}$ and $\theta = \tan^{-1}(y/x)$. Then,

$$w_k = \sqrt{r}e^{i\theta/2+\pi ik}, \qquad k = 0, 1, \tag{1.3.3}$$

or

$$w_0 = \sqrt{r}\left[\cos(\theta/2) + i\sin(\theta/2)\right] \qquad \text{and} \qquad w_1 = -w_0. \tag{1.3.4}$$

Therefore,

$$u_0(x, y) = \sqrt{r}\cos(\theta/2), \qquad v_0(x, y) = \sqrt{r}\sin(\theta/2) \tag{1.3.5}$$

and

$$u_1(x, y) = -\sqrt{r}\cos(\theta/2), \qquad v_1(x, y) = -\sqrt{r}\sin(\theta/2). \tag{1.3.6}$$

Each solution w_0 or w_1 is a *branch* of the multivalued function $\sqrt{z}$. We can make $\sqrt{z}$ single-valued by restricting ourselves to a single branch, say w_0. In that case, the $\text{Re}(w) > 0$ if we restrict $-\pi < \theta < \pi$. Although this is not the only choice that we could have made, it is a popular one. For example, most digital computers use this definition in their complex square root function. The point here is our ability to make a multivalued function single-valued by defining a particular branch.

Although the requirement that a complex function be single-valued is important, it is still too general and would cover all functions of two real variables. To have a useful theory, we must introduce additional constraints. Because an important property associated with most functions is the ability to take their derivative, let us examine the derivative in the complex plane.

Following the definition of a derivative for a single real variable, the derivative of a complex function $w = f(z)$ is defined as

$$\frac{dw}{dz} = \lim_{\Delta z \to 0} \frac{\Delta w}{\Delta z} = \lim_{\Delta z \to 0} \frac{f(z+\Delta z) - f(z)}{\Delta z}. \tag{1.3.7}$$

A function of a complex variable that has a derivative at every point within a region of the complex plane is said to be *analytic* (or *regular* or *holomorphic*) over that region. If the function is analytic everywhere in the complex plane, it is *entire*.

Because the derivative is defined as a limit and limits are well behaved with respect to elementary algebraic operations, the following operations carry over from elementary calculus:

$$\frac{d}{dz}\Big[cf(z)\Big] = cf'(z), \qquad c \text{ a constant} \tag{1.3.8}$$

$$\frac{d}{dz}\left[f(z) \pm g(z)\right] = f'(z) \pm g'(z) \tag{1.3.9}$$

$$\frac{d}{dz}\left[f(z)g(z)\right] = f'(z)g(z) + f(z)g'(z) \tag{1.3.10}$$

$$\frac{d}{dz}\left[\frac{f(z)}{g(z)}\right] = \frac{g(z)f'(z) - g'(z)f(z)}{g^2(z)} \tag{1.3.11}$$

$$\frac{d}{dz}\left\{f[g(z)]\right\} = f'[g(z)]g'(z), \qquad \text{the chain rule.} \tag{1.3.12}$$

Another important property that carries over from real variables is l'Hôspital rule: Let $f(z)$ and $g(z)$ be analytic at z_0, where $f(z)$ has a zero[1] of order m and $g(z)$ has a zero of order n. Then, if $m > n$,

$$\lim_{z\to z_0} \frac{f(z)}{g(z)} = 0; \tag{1.3.13}$$

if $m = n$,

$$\lim_{z\to z_0} \frac{f(z)}{g(z)} = \frac{f^{(m)}(z_0)}{g^{(m)}(z_0)} \tag{1.3.14}$$

and if $m < n$,

$$\lim_{z\to z_0} \frac{f(z)}{g(z)} = \infty. \tag{1.3.15}$$

• **Example 1.3.3**

Let us evaluate $\lim_{z\to i}(z^{10}+1)/(z^6+1)$. From l'Hôspital rule,

$$\lim_{z\to i} \frac{z^{10}+1}{z^6+1} = \lim_{z\to i} \frac{10z^9}{6z^5} = \frac{5}{3}\lim_{z\to i} z^4 = \frac{5}{3}. \tag{1.3.16}$$

So far we have introduced the derivative and some of its properties. But how do we actually know whether a function is analytic or how do we compute its derivative? At this point we must develop some relationships involving the known quantities $u(x,y)$ and $v(x,y)$.

We begin by returning to the definition of the derivative. Because $\Delta z = \Delta x + i\Delta y$, there is an infinite number of different ways of approaching the limit $\Delta z \to 0$. Uniqueness of that limit requires that (1.3.7) must be independent of the manner in which Δz approaches zero. A simple

[1] An analytic function $f(z)$ has a zero of order m at z_0 if and only if $f(z_0) = f'(z_0) = \cdots = f^{(m-1)}(z_0) = 0$ and $f^{(m)}(z_0) \neq 0$.

Figure 1.3.2: Although educated as an engineer, Augustin-Louis Cauchy (1789–1857) would become a mathematician's mathematician, publishing 789 papers and 7 books in the fields of pure and applied mathematics. His greatest writings established the discipline of mathematical analysis as he refined the notions of limit, continuity, function, and convergence. It was this work on analysis that led him to develop complex function theory via the concept of residues. (Portrait courtesy of the Archives de l'Académie des sciences, Paris.)

example is to take Δz in the x-direction so that $\Delta z = \Delta x$; another is to take Δz in the y-direction so that $\Delta z = i\Delta y$. These examples yield

$$\frac{dw}{dz} = \lim_{\Delta z \to 0} \frac{\Delta w}{\Delta z} = \lim_{\Delta x \to 0} \frac{\Delta u + i\Delta v}{\Delta x} = \frac{\partial u}{\partial x} + i\frac{\partial v}{\partial x} \tag{1.3.17}$$

and

$$\frac{dw}{dz} = \lim_{\Delta z \to 0} \frac{\Delta w}{\Delta z} = \lim_{\Delta y \to 0} \frac{\Delta u + i\Delta v}{i\Delta y} = \frac{\partial v}{\partial y} - i\frac{\partial u}{\partial y}. \tag{1.3.18}$$

Figure 1.3.3: Despite his short life, (Georg Friedrich) Bernhard Riemann's (1826–1866) mathematical work contained many imaginative and profound concepts. It was in his doctoral thesis on complex function theory (1851) that he introduced the Cauchy-Riemann differential equations. Riemann's later work dealt with the definition of the integral and the foundations of geometry and non-Euclidean (elliptic) geometry. (Portrait courtesy of Photo AKG, London.)

In both cases we are approaching zero from the positive side. For the limit to be unique and independent of path, (1.3.17) must equal (1.3.18), or

$$\frac{\partial u}{\partial x} = \frac{\partial v}{\partial y} \quad \text{and} \quad \frac{\partial u}{\partial y} = -\frac{\partial v}{\partial x}. \tag{1.3.19}$$

These equations which u and v must both satisfy are the *Cauchy-Riemann* equations. They are necessary but not sufficient to ensure that a function is differentiable. The following example will illustrate this.

• **Example 1.3.4**

Consider the complex function

$$w = \begin{cases} z^5/|z|^4, & z \neq 0 \\ 0, & z = 0. \end{cases} \tag{1.3.20}$$

The derivative at $z = 0$ is given by

$$\frac{dw}{dz} = \lim_{\Delta z \to 0} \frac{(\Delta z)^5/|\Delta z|^4 - 0}{\Delta z} = \lim_{\Delta z \to 0} \frac{(\Delta z)^4}{|\Delta z|^4}, \tag{1.3.21}$$

provided that this limit exists. However, this limit does not exist because, in general, the numerator depends upon the path used to approach zero. For example, if $\Delta z = re^{\pi i/4}$ with $r \to 0$, $dw/dz = -1$. On the other hand, if $\Delta z = re^{\pi i/2}$ with $r \to 0$, $dw/dz = 1$.

Are the Cauchy-Riemann equations satisfied in this case? To check this, we first compute

$$u_x(0,0) = \lim_{\Delta x \to 0} \left(\frac{\Delta x}{|\Delta x|}\right)^4 = 1, \tag{1.3.22}$$

$$v_y(0,0) = \lim_{\Delta y \to 0} \left(\frac{i\Delta y}{|\Delta y|}\right)^4 = 1, \tag{1.3.23}$$

$$u_y(0,0) = \lim_{\Delta y \to 0} \operatorname{Re}\left[\frac{(i\Delta y)^5}{\Delta y|\Delta y|^4}\right] = 0 \tag{1.3.24}$$

and

$$v_x(0,0) = \lim_{\Delta x \to 0} \operatorname{Im}\left[\left(\frac{\Delta x}{|\Delta x|}\right)^4\right] = 0. \tag{1.3.25}$$

Hence, the Cauchy-Riemann equations are satisfied at the origin. Thus, even though the derivative is not uniquely defined, (1.3.21) happens to have the same value for paths taken along the coordinate axes so that the Cauchy-Riemann equations are satisfied.

In summary, if a function is differentiable at a point, the Cauchy-Riemann equations hold. Similarly, if the Cauchy-Riemann equations are not satisfied at a point, then the function is not differentiable at that point. This is one of the important uses of the Cauchy-Riemann equations: the location of nonanalytic points. Isolated nonanalytic points of an otherwise analytic function are called *isolated singularities*. Functions that contain isolated singularities are called *meromorphic*.

The Cauchy-Riemann condition can be modified so that it is a sufficient condition for the derivative to exist. Let us require that u_x, u_y, v_x, and v_y be continuous in some region surrounding a point z_0 and satisfy the Cauchy-Riemann conditions there. Then

$$f(z) - f(z_0) = [u(z) - u(z_0)] + i[v(z) - v(z_0)] \tag{1.3.26}$$

$$\begin{aligned} &= [u_x(z_0)(x - x_0) + u_y(z_0)(y - y_0) \\ &\qquad + \epsilon_1(x - x_0) + \epsilon_2(y - y_0)] \\ &\quad + i[v_x(z_0)(x - x_0) + v_y(z_0)(y - y_0) \\ &\qquad + \epsilon_3(x - x_0) + \epsilon_4(y - y_0)] \end{aligned} \tag{1.3.27}$$

$$\begin{aligned} &= [u_x(z_0) + iv_x(z_0)](z - z_0) \\ &\quad + (\epsilon_1 + i\epsilon_3)(x - x_0) + (\epsilon_2 + i\epsilon_4)(y - y_0), \end{aligned} \tag{1.3.28}$$

where we have used the Cauchy-Riemann equations and $\epsilon_1, \epsilon_2, \epsilon_3, \epsilon_4 \to 0$ as $\Delta x, \Delta y \to 0$. Hence,

$$f'(z_0) = \lim_{\Delta z \to 0} \frac{f(z) - f(z_0)}{\Delta z} = u_x(z_0) + iv_x(z_0), \tag{1.3.29}$$

because $|\Delta x| \le |\Delta z|$ and $|\Delta y| \le |\Delta z|$. Using (1.3.29) and the Cauchy-Riemann equations, we can obtain the derivative from any of the following formulas:

$$\boxed{\frac{dw}{dz} = \frac{\partial u}{\partial x} + i\frac{\partial v}{\partial x} = \frac{\partial v}{\partial y} - i\frac{\partial u}{\partial y}} \tag{1.3.30}$$

and

$$\boxed{\frac{dw}{dz} = \frac{\partial v}{\partial y} + i\frac{\partial v}{\partial x} = \frac{\partial u}{\partial x} - i\frac{\partial u}{\partial y}.} \tag{1.3.31}$$

Furthermore, $f'(z_0)$ is continuous because the partial derivatives are.

• **Example 1.3.5**

Let us show that $\sin(z)$ is an entire function.

$$w = \sin(z) \tag{1.3.32}$$

$$u + iv = \sin(x + iy) = \sin(x)\cos(iy) + \cos(x)\sin(iy) \tag{1.3.33}$$

$$= \sin(x)\cosh(y) + i\cos(x)\sinh(y), \tag{1.3.34}$$

because

$$\cos(iy) = \tfrac{1}{2}\left[e^{i(iy)} + e^{-i(iy)}\right] = \tfrac{1}{2}\left[e^{y} + e^{-y}\right] = \cosh(y) \tag{1.3.35}$$

and

$$\sin(iy) = \tfrac{1}{2i}\left[e^{i(iy)} - e^{-i(iy)}\right] = -\tfrac{1}{2i}\left[e^{y} - e^{-y}\right] = i\sinh(y) \tag{1.3.36}$$

so that

$$u(x,y) = \sin(x)\cosh(y) \quad \text{and} \quad v(x,y) = \cos(x)\sinh(y). \tag{1.3.37}$$

Differentiating both $u(x, y)$ and $v(x, y)$ with respect to x and y, we have that

$$\frac{\partial u}{\partial x} = \cos(x)\cosh(y) \qquad \frac{\partial u}{\partial y} = \sin(x)\sinh(y) \tag{1.3.38}$$

$$\frac{\partial v}{\partial x} = -\sin(x)\sinh(y) \qquad \frac{\partial v}{\partial y} = \cos(x)\cosh(y) \tag{1.3.39}$$

and $u(x, y)$ and $v(x, y)$ satisfy the Cauchy-Riemann equations for all values of x and y. Furthermore, u_x, u_y, v_x, and v_y are continuous for all x and y. Therefore, the function $w = \sin(z)$ is an entire function.

• **Example 1.3.6**

Consider the function $w = 1/z$. Then

$$w = u + iv = \frac{1}{x + iy} = \frac{x}{x^2 + y^2} - \frac{iy}{x^2 + y^2}. \tag{1.3.40}$$

Therefore,

$$u(x,y) = \frac{x}{x^2 + y^2} \qquad \text{and} \qquad v(x,y) = -\frac{y}{x^2 + y^2}. \tag{1.3.41}$$

Now

$$\frac{\partial u}{\partial x} = \frac{(x^2 + y^2) - 2x^2}{(x^2 + y^2)^2} = \frac{y^2 - x^2}{(x^2 + y^2)^2}, \tag{1.3.42}$$

$$\frac{\partial v}{\partial y} = -\frac{(x^2+y^2)-2y^2}{(x^2+y^2)^2} = \frac{y^2-x^2}{(x^2+y^2)^2} = \frac{\partial u}{\partial x}, \tag{1.3.43}$$

$$\frac{\partial v}{\partial x} = -\frac{0-2xy}{(x^2+y^2)^2} = \frac{2xy}{(x^2+y^2)^2} \tag{1.3.44}$$

and

$$\frac{\partial u}{\partial y} = \frac{0-2xy}{(x^2+y^2)^2} = -\frac{2xy}{(x^2+y^2)^2} = -\frac{\partial v}{\partial x}. \tag{1.3.45}$$

The function is analytic at all points except the origin because the function itself ceases to exist when both x and y are zero and the modulus of w becomes infinite.

- **Example 1.3.7**

Let us find the derivative of $\sin(z)$.
Using (1.3.30) and (1.3.34),

$$\frac{d}{dz}\left[\sin(z)\right] = \frac{\partial u}{\partial x} + i\frac{\partial v}{\partial x} \tag{1.3.46}$$

$$= \cos(x)\cosh(y) - i\sin(x)\sinh(y) \tag{1.3.47}$$

$$= \cos(x+iy) = \cos(z). \tag{1.3.48}$$

Similarly,

$$\frac{d}{dz}\left(\frac{1}{z}\right) = \frac{y^2-x^2}{(x^2+y^2)^2} + \frac{2ixy}{(x^2+y^2)^2} \tag{1.3.49}$$

$$= -\frac{1}{(x+iy)^2} = -\frac{1}{z^2}. \tag{1.3.50}$$

The results in the above examples are identical to those for z real. As we showed earlier, the fundamental rules of elementary calculus apply to complex differentiation. Consequently, it is usually simpler to apply those rules to find the derivative rather than breaking $f(z)$ down into its real and imaginary parts, applying either (1.3.30) or (1.3.31), and then putting everything back together.

An additional property of analytic functions follows by cross differentiating the Cauchy-Riemann equations or

$$\frac{\partial^2 u}{\partial x^2} = \frac{\partial^2 v}{\partial x \partial y} = -\frac{\partial^2 u}{\partial y^2} \quad \text{or} \quad \frac{\partial^2 u}{\partial x^2} + \frac{\partial^2 u}{\partial y^2} = 0 \tag{1.3.51}$$

and

$$\frac{\partial^2 v}{\partial x^2} = -\frac{\partial^2 u}{\partial x \partial y} = -\frac{\partial^2 v}{\partial y^2} \quad \text{or} \quad \frac{\partial^2 v}{\partial x^2} + \frac{\partial^2 v}{\partial y^2} = 0. \tag{1.3.52}$$

Any function that has continuous partial derivatives of second order and satisfies Laplace's equation (1.3.51) or (1.3.52) is called a *harmonic function*. Because both $u(x,y)$ and $v(x,y)$ satisfy Laplace's equation if $f(z) = u + iv$ is analytic, $u(x,y)$ and $v(x,y)$ are called *conjugate harmonic functions*.

- **Example 1.3.8**

Given that $u(x,y) = e^{-x}[x\sin(y) - y\cos(y)]$, let us show that u is harmonic and find a conjugate harmonic function $v(x,y)$ such that $f(z) = u + iv$ is analytic.

Because

$$\frac{\partial^2 u}{\partial x^2} = -2e^{-x}\sin(y) + xe^{-x}\sin(y) - ye^{-x}\cos(y) \tag{1.3.53}$$

and

$$\frac{\partial^2 u}{\partial y^2} = -xe^{-x}\sin(y) + 2e^{-x}\sin(y) + ye^{-x}\cos(y), \tag{1.3.54}$$

it follows that $u_{xx} + u_{yy} = 0$. Therefore, $u(x,y)$ is harmonic. From the Cauchy-Riemann equations,

$$\frac{\partial v}{\partial y} = \frac{\partial u}{\partial x} = e^{-x}\sin(y) - xe^{-x}\sin(y) + ye^{-x}\cos(y) \tag{1.3.55}$$

and

$$\frac{\partial v}{\partial x} = -\frac{\partial u}{\partial y} = e^{-x}\cos(y) - xe^{-x}\cos(y) - ye^{-x}\sin(y). \tag{1.3.56}$$

Integrating (1.3.55) with respect to y,

$$v(x,y) = ye^{-x}\sin(y) + xe^{-x}\cos(y) + g(x). \tag{1.3.57}$$

Using (1.3.56),

$$\begin{aligned} v_x = -ye^{-x}\sin(y) &- xe^{-x}\cos(y) + e^{-x}\cos(y) + g'(x) \\ &= e^{-x}\cos(y) - xe^{-x}\cos(y) - ye^{-x}\sin(x). \end{aligned} \tag{1.3.58}$$

Therefore, $g'(x) = 0$ or $g(x) = \text{constant}$. Consequently,

$$v(x,y) = e^{-x}[y\sin(y) + x\cos(y)] + \text{constant}. \tag{1.3.59}$$

Hence, for our real harmonic function $u(x,y)$, there are infinitely many harmonic conjugates $v(x,y)$ which differ from each other by an additive constant.

Problems

Show that the following functions are entire:

1. $f(z) = iz + 2$
2. $f(z) = e^{-z}$
3. $f(z) = z^3$
4. $f(z) = \cosh(z)$

Find the derivative of the following functions:

5. $f(z) = (1 + z^2)^{3/2}$
6. $f(z) = (z + 2z^{1/2})^{1/3}$
7. $f(z) = (1 + 4i)z^2 - 3z - 2$
8. $f(z) = (2z - i)/(z + 2i)$
9. $f(z) = (iz - 1)^{-3}$

Evaluate the following limits:

10. $\lim_{z \to i} \frac{z^2 - 2iz - 1}{z^4 + 2z^2 + 1}$
11. $\lim_{z \to 0} \frac{z - \sin(z)}{z^3}$

12. Show that the function $f(z) = z^*$ is nowhere differentiable.

For each of the following $u(x, y)$, show that it is harmonic and then find a corresponding $v(x, y)$ such that $f(z) = u + iv$ is analytic.

13.
$$u(x, y) = x^2 - y^2$$

14.
$$u(x, y) = x^4 - 6x^2y^2 + y^4 + x$$

15.
$$u(x, y) = x\cos(x)e^{-y} - y\sin(x)e^{-y}$$

16.
$$u(x, y) = (x^2 - y^2)\cos(y)e^x - 2xy\sin(y)e^x$$

1.4 LINE INTEGRALS

So far, we discussed complex numbers, complex functions, and complex differentiation. We are now ready for integration.

Just as we have integrals involving real variables, we can define an integral that involves complex variables. Because the z-plane is two-dimensional there is clearly greater freedom in what we mean by a complex integral. For example, we might ask whether the integral of some function between points A and B depends upon the curve along which we integrate. (In general it does.) Consequently, an important ingredient in any complex integration is the *contour* that we follow during the integration.

The result of a line integral is a complex number or expression. Unlike its counterpart in real variables, there is no physical interpretation for this quantity, such as area under a curve. Generally, integration in the complex plane is an intermediate process with a physically realizable quantity occurring only after we take its real or imaginary part. For example, in potential fluid flow, the lift and drag are found by taking the real and imaginary part of a complex integral, respectively.

How do we compute $\int_C f(z)\,dz$? Let us deal with the definition; we will illustrate the actual method by examples.

A popular method for evaluating complex line integrals consists of breaking everything up into real and imaginary parts. This reduces the integral to line integrals of real-valued functions which we know how to handle. Thus, we write $f(z) = u(x,y) + iv(x,y)$ as usual, and because $z = x + iy$, formally $dz = dx + i\,dy$. Therefore,

$$\int_C f(z)\,dz = \int_C [u(x,y) + iv(x,y)][dx + i\,dy] \tag{1.4.1}$$

$$= \int_C u(x,y)\,dx - v(x,y)\,dy + i\int_C v(x,y)\,dx + u(x,y)\,dy. \tag{1.4.2}$$

The exact method used to evaluate (1.4.2) depends upon the exact path specified.

From the definition of the line integral, we have the following self-evident properties:

$$\int_C f(z)\,dz = -\int_{C'} f(z)\,dz, \tag{1.4.3}$$

where C' is the contour C taken in the opposite direction of C and

$$\int_{C_1+C_2} f(z)\,dz = \int_{C_1} f(z)\,dz + \int_{C_2} f(z)\,dz. \tag{1.4.4}$$

• **Example 1.4.1**

Let us evaluate $\int_C z^*\,dz$ from $z = 0$ to $z = 4+2i$ along two different contours. The first consists of the parametric equation $z = t^2 + it$. The second consists of two "dog legs": the first leg runs along the imaginary

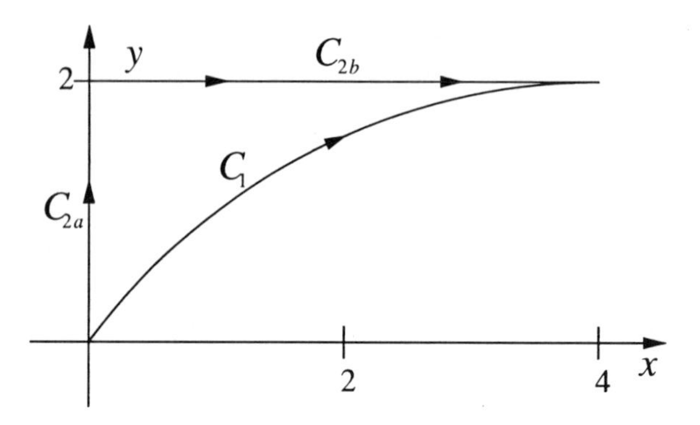

Figure 1.4.1: Contour used in Example 1.4.1.

axis from $z = 0$ to $z = 2i$ and then along a line parallel to the x-axis from $z = 2i$ to $z = 4 + 2i$. See Figure 1.4.1.

For the first case, the points $z = 0$ and $z = 4 + 2i$ on C_1 correspond to $t = 0$ and $t = 2$, respectively. Then the line integral equals

$$\int_{C_1} z^* \, dz = \int_0^2 (t^2 + it)^* \, d(t^2 + it) = \int_0^2 (2t^3 - it^2 + t) \, dt = 10 - \tfrac{8i}{3}. \tag{1.4.5}$$

The line integral for the second contour C_2 equals

$$\int_{C_2} z^* \, dz = \int_{C_{2a}} z^* \, dz + \int_{C_{2b}} z^* \, dz, \tag{1.4.6}$$

where C_{2a} denotes the integration from $z = 0$ to $z = 2i$ while C_{2b} denotes the integration from $z = 2i$ to $z = 4 + 2i$. For the first integral,

$$\int_{C_{2a}} z^* \, dz = \int_{C_{2a}} (x - iy)(dx + i\,dy) = \int_0^2 y\,dy = 2, \tag{1.4.7}$$

because $x = 0$ and $dx = 0$ along C_{2a}. On the other hand, along C_{2b}, $y = 2$ and $dy = 0$ so that

$$\int_{C_{2b}} z^* \, dz = \int_{C_{2b}} (x - iy)(dx + i\,dy) = \int_0^4 x\,dx + i\int_0^4 -2\,dx = 8 - 8i. \tag{1.4.8}$$

Thus the value of entire C_2 contour integral equals the sum of the two parts or $10 - 8i$.

The point here is that integration along two different paths has given us different results even though we integrated from $z = 0$ to $z = 4 + 2i$ both times. This results foreshadows a general result that is extremely important. Because the integrand contains nonanalytic

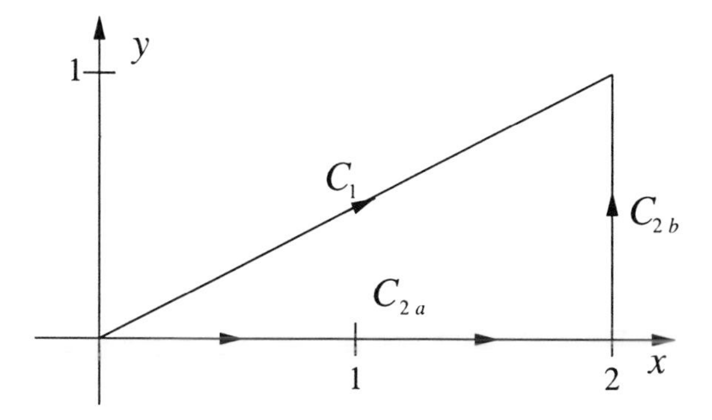

Figure 1.4.2: Contour used in Example 1.4.2.

points along and inside the region enclosed by our two curves, as shown by the Cauchy-Riemann equations, the results depend upon the path taken. Since complex integrations often involve integrands that have nonanalytic points, many line integrations depend upon the contour taken.

• **Example 1.4.2**

Let us integrate the *entire* function $f(z) = z^2$ along the two paths from $z = 0$ to $z = 2 + i$ shown in Figure 1.4.2. For the first integration, $x = 2y$ while along the second path we have two straight paths: $z = 0$ to $z = 2$ and $z = 2$ to $z = 2 + i$.

For the first contour integration,

$$\int_{C_1} z^2 dz = \int_0^1 (2y + iy)^2 (2\,dy + i\,dy) \tag{1.4.9}$$

$$= \int_0^1 (3y^2 + 4y^2 i)(2\,dy + i\,dy) \tag{1.4.10}$$

$$= \int_0^1 6y^2\,dy + 8y^2 i\,dy + 3y^2 i\,dy - 4y^2\,dy \tag{1.4.11}$$

$$= \int_0^1 2y^2\,dy + 11y^2 i\,dy \tag{1.4.12}$$

$$= \tfrac{2}{3}y^3\big|_0^1 + \tfrac{11}{3}iy^3\big|_0^1 = \tfrac{2}{3} + \tfrac{11i}{3}. \tag{1.4.13}$$

For our second integration,

$$\int_{C_2} z^2\,dz = \int_{C_{2a}} z^2\,dz + \int_{C_{2b}} z^2\,dz. \tag{1.4.14}$$

Along C_{2a} we find that $y = dy = 0$ so that

$$\int_{C_{2a}} z^2\,dz = \int_0^2 x^2\,dx = \tfrac{1}{3}x^3\big|_0^2 = \tfrac{8}{3} \tag{1.4.15}$$

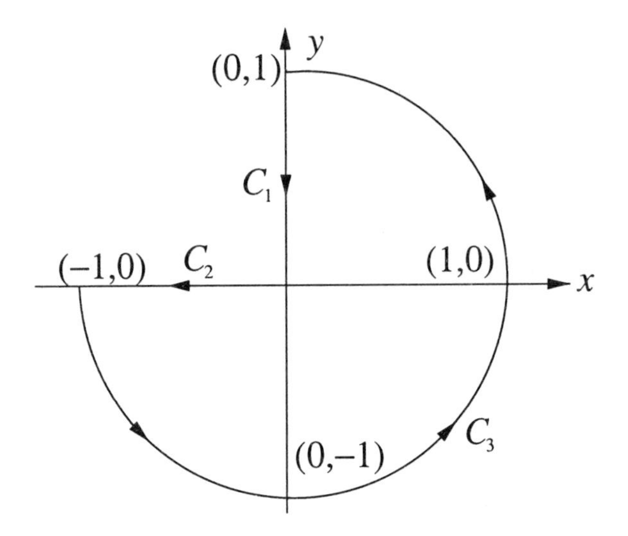

Figure 1.4.3: Contour used in Example 1.4.3.

and

$$\int_{C_{2b}} z^2\,dz = \int_0^1 (2+iy)^2 i\,dy = i\left(4y + 2iy^2 - \frac{y^3}{3}\right)\Big|_0^1 = 4i - 2 - \tfrac{i}{3}, \tag{1.4.16}$$

because $x = 2$ and $dx = 0$. Consequently,

$$\int_{C_2} z^2\,dz = \frac{2}{3} + \frac{11i}{3}. \tag{1.4.17}$$

In this problem we obtained the same results from two different contours of integration. Exploring other contours, we would find that the results are always the same; the integration is path-independent. But what makes these results path-independent while the integration in Example 1.4.1 was not? Perhaps it is the fact that the integrand is analytic everywhere on the complex plane and there are no nonanalytic points. We will explore this later.

Finally, an important class of line integrals involves *closed contours*. We denote this special subclass of line integrals by placing a circle on the integral sign: $\oint$. Consider now the following examples:

• **Example 1.4.3**

Let us integrate $f(z) = z$ around the closed contour shown in Figure 1.4.3.

From Figure 1.4.3,

$$\oint_C z\,dz = \int_{C_1} z\,dz + \int_{C_2} z\,dz + \int_{C_3} z\,dz. \tag{1.4.18}$$

Now

$$\int_{C_1} z\,dz = \int_1^0 iy\,(i\,dy) = -\int_1^0 y\,dy = -\left.\frac{y^2}{2}\right|_1^0 = \frac{1}{2}, \tag{1.4.19}$$

$$\int_{C_2} z\,dz = \int_0^{-1} x\,dx = \left.\frac{x^2}{2}\right|_0^{-1} = \frac{1}{2} \tag{1.4.20}$$

and

$$\int_{C_3} z\,dz = \int_{-\pi}^{\pi/2} e^{\theta i} i e^{\theta i}\,d\theta = \left.\frac{e^{2\theta i}}{2}\right|_{-\pi}^{\pi/2} = -1, \tag{1.4.21}$$

where we have used $z = e^{\theta i}$ around the portion of the unit circle. Therefore, the closed line integral equals zero.

• Example 1.4.4

Let us integrate $f(z) = 1/(z-a)$ around any circle centered on $z = a$. The Cauchy-Riemann equations show that $f(z)$ is a meromorphic function. It is analytic everywhere except at the isolated singularity $z = a$.

If we introduce polar coordinates by letting $z - a = re^{\theta i}$ and $dz = ire^{\theta i}d\theta$,

$$\oint_C \frac{dz}{z-a} = \int_0^{2\pi} \frac{ire^{\theta i}}{re^{\theta i}}\,d\theta = i\int_0^{2\pi} d\theta = 2\pi i. \tag{1.4.22}$$

Note that the integrand becomes undefined at $z = a$. Furthermore, the answer is independent of the size of the circle. Our example suggests that when we have a closed contour integration it is the behavior of the function within the contour rather than the exact shape of the closed contour that is of importance. We will return to this point in later sections.

Problems

1. Evaluate $\oint_C (z^*)^2\,dz$ around the circle $|z| = 1$ taken in the counterclockwise direction.

2. Evaluate $\oint_C |z|^2\,dz$ around the square with vertices at (0,0), (1,0), (1,1), and (0,1) taken in the counterclockwise direction.

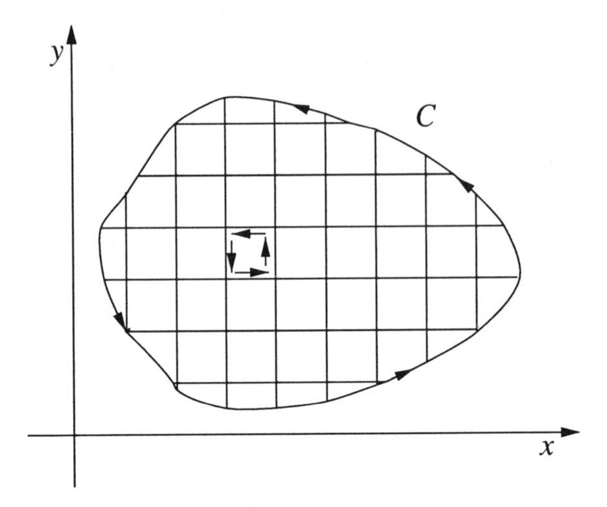

Figure 1.5.1: Diagram used in proving the Cauchy-Goursat theorem.

3. Evaluate $\int_C |z|\, dz$ along the right half of the circle $|z| = 1$ from $z = -i$ to $z = i$.

4. Evaluate $\int_C e^z\, dz$ along the line $y = x$ from $(-1, -1)$ to $(1, 1)$.

5. Evaluate $\int_C (z^*)^2\, dz$ along the line $y = x^2$ from $(0, 0)$ to $(1, 1)$.

6. Evaluate $\int_C z^{-1/2}\, dz$, where C is (a) the upper semicircle $|z| = 1$ and (b) the lower semicircle $|z| = 1$. If $z = re^{\theta i}$, restrict $-\pi < \theta < \pi$. Take both contours in the counterclockwise direction.

1.5 THE CAUCHY-GOURSAT THEOREM

In the previous section we showed how to evaluate line integrations by brute-force reduction to real-valued integrals. In general, this direct approach is quite difficult and we would like to apply some of the deeper properties of complex analysis to work smarter. In the remaining portions of this chapter we will introduce several theorems that will do just that.

If we scan over the examples worked in the previous section, we see considerable differences when the function was analytic inside and on the contour and when it was not. We may formalize this anecdotal evidence into the following theorem:

Cauchy-Goursat theorem[2]: *Let $f(z)$ be analytic in a domain D and*

---

[2] See Goursat, E., 1900: Sur la définition générale des fonctions analytiques, d'après Cauchy. *Trans. Am. Math. Soc.*, **1**, 14–16.

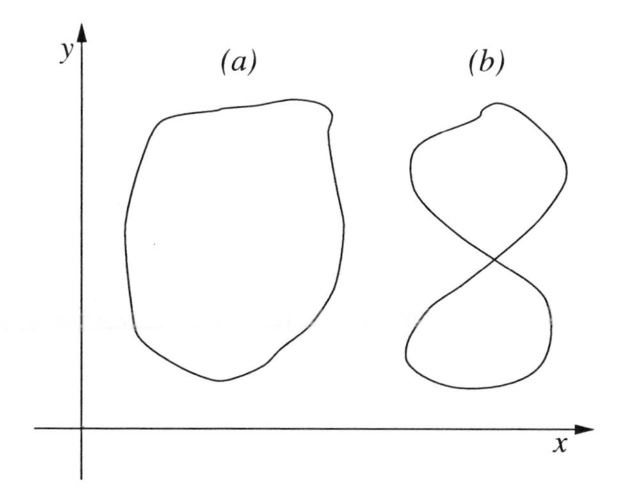

Figure 1.5.2: Examples of a (a) simply closed curve and (b) not simply closed curve.

let C be a simple Jordan curve[3] *inside D so that $f(z)$ is analytic on and inside of C. Then $\oint_C f(z)\,dz = 0$.*

Proof: Let C denote the contour around which we will integrate $w = f(z)$. We divide the region within C into a series of infinitesimal rectangles. See Figure 1.5.1. The integration around each rectangle equals the product of the average value of w on each side and its length,

$$\begin{aligned}\left[w + \frac{\partial w}{\partial x}\frac{dx}{2}\right] dx + \left[w + \frac{\partial w}{\partial x} dx + \frac{\partial w}{\partial(iy)}\frac{d(iy)}{2}\right] d(iy)\\ + \left[w + \frac{\partial w}{\partial x}\frac{dx}{2} + \frac{\partial w}{\partial(iy)} d(iy)\right] (-dx) + \left[w + \frac{\partial w}{\partial(iy)}\frac{d(iy)}{2}\right] d(-iy)\\ = \left(\frac{\partial w}{\partial x} - \frac{\partial w}{i\partial y}\right)(i\,dx\,dy) \qquad \textbf{(1.5.1)}\end{aligned}$$

Substituting $w = u + iv$ into (1.5.1),

$$\frac{\partial w}{\partial x} - \frac{\partial w}{i\,\partial y} = \left(\frac{\partial u}{\partial x} - \frac{\partial v}{\partial y}\right) + i\left(\frac{\partial v}{\partial x} + \frac{\partial u}{\partial y}\right). \qquad \textbf{(1.5.2)}$$

Because the function is analytic, the right side of (1.5.1) and (1.5.2) equals zero. Thus, the integration around each of these rectangles also equals zero.

[3] A Jordan curve is a simply closed curve. It looks like a closed loop that does not cross itself. See Figure 1.5.2.

We note next that in integrating around adjoining rectangles we transverse each side in opposite directions, the net result being equivalent to integrating around the outer curve C. We therefore arrive at the result $\oint_C f(z)\,dz = 0$, where $f(z)$ is analytic within and on the closed contour. □

The Cauchy-Goursat theorem has several useful implications. Suppose we have a domain where $f(z)$ is analytic. Within this domain let us evaluate a line integral from point A to B along two different contours C_1 and C_2. Then, the integral around the closed contour formed by integrating along C_1 and then back along C_2, only in the opposite direction, is

$$\oint_C f(z)\,dz = \int_{C_1} f(z)\,dz - \int_{C_2} f(z)\,dz = 0 \tag{1.5.3}$$

or

$$\int_{C_1} f(z)\,dz = \int_{C_2} f(z)\,dz. \tag{1.5.4}$$

Because C_1 and C_2 are completely arbitrary, we have the result that if, in a domain, $f(z)$ is analytic, the integral between any two points within the domain is *path independent.*

One obvious advantage of path independence is the ability to choose the contour so that the computations are made easier. This obvious choice immediately leads to

The principle of deformation of contours: *The value of a line integral of an analytic function around any simple closed contour remains unchanged if we deform the contour in such a manner that we do not pass over a nonanalytic point.*

• Example 1.5.1

Let us integrate $f(z) = z^{-1}$ around the closed contour C in the counterclockwise direction. This contour consists of a square, centered on the origin, with vertices at $(1, 1)$, $(1, -1)$, $(-1, 1)$, and $(-1, -1)$.

The direct integration of $\oint_C z^{-1} dz$ around the original contour is very cumbersome. However, because the integrand is analytic everywhere except at the origin, we may deform the origin contour into a circle of radius r, centered on the origin. Then, $z = re^{\theta i}$ and $dz = rie^{\theta i} d\theta$ so that

$$\oint_C \frac{dz}{z} = \int_0^{2\pi} \frac{rie^{\theta i}}{re^{\theta i}}\,d\theta = i\int_0^{2\pi} d\theta = 2\pi i. \tag{1.5.5}$$

The point here is that no matter how bizarre the contour is, as long as it encircles the origin and is a simply closed contour, we can deform it into a circle and we will get the same answer for the contour integral. This suggests that it is not the shape of the closed contour that makes the difference but whether we enclose any singularities [points where $f(z)$ becomes undefined] that matters. We shall return to this idea many times in the next few sections.

Finally, suppose that we have a function $f(z)$ such that $f(z)$ is analytic in some domain. Furthermore, let us introduce the analytic function $F(z)$ such that $f(z) = F'(z)$. We would like to evaluate $\int_a^b f(z)\,dz$ in terms of $F(z)$.

We begin by noting that we can represent F, f as $F(z) = U + iV$ and $f(z) = u + iv$. From (1.3.30) we have that $u = U_x$ and $v = V_x$. Therefore,

$$\int_a^b f(z)\,dz = \int_a^b (u + iv)(dx + i\,dy) \tag{1.5.6}$$

$$= \int_a^b U_x\,dx - V_x\,dy + i\int_a^b V_x\,dx + U_x\,dy \tag{1.5.7}$$

$$= \int_a^b U_x\,dx + U_y\,dy + i\int_a^b V_x\,dx + V_y\,dy \tag{1.5.8}$$

$$= \int_a^b dU + i\int_a^b dV = F(b) - F(a) \tag{1.5.9}$$

or

$$\int_a^b f(z)\,dz = F(b) - F(a). \tag{1.5.10}$$

Equation (1.5.10) is the complex variable form of the fundamental theorem of calculus. Thus, if we can find the antiderivative of a function $f(z)$ that is analytic within a specific region, we can evaluate the integral by evaluating the antiderivative at the endpoints for any curves within that region.

• Example 1.5.2

Let us evaluate $\int_0^{\pi i} z\,\sin(z^2)\,dz$.

The integrand $f(z) = z\,\sin(z^2)$ is an entire function and has the antiderivative $-\frac{1}{2}\cos(z^2)$. Therefore,

$$\int_0^{\pi i} z\,\sin(z^2)\,dz = -\tfrac{1}{2}\cos(z^2)\Big|_0^{\pi i} \tag{1.5.11}$$

$$= \tfrac{1}{2}[\cos(0) - \cos(-\pi^2)] \tag{1.5.12}$$

$$= \tfrac{1}{2}[1 - \cos(\pi^2)]. \tag{1.5.13}$$

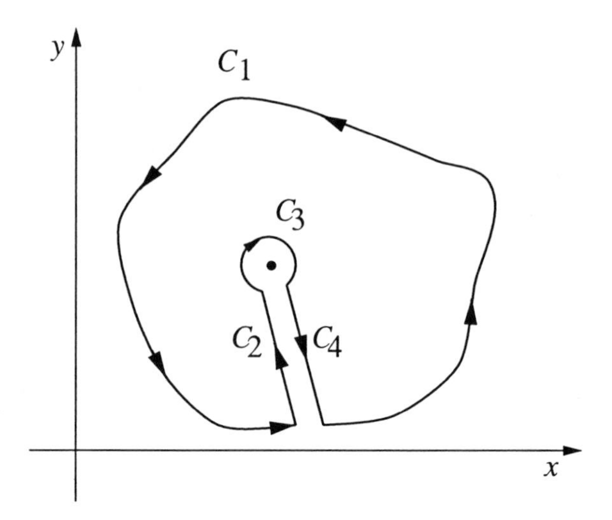

Figure 1.6.1: Diagram used to prove Cauchy's integral formula.

Problems

For the following integrals, show that they are path independent and determine the value of the integral:

1. $\displaystyle\int_{1-\pi i}^{2+3\pi i} e^{-2z}\,dz$
2. $\displaystyle\int_{0}^{2\pi} [e^z - \cos(z)]\,dz$
3. $\displaystyle\int_{0}^{\pi} \sin^2(z)\,dz$
4. $\displaystyle\int_{-i}^{2i} (z+1)\,dz$

1.6 CAUCHY'S INTEGRAL FORMULA

In the previous section, our examples suggested that the presence of a singularity within a contour really determines the value of a closed contour integral. Continuing with this idea, let us consider a class of closed contour integrals that explicitly contain a single singularity within the contour, namely $\oint_C g(z)\,dz$, where $g(z) = f(z)/(z - z_0)$ and $f(z)$ is analytic within and on the contour C. We have closed the contour in the *positive sense* where the enclosed area lies to your left as you move along the contour.

We begin by examining a closed contour integral where the closed contour consists of the C_1, C_2, C_3, and C_4 as shown in Figure 1.6.1. The gap or cut between C_2 and C_4 is very small. Because $g(z)$ is analytic within and on the closed integral, we have that

$$\int_{C_1} \frac{f(z)}{z - z_0}\,dz + \int_{C_2} \frac{f(z)}{z - z_0}\,dz + \int_{C_3} \frac{f(z)}{z - z_0}\,dz + \int_{C_4} \frac{f(z)}{z - z_0}\,dz = 0. \tag{1.6.1}$$

It can be shown that the contribution to the integral from the path C_2 going into the singularity will cancel the contribution from the path C_4 going away from the singularity as the gap between them vanishes. Because $f(z)$ is analytic at z_0, we can approximate its value on C_3 by $f(z) = f(z_0) + \delta(z)$, where δ is a small quantity. Substituting into (1.6.1),

$$\oint_{C_1} \frac{f(z)}{z - z_0}\, dz = -f(z_0) \int_{C_3} \frac{1}{z - z_0}\, dz - \int_{C_3} \frac{\delta(z)}{z - z_0}\, dz. \qquad \textbf{(1.6.2)}$$

Consequently, as the gap between C_2 and C_4 vanishes, the contour C_1 becomes the closed contour C so that (1.6.2) may be written

$$\oint_C \frac{f(z)}{z - z_0}\, dz = 2\pi i f(z_0) + i \int_0^{2\pi} \delta\, d\theta, \qquad \textbf{(1.6.3)}$$

where we have set $z - z_0 = \epsilon e^{\theta i}$ and $dz = i\epsilon e^{\theta i}\, d\theta$.

Let M denote the value of the integral on the right side of (1.6.3) and Δ equal the greatest value of the modulus of δ along the circle. Then

$$|M| < \int_0^{2\pi} |\delta| d\theta \le \int_0^{2\pi} \Delta d\theta = 2\pi\Delta. \qquad \textbf{(1.6.4)}$$

As the radius of the circle diminishes to zero, Δ also diminishes to zero. Therefore, $|M|$, which is positive, becomes less than any finite quantity, however small, and M itself equals zero. Thus, we have that

$$f(z_0) = \frac{1}{2\pi i} \oint_C \frac{f(z)}{z - z_0}\, dz. \qquad \textbf{(1.6.5)}$$

This equation is *Cauchy's integral formula*. By taking n derivatives of (1.6.5), we can extend Cauchy's integral formula[4] to

$$f^{(n)}(z_0) = \frac{n!}{2\pi i} \oint_C \frac{f(z)}{(z - z_0)^{n+1}}\, dz \qquad \textbf{(1.6.6)}$$

[4] See Carrier, G. F., Krook, M., and Pearson, C. E., 1966: *Functions of a Complex Variable: Theory and Technique*, McGraw-Hill, New York, pp. 39–40 for the proof.

for $n = 1, 2, 3, \ldots$. For computing integrals, it is convenient to rewrite (1.6.6) as

$$\oint_C \frac{f(z)}{(z - z_0)^{n+1}}\, dz = \frac{2\pi i}{n!} f^{(n)}(z_0). \tag{1.6.7}$$

• **Example 1.6.1**

Let us find the value of the integral

$$\oint_C \frac{\cos(\pi z)}{(z-1)(z-2)}\, dz, \tag{1.6.8}$$

where C is the circle $|z| = 5$. Using partial fractions,

$$\frac{1}{(z-1)(z-2)} = \frac{1}{z-2} - \frac{1}{z-1} \tag{1.6.9}$$

and

$$\oint_C \frac{\cos(\pi z)}{(z-1)(z-2)}\, dz = \oint_C \frac{\cos(\pi z)}{z-2}\, dz - \oint_C \frac{\cos(\pi z)}{z-1} dz. \tag{1.6.10}$$

By Cauchy's integral formula with $z_0 = 2$ and $z_0 = 1$,

$$\oint_C \frac{\cos(\pi z)}{z-2}\, dz = 2\pi i\ \cos(2\pi) = 2\pi i \tag{1.6.11}$$

and

$$\oint_C \frac{\cos(\pi z)}{z-1}\, dz = 2\pi i \cos(\pi) = -2\pi i, \tag{1.6.12}$$

because $z_0 = 1$ and $z_0 = 2$ lie inside C and $\cos(\pi z)$ is analytic there. Thus the required integral has the value

$$\oint_C \frac{\cos(\pi z)}{(z-1)(z-2)}\, dz = 4\pi i. \tag{1.6.13}$$

• **Example 1.6.2**

Let us use Cauchy's integral formula to evaluate

$$I = \oint_{|z|=2} \frac{e^z}{(z-1)^2(z-3)}\, dz. \tag{1.6.14}$$

We need to convert (1.6.14) into the form (1.6.7). To do this, we rewrite (1.6.14) as

$$\oint_{|z|=2} \frac{e^z}{(z-1)^2(z-3)}\, dz = \oint_{|z|=2} \frac{e^z/(z-3)}{(z-1)^2}\, dz. \tag{1.6.15}$$

Therefore, $f(z) = e^z/(z-3)$, $n = 1$ and $z_0 = 1$. The function $f(z)$ is analytic within the closed contour because the point $z_0 = 3$ lies outside of the contour. Applying Cauchy's integral formula,

$$\oint_{|z|=2} \frac{e^z}{(z-1)^2(z-3)}\,dz = \frac{2\pi i}{1!}\frac{d}{dz}\left(\frac{e^z}{z-3}\right)\Bigg|_{z=1} \tag{1.6.16}$$

$$= 2\pi i\left[\frac{e^z}{z-3} - \frac{e^z}{(z-3)^2}\right]\Bigg|_{z=1} \tag{1.6.17}$$

$$= -\frac{3\pi i e}{2}. \tag{1.6.18}$$

Problems

Use Cauchy's integral formula to evaluate the following integrals. Assume all of the contours are in the positive sense.

1. $\oint_{|z|=1} \frac{\sin^6(z)}{z-\pi/6}\,dz$
2. $\oint_{|z|=1} \frac{\sin^6(z)}{(z-\pi/6)^3}\,dz$
3. $\oint_{|z|=1} \frac{1}{z(z^2+4)}\,dz$
4. $\oint_{|z|=1} \frac{\tan(z)}{z}\,dz$
5. $\oint_{|z-1|=1/2} \frac{1}{(z-1)(z-2)}\,dz$
6. $\oint_{|z|=5} \frac{\exp(z^2)}{z^3}\,dz$
7. $\oint_{|z-1|=1} \frac{z^2+1}{z^2-1}\,dz$
8. $\oint_{|z|=2} \frac{z^2}{(z-1)^4}\,dz$
9. $\oint_{|z|=2} \frac{z^3}{(z+i)^3}\,dz$
10. $\oint_{|z|=1} \frac{\cos(z)}{z^{2n+1}}\,dz$

1.7 TAYLOR AND LAURENT EXPANSIONS AND SINGULARITIES

In the previous section we showed what a crucial role singularities play in complex integration. Before we can find the most general way of computing a closed complex integral, our understanding of singularities must deepen. For this, we employ power series.

One reason why power series are so important is their ability to provide locally a general representation of a function even when its arguments are complex. For example, when we were introduced to trigonometric functions in high school, it was in the context of a right triangle and a real angle. However, when the argument becomes complex this geometrical description disappears and power series provide a formalism for defining the trigonometric functions, regardless of the nature of the argument.

Let us begin our analysis by considering the complex function $f(z)$ which is analytic everywhere on the boundary and the interior of a circle whose center is at $z = z_0$. Then, if z denotes any point within the circle, we have from Cauchy's integral formula that

$$f(z) = \frac{1}{2\pi i} \oint_C \frac{f(\zeta)}{\zeta - z}\, d\zeta = \frac{1}{2\pi i} \oint_C \frac{f(\zeta)}{\zeta - z_0} \left[\frac{1}{1 - (z - z_0)/(\zeta - z_0)} \right] d\zeta, \tag{1.7.1}$$

where C denotes the closed contour. Expanding the bracketed term as a geometric series, we find that

$$f(z) = \frac{1}{2\pi i} \left[\oint_C \frac{f(\zeta)}{\zeta - z_0}\, d\zeta + (z - z_0) \oint_C \frac{f(\zeta)}{(\zeta - z_0)^2}\, d\zeta + \cdots + (z - z_0)^n \oint_C \frac{f(\zeta)}{(\zeta - z_0)^{n+1}}\, d\zeta + \cdots \right]. \tag{1.7.2}$$

Applying Cauchy's integral formula to each integral in (1.7.2), we finally obtain

$$f(z) = f(z_0) + \frac{(z - z_0)}{1!} f'(z_0) + \cdots + \frac{(z - z_0)^n}{n!} f^{(n)}(z_0) + \cdots \tag{1.7.3}$$

or the familiar formula for a Taylor expansion. Consequently, *we can expand any analytic function into a Taylor series.* Interestingly, the radius of convergence[5] of this series may be shown to be the distance between z_0 and the nearest nonanalytic point of $f(z)$.

• Example 1.7.1

Let us find the expansion of $f(z) = \sin(z)$ about the point $z_0 = 0$.

Because $f(z)$ is an entire function, we can construct a Taylor expansion anywhere on the complex plane. For $z_0 = 0$,

$$f(z) = f(0) + \tfrac{1}{1!} f'(0) z + \tfrac{1}{2!} f''(0) z^2 + \tfrac{1}{3!} f'''(0) z^3 + \cdots \tag{1.7.4}$$

Because $f(0) = 0$, $f'(0) = 1$, $f''(0) = 0$, $f'''(0) = -1$ and so forth,

$$f(z) = z - \frac{z^3}{3!} + \frac{z^5}{5!} - \frac{z^7}{7!} + \cdots \tag{1.7.5}$$

Because $\sin(z)$ is an entire function, the radius of convergence is $|z - 0| < \infty$, i.e., all z.

---

[5] A positive number h such that the series diverges for $|z - z_0| > h$ but converges absolutely for $|z - z_0| < h$.

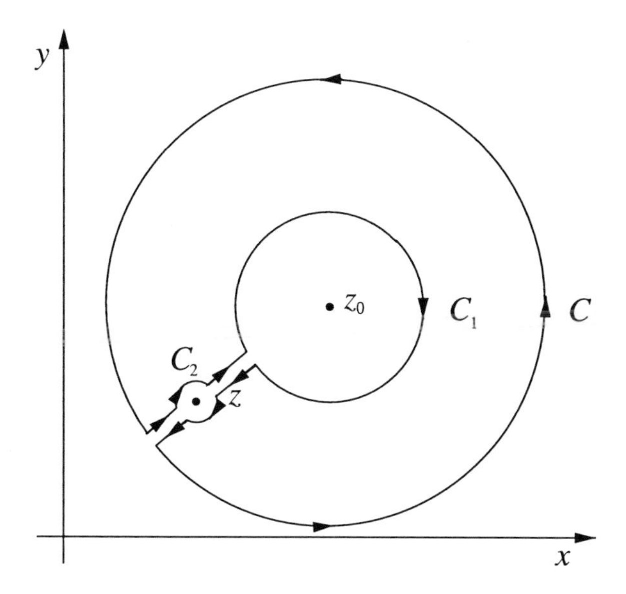

Figure 1.7.1: Contour used in deriving the Laurent expansion.

• **Example 1.7.2**

Let us find the expansion of $f(z) = 1/(1-z)$ about the point $z_0 = 0$. From the formula for a Taylor expansion,

$$f(z) = f(0) + \tfrac{1}{1!}f'(0)z + \tfrac{1}{2!}f''(0)z^2 + \tfrac{1}{3!}f'''(0)z^3 + \cdots \qquad \textbf{(1.7.6)}$$

Because $f^{(n)}(0) = n!$, we find that

$$f(z) = 1 + z + z^2 + z^3 + z^4 + \cdots = \frac{1}{1-z}. \qquad \textbf{(1.7.7)}$$

Equation (1.7.7) is the familiar result for a geometric series. Because the only nonanalytic point is at $z = 1$, the radius of convergence is $|z - 0| < 1$, the unit circle centered at $z = 0$.

Consider now the situation where we draw two concentric circles about some arbitrary point z_0; we denote the outer circle by C while we denote the inner circle by C_1. See Figure 1.7.1. Let us assume that $f(z)$ is analytic inside the annulus between the two circles. Outside of this area, the function may or may not be analytic. Within the annulus we pick a point z and construct a small circle around it, denoting the circle by C_2. As the gap or *cut* in the annulus becomes infinitesimally small, the line integrals that connect the circle C_2 to C_1 and C sum to zero, leaving

$$\oint_C \frac{f(\zeta)}{\zeta - z}\,d\zeta = \oint_{C_1} \frac{f(\zeta)}{\zeta - z}\,d\zeta + \oint_{C_2} \frac{f(\zeta)}{\zeta - z}\,d\zeta. \tag{1.7.8}$$

Because $f(\zeta)$ is analytic everywhere within C_2,

$$2\pi i f(z) = \oint_{C_2} \frac{f(\zeta)}{\zeta - z}\,d\zeta. \tag{1.7.9}$$

Using the relationship:

$$\oint_{C_1} \frac{f(\zeta)}{\zeta - z}\,d\zeta = -\oint_{C_1} \frac{f(\zeta)}{z - \zeta}\,d\zeta, \tag{1.7.10}$$

(1.7.8) becomes

$$f(z) = \frac{1}{2\pi i}\oint_C \frac{f(\zeta)}{\zeta - z}\,d\zeta + \frac{1}{2\pi i}\oint_{C_1} \frac{f(\zeta)}{z - \zeta}\,d\zeta. \tag{1.7.11}$$

Now,

$$\frac{1}{\zeta - z} = \frac{1}{\zeta - z_0 - z + z_0} = \frac{1}{\zeta - z_0}\,\frac{1}{1 - (z - z_0)/(\zeta - z_0)} \tag{1.7.12}$$

$$= \frac{1}{\zeta - z_0}\left[1 + \left(\frac{z - z_0}{\zeta - z_0}\right) + \left(\frac{z - z_0}{\zeta - z_0}\right)^2 + \cdots + \left(\frac{z - z_0}{\zeta - z_0}\right)^n + \cdots\right], \tag{1.7.13}$$

where $|z - z_0|/|\zeta - z_0| < 1$ and

$$\frac{1}{z - \zeta} = \frac{1}{z - z_0 - \zeta + z_0} = \frac{1}{z - z_0}\,\frac{1}{1 - (\zeta - z_0)/(z - z_0)} \tag{1.7.14}$$

$$= \frac{1}{z - z_0}\left[1 + \left(\frac{\zeta - z_0}{z - z_0}\right) + \left(\frac{\zeta - z_0}{z - z_0}\right)^2 + \cdots + \left(\frac{\zeta - z_0}{z - z_0}\right)^n + \cdots\right], \tag{1.7.15}$$

where $|\zeta - z_0|/|z - z_0| < 1$. Upon substituting these expressions into (1.7.11),

$$\begin{aligned}
f(z) = &\left[\frac{1}{2\pi i}\oint_C \frac{f(\zeta)}{\zeta - z_0}\,d\zeta + \frac{z - z_0}{2\pi i}\oint_C \frac{f(\zeta)}{(\zeta - z_0)^2}\,d\zeta + \cdots \right.\\
&\left. + \frac{(z - z_0)^n}{2\pi i}\oint_C \frac{f(\zeta)}{(\zeta - z_0)^{n+1}}\,d\zeta + \cdots\right]\\
&+ \left[\frac{1}{z - z_0}\frac{1}{2\pi i}\oint_{C_1} f(\zeta)\,d\zeta + \frac{1}{(z - z_0)^2}\frac{1}{2\pi i}\oint_{C_1} f(\zeta)(\zeta - z_0)\,d\zeta + \cdots \right.\\
&\left. + \frac{1}{(z - z_0)^n}\frac{1}{2\pi i}\oint_{C_1} f(\zeta)(\zeta - z_0)^{n-1}\,d\zeta + \cdots\right]
\end{aligned} \tag{1.7.16}$$

or

$$f(z) = \frac{a_1}{z - z_0} + \frac{a_2}{(z - z_0)^2} + \cdots + \frac{a_n}{(z - z_0)^n} + \cdots$$
$$+ b_0 + b_1(z - z_0) + \cdots + b_n(z - z_0)^n + \cdots \qquad \textbf{(1.7.17)}$$

Equation (1.7.17) is a *Laurent expansion.*[6] If $f(z)$ is analytic at z_0, then $a_1 = a_2 = \cdots = a_n = \cdots = 0$ and the Laurent expansion reduces to a Taylor expansion. If z_0 is a singularity of $f(z)$, then the Laurent expansion will include both positive and *negative* powers. The coefficient of the $(z - z_0)^{-1}$ term, a_1, is the *residue*, for reasons that will appear in the next section.

Unlike the Taylor series, there is no straightforward method for obtaining a Laurent series. For the remaining portions of this section we will illustrate their construction. These techniques include replacing a function by its appropriate power series, the use of geometric series to expand the denominator, and the use of algebraic tricks to assist in applying the first two method.

• Example 1.7.3

Laurent expansions provide a formalism for the classification of singularities of a function. *Isolated singularities* fall into three types; they are

• *Essential Singularity*: Consider the function $f(z) = \cos(1/z)$. Using the expansion for cosine,

$$\cos\left(\frac{1}{z}\right) = 1 - \frac{1}{2!z^2} + \frac{1}{4!z^4} - \frac{1}{6!z^6} + \cdots \qquad \textbf{(1.7.18)}$$

for $0 < |z| < \infty$. Note that this series never truncates in the inverse powers of z. Essential singularities have Laurent expansions which have an infinite number of inverse powers of $z - z_0$. The value of the residue for this essential singularity at $z = 0$ is zero.

• *Removable Singularity*: Consider the function $f(z) = \sin(z)/z$. This function has a singularity at $z = 0$. Upon applying the expansion for sine,

$$\frac{\sin(z)}{z} = \frac{1}{z}\left(z - \frac{z^3}{3!} + \frac{z^5}{5!} - \frac{z^7}{7!} + \frac{z^9}{9!} - \ldots\right) \qquad \textbf{(1.7.19)}$$

$$= 1 - \frac{z^2}{3!} + \frac{z^4}{5!} - \frac{z^6}{7!} + \frac{z^8}{9!} - \ldots \qquad \textbf{(1.7.20)}$$

[6] See Laurent, M., 1843: Extension du théorème de M. Cauchy relatif à la convergence du développement d'une fonction suivant les puissances ascendantes de la variable x. *C. R. l'Acad. Sci.*, **17**, 938–942.

for all z, if the division is permissible. We have made $f(z)$ analytic by defining it by (1.7.20) and, in the process, removed the singularity. The residue for a removable singularity always equals zero.

• *Pole of order n*: Consider the function

$$f(z) = \frac{1}{(z-1)^3(z+1)}. \tag{1.7.21}$$

This function has two singularities: one at $z = 1$ and the other at $z = -1$. We shall only consider the case $z = 1$. After a little algebra,

$$f(z) = \frac{1}{(z-1)^3} \frac{1}{2+(z-1)} \tag{1.7.22}$$

$$= \frac{1}{2} \frac{1}{(z-1)^3} \frac{1}{1+(z-1)/2} \tag{1.7.23}$$

$$= \frac{1}{2} \frac{1}{(z-1)^3} \left[1 - \frac{z-1}{2} + \frac{(z-1)^2}{4} - \frac{(z-1)^3}{8} + \cdots\right] \tag{1.7.24}$$

$$= \frac{1}{2(z-1)^3} - \frac{1}{4(z-1)^2} + \frac{1}{8(z-1)} - \frac{1}{16} + \cdots \tag{1.7.25}$$

for $0 < |z-1| < 2$. Because the largest inverse (negative) power is three, the singularity at $z = 1$ is a third-order pole; the value of the residue is $1/8$. Generally, we refer to a first-order pole as a *simple* pole.

• **Example 1.7.4**

Let us find the Laurent expansion for

$$f(z) = \frac{z}{(z-1)(z-3)} \tag{1.7.26}$$

about the point $z = 1$.

We begin by rewriting $f(z)$ as

$$f(z) = \frac{1+(z-1)}{(z-1)[-2+(z-1)]} \tag{1.7.27}$$

$$= -\frac{1}{2} \frac{1+(z-1)}{(z-1)[1-\frac{1}{2}(z-1)]} \tag{1.7.28}$$

$$= -\frac{1}{2} \frac{1+(z-1)}{(z-1)}[1+\tfrac{1}{2}(z-1)+\tfrac{1}{4}(z-1)^2+\cdots] \tag{1.7.29}$$

$$= -\frac{1}{2} \frac{1}{z-1} - \frac{3}{4} - \frac{3}{8}(z-1) - \frac{3}{16}(z-1)^2 - \cdots \tag{1.7.30}$$

provided $0 < |z-1| < 2$. Therefore we have a simple pole at $z = 1$ and the value of the residue is $-1/2$. A similar procedure would yield the Laurent expansion about $z = 3$.

For complicated complex functions, it is very difficult to determine the nature of the singularities by finding the complete Laurent expansion and we must try another method. We shall call it "a poor man's Laurent expansion". The idea behind this method is the fact that we generally need only the first few terms of the Laurent expansion to discover its nature. Consequently, we compute these terms through the application of power series where we retain only the leading terms. Consider the following example.

• Example 1.7.5

Let us discover the nature of the singularity at $z = 0$ of the function

$$f(z) = \frac{e^{tz}}{z\sinh(az)}, \tag{1.7.31}$$

where a and t are real.

We begin by replacing the exponential and hyperbolic sine by their Taylor expansion about $z = 0$. Then

$$f(z) = \frac{1 + tz + t^2z^2/2 + \cdots}{z(az - a^3z^3/6 + \cdots)}. \tag{1.7.32}$$

Factoring out az in the denominator,

$$f(z) = \frac{1 + tz + t^2z^2/2 + \cdots}{az^2(1 - a^2z^2/6 + \cdots)}. \tag{1.7.33}$$

Within the parentheses all of the terms except the leading one are small. Therefore, by long division, we formally have that

$$f(z) = \frac{1}{az^2}(1 + tz + t^2z^2/2 + \cdots)(1 + a^2z^2/6 + \cdots) \tag{1.7.34}$$

$$= \frac{1}{az^2}(1 + tz + t^2z^2/2 + a^2z^2/6 + \cdots) \tag{1.7.35}$$

$$= \frac{1}{az^2} + \frac{t}{az} + \frac{3t^2 + a^2}{6a} + \cdots \tag{1.7.36}$$

Thus, we have a second-order pole at $z = 0$ and the residue equals t/a.

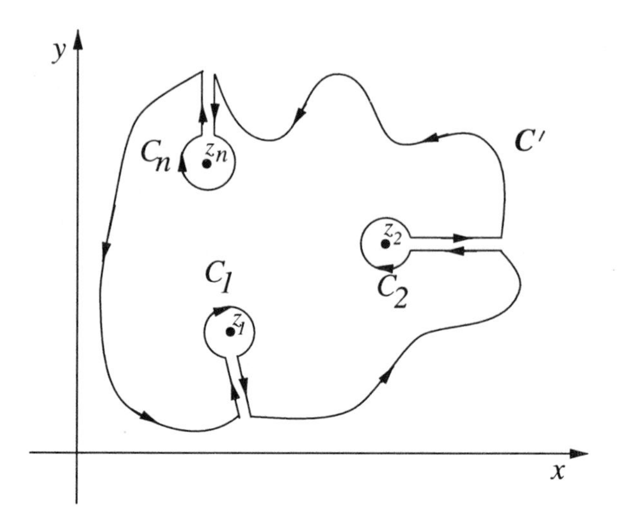

Figure 1.8.1: Contour used in deriving the residue theorem.

Problems

1. Find the Taylor expansion of $f(z) = (1-z)^{-2}$ about the point $z = 0$.

2. Find the Taylor expansion of $f(z) = (z-1)e^z$ about the point $z = 1$. [Hint: Don't find the expansion by taking derivatives.]

By constructing a Laurent expansion, describe the type of singularity and give the residue at z_0 for each of the following functions:

3. $f(z) = z^{10}e^{-1/z}; \quad z_0 = 0$

4. $f(z) = z^{-3}\sin^2(z); \quad z_0 = 0$

5. $f(z) = \dfrac{\cosh(z) - 1}{z^2}; \quad z_0 = 0$

6. $f(z) = \dfrac{z}{(z+2)^2}; \quad z_0 = -2$

7. $f(z) = \dfrac{e^z + 1}{e^{-z} - 1}; \quad z_0 = 0$

8. $f(z) = \dfrac{e^{iz}}{z^2 + b^2}; \quad z_0 = bi$

9. $f(z) = \dfrac{1}{z(z-2)}; \quad z_0 = 2$

10. $f(z) = \dfrac{\exp(z^2)}{z^4}; \quad z_0 = 0$

1.8 THEORY OF RESIDUES

Having shown that around any singularity we may construct a Laurent expansion, we now use this result in the integration of closed complex integrals. Consider a closed contour in which the function $f(z)$ has a number of isolated singularities. As we did in the case of Cauchy's integral formula, we introduce a new contour C' which excludes all of the singularities because they are isolated. See Figure 1.8.1. Therefore,

$$\oint_C f(z)\,dz - \oint_{C_1} f(z)\,dz - \cdots - \oint_{C_n} f(z)\,dz = \oint_{C'} f(z)\,dz = 0. \quad \mathbf{(1.8.1)}$$

Consider now the mth integral, where $1 \leq m \leq n$. Constructing a Laurent expansion for the function $f(z)$ at the isolated singularity $z = z_m$, this integral equals

$$\oint_{C_m} f(z)\,dz = \sum_{k=1}^{\infty} a_k \oint_{C_m} \frac{1}{(z-z_m)^k}\,dz + \sum_{k=0}^{\infty} b_k \oint_{C_m} (z-z_m)^k\,dz. \quad \mathbf{(1.8.2)}$$

Because $(z - z_m)^k$ is an entire function if $k \geq 0$, the integrals equal zero for each term in the second summation. We use Cauchy's integral formula to evaluate the remaining terms. The analytic function in the numerator is 1. Because $d^{k-1}(1)/dz^{k-1} = 0$ if $k > 1$, all of the terms vanish except for $k = 1$. In that case, the integral equals $2\pi i a_1$, where a_1 is the value of the residue for that particular singularity. Applying this approach to each of the singularities, we obtain

Cauchy's residue theorem[7]: *If $f(z)$ is analytic inside and on a closed contour C (taken in the positive sense) except at points $z_1, z_2, \ldots, z_n$ where $f(z)$ has singularities, then*

$$\oint_C f(z)\,dz = 2\pi i \sum_{j=1}^{n} \mathrm{Res}[f(z); z_j], \quad \mathbf{(1.8.3)}$$

where $\mathrm{Res}[f(z); z_j]$ denotes the residue of the jth isolated singularity of $f(z)$ located at $z = z_j$.

• **Example 1.8.1**

Let us compute $\oint_{|z|=2} z^2/(z+1)\,dz$ by the residue theorem, assuming that we take the contour in the positive sense.

Because the contour is a circle of radius 2, centered on the origin, the singularity at $z = -1$ lies within the contour. If the singularity were

[7] See Mitrinović, D. S. and Kečkić, J. D., 1984: *The Cauchy Method of Residues: Theory and Applications*, D. Reidel Publishing, Boston. Section 10.3 gives the historical development of the residue theorem.

not inside the contour, then the integrand would have been analytic inside and on the contour C. In this case, the answer would then be zero by the Cauchy-Goursat theorem.

Returning to the original problem, we construct the Laurent expansion for the integrand around the point $z = 1$ by noting that

$$\frac{z^2}{z+1} = \frac{[(z+1)-1]^2}{z+1} = \frac{1}{z+1} - 2 + (z+1). \tag{1.8.4}$$

The singularity at $z = -1$ is a simple pole and by inspection the value of the residue equals 1. Therefore,

$$\oint_{|z|=2} \frac{z^2}{z+1}\, dz = 2\pi i. \tag{1.8.5}$$

As it presently stands, it would appear that we must always construct a Laurent expansion for each singularity if we wish to use the residue theorem. This becomes increasingly difficult as the structure of the integrand becomes more complicated. In the following paragraphs we will show several techniques that avoid this problem in practice.

We begin by noting that many functions that we will encounter consist of the ratio of two *polynomials*, i.e., rational functions: $f(z) = g(z)/h(z)$. Generally, we can write $h(z)$ as $(z - z_1)^{m_1}(z - z_2)^{m_2} \cdots$. Here we have assumed that we have divided out any common factors between $g(z)$ and $h(z)$ so that $g(z)$ does not vanish at $z_1, z_2, \ldots$. Clearly $z_1, z_2, \ldots,$ are singularities of $f(z)$. Further analysis shows that the nature of the singularities are a pole of order m_1 at $z = z_1$, a pole of order m_2 at $z = z_2$, and so forth.

Having found the nature and location of the singularity, we compute the residue as follows. Suppose we have a pole of order n. Then we know that its Laurent expansion is

$$f(z) = \frac{a_n}{(z-z_0)^n} + \frac{a_{n-1}}{(z-z_0)^{n-1}} + \cdots + b_0 + b_1(z-z_0) + \cdots \tag{1.8.6}$$

Multiplying both sides of (1.8.6) by $(z - z_0)^n$,

$$\begin{aligned} F(z) &= (z-z_0)^n f(z) \\ &= a_n + a_{n-1}(z-z_0) + \cdots + b_0(z-z_0)^n + b_1(z-z_0)^{n+1} + \cdots \end{aligned} \tag{1.8.7}$$

Because $F(z)$ is analytic at $z = z_0$, it has the Taylor expansion

$$F(z) = F(z_0) + F'(z_0)(z-z_0) + \cdots + \frac{F^{(n-1)}(z_0)}{(n-1)!}(z-z_0)^{n-1} + \cdots \tag{1.8.8}$$

Matching powers of $z - z_0$ in (1.8.7) and (1.8.8), the residue equals

$$\operatorname{Res}[f(z); z_0] = a_1 = \frac{F^{(n-1)}(z_0)}{(n-1)!}. \tag{1.8.9}$$

Substituting in $F(z) = (z - z_0)^n f(z)$, we can compute the residue of a pole of order n by

$$\operatorname{Res}[f(z); z_j] = \frac{1}{(n-1)!} \lim_{z \to z_j} \frac{d^{n-1}}{dz^{n-1}} \left[(z - z_j)^n f(z) \right]. \tag{1.8.10}$$

For a simple pole (1.8.10) simplifies to

$$\operatorname{Res}[f(z); z_j] = \lim_{z \to z_j} (z - z_j) f(z). \tag{1.8.11}$$

Quite often, $f(z) = p(z)/q(z)$. From l'Hôspital's rule, it follows that

$$\operatorname{Res}[f(z); z_j] = \frac{p(z_j)}{q'(z_j)}. \tag{1.8.12}$$

Remember that these formulas work only for finite-order poles. For an essential singularity we must compute the residue from its Laurent expansion; however, essential singularities are very rare in applications.

• Example 1.8.2

Let us evaluate

$$\oint_C \frac{e^{iz}}{z^2 + a^2}\, dz, \tag{1.8.13}$$

where C is any contour that includes both $z = \pm ai$ and is in the positive sense.

From Cauchy's residue theorem,

$$\oint_C \frac{e^{iz}}{z^2 + a^2}\, dz = 2\pi i \left[\operatorname{Res}\left(\frac{e^{iz}}{z^2 + a^2}; ai \right) + \operatorname{Res}\left(\frac{e^{iz}}{z^2 + a^2}; -ai \right) \right]. \tag{1.8.14}$$

The singularities at $z = \pm ai$ are simple poles. The corresponding residues are

$$\operatorname{Res}\left(\frac{e^{iz}}{z^2+a^2}; ai\right) = \lim_{z\to ai}(z-ai)\frac{e^{iz}}{(z-ai)(z+ai)} = \frac{e^{-a}}{2ia} \tag{1.8.15}$$

and

$$\operatorname{Res}\left(\frac{e^{iz}}{z^2+a^2}; -ai\right) = \lim_{z\to -ai}(z+ai)\frac{e^{iz}}{(z-ai)(z+ai)} = -\frac{e^{a}}{2ia}. \tag{1.8.16}$$

Consequently,

$$\oint_C \frac{e^{iz}}{z^2+a^2}\,dz = -\frac{2\pi}{2a}\left(e^a - e^{-a}\right) = -\frac{2\pi}{a}\sinh(a). \tag{1.8.17}$$

• **Example 1.8.3**

Let us evaluate

$$\frac{1}{2\pi i}\oint_C \frac{e^{tz}}{z^2(z^2+2z+2)}\,dz, \tag{1.8.18}$$

where C includes all of the singularities and is in the positive sense.

The integrand has a second-order pole at $z = 0$ and two simple poles at $z = -1 \pm i$ which are the roots of $z^2 + 2z + 2 = 0$. Therefore, the residue at $z = 0$ is

$$\operatorname{Res}\left[\frac{e^{tz}}{z^2(z^2+2z+2)}; 0\right] = \lim_{z\to 0}\frac{1}{1!}\frac{d}{dz}\left\{(z-0)^2\left[\frac{e^{tz}}{z^2(z^2+2z+2)}\right]\right\} \tag{1.8.19}$$

$$= \lim_{z\to 0}\left[\frac{te^{tz}}{z^2+2z+2} - \frac{(2z+2)e^{tz}}{(z^2+2z+2)^2}\right] = \frac{t-1}{2}. \tag{1.8.20}$$

The residue at $z = -1+i$ is

$$\operatorname{Res}\left[\frac{e^{tz}}{z^2(z^2+2z+2)}; -1+i\right] = \lim_{z\to -1+i}[z-(-1+i)]\frac{e^{tz}}{z^2(z^2+2z+2)} \tag{1.8.21}$$

$$= \left(\lim_{z\to -1+i}\frac{e^{tz}}{z^2}\right)\left(\lim_{z\to -1+i}\frac{z+1-i}{z^2+2z+2}\right) \tag{1.8.22}$$

$$= \frac{\exp[(-1+i)t]}{2i(-1+i)^2} = \frac{\exp[(-1+i)t]}{4}. \tag{1.8.23}$$

Similarly, the residue at $z = -1 - i$ is

$$\operatorname{Res}\left[\frac{e^{tz}}{z^2(z^2+2z+2)}; -1-i\right] = \lim_{z\to -1-i}[z-(-1-i)]\frac{e^{tz}}{z^2(z^2+2z+2)} \tag{1.8.24}$$

$$= \left(\lim_{z\to -1-i}\frac{e^{tz}}{z^2}\right)\left(\lim_{z\to -1-i}\frac{z+1+i}{z^2+2z+2}\right) \tag{1.8.25}$$

$$= \frac{\exp[(-1-i)t]}{2i(-1-i)^2} = \frac{\exp[(-1-i)t]}{4}. \tag{1.8.26}$$

Then by the residue theorem,

$$\frac{1}{2\pi i}\oint_C \frac{e^{tz}}{z^2(z^2+2z+2)}\,dz = \operatorname{Res}\left[\frac{e^{tz}}{z^2(z^2+2z+2)}; 0\right] + \operatorname{Res}\left[\frac{e^{tz}}{z^2(z^2+2z+2)}; -1+i\right] + \operatorname{Res}\left[\frac{e^{tz}}{z^2(z^2+2z+2)}; -1-i\right] \tag{1.8.27}$$

$$= \frac{t-1}{2} + \frac{\exp[(-1+i)t]}{4} + \frac{\exp[(-1-i)t]}{4} \tag{1.8.28}$$

$$= \tfrac{1}{2}\left[t-1+e^{-t}\cos(t)\right]. \tag{1.8.29}$$

Problems

Assuming that all of the following closed contours are in the positive sense, use the residue theorem to evaluate the following integrals:

1. $\displaystyle\oint_{|z|=1}\frac{z+1}{z^4-2z^3}\,dz$

2. $\displaystyle\oint_{|z|=1}\frac{(z+4)^3}{z^4+5z^3+6z^2}\,dz$

3. $\displaystyle\oint_{|z|=1}\frac{1}{1-e^z}\,dz$

4. $\displaystyle\oint_{|z|=2}\frac{z^2-4}{(z-1)^4}\,dz$

5. $\displaystyle\oint_{|z|=2}\frac{z^3}{z^4-1}\,dz$

6. $\displaystyle\oint_{|z|=1}z^n e^{2/z}\,dz,\quad n>0$

7. $\displaystyle\oint_{|z|=1}e^{1/z}\cos(1/z)\,dz$

8. $\displaystyle\oint_{|z|=2}\frac{2+4\cos(\pi z)}{z(z-1)^2}\,dz$

1.9 EVALUATION OF REAL DEFINITE INTEGRALS

One of the important applications of the theory of residues consists in the evaluation of certain types of real definite integrals. Similar techniques apply when the integrand contains a sine or cosine. See Section 3.4.

• Example 1.9.1

Let us evaluate the integral

$$\int_0^\infty \frac{dx}{x^2+1} = \frac{1}{2}\int_{-\infty}^\infty \frac{dx}{x^2+1}. \tag{1.9.1}$$

This integration occurs along the real axis. In terms of complex variables we can rewrite (1.9.1) as

$$\int_0^\infty \frac{dx}{x^2+1} = \frac{1}{2}\int_{C_1} \frac{dz}{z^2+1}, \tag{1.9.2}$$

where the contour C_1 is the line $\text{Im}(z) = 0$. However, the use of the residue theorem requires an integration along a closed contour. Let us choose the one pictured in Figure 1.9.1. Then

$$\oint_C \frac{dz}{z^2+1} = \int_{C_1} \frac{dz}{z^2+1} + \int_{C_2} \frac{dz}{z^2+1}, \tag{1.9.3}$$

where C denotes the complete closed contour and C_2 denotes the integration path along a semicircle at infinity. Clearly we want the second integral on the right side of (1.9.3) to vanish; otherwise, our choice of the contour C_2 is poor. Because $z = Re^{\theta i}$ and $dz = iRe^{\theta i}\,d\theta$,

$$\left|\int_{C_2} \frac{dz}{z^2+1}\right| = \left|\int_0^\pi \frac{iR\exp(\theta i)\,d\theta}{1+R^2\exp(2\theta i)}\right| \le \int_0^\pi \frac{R\,d\theta}{R^2-1}, \tag{1.9.4}$$

which tends to zero as $R \to \infty$. On the other hand, the residue theorem gives

$$\oint_C \frac{dz}{z^2+1} = 2\pi i \,\text{Res}\left(\frac{1}{z^2+1}; i\right) = 2\pi i \lim_{z\to i} \frac{z-i}{z^2+1} = 2\pi i \times \frac{1}{2i} = \pi. \tag{1.9.5}$$

Therefore,

$$\int_0^\infty \frac{dx}{x^2+1} = \frac{\pi}{2}. \tag{1.9.6}$$

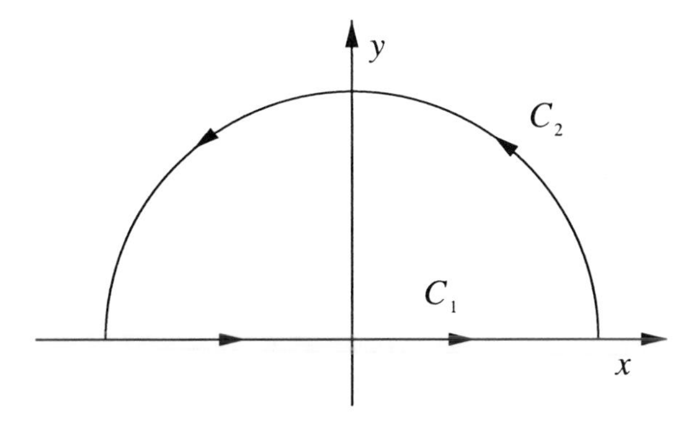

Figure 1.9.1: Contour used in evaluating the integral (1.9.1).

Note that we only evaluated the residue in the upper half-plane because it is the only one inside the contour.

This example illustrates the basic concepts of evaluating definite integrals by the residue theorem. We introduce a closed contour that includes the real axis and an additional contour. We must then evaluate the integral along this additional contour as well as the closed contour integral. If we have properly chosen our closed contour, this additional integral will vanish. For certain classes of general integrals, we shall now show that this additional contour is a circular arc at infinity.

Theorem: *If, on a circular arc C_R with a radius R and center at the origin, $zf(z) \to 0$ uniformly with $|z| \in C_R$ and as $R \to \infty$, then*

$$\lim_{R\to\infty} \int_{C_R} f(z)\, dz = 0. \tag{1.9.7}$$

This follows from the fact that if $|zf(z)| \leq M_R$, then $|f(z)| \leq M_R/R$. Because the length of C_R is αR, where α is the subtended angle,

$$\left| \int_{C_R} f(z)\, dz \right| \leq \frac{M_R}{R}\, \alpha R = \alpha M_R \to 0, \tag{1.9.8}$$

because $M_R \to 0$ as $R \to \infty$. □

• **Example 1.9.2**

A simple illustration of this theorem is the integral

$$\int_{-\infty}^{\infty} \frac{dx}{x^2+x+1} = \int_{C_1} \frac{dz}{z^2+z+1}. \tag{1.9.9}$$

A quick check shows that $z/(z^2+z+1)$ tends to zero uniformly as $R \to \infty$. Therefore, if we use the contour pictured in Figure 1.9.1,

$$\int_{-\infty}^{\infty} \frac{dx}{x^2+x+1} = \oint_C \frac{dz}{z^2+z+1} = 2\pi i \operatorname{Res}\left(\frac{1}{z^2+z+1}; -\tfrac{1}{2}+\tfrac{\sqrt{3}}{2}i\right) \tag{1.9.10}$$

$$= 2\pi i \lim_{z \to -\frac{1}{2}+\frac{\sqrt{3}}{2}i} \left(\frac{1}{2z+1}\right) = \frac{2\pi}{\sqrt{3}}. \tag{1.9.11}$$

• **Example 1.9.3**

Let us evaluate

$$\int_0^{\infty} \frac{dx}{x^6+1}. \tag{1.9.12}$$

In place of an infinite semicircle in the upper half-plane, consider the following integral

$$\oint_C \frac{dz}{z^6+1}, \tag{1.9.13}$$

where we show the closed contour in Figure 1.9.2. We chose this contour for two reasons. First, we only have to evaluate one residue rather than the three enclosed in a traditional upper half-plane contour. Second, the contour integral along C_3 simplifies to a particularly simple and useful form.

Because the only enclosed singularity lies at $z = e^{\pi i/6}$,

$$\oint_C \frac{dz}{z^6+1} = 2\pi i \operatorname{Res}\left(\frac{1}{z^6+1}; e^{\pi i/6}\right) = 2\pi i \lim_{z \to e^{\pi i/6}} \frac{z - e^{\pi i/6}}{z^6+1} \tag{1.9.14}$$

$$= 2\pi i \lim_{z \to e^{\pi i/6}} \frac{1}{6z^5} = -\frac{\pi i}{3} e^{\pi i/6}. \tag{1.9.15}$$

Let us now evaluate (1.9.12) along each of the legs of the contour:

$$\int_{C_1} \frac{dz}{z^6+1} = \int_0^{\infty} \frac{dx}{x^6+1}, \tag{1.9.16}$$

$$\int_{C_2} \frac{dz}{z^6+1} = 0, \tag{1.9.17}$$

because of (1.9.7) and

$$\int_{C_3} \frac{dz}{z^6+1} = \int_{\infty}^{0} \frac{e^{\pi i/3}\,dr}{r^6+1} = -e^{\pi i/3} \int_0^{\infty} \frac{dx}{x^6+1}, \tag{1.9.18}$$

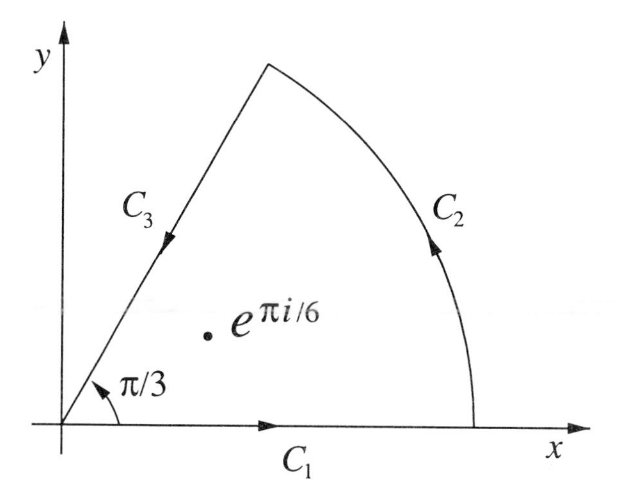

Figure 1.9.2: Contour used in evaluating the integral (1.9.13).

since $z = re^{\pi i/3}$.

Substituting into (1.9.15),

$$\left(1 - e^{\pi i/3}\right) \int_0^\infty \frac{dx}{x^6+1} = -\frac{\pi i}{3} e^{\pi i/6} \tag{1.9.19}$$

or

$$\int_0^\infty \frac{dx}{x^6+1} = \frac{\pi i}{6} \frac{2ie^{\pi i/6}}{e^{\pi i/6}\left(e^{\pi i/6} - e^{-\pi i/6}\right)} = \frac{\pi}{6\sin(\pi/6)} = \frac{\pi}{3}. \tag{1.9.20}$$

Problems

Use the residue theorem to verify the following integral:

1.
$$\int_0^\infty \frac{dx}{x^4+1} = \frac{\pi\sqrt{2}}{4}$$

2.
$$\int_{-\infty}^\infty \frac{dx}{(x^2+4x+5)^2} = \frac{\pi}{2}$$

3.
$$\int_{-\infty}^\infty \frac{x\,dx}{(x^2+1)(x^2+2x+2)} = -\frac{\pi}{5}$$

4.
$$\int_0^\infty \frac{x^2}{x^6+1}\,dx = \frac{\pi}{6}$$

5.

$$\int_0^\infty \frac{dx}{(x^2+1)^2} = \frac{\pi}{4}$$

6.

$$\int_0^\infty \frac{dx}{(x^2+1)(x^2+4)^2} = \frac{5\pi}{288}$$

7.

$$\int_0^\infty \frac{t^2}{(t^2+1)[t^2(a/h+1)+(a/h-1)]}\,dt = \frac{\pi}{4}\left[1-\sqrt{\frac{1-h/a}{1+h/a}}\right],$$

where $h/a < 1$.

8. During an electromagnetic calculation, Strutt[8] needed to prove that

$$\pi\frac{\sinh(\sigma x)}{\cosh(\sigma\pi)} = 2\sigma\sum_{n=0}^{\infty}\frac{\cos\left[\left(n+\frac{1}{2}\right)(x-\pi)\right]}{\sigma^2+\left(n+\frac{1}{2}\right)^2}, \qquad |x|\le\pi.$$

Verify his proof.

Step 1: Using the residue theorem, show that

$$\frac{1}{2\pi i}\oint_{C_N}\pi\frac{\sinh(xz)}{\cosh(\pi z)}\frac{dz}{z-\sigma} = \pi\frac{\sinh(\sigma x)}{\cosh(\sigma\pi)} - \sum_{n=-N-1}^{N}\frac{(-1)^n\sin\left[\left(n+\frac{1}{2}\right)x\right]}{\sigma-i\left(n+\frac{1}{2}\right)},$$

where C_N is a circular contour that includes the poles $z=\sigma$ and $z_n = \pm i\left(n+\frac{1}{2}\right)$, $n=0,1,2,\ldots,N$.

Step 2: Show that in the limit of $N\to\infty$, the contour integral vanishes. [Hint: Examine the behavior of $z\sinh(xz)/[(z-\sigma)\cosh(\pi z)]$ as $|z|\to\infty$. Use (1.9.7) where C_R is the circular contour.]

Step 3: Break the infinite series in Step 1 into two parts and simplify.

In the next chapter we shall show how we can obtain the same series by direct integration.

---

[8] Strutt, M. J. O., 1934: Berechnung des hochfrequenten Feldes einer Kreiszylinderspule in einer konzentrischen leitenden Schirmhülle mit ebenen Deckeln. *Hochfrequenztechn. Elecktroak.*, **43**, 121–123.

Chapter 2
Fourier Series

Fourier series arose during the eighteenth century as a formal solution to the classic wave equation. Later on, it was used to describe physical processes in which events recur in a regular pattern. For example, a musical note usually consists of a simple note, called the fundamental, and a series of auxiliary vibrations, called overtones. Fourier's theorem provides the mathematical language which allows us to precisely describe this complex structure.

2.1 FOURIER SERIES

One of the crowning glories[1] of nineteenth century mathematics

[1] "Fourier's Theorem ... is not only one of the most beautiful results of modern analysis, but may be said to furnish an indispensable instrument in the treatment of nearly every recondite question in modern physics. To mention only sonorous vibrations, the propagation of electric signals along a telegraph wire, and the conduction of heat by the earth's crust, as subjects in their generality intractable without it, is to give but a feeble idea of its importance." (Quote taken from Thomson, W. and Tait, P. G., 1879: *Treatise on Natural Philosophy, Part I,* Cambridge University Press, Cambridge, Section 75.)

was the discovery that the infinite series

$$f(t) = \frac{a_0}{2} + \sum_{n=1}^{\infty} a_n \cos\left(\frac{n\pi t}{L}\right) + b_n \sin\left(\frac{n\pi t}{L}\right) \qquad \textbf{(2.1.1)}$$

can represent a function $f(t)$ under certain general conditions. This series, called a *Fourier series*, converges to the value of the function $f(t)$ at every point in the interval $[-L, L]$ with the possible exceptions of the points at any discontinuities and the endpoints of the interval. Because each term has a period of $2L$, the sum of the series also has the same period. The *fundamental* of the periodic function $f(t)$ is the $n = 1$ term while the *harmonics* are the remaining terms whose frequencies are integer multiples of the fundamental.

We must now find some easy method for computing the a_n's and b_n's for a given function $f(t)$. As a first attempt, we integrate (2.1.1) term by term[2] from $-L$ to L. On the right side, all of the integrals multiplied by a_n and b_n vanish because the average of $\cos(n\pi t/L)$ and $\sin(n\pi t/L)$ is zero. Therefore, we are left with

$$\boxed{a_0 = \frac{1}{L}\int_{-L}^{L} f(t)\,dt.} \qquad \textbf{(2.1.2)}$$

Consequently a_0 is twice the mean value of $f(t)$ over one period.

We next multiply each side of (2.1.1) by $\cos(m\pi t/L)$, where m is a fixed integer. Integrating from $-L$ to L,

$$\begin{aligned}\int_{-L}^{L} f(t)\cos\left(\frac{m\pi t}{L}\right) dt &= \frac{a_0}{2}\int_{-L}^{L} \cos\left(\frac{m\pi t}{L}\right) dt \\ &+ \sum_{n=1}^{\infty} a_n \int_{-L}^{L} \cos\left(\frac{n\pi t}{L}\right) \cos\left(\frac{m\pi t}{L}\right) dt \\ &+ \sum_{n=1}^{\infty} b_n \int_{-L}^{L} \sin\left(\frac{n\pi t}{L}\right) \cos\left(\frac{m\pi t}{L}\right) dt. \end{aligned} \qquad \textbf{(2.1.3)}$$

The a_0 and b_n terms vanish by direct integration. Finally all of the a_n

[2] We assume that the integration of the series can be carried out term by term. This is sometimes difficult to justify but we do it anyway.

integrals vanish when $n \neq m$. Consequently, (2.1.3) simplifies to

$$a_n = \frac{1}{L}\int_{-L}^{L} f(t)\cos\left(\frac{n\pi t}{L}\right)\,dt, \tag{2.1.4}$$

because $\int_{-L}^{L}\cos^2(n\pi t/L)\,dt = L$. Finally, by multiplying both sides of (2.1.1) by $\sin(m\pi t/L)$ (m is again a fixed integer) and integrating from $-L$ to L,

$$b_n = \frac{1}{L}\int_{-L}^{L} f(t)\sin\left(\frac{n\pi t}{L}\right)\,dt. \tag{2.1.5}$$

Although (2.1.2), (2.1.4), and (2.1.5) give us a_0, a_n, and b_n for periodic functions over the interval $[-L, L]$, in certain situations it is convenient to use the interval $[\tau, \tau + 2L]$, where τ is any real number. In that case, (2.1.1) still gives the Fourier series of $f(t)$ and

$$\begin{aligned} a_0 &= \frac{1}{L}\int_{\tau}^{\tau+2L} f(t)\,dt, \\ a_n &= \frac{1}{L}\int_{\tau}^{\tau+2L} f(t)\cos\left(\frac{n\pi t}{L}\right)\,dt, \\ b_n &= \frac{1}{L}\int_{\tau}^{\tau+2L} f(t)\sin\left(\frac{n\pi t}{L}\right)\,dt. \end{aligned} \tag{2.1.6}$$

These results follow when we recall that the function $f(t)$ is a periodic function that extends from minus infinity to plus infinity. The results must remain unchanged, therefore, when we shift from the interval $[-L, L]$ to the new interval $[\tau, \tau + 2L]$.

We now ask the question: what types of functions have Fourier series? Secondly, if a function is discontinuous at a point, what value will the Fourier series give? Dirichlet[3,4] answered these questions in the first half of the nineteenth century. He showed that if any arbitrary function is finite over one period and has a finite number of maxima and minima, then the Fourier series converges. If $f(t)$ is discontinuous at the point t and has two different values at $f(t^-)$ and $f(t^+)$, where t^+ and t^- are points infinitesimally to the right and left of t, the Fourier series converges to the mean value of $[f(t^+) + f(t^-)]/2$. Because *Dirichlet's conditions* are very mild, it is very rare that a convergent Fourier series does not exist for a function that appears in an engineering or scientific problem.

• Example 2.1.1

Let us find the Fourier series for the function

$$f(t) = \begin{cases} 0, & -\pi < t \le 0 \\ t, & 0 \le t < \pi. \end{cases} \tag{2.1.7}$$

We compute the Fourier coefficients a_n and b_n using (2.1.6) by letting $L = \pi$ and $\tau = -\pi$. We then find that

$$a_0 = \frac{1}{\pi}\int_{-\pi}^{\pi} f(t)\,dt = \frac{1}{\pi}\int_0^{\pi} t\,dt = \frac{\pi}{2}, \tag{2.1.8}$$

$$a_n = \frac{1}{\pi}\int_0^{\pi} t\cos(nt)\,dt = \frac{1}{\pi}\left[\frac{t\sin(nt)}{n} + \frac{\cos(nt)}{n^2}\right]\Bigg|_0^{\pi} \tag{2.1.9}$$

$$= \frac{\cos(n\pi) - 1}{n^2\pi} = \frac{(-1)^n - 1}{n^2\pi} \tag{2.1.10}$$

because $\cos(n\pi) = (-1)^n$ and

$$b_n = \frac{1}{\pi}\int_0^{\pi} t\sin(nt)\,dt = \frac{1}{\pi}\left[\frac{-t\cos(nt)}{n} + \frac{\sin(nt)}{n^2}\right]\Bigg|_0^{\pi} \tag{2.1.11}$$

$$= -\frac{\cos(n\pi)}{n} = \frac{(-1)^{n+1}}{n} \tag{2.1.12}$$

---

[3] Dirichlet, P. G. L., 1829: Sur la convergence des séries trigonométriques qui servent à représenter une fonction arbitraire entre des limites données. *J. reine angew. Math.*, **4**, 157–169.

[4] Dirichlet, P. G. L., 1837: Sur l'usage des intégrales définies dans la sommation des séries finies ou infinies. *J. reine angew. Math.*, **17**, 57–67.

Figure 2.1.1: A product of the French Revolution, (Jean Baptiste) Joseph Fourier (1768–1830) held positions within the Napoleonic Empire during his early career. After Napoleon's fall from power, Fourier devoted his talents exclusively to science. Although he won the Institut de France prize in 1811 for his work on heat diffusion, criticism of its mathematical rigor and generality led him to publish the classic book *Théorie analytique de la chaleur* in 1823. Within this book he introduced the world to the series that bears his name. (Portrait courtesy of the Archives de l'Académie des sciences, Paris.)

for $n = 1, 2, 3, \ldots$. Thus, the Fourier series for $f(t)$ is

$$f(t) = \frac{\pi}{4} + \sum_{n=1}^{\infty} \frac{(-1)^n - 1}{n^2\pi} \cos(nt) + \frac{(-1)^{n+1}}{n} \sin(nt) \qquad \textbf{(2.1.13)}$$

$$= \frac{\pi}{4} - \frac{2}{\pi} \sum_{m=1}^{\infty} \frac{\cos[(2m-1)t]}{(2m-1)^2} - \sum_{n=1}^{\infty} \frac{(-1)^n}{n} \sin(nt). \qquad \textbf{(2.1.14)}$$

Figure 2.1.2: Second to Gauss, Peter Gustav Lejeune Dirichlet (1805–1859) was Germany's leading mathematician during the first half of the nineteenth century. Initially drawn to number theory, his later studies in analysis and applied mathematics led him to consider the convergence of Fourier series. These studies eventually produced the modern concept of a function as a correspondence that associates with each real x in an interval some unique value denoted by $f(x)$. (Taken from the frontispiece of Dirichlet, P. G. L., 1889: *Werke.* Druck und Verlag von Georg Reimer, Berlin, 644 pp.)

We note that at the points $t = \pm(2n-1)\pi$, where $n = 1, 2, 3, \ldots$, the function jumps from zero to π. To what value does the Fourier series converge at these points? From Dirichlet's conditions, the series converges to the average of the values of the function just to the right and left of the point of discontinuity, i.e., $(\pi + 0)/2 = \pi/2$. At the remaining points the series converges to $f(t)$.

In Figure 2.1.3 we show how well (2.1.13) approximates the function by graphing various partial sums of (2.1.13) as we include more and more

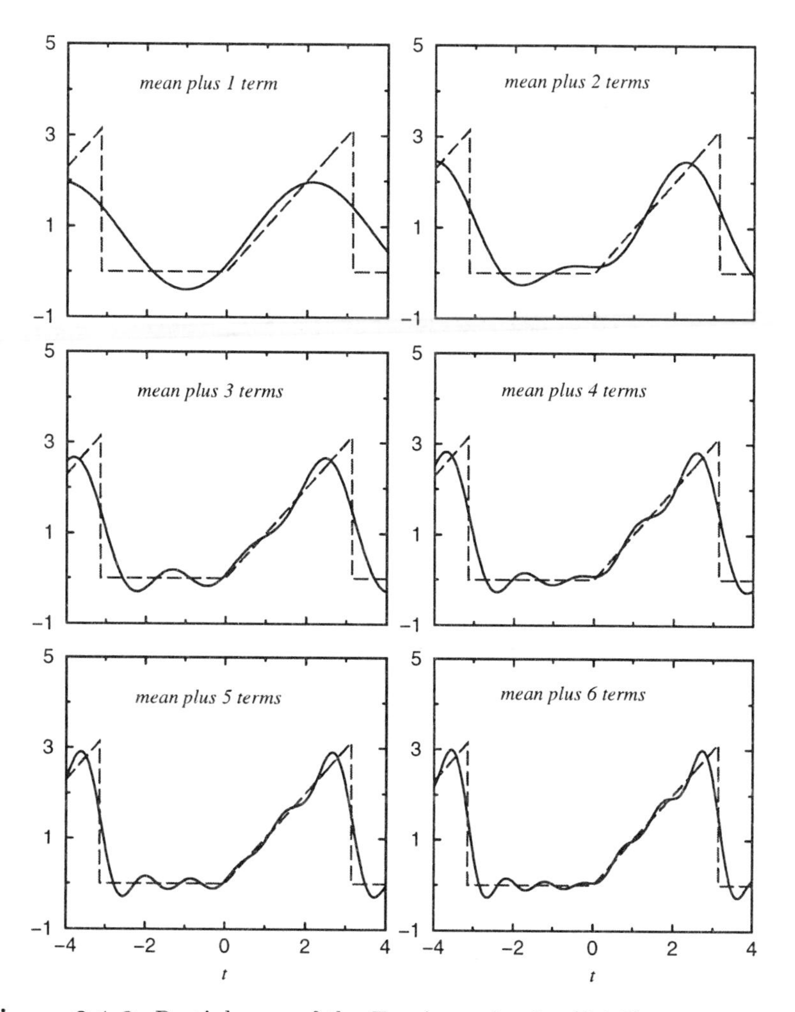

Figure 2.1.3: Partial sum of the Fourier series for (2.1.7).

terms (harmonics). As the figure shows, successive corrections are made to the mean value of the series, $\pi/2$. As each harmonic is added, the Fourier series fits the function better in the sense of least squares:

$$\int_{\tau}^{\tau+2L} [f(x) - f_N(x)]^2 \, dx = \text{ minimum}, \tag{2.1.15}$$

where $f_N(x)$ is the truncated Fourier series of N terms.

• **Example 2.1.2**

Let us calculate the Fourier series of the function $f(t) = |t|$ which is defined over the range $-\pi \leq t \leq \pi$.

From the definition of the Fourier coefficients,

$$a_0 = \frac{1}{\pi}\left[\int_{-\pi}^{0} -t\,dt + \int_0^{\pi} t\,dt\right] = \frac{\pi}{2} + \frac{\pi}{2} = \pi, \tag{2.1.16}$$

$$a_n = \frac{1}{\pi}\left[\int_{-\pi}^{0} -t\cos(nt)\,dt + \int_0^{\pi} t\cos(nt)\,dt\right] \tag{2.1.17}$$

$$= -\left.\frac{nt\sin(nt)+\cos(nt)}{n^2\pi}\right|_{-\pi}^{0} + \left.\frac{nt\sin(nt)+\cos(nt)}{n^2\pi}\right|_{0}^{\pi} \tag{2.1.18}$$

$$= \frac{2}{n^2\pi}[(-1)^n - 1] \tag{2.1.19}$$

and

$$b_n = \frac{1}{\pi}\left[\int_{-\pi}^{0} -t\sin(nt)\,dt + \int_0^{\pi} t\sin(nt)\,dt\right] \tag{2.1.20}$$

$$= \left.\frac{nt\cos(nt)-\sin(nt)}{n^2\pi}\right|_{-\pi}^{0} - \left.\frac{nt\cos(nt)-\sin(nt)}{n^2\pi}\right|_{0}^{\pi} = 0 \tag{2.1.21}$$

for $n = 1, 2, 3, \ldots$. Therefore,

$$|t| = \frac{\pi}{2} + \frac{2}{\pi}\sum_{n=1}^{\infty}\frac{[(-1)^n - 1]}{n^2}\cos(nt) = \frac{\pi}{2} - \frac{4}{\pi}\sum_{m=1}^{\infty}\frac{\cos[(2m-1)t]}{(2m-1)^2} \tag{2.1.22}$$

for $-\pi \le t \le \pi$.

In Figure 2.1.4 we show how well (2.1.22) approximates the function by graphing various partial sums of (2.1.22). As the figure shows, the Fourier series does very well even when we use very few terms. The reason for this rapid convergence is the nature of the function: it does not possess any jump discontinuities.

• Example 2.1.3

Sometimes the function $f(t)$ is an even or odd function.[5] Can we use this property to simplify our work? The answer is yes.

Let $f(t)$ be an even function. Then

$$a_0 = \frac{1}{L}\int_{-L}^{L} f(t)\,dt = \frac{2}{L}\int_0^{L} f(t)\,dt \tag{2.1.23}$$

---

[5] An even function $f_e(t)$ has the property that $f_e(-t) = f_e(t)$; an odd function $f_o(t)$ has the property that $f_o(-t) = -f_o(t)$.

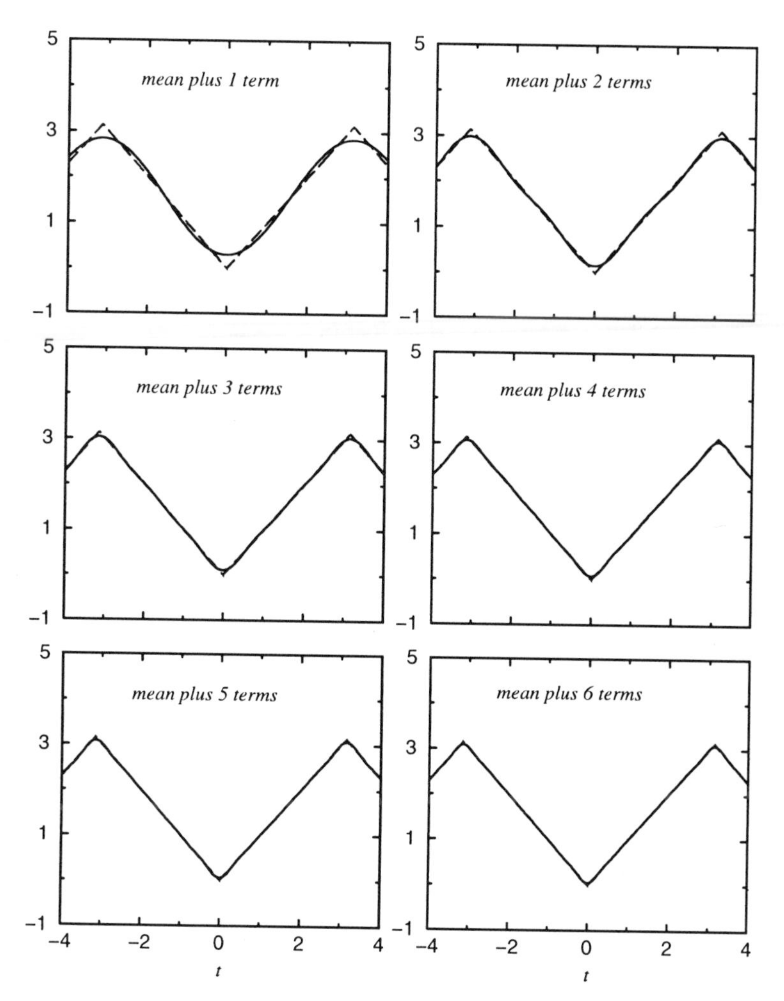

Figure 2.1.4: Partial sum of the Fourier series for $f(t) = |t|$.

and

$$a_n = \frac{1}{L}\int_{-L}^{L} f(t)\cos\left(\frac{n\pi t}{L}\right)\,dt = \frac{2}{L}\int_{0}^{L} f(t)\cos\left(\frac{n\pi t}{L}\right)\,dt \quad \textbf{(2.1.24)}$$

whereas

$$b_n = \frac{1}{L}\int_{-L}^{L} f(t)\sin\left(\frac{n\pi t}{L}\right)\,dt = 0. \quad \textbf{(2.1.25)}$$

Here we have used the properties that $\int_{-L}^{L} f_e(x)\,dx = 2\int_0^L f_e(x)\,dx$ and $\int_{-L}^{L} f_o(x)\,dx = 0$. Thus, if we have an even function, we merely compute a_0 and a_n via (2.1.23)–(2.1.24) and $b_n = 0$. Because the corresponding series contains only cosine terms, it is often called a *Fourier cosine series*.

Similarly, if $f(t)$ is odd, then

$$a_0 = a_n = 0 \quad \text{and} \quad b_n = \frac{2}{L}\int_0^L f(t)\sin\left(\frac{n\pi t}{L}\right)dt. \tag{2.1.26}$$

Thus, if we have an odd function, we merely compute b_n via (2.1.26) and $a_0 = a_n = 0$. Because corresponding series contains only sine terms, it is often called a *Fourier sine series.*

• Example 2.1.4

In the case when $f(x)$ consists of a constant and/or trigonometric functions, it is much easier to find the corresponding Fourier series by inspection rather than by using (2.1.6). For example, let us find the Fourier series for $f(x) = \sin^2(x)$ defined over the range $-\pi \le x \le \pi$.

We begin by rewriting $f(x) = \sin^2(x)$ as $f(x) = \frac{1}{2}[1 - \cos(2x)]$. Next, we note that any function defined over the range $-\pi < x < \pi$ has the Fourier series

$$f(x) = \frac{a_0}{2} + \sum_{n=1}^{\infty} a_n\cos(nx) + b_n\sin(nx) \tag{2.1.27}$$

$$= \frac{a_0}{2} + a_1\cos(x) + b_1\sin(x) + a_2\cos(2x) + b_2\sin(2x) + \cdots \tag{2.1.28}$$

On the other hand,

$$f(x) = \tfrac{1}{2} - \tfrac{1}{2}\cos(2x) \tag{2.1.29}$$

$$= \tfrac{1}{2} + 0\cos(x) + 0\sin(x) - \tfrac{1}{2}\cos(2x) + 0\sin(2x) + \cdots \tag{2.1.30}$$

Consequently, by inspection, we can immediately write that

$$a_0 = 1, a_1 = b_1 = 0, a_2 = -\tfrac{1}{2}, b_2 = 0, a_n = b_n = 0, n \ge 3. \tag{2.1.31}$$

Thus, instead of the usual expansion involving an infinite number of sine and cosine terms, our Fourier series contains only two terms and is simply

$$f(x) = \tfrac{1}{2} - \tfrac{1}{2}\cos(2x), \qquad -\pi \le x \le \pi. \tag{2.1.32}$$

• Example 2.1.5: Quieting snow tires

An application of Fourier series to a problem in industry occurred several years ago, when drivers found that snow tires produced a loud whine[6] on dry pavement. Tire sounds are produced primarily by the dynamic interaction of the tread elements with the road surface.[7] As each tread element passes through the contact patch, it contributes a pulse of acoustic energy to the total sound field radiated by the tire.

For evenly spaced treads we envision that the release of acoustic energy resembles the top of Figure 2.1.5. If we perform a Fourier analysis of this distribution, we find that

$$a_0 = \frac{1}{\pi}\left[\int_{-\pi/2-\epsilon}^{-\pi/2+\epsilon} 1\,dt + \int_{\pi/2-\epsilon}^{\pi/2+\epsilon} 1\,dt\right] = \frac{4\epsilon}{\pi}, \tag{2.1.33}$$

where ϵ is half of the width of the tread and

$$a_n = \frac{1}{\pi}\left[\int_{-\pi/2-\epsilon}^{-\pi/2+\epsilon} \cos(nt)\,dt + \int_{\pi/2-\epsilon}^{\pi/2+\epsilon} \cos(nt)\,dt\right] \tag{2.1.34}$$

$$= \frac{1}{n\pi}\left[\sin(nt)\Big|_{-\pi/2-\epsilon}^{-\pi/2+\epsilon} + \sin(nt)\Big|_{\pi/2-\epsilon}^{\pi/2+\epsilon}\right] \tag{2.1.35}$$

$$= \frac{1}{n\pi}\left[\sin\left(-\frac{n\pi}{2}+n\epsilon\right) - \sin\left(-\frac{n\pi}{2}-n\epsilon\right) + \sin\left(\frac{n\pi}{2}+n\epsilon\right) - \sin\left(\frac{n\pi}{2}-n\epsilon\right)\right] \tag{2.1.36}$$

$$= \frac{1}{n\pi}\left[2\cos\left(-\frac{n\pi}{2}\right) + 2\cos\left(\frac{n\pi}{2}\right)\right]\sin(n\epsilon) \tag{2.1.37}$$

$$= \frac{4}{n\pi}\cos\left(\frac{n\pi}{2}\right)\sin(n\epsilon). \tag{2.1.38}$$

Because $f(t)$ is an even function, $b_n = 0$.

The question now arises of how to best illustrate our Fourier coefficients. In Section 2.4 we will show that any harmonic can be represented as a single wave $A_n \cos(n\pi t/L + \varphi_n)$ or $A_n \sin(n\pi t/L + \psi_n)$, where the amplitude $A_n = \sqrt{a_n^2 + b_n^2}$. At the bottom of Figure 2.1.5, we have plotted this amplitude, usually called the *amplitude* or *frequency spectrum* $\frac{1}{2}\sqrt{a_n^2 + b_n^2}$, as a function of n for an arbitrarily chosen $\epsilon = \pi/12$. Although the value of ϵ will affect the exact shape of the spectrum, the qualitative arguments that we will present remain unchanged. We have added the factor $\frac{1}{2}$ so that our definition of the frequency spectrum is consistent with that for a complex Fourier series stated after (2.5.15). The amplitude spectrum in Figure 2.1.5 shows that the spectrum for periodically placed tire treads has its largest amplitude at small

---

[6] Information based on Varterasian, J. H., 1969: Math quiets rotating machines. *SAE J.*, **77(10)**, 53.

[7] Willett, P. R., 1975: Tire tread pattern sound generation. *Tire Sci. Tech.*, **3**, 252–266.

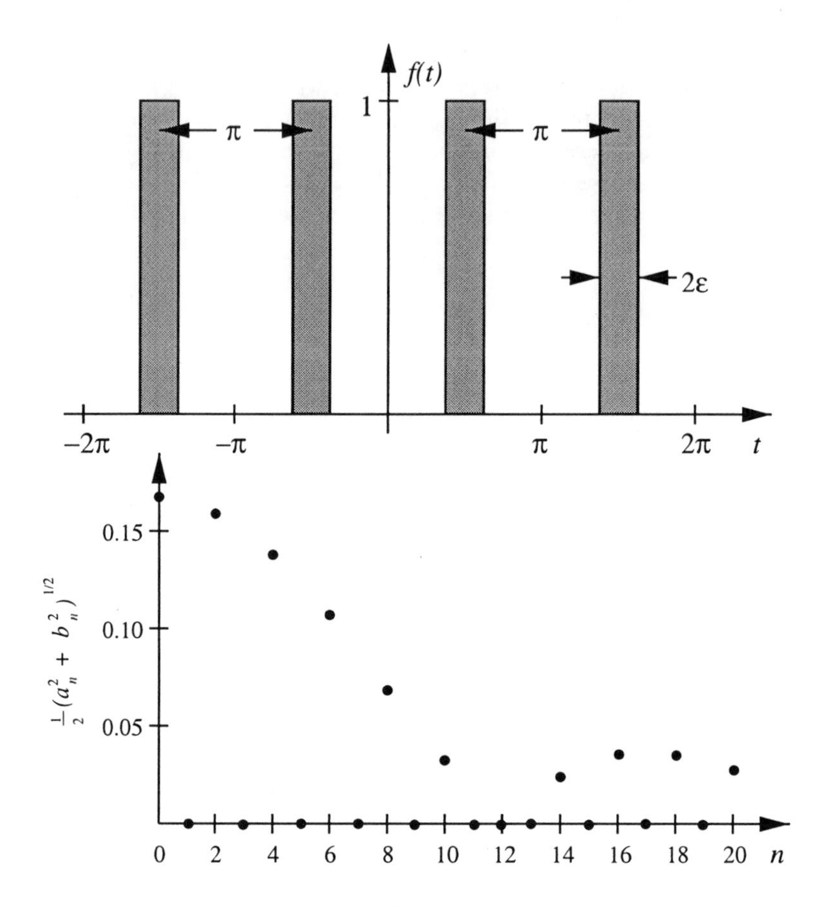

Figure 2.1.5: Temporal spacing (over two periods) and frequency spectrum of a uniformly spaced snow tire.

n. This produces one loud tone plus strong harmonic overtones because the $n = 1$ term (the fundamental) and its overtones are the dominant terms in the Fourier series representation.

Clearly this loud, monotone whine is undesirable. How might we avoid it? Just as soldiers marching in step produce a loud uniform sound, we suspect that our uniform tread pattern is the problem. Therefore, let us now vary the interval between the treads so that the distance between any tread and its nearest neighbor is not equal. Figure 2.1.6 illustrates a simple example. Again we perform a Fourier analysis and obtain that

$$a_0 = \frac{1}{\pi}\left[\int_{-\pi/2-\epsilon}^{-\pi/2+\epsilon} 1\, dt + \int_{\pi/4-\epsilon}^{\pi/4+\epsilon} 1\, dt\right] = \frac{4\epsilon}{\pi}, \tag{2.1.39}$$

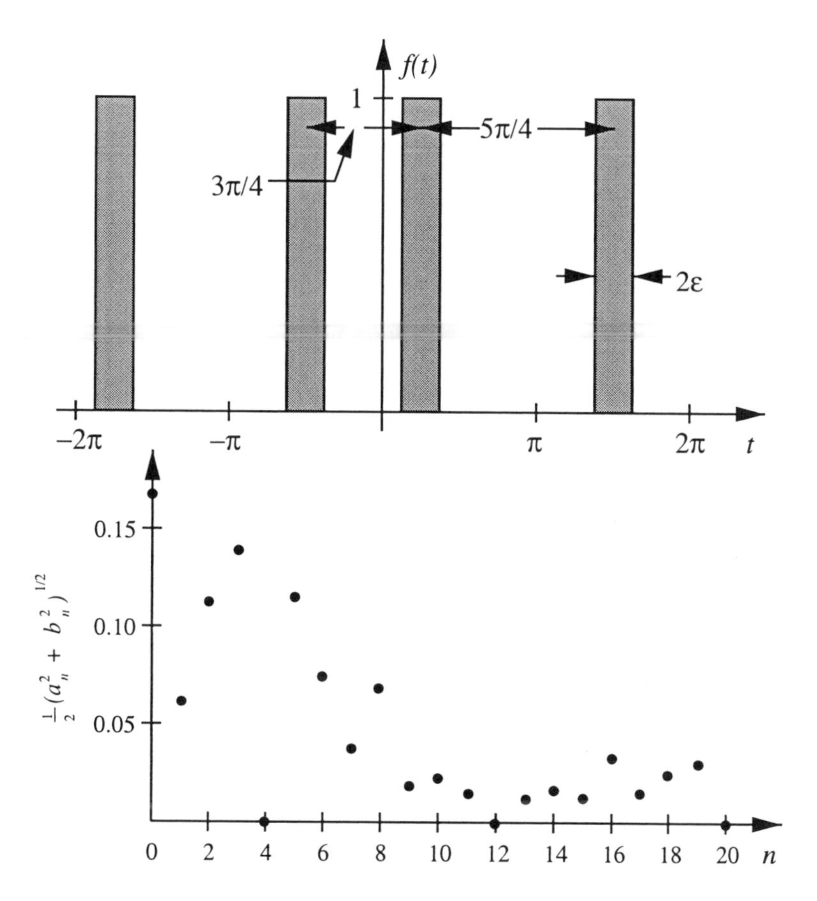

Figure 2.1.6: Temporal spacing and frequency spectrum of a nonuniformly spaced snow tire.

$$a_n = \frac{1}{\pi}\left[\int_{-\pi/2-\epsilon}^{-\pi/2+\epsilon} \cos(nt)\,dt + \int_{\pi/4-\epsilon}^{\pi/4+\epsilon} \cos(nt)\,dt\right] \tag{2.1.40}$$

$$= \frac{1}{n\pi}\sin(nt)\Big|_{-\pi/2-\epsilon}^{-\pi/2+\epsilon} + \frac{1}{n\pi}\sin(nt)\Big|_{\pi/4-\epsilon}^{\pi/4+\epsilon} \tag{2.1.41}$$

$$= -\frac{1}{n\pi}\left[\sin\left(\frac{n\pi}{2}-n\epsilon\right) - \sin\left(\frac{n\pi}{2}+n\epsilon\right)\right] + \frac{1}{n\pi}\left[\sin\left(\frac{n\pi}{4}+n\epsilon\right) - \sin\left(\frac{n\pi}{4}-n\epsilon\right)\right] \tag{2.1.42}$$

$$= \frac{2}{n\pi}\left[\cos\left(\frac{n\pi}{2}\right) + \cos\left(\frac{n\pi}{4}\right)\right]\sin(n\epsilon) \tag{2.1.43}$$

and

$$b_n = \frac{1}{\pi}\left[\int_{-\pi/2-\epsilon}^{-\pi/2+\epsilon} \sin(nt)\,dt + \int_{\pi/4-\epsilon}^{\pi/4+\epsilon} \sin(nt)\,dt\right] \tag{2.1.44}$$

$$= -\frac{1}{n\pi}\left[\cos\left(\frac{n\pi}{2} - n\epsilon\right) - \cos\left(\frac{n\pi}{2} + n\epsilon\right)\right]$$

$$- \frac{1}{n\pi}\left[\cos\left(\frac{n\pi}{4} + n\epsilon\right) - \cos\left(\frac{n\pi}{4} - n\epsilon\right)\right] \tag{2.1.45}$$

$$= \frac{2}{n\pi}\left[\sin\left(\frac{n\pi}{4}\right) - \sin\left(\frac{n\pi}{2}\right)\right]\sin(n\epsilon). \tag{2.1.46}$$

Figure 2.1.6 illustrates the amplitude of each harmonic as a function of n. The important point is that our new choice for the spacing of the treads has reduced or eliminated some of the harmonics compared to the case of equally spaced treads. On the negative side we have excited some of the harmonics that were previously absent. However, the net effect is advantageous because the treads produce less noise at more frequencies rather than a lot of noise at a few select frequencies.

If we were to extend this technique so that the treads occurred at completely random positions, then the treads would produce very little noise at many frequencies and the total noise would be comparable to that generated by other sources within the car. To find the distribution of treads with the whitest noise[8] is a process of trial and error. Assuming a distribution, we can perform a Fourier analysis to obtain the frequency spectrum. If annoying peaks are present in the spectrum, we can then adjust the elements in the tread distribution that may contribute to the peak and analyze the revised distribution. You are finished when no peaks appear.

Problems

Find the Fourier series for the following functions. Plot various partial sums and compare them against the exact function.

1.

$$f(t) = \begin{cases} 1, & -\pi < t < 0 \\ 0, & 0 < t < \pi \end{cases}$$

2.

$$f(t) = \begin{cases} t, & -\pi < t < 0 \\ 0, & 0 < t < \pi \end{cases}$$

[8] White noise is sound that is analogous to white light in that it is uniformly distributed throughout the complete audible sound spectrum.

3.

$$f(t) = \begin{cases} -\pi, & -\pi < t < 0 \\ t, & 0 < t < \pi \end{cases}$$

4.

$$f(t) = \begin{cases} 0, & -\pi \le t \le 0 \\ t, & 0 \le t \le \pi/2 \\ \pi - t, & \pi/2 \le t \le \pi \end{cases}$$

5.

$$f(t) = \begin{cases} \frac{1}{2} + t, & -1 \le t \le 0 \\ \frac{1}{2} - t, & 0 \le t \le 1 \end{cases}$$

6.

$$f(t) = e^{at}, \qquad -L < t < L$$

7.

$$f(t) = \begin{cases} 0, & -\pi \le t \le 0 \\ \sin(t), & 0 \le t \le \pi \end{cases}$$

8.

$$f(t) = t + t^2, \qquad -L < t < L$$

9.

$$f(t) = \begin{cases} t, & -\frac{1}{2} \le t \le \frac{1}{2} \\ 1 - t, & \frac{1}{2} \le t \le \frac{3}{2} \end{cases}$$

10.

$$f(t) = \begin{cases} 0, & -\pi \le t \le -\pi/2 \\ \sin(2t), & -\pi/2 \le t \le \pi/2 \\ 0, & \pi/2 \le t \le \pi \end{cases}$$

11.

$$f(t) = \begin{cases} 0, & -a < t < 0 \\ 2t, & 0 < t < a \end{cases}$$

12.

$$f(t) = \frac{\pi - t}{2}, \qquad 0 < t < 2$$

13.

$$f(t) = t \cos\left(\frac{\pi t}{L}\right), \qquad -L < t < L$$

14.

$$f(t) = \begin{cases} 0, & -\pi < t \le 0 \\ t^2, & 0 \le t < \pi \end{cases}$$

15.

$$f(t) = \sinh\left[a\left(\frac{\pi}{2} - |t|\right)\right], \qquad -\pi \le t \le \pi$$

2.2 PROPERTIES OF FOURIER SERIES

In the previous section we introduced the Fourier series and showed how to compute one given the function $f(t)$. In this section we examine some particular properties of these series.

Differentiation of a Fourier series

In certain instances we only have the Fourier series representation of a function $f(t)$. Can we find the derivative or the integral of $f(t)$ merely by differentiating or integrating the Fourier series term by term? Is this permitted? Let us consider the case of differentiation first.

Consider a function $f(t)$ of period $2L$ which has the derivative $f'(t)$. Let us assume that we can expand $f'(t)$ as a Fourier series. This implies that $f'(t)$ is continuous except for a finite number of discontinuities and $f(t)$ is continuous over an interval that starts at $t = \tau$ and ends at $t = \tau + 2L$. Then

$$f'(t) = \frac{a_0'}{2} + \sum_{n=1}^{\infty} a_n' \cos\left(\frac{n\pi t}{L}\right) + b_n' \sin\left(\frac{n\pi t}{L}\right), \tag{2.2.1}$$

where we have denoted the Fourier coefficients of $f'(t)$ with a prime. Computing the Fourier coefficients,

$$a_0' = \frac{1}{L}\int_{\tau}^{\tau+2L} f'(t)\,dt = \frac{1}{L}[f(\tau+2L) - f(\tau)] = 0, \tag{2.2.2}$$

if $f(\tau + 2L) = f(\tau)$. Similarly, by integrating by parts,

$$a_n' = \frac{1}{L}\int_{\tau}^{\tau+2L} f'(t)\cos\left(\frac{n\pi t}{L}\right)dt \tag{2.2.3}$$

$$= \frac{1}{L}\left[f(t)\cos\left(\frac{n\pi t}{L}\right)\right]\Bigg|_{\tau}^{\tau+2L} + \frac{n\pi}{L^2}\int_{\tau}^{\tau+2L} f(t)\sin\left(\frac{n\pi t}{L}\right)dt \tag{2.2.4}$$

$$= \frac{n\pi b_n}{L} \tag{2.2.5}$$

and

$$b_n' = \frac{1}{L}\int_{\tau}^{\tau+2L} f'(t)\sin\left(\frac{n\pi t}{L}\right)\,dt \tag{2.2.6}$$

$$= \frac{1}{L}\left[f(t)\sin\left(\frac{n\pi t}{L}\right)\right]\Bigg|_{\tau}^{\tau+2L} - \frac{n\pi}{L^2}\int_{\tau}^{\tau+2L} f(t)\cos\left(\frac{n\pi t}{L}\right)\,dt \tag{2.2.7}$$

$$= -\frac{n\pi a_n}{L}. \tag{2.2.8}$$

Consequently, if we have a function $f(t)$ whose derivative $f'(t)$ is continuous except for a finite number of discontinuities and $f(\tau) = f(\tau+2L)$, then

$$f'(t) = \sum_{n=1}^{\infty} \frac{n\pi}{L}\left[b_n\cos\left(\frac{n\pi t}{L}\right) - a_n\sin\left(\frac{n\pi t}{L}\right)\right]. \tag{2.2.9}$$

That is, the derivative of $f(t)$ is given by a term-by-term differentiation of the Fourier series of $f(t)$.

• Example 2.2.1

The Fourier series for the function

$$f(t) = \begin{cases} 0, & -\pi \le t \le 0 \\ t, & 0 \le t \le \pi/2 \\ \pi - t, & \pi/2 \le t \le \pi \end{cases} \tag{2.2.10}$$

is

$$f(t) = \frac{\pi}{8} - \frac{1}{\pi}\sum_{n=1}^{\infty}\frac{\cos[2(2n-1)t]}{(2n-1)^2} - \frac{2}{\pi}\sum_{n=1}^{\infty}\frac{(-1)^n}{(2n-1)^2}\sin[(2n-1)t]. \tag{2.1.11}$$

Because $f(t)$ is continuous over the entire interval $(-\pi, \pi)$ and $f(-\pi) = f(\pi) = 0$, we can find $f'(t)$ by taking the derivative of (2.2.11) term by term:

$$f'(t) = \frac{2}{\pi}\sum_{n=1}^{\infty}\frac{\sin[2(2n-1)t]}{2n-1} - \frac{2}{\pi}\sum_{n=1}^{\infty}\frac{(-1)^n}{2n-1}\cos[(2n-1)t]. \tag{2.2.12}$$

This is the Fourier series that we would have obtained by computing the Fourier series for

$$f'(t) = \begin{cases} 0, & -\pi < t < 0 \\ 1, & 0 < t < \pi/2 \\ -1, & \pi/2 < t < \pi. \end{cases} \tag{2.2.13}$$

Integration of a Fourier series

To determine whether we can find the integral of $f(t)$ by term-by-term integration of its Fourier series, consider a form of the antiderivative of $f(t)$:

$$F(t) = \int_0^t \left[f(\tau) - \frac{a_0}{2}\right] d\tau. \tag{2.2.14}$$

Now

$$F(t+2L) = \int_0^t \left[f(\tau) - \frac{a_0}{2}\right] d\tau + \int_t^{t+2L} \left[f(\tau) - \frac{a_0}{2}\right] d\tau \tag{2.2.15}$$

$$= F(t) + \int_{-L}^{L} \left[f(\tau) - \frac{a_0}{2}\right] d\tau \tag{2.2.16}$$

$$= F(t) + \int_{-L}^{L} f(\tau)\, d\tau - La_0 = F(t), \tag{2.2.17}$$

so that $F(t)$ has a period of $2L$. Consequently we may expand $F(t)$ as the Fourier series

$$F(t) = \frac{A_0}{2} + \sum_{n=1}^{\infty} A_n \cos\left(\frac{n\pi t}{L}\right) + B_n \sin\left(\frac{n\pi t}{L}\right). \tag{2.2.18}$$

For A_n,

$$A_n = \frac{1}{L}\int_{-L}^{L} f(t)\cos\left(\frac{n\pi t}{L}\right) dt \tag{2.2.19}$$

$$= \frac{1}{L}\left[F(t)\frac{\sin(n\pi t/L)}{n\pi/L}\right]\Bigg|_{-L}^{L} - \frac{1}{n\pi}\int_{-L}^{L}\left[f(t) - \frac{a_0}{2}\right]\sin\left(\frac{n\pi t}{L}\right) dt \tag{2.2.20}$$

$$= -\frac{b_n}{n\pi/L}. \tag{2.2.21}$$

Similarly,

$$B_n = \frac{a_n}{n\pi/L}. \tag{2.2.22}$$

Therefore,

$$\int_0^t f(\tau)\, d\tau = \frac{a_0 t}{2} + \frac{A_0}{2} + \sum_{n=1}^{\infty} \frac{a_n \sin(n\pi t/L) - b_n \cos(n\pi t/L)}{n\pi/L}. \tag{2.2.23}$$

This is identical to a term-by-term integration of the Fourier series for $f(t)$. Thus, we can always find the integral of $f(t)$ by a term-by-term integration of its Fourier series.

• Example 2.2.2

The Fourier series for $f(t) = t$ for $-\pi < t < \pi$ is

$$f(t) = -2\sum_{n=1}^{\infty} \frac{(-1)^n}{n}\sin(nt). \tag{2.2.24}$$

To find the Fourier series for $f(t) = t^2$, we integrate (2.2.24) term by term and find that

$$\left.\frac{\tau^2}{2}\right|_0^t = 2\sum_{n=1}^{\infty}\frac{(-1)^n}{n^2}\cos(nt) - 2\sum_{n=1}^{\infty}\frac{(-1)^n}{n^2}. \tag{2.2.25}$$

But $\sum_{n=1}^{\infty}(-1)^n/n^2 = -\pi^2/12$. Substituting and multiplying by 2, we obtain the final result that

$$t^2 = \frac{\pi^2}{3} + 4\sum_{n=1}^{\infty}\frac{(-1)^n}{n^2}\cos(nt). \tag{2.2.26}$$

Parseval's equality

One of the fundamental quantities in engineering is power. The *power content* of a periodic signal $f(t)$ of period $2L$ is $\int_\tau^{\tau+2L} f^2(t)\,dt/L$. This mathematical definition mirrors the power dissipation I^2R that occurs in a resistor of resistance R where I is the root mean square (RMS) of the current. We would like to compute this power content as simply as possible given the coefficients of its Fourier series.

Assume that $f(t)$ has the Fourier series

$$f(t) = \frac{a_0}{2} + \sum_{n=1}^{\infty} a_n\cos\left(\frac{n\pi t}{L}\right) + b_n\sin\left(\frac{n\pi t}{L}\right). \tag{2.2.27}$$

Then,

$$\begin{aligned}\frac{1}{L}\int_\tau^{\tau+2L} f^2(t)\,dt = {} & \frac{a_0}{2L}\int_\tau^{\tau+2L} f(t)\,dt \\ & + \sum_{n=1}^{\infty}\frac{a_n}{L}\int_\tau^{\tau+2L} f(t)\cos\left(\frac{n\pi t}{L}\right)\,dt \\ & + \sum_{n=1}^{\infty}\frac{b_n}{L}\int_\tau^{\tau+2L} f(t)\sin\left(\frac{n\pi t}{L}\right)\,dt \qquad (2.2.28) \\ = {} & \frac{a_0^2}{2} + \sum_{n=1}^{\infty}(a_n^2 + b_n^2). \qquad (2.2.29)\end{aligned}$$

Equation (2.2.29) is *Parseval's equality.*[9] It allows us to sum squares of Fourier coefficients (which we have already computed) rather than performing the integration $\int_{\tau}^{\tau+2L} f^2(t)\,dt$ analytically or numerically.

• **Example 2.2.3**

The Fourier series for $f(t) = t^2$ over the interval $[-\pi, \pi]$ is

$$t^2 = \frac{\pi^2}{3} + 4\sum_{n=1}^{\infty} \frac{(-1)^n}{n^2} \cos(nt). \tag{2.2.30}$$

Then, by Parseval's equality,

$$\frac{1}{\pi}\int_{-\pi}^{\pi} t^4\,dt = \left.\frac{2t^5}{5\pi}\right|_0^{\pi} = \frac{4\pi^4}{18} + 16\sum_{n=1}^{\infty}\frac{1}{n^4} \tag{2.2.31}$$

$$\left(\frac{2}{5} - \frac{4}{18}\right)\pi^4 = 16\sum_{n=1}^{\infty}\frac{1}{n^4} \tag{2.2.32}$$

$$\frac{\pi^4}{90} = \sum_{n=1}^{\infty}\frac{1}{n^4}. \tag{2.2.33}$$

Gibbs phenomena

In the actual application of Fourier series, we cannot sum an infinite number of terms but must be content with N terms. If we denote this partial sum of the Fourier series by $S_N(t)$, we have from the definition of the Fourier series:

$$S_N(t) = \tfrac{1}{2}a_0 + \sum_{n=1}^{N} a_n\cos(nt) + b_n\sin(nt) \tag{2.2.34}$$

$$= \frac{1}{2\pi}\int_0^{2\pi} f(x)\,dx$$

[9] Parseval, M.-A., 1805: Mémoire sur les séries et sur l'intégration complète d'une équation aux différences partielles linéaires du second ordre, à coefficients constants. *Mémoires présentés a l'Institut des sciences, lettres et arts, par divers savans, et lus dans ses assemblées: Sciences mathématiques et Physiques*, **1**, 638–648.

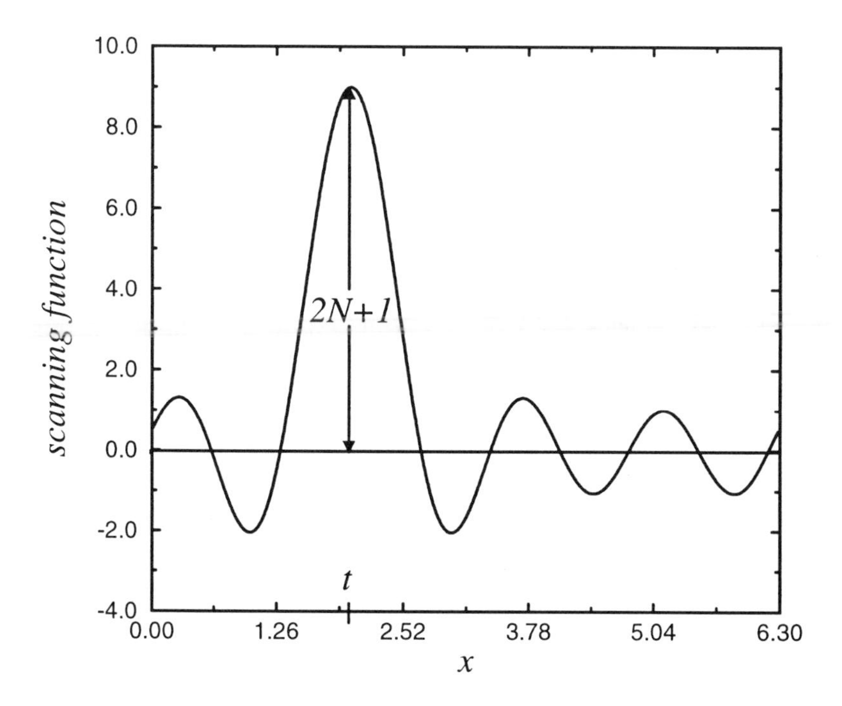

Figure 2.2.1: The scanning function over $0 \leq x \leq 2\pi$ for $N = 5$.

$$+\frac{1}{\pi}\int_0^{2\pi} f(x)\left[\sum_{n=1}^{N}\cos(nt)\cos(nx)+\sin(nt)\sin(nx)\right]dx \tag{2.2.35}$$

$$S_N(t) = \frac{1}{\pi}\int_0^{2\pi} f(x)\left\{\frac{1}{2}+\sum_{n=1}^{N}\cos[n(t-x)]\right\}dx \tag{2.2.36}$$

$$= \frac{1}{2\pi}\int_0^{2\pi} f(x)\frac{\sin[(N+\frac{1}{2})(x-t)]}{\sin[\frac{1}{2}(x-t)]}\,dx. \tag{2.2.37}$$

The quantity $\sin[(N+\frac{1}{2})(x-t)]/\sin[\frac{1}{2}(x-t)]$ is called a *scanning function*. Over the range $0 \leq x \leq 2\pi$ it has a very large peak at $x = t$ where the amplitude equals $2N + 1$. See Figure 2.2.1. On either side of this peak there are oscillations which decrease rapidly with distance from the peak. Consequently, as $N \to \infty$, the scanning function becomes essentially a long narrow slit corresponding to the area under the large peak at $x = t$. If we neglect for the moment the small area under the minor ripples adjacent to this slit, then the integral (2.2.37) essentially equals $f(t)$ times the area of the slit divided by 2π. If $1/2\pi$ times the area of the slit equals unity, then the value of $S_N(t) \approx f(t)$ to a good approximation for large N.

For a relatively small value of N, the scanning function deviates considerably from its ideal form, and the partial sum $S_N(t)$ only crudely approximates the given function $f(t)$. As the partial sum includes more terms and N becomes relatively large, the form of the scanning function improves and so does the degree of approximation between $S_N(t)$ and $f(t)$. The improvement in the scanning function is due to the large hump becoming taller and narrower. At the same time, the adjacent ripples become larger in number and hence also become narrower in the same proportion as the large hump becomes narrower.

The reason why $S_N(t)$ and $f(t)$ will never become identical, even in the limit of $N \to \infty$, is the presence of the positive and negative side lobes near the large peak. Because

$$\frac{\sin[(N+\frac{1}{2})(x-t)]}{\sin[\frac{1}{2}(x-t)]} = 1 + 2\sum_{n=1}^{N} \cos[n(t-x)], \tag{2.2.38}$$

an integration of the scanning function over the interval 0 to 2π shows that the total area under the scanning function equals 2π. However, from Figure 2.2.1 the net area contributed by the ripples is numerically negative so that the area under the large peak must exceed the value of 2π if the area of the entire function equals 2π. Although the exact value depends upon N, it is important to note that this excess does not become zero as $N \to \infty$.

Thus, the presence of these negative side lobes explains the departure of our scanning function from the idealized slit of area 2π. To illustrate this departure, consider the function:

$$f(t) = \begin{cases} 1, & 0 < t < \pi \\ -1, & \pi < t < 2\pi. \end{cases} \tag{2.2.39}$$

Then,

$$S_N(t) = \frac{1}{2\pi}\int_0^{\pi} \frac{\sin[(N+\frac{1}{2})(x-t)]}{\sin[\frac{1}{2}(x-t)]}\,dx - \frac{1}{2\pi}\int_{\pi}^{2\pi} \frac{\sin[(N+\frac{1}{2})(x-t)]}{\sin[\frac{1}{2}(x-t)]}\,dx \tag{2.2.40}$$

$$= \frac{1}{2\pi}\int_0^{\pi} \left\{\frac{\sin[(N+\frac{1}{2})(x-t)]}{\sin[\frac{1}{2}(x-t)]}\,dx + \frac{\sin[(N+\frac{1}{2})(x+t)]}{\sin[\frac{1}{2}(x+t)]}\,dx\right\} \tag{2.2.41}$$

$$= \frac{1}{2\pi}\int_{-t}^{\pi-t} \frac{\sin[(N+\frac{1}{2})\theta]}{\sin(\frac{1}{2}\theta)}\,d\theta - \frac{1}{2\pi}\int_{t}^{\pi+t} \frac{\sin[(N+\frac{1}{2})\theta]}{\sin(\frac{1}{2}\theta)}\,d\theta. \tag{2.2.42}$$

The first integral in (2.2.42) gives the contribution to $S_N(t)$ from the jump discontinuity at $t = 0$ while the second integral gives the contribution from $t = \pi$. In Figure 2.2.2 we have plotted the numerical

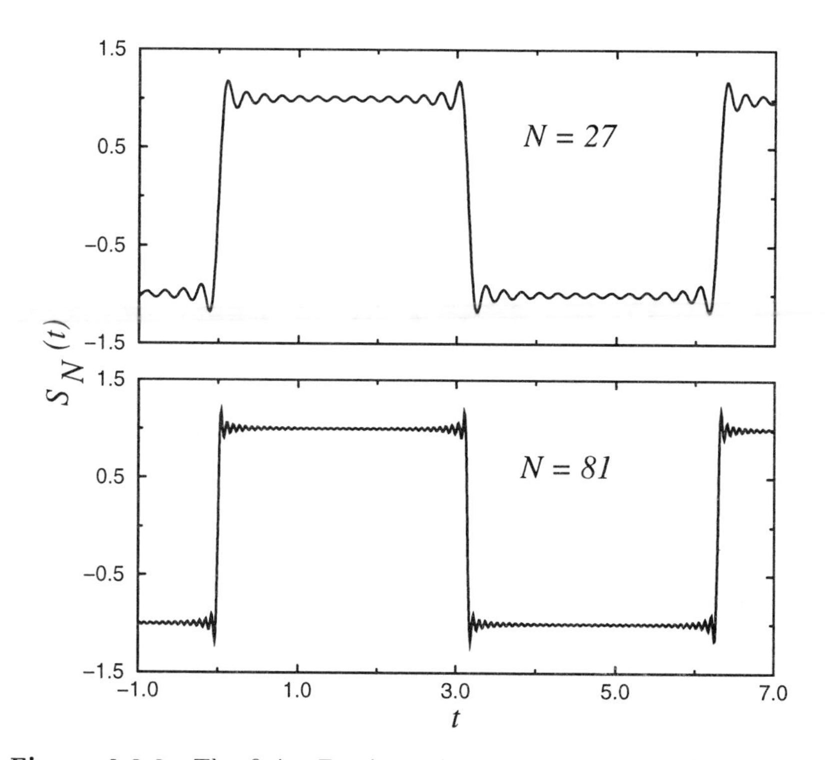

Figure 2.2.2: The finite Fourier series representation $S_N(t)$ for the function (2.2.39) for the range $-1 \leq t \leq 7$ for $N = 27$ and $N = 81$.

integration of (2.2.42) for $N = 27$ and $N = 81$. Residual discrepancies remain even for very large values of N. Indeed, even as N increases this figure changes only in that the ripples in the vicinity of the discontinuity of $f(t)$ show a proportionally increased rate of oscillation as a function of t while their relative magnitude remains the same. As $N \to \infty$ these ripples compress into a single vertical line at the point of discontinuity. True, these oscillations occupy smaller and smaller spaces but they still remain. Thus, we can never approximate a function in the vicinity of a discontinuity by a finite Fourier series without suffering from this over- and undershooting of the series. This peculiarity of Fourier series is called the *Gibbs phenomena*.[10] Gibbs phenomena can only be eliminated by removing the discontinuity.[11]

---

[10] Gibbs, J. W., 1898: Fourier's series. *Nature*, **59**, 200; Gibbs, J. W., 1899: Fourier's series. *Nature*, **59**, 606. For the historical development, see Hewitt, E. and Hewitt, R. E., 1979: The Gibbs-Wilbraham phenomenon: An episode in Fourier analysis. *Arch. Hist. Exact Sci.*, **21**, 129–160.

[11] For a particularly clever method for improving the convergence of

Problems

Additional Fourier series representation can be generated by differentiating or integrating known Fourier series. Work out the following two examples.

1. Given

$$\frac{\pi^2 - 2\pi x}{8} = \sum_{n=0}^{\infty} \frac{\cos[(2n+1)x]}{(2n+1)^2}, \qquad 0 \le x \le \pi,$$

obtain

$$\frac{\pi^2 x - \pi x^2}{8} = \sum_{n=0}^{\infty} \frac{\sin[(2n+1)x]}{(2n+1)^3}, \qquad 0 \le x \le \pi,$$

by term-by-term integration. Could we go the other way, i.e., take the derivative of the second equation to obtain the first? Explain.

2. Given

$$\frac{\pi^2 - 3x^2}{12} = \sum_{n=1}^{\infty} (-1)^{n+1} \frac{\cos(nx)}{n^2}, \qquad -\pi \le x \le \pi,$$

obtain

$$\frac{\pi^2 x - x^3}{12} = \sum_{n=1}^{\infty} (-1)^{n+1} \frac{\sin(nx)}{n^3}, \qquad -\pi \le x \le \pi,$$

by term-by-term integration. Could we go the other way, i.e., take the derivative of the second equation to obtain the first? Explain.

3. (a) Show that the Fourier series for the odd function:

$$f(t) = \begin{cases} 2t + t^2, & -2 < t < 0 \\ 2t - t^2, & 0 < t < 2 \end{cases}$$

is

$$f(t) = \frac{32}{\pi^3} \sum_{n=1}^{\infty} \frac{1}{(2n-1)^3} \sin\left[\frac{(2n-1)\pi t}{2}\right].$$

---

a trigonometric series, see Kantorovich, L. V. and Krylov, V. I., 1964: *Approximate Methods of Higher Analysis*. Interscience, New York, pp. 77–88.

(b) Use Parseval's equality to show that

$$\frac{\pi^6}{960} = \sum_{n=1}^{\infty} \frac{1}{(2n-1)^6}.$$

This series converges very rapidly to $\pi^6/960$ and provides a convenient method for computing π^6.

2.3 HALF-RANGE EXPANSIONS

In certain applications, we will find that we need a Fourier series representation for a function $f(x)$ that applies over the interval $(0, L)$ rather than $(-L, L)$. Because we are completely free to define the function over the interval $(-L, 0)$, it is simplest to have a series that consists only of sines or cosines. In this section we shall show how we can obtain these so-called *half-range expansions.*

Recall in Example 2.1.3 how we saw that if $f(x)$ is an even function $[f_o(x) = 0]$, then $b_n = 0$ for all n. Similarly, if $f(x)$ is an odd function $[f_e(x) = 0]$, then $a_0 = a_n = 0$ for all n. We now use these results to find a Fourier half-range expansion by extending the function defined over the interval $(0, L)$ as either an even or odd function into the interval $(-L, 0)$. If we extend $f(x)$ as an even function, we will get a half-range cosine series; if we extend $f(x)$ as an odd function, we obtain a half-range sine series.

It is important to remember that half-range expansions are a special case of the general Fourier series. For any $f(x)$ we can construct either a Fourier sine or cosine series over the interval $(-L, L)$. Both of these series will give the correct answer over the interval of $(0, L)$. Which one we choose to use depends upon whether we wish to deal with a cosine or sine series.

• Example 2.3.1

Let us find the half-range sine expansion of

$$f(x) = 1, \qquad 0 < x < \pi. \tag{2.3.1}$$

We begin by defining the periodic odd function

$$\tilde{f}(x) = \begin{cases} -1, & -\pi < x < 0 \\ 1, & 0 < x < \pi \end{cases} \tag{2.3.2}$$

with $\tilde{f}(x + 2\pi) = \tilde{f}(x)$. Because $\tilde{f}(x)$ is odd, $a_0 = a_n = 0$ and

$$b_n = \frac{2}{\pi} \int_0^\pi 1 \sin(nx)\, dx = -\frac{2}{n\pi} \cos(nx)\Big|_0^\pi \tag{2.3.3}$$

$$= -\frac{2}{n\pi} [\cos(n\pi) - 1] = -\frac{2}{n\pi} [(-1)^n - 1]. \tag{2.3.4}$$

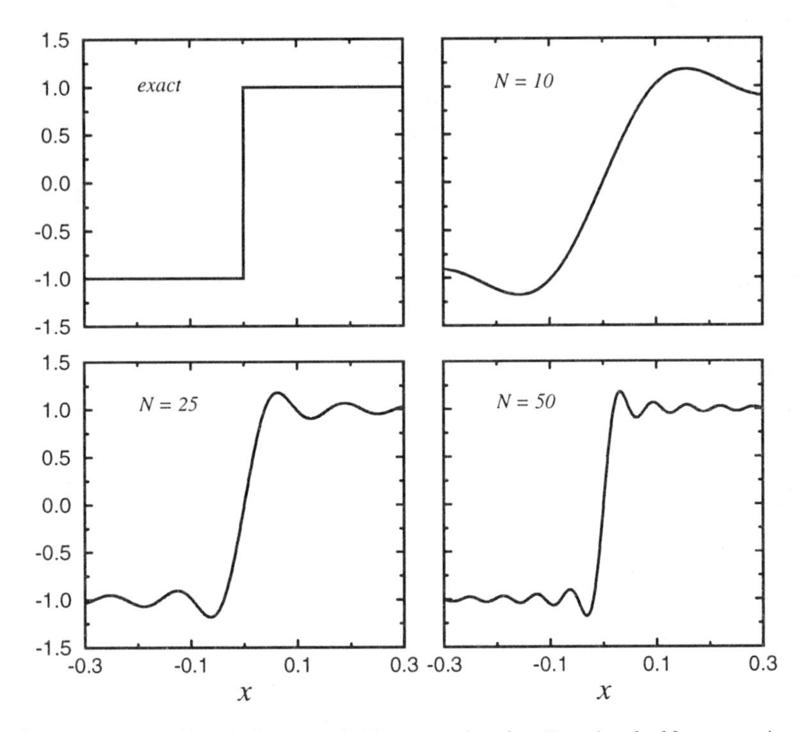

Figure 2.3.1: Partial sum of N terms in the Fourier half-range sine representation of a square wave.

The Fourier half-range sine series expansion of $f(x)$ is therefore

$$f(x) = \frac{2}{\pi}\sum_{n=1}^{\infty}\frac{[1-(-1)^n]}{n}\sin(nx) = \frac{4}{\pi}\sum_{m=1}^{\infty}\frac{\sin[(2m-1)x]}{2m-1}. \tag{2.3.5}$$

As counterpoint, let us find the half-range cosine expansion of $f(x) = 1$, $0 < x < \pi$. Now, we have that $b_n = 0$,

$$a_0 = \frac{2}{\pi}\int_0^{\pi} 1\, dx = 2 \tag{2.3.6}$$

and

$$a_n = \frac{2}{\pi}\int_0^{\pi}\cos(nx)\, dx = \frac{2}{n\pi}\sin(nx)\Big|_0^{\pi} = 0. \tag{2.3.7}$$

Thus, the Fourier half-range cosine expansion equals the single term:

$$f(x) = 1, \qquad 0 < x < \pi. \tag{2.3.8}$$

This is perfectly reasonable. To form a half-range cosine expansion we extend $f(x)$ as an even function into the interval $(-\pi, 0)$. In this case, we would obtain $\tilde{f}(x) = 1$ for $-\pi < x < \pi$. Finally, we note that the Fourier series of a constant is simply that constant.

In practice it is impossible to sum (2.3.5) exactly and we actually sum only the first N terms. Figure 2.3.1 illustrates $f(x)$ when the Fourier series (2.3.5) contains N terms. As seen from the figure, the truncated series tries to achieve the infinite slope at $x = 0$, but in the attempt, it *overshoots* the discontinuity by a certain amount (in this particular case, by 17.9%). This is another example of the Gibbs phenomena. Increasing the number of terms does not remove this peculiarity; it merely shifts it nearer to the discontinuity.

• **Example 2.3.2: Inertial supercharging of an engine**

An important aspect of designing any gasoline engine involves the motion of the fuel, air, and exhaust gas mixture through the engine. Ordinarily an engineer would consider the motion as steady flow; but in the case of a four-stroke, single-cylinder gasoline engine, the closing of the intake valve interrupts the steady flow of the gasoline-air mixture for nearly three quarters of the engine cycle. This periodic interruption sets up standing waves in the intake pipe – waves which can build up an appreciable pressure amplitude just outside the input value.

When one of the harmonics of the engine frequency equals one of the resonance frequencies of the intake pipe, then the pressure fluctuations at the valve will be large. If the intake valve closes during that portion of the cycle when the pressure is less than average, then the waves will reduce the power output. However, if the intake valve closes when the pressure is greater than atmospheric, then the waves will have a supercharging effect and will produce an increase of power. This effect is called *inertia supercharging*.

While studying this problem, Morse et al.[12] found it necessary to express the velocity of the air-gas mixture in the valve, given by

$$f(t) = \begin{cases} 0, & -\pi < \omega t < -\pi/4 \\ \pi\cos(2\omega t)/2, & -\pi/4 < \omega t < \pi/4 \\ 0, & \pi/4 < \omega t < \pi \end{cases} \tag{2.3.9}$$

in terms of a Fourier expansion. The advantage of working with the Fourier series rather than the function itself lies in the ability to write the velocity as a periodic forcing function that highlights the various harmonics that might be resonant with the structure comprising the fuel line.

[12] Morse, P. M., Boden, R. H., and Schecter, H., 1938: Acoustic vibrations and internal combustion engine performance. I. Standing waves in the intake pipe system. *J. Appl. Phys.*, **9**, 16–23.

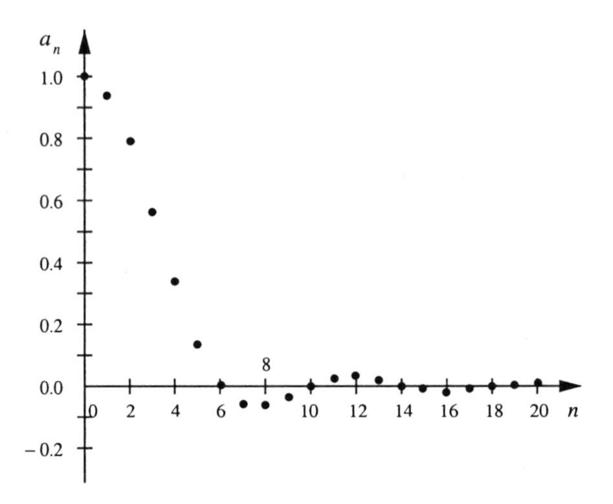

Figure 2.3.2: The spectral coefficients of the Fourier cosine series of the function (2.3.9).

Clearly $f(t)$ is an even function and its Fourier representation will be a cosine series. In this problem $\tau = -\pi/\omega$ and $L = \pi/\omega$. Therefore,

$$a_0 = \frac{2\omega}{\pi} \int_{-\pi/4\omega}^{\pi/4\omega} \frac{\pi}{2} \cos(2\omega t)\, dt = \tfrac{1}{2} \sin(2\omega t)\Big|_{-\pi/4\omega}^{\pi/4\omega} = 1 \tag{2.3.10}$$

and

$$a_n = \frac{2\omega}{\pi} \int_{-\pi/4\omega}^{\pi/4\omega} \frac{\pi}{2} \cos(2\omega t) \cos\left(\frac{n\pi t}{\pi/\omega}\right) dt \tag{2.3.11}$$

$$= \frac{\omega}{2} \int_{-\pi/4\omega}^{\pi/4\omega} \{\cos[(n+2)\omega t] + \cos[(n-2)\omega t]\}\, dt \tag{2.3.12}$$

$$= \begin{cases} \left. \frac{\sin[(n+2)\omega t]}{2(n+2)} + \frac{\sin[(n-2)\omega t]}{2(n-2)} \right|_{-\pi/4\omega}^{\pi/4\omega}, & n \neq 2 \\ \left. \frac{\omega t}{2} + \frac{\sin(4\omega t)}{4} \right|_{-\pi/4\omega}^{\pi/4\omega}, & n = 2 \end{cases} \tag{2.3.13}$$

$$= \begin{cases} -\frac{4}{n^2-4} \cos\left(\frac{n\pi}{4}\right), & n \neq 2 \\ \frac{\pi}{4}, & n = 2. \end{cases} \tag{2.3.14}$$

Because these Fourier coefficients become small rapidly (see Figure 2.3.2), Morse et al. showed that there are only about three resonances where the acoustic properties of the intake pipe can enhance engine performance. These peaks occur when $q = 30c/NL = 3, 4$, or 5, where c is the velocity of sound in the air-gas mixture, L is the effective length

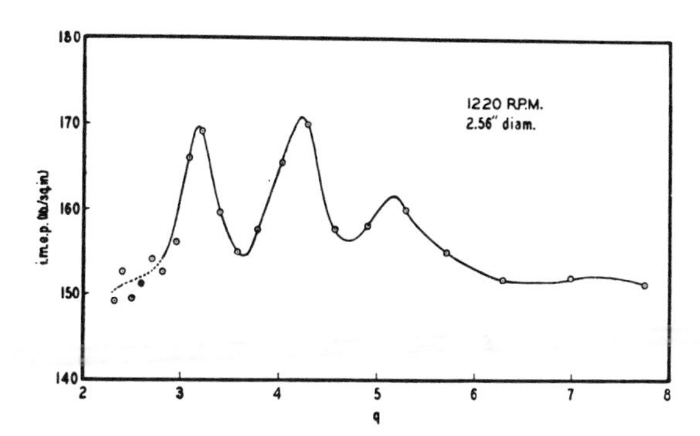

Figure 2.3.3: Experimental verification of the resonance of the $n = 3, 4$, and 5 harmonics of the Fourier representation (2.3.14) of the flow of an air-gas mixture with the intake pipe system. The parameter q is defined in the text. (From Morse, P., Boden, R. H., and Schecter, H., 1938: Acoustic vibrations and internal combustion engine performance. *J. Appl. Phys.*, **9**, 17 with permission.)

of the intake pipe, and N is the engine speed in rpm. See Figure 2.3.3. Subsequent experiments[13] verified these results.

Such analyses are valuable to automotive engineers. Engineers are always seeking ways to optimize a system with little or no additional cost. Our analysis shows that by tuning the length of the intake pipe so that it falls on one of the resonance peaks, we could obtain higher performance from the engine with little or no extra work. Of course, the problem is that no car always performs at some optimal condition.

Problems

Find the Fourier cosine and sine series for the following functions:

1.

$$f(t) = t, \qquad 0 < t < \pi$$

2.

$$f(t) = \pi - t, \qquad 0 < t < \pi$$

3.

$$f(t) = \begin{cases} t, & 0 \le t \le \frac{1}{2} \\ 1 - t, & \frac{1}{2} \le t \le 1 \end{cases}$$

[13] Boden, R. H. and Schecter, H., 1944: Dynamics of the inlet system of a four-stroke engine. *NACA Tech. Note 935.*

4.

$$f(t) = \pi^2 - t^2, \qquad 0 < t < \pi$$

5.

$$f(t) = \begin{cases} t, & 0 \le t \le 1 \\ 1, & 1 \le t < 2 \end{cases}$$

6.

$$f(t) = \begin{cases} 0, & 0 < t < \frac{a}{3} \\ t - \frac{a}{3}, & \frac{a}{3} < t < \frac{2a}{3} \\ \frac{a}{3}, & \frac{2a}{3} < t < a \end{cases}$$

7.

$$f(t) = \begin{cases} \frac{1}{2}, & 0 < t < \frac{a}{2} \\ 1, & \frac{a}{2} < t < a \end{cases}$$

8.

$$f(t) = \begin{cases} \frac{2t}{a}, & 0 < t < \frac{a}{2} \\ \frac{3a-2t}{2a}, & \frac{a}{2} < t < a \end{cases}$$

9.

$$f(t) = \begin{cases} t, & 0 < t < \frac{a}{2} \\ \frac{a}{2}, & \frac{a}{2} < t < a \end{cases}$$

10.

$$f(t) = \frac{a-t}{a}, \qquad 0 < t < a$$

11.

$$f(t) = \begin{cases} 0, & 0 < t < \frac{a}{4} \\ 1, & \frac{a}{4} < t < \frac{3a}{4} \\ 0, & \frac{3a}{4} < t < a \end{cases}$$

12.

$$f(t) = t(a-t), \qquad 0 < t < a$$

13.

$$f(t) = e^{kt}, \qquad 0 < t < a$$

14.

$$f(t) = \begin{cases} 0, & 0 < t < \frac{a}{2} \\ 1, & \frac{a}{2} < t < a \end{cases}$$

15. The function

$$f(t) = 1 - (1+a)\frac{t}{\pi} + (a-1)\frac{t^2}{\pi^2} + (a+1)\frac{t^3}{\pi^3} - a\frac{t^4}{\pi^4}, \quad 0 < t < \pi$$

is a curve fit to the observed pressure trace of an explosion wave in the atmosphere. Because the observed transmission of atmospheric waves depends on the five-fourths power of the frequency, Reed[14] had to re-express this curve fit as a Fourier sine series before he could use the transmission law. He found that

$$\begin{aligned} f(t) = & \frac{1}{\pi}\sum_{n=1}^{\infty}\frac{1}{n}\left[1 - \frac{3(a-1)}{2\pi^2 n^2}\right]\sin(2nt) \\ & + \frac{1}{\pi}\sum_{n=1}^{\infty}\frac{2}{2n-1}\left[1 + \frac{2(a-1)}{\pi^2(2n-1)^2} - \frac{48a}{\pi^4(2n-1)^4}\right]\sin[(2n-1)t]. \end{aligned}$$

Confirm his result.

2.4 FOURIER SERIES WITH PHASE ANGLES

Sometimes it is desirable to rewrite a general Fourier series as a purely cosine or purely sine series with a phase angle. Engineers often like to speak of some quantity leading or lagging another quantity. Re-expressing a Fourier series in terms of amplitude and phase provides a convenient method for determining these phase relationships.

Suppose, for example, that we have a function $f(t)$ of period $2L$, given in the interval $[-L, L]$, whose Fourier series expansion is

$$f(t) = \frac{a_0}{2} + \sum_{n=1}^{\infty} a_n \cos\left(\frac{n\pi t}{L}\right) + b_n \sin\left(\frac{n\pi t}{L}\right). \tag{2.4.1}$$

We wish to replace (2.4.1) by the series:

$$f(t) = \frac{a_0}{2} + \sum_{n=1}^{\infty} B_n \sin\left(\frac{n\pi t}{L} + \varphi_n\right). \tag{2.4.2}$$

[14] From Reed, J. W., 1977: Atmospheric attenuation of explosion waves. *J. Acoust. Soc. Am.*, **61**, 39–47 with permission.

To do this we note that

$$B_n \sin\left(\frac{n\pi t}{L} + \varphi_n\right) = a_n \cos\left(\frac{n\pi t}{L}\right) + b_n \sin\left(\frac{n\pi t}{L}\right) \tag{2.4.3}$$

$$= B_n \sin\left(\frac{n\pi t}{L}\right)\cos(\varphi_n) + B_n \sin(\varphi_n)\cos\left(\frac{n\pi t}{L}\right). \tag{2.4.4}$$

We equate coefficients of $\sin(n\pi t/L)$ and $\cos(n\pi t/L)$ on both sides and obtain

$$a_n = B_n \sin(\varphi_n) \qquad \text{and} \qquad b_n = B_n \cos(\varphi_n). \tag{2.4.5}$$

Hence, upon squaring and adding,

$$B_n = \sqrt{a_n^2 + b_n^2}, \tag{2.4.6}$$

while taking the ratio gives

$$\varphi_n = \tan^{-1}(a_n/b_n). \tag{2.4.7}$$

Similarly we could rewrite (2.4.1) as

$$f(t) = \frac{a_0}{2} + \sum_{n=1}^{\infty} A_n \cos\left(\frac{n\pi t}{L} + \varphi_n\right), \tag{2.4.8}$$

where

$$A_n = \sqrt{a_n^2 + b_n^2} \qquad \text{and} \qquad \varphi_n = \tan^{-1}(-b_n/a_n) \tag{2.4.9}$$

and

$$a_n = A_n \cos(\varphi_n) \qquad \text{and} \qquad b_n = -A_n \sin(\varphi_n). \tag{2.4.10}$$

In both cases, we must be careful in computing φ_n because there are two possible values of φ_n which satisfy (2.4.7) or (2.4.9). These φ_n's must give the correct a_n and b_n using either (2.4.5) or (2.4.10).

• **Example 2.4.1**

The Fourier series for $f(t) = e^t$ over the interval $-L < t < L$ is

$$f(t) = \frac{\sinh(aL)}{aL} + 2\sinh(aL)\sum_{n=1}^{\infty} \frac{aL(-1)^n}{a^2L^2 + n^2\pi^2} \cos\left(\frac{n\pi t}{L}\right)$$
$$- 2\sinh(aL)\sum_{n=1}^{\infty} \frac{n\pi(-1)^n}{a^2L^2 + n^2\pi^2} \sin\left(\frac{n\pi t}{L}\right). \tag{2.4.11}$$

Let us rewrite (2.4.11) as a Fourier series with a phase angle. Regardless of whether we want the new series to contain $\cos(n\pi t/L + \varphi_n)$ or $\sin(n\pi t/L + \varphi_n)$, the amplitude A_n or B_n is the same in both series:

$$A_n = B_n = \sqrt{a_n^2 + b_n^2} = \frac{2\sinh(aL)}{\sqrt{a^2L^2 + n^2\pi^2}}. \tag{2.4.12}$$

If we want our Fourier series to read

$$f(t) = \frac{\sinh(aL)}{aL} + 2\sinh(aL)\sum_{n=1}^{\infty} \frac{\cos(n\pi t/L + \varphi_n)}{\sqrt{a^2L^2 + n^2\pi^2}}, \tag{2.4.13}$$

then

$$\varphi_n = \tan^{-1}\left(-\frac{b_n}{a_n}\right) = \tan^{-1}\left(\frac{n\pi}{aL}\right), \tag{2.4.14}$$

where φ_n lies in the first quadrant if n is even and in the third quadrant if n is odd. This ensures that the sign from the $(-1)^n$ is correct.

On the other hand, if we prefer

$$f(t) = \frac{\sinh(aL)}{aL} + 2\sinh(aL)\sum_{n=1}^{\infty} \frac{\sin(n\pi t/L + \varphi_n)}{\sqrt{a^2L^2 + n^2\pi^2}}, \tag{2.4.15}$$

then

$$\varphi_n = \tan^{-1}\left(\frac{a_n}{b_n}\right) = -\tan^{-1}\left(\frac{aL}{n\pi}\right), \tag{2.4.16}$$

where φ_n lies in the fourth quadrant if n is odd and in the second quadrant if n is even.

Problems

Write the following Fourier series in both the cosine and sine phase angle form:

1.

$$f(t) = \frac{1}{2} + \frac{2}{\pi}\sum_{n=1}^{\infty} \frac{\sin[(2n-1)\pi t]}{2n-1}$$

2.

$$f(t) = \frac{3}{2} + \frac{2}{\pi}\sum_{n=1}^{\infty} \frac{(-1)^n}{2n-1}\cos\left[\frac{(2n-1)\pi t}{2}\right]$$

3.

$$f(t) = -2\sum_{n=1}^{\infty} \frac{(-1)^n}{n}\sin(nt)$$

4.

$$f(t) = \frac{\pi}{2} - \frac{4}{\pi} \sum_{n=1}^{\infty} \frac{\cos[(2n-1)t]}{(2n-1)^2}$$

2.5 COMPLEX FOURIER SERIES

So far in our discussion, we have expressed Fourier series in terms of sines and cosines. We are now ready to reexpress a Fourier series as a series of complex exponentials. There are two reasons for this. First, in certain engineering and scientific applications of Fourier series, the expansion of a function in terms of complex exponentials results in coefficients of considerable simplicity and clarity. Secondly, these complex Fourier series point the way to the development of the Fourier transform in the next chapter.

We begin by introducing the variable

$$\omega_n = \frac{n\pi}{L}, \tag{2.5.1}$$

where $n = 0, \pm 1, \pm 2, \ldots$. Using Euler's formula we can replace the sine and cosine in the Fourier series by exponentials and find that

$$f(t) = \frac{a_0}{2} + \sum_{n=1}^{\infty} \frac{a_n}{2}\left(e^{i\omega_n t} + e^{-i\omega_n t}\right) + \frac{b_n}{2i}\left(e^{i\omega_n t} - e^{-i\omega_n t}\right) \tag{2.5.2}$$

$$= \frac{a_0}{2} + \sum_{n=1}^{\infty} \left(\frac{a_n}{2} - \frac{b_n i}{2}\right) e^{i\omega_n t} + \left(\frac{a_n}{2} + \frac{b_n i}{2}\right) e^{-i\omega_n t}. \tag{2.5.3}$$

If we define c_n as

$$c_n = \tfrac{1}{2}(a_n - ib_n), \tag{2.5.4}$$

then

$$c_n = \tfrac{1}{2}(a_n - ib_n) = \frac{1}{2L} \int_{\tau}^{\tau+2L} f(t)[\cos(\omega_n t) - i\sin(\omega_n t)]\,dt \tag{2.5.5}$$

$$= \frac{1}{2L} \int_{\tau}^{\tau+2L} f(t) e^{-i\omega_n t} dt. \tag{2.5.6}$$

Similarly, the complex conjugate of c_n, c_n^*, equals

$$c_n^* = \tfrac{1}{2}(a_n + ib_n) = \frac{1}{2L} \int_{\tau}^{\tau+2L} f(t) e^{i\omega_n t} dt. \tag{2.5.7}$$

To simplify (2.5.3) we note that

$$\omega_{-n} = \frac{(-n)\pi}{L} = -\frac{n\pi}{L} = -\omega_n, \tag{2.5.8}$$

which yields the result that

$$c_{-n} = \frac{1}{2L}\int_\tau^{\tau+2L} f(t)e^{-i\omega_{-n}t}dt = \frac{1}{2L}\int_\tau^{\tau+2L} f(t)e^{i\omega_n t}dt = c_n^* \quad (2.5.9)$$

so that we can write (2.5.3) as

$$f(t) = \frac{a_0}{2} + \sum_{n=1}^{\infty} c_n e^{i\omega_n t} + c_n^* e^{-i\omega_n t} = \frac{a_0}{2} + \sum_{n=1}^{\infty} c_n e^{i\omega_n t} + c_{-n} e^{-i\omega_n t}. \quad (2.5.10)$$

Letting $n = -m$ in the second summation on the right side of (2.5.10),

$$\sum_{n=1}^{\infty} c_{-n} e^{-i\omega_n t} = \sum_{m=-1}^{-\infty} c_m e^{-i\omega_{-m} t} = \sum_{m=-\infty}^{-1} c_m e^{i\omega_m t} = \sum_{n=-\infty}^{-1} c_n e^{i\omega_n t}, \quad (2.5.11)$$

where we have introduced $m = n$ into the last summation in (2.5.11). Therefore,

$$f(t) = \frac{a_0}{2} + \sum_{n=1}^{\infty} c_n e^{i\omega_n t} + \sum_{n=-\infty}^{-1} c_n e^{i\omega_n t}. \quad (2.5.12)$$

On the other hand,

$$\frac{a_0}{2} = \frac{1}{2L}\int_\tau^{\tau+2L} f(t)\,dt = c_0 = c_0 e^{i\omega_0 t}, \quad (2.5.13)$$

because $\omega_0 = 0\pi/L = 0$. Thus, our final result is

$$\boxed{f(t) = \sum_{n=-\infty}^{\infty} c_n e^{i\omega_n t},} \quad (2.5.14)$$

where

$$\boxed{c_n = \frac{1}{2L}\int_\tau^{\tau+2L} f(t)e^{-i\omega_n t}\,dt} \quad (2.5.15)$$

and $n = 0, \pm 1, \pm 2, \ldots$. Note that even though c_n is generally complex, the summation (2.5.14) always gives a *real-valued* function $f(t)$.

Just as we can represent the function $f(t)$ graphically by a plot of t against $f(t)$, we can plot c_n as a function of n, commonly called the frequency *spectrum*. Because c_n is generally complex, it is necessary to make two plots. Typically the plotted quantities are the amplitude spectra $|c_n|$ and the phase spectra φ_n, where φ_n is the phase of c_n. However, we could just as well plot the real and imaginary parts of c_n. Because n is an integer, these plots consist merely of a series of vertical lines representing the ordinates of the quantity $|c_n|$ or φ_n for each n. For this reason we refer to these plots as the *line spectra*.

Because $2c_n = a_n - ib_n$, the c_n's for an even function will be purely real; the c_n's for an odd function are purely imaginary. It is important to note that we lose the advantage of even and odd functions in the sense that we cannot just integrate over the interval 0 to L and then double the result. In the present case we have a line integral of a complex function along the real axis.

• Example 2.5.1

Let us find the complex Fourier series for

$$f(t) = \begin{cases} 1, & 0 < t < \pi \\ -1, & -\pi < t < 0, \end{cases} \tag{2.5.16}$$

which has the periodicity $f(t + 2\pi) = f(t)$.

With $L = \pi$ and $\tau = -\pi$, $\omega_n = n\pi/L = n$. Therefore,

$$c_n = \frac{1}{2\pi}\int_{-\pi}^{0}(-1)e^{-int}\,dt + \frac{1}{2\pi}\int_{0}^{\pi}(1)e^{-int}\,dt \tag{2.5.17}$$

$$= \frac{1}{2n\pi i}e^{-int}\Big|_{-\pi}^{0} - \frac{1}{2n\pi i}e^{-int}\Big|_{0}^{\pi} \tag{2.5.18}$$

$$= -\frac{i}{2n\pi}\left(1 - e^{n\pi i}\right) + \frac{i}{2n\pi}\left(e^{-n\pi i} - 1\right), \tag{2.5.19}$$

if $n \neq 0$. Because $e^{n\pi i} = \cos(n\pi) + i\sin(n\pi) = (-1)^n$ and $e^{-n\pi i} = \cos(-n\pi) + i\sin(-n\pi) = (-1)^n$, then

$$c_n = -\frac{i}{n\pi}[1 - (-1)^n] = \begin{cases} 0, & n \text{ even} \\ -\frac{2i}{n\pi}, & n \text{ odd} \end{cases} \tag{2.5.20}$$

with

$$f(t) = \sum_{n=-\infty}^{\infty} c_n e^{int}. \tag{2.5.21}$$

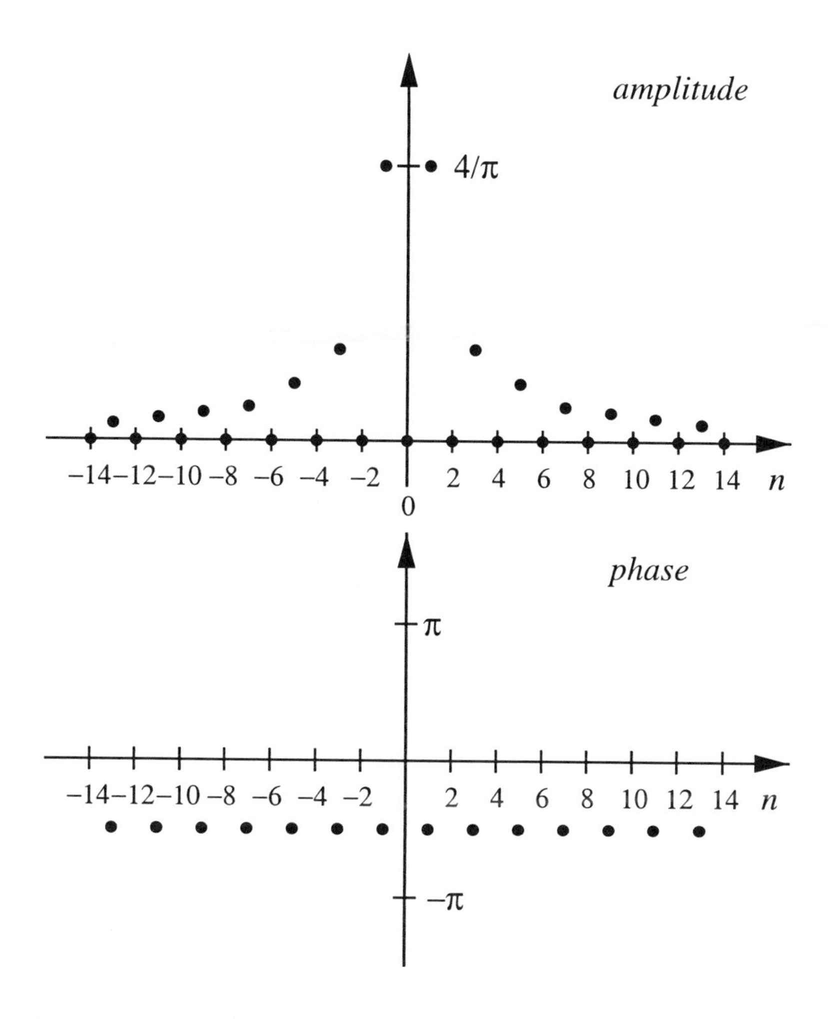

Figure 2.5.1: Amplitude and phase spectra for the function (2.5.16).

In this particular problem we must treat the case $n = 0$ specially because (2.5.18) is undefined for $n = 0$. In that case,

$$c_0 = \frac{1}{2\pi}\int_{-\pi}^{0} (-1)\,dt + \frac{1}{2\pi}\int_{0}^{\pi} (1)\,dt = \frac{1}{2\pi}(-t)\Big|_{-\pi}^{0} + \frac{1}{2\pi}(t)\Big|_{0}^{\pi} = 0. \tag{2.5.22}$$

Because $c_0 = 0$, we can write the expansion:

$$f(t) = -\frac{2i}{\pi}\sum_{m=-\infty}^{\infty} \frac{e^{(2m-1)it}}{2m-1}, \tag{2.5.23}$$

because we can write all odd integers as $2m - 1$, where $m = 0, \pm 1, \pm 2,$

$\pm 3, \ldots$. In Figure 2.5.1 we present the amplitude and phase spectra for the function (2.5.16).

Problems

Find the complex Fourier series for the following functions:

1. $f(t) = |t|, \quad -\pi \le t \le \pi$

2. $f(t) = e^t, \quad 0 < t < 2$

3. $f(t) = t, \quad 0 < t < 2$

4. $f(t) = t^2, \quad -\pi \le t \le \pi$

5. $f(t) = \begin{cases} 0, & -\pi/2 < t < 0 \\ 1, & 0 < t < \pi/2 \end{cases}$

6. $f(t) = t, \quad -1 < t < 1$

2.6 THE USE OF FOURIER SERIES IN THE SOLUTION OF ORDINARY DIFFERENTIAL EQUATIONS

An important application of Fourier series is the solution of ordinary differential equations. Structural engineers especially use this technique because the occupants of buildings and bridges often subject these structures to forcings that are periodic in nature.[15]

• Example 2.6.1

Let us find the general solution to the ordinary differential equation

$$y'' + 9y = f(t), \tag{2.6.1}$$

where the forcing is

$$f(t) = |t|, \qquad -\pi \le t \le \pi, \qquad f(t + 2\pi) = f(t). \tag{2.6.2}$$

This equation represents an oscillator forced by a driver whose displacement is the saw-tooth function.

We begin by replacing the function $f(t)$ by its Fourier series representation because the forcing function is periodic. The advantage of expressing $f(t)$ as a Fourier series is its validity for any time t. The alternative would have been to construct a solution over each interval $n\pi < t < (n+1)\pi$ and then piece together the final solution assuming that the solution and its first derivative is continuous at each junction

---

[15] Timoshenko, S. P., 1943: Theory of suspension bridges. Part II. *J. Franklin Inst.*, **235**, 327–349; Inglis, C. E., 1934: *A Mathematical Treatise on Vibrations in Railway Bridges*, Cambridge University Press, Cambridge.

$t = n\pi$. Because the function is an even function, all of the sine terms vanish and the Fourier series is

$$|t| = \frac{\pi}{2} - \frac{4}{\pi} \sum_{n=1}^{\infty} \frac{\cos[(2n-1)t]}{(2n-1)^2}. \tag{2.6.3}$$

Next, we note that the general solution consists of the complementary solution, which equals

$$y_H(t) = A\cos(3t) + B\sin(3t), \tag{2.6.4}$$

and the particular solution $y_p(t)$ which satisfies the differential equation

$$y_p'' + 9y_p = \frac{\pi}{2} - \frac{4}{\pi} \sum_{n=1}^{\infty} \frac{\cos[(2n-1)t]}{(2n-1)^2}. \tag{2.6.5}$$

To determine this particular solution, we write (2.6.5) as

$$y_p'' + 9y_p = \frac{\pi}{2} - \frac{4}{\pi}\cos(t) - \frac{4}{9\pi}\cos(3t) - \frac{4}{25\pi}\cos(5t) - \cdots \tag{2.6.6}$$

By the method of undetermined coefficients, we would have guessed the particular solution:

$$y_p(t) = \frac{a_0}{2} + a_1\cos(t) + b_1\sin(t) + a_2\cos(3t) + b_2\sin(3t) + \cdots \tag{2.6.7}$$

or

$$y_p(t) = \tfrac{1}{2}a_0 + \sum_{n=1}^{\infty} a_n\cos[(2n-1)t] + b_n\sin[(2n-1)t]. \tag{2.6.8}$$

Because

$$y_p''(t) = \sum_{n=1}^{\infty} -(2n-1)^2\{a_n\cos[(2n-1)t] + b_n\sin[(2n-1)t]\}, \tag{2.6.9}$$

$$\begin{aligned}\sum_{n=1}^{\infty} -(2n-1)^2\{a_n\cos[(2n-1)t] + b_n\sin[(2n-1)t]\}&\\ + \tfrac{9}{2}a_0 + 9\sum_{n=1}^{\infty} a_n\cos[(2n-1)t] + b_n\sin[(2n-1)t]&\\ = \frac{\pi}{2} - \frac{4}{\pi}\sum_{n=1}^{\infty}\frac{\cos[(2n-1)t]}{(2n-1)^2}&\end{aligned} \tag{2.6.10}$$

or

$$\frac{9}{2}a_0 - \frac{\pi}{2} + \sum_{n=1}^{\infty}\left\{[9-(2n-1)^2]a_n + \frac{4}{\pi(2n-1)^2}\right\}\cos[(2n-1)t]$$
$$+\sum_{n=1}^{\infty}[9-(2n-1)^2]b_n\sin[(2n-1)t] = 0. \tag{2.6.11}$$

Because (2.6.11) must hold true for any time, each harmonic must vanish separately:

$$a_0 = \frac{\pi}{9}, \qquad a_n = -\frac{4}{\pi(2n-1)^2[9-(2n-1)^2]} \tag{2.6.12}$$

and $b_n = 0$. All of the a_n's are finite except for $n = 2$, where a_2 becomes undefined. The coefficient a_2 is undefined because the harmonic $\cos(3t)$ in the forcing function is resonating with the natural mode of the system.

Let us review our analysis to date. We found that each harmonic in the forcing function yields a corresponding harmonic in the particular solution (2.6.8). The only difficulty arises with the harmonic $n = 2$. Although our particular solution is not correct because it contains $\cos(3t)$, we suspect that if we remove that term then the remaining harmonic solutions are correct. The problem is linear, and difficulties with one harmonic term should not affect other harmonics. But how shall we deal with the $\cos(3t)$ term in the forcing function? Let us denote the particular solution for that harmonic by $Y(t)$ and modify our particular solution as follows:

$$y_p(t) = \tfrac{1}{2}a_0 + a_1\cos(t) + Y(t) + a_3\cos(5t) + \cdots \tag{2.6.13}$$

Substituting this solution into the differential equation and simplifying, everything cancels except

$$Y'' + 9Y = -\frac{4}{9\pi}\cos(3t). \tag{2.6.14}$$

The solution of this equation by the method of undetermined coefficients is

$$Y(t) = -\frac{2}{27\pi}t\sin(3t). \tag{2.6.15}$$

This term, called a *secular term*, is the most important one in the solution. While the other terms merely represent simple oscillatory motion, the term $t\sin(3t)$ grows linearly with time and eventually becomes the dominant term in the series. Consequently, the general solution equals the complementary plus the particular solution:

$$y(t) = A\cos(3t) + B\sin(3t)$$
$$+\frac{\pi}{18} - \frac{2}{27\pi}t\sin(3t) - \frac{4}{\pi}\sum_{\substack{n=1\\n\neq 2}}^{\infty}\frac{\cos[(2n-1)t]}{(2n-1)^2[9-(2n-1)^2]}. \tag{2.6.16}$$

• **Example 2.6.2**

Let us redo the previous problem only using complex Fourier series. That is, let us find the general solution to the ordinary differential equation

$$y'' + 9y = \frac{\pi}{2} - \frac{2}{\pi} \sum_{n=-\infty}^{\infty} \frac{e^{i(2n-1)t}}{(2n-1)^2}. \tag{2.6.17}$$

From the method of undetermined coefficients we guess the particular solution for (2.6.17) to be

$$y_p(t) = c_0 + \sum_{n=-\infty}^{\infty} c_n e^{i(2n-1)t}. \tag{2.6.18}$$

Then

$$y_p''(t) = \sum_{n=-\infty}^{\infty} -(2n-1)^2 c_n e^{i(2n-1)t}. \tag{2.6.19}$$

Substituting (2.6.18) and (2.6.19) into (2.6.17),

$$9c_0 + \sum_{n=-\infty}^{\infty} [9-(2n-1)^2] c_n e^{i(2n-1)t} = \frac{\pi}{2} - \frac{2}{\pi} \sum_{n=-\infty}^{\infty} \frac{e^{i(2n-1)t}}{(2n-1)^2}. \tag{2.6.20}$$

Because (2.6.20) must be true for any t,

$$c_0 = \frac{\pi}{18} \quad \text{and} \quad c_n = \frac{2}{\pi(2n-1)^2[(2n-1)^2-9]}. \tag{2.6.21}$$

Therefore,

$$y_p(t) = \frac{\pi}{18} + \frac{2}{\pi} \sum_{n=-\infty}^{\infty} \frac{e^{i(2n-1)t}}{(2n-1)^2[(2n-1)^2-9]} e^{i(2n-1)t}. \tag{2.6.22}$$

However, there is a problem when $n = -1$ and $n = 2$. Therefore, we modify (2.6.22) to read

$$\begin{aligned} y_p(t) = {} & \frac{\pi}{18} + c_2 t e^{3it} + c_{-1} t e^{-3it} \\ & + \frac{2}{\pi} \sum_{\substack{n=-\infty \\ n \neq -1,2}}^{\infty} \frac{e^{i(2n-1)t}}{(2n-1)^2[(2n-1)^2-9]} e^{i(2n-1)t}. \end{aligned} \tag{2.6.23}$$

Substituting (2.6.23) into (2.6.17) and simplifying,

$$c_2 = -\frac{1}{27\pi i} \quad \text{and} \quad c_{-1} = -\frac{1}{27\pi i}. \tag{2.6.24}$$

The general solution is then

$$
\begin{aligned}
y_p(t) = {} & Ae^{3it} + Be^{-3it} + \frac{\pi}{18} - \frac{te^{3it}}{27\pi i} + \frac{te^{-3it}}{27\pi i} \\
& + \frac{2}{\pi} \sum_{\substack{n=-\infty \\ n\neq -1,2}}^{\infty} \frac{e^{i(2n-1)t}}{(2n-1)^2[(2n-1)^2-9]}.
\end{aligned} \tag{2.6.25}
$$

The first two terms on the right side of (2.6.25) represent the complementary solution. Although (2.6.25) is equivalent to (2.6.16), we have all of the advantages of dealing with exponentials rather than sines and cosines. These advantages include ease of differentiation and integration, and writing the series in terms of amplitude and phase.

• **Example 2.6.3: Temperature within a spinning satellite**

In the design of artificial satellites, it is important to determine the temperature distribution on the spacecraft's surface. An interesting special case is the temperature fluctuation in the skin due to the spinning of the vehicle. If the craft is thin-walled so that there is no radial dependence, Hrycak[16] showed that he could approximate the nondimensional temperature field at the equator of the rotating satellite by

$$
\frac{d^2T}{d\eta^2} + b\frac{dT}{d\eta} - c\left(T - \frac{3}{4}\right) = -\frac{\pi c}{4}\frac{F(\eta) + \beta/4}{1 + \pi\beta/4}, \tag{2.6.26}
$$

where

$$
b = 4\pi^2 r^2 f/a, \qquad c = \frac{16\pi S}{\gamma T_\infty}\left(1 + \frac{\pi\beta}{4}\right), \tag{2.6.27}
$$

$$
F(\eta) = \begin{cases} \cos(2\pi\eta), & 0 < \eta < \frac{1}{4} \\ 0, & \frac{1}{4} < \eta < \frac{3}{4} \\ \cos(2\pi\eta), & \frac{3}{4} < \eta < 1, \end{cases} \tag{2.6.28}
$$

$$
T_\infty = \left(\frac{S}{\pi\sigma\epsilon}\right)^{1/4}\left(\frac{1 + \pi\beta/4}{1 + \beta}\right)^{1/4}, \tag{2.6.29}
$$

a is the thermal diffusivity of the shell, f is the rate of spin, r is the radius of the spacecraft, S is the net direct solar heating, β is the ratio of the emissivity of the interior shell to the emissivity of the exterior surface, ϵ is the overall emissivity of the exterior surface, γ is the satellite's skin conductance, and σ is the Stefan-Boltzmann constant. The independent variable η is the longitude along the equator with the effect of rotation

---

[16] Hrycak, P., 1963: Temperature distribution in a spinning spherical space vehicle. *AIAA J.*, **1**, 96–99.

subtracted out ($2\pi\eta = \varphi - 2\pi ft$). The reference temperature T_∞ equals the temperature that the spacecraft would have if it spun with infinite angular speed so that the solar heating would be uniform around the craft. We have nondimensionalized the temperature with respect to T_∞.

We begin our analysis by introducing the new variables

$$y = T - \frac{3}{4} - \frac{\pi\beta}{16 + 4\pi\beta}, \quad \nu_0 = \frac{2\pi^2 r^2 f}{a\rho_0}, \quad A_0 = -\frac{\pi\rho^2}{4 + \pi\beta} \tag{2.6.30}$$

and $\rho_0^2 = c$ so that

$$\frac{d^2y}{d\eta^2} + 2\rho_0\nu_0\frac{dy}{d\eta} - \rho_0^2 y = A_0 F(\eta). \tag{2.6.31}$$

Next, we expand $F(\eta)$ as a Fourier series because it is a periodic function of period 1. Because it is an even function,

$$f(\eta) = \tfrac{1}{2}a_0 + \sum_{n=1}^{\infty} a_n \cos(2n\pi\eta), \tag{2.6.32}$$

where

$$a_0 = \frac{1}{1/2}\int_0^{1/4} \cos(2\pi x)\,dx + \frac{1}{1/2}\int_{3/4}^{1} \cos(2\pi x)\,dx = \frac{2}{\pi}, \tag{2.6.33}$$

$$a_1 = \frac{1}{1/2}\int_0^{1/4} \cos^2(2\pi x)\,dx + \frac{1}{1/2}\int_{3/4}^{1} \cos^2(2\pi x)\,dx = \frac{1}{2} \tag{2.6.34}$$

and

$$a_n = \frac{1}{1/2}\int_0^{1/4} \cos(2\pi x)\cos(2n\pi x)\,dx + \frac{1}{1/2}\int_{3/4}^{1} \cos(2\pi x)\cos(2n\pi x)\,dx \tag{2.6.35}$$

$$= -\frac{2(-1)^n}{\pi(n^2-1)}\cos\left(\frac{n\pi}{2}\right), \tag{2.6.36}$$

if $n \geq 2$. Therefore,

$$f(\eta) = \frac{1}{\pi} + \frac{1}{2}\cos(2\pi\eta) - \frac{2}{\pi}\sum_{n=1}^{\infty}\frac{(-1)^n}{4n^2-1}\cos(4n\pi\eta). \tag{2.6.37}$$

From the method of undetermined coefficients, the particular solution is

$$y_p(\eta) = \tfrac{1}{2}a_0 + a_1\cos(2\pi\eta) + b_1\sin(2\pi\eta) + \sum_{n=1}^{\infty} a_{2n}\cos(4n\pi\eta) + b_{2n}\sin(4n\pi\eta), \tag{2.6.38}$$

which yields

$$y_p'(\eta) = -2\pi a_1 \sin(2\pi\eta) + 2\pi b_1 \cos(2\pi\eta) + \sum_{n=1}^{\infty}[-4n\pi a_{2n} \sin(4n\pi\eta) + 4n\pi b_{2n} \cos(4n\pi\eta)] \quad (2.6.39)$$

and

$$y_p''(\eta) = -4\pi^2 a_1 \cos(2\pi\eta) - 4\pi^2 b_1 \sin(2\pi\eta) + \sum_{n=1}^{\infty}[-16n^2\pi^2 a_{2n} \cos(4n\pi\eta) - 16n^2\pi^2 b_{2n} \sin(4n\pi\eta)]. \quad (2.6.40)$$

Substituting into (2.6.31),

$$-\frac{1}{2}\rho_0^2 a_0 - \frac{A_0}{\pi} + \left(-4\pi^2 a_1 + 4\pi\rho_0\nu_0 b_1 - \rho_0^2 a_1 - \frac{A_0}{2}\right)\cos(2\pi\eta) + \left(-4\pi^2 b_1 - 4\pi\rho_0\nu_0 a_1 - \rho_0^2 b_1\right)\sin(2\pi\eta) + \sum_{n=1}^{\infty}\left[-16n^2\pi^2 a_{2n} + 8n\pi\rho_0\nu_0 b_{2n} - \rho_0^2 a_{2n} + \frac{2A_0(-1)^n}{\pi(4n^2-1)}\right]\cos(4n\pi\eta) + \sum_{n=1}^{\infty}\left(-16n^2\pi^2 b_{2n} - 8n\pi\rho_0\nu_0 a_{2n} - \rho_0^2 b_{2n}\right)\sin(4n\pi\eta) = 0. \quad (2.6.41)$$

In order to satisfy (2.6.41) for any η, we set

$$a_0 = -\frac{2A_0}{\pi\rho_0^2}, \quad (2.6.42)$$

$$-(4\pi^2 + \rho_0^2)a_1 + 4\pi\rho_0\nu_0 b_1 = \frac{A_0}{2}, \quad (2.6.43)$$

$$4\pi\rho_0\nu_0 a_1 + (4\pi^2 + \rho_0^2)b_1 = 0, \quad (2.6.44)$$

$$(16n^2\pi^2 + \rho_0^2)a_{2n} - 8n\pi\rho_0\nu_0 b_{2n} = \frac{2A_0(-1)^n}{\pi(4n^2-1)} \quad (2.6.45)$$

and

$$8n\pi\rho_0\nu_0 a_{2n} + (16n^2\pi^2 + \rho_0^2)b_{2n} = 0 \quad (2.6.46)$$

or

$$[16\pi^2\rho_0^2\nu_0^2 + (4\pi^2 + \rho_0^2)^2]a_1 = -\frac{(4\pi^2 + \rho_0^2)A_0}{2}, \quad (2.6.47)$$

$$[16\pi^2\rho_0^2\nu_0^2 + (4\pi^2 + \rho_0^2)^2]b_1 = 2\pi\rho_0\nu_0 A_0, \quad (2.6.48)$$

$$[64n^2\pi^2\rho_0^2\nu_0^2 + (16n^2\pi^2 + \rho_0^2)^2]a_{2n} = \frac{2A_0(-1)^n(16n^2\pi^2 + \rho_0^2)}{\pi(4n^2-1)} \quad (2.6.49)$$

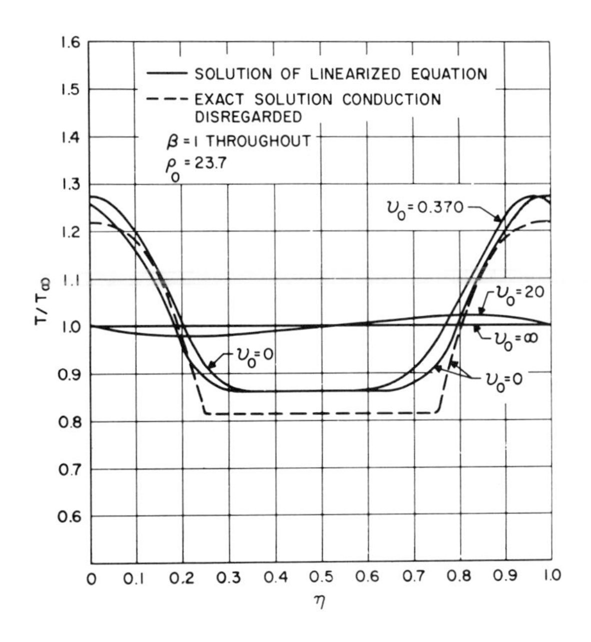

Figure 2.6.1: Temperature distribution along the equator of a spinning spherical satellite. (From Hrycak, P., 1963: Temperature distribution in a spinning spherical space vehicle. *AIAA J.*, **1**, 97. ©1963 AIAA, reprinted with permission.)

and

$$[64n^2\pi^2\rho_0^2\nu_0^2 + (16n^2\pi^2 + \rho_0^2)^2]b_{2n} = -\frac{16(-1)^n\rho_0\nu_0 nA_0}{4n^2-1}. \qquad \mathbf{(2.6.50)}$$

Substituting for a_0, a_1, b_1, a_{2n}, and b_{2n}, the particular solution is

$$\begin{aligned} y(\eta) = &-\frac{A_0}{\pi\rho_0^2} - \frac{(4\pi^2+\rho_0^2)A_0\cos(2\pi\eta)}{2[(4\pi^2+\rho_0^2)^2+16\pi^2\rho_0^2\nu_0^2]} + \frac{2\pi\rho_0\nu_0A_0\sin(2\pi\eta)}{(4\pi^2+\rho_0^2)^2+16\pi^2\rho_0^2\nu_0^2} \\ &+ \frac{2A_0}{\pi}\sum_{n=1}^{\infty}\frac{(-1)^n(16n^2\pi^2+\rho_0^2)\cos(2n\pi\eta)}{(4n^2-1)[64n^2\pi^2\rho_0^2\nu_0^2+(16n^2\pi^2+\rho_0^2)^2]} \\ &- 16\rho_0\nu_0A_0\sum_{n=1}^{\infty}\frac{(-1)^n n\sin(2n\pi\eta)}{(4n^2-1)[64n^2\pi^2\rho_0^2\nu_0^2+(16n^2\pi^2+\rho_0^2)^2]}. \end{aligned} \qquad \mathbf{(2.6.51)}$$

In Figure 2.6.1 we reproduce a figure from Hrycak's paper showing the variation of the nondimensional temperature as a function of η for the spinning rate ν_0. The other parameters are typical of a satellite with aluminum skin and fully covered with glass-protected solar cells.

As a check on the solution, we show the temperature field (the dashed line) of a nonrotating satellite where we neglect the effects of conduction and only radiation occurs. The difference between the $\nu_0 = 0$ solid and dashed lines arises primarily due to the *linearization* of the nonlinear radiation boundary condition during the derivation of the governing equations.

Problems

Solve the following ordinary differential equations by Fourier series if the forcing is by the periodic function

$$f(t) = \begin{cases} 1, & 0 < t < \pi \\ 0, & \pi < t < 2\pi \end{cases}$$

and $f(t) = f(t + 2\pi)$:

1. $y'' - y = f(t)$, 2. $y'' + y = f(t)$, 3. $y'' - 3y' + 2y = f(t)$.

Solve the following ordinary differential equations by *complex* Fourier series if the forcing is by the periodic function

$$f(t) = |t|, \qquad -\pi < t < \pi,$$

and $f(t) = f(t + 2\pi)$:

4. $y'' - y = f(t)$, 5. $y'' + 4y = f(t)$.

6. An object radiating into its nocturnal surrounding has a temperature $y(t)$ governed by the equation[17]:

$$\frac{dy}{dt} + ay = A_0 + \sum_{n=1}^{\infty} A_n \cos(n\omega t) + B_n \sin(n\omega t),$$

where the constant a is the heat loss coefficient and the Fourier series describes the temporal variation of the atmospheric air temperature and the effective sky temperature. If $y(0) = T_0$, find $y(t)$.

7. The equation that governs the charge q on the capacitor of an LRC electrical circuit is

$$q'' + 2\alpha q' + \omega^2 q = \omega^2 E,$$

[17] Reprinted from *Solar Energy*, **28**, Sodha, M. S., Transient radiative cooling, 541, ©1982, with the kind permission from Elsevier Science Ltd, The Boulevard, Langford Lane, Kidlington, OX5 1GB, UK.

where $\alpha = R/2L$, $\omega^2 = 1/LC$, R denotes resistance, C denotes capacitance, L denotes the inductance, and E is the electromotive force driving the circuit. If E is given by

$$E = \sum_{n=-\infty}^{\infty} \varphi_n e^{in\omega_0 t},$$

find $q(t)$.

2.7 FINITE FOURIER SERIES

In many applications we must construct a Fourier series from values given by data or a graph. Unlike the situation for an analytic formula where we have an infinite number of data points and, consequently, an infinite number of terms in the Fourier series, the Fourier series contains a finite number of sine and cosines. This number is controlled by the number of data points; there must be at least two points (one for the crest, the other for the trough) to resolve the highest harmonic.

Assuming that these series are useful, the next question is how do we find the Fourier coefficients? We could compute them by numerically integrating (2.1.6). However, the results would suffer from the truncation errors that afflict all numerical schemes. On the other hand, we can avoid this problem if we again employ the orthogonality properties of sines and cosines, now in their discrete form. Just as in the case of conventional Fourier series, we can use these properties to derive formulas for computing the Fourier coefficients. These results will be *exact* except for roundoff errors.

We begin our analysis by deriving some preliminary results. Let us define $x_m = mP/(2N)$. Then, if k is an integer,

$$\sum_{m=0}^{2N-1} \exp\left(\frac{2\pi i k x_m}{P}\right) = \sum_{m=0}^{2N-1} \exp\left(\frac{km\pi i}{N}\right) = \sum_{m=0}^{2N-1} r^m \quad \textbf{(2.7.1)}$$

$$= \begin{cases} \frac{1-r^{2N}}{1-r} = 0, & r \neq 1 \\ 2N, & r = 1 \end{cases} \quad \textbf{(2.7.2)}$$

because $r^{2N} = \exp(2\pi k i) = 1$ if $r \neq 1$. If $r = 1$, then the sum consists of $2N$ terms, each of which equals one. The condition $r = 1$ corresponds to $k = 0, \pm 2N, \pm 4N, \ldots$. Taking the real and imaginary part of (2.7.2),

$$\sum_{m=0}^{2N-1} \cos\left(\frac{2\pi k x_m}{P}\right) = \begin{cases} 0, & k \neq 0, \pm 2N, \pm 4N, \ldots \\ 2N, & k = 0, \pm 2N, \pm 4N, \ldots \end{cases} \quad \textbf{(2.7.3)}$$

and

$$\sum_{m=0}^{2N-1} \sin\left(\frac{2\pi k x_m}{P}\right) = 0 \tag{2.7.4}$$

for all k.

Consider now the following sum:

$$\sum_{m=0}^{2N-1} \cos\left(\frac{2\pi k x_m}{P}\right) \cos\left(\frac{2\pi j x_m}{P}\right)$$

$$= \frac{1}{2} \sum_{m=0}^{2N-1} \left\{ \cos\left[\frac{2\pi(k+j)x_m}{P}\right] + \cos\left[\frac{2\pi(k-j)x_m}{P}\right] \right\} \tag{2.7.5}$$

$$= \begin{cases} 0, & |k-j| \text{ and } |k+m| \neq 0, 2N, 4N, \ldots \\ N, & |k-j| \text{ or } |k+m| \neq 0, 2N, 4N, \ldots \\ 2N, & |k-j| \text{ and } |k+m| = 0, 2N, 4N, \ldots \end{cases} \tag{2.7.6}$$

Let us simplify the right side of (2.7.6) by restricting ourselves to $k+j$ lying between 0 to $2N$. This is permissible because of the periodic nature of (2.7.5). If $k+j=0$, $k=j=0$; if $k+j=2N$, $k=j=N$. In either case, $k-j=0$ and the right side of (2.7.6) equals $2N$. Consider now the case $k \neq j$. Then $k+j \neq 0$ or $2N$ and $k-j \neq 0$ or $2N$. The right side of (2.7.6) must equal 0. Finally, if $k=j \neq 0$ or N, then $k+j \neq 0$ or $2N$ but $k-j=0$ and the right side of (2.7.6) equals N. In summary,

$$\sum_{m=0}^{2N-1} \cos\left(\frac{2\pi k x_m}{P}\right) \cos\left(\frac{2\pi j x_m}{P}\right) = \begin{cases} 0, & k \neq j \\ N, & k = j \neq 0, N \\ 2N, & k = j = 0, N. \end{cases} \tag{2.7.7}$$

In a similar manner,

$$\sum_{m=0}^{2N-1} \cos\left(\frac{2\pi k x_m}{P}\right) \sin\left(\frac{2\pi j x_m}{P}\right) = 0 \tag{2.7.8}$$

for all k and j and

$$\sum_{m=0}^{2N-1} \sin\left(\frac{2\pi k x_m}{P}\right) \sin\left(\frac{2\pi j x_m}{P}\right) = \begin{cases} 0, & k \neq j \\ N, & k = j \neq 0, N \\ 0, & k = j = 0, N. \end{cases} \tag{2.7.9}$$

Armed with (2.7.7)–(2.7.9) we are ready to find the coefficients A_n and B_n of the finite Fourier series,

$$f(x) = \frac{A_0}{2} + \sum_{k=1}^{N-1} \left[A_k \cos\left(\frac{2\pi k x}{P}\right) + B_k \sin\left(\frac{2\pi k x}{P}\right) \right]$$

$$+ \frac{A_N}{2} \cos\left(\frac{2\pi N x}{P}\right), \tag{2.7.10}$$

where we have $2N$ data points and we now define P as the period of the function.

To find A_k we proceed as before and multiply (2.7.10) by $\cos(2\pi jx/P)$ (j may take on values from 0 to N) and sum from 0 to $2N-1$. At the point $x = x_m$,

$$\sum_{m=0}^{2N-1} f(x_m)\cos\left(\frac{2\pi j}{P}x_m\right) = \frac{A_0}{2}\sum_{m=0}^{2N-1}\cos\left(\frac{2\pi j}{P}x_m\right) + \sum_{k=1}^{N-1} A_k \sum_{m=0}^{2N-1}\cos\left(\frac{2\pi k}{P}x_m\right)\cos\left(\frac{2\pi j}{P}x_m\right) + \sum_{k=1}^{N-1} B_k \sum_{m=0}^{2N-1}\sin\left(\frac{2\pi k}{P}x_m\right)\cos\left(\frac{2\pi j}{P}x_m\right) + \frac{A_N}{2}\sum_{m=0}^{2N-1}\cos\left(\frac{2\pi N}{P}x_m\right)\cos\left(\frac{2\pi j}{P}x_m\right). \quad \textbf{(2.7.11)}$$

If $j \neq 0$ or N, then the first summation on the right side vanishes by (2.7.3), the third by (2.7.9), and the fourth by (2.7.7). The second summation does *not* vanish if $k = j$ and equals N. Similar considerations lead to the formulas for the calculation of A_k and B_k:

$$A_k = \frac{1}{N}\sum_{m=0}^{2N-1} f(x_m)\cos\left(\frac{2\pi k}{P}x_m\right), \qquad k = 0, 1, 2, \ldots, N \quad \textbf{(2.7.12)}$$

and

$$B_k = \frac{1}{N}\sum_{m=0}^{2N-1} f(x_m)\sin\left(\frac{2\pi k}{P}x_m\right), \qquad k = 1, 2, \ldots, N-1. \quad \textbf{(2.7.13)}$$

If there are $2N+1$ data points and $f(x_0) = f(x_{2N})$, then (2.7.12)–(2.7.13) is still valid and we need only consider the first $2N$ points. If $f(x_0) \neq f(x_{2N})$, we can still use our formulas if we require that the endpoints have the value of $[f(x_0)+f(x_{2N})]/2$. In this case the formulas for the coefficients A_k and B_k are

$$A_k = \frac{1}{N}\left[\frac{f(x_0)+f(x_{2N})}{2} + \sum_{m=1}^{2N-1} f(x_m)\cos\left(\frac{2\pi k}{P}x_m\right)\right], \quad \textbf{(2.7.14)}$$

where $k = 0, 1, 2, \ldots, N$ and

$$B_k = \frac{1}{N}\sum_{m=1}^{2N-1} f(x_m)\sin\left(\frac{2\pi k}{P}x_m\right), \quad \textbf{(2.7.15)}$$

Table 2.7.1: The Depth of Water in the Harbor at Buffalo, NY (Minus the Low-Water Datum of 568.8 ft) on the 15th Day of Each Month During 1977.

mo	n	depth	mo	n	depth	mo	n	depth
Jan	1	1.61	May	5	3.16	Sep	9	2.42
Feb	2	1.57	Jun	6	2.95	Oct	10	2.95
Mar	3	2.01	Jul	7	3.10	Nov	11	2.74
Apr	4	2.68	Aug	8	2.90	Dec	12	2.63

where $k = 1, 2, \ldots, N-1$.

It is important to note that $2N$ data points yield $2N$ Fourier coefficients A_k and B_k. Consequently our sampling frequency will always limit the amount of information, whether in the form of data points or Fourier coefficients. It might be argued that from the Fourier series representation of $f(t)$ we could find the value of $f(t)$ for any given t, which is more than we can do with the data alone. This is not true. Although we can calculate a value for $f(t)$ at any t using the finite Fourier series, we simply do not know whether those values are correct or not. They are simply those given by a finite Fourier series which fit the given data points. Despite this, the Fourier analysis of finite data sets yields valuable physical insights into the processes governing many physical systems.

• **Example 2.7.1: Water depth at Buffalo, NY**

Each entry[18] in Table 2.7.1 gives the observed depth of water at Buffalo, NY (minus the low-water datum of 568.6 ft) on the 15th of the corresponding month during 1977. Assuming that the water level is a periodic function of 1 year, and that we took the observations at equal intervals, we want to construct a finite Fourier series from these data. This corresponds to computing the Fourier coefficients $A_0, A_1, \ldots, A_6, B_1, \ldots, B_5$, which give the mean level and harmonic fluctuations of the depth of water, the harmonics having the periods 12 months, 6 months, 4 months, and so forth.

In this problem, P equals 12 months, $N = P/2 = 6$ and $x_m = mP/(2N) = m(12 \text{ mo})/12 = m$ mo. That is, there should be a data

[18] National Ocean Survey, 1977: Great Lakes Water Level, 1977, Daily and Monthly Average Water Surface Elevations, National Oceanic and Atmospheric Administration, Rockville, MD.

point for each month. From (2.7.12) and (2.7.13),

$$A_k = \frac{1}{6} \sum_{m=0}^{11} f(x_m) \cos\left(\frac{mk\pi}{6}\right), \quad k = 0, 1, 2, 3, 4, 5, 6 \qquad \textbf{(2.7.16)}$$

and

$$B_k = \frac{1}{6} \sum_{m=0}^{11} f(x_m) \sin\left(\frac{mk\pi}{6}\right), \quad k = 1, 2, 3, 4, 5. \qquad \textbf{(2.7.17)}$$

Substituting the data into (2.7.16)–(2.7.17) yields

A_0	= twice the mean level		= +5.120 ft
A_1	= harmonic component with a period of	12 mo	= −0.566 ft
B_1	= harmonic component with a period of	12 mo	= −0.128 ft
A_2	= harmonic component with a period of	6 mo	= −0.177 ft
B_2	= harmonic component with a period of	6 mo	= −0.372 ft
A_3	= harmonic component with a period of	4 mo	= −0.110 ft
B_3	= harmonic component with a period of	4 mo	= −0.123 ft
A_4	= harmonic component with a period of	3 mo	= +0.025 ft
B_4	= harmonic component with a period of	3 mo	= +0.052 ft
A_5	= harmonic component with a period of	2.4 mo	= −0.079 ft
B_5	= harmonic component with a period of	2.4 mo	= −0.131 ft
A_6	= harmonic component with a period of	2 mo	= −0.107 ft

Figure 2.7.1 is a plot of our results using (2.7.10). Note that when we include all of the harmonic terms, the finite Fourier series fits the data points exactly. The values given by the series at points between the data points may be right or they may not. To illustrate this, we also plotted the values for the first of each month. Sometimes the values given by the Fourier series and these intermediate data points are quite different.

Let us now examine our results in terms of various physical processes. In the long run the depth of water in the harbor at Buffalo, NY depends upon the three-way balance between precipitation, evaporation, and inflow-outflow of any rivers. Because the inflow and outflow of the rivers depends strongly upon precipitation, and evaporation is of secondary importance, the water level should correlate with the precipitation rate. It is well known that more precipitation falls during the warmer months rather than the colder months. The large amplitude of the Fourier coefficient A_1 and B_1, corresponding to the annual cycle ($k = 1$), reflects this.

Another important term in the harmonic analysis corresponds to the semiannual cycle ($k = 2$). During the winter months around Lake

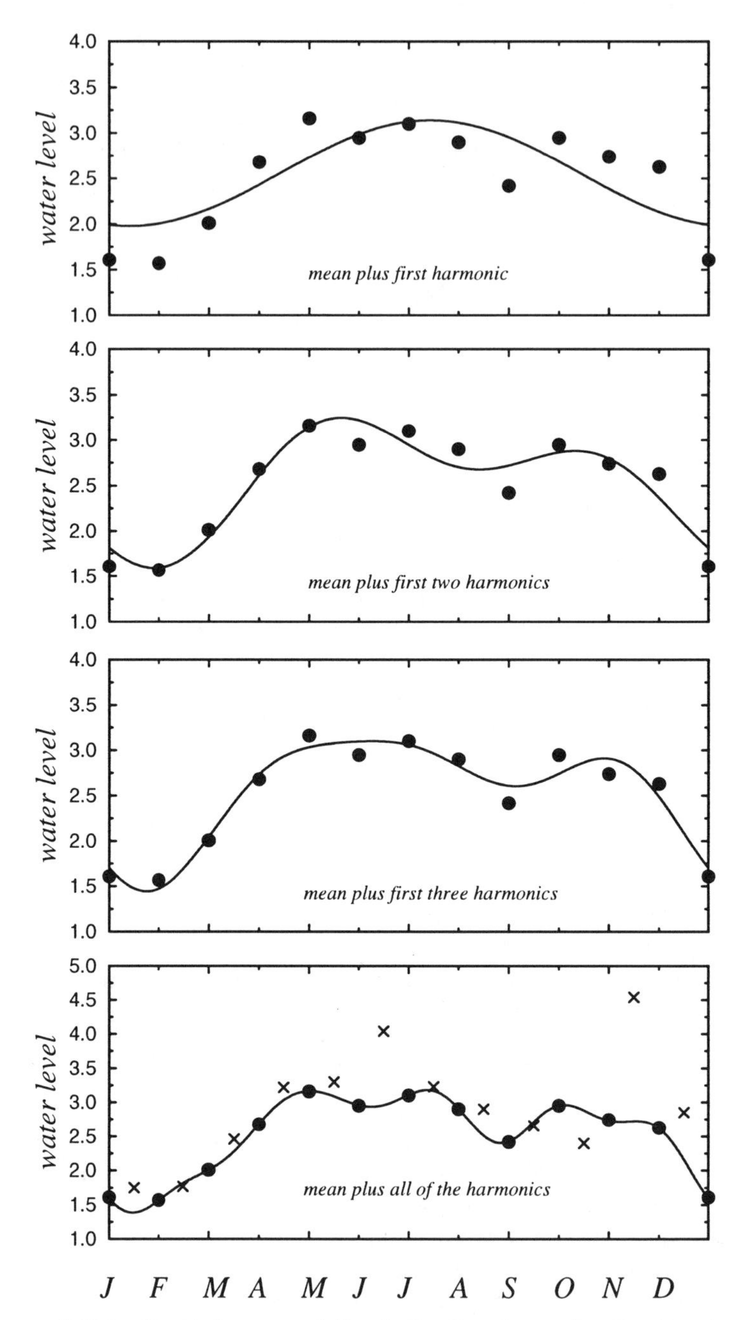

Figure 2.7.1: Partial sums of the finite Fourier series for the depth of water in the harbor of Buffalo, NY during 1977. Circles indicate observations on the 15th of the month; crosses are observations on the first.

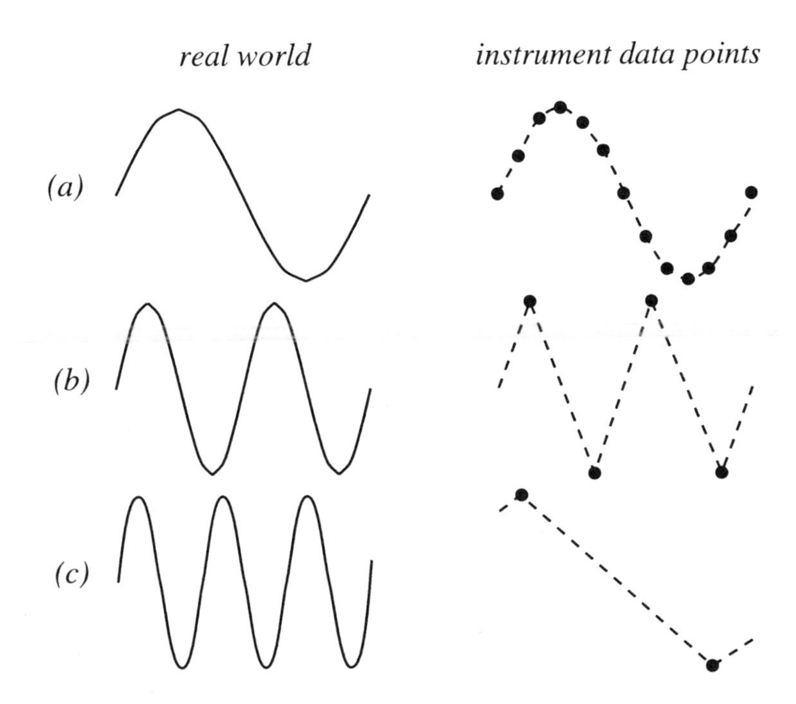

Figure 2.7.2: The effect of sampling in the representation of periodic functions.

Ontario, precipitation falls as snow. Therefore, the inflow from rivers is greatly reduced. When spring comes, the snow and ice melt and a jump in the water level occurs. Because the second harmonic gives periodic variations associated with seasonal variations, this harmonic is absolutely necessary if we want to get the correct answer while the higher harmonics do not represent any specific physical process.

• **Example 2.7.2: Aliasing**

In the previous example, we could only resolve phenomena with a period of 2 months or greater although we had data for each of the 12 months. This is an example of *Nyquist's sampling criteria*[19]: At least two samples are required to resolve the highest frequency in a periodically sampled record.

Figure 2.7.2 will help explain this phenomenon. In case (a) we have quite a few data points over one cycle. Consequently our picture, constructed from data, is fairly good. In case (b), we have only taken samples at the ridges and troughs of the wave. Although our picture

[19] Nyquist, H., 1928: Certain topics in telegraph transmission theory. *AIEE Trans.*, **47**, 617–644.

of the real phenomenon is poor, at least we know that there is a wave. From this picture we see that even if we are lucky enough to take our observations at the ridges and troughs of a wave, we need at least two data points per cycle (one for the ridge, the other for the trough) to resolve the highest-frequency wave.

In case (c) we have made a big mistake. We have taken a wave of frequency N Hz and misrepresented it as a wave of frequency $N/2$ Hz. This mispresentation of a high-frequency wave by a lower-frequency wave is called *aliasing*. It arises because we are sampling a continuous signal at equal intervals. By comparing cases (b) and (c), we see that there is a cutoff between aliased and nonaliased frequencies. This frequency is called the *Nyquist* or *folding* frequency. It corresponds to the highest frequency resolved by our finite Fourier analysis.

Because most periodic functions require an infinite number of harmonics for their representation, aliasing of signals is a common problem. Thus the question is not "can I avoid aliasing?" but "can I live with it?" Quite often, we can construct our experiments to say yes. An example where aliasing is unavoidable occurs in a Western at the movies when we see the rapidly rotating spokes of the stagecoach's wheel. A movie is a sampling of continuous motion where we present the data as a succession of pictures. Consequently, a film aliases the high rate of revolution of the stagecoach's wheel in such a manner so that it appears to be stationary or rotating very slowly.

• Example 2.7.3: Spectrum of the Chesapeake Bay

For our final example we will perform a Fourier analysis of hourly sea-level measurements taken at the mouth of the Chesapeake Bay during the 2000 days from 9 April 1985 to 29 June 1990. Figure 2.7.3 shows 200 days of this record, starting from 1 July 1985. As this figure shows, the measurements contain a wide range of oscillations. In particular, note the large peak near day 90 which corresponds to the passage of hurricane Gloria during the early hours of 27 September 1985.

Utilizing the entire 2000 days, we have plotted the amplitude of the Fourier coefficients as a function of period in Figure 2.7.4. We see a general rise of the amplitude as the period increases. Especially noteworthy are the sharp peaks near periods of 12 and 24 hours. The largest peak is at 12.417 hours and corresponds to the semidiurnal tide. Thus, our Fourier analysis has shown that the dominant oscillations at the mouth of the Chesapeake Bay are the tides. A similar situation occurs in Baltimore harbor. Furthermore, with this spectral information we could predict high and low tides very accurately.

Although the tides are of great interest to many, they are a nuisance to others because they mask other physical processes that might be occurring. For that reason we would like to remove them from the tidal

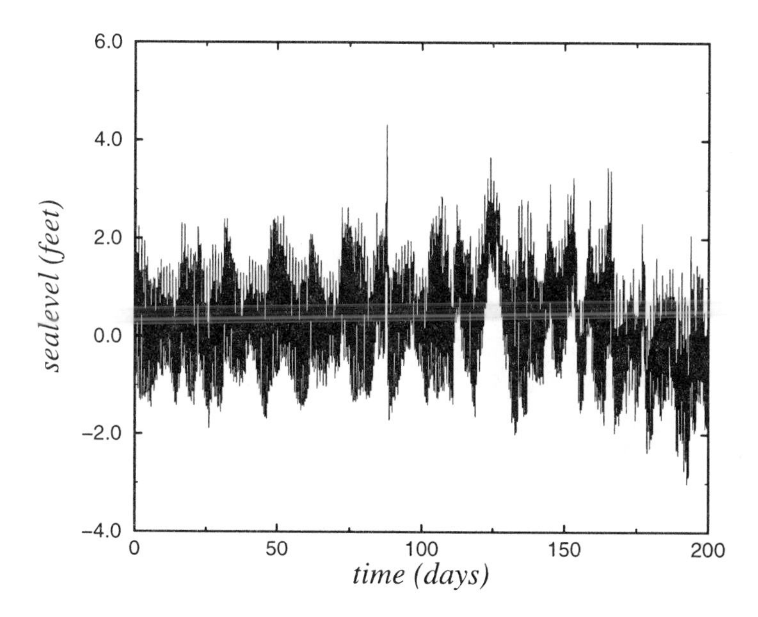

Figure 2.7.3: The sea elevation at the mouth of the Chesapeake Bay from its average depth as a function of time after 1 July 1985.

gauge history and see what is left. One way would be to zero out the Fourier coefficients corresponding to the tidal components and then plot the resulting Fourier series. Another method is to replace each hourly report with an average of hourly reports that occurred 24 hours ahead and behind of a particular report. We will construct this average in such a manner that waves with periods of the tides sum to zero.[20] Such a *filter* is a popular method for eliminating unwanted waves from a record. Filters play an important role in the analysis of data. We have plotted the filtered sea level data in Figure 2.7.5. Note that summertime (0–50 days) produces little variation in the sea level compared to wintertime (100–150 days) when intense coastal storms occur.

Problems

Find the finite Fourier series for the following pieces of data:

1. $x(0) = 0$, $x(1) = 1$, $x(2) = 2$, $x(3) = 3$ and $N = 2$.

2. $x(0) = 1$, $x(1) = 1$, $x(2) = -1$, $x(3) = -1$ and $N = 2$.

---

[20] See Godin, G., 1972: *The Analysis of Tides*, University of Toronto Press, Toronto, Section 2.1.

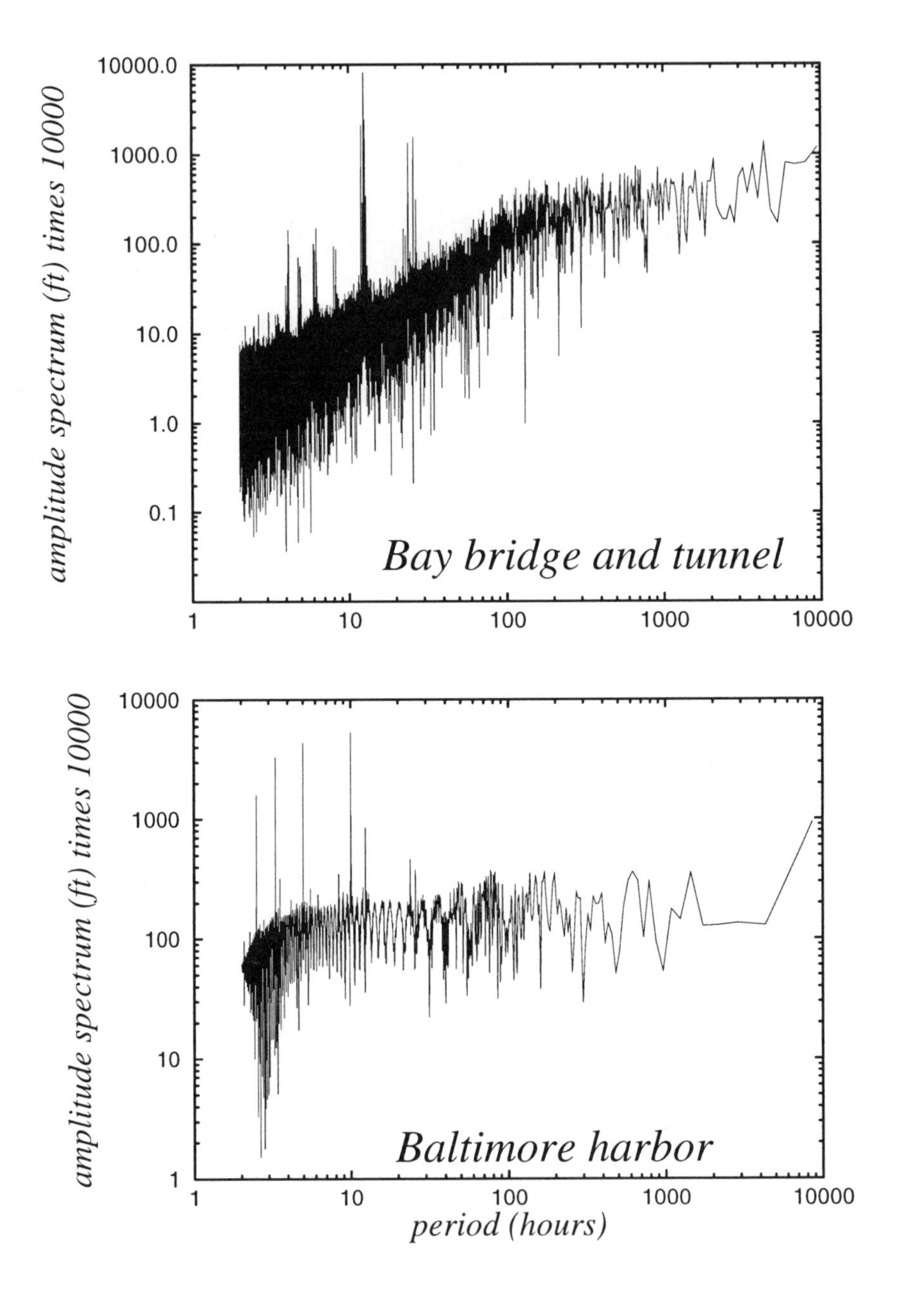

Figure 2.7.4: The amplitude of the Fourier coefficients for the sea elevation at the Chesapeake Bay bridge and tunnel (top) and Baltimore harbor (bottom) as a function of period.

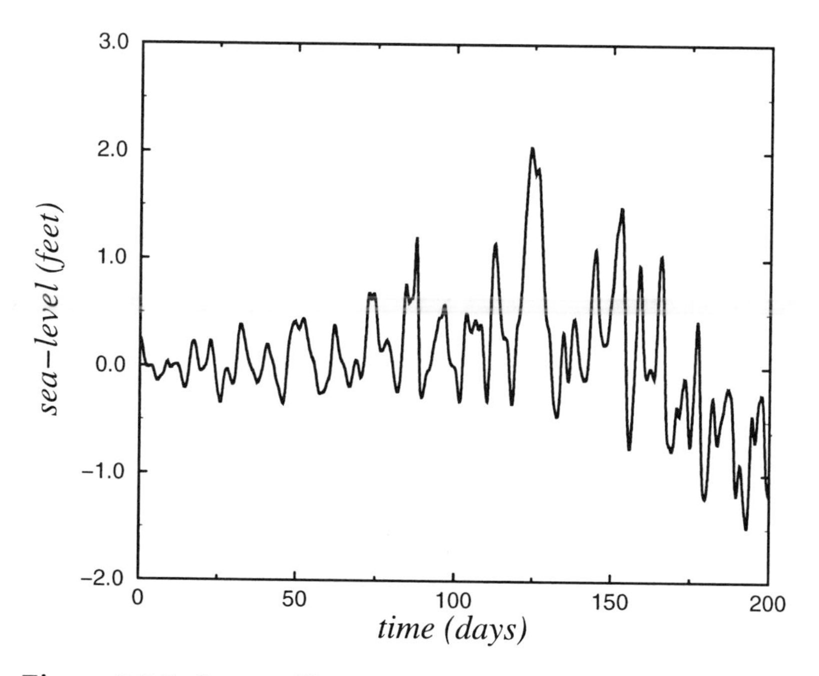

Figure 2.7.5: Same as Figure 2.7.3 but with the tides removed.

Project: Spectrum of the Earth's Orography

Table 2.7.3 gives the orographic height of the earth's surface used in an atmospheric general circulation model (GCM) at a resolution of 2.5° longitude along the latitude belts of 28°S, 36°N, and 66°N. In this project you will find the spectrum of this orographic field along the various latitude belts.

Step 1: Write code to read in the data and find A_n and B_n. Although you could code (2.7.12)–(2.7.13), no one does Fourier analysis that way any more. They use a fast Fourier transform (FFT) that is available as a system's routine on their computer or use one that is given in various computer books.[21] Many of these routines deal with finite Fourier series in its complex form. The only way that you can be confident of your results is to first create a data set with a known Fourier series, for example:

$$f(x) = 5 + \cos\left(\frac{2\pi x}{2N}\right) + 3\sin\left(\frac{2\pi x}{2N}\right) + 6\cos\left(\frac{6\pi x}{2N}\right),$$

[21] For example, Press, W. H., Flannery, B. P., Teukolsky, S. A., and Vetterling, W. T., 1986: *Numerical Recipes: The Art of Scientific Computing*, Cambridge University Press, New York, chap. 12.

Table 2.7.2: The Fourier Coefficients Generated by the IMSL Subroutine FFTRF with $N = 8$ for the Test Case Given in Step 1 of the Project.

x	$f(x)$	Fourier coefficient	Value of Fourier coefficient
0.00000	12.00000	$2NA_0$	80.00001
1.00000	9.36803	NA_1	8.00001
2.00000	3.58579	$-NB_1$	−24.00000
3.00000	2.61104	NA_2	0.00000
4.00000	8.00000	$-NB_2$	0.00001
5.00000	12.93223	NA_3	48.00000
6.00000	10.65685	$-NB_3$	0.00001
7.00000	2.92807	NA_4	−0.00001
8.00000	−2.00000	$-NB_4$	0.00000
9.00000	0.63197	NA_5	0.00000
10.00000	6.41421	$-NB_5$	0.00000
11.00000	7.38895	NA_6	0.00000
12.00000	2.00000	$-NB_6$	0.00000
13.00000	−2.93223	NA_7	−0.00001
14.00000	−0.65685	$-NB_7$	0.00000
15.00000	7.07193	$2NA_8$	0.00001

and then find the Fourier coefficients given by the subroutine. In Table 2.7.3 we show the results from using the IMSL routine FFTRF. From these results, you see that the Fourier coefficients given by the subroutine are multiplied by N and the B_ns are of opposite sign.

Step 2: Construct several spectra by using every data point, every other data point, etc. How do the magnitudes of the Fourier coefficient change? You might like to read about *leakage* from a book on harmonic analysis.[22]

Step 3: Compare and contrast the spectra from the various latitude belts. How do the magnitudes of the Fourier coefficients decrease with n? Why are there these differences?

Step 4: You may have noted that some of the heights are negative, even in the middle of the ocean! Take the original data (for any latitude belt) and zero out all of the negative heights. Find the spectra for this

[22] For example, Bloomfield, P., 1976: *Fourier Analysis of Time Series: An Introduction*, John Wiley & Sons, New York.

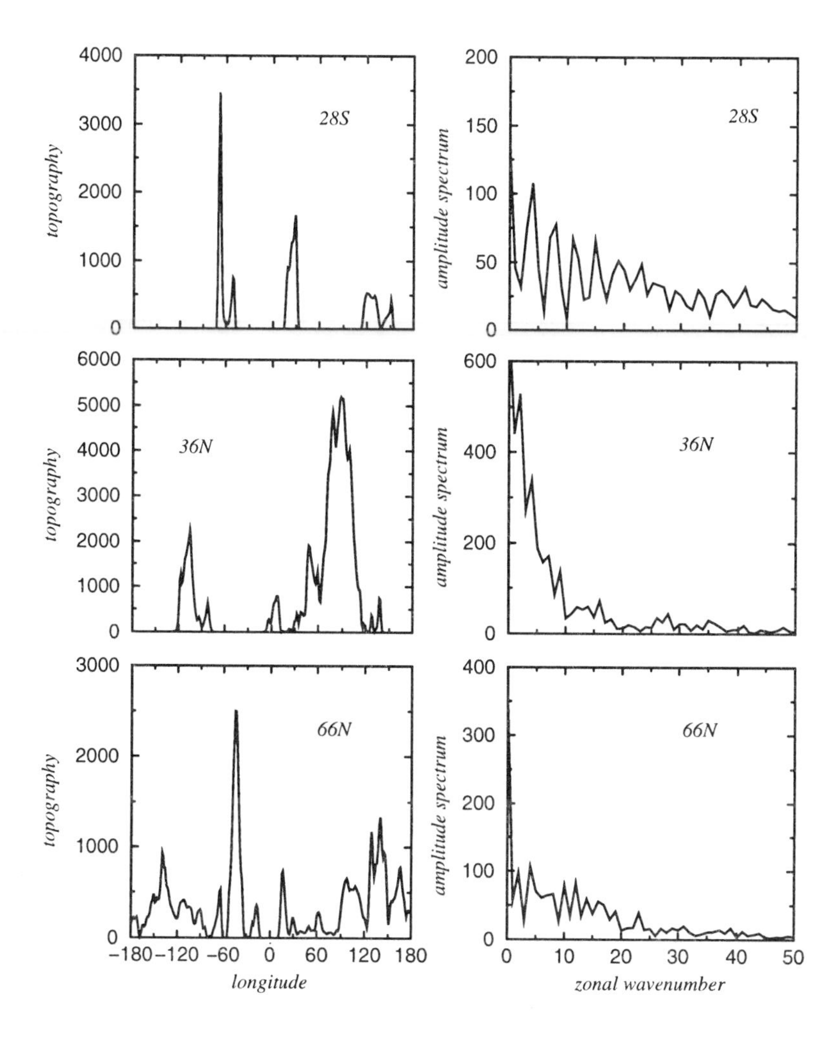

Figure 2.7.6: The orography of the earth and its spectrum in meters along three latitude belts.

new data set. How has the spectra changed? Is there a reason why the negative heights were introduced?

Table 2.7.3: Orographic Heights (in m) Along Three Latitude Belts.

Longitude	28°S	36°N	66°N	Longitude	28°S	36°N	66°N
−180.0	4.	3.	2532.	−82.5	36.	4047.	737.
−177.5	1.	−2.	1665.	−80.0	−64.	3938.	185.
−175.0	1.	2.	1432.	−77.5	138.	1669.	71.
−172.5	1.	−3.	1213.	−75.0	−363.	236.	160.
−170.0	1.	1.	501.	−72.5	4692.	31.	823.
−167.5	1.	−3.	367.	−70.0	19317.	−8.	1830.
−165.0	1.	1.	963.	−67.5	21681.	0.	3000.
−162.5	0.	0.	1814.	−65.0	9222.	−2.	3668.
−160.0	−1.	6.	2562.	−62.5	1949.	−2.	2147.
−157.5	0.	1.	3150.	−60.0	774.	0.	391.
−155.0	0.	3.	4008.	−57.5	955.	5.	−77.
−152.5	1.	−2.	4980.	−55.0	2268.	6.	601.
−150.0	−1.	4.	6011.	−52.5	4636.	−1.	3266.
−147.5	6.	−1.	6273.	−50.0	4621.	2.	9128.
−145.0	14.	3.	5928.	−47.5	1300.	−4.	17808.
−142.5	6.	−1.	6509.	−45.0	−91.	1.	22960.
−140.0	−2.	6.	7865.	−42.5	57.	−1.	20559.
−137.5	0.	3.	7752.	−40.0	−25.	4.	14296.
−135.0	−2.	5.	6817.	−37.5	13.	−1.	9783.
−132.5	1.	−2.	6272.	−35.0	−10.	6.	5969.
−130.0	−2.	0.	5582.	−32.5	8.	2.	1972.
−127.5	0.	5.	4412.	−30.0	−4.	22.	640.
−125.0	−2.	423.	3206.	−27.5	6.	33.	379.
−122.5	1.	3688.	2653.	−25.0	−2.	39.	286.
−120.0	−3.	10919.	2702.	−22.5	3.	2.	981.
−117.5	2.	16148.	3062.	−20.0	−3.	11.	1971.
−115.0	−3.	17624.	3344.	−17.5	1.	−6.	2576.
−112.5	7.	18132.	3444.	−15.0	−1.	19.	1692.
−110.0	12.	19511.	3262.	−12.5	0.	−18.	357.
−107.5	9.	22619.	3001.	−10.0	−1.	490.	−21.
−105.0	−5.	20273.	2931.	−7.5	0.	2164.	−5.
−102.5	3.	12914.	2633.	−5.0	1.	4728.	−10.
−100.0	−5.	7434.	1933.	−2.5	0.	5347.	0.
−97.5	6.	4311.	1473.	0.0	4.	2667.	−6.
−95.0	−8.	2933.	1689.	2.5	−5.	1213.	−1.
−92.5	8.	2404.	2318.	5.0	7.	1612.	−31.
−90.0	−12.	1721.	2285.	7.5	−13.	1744.	−58.
−87.5	18.	1681.	1561.	10.0	28.	1153.	381.
−85.0	−23.	2666.	1199.	12.5	107.	838.	2472.

Table 2.7.3, contd.: Orographic Heights (in m) Along Three Latitude Belts.

Longitude	28°S	36°N	66°N	Longitude	28°S	36°N	66°N
15.0	2208.	1313.	5263.	97.5	0.	35538.	6222.
17.5	6566.	862.	5646.	100.0	−2.	31985.	5523.
20.0	9091.	1509.	3672.	102.5	0.	23246.	4823.
22.5	10690.	2483.	1628.	105.0	−4.	17363.	4689.
25.0	12715.	1697.	889.	107.5	2.	14315.	4698.
27.5	14583.	3377.	1366.	110.0	−17.	12639.	4674.
30.0	11351.	7682.	1857.	112.5	302.	10543.	4435.
32.5	3370.	9663.	1534.	115.0	1874.	4967.	3646.
35.0	15.	10197.	993.	117.5	4005.	1119.	2655.
37.5	49.	10792.	863.	120.0	4989.	696.	2065.
40.0	−31.	11322.	756.	122.5	4887.	475.	1583.
42.5	20.	13321.	620.	125.0	4445.	1631.	3072.
45.0	−17.	15414.	626.	127.5	4362.	2933.	7290.
47.5	−19.	12873.	836.	130.0	4368.	1329.	8541.
50.0	−18.	6114.	1029.	132.5	3485.	88.	7078.
52.5	6.	2962.	946.	135.0	1921.	598.	7322.
55.0	−2.	4913.	828.	137.5	670.	1983.	9445.
57.5	3.	6600.	1247.	140.0	666.	2511.	10692.
60.0	−3.	4885.	2091.	142.5	1275.	866.	9280.
62.5	2.	3380.	2276.	145.0	1865.	13.	8372.
65.0	−1.	5842.	1870.	147.5	2452.	11.	6624.
67.5	2.	12106.	1215.	150.0	3160.	−4.	3617.
70.0	0.	23032.	680.	152.5	2676.	−1.	2717.
72.5	2.	35376.	531.	155.0	697.	0.	3474.
75.0	−1.	36415.	539.	157.5	−67.	−3.	4337.
77.5	1.	26544.	579.	160.0	25.	3.	4824.
80.0	0.	19363.	554.	162.5	−12.	−1.	5525.
82.5	1.	17915.	632.	165.0	10.	4.	6323.
85.0	−2.	22260.	791.	167.5	−5.	−2.	5899.
87.5	−1.	30442.	1455.	170.0	0.	1.	4330.
90.0	−3.	33601.	3194.	172.5	0.	−4.	3338.
92.5	−1.	30873.	4878.	175.0	4.	3.	3408.
95.0	0.	31865.	5903.	177.5	3.	−1.	3407.

Chapter 3

The Fourier Transform

In the previous chapter we showed how we could expand a periodic function in terms of an infinite sum of sines and cosines. However, most functions encountered in engineering are aperiodic. As we shall see, the extension of Fourier series to these functions leads to the Fourier transform.

3.1 FOURIER TRANSFORMS

The Fourier transform is the natural extension of Fourier series to a function $f(t)$ of infinite period. To show this, consider a periodic function $f(t)$ of period $2T$ that satisfies the so-called Dirichlet's conditions.[1] If the integral $\int_a^b |f(t)|\,dt$ exists, this function has the complex Fourier series:

$$f(t) = \sum_{n=-\infty}^{\infty} c_n e^{in\pi t/T}, \tag{3.1.1}$$

[1] A function $f(t)$ satisfies Dirichlet's conditions in the interval (a, b) if (1) it is bounded in (a, b), and (2) it has at most a finite number of discontinuities and a finite number of maxima and minima in the interval (a, b).

where

$$c_n = \frac{1}{2T} \int_{-T}^{T} f(t) e^{-in\pi t/T}\, dt. \tag{3.1.2}$$

Equation (3.1.1) applies only if $f(t)$ is continuous at t; if $f(t)$ suffers from a jump discontinuity at t, then the left side of (3.1.1) equals $\frac{1}{2}[f(t^+) + f(t^-)]$, where $f(t^+) = \lim_{x \to t^+} f(x)$ and $f(t^-) = \lim_{x \to t^-} f(x)$. Substituting (3.1.2) into (3.1.1),

$$f(t) = \frac{1}{2T} \sum_{n=-\infty}^{\infty} e^{in\pi t/T} \int_{-T}^{T} f(x) e^{-in\pi x/T}\, dx. \tag{3.1.3}$$

Let us now introduce the notation $\omega_n = n\pi/T$ so that $\Delta\omega_n = \omega_{n+1} - \omega_n = \pi/T$. Then,

$$f(t) = \frac{1}{2\pi} \sum_{n=-\infty}^{\infty} F(\omega_n) e^{i\omega_n t} \Delta\omega_n, \tag{3.1.4}$$

where

$$F(\omega_n) = \int_{-T}^{T} f(x) e^{-i\omega_n x}\, dx. \tag{3.1.5}$$

As $T \to \infty$, ω_n approaches a continuous variable ω and $\Delta\omega_n$ may be interpreted as the infinitesimal $d\omega$. Therefore, ignoring any possible difficulties.[2]

$$\boxed{f(t) = \frac{1}{2\pi} \int_{-\infty}^{\infty} F(\omega) e^{i\omega t}\, d\omega} \tag{3.1.6}$$

and

$$\boxed{F(\omega) = \int_{-\infty}^{\infty} f(t) e^{-i\omega t}\, dt.} \tag{3.1.7}$$

---

[2] For a rigorous derivation, see Titchmarsh, E. C., 1948: *Introduction to the Theory of Fourier Integrals*, Clarendon Press, Oxford, chap. 1.

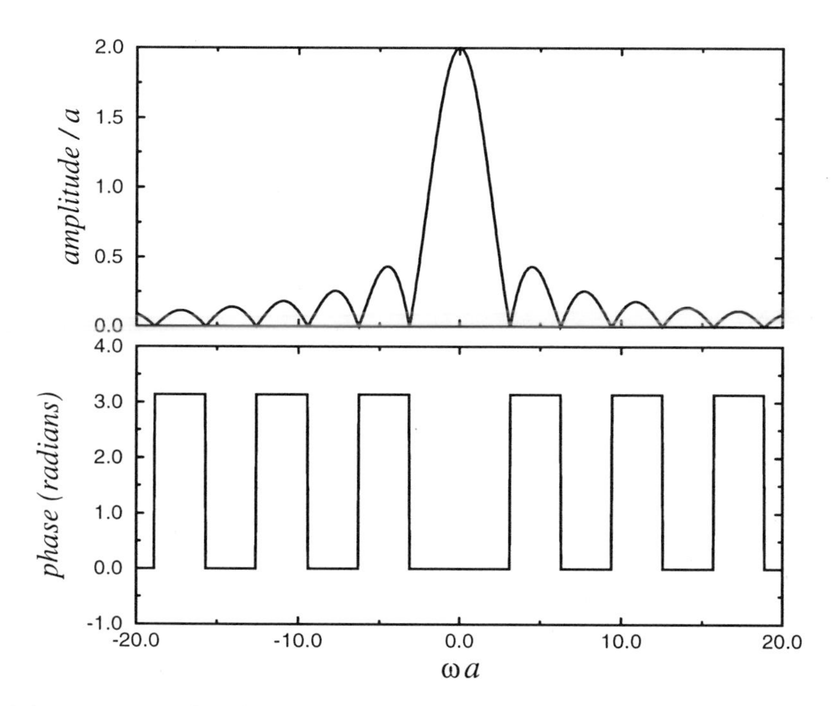

Figure 3.1.1: Graph of the Fourier transform for (3.1.9).

Equation (3.1.7) is the *Fourier transform* of $f(t)$ while (3.1.6) is the *inverse Fourier transform* which converts a Fourier transform back to $f(t)$. Alternatively, we may combine (3.1.6)–(3.1.7) to yield the equivalent real form:

$$f(t) = \frac{1}{\pi}\int_0^\infty \left\{\int_{-\infty}^\infty f(x)\cos[\omega(t-x)]\,dx\right\} d\omega. \tag{3.1.8}$$

Hamming[3] has suggested the following analog in understanding the Fourier transform. Let us imagine that $f(t)$ is a light beam. Then the Fourier transform, like a glass prism, breaks up the function into its component frequencies ω, each of intensity $F(\omega)$. In optics, the various frequencies are called colors; by analogy the Fourier transform gives us the color spectrum of a function. On the other hand, the inverse Fourier transform blends a function's spectrum to give back the original function.

Most signals encountered in practice have Fourier transforms because they are absolutely integrable since they are bounded and of finite duration. However, there are some notable exceptions. Examples include the trigonometric functions sine and cosine.

[3] Hamming, R. W., 1977: *Digital Filters*, Prentice-Hall, Englewood Cliffs, NJ, p. 136.

- **Example 3.1.1**

Let us find the Fourier transform for

$$f(t) = \begin{cases} 1, & |t| < a \\ 0, & |t| > a. \end{cases} \tag{3.1.9}$$

From the definition of the Fourier transform,

$$F(\omega) = \int_{-\infty}^{-a} 0\, e^{-i\omega t} dt + \int_{-a}^{a} 1\, e^{-i\omega t} dt + \int_{a}^{\infty} 0\, e^{-i\omega t} dt \tag{3.1.10}$$

$$= \frac{e^{\omega a i} - e^{-\omega a i}}{\omega i} = \frac{2\sin(\omega a)}{\omega} = 2a\,\mathrm{sinc}(\omega a), \tag{3.1.11}$$

where $\mathrm{sinc}(x) = \sin(x)/x$ is the *sinc function.*

Although this particular example does not show it, the Fourier transform is, in general, a complex function. The most common method of displaying it is to plot its amplitude and phase on two separate graphs for all values of ω. See Figure 3.1.1. Of these two quantities, the amplitude is by far the more popular one and is given the special name of *frequency spectrum.*

From the definition of the inverse Fourier transform,

$$f(t) = \frac{1}{\pi} \int_{-\infty}^{\infty} \frac{\sin(\omega a)}{\omega} e^{i\omega t} d\omega = \begin{cases} 1, & |t| < a \\ 0, & |t| > a. \end{cases} \tag{3.1.12}$$

An important question is what value does $f(t)$ converge to in the limit as $t \to a$ and $t \to -a$? Because Fourier transforms are an extension of Fourier series, the behavior at a jump is the same as that for a Fourier series. For that reason, $f(a) = \frac{1}{2}[f(a^+) + f(a^-)] = \frac{1}{2}$ and $f(-a) = \frac{1}{2}[f(-a^+) + f(-a^-)] = \frac{1}{2}$.

- **Example 3.1.2: Dirac delta function**

Of the many functions that have a Fourier transform, a particularly important one is the *(Dirac) delta function.*[4] For example, in Section 3.6 we will use it to solve differential equations. We *define* it as the inverse of the Fourier transform $F(\omega) = 1$. Therefore,

$$\delta(t) = \frac{1}{2\pi} \int_{-\infty}^{\infty} e^{i\omega t} d\omega. \tag{3.1.13}$$

[4] Dirac, P. A. M., 1947: *The Principles of Quantum Mechanics*, Clarendon Press, Oxford, Section 15.

Table 3.1.1: The Fourier Transforms of Some Commonly Encountered Functions. The Heaviside Step Function $H(t)$ Is Defined by (3.2.16).

	$\mathbf{f(t)},\ \mathbf{|t|<\infty}$	$\mathbf{F(\omega)}$
1.	$e^{-at}H(t),\quad a>0$	$\dfrac{1}{a+\omega i}$
2.	$e^{at}H(-t),\quad a>0$	$\dfrac{1}{a-\omega i}$
3.	$te^{-at}H(t),\quad a>0$	$\dfrac{1}{(a+\omega i)^2}$
4.	$te^{at}H(-t),\quad a>0$	$\dfrac{-1}{(a-\omega i)^2}$
5.	$t^n e^{-at}H(t),\ \mathrm{Re}(a)>0,\ n=1,2,\ldots$	$\dfrac{n!}{(a+\omega i)^{n+1}}$
6.	$e^{-a\|t\|},\quad a>0$	$\dfrac{2a}{\omega^2+a^2}$
7.	$te^{-a\|t\|},\quad a>0$	$\dfrac{-4a\omega i}{(\omega^2+a^2)^2}$
8.	$\dfrac{1}{1+a^2t^2}$	$\dfrac{\pi}{\|a\|}e^{-\|\omega/a\|}$
9.	$\dfrac{\cos(at)}{1+t^2}$	$\frac{\pi}{2}\left(e^{-\|\omega-a\|}+e^{-\|\omega+a\|}\right)$
10.	$\dfrac{\sin(at)}{1+t^2}$	$\frac{\pi}{2i}\left(e^{-\|\omega-a\|}-e^{-\|\omega+a\|}\right)$
11.	$\begin{cases}1, & \|t\|<a\\ 0, & \|t\|>a\end{cases}$	$\dfrac{2\sin(\omega a)}{\omega}$
12.	$\dfrac{\sin(at)}{at}$	$\begin{cases}\pi/a, & \|\omega\|<a\\ 0, & \|\omega\|>a\end{cases}$

To give some insight into the nature of the delta function, consider another band-limited transform:

$$F_\Omega(\omega)=\begin{cases}1, & |\omega|<\Omega\\ 0, & |\omega|>\Omega,\end{cases} \tag{3.1.14}$$

where Ω is real and positive. Then,

$$f_\Omega(t)=\frac{1}{2\pi}\int_{-\Omega}^{\Omega}e^{i\omega t}\,d\omega=\frac{\Omega}{\pi}\frac{\sin(\Omega t)}{\Omega t}. \tag{3.1.15}$$

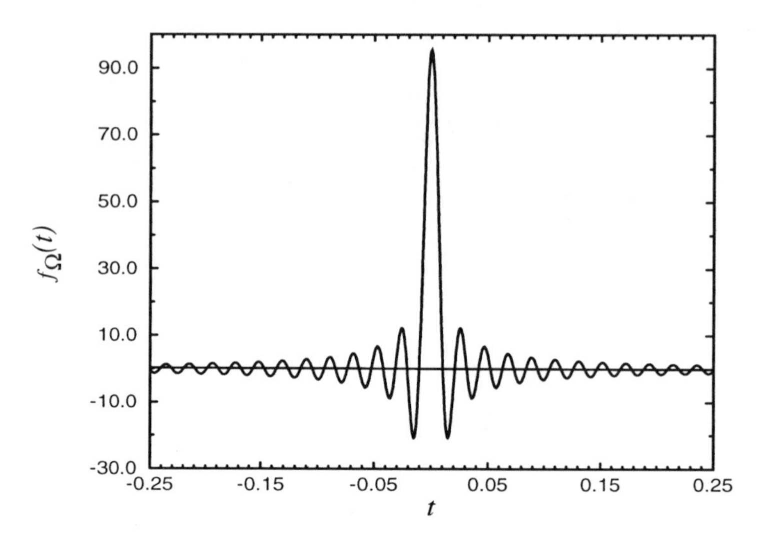

Figure 3.1.2: Graph of the function given in (3.1.15) for $\Omega = 300$.

Figure 3.1.2 illustrates $f_\Omega(t)$ for a large value of Ω. We observe that as $\Omega \to \infty$, $f_\Omega(t)$ becomes very large near $t = 0$ as well as very narrow. On the other hand, $f_\Omega(t)$ rapidly approaches zero as $|t|$ increases. Therefore, we may consider the delta function as the limit:

$$\delta(t) = \lim_{\Omega \to \infty} \frac{\sin(\Omega t)}{\pi t} \tag{3.1.16}$$

or

$$\delta(t) = \begin{cases} \infty, & t = 0 \\ 0, & t \neq 0. \end{cases} \tag{3.1.17}$$

Because the Fourier transform of the delta function equals one,

$$\int_{-\infty}^{\infty} \delta(t) e^{-i\omega t} \, dt = 1. \tag{3.1.18}$$

Since (3.1.18) must hold for any ω, we take $\omega = 0$ and find that

$$\int_{-\infty}^{\infty} \delta(t) \, dt = 1. \tag{3.1.19}$$

Thus, the area under the delta function equals unity. Taking (3.1.17) into account, we can also write (3.1.19) as

$$\int_{-a}^{b} \delta(t) \, dt = 1, \qquad a, b > 0. \tag{3.1.20}$$

Finally,

$$\int_{a}^{b} f(t) \delta(t - t_0) \, dt = f(t_0), \tag{3.1.21}$$

if $a < t_0 < b$. This follows from the law of the mean of integrals.

We may also use several other functions with equal validity to represent the delta function. These include the limiting case of the following rectangular or triangular distributions:

$$\delta(t) = \lim_{\epsilon \to 0} \begin{cases} \frac{1}{\epsilon}, & |t| < \frac{\epsilon}{2} \\ 0, & |t| > \frac{\epsilon}{2} \end{cases} \tag{3.1.22}$$

or

$$\delta(t) = \lim_{\epsilon \to 0} \begin{cases} \frac{1}{\epsilon}\left(1 - \frac{|t|}{\epsilon}\right), & |t| < \epsilon \\ 0, & |t| > \epsilon \end{cases} \tag{3.1.23}$$

and the Gaussian function:

$$\delta(t) = \lim_{\epsilon \to 0} \frac{\exp(-\pi t^2/\epsilon)}{\sqrt{\epsilon}}. \tag{3.1.24}$$

Note that the delta function is an even function.

Problems

1. Show that the Fourier transform of

$$f(t) = e^{-a|t|}, \qquad a > 0,$$

is

$$F(\omega) = \frac{2a}{\omega^2 + a^2}.$$

Now plot the amplitude and phase spectra for this transform.

2. Show that the Fourier transform of

$$f(t) = te^{-a|t|}, \qquad a > 0,$$

is

$$F(\omega) = -\frac{4a\omega i}{(\omega^2 + a^2)^2}.$$

Now plot the amplitude and phase spectra for this transform.

3. Show that the Fourier transform of

$$f(t) = \begin{cases} e^{2t}, & t < 0 \\ e^{-t}, & t > 0 \end{cases}$$

is

$$F(\omega) = \frac{3}{(2 - i\omega)(1 + i\omega)}.$$

Now plot the amplitude and phase spectra for this transform.

4. Show that the Fourier transform of

$$f(t) = \begin{cases} e^{-(1+i)t}, & t > 0 \\ -e^{(1-i)t}, & t < 0 \end{cases}$$

is

$$F(\omega) = \frac{-2i(\omega + 1)}{(\omega + 1)^2 + 1}.$$

Now plot the amplitude and phase spectra for this transform.

5. Show that the Fourier transform of

$$f(t) = \begin{cases} \cos(at), & |t| < 1 \\ 0, & |t| > 1 \end{cases}$$

is

$$F(\omega) = \frac{\sin(\omega - a)}{\omega - a} + \frac{\sin(\omega + a)}{\omega + a}.$$

Now plot the amplitude and phase spectra for this transform.

6. Show that the Fourier transform of

$$f(t) = \begin{cases} \sin(t), & 0 \leq t < 1 \\ 0, & \text{otherwise} \end{cases}$$

is

$$\begin{aligned} F(\omega) = &-\frac{1}{2}\left[\frac{1 - \cos(\omega - 1)}{\omega - 1} + \frac{\cos(\omega + 1) - 1}{\omega + 1}\right] \\ &- \frac{i}{2}\left[\frac{\sin(\omega - 1)}{\omega - 1} - \frac{\sin(\omega + 1)}{\omega + 1}\right]. \end{aligned}$$

Now plot the amplitude and phase spectra for this transform.

7. Show that the Fourier transform of

$$f(t) = \begin{cases} 1 - t/\tau, & 0 \leq t < 2\tau \\ 0, & \text{otherwise} \end{cases}$$

is

$$F(\omega) = \frac{2e^{-i\omega t}}{i\omega}\left[\frac{\sin(\omega\tau)}{\omega\tau} - \cos(\omega\tau)\right]$$

Now plot the amplitude and phase spectra for this transform.

8. The integral representation[5] of the modified Bessel function $K_\nu(\ )$ is

$$K_\nu\left(a|\omega|\right) = \frac{\Gamma\left(\nu+\frac{1}{2}\right)(2a)^\nu}{|\omega|^\nu\Gamma\left(\frac{1}{2}\right)} \int_0^\infty \frac{\cos(\omega t)}{(t^2+a^2)^{\nu+1/2}}\,dt,$$

where $\Gamma(\ )$ is the gamma function, $\nu \geq 0$ and $a > 0$. Use this relationship to show that

$$\mathcal{F}\left[\frac{1}{(t^2+a^2)^{\nu+1/2}}\right] = \frac{2|\omega|^\nu\Gamma\left(\frac{1}{2}\right)K_\nu\left(a|\omega|\right)}{\Gamma\left(\nu+\frac{1}{2}\right)(2a)^\nu}.$$

9. Show that the Fourier transform of a constant K is $2\pi\delta(\omega)K$.

3.2 FOURIER TRANSFORMS CONTAINING THE DELTA FUNCTION

In the previous section we stressed the fact that such simple functions as cosine and sine are not absolutely integrable. Does this mean that these functions do not possess a Fourier transform? In this section we shall show that certain functions can still have a Fourier transform even though we cannot compute them directly.

The reason why we can find the Fourier transform of certain functions that are not absolutely integrable lies with the introduction of the delta function because

$$\int_{-\infty}^{\infty} \delta(\omega-\omega_0)e^{it\omega}\,d\omega = e^{i\omega_0 t} \tag{3.2.1}$$

for all t. Thus, the inverse of the Fourier transform $\delta(\omega-\omega_0)$ is the complex exponential $e^{i\omega_0 t}/2\pi$ or

$$\mathcal{F}\left(e^{i\omega_0 t}\right) = 2\pi\delta(\omega-\omega_0). \tag{3.2.2}$$

This yields immediately the result that

$$\mathcal{F}(1) = 2\pi\delta(\omega), \tag{3.2.3}$$

if we set $\omega_0 = 0$. Thus, the Fourier transform of 1 is an impulse at $\omega = 0$ with weight 2π. Because the Fourier transform equals zero for all $\omega \neq 0$, $f(t) = 1$ does not contain a nonzero frequency and is consequently a DC signal.

---

[5] Watson, G. N., 1966: *A Treatise on the Theory of Bessel Functions*, Cambridge University Press, Cambridge, p. 185.

Another set of transforms arises from Euler's formula because we have that

$$\mathcal{F}[\sin(\omega_0 t)] = \tfrac{1}{2i}\left[\mathcal{F}\left(e^{i\omega_0 t}\right) - \mathcal{F}\left(e^{-i\omega_0 t}\right)\right] \tag{3.2.4}$$

$$= \tfrac{\pi}{i}\left[\delta(\omega - \omega_0) - \delta(\omega + \omega_0)\right] \tag{3.2.5}$$

$$= -\pi i\delta(\omega - \omega_0) + \pi i\delta(\omega + \omega_0) \tag{3.2.6}$$

and

$$\mathcal{F}[\cos(\omega_0 t)] = \tfrac{1}{2}\left[\mathcal{F}\left(e^{i\omega_0 t}\right) + \mathcal{F}\left(e^{-i\omega_0 t}\right)\right] \tag{3.2.7}$$

$$= \pi\left[\delta(\omega - \omega_0) + \delta(\omega + \omega_0)\right]. \tag{3.2.8}$$

Note that although the amplitude spectra of $\sin(\omega_0 t)$ and $\cos(\omega_0 t)$ are the same, their phase spectra are different.

Let us consider the Fourier transform of any arbitrary periodic function. Recall that any such function $f(t)$ with period $2L$ can be rewritten as the complex Fourier series:

$$f(t) = \sum_{n=-\infty}^{\infty} c_n e^{in\omega_0 t}, \tag{3.2.9}$$

where $\omega_0 = \pi/L$. The Fourier transform of $f(t)$ is

$$F(\omega) = \mathcal{F}[f(t)] = \sum_{n=-\infty}^{\infty} 2\pi c_n \delta(\omega - n\omega_0). \tag{3.2.10}$$

Therefore, the Fourier transform of any arbitrary periodic function is a sequence of impulses with weight $2\pi c_n$ located at $\omega = n\omega_0$ with $n = 0, \pm 1, \pm 2, \ldots$. Thus, the Fourier series and transform of a periodic function are closely related.

• **Example 3.2.1: Fourier transform of the sign function**

Consider the sign function

$$\operatorname{sgn}(t) = \begin{cases} 1, & t > 0 \\ 0, & t = 0 \\ -1, & t < 0. \end{cases} \tag{3.2.11}$$

The function is not absolutely integrable. However, let us approximate it by $e^{-\epsilon|t|}\operatorname{sgn}(t)$, where ϵ is a small positive number. This new function is absolutely integrable and we have that

$$\mathcal{F}[\operatorname{sgn}(t)] = \lim_{\epsilon\to 0}\left[-\int_{-\infty}^{0} e^{\epsilon t} e^{-i\omega t}\,dt + \int_0^{\infty} e^{-\epsilon t} e^{-i\omega t}\,dt\right] \tag{3.2.12}$$

$$= \lim_{\epsilon\to 0}\left(\frac{-1}{\epsilon - i\omega} + \frac{1}{\epsilon + i\omega}\right). \tag{3.2.13}$$

If $\omega \neq 0$, (3.2.13) equals $2/i\omega$. If $\omega = 0$, (3.2.13) equals 0 because

$$\lim_{\epsilon \to 0} \left(\frac{-1}{\epsilon} + \frac{1}{\epsilon} \right) = 0. \tag{3.2.14}$$

Thus, we conclude that

$$\mathcal{F}[\text{sgn}(t)] = \begin{cases} 2/i\omega, & \omega \neq 0 \\ 0, & \omega = 0. \end{cases} \tag{3.2.15}$$

• **Example 3.2.2: Fourier transform of the step function**

An important function in transform methods is the *(Heaviside) step function*:

$$H(t) = \begin{cases} 1, & t > 0 \\ 0, & t < 0. \end{cases} \tag{3.2.16}$$

In terms of the sign function it can be written

$$H(t) = \tfrac{1}{2} + \tfrac{1}{2}\text{sgn}(t). \tag{3.2.17}$$

Because the Fourier transforms of 1 and $\text{sgn}(t)$ are $2\pi\delta(\omega)$ and $2/i\omega$, respectively, we have that

$$\mathcal{F}[H(t)] = \pi\delta(\omega) + \frac{1}{i\omega}. \tag{3.2.18}$$

These transforms are used in engineering but the presence of the delta function requires extra care to ensure their proper use.

Problems

1. Verify that

$$\mathcal{F}[\sin(\omega_0 t)H(t)] = \frac{\omega_0}{\omega_0^2 - \omega^2} + \frac{\pi i}{2}[\delta(\omega + \omega_0) - \delta(\omega - \omega_0)].$$

2. Verify that

$$\mathcal{F}[\cos(\omega_0 t)H(t)] = \frac{i\omega}{\omega_0^2 - \omega^2} + \frac{\pi}{2}[\delta(\omega + \omega_0) + \delta(\omega - \omega_0)].$$

3.3 PROPERTIES OF FOURIER TRANSFORMS

In principle we can compute any Fourier transform from the definition. However, it is far more efficient to derive some simple relationships that relate transforms to each other. This is the purpose of this section.

Linearity

If $f(t)$ and $g(t)$ are functions with Fourier transforms $F(\omega)$ and $G(\omega)$, respectively, then

$$\mathcal{F}[c_1 f(t) + c_2 g(t)] = c_1 F(\omega) + c_2 G(\omega), \tag{3.3.1}$$

where c_1 and c_2 are (real or complex) constants.

This result follows from the integral definition:

$$\mathcal{F}[c_1 f(t) + c_2 g(t)] = \int_{-\infty}^{\infty} [c_1 f(t) + c_2 g(t)] e^{-i\omega t}\, dt \tag{3.3.2}$$

$$= c_1 \int_{-\infty}^{\infty} f(t) e^{-i\omega t}\, dt + c_2 \int_{-\infty}^{\infty} g(t) e^{-i\omega t}\, dt \tag{3.3.3}$$

$$= c_1 F(\omega) + c_2 G(\omega). \tag{3.3.4}$$

Time shifting

If $f(t)$ is a function with a Fourier transform $F(\omega)$, then $\mathcal{F}[f(t - \tau)] = e^{-i\omega\tau} F(\omega)$.

This follows from the definition of the Fourier transform:

$$\mathcal{F}[f(t-\tau)] = \int_{-\infty}^{\infty} f(t-\tau) e^{-i\omega t}\, dt = \int_{-\infty}^{\infty} f(x) e^{-i\omega(x+\tau)}\, dx \tag{3.3.5}$$

$$= e^{-i\omega\tau} \int_{-\infty}^{\infty} f(x) e^{-i\omega x}\, dx = e^{-i\omega\tau} F(\omega). \tag{3.3.6}$$

• Example 3.3.1

The Fourier transform of $f(t) = \cos(at)H(t)$ is $F(\omega) = i\omega/(a^2 - \omega^2) + \pi[\delta(\omega + a) + \delta(\omega - a)]/2$. Therefore,

$$\mathcal{F}\{\cos[a(t-k)]H(t-k)\} = e^{-ik\omega} \mathcal{F}[\cos(at)H(t)] \tag{3.3.7}$$

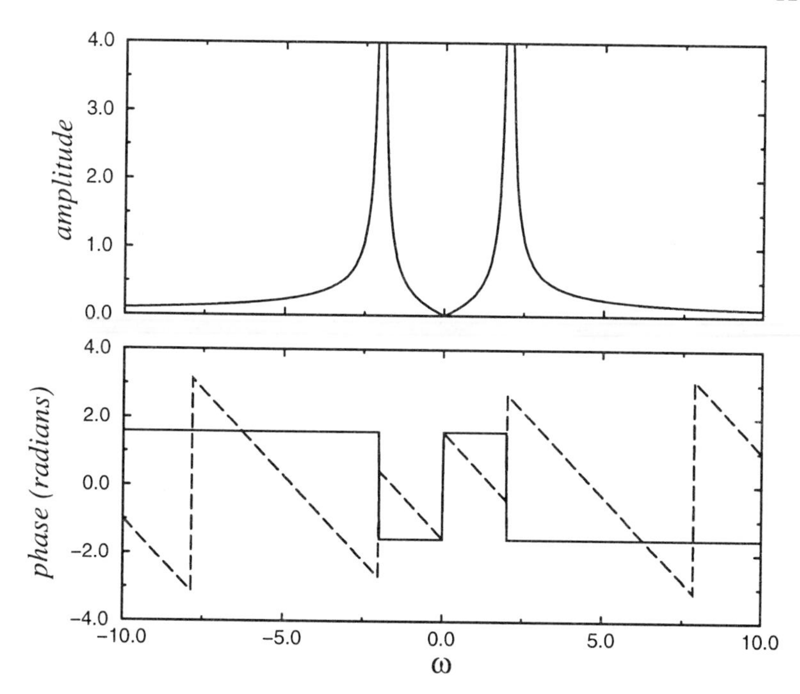

Figure 3.3.1: The amplitude and phase spectra of the Fourier transform for $\cos(2t)H(t)$ (solid line) and $\cos[2(t-1)]H(t-1)$ (dashed line). The amplitude becomes infinite at $\omega = \pm 2$.

or

$$\mathcal{F}\{\cos[a(t-k)]H(t-k)\} = \frac{i\omega e^{-ik\omega}}{a^2-\omega^2} + \frac{\pi}{2}e^{-ik\omega}[\delta(\omega+a)+\delta(\omega-a)]. \tag{3.3.8}$$

In Figure 3.3.1 we present the amplitude and phase spectra for $\cos(2t)$ $H(t)$ (the solid line) while the dashed line gives these spectra for $\cos[2(t-1)]H(t-1)$. This figure shows that the amplitude spectra are identical (why?) while the phase spectra are considerably different.

Scaling factor

Let $f(t)$ be a function with a Fourier transform $F(\omega)$ and k be a real, nonzero constant. Then $\mathcal{F}[f(kt)] = F(\omega/k)/|k|$.

From the definition of the Fourier transform:

$$\mathcal{F}[f(kt)] = \int_{-\infty}^{\infty} f(kt)e^{-i\omega t}\,dt = \frac{1}{|k|}\int_{-\infty}^{\infty} f(x)e^{-i(\omega/k)x}\,dx = \frac{1}{|k|}F\left(\frac{\omega}{k}\right). \tag{3.3.9}$$

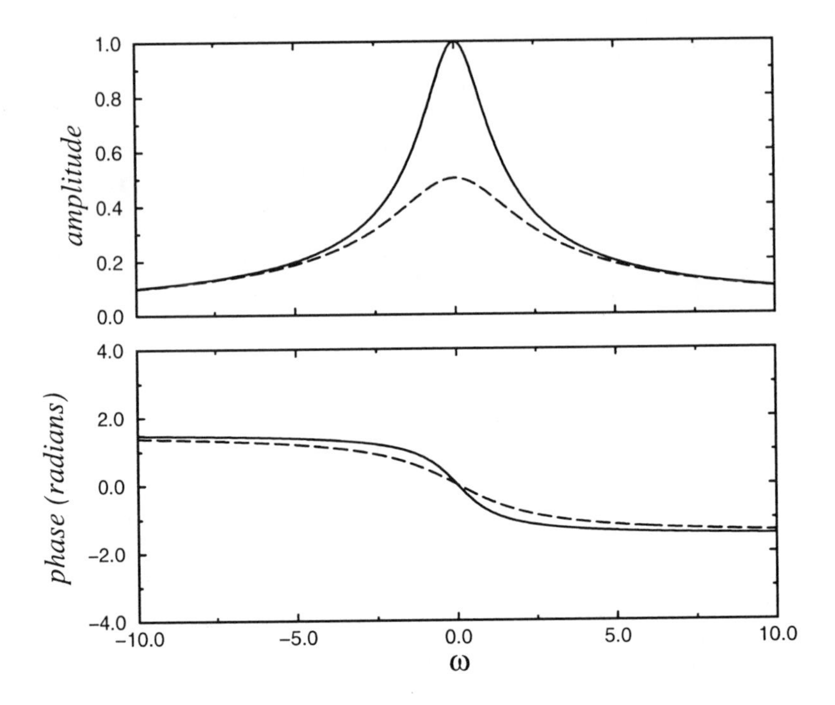

Figure 3.3.2: The amplitude and phase spectra of the Fourier transform for $e^{-t}H(t)$ (solid line) and $e^{-2t}H(t)$ (dashed line).

• **Example 3.3.2**

The Fourier transform of $f(t) = e^{-t}H(t)$ is $F(\omega) = 1/(1+\omega i)$. Therefore, the Fourier transform for $f(at) = e^{-at}H(t)$, $a > 0$, is

$$\mathcal{F}[f(at)] = \left(\frac{1}{a}\right)\left(\frac{1}{1+i\omega/a}\right) = \frac{1}{a+\omega i}. \tag{3.3.10}$$

In Figure 3.3.2 we present the amplitude and phase spectra for $e^{-t}H(t)$ (solid line) while the dashed line gives these spectra for $e^{-2t}H(t)$. This figure shows that the amplitude spectra has decreased by a factor of two for $e^{-2t}H(t)$ compared to $e^{-t}H(t)$ while the differences in the phase are smaller.

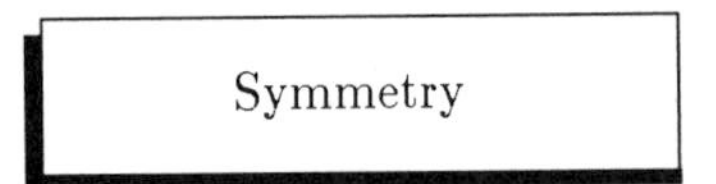

If the function $f(t)$ has the Fourier transform $F(\omega)$, then $\mathcal{F}[F(t)] = 2\pi f(-\omega)$.

From the definition of the inverse Fourier transform,

$$f(t) = \frac{1}{2\pi} \int_{-\infty}^{\infty} F(\omega) e^{i\omega t} d\omega = \frac{1}{2\pi} \int_{-\infty}^{\infty} F(x) e^{ixt} dx. \tag{3.3.11}$$

Then

$$2\pi f(-\omega) = \int_{-\infty}^{\infty} F(x) e^{-i\omega x} dx = \int_{-\infty}^{\infty} F(t) e^{-i\omega t} dt = \mathcal{F}[F(t)]. \tag{3.3.12}$$

- **Example 3.3.3**

The Fourier transform of $1/(1+t^2)$ is $\pi e^{-|\omega|}$. Therefore,

$$\mathcal{F}\left(\pi e^{-|t|}\right) = \frac{2\pi}{1+\omega^2} \tag{3.3.13}$$

or

$$\mathcal{F}\left(e^{-|t|}\right) = \frac{2}{1+\omega^2}. \tag{3.3.14}$$

Derivatives of functions

Let $f^{(k)}(t), k = 0, 1, 2, \ldots, n-1$, be continuous and $f^{(n)}(t)$ be piecewise continuous. Let $|f^{(k)}(t)| \le Ke^{-bt}, b > 0, 0 \le t < \infty; |f^{(k)}(t)| \le Me^{at}, a > 0, -\infty < t \le 0, k = 0, 1, .., n$. Then, $\mathcal{F}[f^{(n)}(t)] = (i\omega)^n F(\omega)$.

We begin by noting that if the transform $\mathcal{F}[f'(t)]$ exists, then

$$\mathcal{F}[f'(t)] = \int_{-\infty}^{\infty} f'(t) e^{-i\omega t} dt \tag{3.3.15}$$

$$= \int_{-\infty}^{\infty} f'(t) e^{\omega_i t} [\cos(\omega_r t) - i \sin(\omega_r t)] \, dt \tag{3.3.16}$$

$$= (-\omega_i + i\omega_r) \int_{-\infty}^{\infty} f(t) e^{\omega_i t} [\cos(\omega_r t) - i \sin(\omega_r t)] \, dt \tag{3.3.17}$$

$$= i\omega \int_{-\infty}^{\infty} f(t) e^{-i\omega t} dt = i\omega F(\omega). \tag{3.3.18}$$

Finally,

$$\mathcal{F}[f^{(n)}(t)] = i\omega \mathcal{F}[f^{(n-1)}(t)] = (i\omega)^2 \mathcal{F}[f^{(n-2)}(t)] = \cdots = (i\omega)^n F(\omega). \tag{3.3.19}$$

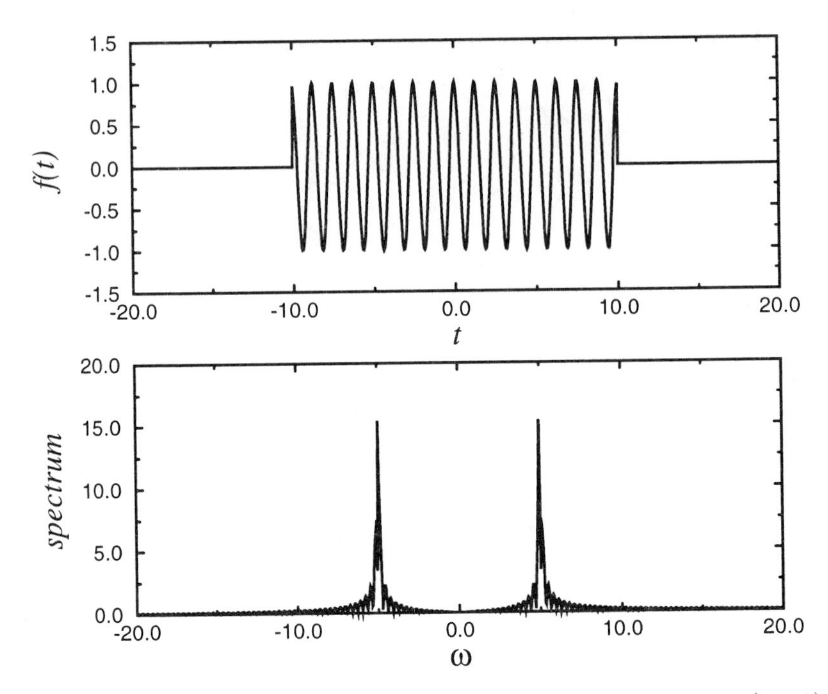

Figure 3.3.3: The (amplitude) spectrum of a rectangular pulse (3.1.9) with a half width $a = 10$ that has been modulated with $\cos(5t)$.

• **Example 3.3.4**

The Fourier transform of $f(t) = 1/(1+t^2)$ is $F(\omega) = \pi e^{-|\omega|}$. Therefore,

$$\mathcal{F}\left[-\frac{2t}{(1+t^2)^2}\right] = i\omega\pi e^{-|\omega|} \tag{3.3.20}$$

or

$$\mathcal{F}\left[\frac{t}{(1+t^2)^2}\right] = -\frac{i\omega\pi}{2} e^{-|\omega|}. \tag{3.3.21}$$

Modulation

In communications a popular method of transmitting information is by *amplitude modulation* (AM). In this process some signal is carried according to the expression $f(t)e^{i\omega_0 t}$, where ω_0 is the *carrier frequency* and $f(t)$ is some arbitrary function of time whose amplitude spectrum peaks at some frequency that is usually small compared to ω_0. We now

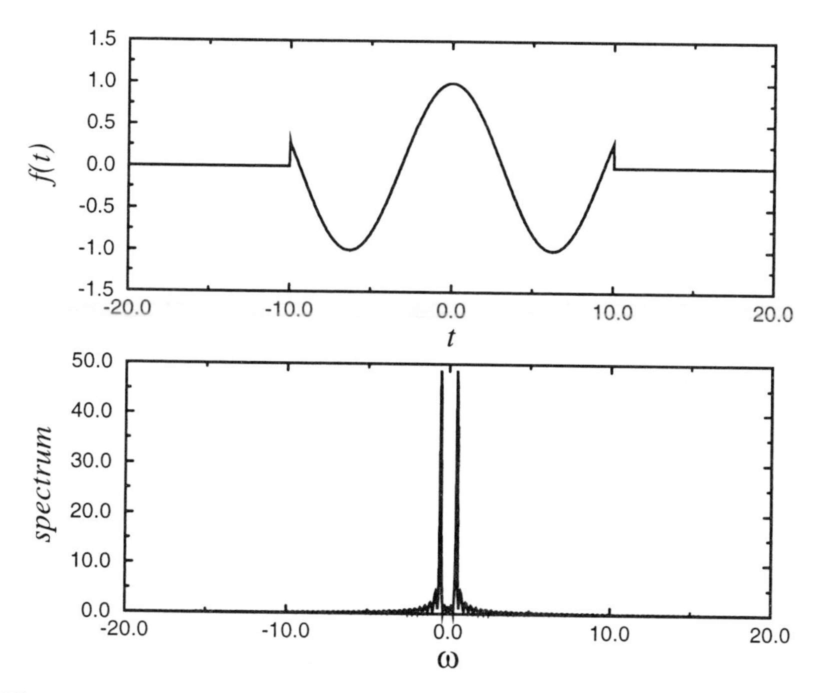

Figure 3.3.4: The (amplitude) spectrum of a rectangular pulse (3.1.9) with a half width $a = 10$ that has been modulated with $\cos(t/2)$.

want to show that the Fourier transform of $f(t)e^{i\omega_0 t}$ is $F(\omega - \omega_0)$, where $F(\omega)$ is the Fourier transform of $f(t)$.

We begin by using the definition of the Fourier transform:

$$\mathcal{F}[f(t)e^{i\omega_0 t}] = \int_{-\infty}^{\infty} f(t)e^{i\omega_0 t}e^{-i\omega t}\,dt = \int_{-\infty}^{\infty} f(t)e^{-i(\omega-\omega_0)t}\,dt \quad (3.3.22)$$

$$= F(\omega - \omega_0). \quad (3.3.23)$$

Therefore, if we have the spectrum of a particular function $f(t)$, then the Fourier transform of the modulated function $f(t)e^{i\omega_0 t}$ is the same as that for $f(t)$ except that it is now centered on the frequency ω_0 rather than on the zero frequency.

• Example 3.3.5

Let us determine the Fourier transform of a square pulse modulated by a cosine wave as shown in Figures 3.3.3 and 3.3.4. Because $\cos(\omega_0 t) = \frac{1}{2}[e^{i\omega_0 t} + e^{-i\omega_0 t}]$ and the Fourier transform of a square pulse is $F(\omega) = 2\sin(\omega a)/\omega$,

$$\mathcal{F}[f(t)\cos(\omega_0 t)] = \frac{\sin[(\omega - \omega_0)a]}{\omega - \omega_0} + \frac{\sin[(\omega + \omega_0)a]}{\omega + \omega_0}. \quad (3.3.24)$$

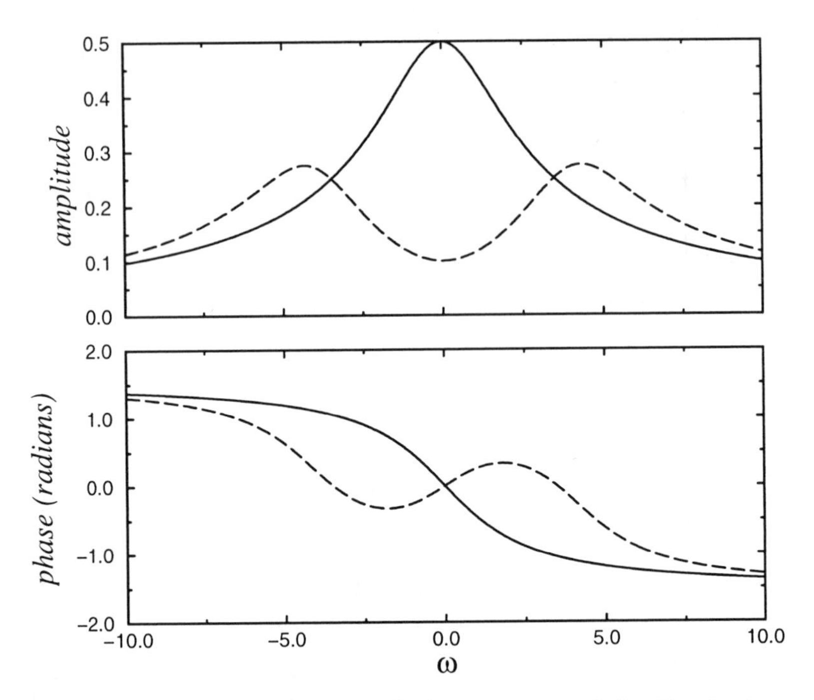

Figure 3.3.5: The amplitude and phase spectra of the Fourier transform for $e^{-2t}H(t)$ (solid line) and $e^{-2t}\cos(4t)H(t)$ (dashed line).

Therefore, the Fourier transform of the modulated pulse equals one half of the sum of the pulse centered on ω_0 and the other that of the pulse centered on $-\omega_0$. See Figures 3.3.3 and 3.3.4.

In many practical situations, $\omega_0 \gg \pi/a$. In this case we may consider that the two terms are completely independent from each other and the contribution from the peak at $\omega = \omega_0$ has a negligible effect on the peak at $\omega = -\omega_0$.

• **Example 3.3.6**

The Fourier transform of $f(t) = e^{-bt}H(t)$ is $F(\omega) = 1/(b + i\omega)$. Therefore,

$$\mathcal{F}[e^{-bt}\cos(at)H(t)] = \tfrac{1}{2}\mathcal{F}\left(e^{iat}e^{-bt} + e^{-iat}e^{-bt}\right) \tag{3.3.25}$$

$$= \frac{1}{2}\left(\frac{1}{b+i\omega'}\bigg|_{\omega'=\omega-a} + \frac{1}{b+i\omega'}\bigg|_{\omega'=\omega+a}\right) \tag{3.3.26}$$

$$\mathcal{F}[e^{-bt}\cos(at)H(t)] = \frac{1}{2}\left[\frac{1}{(b+i\omega)-ai} + \frac{1}{(b+i\omega)+ai}\right] \tag{3.3.27}$$

$$= \frac{b+i\omega}{(b+i\omega)^2+a^2}. \tag{3.3.28}$$

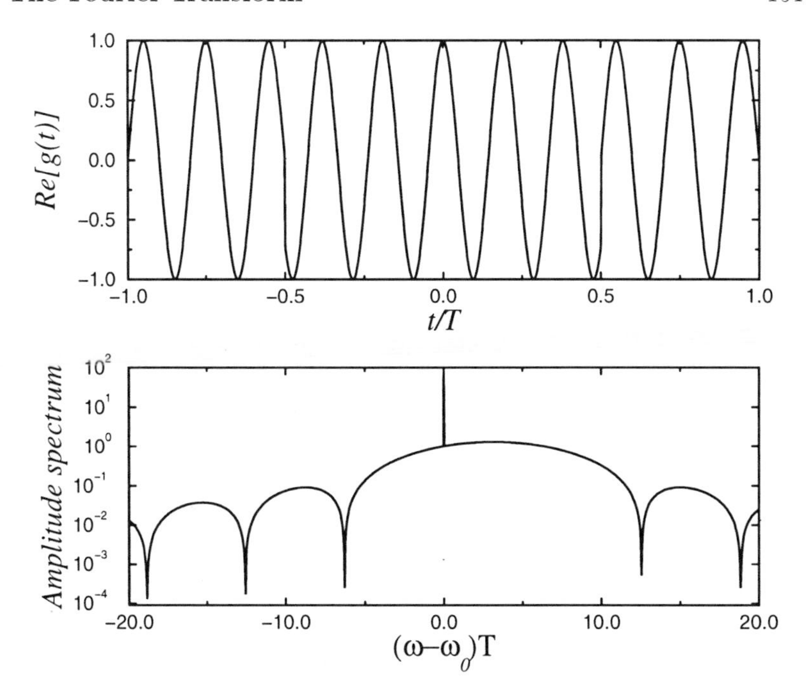

Figure 3.3.6: The (amplitude) spectrum $|G(\omega)|/T$ of a frequency-modulated signal (shown top) using the parameters $\omega_1 T = 2\pi$ and $\omega_0 T = 10\pi$. The transform becomes undefined at $\omega = \omega_0$.

We have illustrated this result using $e^{-2t}H(t)$ and $e^{-2t}\cos(4t)H(t)$ in Figure 3.3.5.

• **Example 3.3.7: Frequency modulation**

In contrast to amplitude modulation, *frequency modulation* (FM) transmits information by instantaneous variations of the carrier frequency. It may be expressed mathematically as $\exp\left[i\int_{-\infty}^{t} f(\tau)\,d\tau + iC\right] e^{i\omega_0 t}$, where C is a constant. To illustrate this concept, let us find the Fourier transform of a simple frequency modulation:

$$f(t) = \begin{cases} \omega_1, & |t| < T/2 \\ 0, & |t| > T/2 \end{cases} \tag{3.3.29}$$

and $C = -\omega_1 T/2$. In this case, the signal in the time domain is

$$g(t) = \exp\left[i\int_{-\infty}^{t} f(\tau)\,d\tau + iC\right] e^{i\omega_0 t} \tag{3.3.30}$$

$$= \begin{cases} e^{-i\omega_1 T/2} e^{i\omega_0 t}, & t < -T/2 \\ e^{i\omega_1 t} e^{i\omega_0 t}, & -T/2 < t < T/2 \\ e^{i\omega_1 T/2} e^{i\omega_0 t}, & t > T/2. \end{cases} \tag{3.3.31}$$

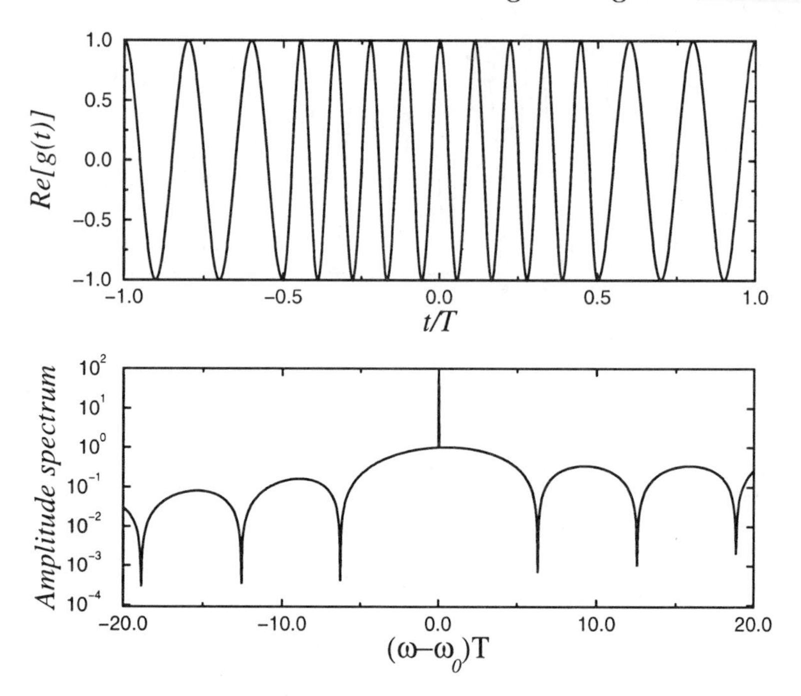

Figure 3.3.7: The (amplitude) spectrum $|G(\omega)|/T$ of a frequency-modulated signal (shown top) using the parameters $\omega_1 T = 8\pi$ and $\omega_0 T = 10\pi$. The transform becomes undefined at $\omega = \omega_0$.

We have illustrated this signal in Figures 3.3.6 and 3.3.7.

The Fourier transform of the signal $G(\omega)$ equals

$$G(\omega) = e^{-i\omega_1 T/2} \int_{-\infty}^{-T/2} e^{i(\omega_0-\omega)t}\,dt + \int_{-T/2}^{T/2} e^{i(\omega_0+\omega_1-\omega)t}\,dt + e^{i\omega_1 T/2} \int_{T/2}^{\infty} e^{i(\omega_0-\omega)t}\,dt \tag{3.3.32}$$

$$= e^{-i\omega_1 T/2} \int_{-\infty}^{0} e^{i(\omega_0-\omega)t}\,dt + e^{i\omega_1 T/2} \int_{0}^{\infty} e^{i(\omega_0-\omega)t}\,dt - e^{-i\omega_1 T/2} \int_{-T/2}^{0} e^{i(\omega_0-\omega)t}\,dt + \int_{-T/2}^{T/2} e^{i(\omega_0+\omega_1-\omega)t}\,dt - e^{i\omega_1 T/2} \int_{0}^{T/2} e^{i(\omega_0-\omega)t}\,dt. \tag{3.3.33}$$

Applying the fact that

$$\int_0^\infty e^{\pm i\alpha t}\,dt = \pi\delta(\alpha) \pm \frac{i}{\alpha}, \tag{3.3.34}$$

$$G(\omega) = \pi\delta(\omega-\omega_0)\left[e^{i\omega_1 T/2} + e^{-i\omega_1 T/2}\right] + \frac{\left[e^{i(\omega_0+\omega_1-\omega)T/2} - e^{-i(\omega_0+\omega_1-\omega)T/2}\right]}{i(\omega_0+\omega_1-\omega)} - \frac{\left[e^{i(\omega_0+\omega_1-\omega)T/2} - e^{-i(\omega_0+\omega_1-\omega)T/2}\right]}{i(\omega_0-\omega)} \tag{3.3.35}$$

$$= 2\pi\delta(\omega-\omega_0)\cos(\omega_1 T/2) + \frac{2\omega_1 \sin[(\omega-\omega_0-\omega_1)T/2]}{(\omega-\omega_0)(\omega-\omega_0-\omega_1)}. \tag{3.3.36}$$

In Figures 3.3.6 and 3.3.7 we have illustrated the amplitude spectrum for various parameters. In general, the transform is not symmetric, with an increasing number of humped curves as $\omega_1 T$ increases.

Parseval's equality

In applying Fourier methods to practical problems we may encounter a situation where we are interested in computing the energy of a system. Energy is usually expressed by the integral $\int_{-\infty}^{\infty} |f(t)|^2\, dt$. Can we compute this integral if we only have the Fourier transform of $F(\omega)$?

From the definition of the inverse Fourier transform

$$f(t) = \frac{1}{2\pi}\int_{-\infty}^{\infty} F(\omega)e^{i\omega t}\,d\omega, \tag{3.3.37}$$

we have that

$$\int_{-\infty}^{\infty} |f(t)|^2\, dt = \frac{1}{2\pi}\int_{-\infty}^{\infty} f(t)\left[\int_{-\infty}^{\infty} F(\omega)e^{i\omega t}\,d\omega\right]dt. \tag{3.3.38}$$

Interchanging the order of integration on the right side of (3.3.38),

$$\int_{-\infty}^{\infty} |f(t)|^2\, dt = \frac{1}{2\pi}\int_{-\infty}^{\infty} F(\omega)\left[\int_{-\infty}^{\infty} f(t)e^{i\omega t}\,dt\right]d\omega. \tag{3.3.39}$$

However,

$$F^*(\omega) = \int_{-\infty}^{\infty} f(t)e^{i\omega t}\,dt. \tag{3.3.40}$$

Therefore,

$$\int_{-\infty}^{\infty} |f(t)|^2\, dt = \frac{1}{2\pi}\int_{-\infty}^{\infty} |F(\omega)|^2\, d\omega. \tag{3.3.41}$$

This is *Parseval's equality*[6] as it applies to Fourier transforms. The quantity $|F(\omega)|^2$ is called the *power spectrum.*

• Example 3.3.8

In Example 3.1.1, we showed that the Fourier transform for a unit rectangular pulse between $-a < t < a$ is $2\sin(\omega a)/\omega$. Therefore, by Parseval's equality,

$$\frac{2}{\pi}\int_{-\infty}^{\infty} \frac{\sin^2(\omega a)}{\omega^2}\, d\omega = \int_{-a}^{a} 1^2\, dt = 2a \tag{3.3.42}$$

or

$$\int_{-\infty}^{\infty} \frac{\sin^2(\omega a)}{\omega^2}\, d\omega = \pi a. \tag{3.3.43}$$

Poisson's summation formula

If $f(x)$ is integrable over $(-\infty, \infty)$, there exists a relationship between the function and its Fourier transform, commonly called *Poisson's summation formula.*[7]

We begin by inventing a periodic function $g(x)$ defined by

$$g(x) = \sum_{k=-\infty}^{\infty} f(x + 2\pi k). \tag{3.3.44}$$

Because $g(x)$ is a periodic function of 2π, it can be represented by the complex Fourier series:

$$g(x) = \sum_{n=-\infty}^{\infty} c_n e^{inx} \tag{3.3.45}$$

or

$$g(0) = \sum_{k=-\infty}^{\infty} f(2\pi k) = \sum_{n=-\infty}^{\infty} c_n. \tag{3.3.46}$$

---

[6] Apparently first derived by Rayleigh, J. W., 1889: On the character of the complete radiation at a given temperature. *Philos. Mag., Ser. 5*, **27**, 460–469.

[7] Poisson, S. D., 1823: Suite du mémoire sur les intégrales définies et sur la sommation des séries. *J. École Polytech.*, **19**, 404–509. See page 451.

Computing c_n, we find that

$$c_n = \frac{1}{2\pi}\int_{-\pi}^{\pi} g(x)e^{-inx}\,dx = \frac{1}{2\pi}\int_{-\pi}^{\pi}\sum_{k=-\infty}^{\infty} f(x+2k\pi)e^{-inx}\,dx \tag{3.3.47}$$

$$= \frac{1}{2\pi}\sum_{k=-\infty}^{\infty}\int_{-\pi}^{\pi} f(x+2k\pi)e^{-inx}\,dx = \frac{1}{2\pi}\int_{-\infty}^{\infty} f(x)e^{-inx}\,dx \tag{3.3.48}$$

$$= \frac{F(n)}{2\pi}, \tag{3.3.49}$$

where $F(\omega)$ is the Fourier transform of $f(x)$. Substituting (3.3.49) into (3.3.46), we obtain

$$\sum_{k=-\infty}^{\infty} f(2\pi k) = \frac{1}{2\pi}\sum_{n=-\infty}^{\infty} F(n) \tag{3.3.50}$$

or

$$\sum_{k=-\infty}^{\infty} f(\alpha k) = \frac{1}{\alpha}\sum_{n=-\infty}^{\infty} F\left(\frac{2\pi n}{\alpha}\right). \tag{3.3.51}$$

• Example 3.3.9

One of the popular uses of Poisson's summation formula is the evaluation of infinite series. For example, let $f(x) = 1/(a^2+x^2)$ with a real and nonzero. Then, $F(\omega) = \pi e^{-|a\omega|}/|a|$ and

$$\sum_{k=-\infty}^{\infty}\frac{1}{a^2+(2\pi k)^2} = \frac{1}{2}\sum_{n=-\infty}^{\infty}\frac{1}{|a|}e^{-|an|} = \frac{1}{2|a|}\left(1+2\sum_{n=1}^{\infty}e^{-|a|n}\right) \tag{3.3.52}$$

$$= \frac{1}{2|a|}\left(-1+\frac{2}{1-e^{-|a|}}\right) = \frac{1}{2|a|}\coth\left(\frac{|a|}{2}\right). \tag{3.3.53}$$

Problems

1. Find the Fourier transform of $1/(1+a^2t^2)$, where a is real, given that $\mathcal{F}[1/(1+t^2)] = \pi e^{-|\omega|}$.

2. Find the Fourier transform of $\cos(at)/(1+t^2)$, where a is real, given that $\mathcal{F}[1/(1+t^2)] = \pi e^{-|\omega|}$.

3. Use the fact that $\mathcal{F}[e^{-at}H(t)] = 1/(a+i\omega)$ with $a > 0$ and Parseval's equality to show that

$$\int_{-\infty}^{\infty} \frac{dx}{x^2+a^2} = \frac{\pi}{a}.$$

4. Use the fact that $\mathcal{F}[1/(1+t^2)] = \pi e^{-|\omega|}$ and Parseval's equality to show that

$$\int_{-\infty}^{\infty} \frac{dx}{(x^2+1)^2} = \frac{\pi}{2}.$$

5. Use the function $f(t) = e^{-at}\sin(bt)H(t)$ with $a > 0$ and Parseval's equality to show that

$$2\int_0^{\infty} \frac{dx}{(x^2+a^2-b^2)^2+4a^2b^2} = \int_{-\infty}^{\infty} \frac{dx}{(x^2+a^2-b^2)^2+4a^2b^2} = \frac{\pi}{2a(a^2+b^2)}.$$

6. Using the modulation property and $\mathcal{F}[e^{-bt}H(t)] = 1/(b+i\omega)$, show that

$$\mathcal{F}\left[e^{-bt}\sin(at)H(t)\right] = \frac{a}{(b+i\omega)^2+a^2}.$$

Plot and compare the amplitude and phase spectra for $e^{-t}H(t)$ and $e^{-t}\sin(2t)\ H(t)$.

7. Use Poisson's summation formula to prove that

$$\sum_{n=-\infty}^{\infty} e^{-ianT} = \frac{2\pi}{T}\sum_{n=-\infty}^{\infty} \delta\left(\frac{2\pi n}{T} - a\right),$$

where $\delta(\ )$ is the Dirac delta function.

3.4 INVERSION OF FOURIER TRANSFORMS

Having focused on the Fourier transform in the previous sections, we consider in detail the inverse Fourier transform in this section. Recall that the improper integral (3.1.6) defines the inverse. Consequently one method of inversion is direct integration.

• Example 3.4.1

Let us find the inverse of $F(\omega) = \pi e^{-|\omega|}$.
From the definition of the inverse Fourier transform,

$$f(t) = \frac{1}{2\pi}\int_{-\infty}^{\infty} \pi e^{-|\omega|} e^{i\omega t}\, d\omega \tag{3.4.1}$$

$$= \frac{1}{2}\int_{-\infty}^{0} e^{(1+it)\omega}\, d\omega + \frac{1}{2}\int_{0}^{\infty} e^{(-1+it)\omega}\, d\omega \tag{3.4.2}$$

$$= \frac{1}{2}\left[\left.\frac{e^{(1+it)\omega}}{1+it}\right|_{-\infty}^{0} + \left.\frac{e^{(-1+it)\omega}}{-1+it}\right|_{0}^{\infty} \right] \tag{3.4.3}$$

$$= \frac{1}{2}\left[\frac{1}{1+it} - \frac{1}{-1+it}\right] = \frac{1}{1+t^2}. \tag{3.4.4}$$

Another method for inverting Fourier transforms is rewriting the Fourier transform using partial fractions so that we can use transform tables. The following example illustrates this technique.

• Example 3.4.2

Let us invert the transform

$$F(\omega) = \frac{1}{(1+i\omega)(1-2i\omega)^2}. \tag{3.4.5}$$

We begin by rewriting (3.4.5) as

$$F(\omega) = \frac{1}{9}\left[\frac{1}{1+i\omega} + \frac{2}{1-2i\omega} + \frac{6}{(1-2i\omega)^2}\right] \tag{3.4.6}$$

$$= \frac{1}{9(1+i\omega)} + \frac{1}{\frac{1}{2}-i\omega} + \frac{1}{6(\frac{1}{2}-i\omega)^2}. \tag{3.4.7}$$

Using Table 3.1.1, we invert (3.4.7) term by term and find that

$$f(t) = \tfrac{1}{9}e^{-t}H(t) + \tfrac{1}{9}e^{t/2}H(-t) - \tfrac{1}{6}te^{t/2}H(-t). \tag{3.4.8}$$

Although we may find the inverse by direct integration or partial fractions, in many instances the Fourier transform does not lend itself to these techniques. On the other hand, if we view the inverse Fourier transform as a line integral along the real axis in the complex ω-plane, then perhaps some of the techniques that we developed in Chapter 1 might be applicable to this problem. To this end, we rewrite the inversion integral (3.1.6) as

$$f(t) = \frac{1}{2\pi} \int_{-\infty}^{\infty} F(\omega) e^{it\omega}\, d\omega = \frac{1}{2\pi} \oint_C F(z) e^{itz}\, dz - \frac{1}{2\pi} \int_{C_R} F(z) e^{itz}\, dz, \tag{3.4.9}$$

where C denotes a closed contour consisting of the entire real axis plus a new contour C_R that joins the point $(\infty, 0)$ to $(-\infty, 0)$. There are countless possibilities for C_R. For example, it could be the loop $(\infty, 0)$ to (∞, R) to $(-\infty, R)$ to $(-\infty, 0)$ with $R > 0$. However, any choice of C_R must be such that we can compute $\int_{C_R} F(z) e^{itz}\, dz$. When we take that constraint into account, the number of acceptable contours decrease to just a few. The best is given by *Jordan's lemma*:[8]

Jordan's lemma: *Suppose that, on a circular arc C_R with radius R and center at the origin, $f(z) \to 0$ uniformly as $R \to \infty$. Then*

(1) $$\lim_{R\to\infty} \int_{C_R} f(z) e^{imz}\, dz = 0, \qquad (m > 0) \tag{3.4.10}$$

if C_R lies in the first and/or second quadrant;

(2) $$\lim_{R\to\infty} \int_{C_R} f(z) e^{-imz}\, dz = 0, \qquad (m > 0) \tag{3.4.11}$$

if C_R lies in the third and/or fourth quadrant;

(3) $$\lim_{R\to\infty} \int_{C_R} f(z) e^{mz}\, dz = 0, \qquad (m > 0) \tag{3.4.12}$$

if C_R lies in the second and/or third quadrant; and

(4) $$\lim_{R\to\infty} \int_{C_R} f(z) e^{-mz}\, dz = 0, \qquad (m > 0) \tag{3.4.13}$$

if C_R lies in the first and/or fourth quadrant.

---

[8] Jordan, C., 1894: *Cours D'Analyse de l'École Polytechnique. Vol. 2*, Gauthier-Villars, Paris, pp. 285–286. See also Whittaker, E. T. and Watson, G. N., 1963: *A Course of Modern Analysis*, Cambridge University Press, Cambridge, p. 115.

Technically, only (1) is actually Jordan's lemma while the remaining points are variations.

Proof: We shall prove the first part; the remaining portions follow by analog. We begin by noting that

$$|I_R| = \left| \int_{C_R} f(z) e^{imz}\, dz \right| \leq \int_{C_R} |f(z)| \left| e^{imz} \right| |dz|. \tag{3.4.14}$$

Now

$$|dz| = R\, d\theta, \quad |f(z)| \leq M_R, \tag{3.4.15}$$

$$\left| e^{imz} \right| = \left| \exp(imRe^{\theta i}) \right| = |\exp\{imR[\cos(\theta) + i\sin(\theta)]\}| = e^{-mR\sin(\theta)}. \tag{3.4.16}$$

Therefore,

$$|I_R| \leq RM_R \int_{\theta_0}^{\theta_1} \exp[-mR\sin(\theta)]\, d\theta, \tag{3.4.17}$$

where $0 \leq \theta_0 < \theta_1 \leq \pi$. Because the integrand is positive, the right side of (3.4.17) is largest if we take $\theta_0 = 0$ and $\theta_1 = \pi$. Then

$$|I_R| \leq RM_R \int_0^{\pi} e^{-mR\sin(\theta)} d\theta = 2RM_R \int_0^{\pi/2} e^{-mR\sin(\theta)} d\theta. \tag{3.4.18}$$

We cannot evaluate the integrals in (3.4.18) as they stand. However, because $\sin(\theta) \geq 2\theta/\pi$ if $0 \leq \theta \leq \pi/2$, we can bound the value of the integral by

$$|I_R| \leq 2RM_R \int_0^{\pi/2} e^{-2mR\theta/\pi} d\theta = \frac{\pi}{m} M_R \left(1 - e^{-mR}\right). \tag{3.4.19}$$

If $m > 0$, $|I_R|$ tends to zero with M_R as $R \to \infty$. □

Consider now the following inversions of Fourier transforms:

• Example 3.4.3

For our first example we find the inverse for

$$F(\omega) = \frac{1}{\omega^2 - 2ib\omega - a^2 - b^2}. \tag{3.4.20}$$

From the inversion integral,

$$f(t) = \frac{1}{2\pi} \int_{-\infty}^{\infty} \frac{e^{it\omega}}{\omega^2 - 2ib\omega - a^2 - b^2}\, d\omega \tag{3.4.21}$$

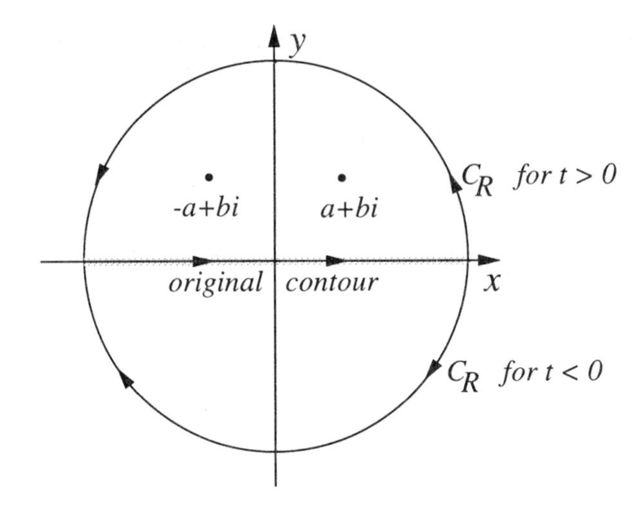

Figure 3.4.1: Contour used to find the inverse of the Fourier transform (3.4.20). The contour C consists of the line integral along the real axis plus C_R.

or

$$f(t) = \frac{1}{2\pi} \oint_C \frac{e^{itz}}{z^2 - 2ibz - a^2 - b^2} \, dz - \frac{1}{2\pi} \int_{C_R} \frac{e^{itz}}{z^2 - 2ibz - a^2 - b^2} \, dz, \tag{3.4.22}$$

where C denotes a closed contour consisting of the entire real axis plus C_R. Because $f(z) = 1/(z^2 - 2ibz - a^2 - b^2)$ tends to zero uniformly as $|z| \to \infty$ and $m = t$, the second integral in (3.4.22) will vanish by Jordan's lemma if C_R is a semicircle of infinite radius in the upper half of the z-plane when $t > 0$ and a semicircle in the lower half of the z-plane when $t < 0$.

Next we must find the location and nature of the singularities. They are located at

$$z^2 - 2ibz - a^2 - b^2 = 0 \tag{3.4.23}$$

or

$$z = \pm a + bi. \tag{3.4.24}$$

Therefore we can rewrite (3.4.22) as

$$f(t) = \frac{1}{2\pi} \oint_C \frac{e^{itz}}{(z - a - bi)(z + a - bi)} \, dz. \tag{3.4.25}$$

Thus, all of the singularities are simple poles.

Consider now $t > 0$. As stated earlier, we close the line integral with an infinite semicircle in the upper half-plane. See Figure 3.4.1.

Inside this closed contour there are two singularities: $z = \pm a + bi$. For these poles,

$$\begin{aligned}
&\text{Res}\left(\frac{e^{itz}}{z^2 - 2ibz - a^2 - b^2}; a + bi\right) \\
&= \lim_{z \to a+bi} (z - a - bi) \frac{e^{itz}}{(z - a - bi)(z + a - bi)} \qquad (3.4.26) \\
&= \frac{e^{iat}e^{-bt}}{2a} = \frac{1}{2a}e^{-bt}[\cos(at) + i\sin(at)], \qquad (3.4.27)
\end{aligned}$$

where we have used Euler's formula to eliminate e^{iat}. Similarly,

$$\text{Res}\left(\frac{e^{itz}}{z^2 - 2ibz - a^2 - b^2}; -a + bi\right) = -\frac{1}{2a}e^{-bt}[\cos(at) - i\sin(at)]. \tag{3.4.28}$$

Consequently the inverse Fourier transform follows from (3.4.25) after applying the residue theorem and equals

$$f(t) = -\frac{1}{2a}e^{-bt}\sin(at) \tag{3.4.29}$$

for $t > 0$.

For $t < 0$ the semicircle is in the lower half-plane because the contribution from the semicircle vanishes as $R \to \infty$. Because there are no singularities within the closed contour, $f(t) = 0$. Therefore, we can write in general that

$$f(t) = -\frac{1}{2a}e^{-bt}\sin(at)H(t). \tag{3.4.30}$$

• **Example 3.4.4**

Let us find the inverse of the Fourier transform

$$F(\omega) = \frac{e^{-\omega i}}{\omega^2 + a^2}, \tag{3.4.31}$$

where a is real and positive.

From the inversion integral,

$$\begin{aligned}
f(t) &= \frac{1}{2\pi}\int_{-\infty}^{\infty} \frac{e^{i(t-1)\omega}}{\omega^2 + a^2}\,d\omega \qquad (3.4.32) \\
&= \frac{1}{2\pi}\oint_C \frac{e^{i(t-1)z}}{z^2 + a^2}\,dz - \frac{1}{2\pi}\int_{C_R} \frac{e^{i(t-1)z}}{z^2 + a^2}\,dz, \qquad (3.4.33)
\end{aligned}$$

where C denotes a closed contour consisting of the entire real axis plus C_R. The contour C_R is determined by Jordan's lemma because $1/(z^2 + a^2) \to 0$ uniformly as $|z| \to \infty$. Since $m = t - 1$, the semicircle C_R of infinite radius lies in the upper half-plane if $t > 1$ and in the lower half-plane if $t < 1$. Thus, if $t > 1$,

$$f(t) = \frac{1}{2\pi}(2\pi i)\mathrm{Res}\left[\frac{e^{i(t-1)z}}{z^2 + a^2}; ai\right] = \frac{e^{-a(t-1)}}{2a}, \tag{3.4.34}$$

whereas for $t < 1$,

$$f(t) = \frac{1}{2\pi}(-2\pi i)\mathrm{Res}\left[\frac{e^{i(t-1)z}}{z^2 + a^2}; -ai\right] = \frac{e^{a(t-1)}}{2a}. \tag{3.4.35}$$

The minus sign in front of the $2\pi i$ arises from the negative sense of the contour. We may write the inverse as the single expression:

$$f(t) = \frac{1}{2a}e^{-a|t-1|}. \tag{3.4.36}$$

• **Example 3.4.5**

Let us evaluate the integral

$$\int_{-\infty}^{\infty} \frac{\cos(kx)}{x^2 + a^2}\, dx, \tag{3.4.37}$$

where $a, k > 0$.

We begin by noting that

$$\int_{-\infty}^{\infty} \frac{\cos(kx)}{x^2 + a^2}\, dx = \mathrm{Re}\left(\int_{-\infty}^{\infty} \frac{e^{ikx}}{x^2 + a^2}\, dx\right) = \mathrm{Re}\left(\oint_{C_1} \frac{e^{ikz}}{z^2 + a^2}\, dz\right), \tag{3.4.38}$$

where C_1 denotes a line integral along the real axis from $-\infty$ to ∞. A quick check shows that the integrand of the right side of (3.4.38) satisfies Jordan's lemma. Therefore,

$$\int_{-\infty}^{\infty} \frac{e^{ikx}}{x^2 + a^2}\, dx = \oint_C \frac{e^{ikz}}{z^2 + a^2}\, dz = 2\pi i \ \mathrm{Res}\left(\frac{e^{ikz}}{z^2 + a^2}; ai\right) \tag{3.4.39}$$

$$= 2\pi i \lim_{z \to ai} \frac{(z - ai)e^{ikz}}{z^2 + a^2} = \frac{\pi}{a}e^{-ka}, \tag{3.4.40}$$

where C denotes the closed infinite semicircle in the upper half-plane. Taking the real and imaginary parts of (3.4.40),

$$\int_{-\infty}^{\infty} \frac{\cos(kx)}{x^2 + a^2}\, dx = \frac{\pi}{a}e^{-ka} \tag{3.4.41}$$

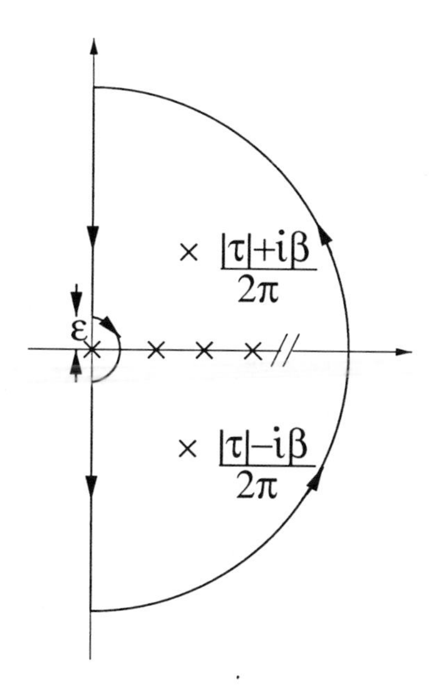

Figure 3.4.2: Contour used in Example 3.4.6.

and

$$\int_{-\infty}^{\infty} \frac{\sin(kx)}{x^2 + a^2}\, dx = 0. \tag{3.4.42}$$

• **Example 3.4.6**

So far we have used only the first two points of Jordan's lemma. In this example[9] we illustrate how the remaining two points may be applied.

Consider the contour integral

$$\oint_C \cot(\pi z) \left[\frac{e^{-cz}}{(\tau + 2\pi z)^2 + \beta^2} + \frac{e^{-cz}}{(\tau - 2\pi z)^2 + \beta^2} \right] dz,$$

where $c > 0$ and β, τ are real. Let us evaluate this contour integral where the contour is shown in Figure 3.4.2.

---

[9] Reprinted from *Int. J. Heat Mass Transfer*, **15**, Hsieh, T. C., and R. Greif, Theoretical determination of the absorption coefficient and the total band absorptance including a specific application to carbon monoxide, 1477–1487, ©1972, with kind permission from Elsevier Science Ltd., The Boulevard, Langford Lane, Kidlington OX5 1GB, UK.

From the residue theorem,

$$\oint_C \cot(\pi z)\left[\frac{e^{-cz}}{(\tau+2\pi z)^2+\beta^2}+\frac{e^{-cz}}{(\tau-2\pi z)^2+\beta^2}\right]dz$$
$$= 2\pi i\sum_{n=1}^{\infty}\operatorname{Res}\left\{\cot(\pi z)\left[\frac{e^{-cz}}{(\tau+2\pi z)^2+\beta^2}+\frac{e^{-cz}}{(\tau-2\pi z)^2+\beta^2}\right];n\right\}$$
$$+ 2\pi i\operatorname{Res}\left\{\cot(\pi z)\left[\frac{e^{-cz}}{(\tau+2\pi z)^2+\beta^2}+\frac{e^{-cz}}{(\tau-2\pi z)^2+\beta^2}\right];\frac{|\tau|+\beta i}{2\pi}\right\}$$
$$+ 2\pi i\operatorname{Res}\left\{\cot(\pi z)\left[\frac{e^{-cz}}{(\tau+2\pi z)^2+\beta^2}+\frac{e^{-cz}}{(\tau-2\pi z)^2+\beta^2}\right];\frac{|\tau|-\beta i}{2\pi}\right\}. \quad \textbf{(3.4.43)}$$

Now

$$\operatorname{Res}\left\{\cot(\pi z)\left[\frac{e^{-cz}}{(\tau+2\pi z)^2+\beta^2}+\frac{e^{-cz}}{(\tau-2\pi z)^2+\beta^2}\right];n\right\}$$
$$= \lim_{z\to n}\frac{(z-n)\cos(\pi z)}{\sin(\pi z)}\lim_{z\to n}\left[\frac{e^{-cz}}{(\tau+2\pi z)^2+\beta^2}+\frac{e^{-cz}}{(\tau-2\pi z)^2+\beta^2}\right] \quad \textbf{(3.4.44)}$$
$$= \frac{1}{\pi}\left[\frac{e^{-nc}}{(\tau+2n\pi)^2+\beta^2}+\frac{e^{-nc}}{(\tau-2n\pi)^2+\beta^2}\right], \quad \textbf{(3.4.45)}$$

$$\operatorname{Res}\left\{\cot(\pi z)\left[\frac{e^{-cz}}{(\tau+2\pi z)^2+\beta^2}+\frac{e^{-cz}}{(\tau-2\pi z)^2+\beta^2}\right];\frac{|\tau|+\beta i}{2\pi}\right\}$$
$$= \lim_{z\to(|\tau|+\beta i)/2\pi}\frac{\cot(\pi z)}{4\pi^2}$$
$$\times\left[\frac{(z-|\tau|-\beta i)e^{-cz}}{(z+\tau/2\pi)^2+\beta^2/4\pi^2}+\frac{(z-|\tau|-\beta i)e^{-cz}}{(z-\tau/2\pi)^2+\beta^2/4\pi^2}\right] \quad \textbf{(3.4.46)}$$
$$= \frac{\cot(|\tau|/2+\beta i/2)\exp(-c|\tau|/2\pi)[\cos(c\beta/2\pi)-i\sin(c\beta/2\pi)]}{4\pi\beta i} \quad \textbf{(3.4.47)}$$

and

$$\operatorname{Res}\left\{\cot(\pi z)\left[\frac{e^{-cz}}{(\tau+2\pi z)^2+\beta^2}+\frac{e^{-cz}}{(\tau-2\pi z)^2+\beta^2}\right];\frac{|\tau|-\beta i}{2\pi}\right\}$$
$$= \lim_{z\to(|\tau|-\beta i)/2\pi}\frac{\cot(\pi z)}{4\pi^2}$$
$$\times\left[\frac{(z-|\tau|+\beta i)e^{-cz}}{(z+\tau/2\pi)^2+\beta^2/4\pi^2}+\frac{(z-|\tau|+\beta i)e^{-cz}}{(z-\tau/2\pi)^2+\beta^2/4\pi^2}\right] \quad \textbf{(3.4.48)}$$
$$= \frac{\cot(|\tau|/2-\beta i/2)\exp(-c|\tau|/2\pi)[\cos(c\beta/2\pi)+i\sin(c\beta/2\pi)]}{-4\pi\beta i}. \quad \textbf{(3.4.49)}$$

Therefore,

$$\oint_C \cot(\pi z)\left[\frac{e^{-cz}}{(\tau+2\pi z)^2+\beta^2}+\frac{e^{-cz}}{(\tau-2\pi z)^2+\beta^2}\right]dz$$

$$= 2i\sum_{n=1}^{\infty}\left[\frac{e^{-nc}}{(\tau+2n\pi)^2+\beta^2}+\frac{e^{-nc}}{(\tau-2n\pi)^2+\beta^2}\right]$$

$$+\frac{i}{2\beta}\frac{e^{i|\tau|}+e^{\beta}}{e^{i|\tau|}-e^{\beta}}e^{-c|\tau|/2\pi}[\cos(c\beta/2\pi)-i\sin(c\beta/2\pi)]$$

$$-\frac{i}{2\beta}\frac{e^{i|\tau|}+e^{-\beta}}{e^{i|\tau|}-e^{-\beta}}e^{-c|\tau|/2\pi}[\cos(c\beta/2\pi)+i\sin(c\beta/2\pi)] \quad (3.4.50)$$

$$= 2i\sum_{n=1}^{\infty}\left[\frac{e^{-nc}}{(\tau+2n\pi)^2+\beta^2}+\frac{e^{-nc}}{(\tau-2n\pi)^2+\beta^2}\right]$$

$$-\frac{i}{\beta}\frac{\sinh(\beta)\cos(c\beta/2\pi)+\sin(|\tau|)\sin(c\beta/2\pi)}{\cosh(\beta)-\cos(\tau)}e^{-c|\tau|/2\pi}, \quad (3.4.51)$$

where $\cot(\alpha)=i(e^{2i\alpha}+1)/(e^{2i\alpha}-1)$ and we have made extensive use of Euler's formula.

Let us now evaluate the contour integral by direct integration. The contribution from the integration along the semicircle at infinity vanishes according to Jordan's lemma. Indeed that is why this particular contour was chosen. Therefore,

$$\oint_C \cot(\pi z)\left[\frac{e^{-cz}}{(\tau+2\pi z)^2+\beta^2}+\frac{e^{-cz}}{(\tau-2\pi z)^2+\beta^2}\right]dz$$

$$= \int_{i\infty}^{i\epsilon}\cot(\pi z)\left[\frac{e^{-cz}}{(\tau+2\pi z)^2+\beta^2}+\frac{e^{-cz}}{(\tau-2\pi z)^2+\beta^2}\right]dz$$

$$+\int_{C_\epsilon}\cot(\pi z)\left[\frac{e^{-cz}}{(\tau+2\pi z)^2+\beta^2}+\frac{e^{-cz}}{(\tau-2\pi z)^2+\beta^2}\right]dz$$

$$+\int_{-i\epsilon}^{-i\infty}\cot(\pi z)\left[\frac{e^{-cz}}{(\tau+2\pi z)^2+\beta^2}+\frac{e^{-cz}}{(\tau-2\pi z)^2+\beta^2}\right]dz. \quad (3.4.52)$$

Now, because $z = iy$,

$$\int_{i\infty}^{i\epsilon}\cot(\pi z)\left[\frac{e^{-cz}}{(\tau+2\pi z)^2+\beta^2}+\frac{e^{-cz}}{(\tau-2\pi z)^2+\beta^2}\right]dz$$

$$= \int_{\infty}^{\epsilon}\coth(\pi y)\left[\frac{e^{-icy}}{(\tau+2\pi iy)^2+\beta^2}+\frac{e^{-icy}}{(\tau-2\pi iy)^2+\beta^2}\right]dy \quad (3.4.53)$$

$$= -2\int_{\epsilon}^{\infty}\frac{\coth(\pi y)(\tau^2+\beta^2-4\pi^2y^2)e^{-icy}}{(\tau^2+\beta^2-4\pi^2y^2)^2+16\pi^2\tau^2y^2}dy, \quad (3.4.54)$$

$$\int_{-i\epsilon}^{-i\infty} \cot(\pi z) \left[\frac{e^{-cz}}{(\tau + 2\pi z)^2 + \beta^2} + \frac{e^{-cz}}{(\tau - 2\pi z)^2 + \beta^2} \right] dz$$

$$= \int_{-\epsilon}^{-\infty} \coth(\pi y) \left[\frac{e^{-icy}}{(\tau + 2\pi iy)^2 + \beta^2} + \frac{e^{-icy}}{(\tau - 2\pi iy)^2 + \beta^2} \right] dy \tag{3.4.55}$$

$$= 2 \int_{\epsilon}^{\infty} \frac{\coth(\pi y)(\tau^2 + \beta^2 - 4\pi^2 y^2) e^{icy}}{(\tau^2 + \beta^2 - 4\pi^2 y^2)^2 + 16\pi^2 \tau^2 y^2}\, dy \tag{3.4.56}$$

and

$$\int_{C_\epsilon} \cot(\pi z) \left[\frac{e^{-cz}}{(\tau + 2\pi z)^2 + \beta^2} + \frac{e^{-cz}}{(\tau - 2\pi z)^2 + \beta^2} \right] dz$$

$$= \int_{\pi/2}^{-\pi/2} \left[\frac{1}{\pi \epsilon e^{\theta i}} - \frac{\pi \epsilon e^{\theta i}}{3} - \cdots \right] \epsilon i e^{\theta i}\, d\theta$$

$$\times \left[\frac{\exp(-c\epsilon e^{\theta i})}{(\tau + 2\pi \epsilon e^{\theta i})^2 + \beta^2} + \frac{\exp(-c\epsilon e^{\theta i})}{(\tau - 2\pi \epsilon e^{\theta i})^2 + \beta^2} \right]. \tag{3.4.57}$$

In the limit of $\epsilon \to 0$,

$$\oint_C \cot(\pi z) \left[\frac{e^{-cz}}{(\tau + 2\pi z)^2 + \beta^2} + \frac{e^{-cz}}{(\tau - 2\pi z)^2 + \beta^2} \right] dz$$

$$= 4i \int_0^{\infty} \frac{\coth(\pi y)(\tau^2 + \beta^2 - 4\pi^2 y^2) \sin(cy)}{(\tau^2 + \beta^2 - 4\pi^2 y^2)^2 + 16\pi^2 \tau^2 y^2}\, dy - \frac{2i}{\tau^2 + \beta^2} \tag{3.4.58}$$

$$= 2i \sum_{n=1}^{\infty} \left[\frac{e^{-nc}}{(\tau + 2n\pi)^2 + \beta^2} + \frac{e^{-nc}}{(\tau - 2n\pi)^2 + \beta^2} \right]$$

$$- \frac{i}{\beta} \frac{\sinh(\beta)\cos(c\beta/2\pi) + \sin(|\tau|)\sin(c\beta/2\pi)}{\cosh(\beta) - \cos(\tau)} e^{-c|\tau|/2\pi} \tag{3.4.59}$$

or

$$4 \int_0^{\infty} \frac{\coth(\pi y)(\tau^2 + \beta^2 - 4\pi^2 y^2) \sin(cy)}{(\tau^2 + \beta^2 - 4\pi^2 y^2)^2 + 16\pi^2 \tau^2 y^2}\, dy$$

$$= 2 \sum_{n=1}^{\infty} \left[\frac{e^{-nc}}{(\tau + 2n\pi)^2 + \beta^2} + \frac{e^{-nc}}{(\tau - 2n\pi)^2 + \beta^2} \right]$$

$$- \frac{1}{\beta} \frac{\sinh(\beta)\cos(c\beta/2\pi) + \sin(|\tau|)\sin(c\beta/2\pi)}{\cosh(\beta) - \cos(\tau)} e^{-c|\tau|/2\pi}$$

$$+ \frac{2}{\tau^2 + \beta^2}. \tag{3.4.60}$$

If we let $y = x/2\pi$,

$$\frac{\beta}{\pi} \int_0^{\infty} \frac{\coth(x/2)(\tau^2 + \beta^2 - x^2) \sin(cx/2\pi)}{(\tau^2 + \beta^2 - x^2)^2 + 4\tau^2 x^2}\, dx$$

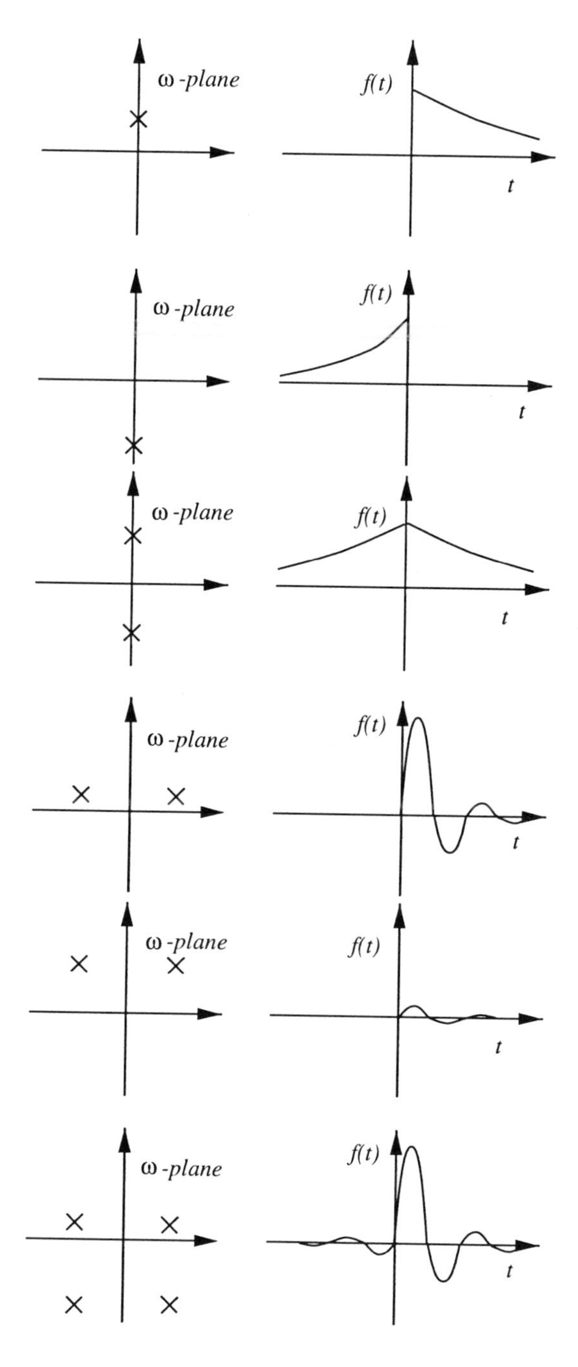

Figure 3.4.3: The correspondence between the location of the simple poles of the Fourier transform $F(\omega)$ and the behavior of $f(t)$.

$$= 2\beta \sum_{n=1}^{\infty} \left[\frac{e^{-nc}}{(\tau + 2n\pi)^2 + \beta^2} + \frac{e^{-nc}}{(\tau - 2n\pi)^2 + \beta^2} \right]$$
$$- \frac{\sinh(\beta)\cos(c\beta/2\pi) + \sin(|\tau|)\sin(c\beta/2\pi)}{\cosh(\beta) - \cos(\tau)} e^{-c|\tau|/2\pi}$$
$$+ \frac{2\beta}{\tau^2 + \beta^2}. \tag{3.4.61}$$

- **Example 3.4.7**

An additional benefit of understanding inversion by the residue method is the ability to qualitatively anticipate the inverse by knowing the location of the poles of $F(\omega)$. This intuition is important because many engineering analyses discuss stability and performance entirely in terms of the properties of the system's Fourier transform. In Figure 3.4.3 we have graphed the location of the poles of $F(\omega)$ and the corresponding $f(t)$. The student should go through the mental exercise of connecting the two pictures.

Problems

1. Use direct integration to find the inverse of the Fourier transform

$$F(\omega) = \frac{i\omega\pi}{2} e^{-|\omega|}.$$

Use partial fractions to invert the following Fourier transforms:

2. $\dfrac{1}{(1+i\omega)(1+2i\omega)}$

3. $\dfrac{1}{(1+i\omega)(1-i\omega)}$

4. $\dfrac{i\omega}{(1+i\omega)(1+2i\omega)}$

5. $\dfrac{1}{(1+i\omega)(1+2i\omega)^2}$

By taking the appropriate closed contour, find the inverse of the following Fourier transform by contour integration. The parameter a is real and positive.

6. $\dfrac{1}{\omega^2 + a^2}$

7. $\dfrac{\omega}{\omega^2 + a^2}$

8. $\dfrac{\omega}{(\omega^2 + a^2)^2}$

9. $\dfrac{\omega^2}{(\omega^2 + a^2)^2}$

10. $\dfrac{1}{\omega^2 - 3i\omega - 3}$

11. $\dfrac{1}{(\omega - ia)^{2n+2}}$

12. $\dfrac{\omega^2}{(\omega^2 - 1)^2 + 4a^2\omega^2}$

13. $\dfrac{3}{(2 - \omega i)(1 + \omega i)}$

14. Find the inverse of $F(\omega) = \cos(\omega)/(\omega^2+a^2)$, $a > 0$, by first rewriting the transform as

$$F(\omega) = \frac{e^{i\omega}}{2(\omega^2 + a^2)} + \frac{e^{-i\omega}}{2(\omega^2 + a^2)}$$

and then using the residue theorem on each term.

15. As we shall show shortly, Fourier transforms can be used to solve differential equations. During the solution of the heat equation, Taitel et al.[10] had to invert the Fourier transform

$$F(\omega) = \frac{\cosh(y\sqrt{\omega^2+1}\,)}{\sqrt{\omega^2+1}\,\sinh(p\sqrt{\omega^2+1}/2)},$$

where y and p are real. Show that they should have found

$$f(t) = \frac{e^{-|t|}}{p} + \frac{2}{p}\sum_{n=1}^{\infty} \frac{(-1)^n}{\sqrt{1+4n^2\pi^2/p^2}} \cos\left(\frac{2n\pi y}{p}\right) e^{-\sqrt{1+4n^2\pi^2/p^2}\,|t|}.$$

In this case, our time variable t was their spatial variable $x - \xi$.

16. Find the inverse of the Fourier transform

$$F(\omega) = \left[\cos\left\{\frac{\omega L}{\beta[1 + i\gamma\ \mathrm{sgn}(\omega)]}\right\}\right]^{-1},$$

where L, β, and γ are real and positive and $\mathrm{sgn}(z) = 1$ if $\mathrm{Re}(z) > 0$ and -1 if $\mathrm{Re}(z) < 0$.

Use the residue theorem to verify the following integrals:

17.

$$\int_{-\infty}^{\infty} \frac{\sin(x)}{x^2 + 4x + 5}\,dx = -\frac{\pi}{e}\sin(2)$$

18.

$$\int_0^{\infty} \frac{\cos(x)}{(x^2 + 1)^2}\,dx = \frac{\pi}{2e}$$

---

[10] Reprinted from *Int. J. Heat Mass Transfer*, **16**, Taitel, Y., M. Bentwich and A. Tamir, Effects of upstream and downstream boundary conditions on heat (mass) transfer with axial diffusion, 359–369, ©1973, with kind permission from Elsevier Science Ltd., The Boulevard, Langford Lane, Kidlington OX5 1GB, UK.

19.

$$\int_{-\infty}^{\infty} \frac{x\sin(ax)}{x^2+4}\, dx = \pi e^{-2a}, \qquad a > 0$$

20. The concept of forced convection is normally associated with heat streaming through a duct or past an obstacle. Bentwich[11] wanted to show a similar transport can exist when convection results from a wave traveling through an essentially stagnant fluid. In the process of computing the amount of heating he had to prove the following identity:

$$\int_{-\infty}^{\infty} \frac{\cosh(hx)-1}{x\sinh(hx)} \cos(ax)\, dx = \ln[\coth(|a|\pi/h)], \qquad h > 0.$$

Confirm his result.

3.5 CONVOLUTION

The most important property of Fourier transforms is convolution. We shall use it extensively in the solution of differential equations and the design of filters because it yields in time or space the effect of multiplying two transforms together.

The convolution operation is

$$f(t) * g(t) = \int_{-\infty}^{\infty} f(x)g(t-x)\, dx = \int_{-\infty}^{\infty} f(t-x)g(x)\, dx. \qquad (3.5.1)$$

Then,

$$\mathcal{F}[f(t) * g(t)] = \int_{-\infty}^{\infty} f(x)e^{-i\omega x}\left[\int_{-\infty}^{\infty} g(t-x)e^{-i\omega(t-x)} dt\right] dx \qquad (3.5.2)$$

$$= \int_{-\infty}^{\infty} f(x)G(\omega)e^{-i\omega x}\, dx = F(\omega)G(\omega). \qquad (3.5.3)$$

Thus, the Fourier transform of the convolution of two functions equals the product of the Fourier transforms of each of the functions.

• **Example 3.5.1**

Verify the convolution theorem using the functions $f(t) = H(t+a) - H(t-a)$ and $g(t) = e^{-t}H(t)$, where $a > 0$.

---

[11] Reprinted from *Int. J. Heat Mass Transfer*, **9**, Bentwich, M., Convection enforced by surface and tidal waves, 663–670, ©1966, with kind permission from Elsevier Science Ltd., The Boulevard, Langford Lane, Kidlington OX5 1GB, UK.

The convolution of $f(t)$ with $g(t)$ is

$$f(t) * g(t) = \int_{-\infty}^{\infty} e^{-(t-x)} H(t-x) \left[H(x+a) - H(x-a)\right] dx \tag{3.5.4}$$

$$= e^{-t} \int_{-a}^{a} e^{x} H(t-x)\, dx. \tag{3.5.5}$$

If $t < -a$, then the integrand of (3.5.5) is always zero and $f(t)*g(t) = 0$. If $t > a$,

$$f(t) * g(t) = e^{-t} \int_{-a}^{a} e^{x}\, dx = e^{-(t-a)} - e^{-(t+a)}. \tag{3.5.6}$$

Finally, for $-a < t < a$,

$$f(t) * g(t) = e^{-t} \int_{-a}^{t} e^{x}\, dx = 1 - e^{-(t+a)}. \tag{3.5.7}$$

In summary,

$$f(t) * g(t) = \begin{cases} 0, & t < -a \\ 1 - e^{-(t+a)}, & -a < t < a \\ e^{-(t-a)} - e^{-(t+a)}, & t > a. \end{cases} \tag{3.5.8}$$

The Fourier transform of $f(t) * g(t)$ is

$$\mathcal{F}[f(t) * g(t)] = \int_{-a}^{a} \left[1 - e^{-(t+a)}\right] e^{-i\omega t} dt + \int_{a}^{\infty} \left[e^{-(t-a)} - e^{-(t+a)}\right] e^{-i\omega t} dt \tag{3.5.9}$$

$$= \frac{2\sin(\omega a)}{\omega} - \frac{2i\sin(\omega a)}{1+\omega i} \tag{3.5.10}$$

$$= \frac{2\sin(\omega a)}{\omega} \left(\frac{1}{1+\omega i}\right) = F(\omega)G(\omega) \tag{3.5.11}$$

and the convolution theorem is true for this special case.

• **Example 3.5.2**

Let us consider the convolution of $f(t) = f_+(t)H(t)$ with $g(t) = g_+H(t)$. Note that both of the functions are nonzero only for $t > 0$.

From the definition of convolution,

$$f(t) * g(t) = \int_{-\infty}^{\infty} f_+(t-x)H(t-x)g_+(x)H(x)\,dx \tag{3.5.12}$$

$$= \int_0^{\infty} f_+(t-x)H(t-x)g_+(x)\,dx. \tag{3.5.13}$$

For $t < 0$, the integrand is always zero and $f(t) * g(t) = 0$. For $t > 0$,

$$f(t) * g(t) = \int_0^t f_+(t-x)g_+(x)\,dx. \tag{3.5.14}$$

Therefore, in general,

$$f(t) * g(t) = \left[\int_0^t f_+(t-x)g_+(x)\,dx\right] H(t). \tag{3.5.15}$$

This is the definition of convolution that we will use for Laplace transforms where all of the functions equal zero for $t < 0$.

Problems

1. Show that

$$e^{-t}H(t) * e^{-t}H(t) = te^{-t}H(t).$$

2. Show that

$$e^{-t}H(t) * e^{t}H(-t) = \tfrac{1}{2}e^{-|t|}.$$

3. Show that

$$e^{-t}H(t) * e^{-2t}H(t) = \left(e^{-t} - e^{-2t}\right)H(t).$$

4. Show that

$$e^{t}H(-t) * [H(t) - H(t-2)] = \begin{cases} e^{t} - e^{t-2}, & t < 0 \\ 1 - e^{t-2}, & 0 < t < 2 \\ 0, & t > 2. \end{cases}$$

5. Show that

$$[H(t) - H(t-2)] * [H(t) - H(t-2)] = \begin{cases} 0, & t < 0 \\ t, & 0 < t < 2 \\ 4 - t, & 2 < t < 4 \\ 0, & t > 4. \end{cases}$$

6. Show that

$$e^{-|t|} * e^{-|t|} = (1 + |t|)e^{-|t|}.$$

7. Prove that the convolution of two Dirac delta functions is a Dirac delta function.

3.6 SOLUTION OF ORDINARY DIFFERENTIAL EQUATIONS BY FOURIER TRANSFORMS

As with Laplace transforms, we may use Fourier transforms to solve ordinary differential equations. However, this method gives only the particular solution and we must find the complementary solution separately.

Consider the differential equations

$$y' + y = \tfrac{1}{2}e^{-|t|}, \qquad -\infty < t < \infty. \tag{3.6.1}$$

Taking the Fourier transform of both sides of (3.6.1),

$$i\omega Y(\omega) + Y(\omega) = \frac{1}{\omega^2 + 1}, \tag{3.6.2}$$

where we have used the derivative rule (3.3.19) to obtain the transform of y' and $Y(\omega) = \mathcal{F}[y(t)]$. Therefore,

$$Y(\omega) = \frac{1}{(\omega^2 + 1)(1 + \omega i)}. \tag{3.6.3}$$

Applying the inversion integral to (3.6.3),

$$y(t) = \frac{1}{2\pi}\int_{-\infty}^{\infty} \frac{e^{it\omega}}{(\omega^2 + 1)(1 + \omega i)}\, d\omega. \tag{3.6.4}$$

We evaluate (3.6.4) by contour integration. For $t > 0$ we close the line integral with an infinite semicircle in the upper half of the ω-plane. The integration along this arc equals zero by Jordan's lemma. Within this closed contour we have a second-order pole at $z = i$. Therefore,

$$\operatorname{Res}\left[\frac{e^{itz}}{(z^2 + 1)(1 + zi)}; i\right] = \lim_{z \to i} \frac{d}{dz}\left[(z - i)^2 \frac{e^{itz}}{i(z - i)^2(z + i)}\right] \tag{3.6.5}$$

$$= \frac{te^{-t}}{2i} + \frac{e^{-t}}{4i} \tag{3.6.6}$$

and

$$y(t) = \frac{1}{2\pi}(2\pi i)\left[\frac{te^{-t}}{2i} + \frac{e^{-t}}{4i}\right] = \frac{e^{-t}}{4}(2t + 1). \tag{3.6.7}$$

For $t < 0$, we again close the line integral with an infinite semicircle but this time it is in the lower half of the ω-plane. The contribution from the line integral along the arc vanishes by Jordan's lemma. Within the contour, we have a simple pole at $z = -i$. Therefore,

$$\text{Res}\left[\frac{e^{itz}}{(z^2+1)(1+zi)}; -i\right] = \lim_{z\to -i}(z+i)\frac{e^{itz}}{i(z+i)(z-i)^2} = -\frac{e^t}{4i} \tag{3.6.8}$$

and

$$y(t) = \frac{1}{2\pi}(-2\pi i)\left(-\frac{e^t}{4i}\right) = \frac{e^t}{4}. \tag{3.6.9}$$

The minus sign in front of the $2\pi i$ results from the contour being taken in the negative sense. Using the step function, we can combine (3.6.7) and (3.6.9) into the single expression

$$y(t) = \tfrac{1}{4}e^{-|t|} + \tfrac{1}{2}te^{-t}H(t). \tag{3.6.10}$$

Note that we have only found the particular or forced solution to (3.6.1). The most general solution therefore requires that we add the complementary solution Ae^{-t}, yielding

$$y(t) = Ae^{-t} + \tfrac{1}{4}e^{-|t|} + \tfrac{1}{2}te^{-t}H(t). \tag{3.6.11}$$

The arbitrary constant A would be determined by the initial condition which we have not specified.

Consider now a more general problem of

$$y' + y = f(t), \qquad -\infty < t < \infty, \tag{3.6.12}$$

where we assume that $f(t)$ has the Fourier transform $F(\omega)$. Then the Fourier-transformed solution to (3.6.12) is

$$Y(\omega) = \frac{1}{1+\omega i}F(\omega) = G(\omega)F(\omega) \tag{3.6.13}$$

or

$$y(t) = g(t) * f(t), \tag{3.6.14}$$

where $g(t) = \mathcal{F}^{-1}[1/(1+\omega i)] = e^{-t}H(t)$. Thus, we can obtain our solution in one of two ways. First, we can take the Fourier transform of $f(t)$, multiply this transform by $G(\omega)$, and finally compute the inverse. The second method requires a convolution of $f(t)$ with $g(t)$. Which method is easiest depends upon $f(t)$ and $g(t)$.

The function $g(t)$ may also be viewed as the particular solution of (3.6.12) resulting from the forcing function $\delta(t)$, the Dirac delta function, because $\mathcal{F}[\delta(t)] = 1$. Traditionally this forced solution $g(t)$ is called the

Green's function and $G(\omega)$ is called the *frequency response* or *steady-state transfer function* of our system. Engineers often extensively study the frequency response in their analysis rather than the Green's function because the frequency response is easier to obtain experimentally and the output from a linear system is just the product of two transforms [see (3.6.13)] rather than an integration.

In summary, we may use Fourier transforms to find particular solutions to differential equations. The complete solution consists of this particular solution plus any homogeneous solution that we need to satisfy the initial conditions. Convolution of the Green's function with the forcing function also gives the particular solution.

• Example 3.6.1: Spectrum of a damped harmonic oscillator

Second-order differential equations are ubiquitous in engineering. In electrical engineering many electrical circuits are governed by second-order, linear ordinary differential equations. In mechanical engineering they arise during the application of Newton's second law. For example, in mechanics the damped oscillations of a mass m attached to a spring with a spring constant k and damped with a velocity dependent resistance is govern by the equation

$$my'' + cy' + ky = f(t), \tag{3.6.15}$$

where $y(t)$ denotes the displacement of the oscillator from its equilibrium position, c denotes the damping coefficient and $f(t)$ denotes the forcing.

Assuming that both $f(t)$ and $y(t)$ have Fourier transforms, let us analyze this system by finding its frequency response. We begin our analysis by solving for the Green's function $g(t)$ which is given by

$$mg'' + cg' + kg = \delta(t), \tag{3.6.16}$$

because the Green's function is the response of a system to a delta function forcing. Taking the Fourier transform of both sides of (3.6.16), the frequency response is

$$G(\omega) = \frac{1}{k + ic\omega - m\omega^2} = \frac{1/m}{\omega_0^2 + ic\omega/m - \omega^2}, \tag{3.6.17}$$

where $\omega_0^2 = k/m$ is the natural frequency of the system. The most useful quantity to plot is the frequency response or

$$|G(\omega)| = \frac{\omega_0^2}{k\sqrt{(\omega^2 - \omega_0^2)^2 + \omega^2\omega_0^2(c^2/km)}} \tag{3.6.18}$$

$$= \frac{1}{k\sqrt{[(\omega/\omega_0)^2 - 1]^2 + (c^2/km)(\omega/\omega_0)^2}}. \tag{3.6.19}$$

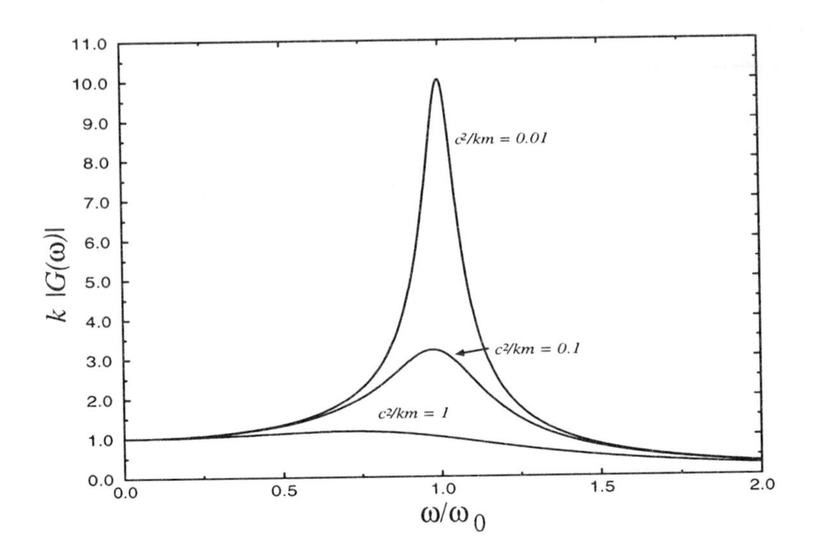

Figure 3.6.1: The variation of the frequency response for a damped harmonic oscillator as a function of driving frequency ω. See the text for the definition of the parameters.

In Figure 3.6.1 we have plotted with frequence response for different c^2/km's. Note that as the damping becomes larger, the sharp peak at $\omega = \omega_0$ essentially vanishes. As $c^2/km \to 0$, we obtain a very finely tuned response curve.

Let us now find the Green's function. From the definition of the inverse Fourier transform,

$$mg(t) = -\frac{1}{2\pi}\int_{-\infty}^{\infty} \frac{e^{i\omega t}}{\omega^2 - ic\omega/m - \omega_0^2}\, d\omega \tag{3.6.20}$$

$$= -\frac{1}{2\pi}\int_{-\infty}^{\infty} \frac{e^{i\omega t}}{(\omega-\omega_1)(\omega-\omega_2)}\, d\omega, \tag{3.6.21}$$

where

$$\omega_{1,2} = \pm\sqrt{\omega_0^2 - \gamma^2} + \gamma i \tag{3.6.22}$$

and $\gamma = c/2m > 0$. We can evaluate (3.6.21) by residues. Clearly the poles always lie in the upper half of the ω-plane. Thus, if $t < 0$ in (3.6.21) we can close the line integration along the real axis with a semicircle of infinite radius in the lower half of the ω-plane by Jordan's lemma. Because the integrand is analytic within the closed contour, $g(t) = 0$ for $t < 0$. This is simply the causality condition,[12] the impulse

[12] The principle stating that an event cannot precede its cause.

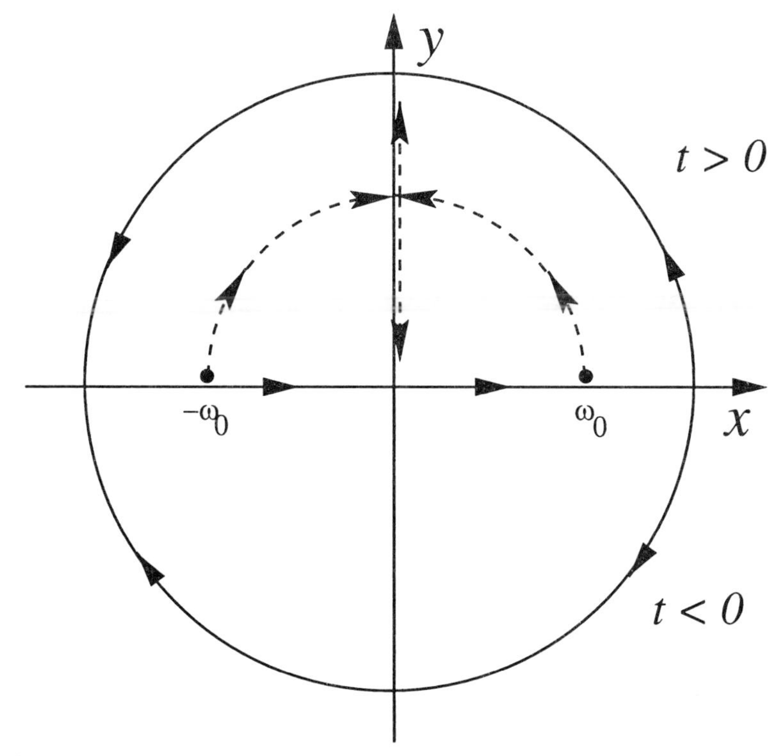

Figure 3.6.2: The migration of the poles of the frequency response of a damped harmonic oscillator as a function of γ.

forcing being the cause of the excitation. Clearly, causality is closely connected with the analyticity of the frequency response in the lower half of the ω-plane.

If $t > 0$, we close the line integration along the real axis with a semicircle of infinite radius in the upper half of the ω-plane and obtain

$$mg(t) = 2\pi i\left(-\frac{1}{2\pi}\right)\left\{\operatorname{Res}\left[\frac{e^{izt}}{(z-\omega_1)(z-\omega_2)};\omega_1\right]\right.$$
$$\left.+\operatorname{Res}\left[\frac{e^{izt}}{(z-\omega_1)(z-\omega_2)};\omega_2\right]\right\} \quad (3.6.23)$$

$$= \frac{-i}{\omega_1-\omega_2}\left(e^{i\omega_1 t}-e^{i\omega_2 t}\right) \quad (3.6.24)$$

$$= \frac{e^{-\gamma t}\sin\left(t\sqrt{\omega_0^2-\gamma^2}\right)}{\sqrt{\omega_0^2-\gamma^2}}H(t). \quad (3.6.25)$$

Let us now examine the damped harmonic oscillator by describing the migration of the poles $\omega_{1,2}$ in the complex ω-plane as γ increase from 0 to ∞. See Figure 3.6.2. For $\gamma \ll \omega_0$ (weak damping), the poles $\omega_{1,2}$ are very near to the real axis, above the points $\pm\omega_0$, respectively. This corresponds to the narrow resonance band discussed earlier and we have an underdamped harmonic oscillator. As γ increases from 0 to ω_0, the poles approach the positive imaginary axis, moving along a semicircle of radius ω_0 centered at the origin. They coalesce at the point $i\omega_0$ for $\gamma = \omega_0$, yielding repeated roots, and we have a critically damped oscillator. For $\gamma > \omega_0$, the poles move in opposite directions along the positive imaginary axis; one of them approaches the origin, while the other tends to $i\infty$ as $\gamma \to \infty$. The solution then has two purely decaying, overdamped solutions.

During the early 1950s, a similar diagram was invented by Evans[13] where the movement of closed-loop poles is plotted for all values of a system parameter, usually the gain. This *root-locus method* is very popular in system control theory for two reasons. First, the investigator can easily determine the contribution of a particular closed-loop pole to the transient response. Second, he may determine the manner in which open-loop poles or zeros should be introduced or their location modified so that he will achieve a desired performance characteristic for his system.

• Example 3.6.2: Low frequency filter

Consider the ordinary differential equation

$$Ry' + \frac{1}{C}y = f(t), \tag{3.6.26}$$

where R and C are real, positive constants. If $y(t)$ denotes current, then (3.6.26) would be the equation that gives the voltage across a capacitor in a RC circuit. Let us find the frequency response and Green's function for this system.

We begin by writing (3.6.26) as

$$Rg' + \frac{1}{C}g = \delta(t), \tag{3.6.27}$$

where $g(t)$ denotes the Green's function. If the Fourier transform of $g(t)$ is $G(\omega)$, the frequency response $G(\omega)$ is given by

$$i\omega RG(\omega) + \frac{G(\omega)}{C} = 1 \tag{3.6.28}$$

[13] Evans, W. R., 1948: Graphical analysis of control systems. *Trans. AIEE*, **67**, 547–551; Evans, W. R., 1954: *Control-System Dynamics*, McGraw-Hill, New York.

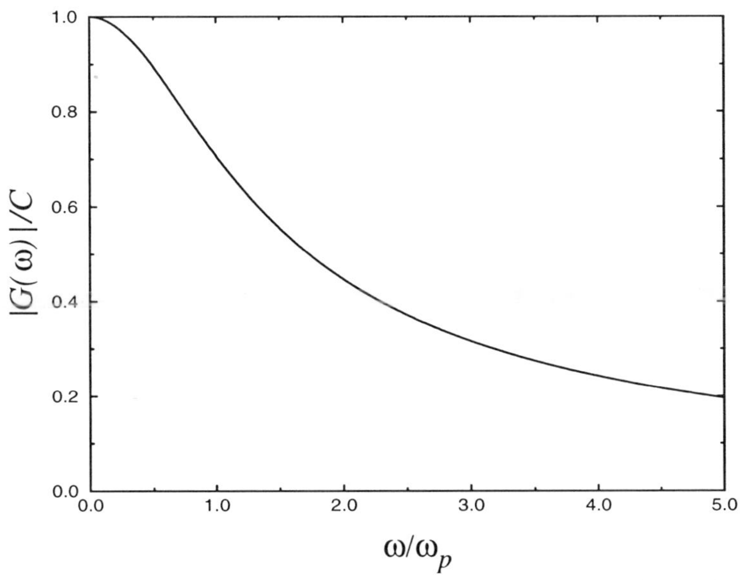

Figure 3.6.3: The variation of the frequency response (3.6.30) as a function of driving frequency ω. See the text for the definition of the parameters.

or

$$G(\omega) = \frac{1}{i\omega R + 1/C} = \frac{C}{1 + i\omega RC} \tag{3.6.29}$$

and

$$|G(\omega)| = \frac{C}{\sqrt{1 + \omega^2 R^2 C^2}} = \frac{C}{\sqrt{1 + \omega^2/\omega_p^2}}, \tag{3.6.30}$$

where $\omega_p = 1/(RC)$ is an intrinsic constant of the system. In Figure 3.6.3 we have plotted $|G(\omega)|$ as a function of ω. From this figure, we see that the response is largest for small ω and decreases as ω increases.

This is an example of a *low frequency filter* because relatively more signal passes through at lower frequencies than at higher frequencies. To understand this, let us drive the system with a forcing function that has the Fourier transform $F(\omega)$. The response of the system will be $G(\omega)F(\omega)$. Thus, that portion of the forcing function's spectrum at the lower frequencies will be relatively unaffected because $|G(\omega)|$ is near unity. However, at higher frequencies where $|G(\omega)|$ is smaller, the magnitude of the output will be greatly reduced.

Problems

Find the particular solutions for the following differential equations:

1. $y'' + 3y' + 2y = e^{-t}H(t)$

2. $y'' + 4y' + 4y = \frac{1}{2}e^{-|t|}$

3. $y'' - 4y' + 4y = e^{-t}H(t)$ 4. $y^{iv} - \lambda^4 y = \delta(x)$,
where λ has a positive real part and a negative imaginary part.

Chapter 4
The Laplace Transform

The previous chapter introduced the concept of the Fourier integral. If the function is nonzero only when $t > 0$, a similar transform, the *Laplace transform*,[1] exists. It is particularly useful in solving initial-value problems involving linear, constant coefficient, ordinary and partial differential equations. The present chapter develops the general properties and techniques of Laplace transforms.

4.1 DEFINITION AND ELEMENTARY PROPERTIES

Consider a function $f(t)$ such that $f(t) = 0$ for $t < 0$. Then the *Laplace integral*

$$\mathcal{L}[f(t)] = F(s) = \int_0^\infty f(t)e^{-st}\,dt \tag{4.1.1}$$

[1] The standard reference for Laplace transforms is Doetsch, G., 1950: *Handbuch der Laplace-Transformation. Band 1. Theorie der Laplace-Transformation*, Birkhäuser Verlag, 581 pp.; Doetsch, G., 1955: *Handbuch der Laplace-Transformation. Band 2. Anwendungen der Laplace-Transformation. 1. Abteilung*, Birkhäuser Verlag, 433 pp.; Doetsch, G., 1956: *Handbuch der Laplace-Transformation. Band 3. Anwendungen der Laplace-Transformation. 2. Abteilung*, Birkhäuser Verlag, 298 pp.

defines the Laplace transform of $f(t)$, which we shall write $\mathcal{L}[f(t)]$ or $F(s)$. The Laplace transform converts a function of t into a function of the transform variable s.

Not all functions have a Laplace transform because the integral (4.1.1) may fail to exist. For example, the function may have infinite discontinuities. For this reason, $f(t) = \tan(t)$ does *not* have a Laplace transform. We may avoid this difficulty by requiring that $f(t)$ be *piece-wise continuous.* That is, we can divide a finite range into a finite number of intervals in such a manner that $f(t)$ is continuous inside each interval and approaches finite values as we approach either end of any interval from the interior.

Another unacceptable function is $f(t) = 1/t$ because the integral (4.1.1) fails to exist. This leads to the requirement that the product $t^n|f(t)|$ is bounded near $t = 0$ for some number $n < 1$.

Finally $|f(t)|$ cannot grow too rapidly or it could overwhelm the e^{-st} term. To express this, we introduce the concept of functions of *exponential order.* By exponential order we mean that there exists some constants, M and k, for which

$$|f(t)| \le Me^{kt} \tag{4.1.2}$$

for all $t > 0$. Then, the Laplace transform of $f(t)$ exists if s, or just the real part of s, is greater than k.

In summary, the Laplace transform of $f(t)$ exists, for sufficiently large s, provided $f(t)$ satisfies the following conditions:

- $f(t) = 0$ for $t < 0$,
- $f(t)$ is continuous or piece-wise continuous in every interval,
- $t^n|f(t)| < \infty$ as $t \to 0$ for some number n, where $n < 1$,
- $e^{-s_0 t}|f(t)| < \infty$ as $t \to \infty$, for some number s_0. The quantity s_0 is called the *abscissa of convergence.*

• Example 4.1.1

Let us find the Laplace transform of 1, e^{at}, $\sin(at)$, $\cos(at)$, and t^n from the definition of the Laplace transform. From (4.1.1), direct integration yields:

$$\mathcal{L}(1) = \int_0^\infty e^{-st}\,dt = -\left.\frac{e^{-st}}{s}\right|_0^\infty = \frac{1}{s}, \qquad s > 0, \tag{4.1.3}$$

$$\mathcal{L}(e^{at}) = \int_0^\infty e^{at}e^{-st}\,dt = \int_0^\infty e^{-(s-a)t}\,dt \tag{4.1.4}$$

$$= -\left.\frac{e^{-(s-a)t}}{s-a}\right|_0^\infty = \frac{1}{s-a}, \qquad s > a, \tag{4.1.5}$$

$$\mathcal{L}[\sin(at)] = \int_0^\infty \sin(at)e^{-st}dt = -\left.\frac{e^{-st}}{s^2+a^2}[s\sin(at) + a\cos(at)]\right|_0^\infty \tag{4.1.6}$$

$$= \frac{a}{s^2+a^2}, \qquad s > 0, \tag{4.1.7}$$

$$\mathcal{L}[\cos(at)] = \int_0^\infty \cos(at)e^{-st}dt = \left.\frac{e^{-st}}{s^2+a^2}[-s\cos(at) + a\sin(at)]\right|_0^\infty \tag{4.1.8}$$

$$= \frac{s}{s^2+a^2}, \qquad s > 0 \tag{4.1.9}$$

and

$$\mathcal{L}(t^n) = \int_0^\infty t^n e^{-st}dt = n!e^{-st}\sum_{m=0}^{n}\left.\frac{t^{n-m}}{(n-m)!s^{m+1}}\right|_0^\infty = \frac{n!}{s^{n+1}}, \qquad s > 0, \tag{4.1.10}$$

where n is a positive integer.

The Laplace transform inherits two important properties from its integral definition. First, the transform of a sum equals the sum of the transforms:

$$\mathcal{L}[c_1 f(t) + c_2 g(t)] = c_1\mathcal{L}[f(t)] + c_2\mathcal{L}[g(t)]. \tag{4.1.11}$$

This linearity property holds with complex numbers and functions as well.

• Example 4.1.2

Success with Laplace transforms often rests with the ability to manipulate a given transform into a form which you can invert by inspection. Consider the following examples.

Given $F(s) = 4/s^3$, then

$$F(s) = 2 \times \frac{2}{s^3} \quad \text{and} \quad f(t) = 2t^2 \tag{4.1.12}$$

from (4.1.10).

Table 4.1.1: The Laplace Transforms of Some Commonly Encountered Functions.

	$\mathbf{f(t),\ t \geq 0}$	$\mathbf{F(s)}$
1.	1	$\frac{1}{s}$
2.	e^{-at}	$\frac{1}{s+a}$
3.	$\frac{1}{a}(1-e^{-at})$	$\frac{1}{s(s+a)}$
4.	$\frac{1}{a-b}(e^{-bt}-e^{-at})$	$\frac{1}{(s+a)(s+b)}$
5.	$\frac{1}{b-a}(be^{-bt}-ae^{-at})$	$\frac{s}{(s+a)(s+b)}$
6.	$\sin(at)$	$\frac{a}{s^2+a^2}$
7.	$\cos(at)$	$\frac{s}{s^2+a^2}$
8.	$\sinh(at)$	$\frac{a}{s^2-a^2}$
9.	$\cosh(at)$	$\frac{s}{s^2-a^2}$
10.	$t\sin(at)$	$\frac{2as}{(s^2+a^2)^2}$
11.	$1-\cos(at)$	$\frac{a^2}{s(s^2+a^2)}$
12.	$at-\sin(at)$	$\frac{a^3}{s^2(s^2+a^2)}$
13.	$t\cos(at)$	$\frac{s^2-a^2}{(s^2+a^2)^2}$
14.	$\sin(at)-at\cos(at)$	$\frac{2a^3}{(s^2+a^2)^2}$
15.	$t\sinh(at)$	$\frac{2as}{(s^2-a^2)^2}$
16.	$t\cosh(at)$	$\frac{s^2+a^2}{(s^2-a^2)^2}$
17.	$at\cosh(at)-\sinh(at)$	$\frac{2a^3}{(s^2-a^2)^2}$

Table 4.1.1 (contd.): The Laplace Transforms of Some Commonly Encountered Functions.

	$\mathbf{f(t),\ t \geq 0}$	$\mathbf{F(s)}$
18.	$e^{-bt}\sin(at)$	$\dfrac{a}{(s+b)^2+a^2}$
19.	$e^{-bt}\cos(at)$	$\dfrac{s+b}{(s+b)^2+a^2}$
20.	$(1+a^2t^2)\sin(at)-\cos(at)$	$\dfrac{8a^3s^2}{(s^2+a^2)^3}$
21.	$\sin(at)\cosh(at)-\cos(at)\sinh(at)$	$\dfrac{4a^3}{s^4+4a^4}$
22.	$\sin(at)\sinh(at)$	$\dfrac{2a^2s}{s^4+4a^4}$
23.	$\sinh(at)-\sin(at)$	$\dfrac{2a^3}{s^4-a^4}$
24.	$\cosh(at)-\cos(at)$	$\dfrac{2a^2s}{s^4-a^4}$
25.	$\dfrac{a\sin(at)-b\sin(bt)}{a^2-b^2}, a^2 \neq b^2$	$\dfrac{s^2}{(s^2+a^2)(s^2+b^2)}$
26.	$\dfrac{b\sin(at)-a\sin(bt)}{ab(b^2-a^2)}, a^2 \neq b^2$	$\dfrac{1}{(s^2+a^2)(s^2+b^2)}$
27.	$\dfrac{\cos(at)-\cos(bt)}{b^2-a^2}, a^2 \neq b^2$	$\dfrac{s}{(s^2+a^2)(s^2+b^2)}$
28.	$t^n, n \geq 0$	$\dfrac{n!}{s^{n+1}}$
29.	$\dfrac{t^{n-1}e^{-at}}{(n-1)!}, n > 0$	$\dfrac{1}{(s+a)^n}$
30.	$\dfrac{(n-1)-at}{(n-1)!}t^{n-2}e^{-at}, n > 1$	$\dfrac{s}{(s+a)^n}$
31.	$t^n e^{-at}, n \geq 0$	$\dfrac{n!}{(s+a)^{n+1}}$
32.	$\dfrac{2^n t^{n-(1/2)}}{1\cdot 3\cdot 5\cdots(2n-1)\sqrt{\pi}}, n \geq 1$	$s^{-[n+(1/2)]}$
33.	$J_0(at)$	$\dfrac{1}{\sqrt{s^2+a^2}}$

Table 4.1.1 (contd.): The Laplace Transforms of Some Commonly Encountered Functions.

	$\mathbf{f(t),\ t \geq 0}$	$\mathbf{F(s)}$
34.	$I_0(at)$	$\dfrac{1}{\sqrt{s^2-a^2}}$
35.	$\dfrac{1}{\sqrt{a}}\operatorname{erf}(\sqrt{at}\,)$	$\dfrac{1}{s\sqrt{s+a}}$
36.	$\dfrac{1}{\sqrt{\pi t}}e^{-at}+\sqrt{a}\operatorname{erf}(\sqrt{at}\,)$	$\dfrac{\sqrt{s+a}}{s}$
37.	$\dfrac{1}{\sqrt{\pi t}}-ae^{a^2t}\operatorname{erfc}(a\sqrt{t}\,)$	$\dfrac{1}{a+\sqrt{s}}$
38.	$e^{at}\operatorname{erfc}(\sqrt{at}\,)$	$\dfrac{1}{s+\sqrt{as}}$
39.	$\dfrac{1}{2\sqrt{\pi t^3}}\left(e^{bt}-e^{at}\right)$	$\sqrt{s-a}-\sqrt{s-b}$
40.	$\dfrac{1}{\sqrt{\pi t}}+ae^{a^2t}\operatorname{erf}(a\sqrt{t}\,)$	$\dfrac{\sqrt{s}}{s-a^2}$
41.	$\dfrac{1}{\sqrt{\pi t}}e^{at}(1+2at)$	$\dfrac{s}{(s-a)\sqrt{s-a}}$
42.	$\dfrac{1}{a}e^{a^2t}\operatorname{erf}(a\sqrt{t}\,)$	$\dfrac{1}{(s-a^2)\sqrt{s}}$
43.	$\sqrt{\dfrac{a}{\pi t^3}}e^{-a/t},\ a>0$	$e^{-2\sqrt{as}}$
44.	$\dfrac{1}{\sqrt{\pi t}}e^{-a/t},\ a\geq 0$	$\dfrac{1}{\sqrt{s}}e^{-2\sqrt{as}}$
45.	$\operatorname{erfc}\left(\sqrt{\dfrac{a}{t}}\right),\ a\geq 0$	$\dfrac{1}{s}e^{-2\sqrt{as}}$
46.	$2\sqrt{\dfrac{t}{\pi}}\exp\left(-\dfrac{a^2}{4t}\right)-a\operatorname{erfc}\left(\dfrac{a}{2\sqrt{t}}\right),\ a\geq 0$	$\dfrac{e^{-a\sqrt{s}}}{s\sqrt{s}}$
47.	$-e^{b^2t+ab}\operatorname{erfc}\left(b\sqrt{t}+\frac{a}{2\sqrt{t}}\right)+\operatorname{erfc}\left(\frac{a}{2\sqrt{t}}\right),\ a\geq 0$	$\dfrac{be^{-a\sqrt{s}}}{s(b+\sqrt{s}\,)}$
48.	$e^{ab}e^{b^2t}\operatorname{erfc}\left(b\sqrt{t}+\dfrac{a}{2\sqrt{t}}\right),\ a\geq 0$	$\dfrac{e^{-a\sqrt{s}}}{\sqrt{s}\,(b+\sqrt{s}\,)}$

Notes: Error function: $\operatorname{erf}(x)=\dfrac{2}{\pi}\displaystyle\int_0^x e^{-y^2}\,dy$

Complementary error function: $\operatorname{erfc}(x)=1-\operatorname{erf}(x)$

Given

$$F(s) = \frac{s+2}{s^2+1} = \frac{s}{s^2+1} + \frac{2}{s^2+1}, \tag{4.1.13}$$

then

$$f(t) = \cos(t) + 2\sin(t) \tag{4.1.14}$$

by (4.1.7), (4.1.9), and (4.1.11).

Because

$$F(s) = \frac{1}{s(s-1)} = \frac{1}{s-1} - \frac{1}{s} \tag{4.1.15}$$

by partial fractions, then

$$f(t) = e^t - 1 \tag{4.1.16}$$

by (4.1.3), (4.1.5), and (4.1.11).

The second important property deals with derivatives. Suppose $f(t)$ is continuous and has a piece-wise continuous derivative $f'(t)$. Then

$$\mathcal{L}[f'(t)] = \int_0^\infty f'(t)e^{-st}dt = e^{-st}f(t)\Big|_0^\infty + s\int_0^\infty f(t)e^{-st}dt \tag{4.1.17}$$

by integration by parts. If $f(t)$ is of exponential order, $e^{-st}f(t)$ tends to zero as $t \to \infty$, for large enough s, so that

$$\mathcal{L}[f'(t)] = sF(s) - f(0). \tag{4.1.18}$$

Similarly, if $f(t)$ and $f'(t)$ are continuous, $f''(t)$ is piece-wise continuous, and all three functions are of exponential order, then

$$\mathcal{L}[f''(t)] = s\mathcal{L}[f'(t)] - f'(0) = s^2F(s) - sf(0) - f'(0). \tag{4.1.19}$$

In general,

$$\boxed{\mathcal{L}[f^{(n)}(t)] = s^nF(s) - s^{n-1}f(0) - \cdots - sf^{(n-2)}(0) - f^{(n-1)}(0)} \tag{4.1.20}$$

on the assumption that $f(t)$ and its first $n-1$ derivatives are continuous, $f^{(n)}(t)$ is piece-wise continuous, and all are of exponential order so that the Laplace transform exists.

The converse of (4.1.20) is also of some importance. If

$$u(t) = \int_0^t f(\tau)\, d\tau, \tag{4.1.21}$$

then

$$\mathcal{L}[u(t)] = \int_0^\infty e^{-st} \left[\int_0^t f(\tau)\, d\tau \right] dt \tag{4.1.22}$$

$$= -\frac{e^{-st}}{s} \int_0^t f(\tau)\, d\tau \Big|_0^\infty + \frac{1}{s} \int_0^\infty f(t) e^{-st}\, dt \tag{4.1.23}$$

and

$$\mathcal{L}\left[\int_0^t f(\tau)\, d\tau \right] = \frac{F(s)}{s}, \tag{4.1.24}$$

where $u(0) = 0$.

Problems

Using the definition of the Laplace transform, find the Laplace transform of the following function:

1. $f(t) = \cosh(at)$

2. $f(t) = \cos^2(at)$

3. $f(t) = (t+1)^2$

4. $f(t) = (t+1)e^{-at}$

5. $f(t) = \begin{cases} e^t, & 0 < t < 2 \\ 0, & t > 2 \end{cases}$

6. $f(t) = \begin{cases} \sin(t), & 0 < t < \pi \\ 0, & t > \pi \end{cases}$

Using your knowledge of the transform for 1, e^{at}, $\sin(at)$, $\cos(at)$, and t^n, find the Laplace transform of

7. $f(t) = 2\sin(t) - \cos(2t) + \cos(3) - t$

8. $f(t) = t - 2 + e^{-5t} - \sin(5t) + \cos(2)$.

Find the inverse of the following transforms:

9. $F(s) = 1/(s+3)$

10. $F(s) = 1/s^4$

11. $F(s) = 1/(s^2+9)$

12. $F(s) = (2s+3)/(s^2+9)$

13. $F(s) = 2/(s^2+1) - 15/s^3 + 2/(s+1) - 6s/(s^2+4)$

14. $F(s) = 3/s + 15/s^3 + (s+5)/(s^2+1) - 6/(s-2)$.

15. Verify the derivative rule for Laplace transforms using the function $f(t) = \sin(at)$.

16. Show that $\mathcal{L}[f(at)] = F(s/a)/a$, where $F(s) = \mathcal{L}[f(t)]$.

17. Using the trigonometric identity $\sin^2(x) = [1 - \cos(2x)]/2$, find the Laplace transform of $f(t) = \sin^2[\pi t/(2T)]$.

4.2 THE HEAVISIDE STEP AND DIRAC DELTA FUNCTIONS

Change can occur abruptly. We throw a switch and electricity suddenly flows. In this section we introduce two functions, the Heaviside step and Dirac delta, that will give us the ability to construct complicated discontinuous functions to express these changes.

Heaviside step function

We define the *Heaviside step function* as

$$H(t-a) = \begin{cases} 1, & t > a \\ 0, & t < a, \end{cases} \tag{4.2.1}$$

where $a \geq 0$. From this definition,

$$\mathcal{L}[H(t-a)] = \int_a^\infty e^{-st}\,dt = \frac{e^{-as}}{s}, \qquad s > 0. \tag{4.2.2}$$

Note that this transform is identical to that for $f(t) = 1$ if $a = 0$. This should not surprise us. As pointed out earlier, the function $f(t)$ is zero for all $t < 0$ by definition. Thus, when dealing with Laplace transforms $f(t) = 1$ and $H(t)$ are identical. Generally we will take 1 rather than $H(t)$ as the inverse of $1/s$.

The Heaviside step function is essentially a bookkeeping device that gives us the ability to "switch on" and "switch off" a given function. For example, if we want a function $f(t)$ to become nonzero at time $t = a$, we represent this process by the product $f(t)H(t-a)$. On the other hand, if we only want the function to be "turned on" when $a < t < b$, the desired expression is then $f(t)[H(t-a) - H(t-b)]$. For $t < a$, both step

Figure 4.2.1: Largely a self-educated man, Oliver Heaviside (1850–1925) lived the life of a recluse. It was during his studies of the implications of Maxwell's theory of electricity and magnetism that he re-invented Laplace transforms. Initially rejected, it would require the work of Bromwich to justify its use. (Portrait courtesy of the Institution of Electrical Engineers, London.)

functions in the brackets have the value of zero. For $a < t < b$, the first step function has the value of unity and the second step function has the value of zero, so that we have $f(t)$. For $t > b$, both step functions equal unity so that their difference is zero.

• **Example 4.2.1**

Quite often we need to express the graphical representation of a function by a mathematical equation. We can conveniently do this through the use of step functions in a two-step procedure. The following example illustrates this procedure.

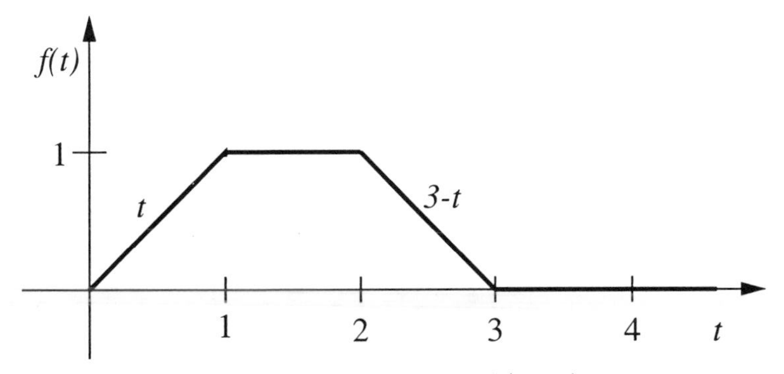

Figure 4.2.2: Graphical representation of (4.2.5).

Consider Figure 4.2.2. We would like to express this graph in terms of Heaviside step functions. We begin by introducing step functions at each point where there is a kink (discontinuity in the first derivative) or jump in the graph – in the present case at $t = 0$, $t = 1$, $t = 2$, and $t = 3$. Thus,

$$f(t) = a_0(t)H(t) + a_1(t)H(t-1) + a_2(t)H(t-2) + a_3(t)H(t-3), \quad (4.2.3)$$

where the coefficients $a_0(t), a_1(t), \ldots$ are yet to be determined. Proceeding from left to right in Figure 4.2.2, the coefficient of each step function equals the mathematical expression that we want after the kink or jump minus the expression before the kink or jump. Thus, in the present example,

$$f(t) = (t-0)H(t) + (1-t)H(t-1) + [(3-t)-1]H(t-2) + [0-(3-t)]H(t-3) \quad (4.2.4)$$

or

$$f(t) = tH(t) - (t-1)H(t-1) - (t-2)H(t-2) + (t-3)H(t-3). \quad (4.2.5)$$

We can easily find the Laplace transform of (4.2.5) by the "second shifting" theorem introduced in the next section.

• Example 4.2.2

Laplace transforms are particularly useful in solving initial-value problems involving linear, constant coefficient, ordinary differential equations where the nonhomogeneous term is discontinuous. As we shall show in the next section, we must first rewrite the nonhomogeneous term using the Heaviside step function before we can use Laplace transforms. For example, given the nonhomogeneous ordinary differential equation:

$$y'' + 3y' + 2y = \begin{cases} t, & 0 < t < 1 \\ 0, & t > 1, \end{cases} \quad (4.2.6)$$

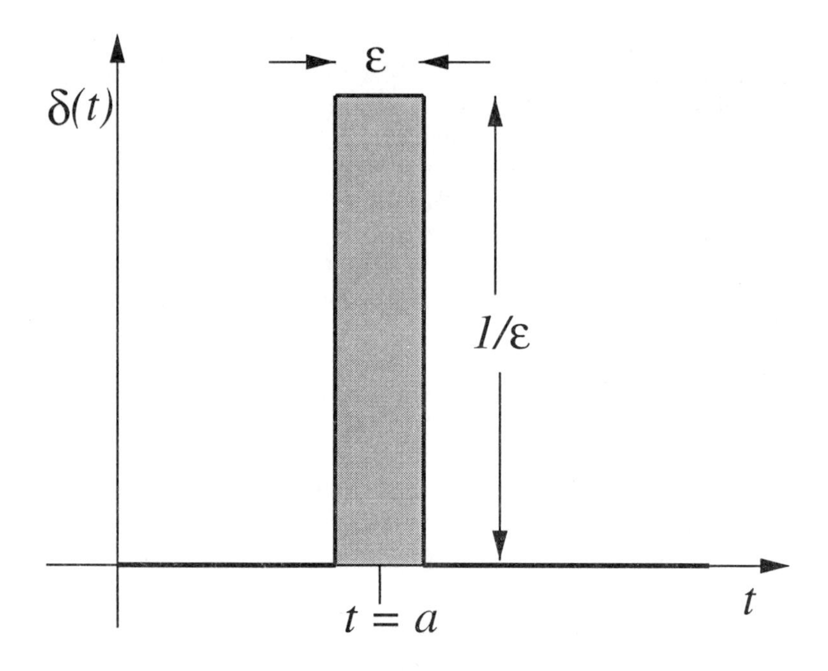

Figure 4.2.3: The Dirac delta function.

we can rewrite the right side of (4.2.6) as

$$y'' + 3y' + 2y = t - tH(t-1) \tag{4.2.7}$$
$$= t - (t-1)H(t-1) - H(t-1). \tag{4.2.8}$$

In Section 4.8 we will show how to solve this type of ordinary differential equation using Laplace transforms.

Dirac delta function

The second special function is the *Dirac delta function* or *impulse function*. We define it by

$$\delta(t-a) = \begin{cases} \infty, & t = a \\ 0, & t \neq a, \end{cases} \qquad \int_0^\infty \delta(t-a)\, dt = 1, \tag{4.2.9}$$

where $a \geq 0$.

A popular way of visualizing the delta function is as a very narrow rectangular pulse:

$$\delta(t-a) = \lim_{\epsilon \to 0} \begin{cases} 1/\epsilon, & 0 < |t-a| < \epsilon/2 \\ 0, & |t-a| > \epsilon/2, \end{cases} \tag{4.2.10}$$

where $\epsilon > 0$ is some small number and $a > 0$. This pulse has a width ϵ, height $1/\epsilon$, and centered at $t = a$ so that its area is unity. Now as this pulse shrinks in width ($\epsilon \to 0$), its height increases so that it remains centered at $t = a$ and its area equals unity. If we continue this process, always keeping the area unity and the pulse symmetric about $t = a$, eventually we obtain an extremely narrow, very large amplitude pulse at $t = a$. If we proceed to the limit, where the width approaches zero and the height approaches infinity (but still with unit area), we obtain the delta function $\delta(t - a)$.

The delta function was introduced earlier during our study of Fourier transforms. So what is the difference between the delta function introduced then and the delta function now? Simply put, the delta function can now only be used on the interval $[0, \infty)$. Outside of that, we shall use it very much as we did with Fourier transforms.

Using (4.2.10), the Laplace transform of the delta function is

$$\mathcal{L}[\delta(t-a)] = \int_0^\infty \delta(t-a) e^{-st}\, dt = \lim_{\epsilon \to 0} \frac{1}{\epsilon} \int_{a-\epsilon/2}^{a+\epsilon/2} e^{-st}\, dt \tag{4.2.11}$$

$$= \lim_{\epsilon \to 0} \frac{1}{\epsilon s} \left(e^{-as+\epsilon s/2} - e^{-as-\epsilon s/2} \right) \tag{4.2.12}$$

$$= \lim_{\epsilon \to 0} \frac{e^{-as}}{\epsilon s} \left(1 + \frac{\epsilon s}{2} + \frac{\epsilon^2 s^2}{8} + \cdots - 1 + \frac{\epsilon s}{2} - \frac{\epsilon^2 s^2}{8} + \cdots \right) \tag{4.2.13}$$

$$= e^{-as}. \tag{4.2.14}$$

In the special case when $a = 0$, $\mathcal{L}[\delta(t)] = 1$, a property that we will use in Section 4.9. Note that this is exactly the result that we obtained for the Fourier transform of the delta function.

If we integrate the impulse function,

$$\int_0^t \delta(\tau - a)\, d\tau = \begin{cases} 0, & t < a \\ 1, & t > a, \end{cases} \tag{4.2.15}$$

according to whether the impulse does or does not come within the range of integration. This integral gives a result that is precisely the definition of the Heaviside step function so that we can rewrite (4.2.15)

$$\int_0^t \delta(\tau - a)\, d\tau = H(t-a). \tag{4.2.16}$$

Consequently the delta function behaves like the derivative of the step function or

$$\frac{d}{dt}\Big[H(t-a) \Big] = \delta(t-a). \tag{4.2.17}$$

Because the conventional derivative does not exist at a point of discontinuity, we can only make sense of (4.2.17) if we extend the definition of the derivative. Here we have extended the definition formally, but a richer and deeper understanding arises from the theory of generalized functions.[2]

Problems

Sketch the following functions and express them in terms of Heaviside's step functions:

1. $$f(t) = \begin{cases} 0, & 0 \le t \le 2 \\ t-2, & 2 \le t < 3 \\ 0, & t > 3 \end{cases}$$

2. $$f(t) = \begin{cases} 0, & 0 < t < a \\ 1, & a < t < 2a \\ -1, & 2a < t < 3a \\ 0, & t > 3a \end{cases}$$

Rewrite the following nonhomogeneous ordinary differential equations using Heaviside's step functions.

3. $$y'' + 3y' + 2y = \begin{cases} 0, & 0 < t < 1 \\ 1, & t > 1 \end{cases}$$

4. $$y'' + 4y = \begin{cases} 0, & 0 < t < 4 \\ 3, & t > 4 \end{cases}$$

5. $$y'' + 4y' + 4y = \begin{cases} 0, & 0 < t < 2 \\ t, & t > 2 \end{cases}$$

6. $$y'' + 3y' + 2y = \begin{cases} 0, & 0 < t < 1 \\ e^t, & t > 1 \end{cases}$$

7. $$y'' - 3y' + 2y = \begin{cases} 0, & 0 < t < 2 \\ e^{-t}, & t > 2 \end{cases}$$

8. $$y'' - 3y' + 2y = \begin{cases} 0, & 0 < t < 1 \\ t^2, & t > 1 \end{cases}$$

---

[2] The generalization of the definition of a function so that it can express in a mathematically correct form such idealized concepts as the density of a material point, a point charge or point dipole, the space charge of a simple or double layer, the intensity of an instantaneous source, etc.

9. $y'' + y = \begin{cases} \sin(t), & 0 \le t \le \pi \\ 0, & t \ge \pi \end{cases}$

10. $y'' + 3y' + 2y = \begin{cases} t, & 0 \le t \le a \\ ae^{-(t-a)}, & t \ge a \end{cases}$

4.3 SOME USEFUL THEOREMS

Although at first sight there would appear to be a bewildering number of transforms to either memorize or tabulate, there are several useful theorems which can extend the applicability of a given transform.

First shifting theorem

Consider the transform of the function $e^{-at}f(t)$, where a is any real number. Then, by definition,

$$\mathcal{L}\left[e^{-at}f(t)\right] = \int_0^\infty e^{-st}e^{-at}f(t)\,dt = \int_0^\infty e^{-(s+a)t}f(t)\,dt, \qquad (4.3.1)$$

or

$$\mathcal{L}\left[e^{-at}f(t)\right] = F(s+a). \qquad (4.3.2)$$

That is, if $F(s)$ is the transform of $f(t)$ and a is a constant, then $F(s+a)$ is the transform of $e^{-at}f(t)$.

• Example 4.3.1

Let us find the Laplace transform of $f(t) = e^{-at}\sin(bt)$. Because the Laplace transform of $\sin(bt)$ is $b/(s^2+b^2)$,

$$\mathcal{L}\left[e^{-at}\sin(bt)\right] = \frac{b}{(s+a)^2+b^2}, \qquad (4.3.3)$$

where we have simply replaced s by $s+a$ in the transform for $\sin(bt)$.

• Example 4.3.2

Let us find the inverse of the Laplace transform

$$F(s) = \frac{s+2}{s^2+6s+1}. \qquad (4.3.4)$$

Rearranging terms,

$$F(s) = \frac{s+2}{s^2+6s+1} = \frac{s+2}{(s+3)^2-8} \tag{4.3.5}$$

$$= \frac{s+3}{(s+3)^2-8} - \frac{1}{2\sqrt{2}}\frac{2\sqrt{2}}{(s+3)^2-8}. \tag{4.3.6}$$

Immediately, from the first shifting theorem,

$$f(t) = e^{-3t}\cosh(2\sqrt{2}t) - \frac{1}{2\sqrt{2}}e^{-3t}\sinh(2\sqrt{2}t). \tag{4.3.7}$$

Second shifting theorem

The *second shifting theorem* states that if $F(s)$ is the transform of $f(t)$, then $e^{-bs}F(s)$ is the transform of $f(t-b)H(t-b)$, where b is real and positive. To show this, consider the Laplace transform of $f(t-b)H(t-b)$. Then, from the definition,

$$\mathcal{L}[f(t-b)H(t-b)] = \int_0^\infty f(t-b)H(t-b)e^{-st}\,dt \tag{4.3.8}$$

$$= \int_b^\infty f(t-b)e^{-st}\,dt = \int_0^\infty e^{-bs}e^{-sx}f(x)\,dx \tag{4.3.9}$$

$$= e^{-bs}\int_0^\infty e^{-sx}f(x)\,dx \tag{4.3.10}$$

or

$$\mathcal{L}[f(t-b)H(t-b)] = e^{-bs}F(s), \tag{4.3.11}$$

where we have set $x = t-b$. This theorem is of fundamental importance because it allows us to write down the transforms for "delayed" time functions. That is, functions which "turn on" b units after the initial time.

• **Example 4.3.3**

Let us find the inverse of the transform $(1-e^{-s})/s$. Since

$$\frac{1-e^{-s}}{s} = \frac{1}{s} - \frac{e^{-s}}{s}, \tag{4.3.12}$$

$$\mathcal{L}^{-1}\left(\frac{1}{s} - \frac{e^{-s}}{s}\right) = \mathcal{L}^{-1}\left(\frac{1}{s}\right) - \mathcal{L}^{-1}\left(\frac{e^{-s}}{s}\right) = H(t) - H(t-1), \tag{4.3.13}$$

because $\mathcal{L}^{-1}(1/s) = f(t) = 1$ and $f(t-1) = 1$.

- **Example 4.3.4**

Let us find the Laplace transform of $f(t) = (t^2 - 1)H(t-1)$. We begin by noting that

$$(t^2 - 1)H(t-1) = [(t-1+1)^2 - 1]H(t-1) \tag{4.3.14}$$

$$= [(t-1)^2 + 2(t-1)]H(t-1) \tag{4.3.15}$$

$$= (t-1)^2 H(t-1) + 2(t-1)H(t-1). \tag{4.3.16}$$

A direct application of the second shifting theorem leads then to

$$\mathcal{L}[(t^2-1)H(t-1)] = \frac{2e^{-s}}{s^3} + \frac{2e^{-s}}{s^2}. \tag{4.3.17}$$

- **Example 4.3.5**

In Example 4.2.2 we discussed the use of Laplace transforms in solving ordinary differential equations. One further step along the road consists of finding $Y(s) = \mathcal{L}[y(t)]$. Now that we have the second shifting theorem, let us do this.

Continuing Example 4.2.2 with $y(0) = 0$ and $y'(0) = 1$, let us take the Laplace transform of (4.2.8). Employing the second shifting theorem and (4.1.20), we find that

$$s^2Y(s) - sy(0) - y'(0) + 3sY(s) - 3y(0) + 2Y(s) = \frac{1}{s^2} - \frac{e^{-s}}{s^2} - \frac{e^{-s}}{s}. \tag{4.3.18}$$

Substituting in the initial conditions and solving for $Y(s)$, we finally obtain

$$Y(s) = \frac{1}{(s+2)(s+1)} + \frac{1}{s^2(s+2)(s+1)} + \frac{e^{-s}}{s^2(s+2)(s+1)} + \frac{e^{-s}}{s(s+2)(s+1)}. \tag{4.3.19}$$

Laplace transform of $t^n f(t)$

In addition to the shifting theorems, there are two other particularly useful theorems that involve the derivative and integral of the transform $F(s)$. For example, if we write

$$F(s) = \mathcal{L}[f(t)] = \int_0^\infty f(t)e^{-st}\,dt \tag{4.3.20}$$

and differentiate with respect to s, then

$$F'(s) = \int_0^\infty -tf(t)e^{-st}\,dt = -\mathcal{L}[tf(t)]. \tag{4.3.21}$$

In general, we have that

$$F^{(n)}(s) = (-1)^n \mathcal{L}[t^n f(t)]. \tag{4.3.22}$$

Laplace transform of $f(t)/t$

Consider the following integration of the Laplace transform $F(s)$:

$$\int_s^\infty F(z)\,dz = \int_s^\infty \left[\int_0^\infty f(t)e^{-zt}\,dt\right] dz. \tag{4.3.23}$$

Upon interchanging the order of integration, we find that

$$\int_s^\infty F(z)\,dz = \int_0^\infty f(t)\left[\int_s^\infty e^{-zt}\,dz\right] dt \tag{4.3.24}$$

$$= -\int_0^\infty f(t)\left.\frac{e^{-zt}}{t}\right|_s^\infty dt = \int_0^\infty \frac{f(t)}{t}e^{-st}\,dt. \tag{4.3.25}$$

Therefore,

$$\int_s^\infty F(z)\,dz = \mathcal{L}\left[\frac{f(t)}{t}\right]. \tag{4.3.26}$$

- **Example 4.3.6**

Let us find the transform of $t\sin(at)$. From (4.3.21),

$$\mathcal{L}[t\sin(at)] = -\frac{d}{ds}\left\{\mathcal{L}[\sin(at)]\right\} = -\frac{d}{ds}\left[\frac{a}{s^2+a^2}\right] \tag{4.3.27}$$

$$= \frac{2as}{(s^2+a^2)^2}. \tag{4.3.28}$$

• **Example 4.3.7**

Let us find the transform of $[1-\cos(at)]/t$. To solve this problem, we apply (4.3.26) and find that

$$\mathcal{L}\left[\frac{1-\cos(at)}{t}\right] = \int_s^\infty \mathcal{L}[1-\cos(at)]\Big|_{s=z} dz = \int_s^\infty \left(\frac{1}{z} - \frac{z}{z^2+a^2}\right) dz \tag{4.3.29}$$

$$= \ln(z) - \tfrac{1}{2}\ln(z^2+a^2)\Big|_s^\infty = \ln\left(\frac{z}{\sqrt{z^2+a^2}}\right)\Big|_s^\infty \tag{4.3.30}$$

$$= \ln(1) - \ln\left(\frac{s}{\sqrt{s^2+a^2}}\right) = -\ln\left(\frac{s}{\sqrt{s^2+a^2}}\right). \tag{4.3.31}$$

Initial-value theorem

Let $f(t)$ and $f'(t)$ possess Laplace transforms. Then, from the definition of the Laplace transform,

$$\int_0^\infty f'(t)e^{-st}dt = sF(s) - f(0). \tag{4.3.32}$$

Because s is a parameter in (4.3.32) and the existence of the integral is implied by the derivative rule, we can let $s \to \infty$ before we integrate. In that case, the left side of (4.3.32) vanishes to zero, which leads to

$$\lim_{s\to\infty} sF(s) = f(0). \tag{4.3.33}$$

This is the *initial-value theorem.*

• **Example 4.3.8**

Let us verify the initial-value theorem using $f(t) = e^{3t}$. Because $F(s) = 1/(s-3)$, $\lim_{s\to\infty} s/(s-3) = 1$. This agrees with $f(0) = 1$.

Final-value theorem

Let $f(t)$ and $f'(t)$ possess Laplace transforms. Then, in the limit of $s \to 0$, (4.3.32) becomes

$$\int_0^\infty f'(t)\,dt = \lim_{t\to\infty}\int_0^t f'(\tau)\,d\tau = \lim_{t\to\infty} f(t) - f(0) = \lim_{s\to 0} sF(s) - f(0). \tag{4.3.34}$$

Because $f(0)$ is not a function of t or s, the quantity $f(0)$ cancels from the (4.3.34), leaving

$$\lim_{t\to\infty} f(t) = \lim_{s\to 0} sF(s). \tag{4.3.35}$$

Equation (4.3.35) is the *final-value theorem*. It should be noted that this theorem assumes that $\lim_{t\to\infty} f(t)$ exists. For example, it does not apply to sinusoidal functions. Thus, we must restrict ourselves to Laplace transforms that have singularities in the left half of the s-plane unless they occur at the origin.

• Example 4.3.9

Let us verify the final-value theorem using $f(t) = t$. Because $F(s) = 1/s^2$, $\lim_{s\to 0} sF(s) = \lim_{s\to 0} 1/s = \infty$. The limit of $f(t)$ as $t \to \infty$ is also undefined.

Problems

Find the Laplace transform of the following functions:

1. $f(t) = e^{-t}\sin(2t)$ 2. $f(t) = e^{-2t}\cos(2t)$

3. $f(t) = te^t + \sin(3t)e^t + \cos(5t)e^{2t}$

4. $f(t) = t^4e^{-2t} + \sin(3t)e^t + \cos(4t)e^{2t}$

5. $f(t) = t^2e^{-t} + \sin(2t)e^t + \cos(3t)e^{-3t}$

6. $f(t) = t^2H(t-1)$

7. $f(t) = e^{2t}H(t-3)$ 8. $f(t) = t^2H(t-1) + e^tH(t-2)$

9. $f(t) = (t^2+2)H(t-1) + H(t-2)$

10. $f(t) = (t+1)^2H(t-1) + e^tH(t-2)$

11. $f(t) = \begin{cases} \sin(t), & 0 \le t \le \pi \\ 0, & t > \pi \end{cases}$

12. $f(t) = \begin{cases} t, & 0 \le t \le 2 \\ 2, & t \ge 2 \end{cases}$

13. $f(t) = te^{-3t}\sin(2t)$

Find the inverse of the following Laplace transforms:

14. $F(s) = 1/(s+2)^4$

15. $F(s) = s/(s+2)^4$

16. $F(s) = s/(s^2+2s+2)$

17. $F(s) = (s+3)/(s^2+2s+2)$

18. $F(s) = s/(s+1)^3 + (s+1)/(s^2+2s+2)$

19. $F(s) = s/(s+2)^2 + (s+2)/(s^2+2s+2)$

20. $F(s) = s/(s+2)^3 + (s+4)/(s^2+4s+5)$

21. $F(s) = e^{-3s}/(s-1)$

22. $F(s) = e^{-2s}/(s+1)^2$

23. $F(s) = se^{-s}/(s^2+2s+2)$

24. $F(s) = e^{-4s}/(s^2+4s+5)$

25. $F(s) = se^{-s}/(s^2+4) + e^{-3s}/(s-2)^4$

26. $F(s) = e^{-s}/(s^2+4) + (s-1)e^{-3s}/s^4$

27. $F(s) = (s+1)e^{-s}/(s^2+4) + e^{-3s}/s^4$

28. Find the Laplace transform of $f(t) = te^t[H(t-1)-H(t-2)]$ by using (a) the definition of the Laplace transform, and (b) a joint application of the first and second shifting theorems.

29. Write the function

$$f(t) = \begin{cases} t, & 0 < t < a \\ 0, & t > a \end{cases}$$

in terms of Heaviside's step functions. Then find its transform using (a) the definition of the Laplace transform, and (b) the second shifting theorem.

In problems 30–33, write the function $f(t)$ in terms of Heaviside's step functions and then find its transform using the second shifting theorem.

30.

$$f(t) = \begin{cases} t/2, & 0 \le t < 2 \\ 0, & t > 2 \end{cases}$$

31.

$$f(t) = \begin{cases} t, & 0 \le t \le 1 \\ 1, & 1 \le t < 2 \\ 0, & t > 2 \end{cases}$$

32.

$$f(t) = \begin{cases} t, & 0 \le t \le 2 \\ 4 - t, & 2 \le t \le 4 \\ 0, & t \ge 4 \end{cases}$$

33.

$$f(t) = \begin{cases} 0, & 0 \le t \le 1 \\ t - 1, & 1 \le t \le 2 \\ 1, & 2 \le t < 3 \\ 0, & t > 3 \end{cases}$$

Find $Y(s)$ for the following ordinary differential equations:

34. $y'' + 3y' + 2y = H(t-1); \qquad y(0) = y'(0) = 0$

35. $y'' + 4y = 3H(t-4); \qquad y(0) = 1,\ y'(0) = 0$

36. $y'' + 4y' + 4y = tH(t-2); \qquad y(0) = 0,\ y'(0) = 2$

37. $y'' + 3y' + 2y = e^t H(t-1); \qquad y(0) = y'(0) = 0$

38. $y'' - 3y' + 2y = e^{-t} H(t-2); \qquad y(0) = 2,\ y'(0) = 0$

39. $y'' - 3y' + 2y = t^2 H(t-1); \qquad y(0) = 0,\ y'(0) = 5$

40. $y'' + y = \sin(t)[1 - H(t-\pi)]; \qquad y(0) = y'(0) = 0$

41. $y'' + 3y' + 2y = t + \left[ae^{-(t-a)} - t\right] H(t-a); \qquad y(0) = y'(0) = 0.$

For each of the following functions, find its value at $t = 0$. Then check your answer using the initial-value theorem.

42. $f(t) = t$

43. $f(t) = \cos(at)$

44. $f(t) = te^{-t}$

45. $f(t) = e^t \sin(3t)$

For each of the following Laplace transforms, state whether you can or cannot apply the final-value theorem. If you can, find the final value. Check your result by finding the inverse and finding the limit as $t \to \infty$.

46. $F(s) = \dfrac{1}{s-1}$

47. $F(s) = \dfrac{1}{s}$

48. $F(s) = \dfrac{1}{s+1}$

49. $F(s) = \dfrac{s}{s^2+1}$

50. $F(s) = \dfrac{2}{s(s^2+3s+2)}$

51. $F(s) = \dfrac{2}{s(s^2-3s+2)}$

4.4 THE LAPLACE TRANSFORM OF A PERIODIC FUNCTION

Periodic functions frequently occur in engineering problems and we shall now show how to calculate their transform. They possess the property that $f(t+T) = f(t)$ for $t > 0$ and equal zero for $t < 0$, where T is the period of the function.

For convenience let us define a function $x(t)$ which equals zero except over the interval $(0, T)$ where it equals $f(t)$:

$$x(t) = \begin{cases} f(t), & 0 < t < T \\ 0, & t > T. \end{cases} \tag{4.4.1}$$

By definition

$$F(s) = \int_0^\infty f(t)e^{-st}\,dt \tag{4.4.2}$$

$$= \int_0^T f(t)e^{-st}\,dt + \int_T^{2T} f(t)e^{-st}\,dt + \cdots + \int_{kT}^{(k+1)T} f(t)e^{-st}\,dt + \cdots \tag{4.4.3}$$

Now let $z = t - kT$, where $k = 0, 1, 2, \ldots$, in the kth integral and $F(s)$ becomes

$$F(s) = \int_0^T f(z)e^{-sz}\,dz + \int_0^T f(z+T)e^{-s(z+T)}\,dz + \cdots$$
$$+ \int_0^T f(z+kT)e^{-s(z+kT)}\,dz + \cdots \tag{4.4.4}$$

However,

$$x(z) = f(z) = f(z+T) = \ldots = f(z+kT) = \ldots, \tag{4.4.5}$$

because the range of integration in each integral is from 0 to T. Thus, $F(s)$ becomes

$$F(s) = \int_0^T x(z)e^{-sz}\,dz + e^{-sT}\int_0^T x(z)e^{-sz}\,dz + \cdots$$
$$+ e^{-ksT}\int_0^T x(z)e^{-sz}\,dz + \cdots \tag{4.4.6}$$

or

$$F(s) = \left(1 + e^{-sT} + e^{-2sT} + \cdots + e^{-ksT} + \cdots\right)X(s). \tag{4.4.7}$$

The first term on the right side of (4.4.7) is a geometric series with common ratio e^{-sT}. If $|e^{-sT}| < 1$, then the series converges and

$$F(s) = \frac{X(s)}{1 - e^{-sT}}. \tag{4.4.8}$$

• Example 4.4.1

Let us find the Laplace transform of the square wave with period T:

$$f(t) = \begin{cases} h, & 0 < t < T/2 \\ -h, & T/2 < t < T. \end{cases} \tag{4.4.9}$$

By definition $x(t)$ is

$$x(t) = \begin{cases} h, & 0 < t < T/2 \\ -h, & T/2 < t < T \\ 0, & t > T. \end{cases} \tag{4.4.10}$$

Then

$$X(s) = \int_0^\infty x(t)e^{-st}\,dt = \int_0^{T/2} h\,e^{-st}\,dt + \int_{T/2}^{T} (-h)\,e^{-st}\,dt \tag{4.4.11}$$

$$= \frac{h}{s}\left(1 - 2e^{-sT/2} + e^{-sT}\right) = \frac{h}{s}\left(1 - e^{-sT/2}\right)^2 \tag{4.4.12}$$

and

$$F(s) = \frac{h\left(1 - e^{-sT/2}\right)^2}{s\left(1 - e^{-sT}\right)} = \frac{h\left(1 - e^{-sT/2}\right)}{s\left(1 + e^{-sT/2}\right)}. \tag{4.4.13}$$

If we multiply numerator and denominator by $\exp(sT/4)$ and recall that $\tanh(u) = (e^u - e^{-u})/(e^u + e^{-u})$, we have that

$$F(s) = \frac{h}{s}\tanh\left(\frac{sT}{4}\right). \tag{4.4.14}$$

- **Example 4.4.2**

Let us find the Laplace transform of the periodic function

$$f(t) = \begin{cases} \sin(2\pi t/T), & 0 < t < T/2 \\ 0, & T/2 < t < T. \end{cases} \tag{4.4.15}$$

By definition $x(t)$ is

$$x(t) = \begin{cases} \sin(2\pi t/T), & 0 < t < T/2 \\ 0, & t > T/2. \end{cases} \tag{4.4.16}$$

Then

$$X(s) = \int_0^{T/2} \sin\left(\frac{2\pi t}{T}\right) e^{-st}\, dt = \frac{2\pi T}{s^2T^2 + 4\pi^2}\left(1 + e^{-sT/2}\right). \tag{4.4.17}$$

Hence,

$$F(s) = \frac{X(s)}{1 - e^{-sT}} = \frac{2\pi T}{s^2T^2 + 4\pi^2} \times \frac{1 + e^{-sT/2}}{1 - e^{-sT}} \tag{4.4.18}$$

$$= \frac{2\pi T}{s^2T^2 + 4\pi^2} \times \frac{1}{1 - e^{-sT/2}}. \tag{4.4.19}$$

Problems

Find the Laplace transform for the following periodic functions:

1. $f(t) = \sin(t), \ 0 \le t \le \pi, \qquad f(t) = f(t + \pi)$

2. $f(t) = \begin{cases} \sin(t), & 0 \le t \le \pi \\ 0, & \pi \le t \le 2\pi, \end{cases} \qquad f(t) = f(t + 2\pi)$

3. $f(t) = \begin{cases} t, & 0 \le t < a \\ 0, & a < t \le 2a, \end{cases} \qquad f(t) = f(t + 2a)$

4. $f(t) = \begin{cases} 1, & 0 < t < a \\ 0, & a < t < 2a \\ -1, & 2a < t < 3a \\ 0, & 3a < t < 4a, \end{cases} \qquad f(t) = f(t + 4a).$

4.5 INVERSION BY PARTIAL FRACTIONS: HEAVISIDE'S EXPANSION THEOREM

In the previous sections, we have devoted our efforts to calculating the Laplace transform of a given function. Obviously we must have a method for going the other way. Given a transform, we must find the corresponding function. This is often a very formidable task. In the next few sections we shall present some general techniques for the inversion of a Laplace transform.

The first technique involves transforms that we can express as the ratio of two polynomials: $F(s) = q(s)/p(s)$. We shall assume that the order of $q(s)$ is *less* than $p(s)$ and we have divided out any common factor between them. In principle we know that $p(s)$ has n zeros, where n is the order of the $p(s)$ polynomial. Some of the zeros may be complex, some of them may be real, and some of them may be duplicates of other zeros. In the case when $p(s)$ has n simple zeros (nonrepeating roots), a simple method exists for inverting the transform.

We want to rewrite $F(s)$ in the form:

$$F(s) = \frac{a_1}{s-s_1} + \frac{a_2}{s-s_2} + \cdots + \frac{a_n}{s-s_n} = \frac{q(s)}{p(s)}, \tag{4.5.1}$$

where $s_1, s_2, \ldots, s_n$ are the n simple zeros of $p(s)$. We now multiply both sides of (4.5.1) by $s - s_1$ so that

$$\frac{(s-s_1)q(s)}{p(s)} = a_1 + \frac{(s-s_1)a_2}{s-s_2} + \cdots + \frac{(s-s_1)a_n}{s-s_n}. \tag{4.5.2}$$

If we set $s = s_1$, the right side of (4.5.2) becomes simply a_1. The left side takes the form 0/0 and there are two cases. If $p(s) = (s - s_1)g(s)$, then $a_1 = q(s_1)/g(s_1)$. If we cannot explicitly factor out $s - s_1$, l'Hôspital's rule gives

$$a_1 = \lim_{s\to s_1} \frac{(s-s_1)q(s)}{p(s)} = \lim_{s\to s_1} \frac{(s-s_1)q'(s) + q(s)}{p'(s)} = \frac{q(s_1)}{p'(s_1)}. \tag{4.5.3}$$

In a similar manner, we can compute all of the a_k's, where $k = 1, 2, \ldots, n$. Therefore,

$$\mathcal{L}^{-1}[F(s)] = \mathcal{L}^{-1}\left[\frac{q(s)}{p(s)}\right] = \mathcal{L}^{-1}\left(\frac{a_1}{s-s_1} + \frac{a_2}{s-s_2} + \cdots + \frac{a_n}{s-s_n}\right) \tag{4.5.4}$$

$$= a_1 e^{s_1 t} + a_2 e^{s_2 t} + \cdots + a_n e^{s_n t}. \tag{4.5.5}$$

This is *Heaviside's expansion theorem*, applicable when $p(s)$ has only simple poles.

• Example 4.5.1

Let us invert the transform $s/[(s+2)(s^2+1)]$. It has three simple poles at $s = -2$ and $s = \pm i$. From our earlier discussion, $q(s) = s$, $p(s) = (s+2)(s^2+1)$, and $p'(s) = 3s^2 + 4s + 1$. Therefore,

$$\mathcal{L}^{-1}\left[\frac{s}{(s+2)(s^2+1)}\right] = \frac{-2}{12-8+1}e^{-2t} + \frac{i}{-3+4i+1}e^{it} + \frac{-i}{-3-4i+1}e^{-it} \tag{4.5.6}$$

$$= -\frac{2}{5}e^{-2t} + \frac{i}{-2+4i}e^{it} - \frac{i}{-2-4i}e^{-it} \tag{4.5.7}$$

$$= -\frac{2}{5}e^{-2t} + i\frac{-2-4i}{4+16}e^{it} - i\frac{-2+4i}{4+16}e^{-it} \tag{4.5.8}$$

$$= -\frac{2}{5}e^{-2t} + \frac{1}{5}\sin(t) + \frac{2}{5}\cos(t), \tag{4.5.9}$$

where we have used $\sin(t) = \frac{1}{2i}(e^{it} - e^{-it})$ and $\cos(t) = \frac{1}{2}(e^{it} + e^{-it})$.

• Example 4.5.2

Let us invert the transform $1/[(s-1)(s-2)(s-3)]$. There are three simple poles at $s_1 = 1$, $s_2 = 2$, and $s_3 = 3$. In this case, the easiest method for computing a_1, a_2, and a_3 is

$$a_1 = \lim_{s\to 1} \frac{s-1}{(s-1)(s-2)(s-3)} = \frac{1}{2}, \tag{4.5.10}$$

$$a_2 = \lim_{s\to 2} \frac{s-2}{(s-1)(s-2)(s-3)} = -1 \tag{4.5.11}$$

and

$$a_3 = \lim_{s\to 3} \frac{s-3}{(s-1)(s-2)(s-3)} = \frac{1}{2}. \tag{4.5.12}$$

Therefore,

$$\mathcal{L}^{-1}\left[\frac{1}{(s-1)(s-2)(s-3)}\right] = \mathcal{L}^{-1}\left[\frac{a_1}{s-1} + \frac{a_2}{s-2} + \frac{a_3}{s-3}\right] = \tfrac{1}{2}e^{t} - e^{2t} + \tfrac{1}{2}e^{3t}. \tag{4.5.13}$$

Note that for inverting transforms of the form $F(s)e^{-as}$ with $a > 0$, you should use Heaviside's expansion theorem to first invert $F(s)$ and then apply the second shifting theorem.

Let us now find the expansion when we have multiple roots, namely

$$F(s) = \frac{q(s)}{p(s)} = \frac{q(s)}{(s-s_1)^{m_1}(s-s_2)^{m_2}\cdots(s-s_n)^{m_n}}, \tag{4.5.14}$$

where the order of the denominator, $m_1+m_2+\cdots+m_n$, is greater than that for the numerator. Once again we have eliminated any common factor between the numerator and denominator. Now we can write $F(s)$ as

$$F(s) = \sum_{k=1}^{n}\sum_{j=1}^{m_k} \frac{a_{kj}}{(s-s_k)^{m_k-j+1}}. \tag{4.5.15}$$

Multiplying (4.5.15) by $(s-s_k)^{m_k}$,

$$\begin{aligned}\frac{(s-s_k)^{m_k}q(s)}{p(s)} &= a_{k1} + a_{k2}(s-s_k) + \cdots + a_{km_k}(s-s_k)^{m_k-1}\\ &\quad + (s-s_k)^{m_k}\left[\frac{a_{11}}{(s-s_1)^{m_1}} + \cdots + \frac{a_{nm_n}}{s-s_n}\right],\end{aligned} \tag{4.5.16}$$

where we have grouped together into the square-bracketed term all of the terms except for those with a_{kj} coefficients. Taking the limit as $s \to s_k$,

$$a_{k1} = \lim_{s\to s_k} \frac{(s-s_k)^{m_k}q(s)}{p(s)}. \tag{4.5.17}$$

Let us now take the derivative of (4.5.16),

$$\begin{aligned}\frac{d}{ds}&\left[\frac{(s-s_k)^{m_k}q(s)}{p(s)}\right]\\ &= a_{k2} + 2a_{k3}(s-s_k) + \cdots + (m_k-1)a_{km_k}(s-s_k)^{m_k-2}\\ &+ \frac{d}{ds}\left\{(s-s_k)^{m_k}\left[\frac{a_{11}}{(s-s_1)^{m_1}} + \cdots + \frac{a_{nm_n}}{s-s_n}\right]\right\}.\end{aligned} \tag{4.5.18}$$

Taking the limit as $s \to s_k$,

$$a_{k2} = \lim_{s\to s_k} \frac{d}{ds}\left[\frac{(s-s_k)^{m_k}q(s)}{p(s)}\right]. \tag{4.5.19}$$

In general,

$$a_{kj} = \lim_{s\to s_k} \frac{1}{(j-1)!}\frac{d^{j-1}}{ds^{j-1}}\left[\frac{(s-s_k)^{m_k}q(s)}{p(s)}\right] \tag{4.5.20}$$

and by direct inversion,

$$f(t) = \sum_{k=1}^{n}\sum_{j=1}^{m_k} \frac{a_{kj}}{(m_k-j)!}t^{m_k-j}e^{s_k t}. \tag{4.5.21}$$

• **Example 4.5.3**

Let us find the inverse of

$$F(s) = \frac{s}{(s+2)^2(s^2+1)}. \tag{4.5.22}$$

We first note that the denominator has simple zeros at $s = \pm i$ and a repeated root at $s = -2$. Therefore,

$$F(s) = \frac{A}{s-i} + \frac{B}{s+i} + \frac{C}{s+2} + \frac{D}{(s+2)^2}, \tag{4.5.23}$$

where

$$A = \lim_{s\to i}(s-i)F(s) = \tfrac{1}{6+8i}, \tag{4.5.24}$$

$$B = \lim_{s\to -i}(s+i)F(s) = \tfrac{1}{6-8i}, \tag{4.5.25}$$

$$C = \lim_{s\to -2}\frac{d}{ds}\left[(s+2)^2F(s)\right] = \lim_{s\to -2}\frac{d}{ds}\left[\frac{s}{s^2+1}\right] = -\tfrac{3}{25} \tag{4.5.26}$$

and

$$D = \lim_{s\to -2}(s+2)^2F(s) = -\tfrac{2}{5}. \tag{4.5.27}$$

Thus,

$$f(t) = \tfrac{1}{6+8i}e^{it} + \tfrac{1}{6-8i}e^{-it} - \tfrac{3}{25}e^{-2t} - \tfrac{2}{5}te^{-2t} \tag{4.5.28}$$

$$= \tfrac{3}{25}\cos(t) + \tfrac{4}{25}\sin(t) - \tfrac{3}{25}e^{-2t} - \tfrac{10}{25}te^{-2t}. \tag{4.5.29}$$

In Section 4.10 we shall see that we can invert transforms just as easily with the residue theorem.

Let us now find the inverse of

$$F(s) = \frac{cs+(ca-\omega d)}{(s+a)^2+\omega^2} = \frac{cs+(ca-\omega d)}{(s+a-\omega i)(s+a+\omega i)} \tag{4.5.30}$$

by Heaviside's expansion theorem. Then

$$F(s) = \frac{c+di}{2(s+a-\omega i)} + \frac{c-di}{2(s+a+\omega i)} \tag{4.5.31}$$

$$= \frac{\sqrt{c^2+d^2}e^{\theta i}}{2(s+a-\omega i)} + \frac{\sqrt{c^2+d^2}e^{-\theta i}}{2(s+a+\omega i)}, \tag{4.5.32}$$

where $\theta = \tan^{-1}(d/c)$. Note that we must choose θ so that it gives the correct sign for c and d.

Taking the inverse of (4.5.32),

$$f(t) = \tfrac{1}{2}\sqrt{c^2+d^2}e^{-at+\omega ti+\theta i} + \tfrac{1}{2}\sqrt{c^2+d^2}e^{-at-\omega ti-\theta i} \tag{4.5.33}$$

$$= \sqrt{c^2+d^2}e^{-at}\cos(\omega t+\theta). \tag{4.5.34}$$

Equation (4.5.34) is the amplitude/phase form of the inverse of (4.5.30). It is particularly popular with electrical engineers.

• **Example 4.5.4**

Let us express the inverse of

$$F(s) = \frac{8s-3}{s^2+4s+13} \tag{4.5.35}$$

in the amplitude/phase form.

Starting with

$$F(s) = \frac{8s-3}{(s+2-3i)(s+2+3i)} \tag{4.5.36}$$

$$= \frac{4+19i/6}{s+2-3i} + \frac{4-19i/6}{s+2+3i} \tag{4.5.37}$$

$$= \frac{5.1017e^{38.3675^\circ i}}{s+2-3i} + \frac{5.1017e^{-38.3675^\circ i}}{s+2+3i} \tag{4.5.38}$$

or

$$f(t) = 5.1017e^{-2t+3it+38.3675^\circ i} + 5.1017e^{-2t-3it-38.3675^\circ i} \tag{4.5.39}$$

$$= 10.2034e^{-2t}\cos(3t+38.3675^\circ). \tag{4.5.40}$$

• **Example 4.5.5: The design of film projectors**

For our final example we anticipate future work. The primary use of Laplace transforms is the solution of differential equations. In this example we illustrate this technique that includes Heaviside's expansion theorem in the form of amplitude and phase.

This problem[3] arose in the design of projectors for motion pictures. An early problem was ensuring that the speed at which the film passed the electric eye remained essentially constant; otherwise, a frequency modulation of the reproduced sound resulted. Figure 4.5.1(A) shows a diagram of the projector. Many will remember this design from their

[3] Cook, E. D., 1935: The technical aspects of the high-fidelity reproducer. *J. Soc. Motion Pict. Eng.*, **25**, 289–312.

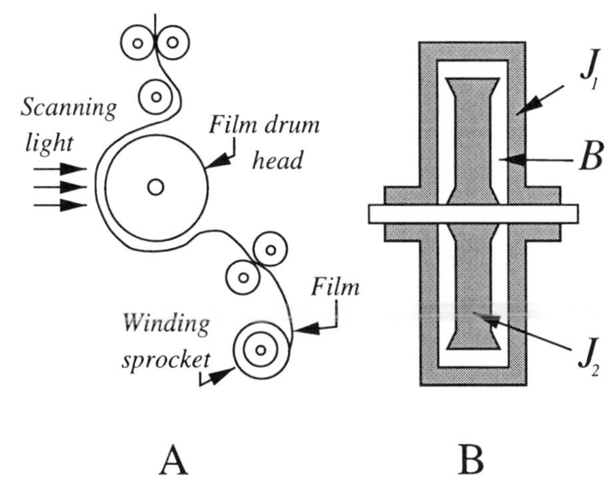

Figure 4.5.1: (A) The schematic for the scanning light in a motion-picture projector and (B) interior of the film drum head.

days as a school projectist. In this section we shall show that this particular design filters out variations in the film speed caused by irregularities either in the driving-gear trains or in the engagement of the sprocket teeth with the holes in the film.

Let us now focus on the film head – a hollow drum of small moment of inertia J_1. See Figure 4.5.1(B). Within it there is a concentric inner flywheel of moment of inertia J_2, where $J_2 \gg J_1$. The remainder of the space within the drum is filled with oil. The inner flywheel rotates on precision ball bearings on the drum shaft. The only coupling between the drum and flywheel is through fluid friction and the very small friction in the ball bearings. The flection of the film loops between the drum head and idler pulleys provides the spring restoring force for the system as the film runs rapidly through the system.

From Figure 4.5.1 the dynamical equations governing the outer case and inner flywheel are (1) the rate of change of the outer casing of the film head equals the frictional torque given to the casing from the inner flywheel plus the restoring torque due to the flection of the film, and (2) the rate of change of the inner flywheel equals the negative of the frictional torque given to the outer casing by the inner flywheel.

Assuming that the frictional torque between the two flywheels is proportional to the difference in their angular velocities, the frictional torque given to the casing from the inner flywheel is $B(\omega_2 - \omega_1)$, where B is the frictional resistance, ω_1 and ω_2 are the deviations of the drum and inner flywheel from their normal angular velocities, respectively. If r is the ratio of the diameter of the winding sprocket to the diameter of the drum, the restoring torque due to the flection of the film and its corresponding angular twist equals $K \int_0^t (r\omega_0 - \omega_1)\, d\tau$, where K is the rotational stiffness and ω_0 is the deviation of the winding sprocket from its normal angular velocity. The quantity $r\omega_0$ gives the angular velocity at which the film is running through the projector because the winding sprocket is the mechanism that pulls the film. Consequently the equations governing this mechanical system are

$$J_1 \frac{d\omega_1}{dt} = K \int_0^t (r\omega_0 - \omega_1)\, d\tau + B(\omega_2 - \omega_1) \tag{4.5.41}$$

and

$$J_2 \frac{d\omega_2}{dt} = -B(\omega_2 - \omega_1). \tag{4.5.42}$$

With the winding sprocket, the drum, and the flywheel running at their normal uniform angular velocities, let us assume that the winding sprocket introduces a disturbance equivalent to an unit increase in its angular velocity for 0.15 seconds, followed by the resumption of its normal velocity. It is assumed that the film in contact with the drum cannot slip. The initial conditions are $\omega_1(0) = \omega_2(0) = 0$.

Taking the Laplace transform of (4.5.41)–(4.5.42) using (4.1.18),

$$\left(J_1 s + B + \frac{K}{s}\right) \Omega_1(s) - B\Omega_2(s) = \frac{rK}{s}\Omega_0(s) = rK\mathcal{L}\left[\int_0^t \omega_0(\tau)\, d\tau\right] \tag{4.5.43}$$

and

$$-B\Omega_1(s) + (J_2 s + B)\Omega_2(s) = 0. \tag{4.5.44}$$

The solution of (4.5.43)–(4.5.44) for $\Omega_1(s)$ is

$$\Omega_1(s) = \frac{rK}{J_1} \quad \frac{(s + a_0)\Omega_0(s)}{s^3 + b_2 s^2 + b_1 s + b_0}, \tag{4.5.45}$$

where typical values[4] are

$$\frac{rK}{J_1} = 90.8, \quad a_0 = \frac{B}{J_2} = 1.47, \quad b_0 = \frac{BK}{J_1 J_2} = 231, \tag{4.5.46}$$

$$b_1 = \frac{K}{J_1} = 157 \quad \text{and} \quad b_2 = \frac{B(J_1 + J_2)}{J_1 J_2} = 8.20. \tag{4.5.47}$$

The transform $\Omega_1(s)$ has three simple poles located at $s_1 = -1.58$, $s_2 = -3.32 + 11.6i$, and $s_3 = -3.32 - 11.6i$.

---

[4] $J_1 = 1.84 \times 10^4$ dyne cm sec^2 per radian, $J_2 = 8.43 \times 10^4$ dyne cm sec^2 per radian, $B = 12.4 \times 10^4$ dyne cm sec per radian, $K = 2.89 \times 10^6$ dyne cm per radian, and $r = 0.578$

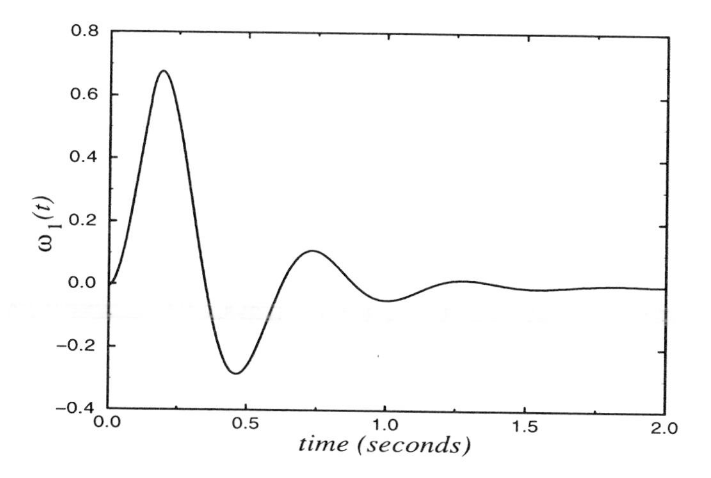

Figure 4.5.2: The deviation ω_1 of a film drum head from its uniform angular velocity when the sprocket angular velocity is perturbed by a unit amount for the duration of 0.15 seconds.

Because the sprocket angular velocity deviation $\omega_0(t)$ is a pulse of unit amplitude and 0.15 second duration, we express it as the difference of two Heaviside step functions:

$$\omega_0(t) = H(t) - H(t - 0.15). \tag{4.5.48}$$

Its Laplace transform is

$$\Omega_0(s) = \frac{1}{s} - \frac{1}{s}e^{-0.15s} \tag{4.5.49}$$

so that (4.5.45) becomes

$$\Omega_1(s) = \frac{rK}{J_1}\frac{(s + a_0)}{s(s - s_1)(s - s_2)(s - s_3)}\left(1 - e^{-0.15s}\right). \tag{4.5.50}$$

The inversion of (4.5.50) follows directly from the second shifting theorem and Heaviside's expansion theorem:

$$\begin{aligned}\omega_1(t) = {} & K_0 + K_1e^{s_1t} + K_2e^{s_2t} + K_3e^{s_3t} \\ & - [K_0 + K_1e^{s_1(t-0.15)} + K_2e^{s_2(t-0.15)} + K_3e^{s_3(t-0.15)}]H(t - 0.15),\end{aligned} \tag{4.5.51}$$

where

$$K_0 = \frac{rK}{J_1}\left.\frac{s + a_0}{(s - s_1)(s - s_2)(s - s_3)}\right|_{s=0} = 0.578, \tag{4.5.52}$$

$$K_1 = \frac{rK}{J_1} \left. \frac{s + a_0}{s(s - s_2)(s - s_3)} \right|_{s=s_1} = 0.046, \tag{4.5.53}$$

$$K_2 = \frac{rK}{J_1} \left. \frac{s + a_0}{s(s - s_1)(s - s_3)} \right|_{s=s_2} = 0.326e^{165°i} \tag{4.5.54}$$

and

$$K_3 = \frac{rK}{J_1} \left. \frac{s + a_0}{s(s - s_1)(s - s_2)} \right|_{s=s_3} = 0.326e^{-165°i}. \tag{4.5.55}$$

Using Euler's identity $\cos(t) = (e^{it} + e^{-it})/2$, we can write (4.5.51) as

$$\begin{aligned} \omega_1(t) &= 0.578 + 0.046e^{-1.58t} + 0.652e^{-3.32t}\cos(11.6t + 165°) \\ &- \{0.578 + 0.046e^{-1.58(t-0.15)} + 0.652e^{-3.32(t-0.15)} \\ &\times \cos[11.6(t - 0.15) + 165°]\}H(t - 0.15). \end{aligned} \tag{4.5.56}$$

Equation (4.5.56) is plotted in Figure 4.5.2. Note that fluctuations in $\omega_1(t)$ are damped out by the particular design of this film projector. Because this mechanical device dampens unwanted fluctuations (or noise) in the motion-picture projector, this particular device is an example of a *mechanical filter.*

Problems

Use Heaviside's expansion theorem to find the inverse of the following Laplace transforms:

1. $F(s) = \dfrac{1}{s^2 + 3s + 2}$

2. $F(s) = \dfrac{s + 3}{(s + 4)(s - 2)}$

3. $F(s) = \dfrac{s - 4}{(s + 2)(s + 1)(s - 3)}$

4. $F(s) = \dfrac{s - 3}{(s^2 + 4)(s + 1)}$.

Find the inverse of the following transforms and express them in amplitude/phase form:

5. $F(s) = \dfrac{1}{s^2 + 4s + 5}$

6. $F(s) = \dfrac{1}{s^2 + 6s + 13}$

7. $F(s) = \dfrac{2s - 5}{s^2 + 16}$

8. $F(s) = \dfrac{1}{s(s^2 + 2s + 2)}$

9. $F(s) = \dfrac{s + 2}{s(s^2 + 4)}$

4.6 CONVOLUTION

In this section we turn to a fundamental concept in Laplace transforms: convolution. We shall restrict ourselves to its use in finding the inverse of a transform when that transform consists of the *product* of two simpler transforms. In subsequent sections we will use it to solve ordinary differential equations.

We begin by formally introducing the mathematical operation of the *convolution product*:

$$f(t) * g(t) = \int_0^t f(t-x)g(x)\,dx = \int_0^t f(x)g(t-x)\,dx. \qquad (4.6.1)$$

In most cases the operations required by (4.6.1) are straightforward.

• **Example 4.6.1**

Let us find the convolution between $\cos(t)$ and $\sin(t)$.

$$\cos(t) * \sin(t) = \int_0^t \sin(t-x)\cos(x)\,dx \qquad (4.6.2)$$

$$= \tfrac{1}{2}\int_0^t [\sin(t) + \sin(t-2x)]\,dx \qquad (4.6.3)$$

$$= \tfrac{1}{2}\int_0^t \sin(t)\,dx + \tfrac{1}{2}\int_0^t \sin(t-2x)\,dx \qquad (4.6.4)$$

$$= \tfrac{1}{2}\sin(t)\,x\Big|_0^t + \tfrac{1}{4}\cos(t-2x)\Big|_0^t = \tfrac{1}{2}t\sin(t). \qquad (4.6.5)$$

• **Example 4.6.2**

Similarly, the convolution between t^2 and $\sin(t)$ is

$$t^2 * \sin(t) = \int_0^t (t-x)^2\sin(x)\,dx \qquad (4.6.6)$$

$$= -(t-x)^2\cos(x)\Big|_0^t - 2\int_0^t (t-x)\cos(x)\,dx \qquad (4.6.7)$$

$$= t^2 - 2(t-x)\sin(x)\Big|_0^t - 2\int_0^t \sin(x)\,dx \qquad (4.6.8)$$

$$= t^2 + 2\cos(t) - 2 \qquad (4.6.9)$$

by integration by parts.

• **Example 4.6.3**

Consider now the convolution between e^t and the discontinuous function $H(t-1) - H(t-2)$:

$$e^t * [H(t-1) - H(t-2)] = \int_0^t e^{t-x}[H(x-1) - H(x-2)]\,dx \quad \mathbf{(4.6.10)}$$

$$= e^t \int_0^t e^{-x}[H(x-1) - H(x-2)]\,dx. \quad \mathbf{(4.6.11)}$$

In order to evaluate the integral (4.6.11) we must examine various cases. If $t < 1$, then both of the step functions equal zero and the convolution equals zero. However, when $1 < t < 2$, the first step function equals one while the second equals zero. Therefore,

$$e^t * [H(t-1) - H(t-2)] = e^t \int_1^t e^{-x}\,dx = e^{t-1} - 1, \quad \mathbf{(4.6.12)}$$

because the portion of the integral from zero to one equals zero. Finally, when $t > 2$, the integrand is only nonzero for that portion of the integration when $1 < x < 2$. Consequently,

$$e^t * [H(t-1) - H(t-2)] = e^t \int_1^2 e^{-x}\,dx = e^{t-1} - e^{t-2}. \quad \mathbf{(4.6.13)}$$

Thus, the convolution of e^t with the pulse $H(t-1) - H(t-2)$ is

$$e^t * [H(t-1) - H(t-2)] = \begin{cases} 0, & 0 < t < 1 \\ e^{t-1} - 1, & 1 < t < 2 \\ e^{t-1} - e^{t-2}, & t > 2. \end{cases} \quad \mathbf{(4.6.14)}$$

The reason why we have introduced convolution follows from the following fundamental theorem (often called *Borel's theorem*[5]). If

$$w(t) = u(t) * v(t) \quad \mathbf{(4.6.15)}$$

then

$$W(s) = U(s)V(s). \quad \mathbf{(4.6.16)}$$

[5] Borel, É., 1901: *Leçons sur les séries divergentes.* Gauthier-Villars, Paris, p. 104.

In other words, we can invert a complicated transform by convoluting the inverses to two simpler functions. The proof is as follows:

$$W(s) = \int_0^\infty \left[\int_0^t u(x)v(t-x)\,dx\right] e^{-st} dt \tag{4.6.17}$$

$$= \int_0^\infty \left[\int_x^\infty u(x)v(t-x)e^{-st} dt\right] dx \tag{4.6.18}$$

$$= \int_0^\infty u(x) \left[\int_0^\infty v(r)e^{-s(r+x)} dr\right] dx \tag{4.6.19}$$

$$= \left[\int_0^\infty u(x)e^{-sx} dx\right] \left[\int_0^\infty v(r)e^{-sr} dr\right] = U(s)V(s), \tag{4.6.20}$$

where $t = r + x$. □

- **Example 4.6.4**

Let us find the inverse of the transform

$$\frac{s}{(s^2+1)^2} = \frac{s}{s^2+1} \times \frac{1}{s^2+1} = \mathcal{L}[\cos(t)]\mathcal{L}[\sin(t)] \tag{4.6.21}$$

$$= \mathcal{L}[\cos(t) * \sin(t)] = \mathcal{L}[\tfrac{1}{2} t \sin(t)] \tag{4.6.22}$$

from Example 4.6.1.

- **Example 4.6.5**

Let us find the inverse of the transform

$$\frac{1}{(s^2+a^2)^2} = \frac{1}{a^2}\left(\frac{a}{s^2+a^2} \times \frac{a}{s^2+a^2}\right) \tag{4.6.23}$$

$$= \frac{1}{a^2}\mathcal{L}[\sin(at)]\mathcal{L}[\sin(at)]. \tag{4.6.24}$$

Therefore,

$$\mathcal{L}^{-1}\left[\frac{1}{(s^2+a^2)^2}\right] = \frac{1}{a^2}\int_0^t \sin[a(t-x)]\sin(ax)\,dx \tag{4.6.25}$$

$$= \frac{1}{2a^2}\int_0^t \cos[a(t-2x)]\,dx - \frac{1}{2a^2}\int_0^t \cos(at)\,dx \tag{4.6.26}$$

$$= -\frac{1}{4a^3}\sin[a(t-2x)]\Big|_0^t - \frac{1}{2a^2}\cos(at)\,x\Big|_0^t \tag{4.6.27}$$

$$= \frac{1}{2a^3}[\sin(at) - at\cos(at)]. \tag{4.6.28}$$

• **Example 4.6.6**

Let us use the results from Example 4.6.3 to verify the convolution theorem.

We begin by rewriting (4.6.14) in terms of Heaviside's step functions. Using the method outline in Example 4.2.1,

$$f(t) * g(t) = \left(e^{t-1} - 1\right) H(t-1) + \left(1 - e^{t-2}\right) H(t-2). \quad \textbf{(4.6.29)}$$

Employing the second shifting theorem,

$$\mathcal{L}[f * g] = \frac{e^{-s}}{s-1} - \frac{e^{-s}}{s} + \frac{e^{-2s}}{s} - \frac{e^{-2s}}{s-1} \quad \textbf{(4.6.30)}$$

$$= \frac{e^{-s}}{s(s-1)} - \frac{e^{-2s}}{s(s-1)} = \frac{1}{s-1}\left(\frac{e^{-s}}{s} - \frac{e^{-2s}}{s}\right) \quad \textbf{(4.6.31)}$$

$$= \mathcal{L}[e^t]\mathcal{L}[H(t-1) - H(t-2)] \quad \textbf{(4.6.32)}$$

and the convolution theorem holds true. If we had not rewritten (4.6.14) in terms of step functions, we could still have found $\mathcal{L}[f * g]$ from the definition of the Laplace transform.

Problems

Verify the following convolutions and then show that the convolution theorem is true.

1. $1 * 1 = t$
2. $1 * \cos(at) = \sin(at)/a$
3. $1 * e^t = e^t - 1$
4. $t * t = t^3/6$
5. $t * \sin(t) = t - \sin(t)$
6. $t * e^t = e^t - t - 1$

7.
$$t^2 * \sin(at) = \frac{t^2}{a} - \frac{4}{a^3}\sin^2\left(\frac{at}{2}\right)$$

8.
$$t * H(t-1) = \tfrac{1}{2}(t-1)^2 H(t-1)$$

9.
$$H(t-a) * H(t-b) = (t-a-b)H(t-a-b)$$

10.
$$t * [H(t) - H(t-2)] = \frac{t^2}{2} - \frac{(t-2)^2}{2}H(t-2)$$

Use the convolution theorem to invert the following functions:

11.

$$F(s) = \frac{1}{s^2(s-1)}$$

12.

$$F(s) = \frac{1}{s^2(s+a)^2}$$

13. Prove that the convolution of two Dirac delta functions is a Dirac delta function.

4.7 INTEGRAL EQUATIONS

An *integral equation* contains the dependent variable under an integral sign. The convolution theorem provides an excellent tool for solving a very special class of these equations, *Volterra equation of the second kind* :[6]

$$f(t) - \int_0^t K[t, x, f(x)]\, dx = g(t), \qquad 0 \le t \le T. \tag{4.7.1}$$

These equations appear in history-dependent problems, such as epidemics,[7] vibration problems,[8] and viscoelasticity.[9]

• **Example 4.7.1**

Let us find $f(t)$ from the integral equation

$$f(t) = 4t - 3\int_0^t f(x)\sin(t-x)\, dx. \tag{4.7.2}$$

[6] Fock, V., 1924: Über eine Klasse von Integralgleichungen. *Math. Z.*, **21**, 161–173; Koizumi, S., 1931: On Heaviside's operational solution of a Volterra's integral equation when its nucleus is a function of $(x-\xi)$. *Philos. Mag., Ser. 7*, **11**, 432–441.

[7] Wang, F. J. S., 1978: Asymptotic behavior of some deterministic epidemic models. *SIAM J. Math. Anal.*, **9**, 529–534.

[8] Lin, S. P., 1975: Damped vibration of a string. *J. Fluid Mech.*, **72**, 787–797.

[9] Rogers, T. G. and Lee, E. H., 1964: The cylinder problem in viscoelastic stress analysis. *Q. Appl. Math.*, **22**, 117–131.

The integral in (4.7.2) is such that we can use the convolution theorem to find its Laplace transform. Then, because $\mathcal{L}[\sin(t)] = 1/(s^2+1)$, the convolution theorem yields

$$\mathcal{L}\left[\int_0^t f(x)\sin(t-x)\,dx\right] = \frac{F(s)}{s^2+1}. \tag{4.7.3}$$

Therefore, the Laplace transform converts (4.7.2) into

$$F(s) = \frac{4}{s^2} - \frac{3F(s)}{s^2+1}. \tag{4.7.4}$$

Solving for $F(s)$,

$$F(s) = \frac{4(s^2+1)}{s^2(s^2+4)}. \tag{4.7.5}$$

By partial fractions, or by inspection,

$$F(s) = \frac{1}{s^2} + \frac{3}{s^2+4}. \tag{4.7.6}$$

Therefore, inverting term by term,

$$f(t) = t + \tfrac{3}{2}\sin(2t). \tag{4.7.7}$$

Note that the integral equation

$$f(t) = 4t - 3\int_0^t f(t-x)\sin(x)\,dx \tag{4.7.8}$$

also has the same solution.

- **Example 4.7.2**

Let us solve the equation

$$g(t) = \frac{t^2}{2} - \int_0^t (t-x)g(x)\,dx. \tag{4.7.9}$$

Again the integral is one of the convolution type. Taking the Laplace transform of (4.7.9),

$$G(s) = \frac{1}{s^3} - \frac{G(s)}{s^2}, \tag{4.7.10}$$

which yields

$$\left(1+\frac{1}{s^2}\right)G(s) = \frac{1}{s^3} \tag{4.7.11}$$

or

$$G(s) = \frac{1}{s(s^2+1)} = \frac{1}{s} - \frac{s}{s^2+1}. \tag{4.7.12}$$

Then

$$g(t) = 1 - \cos(t). \tag{4.7.13}$$

Problems

Solve the following integral equations:

1. $$f(t) = 1 + 2\int_0^t f(t-x)e^{-2x}\,dx$$

2. $$f(t) = 1 + \int_0^t f(x)\sin(t-x)\,dx$$

3. $$f(t) = t + \int_0^t f(t-x)e^{-x}\,dx$$

4. $$f(t) = 4t^2 - \int_0^t f(t-x)e^{-x}\,dx$$

5. $$f(t) = t^3 + \int_0^t f(x)\sin(t-x)\,dx$$

6. $$f(t) = 8t^2 - 3\int_0^t f(x)\sin(t-x)\,dx$$

7. $$f(t) = t^2 - 2\int_0^t f(t-x)\sinh(2x)\,dx$$

8. $$f(t) = 1 + 2\int_0^t f(t-x)\cos(x)\,dx$$

9. $$f(t) = e^{2t} - 2\int_0^t f(t-x)\cos(x)\,dx$$

10. $$f(t) = t^2 + \int_0^t f(x)\sin(t-x)\,dx$$

11.

$$f(t) = e^{-t} - 2\int_0^t f(x)\cos(t-x)\,dx$$

12.

$$f(t) = 6t + 4\int_0^t f(x)(x-t)^2\,dx$$

13. Solve the following equation for $f(t)$ with the condition that $f(0) = 4$:

$$f'(t) = t + \int_0^t f(t-x)\cos(x)\,dx.$$

14. Solve the following equation for $f(t)$ with the condition that $f(0) = 0$:

$$f'(t) = \sin(t) + \int_0^t f(t-x)\cos(x)\,dx.$$

15. During a study of nucleation involving idealized active sites along a boiling surface, Marto and Rohsenow[10] had to solve the integral equation

$$A = B\sqrt{t} + C\int_0^t \frac{x'(\tau)}{\sqrt{t-\tau}}\,d\tau$$

to find the position $x(t)$ of the liquid/vapor interface. If A, B, and C are constants and $x(0) = 0$, find the solution for them.

16. Solve the following equation for $x(t)$ with the condition that $x(0) = 0$:

$$x(t) + t = \frac{1}{c\sqrt{\pi}}\int_0^t \frac{x'(\tau)}{\sqrt{t-\tau}}\,d\tau,$$

where c is constant.

17. During a study of the temperature $f(t)$ of a heat reservoir attached to a semi-infinite heat-conducting rod, Huber[11] had to solve the integral equation

$$f'(t) = \alpha - \frac{\beta}{\sqrt{\pi}}\int_0^t \frac{f'(\tau)}{\sqrt{t-\tau}}\,d\tau,$$

where α and β are constants and $f(0) = 0$. Find $f(t)$ for him. Hint:

$$\frac{\alpha}{s^{3/2}(s^{1/2}+\beta)} = \frac{\alpha}{s(s-\beta^2)} - \frac{\alpha\beta}{s^{3/2}(s-\beta^2)}.$$

---

[10] From Marto, P. J. and Rohsenow, W. M., 1966: Nucleate boiling instability of alkali metals. *J. Heat Transfer*, **88**, 183–193 with permission.

[11] From Huber, A., 1934: Eine Methode zur Bestimmung der Wärme- und Temperaturleitfähigkeit. *Monatsh. Math. Phys.*, **41**, 35–42.

18. During the solution of a diffusion problem, Zhdanov, Chikhachev, and Yavlinskii[12] solved an integral equation similar to

$$\int_0^t f(\tau)\left[1 - \operatorname{erf}\left(a\sqrt{t-\tau}\right)\right] d\tau = at,$$

where $\operatorname{erf}(x) = \dfrac{2}{\sqrt{\pi}} \displaystyle\int_0^x e^{-y^2}\, dy$ is the error function. What should have they found? Hint: You will need to prove that

$$\mathcal{L}\left[t \operatorname{erf}(a\sqrt{t}\,) - \frac{1}{2a^2}\operatorname{erf}(a\sqrt{t}\,) + \frac{\sqrt{t}}{a\sqrt{\pi}} e^{-a^2 t}\right] = \frac{a}{s^2\sqrt{s+a^2}}.$$

4.8 SOLUTION OF LINEAR DIFFERENTIAL EQUATIONS WITH CONSTANT COEFFICIENTS

For the engineer, as it was for Oliver Heaviside, the primary use of Laplace transforms is the solution of ordinary, constant coefficient, linear differential equations. These equations are important not only because they appear in many engineering problems but also because they may serve as approximations, even if locally, to ordinary differential equations with nonconstant coefficients or to nonlinear ordinary differential equations.

For all of these reasons, we wish to solve the *initial-value problem*

$$\frac{d^n y}{dt^n} + a_1 \frac{d^{n-1} y}{dt^{n-1}} + \cdots + a_{n-1}\frac{dy}{dt} + a_n y = f(t), \quad t > 0 \qquad \textbf{(4.8.1)}$$

by Laplace transforms, where $a_1, a_2, \ldots$ are constants and we know the value of $y, y', \ldots, y^{(n-1)}$ at $t = 0$. The procedure is as follows. Applying the derivative rule (4.1.20) to (4.8.1), we reduce the *differential* equation to an *algebraic* one involving the constants $a_1, a_2, \ldots, a_n$, the parameter s, the Laplace transform of $f(t)$, and the values of the initial conditions. We then solve for the Laplace transform of $y(t)$, $Y(s)$. Finally, we apply one of the many techniques of inverting a Laplace transform to find $y(t)$.

Similar considerations hold with *systems* of ordinary differential equations. The Laplace transform of the system of ordinary differential equations results in an algebraic set of equations containing $Y_1(s), Y_2(s), \ldots, Y_n(s)$. By some method we solve this set of equations and invert each transform $Y_1(s), Y_2(s), \ldots, Y_n(s)$ in turn to give $y_1(t), y_2(t), \ldots, y_n(t)$.

---

[12] Zhdanov, S. K., Chikhachev, A. S., and Yavlinskii, Yu. N., 1976: Diffusion boundary-value problem for regions with moving boundaries and conservation of particles. *Sov. Phys. Tech. Phys.*, **21**, 883–884.

The following examples will illustrate the details of the process.

• Example 4.8.1

Let us solve the ordinary differential equation

$$y'' + 2y' = 8t \tag{4.8.2}$$

subject to the initial conditions that $y'(0) = y(0) = 0$. Taking the Laplace transform of both sides of (4.8.2),

$$\mathcal{L}(y'') + 2\mathcal{L}(y') = 8\mathcal{L}(t) \tag{4.8.3}$$

or

$$s^2Y(s) - sy(0) - y'(0) + 2sY(s) - 2y(0) = \frac{8}{s^2}, \tag{4.8.4}$$

where $Y(s) = \mathcal{L}[y(t)]$. Substituting the initial conditions into (4.8.4) and solving for $Y(s)$,

$$Y(s) = \frac{8}{s^3(s+2)} = \frac{A}{s^3} + \frac{B}{s^2} + \frac{C}{s} + \frac{D}{s+2} \tag{4.8.5}$$

$$= \frac{8}{s^3(s+2)} = \frac{(s+2)A + s(s+2)B + s^2(s+2)C + s^3D}{s^3(s+2)}. \tag{4.8.6}$$

Matching powers of s in the numerators of (4.8.6), $C+D=0$, $B+2C=0$, $A+2B=0$, and $2A=8$ or $A=4$, $B=-2$, $C=1$, and $D=-1$. Therefore,

$$Y(s) = \frac{4}{s^3} - \frac{2}{s^2} + \frac{1}{s} - \frac{1}{s+2}. \tag{4.8.7}$$

Finally, performing term-by-term inversion of (4.8.7), the final solution is

$$y(t) = 2t^2 - 2t + 1 - e^{-2t}. \tag{4.8.8}$$

• Example 4.8.2

Let us solve the ordinary differential equation

$$y'' + y = H(t) - H(t-1) \tag{4.8.9}$$

with the initial conditions that $y'(0) = y(0) = 0$. Taking the Laplace transform of both sides of (4.8.9),

$$s^2Y(s) - sy(0) - y'(0) + Y(s) = \frac{1}{s} - \frac{e^{-s}}{s}, \tag{4.8.10}$$

where $Y(s) = \mathcal{L}[y(t)]$. Substituting the initial conditions into (4.8.10) and solving for $Y(s)$,

$$Y(s) = \left(\frac{1}{s} - \frac{s}{s^2+1}\right) - \left(\frac{1}{s} - \frac{s}{s^2+1}\right)e^{-s}. \tag{4.8.11}$$

Using the second shifting theorem, the final solution is

$$y(t) = 1 - \cos(t) - [1 - \cos(t-1)]H(t-1). \tag{4.8.12}$$

• **Example 4.8.3**

Let us solve the ordinary differential equation

$$y'' + 2y' + y = f(t) \tag{4.8.13}$$

with the initial conditions that $y'(0) = y(0) = 0$, where $f(t)$ is an unknown function whose Laplace transform exists. Taking the Laplace transform of both sides of (4.8.13),

$$s^2Y(s) - sy(0) - y'(0) + 2sY(s) - 2y(0) + Y(s) = F(s), \tag{4.8.14}$$

where $Y(s) = \mathcal{L}[y(t)]$. Substituting the initial conditions into (4.8.14) and solving for $Y(s)$,

$$Y(s) = \frac{1}{(s+1)^2}F(s). \tag{4.8.15}$$

We have written (4.8.15) in this form because the transform $Y(s)$ equals the product of two transforms $1/(s+1)^2$ and $F(s)$. Therefore, by the convolution theorem we can immediately write

$$y(t) = te^{-t} * f(t) = \int_0^t xe^{-x} f(t-x)\, dx. \tag{4.8.16}$$

Without knowing $f(t)$, this is as far as we can go.

• **Example 4.8.4: Forced harmonic oscillator**

Let us solve the *simple harmonic oscillator* forced by a harmonic forcing:

$$y'' + \omega^2 y = \cos(\omega t) \tag{4.8.17}$$

subject to the initial conditions that $y'(0) = y(0) = 0$. Although the complete solution could be found by summing the complementary solution and a particular solution obtained, say, from the method of undetermined coefficients, we will now illustrate how we can use Laplace transforms to solve this problem.

Taking the Laplace transform of both sides of (4.8.17), substituting in the initial conditions, and solving for $Y(s)$,

$$Y(s) = \frac{s}{(s^2+\omega^2)^2} \tag{4.8.18}$$

and

$$y(t) = \frac{1}{\omega}\sin(\omega t) * \cos(\omega t) = \frac{t}{2\omega}\sin(\omega t). \tag{4.8.19}$$

Equation (4.8.19) gives an oscillation that grows linearly with time although the forcing function is simply periodic. Why does this occur? Recall that our simple harmonic oscillator has the natural frequency ω. But that is exactly the frequency at which we drive the system. Consequently, our choice of forcing has resulted in *resonance* where energy continuously feeds into the oscillator.

• Example 4.8.5

Let us solve the *system* of ordinary differential equations:

$$2x' + y = \cos(t) \tag{4.8.20}$$

and

$$y' - 2x = \sin(t) \tag{4.8.21}$$

subject to the initial conditions that $x(0) = 0$ and $y(0) = 1$. Taking the Laplace transform of (4.8.20) and (4.8.21),

$$2sX(s) + Y(s) = \frac{s}{s^2+1} \tag{4.8.22}$$

and

$$-2X(s) + sY(s) = 1 + \frac{1}{s^2+1}, \tag{4.8.23}$$

after introducing the initial conditions. Solving for $X(s)$ and $Y(s)$,

$$X(s) = -\frac{1}{(s^2+1)^2} \tag{4.8.24}$$

and

$$Y(s) = \frac{s}{s^2+1} + \frac{2s}{(s^2+1)^2}. \tag{4.8.25}$$

Taking the inverse of (4.8.24)–(4.8.25) term by term,

$$x(t) = \tfrac{1}{2}[t\cos(t) - \sin(t)] \tag{4.8.26}$$

and

$$y(t) = t\sin(t) + \cos(t). \tag{4.8.27}$$

• Example 4.8.6

Let us determine the displacement of a mass m attached to a spring and excited by the driving force:

$$F(t) = mA\left(1 - \frac{t}{T}\right)e^{-t/T}. \tag{4.8.28}$$

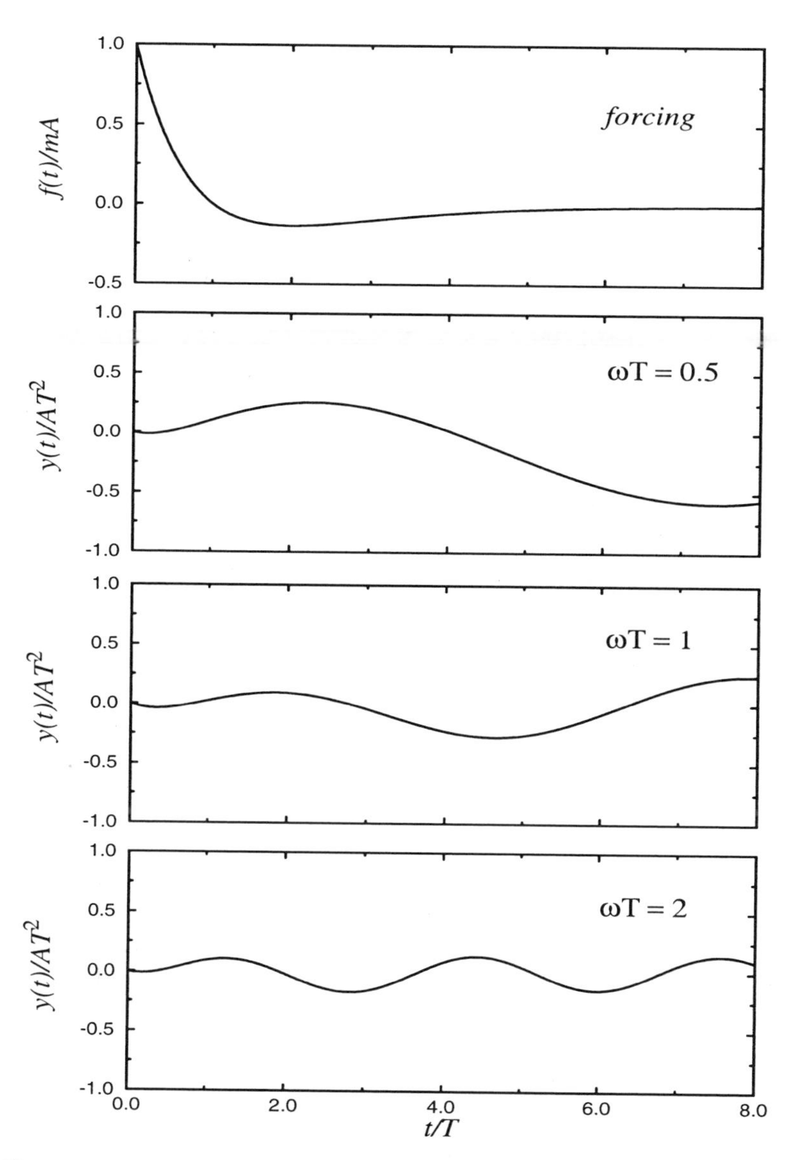

Figure 4.8.1: Displacement of a simple harmonic oscillator with nondimensional frequency ωT as a function of time t/T. The top frame shows the forcing function.

The dynamical equation governing this system is

$$y'' + \omega^2 y = A\left(1 - \frac{t}{T}\right) e^{-t/T}, \tag{4.8.29}$$

where $\omega^2 = k/m$ and k is the spring constant. Assuming that the system is initially at rest, the Laplace transform of the dynamical system is

$$(s^2 + \omega^2)Y(s) = \frac{A}{s + 1/T} - \frac{A}{T(s + 1/T)^2} \tag{4.8.30}$$

or

$$Y(s) = \frac{A}{(s^2 + \omega^2)(s + 1/T)} - \frac{A}{T(s^2 + \omega^2)(s + 1/T)^2}. \tag{4.8.31}$$

Partial fractions yield

$$\begin{aligned} Y(s) &= \frac{A}{\omega^2 + 1/T^2}\left(\frac{1}{s + 1/T} - \frac{s - 1/T}{s^2 + \omega^2}\right) - \frac{A}{T(\omega^2 + 1/T^2)^2} \\ &\times \left[\frac{1/T^2 - \omega^2}{s^2 + \omega^2} - \frac{2s/T}{s^2 + \omega^2} + \frac{\omega^2 + 1/T^2}{(s + 1/T)^2} + \frac{2/T}{s + 1/T}\right]. \end{aligned} \tag{4.8.32}$$

Inverting (4.8.32) term by term,

$$\begin{aligned} y(t) &= \frac{AT^2}{1 + \omega^2 T^2}\left[e^{-t/T} - \cos(\omega t) + \frac{\sin(\omega t)}{\omega T}\right] \\ &- \frac{AT^2}{(1 + \omega^2 T^2)^2}\left\{(1 - \omega^2 T^2)\frac{\sin(\omega t)}{\omega T} + 2\left[e^{-t/T} - \cos(\omega t)\right]\right. \\ &\left. + (1 + \omega^2 T^2)(t/T)e^{-t/T}\right\}. \end{aligned} \tag{4.8.33}$$

The solution to this problem consists of two parts. The exponential terms result from the forcing and will die away with time. This is the *transient* portion of the solution. The sinusoidal terms are those natural oscillations that are necessary so that the solution satisfies the initial conditions. They are the *steady-state* portion of the solution. They endure forever. Figure 4.8.1 illustrates the solution when $\omega T = 0.1$, 1, and 2. Note that the displacement decreases in magnitude as the nondimensional frequency of the oscillator increases.

• **Example 4.8.7**

Let us solve the equation

$$y'' + 16y = \delta(t - \pi/4) \tag{4.8.34}$$

with the initial conditions that $y(0) = 1$ and $y'(0) = 0$.

Taking the Laplace transform of (4.8.34) and inserting the initial conditions,

$$(s^2 + 16)Y(s) = s + e^{-s\pi/4} \tag{4.8.35}$$

or

$$Y(s) = \frac{s}{s^2+16} + \frac{e^{-s\pi/4}}{s^2+16}. \tag{4.8.36}$$

Applying the second shifting theorem,

$$y(t) = \cos(4t) + \tfrac{1}{4}\sin[4(t-\pi/4)]H(t-\pi/4) \tag{4.8.37}$$

$$= \cos(4t) - \tfrac{1}{4}\sin(4t)H(t-\pi/4). \tag{4.8.38}$$

• Example 4.8.8: Oscillations in electric circuits

During the middle of the nineteenth century, Lord Kelvin[13] analyzed the LCR electrical circuit shown in Figure 4.8.2 which contains resistance R, capacitance C, and inductance L. For reasons that we shall shortly show, this LCR circuit has become one of the quintessential circuits for electrical engineers. In this example, we shall solve the problem by Laplace transforms.

Because we can add the potential differences across the elements, the equation governing the LCR circuit is

$$L\frac{dI}{dt} + RI + \frac{1}{C}\int_0^t I\,d\tau = E(t), \tag{4.8.39}$$

where I denotes the current in the circuit. Let us solve (4.8.39) when we close the circuit and the initial conditions are $I(0) = 0$ and $Q(0) = -Q_0$. Taking the Laplace transform of (4.8.39),

$$\left(Ls + R + \frac{1}{Cs}\right)\overline{I}(s) = LI(0) - \frac{Q(0)}{Cs}. \tag{4.8.40}$$

Solving for $\overline{I}(s)$,

$$\overline{I}(s) = \frac{Q_0}{Cs(Ls+R+1/Cs)} = \frac{\omega_0^2 Q_0}{s^2+2\alpha s+\omega_0^2} \tag{4.8.41}$$

$$= \frac{\omega_0^2 Q_0}{(s+\alpha)^2+\omega_0^2-\alpha^2}, \tag{4.8.42}$$

where $\alpha = R/2L$ and $\omega_0^2 = 1/(LC)$. From the first shifting theorem,

$$I(t) = \frac{\omega_0^2 Q_0}{\omega} e^{-\alpha t}\sin(\omega t), \tag{4.8.43}$$

---

[13] Thomson, W., 1853: On transient electric currents. *Philos. Mag., Ser. 4*, **5**, 393–405.

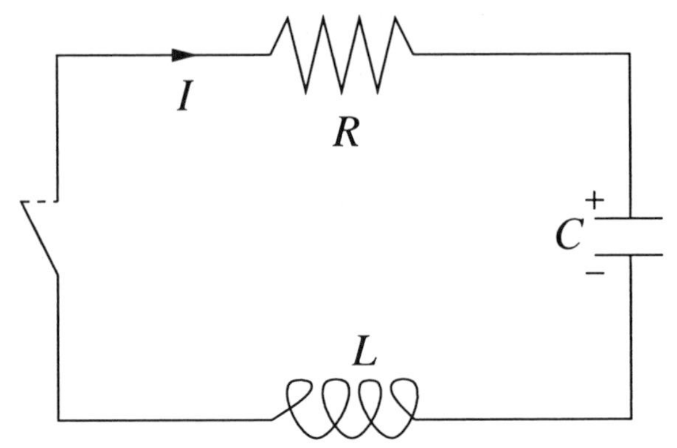

Figure 4.8.2: Schematic of a LCR circuit.

where $\omega^2 = \omega_0^2 - \alpha^2 > 0$. The quantity ω is the natural frequency of the circuit, which is lower than the free frequency ω_0 of a circuit formed by a condenser and coil. Most importantly, the solution decays in amplitude with time.

Although Kelvin's solution was of academic interest when he originally published it, this radically changed with the advent of radio telegraphy[14] because the LCR circuit described the fundamental physical properties of wireless transmitters and receivers.[15] The inescapable conclusion from this analysis was that no matter how clever the receiver was designed, eventually the resistance in the circuit would rapidly dampen the electrical oscillations and thus limit the strength of the received signal.

This technical problem was overcome by Armstrong[16] who invented an electrical circuit that used De Forest's audion (the first vacuum tube) for generating electrical oscillations and for amplifying externally impressed oscillations by "regenerative action". The effect of adding the "thermionic amplifier" is seen by again considering the LRC circuit as shown in Figure 4.8.3 with the modification suggested by Armstrong.[17]

The governing equations of this new circuit are

$$L_1 \frac{dI}{dt} + RI + \frac{1}{C} \int_0^t I \, d\tau + M \frac{dI_p}{dt} = 0 \qquad \textbf{(4.8.44)}$$

[14] Stone, J S., 1914: The resistance of the spark and its effect on the oscillations of electrical oscillators. *Proc. IRE*, **2**, 307–324.

[15] See Hogan, J. L., 1916: Physical aspects of radio telegraphy. *Proc. IRE*, **4**, 397–420.

[16] Armstrong, E. H., 1915: Some recent developments in the audion receiver. *Proc. IRE*, **3**, 215–247.

[17] From Ballantine, S., 1919: The operational characteristics of thermionic amplifiers. *Proc. IRE*, **7**, 129–161. ©IRE (now IEEE).

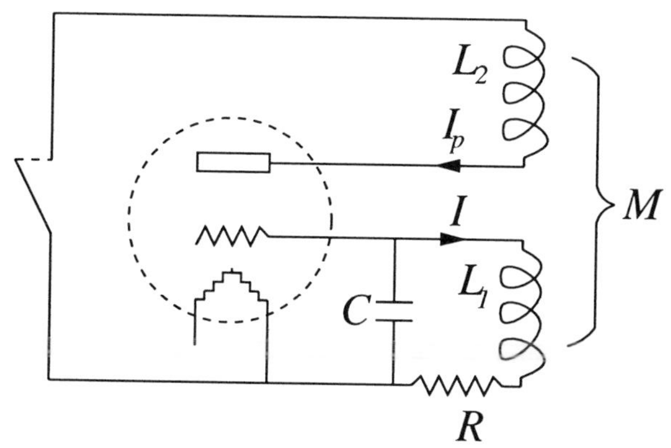

Figure 4.8.3: Schematic of a LCR circuit with the addition of a thermionic amplifier. [From Ballantine, S., 1919: The operational characteristics of thermionic amplifiers. *Proc. IRE*, **7**, 155. ©IRE (now IEEE).]

and

$$L_2 \frac{dI_p}{dt} + R_0 I_p + M \frac{dI}{dt} + \frac{\mu}{C} \int_0^t I \, d\tau = 0, \qquad \textbf{(4.8.45)}$$

where the plate circuit has the current I_p, the resistance R_0, the inductance L_2, and the electromotive force (emf) of $\mu \int_0^t I \, d\tau / C$. The mutual inductance between the two circuits is given by M. Taking the Laplace transform of (4.8.44)–(4.8.45),

$$L_1 s \overline{I}(s) + R\overline{I}(s) + \frac{\overline{I}(s)}{sC} + Ms\overline{I}_p(s) = \frac{Q_0}{sC} \qquad \textbf{(4.8.46)}$$

and

$$L_2 s \overline{I}_p(s) + R_0 \overline{I}_p(s) + Ms\overline{I}(s) + \frac{\mu}{sC}\overline{I}(s) = 0. \qquad \textbf{(4.8.47)}$$

Eliminating $\overline{I}_p(s)$ between (4.8.46)–(4.8.47) and solving for $\overline{I}(s)$,

$$\overline{I}(s) = \frac{(L_2 s + R_0)Q_0}{(L_1 L_2 - M^2)Cs^3 + (RL_2 + R_0 L_1)Cs^2 + (L_2 + CRR_0 - \mu M)s + R_0}. \qquad \textbf{(4.8.48)}$$

For high-frequency radio circuits, we can approximate the roots of the denominator of (4.8.48) as

$$s_1 \approx -\frac{R_0}{L_2 + CRR_0 - \mu M} \qquad \textbf{(4.8.49)}$$

and

$$s_{2,3} \approx \frac{R_0}{2(L_2 + CRR_0 - \mu M)} - \frac{R_0 L_1 + RL_2}{2(L_1 L_2 - M^2)} \pm i\omega. \qquad \textbf{(4.8.50)}$$

In the limit of M and R_0 vanishing, we recover our previous result for the LRC circuit. However, in reality, R_0 is very large and our solution has three terms. The term associated with s_1 is a rapidly decaying transient while the s_2 and s_3 roots yield oscillatory solutions with a *slight* amount of damping. Thus, our analysis has shown that in the ordinary regenerative circuit, the tube effectively introduces sufficient "negative" resistance so that the resultant positive resistance of the equivalent LCR circuit is relatively low, and the response of an applied signal voltage at the resonant frequency of the circuit is therefore relatively great. Later, Armstrong[18] extended his work on regeneration by introducing an electrical circuit – the superregenerative circuit – where the regeneration is made large enough so that the resultant resistance is negative, and self-sustained oscillations can occur.[19] It was this circuit[20] which led to the explosive development of radio in the 1920s and 1930s.

• **Example 4.8.9: Resonance transformer circuit**

One of the fundamental electrical circuits of early radio telegraphy[21] is the resonance transformer circuit shown in Figure 4.8.4. Its development gave transmitters and receivers the ability to tune to each other.

The governing equations follow from Kirchhoff's law and are

$$L_1\frac{dI_1}{dt} + M\frac{dI_2}{dt} + \frac{1}{C_1}\int_0^t I_1\,d\tau = E(t) \tag{4.8.51}$$

and

$$M\frac{dI_1}{dt} + L_2\frac{dI_2}{dt} + RI_2 + \frac{1}{C_2}\int_0^t I_2\,d\tau = 0. \tag{4.8.52}$$

Let us examine the oscillations generated if initially the system has no currents or charges and the forcing function is $E(t) = \delta(t)$.

Taking the Laplace transform of (4.8.51)–(4.8.52),

$$L_1 s\overline{I}_1 + Ms\overline{I}_2 + \frac{\overline{I}_2}{sC_1} = 1 \tag{4.8.53}$$

---

[18] Armstrong, E. H., 1922: Some recent developments of regenerative circuits. *Proc. IRE*, **10**, 244–260.

[19] See Frink, F. W., 1938: The basic principles of superregenerative reception. *Proc. IRE*, **26**, 76–106.

[20] Lewis, T., 1991: *Empire of the Air: The Men Who Made Radio*, HarperCollins Publishers, New York.

[21] Fleming, J. A., 1919: *The Principles of Electric Wave Telegraphy and Telephony*, Longmans, Green, Chicago.

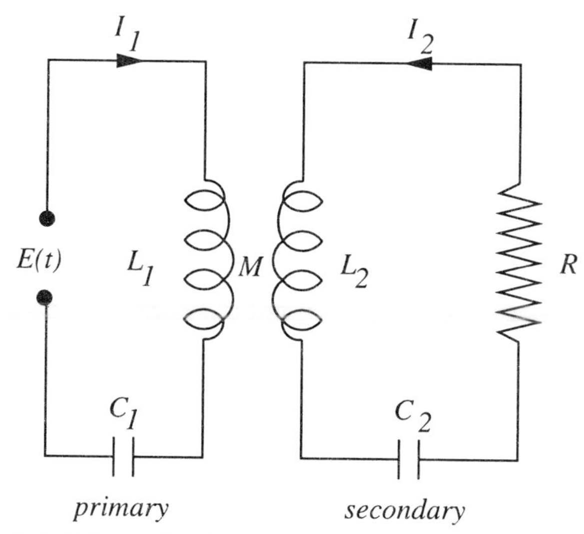

Figure 4.8.4: Schematic of a resonance transformer circuit.

and

$$Ms\overline{I}_1 + L_2 s\overline{I}_2 + R\overline{I}_2 + \frac{\overline{I}_2}{sC_2} = 0. \qquad (4.8.54)$$

Because the current in the second circuit is of greater interest, we solve for $\overline{I}_2$ and find that

$$\overline{I}_2(s) = -\frac{Ms^3}{L_1L_2[(1-k^2)s^4 + 2\alpha\omega_2^2 s^3 + (\omega_1^2+\omega_2^2)s^2 + 2\alpha\omega_1^2 s + \omega_1^2\omega_2^2]}, \qquad (4.8.55)$$

where $\alpha = R/2L_2$, $\omega_1^2 = 1/L_1C_1$, $\omega_2^2 = 1/L_2C_2$, and $k^2 = M^2/L_1L_2$, the so-called coefficient of coupling.

We can obtain analytic solutions if we assume that the coupling is weak ($k^2 \ll 1$). Equation (4.8.55) becomes

$$\overline{I}_2 = -\frac{Ms^3}{L_1L_2(s^2+\omega_1^2)(s^2+2\alpha s+\omega_2^2)}. \qquad (4.8.56)$$

Using partial fractions and inverting term by term, we find that

$$\begin{aligned} I_2(t) = \frac{M}{L_1L_2}\Bigg[& \frac{2\alpha\omega_1^3\sin(\omega_1 t)}{(\omega_2^2-\omega_1^2)^2+4\alpha^2\omega_1^2} + \frac{\omega_1^2(\omega_2^2-\omega_1^2)\cos(\omega_1 t)}{(\omega_2^2-\omega_1^2)^2+4\alpha^2\omega_1^2} \\ & + \frac{\alpha\omega_2^4 - 3\alpha\omega_1^2\omega_2^2 + 4\alpha^3\omega_1^2}{(\omega_2^2-\omega_1^2)^2+4\alpha^2\omega_1^2} e^{-\alpha t}\frac{\sin(\omega t)}{\omega} \\ & - \frac{\omega_2^2(\omega_2^2-\omega_1^2)+4\alpha^2\omega_1^2}{(\omega_2^2-\omega_1^2)^2+4\alpha^2\omega_1^2} e^{-\alpha t}\cos(\omega t)\Bigg], \end{aligned} \qquad (4.8.57)$$

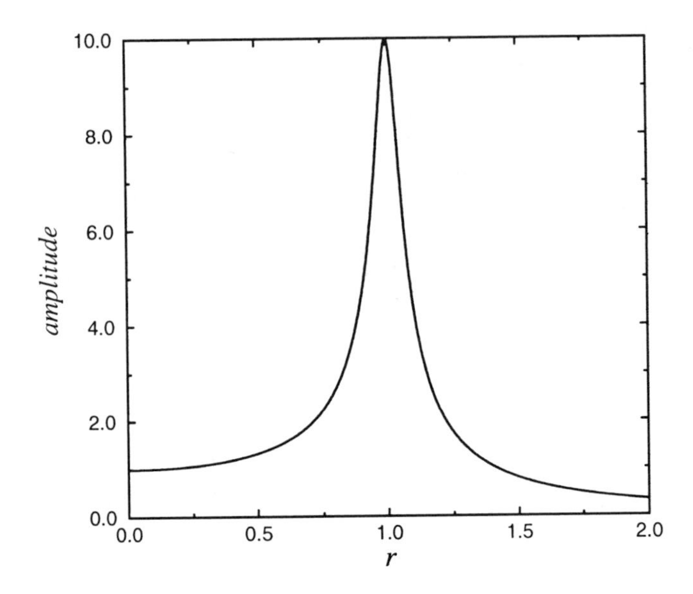

Figure 4.8.5: The resonance curve $1/\sqrt{(r^2-1)^2+0.01}$ for a resonance transformer circuit with $r = \omega_2/\omega_1$.

where $\omega^2 = \omega_2^2 - \alpha^2$.

The exponentially damped solutions will eventually disappear, leaving only the steady-state oscillations which vibrate with the angular frequency ω_1, the natural frequency of the primary circuit. If we rewrite this steady-state solution in amplitude/phase form, the amplitude is

$$\frac{M}{L_1 L_2 \sqrt{(r^2-1)^2 + 4\alpha^2/\omega_1^2}}, \qquad (4.8.58)$$

where $r = \omega_2/\omega_1$. As Figure 4.8.5 shows, as r increases from zero to two, the amplitude rises until a very sharp peak occurs at $r = 1$ and then decreases just as rapidly as we approach $r = 2$. Thus, the resonance transformer circuit provides a convenient way to tune a transmitter or receiver to the frequency ω_1.

Problems

Solve the following ordinary differential equations by Laplace transforms:

1. $y' - 2y = 1 - t; \quad y(0) = 1$

2. $y'' - 4y' + 3y = e^t; \quad y(0) = 0, y'(0) = 0$

3. $y'' - 4y' + 3y = e^{2t}; \quad y(0) = 0, y'(0) = 1$

4. $y'' - 6y' + 8y = e^{t}; \quad y(0) = 3, y'(0) = 9$

5. $y'' + 4y' + 3y = e^{-t}; \quad y(0) = 1, y'(0) = 1$

6. $y'' + y = t; \quad y(0) = 1, y'(0) = 0$

7. $y'' + 4y' + 3y = e^{t}; \quad y(0) = 0, y'(0) = 2$

8. $y'' - 4y' + 5y = 0; \quad y(0) = 2, y'(0) = 4$

9. $y' + y = tH(t-1); \quad y(0) = 0$

10. $y'' + 3y' + 2y = H(t-1); \quad y(0) = 0, y'(0) = 1$

11. $y'' - 3y' + 2y = H(t-1); \quad y(0) = 0, y'(0) = 1$

12. $y'' + 4y = 3H(t-4); \quad y(0) = 1, y'(0) = 0$

13. $y'' + 4y' + 4y = 4H(t-2); \quad y(0) = 0, y'(0) = 0$

14. $y'' + 3y' + 2y = e^{t-1}H(t-1); \quad y(0) = 0, y'(0) = 1$

15. $y'' - 3y' + 2y = e^{-(t-2)}H(t-2); \quad y(0) = 0, y'(0) = 0$

16. $y'' - 3y' + 2y = H(t-1) - H(t-2); \quad y(0) = 0, y'(0) = 0$

17. $y'' + y = 1 - H(t-T); \quad y(0) = 0, y'(0) = 0$

18. $y'' + y = \begin{cases} \sin(t), & 0 \leq t \leq \pi \\ 0, & t \geq \pi; \end{cases} \quad y(0) = 0, y'(0) = 0$

19. $y'' + 3y' + 2y = \begin{cases} t, & 0 \leq t \leq a \\ ae^{-(t-a)}, & t \geq a; \end{cases} \quad y(0) = 0, y'(0) = 0$

20. $y'' + \omega^2 y = \begin{cases} t/a, & 0 \leq t \leq a \\ 1 - (t-a)/(b-a), & a \leq t \leq b \\ 0, & t \geq b; \end{cases}$
$y(0) = 0, y'(0) = 0$

21. $y'' - 2y' + y = 3\delta(t-2); \quad y(0) = 0, y'(0) = 1$

22. $y'' - 5y' + 4y = \delta(t-1); \quad y(0) = 0, y'(0) = 0$

23. $y'' + 5y' + 6y = 3\delta(t-2) - 4\delta(t-5); \quad y(0) = y'(0) = 0$

24. $x' - 2x + y = 0, y' - 3x - 4y = 0; \quad x(0) = 1, y(0) = 0$

25. $x' - 2y' = 1, x' + y - x = 0; \quad x(0) = y(0) = 0$

26. $x' + 2x - y' = 0, x' + y + x = t^2; \quad x(0) = y(0) = 0$

27. $x' + 3x - y = 1, x' + y' + 3x = 0; \quad x(0) = 2, y(0) = 0$

28. Forster, Escobal, and Lieske[22] used Laplace transforms to solve the linearized equations of motion of a vehicle in a gravitational field created by two other bodies. A simplified form of this problem involves solving the following system of ordinary differential equations:

$$x'' - 2y' = F_1 + x + 2y, \qquad 2x' + y'' = F_2 + 2x + 3y$$

subject to the initial conditions that $x(0) = y(0) = x'(0) = y'(0) = 0$. Find the solution to this system.

4.9 TRANSFER FUNCTIONS, GREEN'S FUNCTION, AND INDICIAL ADMITTANCE

One of the drawbacks of using Laplace transforms to solve ordinary differential equations with a forcing term is its lack of generality. Each new forcing function requires a repetition of the entire process. In this section we give some methods for finding the solution in a somewhat more general manner for stationary systems where the forcing, not any initially stored energy (i.e., nonzero initial conditions), produces the total output. Unfortunately, the solution must be written as an integral.

In Example 4.8.3 we solved the linear differential equation

$$y'' + 2y' + y = f(t) \tag{4.9.1}$$

subject to the initial conditions $y(0) = y'(0) = 0$. At that time we wrote the Laplace transform of $y(t)$, $Y(s)$, as the product of two Laplace transforms:

$$Y(s) = \frac{1}{(s+1)^2} F(s). \tag{4.9.2}$$

---

[22] Reprinted from *Astronaut. Acta*, **14**, Forster, K., P. R. Escobal and H. A. Lieske, Motion of a vehicle in the transition region of the three-body problem, 1–10, ©1968, with kind permission from Elsevier Science Ltd, The Boulevard, Langford Lane, Kidlington OX5 1GB, UK.

One drawback in using (4.9.2) is its dependence upon an unspecified Laplace transform $F(s)$. Is there a way to eliminate this dependence and yet retain the essence of the solution?

One way of obtaining a quantity that is independent of the forcing is to consider the ratio:

$$\frac{Y(s)}{F(s)} = G(s) = \frac{1}{(s+1)^2}. \tag{4.9.3}$$

This ratio is called the ***transfer function*** because we can transfer the input $F(s)$ into the output $Y(s)$ by multiplying $F(s)$ by $G(s)$. It depends only upon the properties of the system.

Let us now consider a related problem to (4.9.1), namely

$$g'' + 2g' + g = \delta(t), \qquad t > 0 \tag{4.9.4}$$

with $g(0) = g'(0) = 0$. Because the forcing equals the Dirac delta function, $g(t)$ is called the ***impulse response*** or ***Green's function***.[23] Computing $G(s)$,

$$G(s) = \frac{1}{(s+1)^2}. \tag{4.9.5}$$

From (4.9.3) we see that $G(s)$ is also the transfer function. Thus, an alternative method for computing the transfer function is to subject the system to impulse forcing and the Laplace transform of the response is the transfer function.

From (4.9.3),

$$Y(s) = G(s)F(s) \tag{4.9.6}$$

or

$$y(t) = g(t) * f(t). \tag{4.9.7}$$

That is, the convolution of the impulse response with the particular forcing gives the response of the system. Thus, we may describe a stationary system in one of two ways: (1) in the transform domain we have the transfer function, and (2) in the time domain there is the impulse response.

Despite the fundamental importance of the impulse response or Green's function for a given linear system, it is often quite difficult to determine, especially experimentally, and a more convenient practice is to deal with the response to the unit step $H(t)$. This response is called the ***indicial admittance*** or ***step response***, which we shall denote by $a(t)$.

---

[23] For the origin of the Green's function, see Farina, J. E. G., 1976: The work and significance of George Green, the miller mathematician, 1793–1841. *Bull. Inst. Math. Appl.*, **12**, 98–105.

Because $\mathcal{L}[H(t)] = 1/s$, we can determine the transfer function from the indicial admittance because $\mathcal{L}[a(t)] = G(s)\mathcal{L}[H(t)]$ or $sA(s) = G(s)$. Furthermore, because

$$\mathcal{L}[g(t)] = G(s) = \frac{\mathcal{L}[a(t)]}{\mathcal{L}[H(t)]}, \tag{4.9.8}$$

then

$$g(t) = \frac{da(t)}{dt} \tag{4.9.9}$$

from (4.1.18).

• Example 4.9.1

Let us find the transfer function, impulse response, and step response for the system

$$y'' - 3y' + 2y = f(t) \tag{4.9.10}$$

with $y(0) = y'(0) = 0$. To find the impulse response, we solve

$$g'' - 3g' + 2g = \delta(t) \tag{4.9.11}$$

with $g(0) = g'(0) = 0$. Taking the Laplace transform of (4.9.11), we find that

$$G(s) = \frac{1}{s^2 - 3s + 2}, \tag{4.9.12}$$

which is the transfer function for this system. The impulse response equals the inverse of $G(s)$ or

$$g(t) = e^{2t} - e^t. \tag{4.9.13}$$

To find the step response, we solve

$$a'' - 3a' + 2a = H(t) \tag{4.9.14}$$

with $a(0) = a'(0) = 0$. Taking the Laplace transform of (4.9.14),

$$A(s) = \frac{1}{s(s-1)(s-2)} \tag{4.9.15}$$

or

$$a(t) = \tfrac{1}{2} + \tfrac{1}{2}e^{2t} - e^t. \tag{4.9.16}$$

Note that $a'(t) = g(t)$.

• **Example 4.9.2**

There is an old joke about a man who took his car into a garage because of a terrible knocking sound. Upon his arrival the mechanic took one look at it and gave it a hefty kick.[24] Then, without a moment's hesitation he opened the hood, bent over, and tightened up a loose bolt. Turning to the owner, he said, "Your car is fine. That'll be $50." The owner felt that the charge was somewhat excessive, and demanded an itemized account. The mechanic said, "The kicking of the car and tightening one bolt, cost you a buck. The remaining $49 comes from knowing where to kick the car and finding the loose bolt."

Although the moral of the story may be about expertise as a marketable commodity, it also illustrates the concept of transfer function.[25] Let us model the car as a linear system where the equation

$$a_n \frac{d^n y}{dt^n} + a_{n-1} \frac{d^{n-1} y}{dt^{n-1}} + \cdots + a_1 \frac{dy}{dt} + a_0 y = f(t) \tag{4.9.17}$$

governs the response $y(t)$ to a forcing $f(t)$. Assuming that the car has been sitting still, the initial conditions are zero and the Laplace transform of (4.9.17) is

$$K(s)Y(s) = F(s), \tag{4.9.18}$$

where

$$K(s) = a_n s^n + a_{n-1} s^{n-1} + \cdots + a_1 s + a_0. \tag{4.9.19}$$

Hence

$$Y(s) = \frac{F(s)}{K(s)} = G(s)F(s), \tag{4.9.20}$$

where the transfer function $G(s)$ clearly depends only on the internal workings of the car. So if we know the transfer function, we understand how the car vibrates because

$$y(t) = \int_0^t g(t-x) f(x)\, dx. \tag{4.9.21}$$

But what does this have to do with our mechanic? He realized that a short sharp kick mimics an impulse forcing with $f(t) = \delta(t)$ and

[24] This is obviously a very old joke.

[25] Originally suggested by Stern, M. D., 1987: Why the mechanic kicked the car – A teaching aid for transfer functions. *Math. Gaz.*, **71**, 62–64.

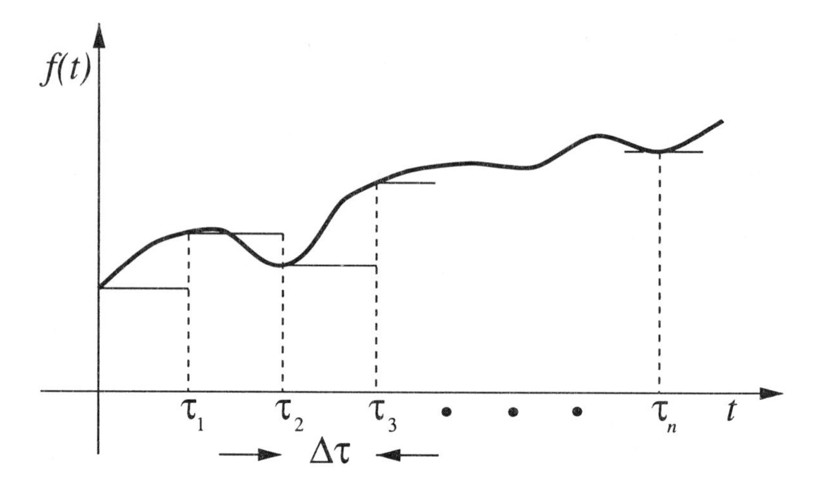

Figure 4.9.1: Diagram used in the derivation of Duhamel's integral.

$y(t) = g(t)$. Therefore, by observing the response of the car to his kick, he diagnosed the loose bolt and fixed the car.

In this section we have shown how the response of any system may be expressed in terms of its Green's function and the arbitrary forcing. Can we also determine the response using the indicial admittance $a(t)$?

Consider first a system that is dormant until a certain time $t = \tau_1$. At that instant we subject the system to a forcing $H(t - \tau_1)$. Then the response will be zero if $t < \tau_1$ and will equal the indicial admittance $a(t - \tau_1)$ when $t > \tau_1$ because the indicial admittance is the response of a system to the step function. Here $t - \tau_1$ is the time measured from the instant of change.

Next, suppose that we now force the system with the value $f(0)$ when $t = 0$ and hold that value until $t = \tau_1$. We then abruptly change the forcing by an amount $f(\tau_1) - f(0)$ to the value $f(\tau_1)$ at the time τ_1 and hold it at that value until $t = \tau_2$. Then we again abruptly change the forcing by an amount $f(\tau_2) - f(\tau_1)$ at the time τ_2, and so forth (see Figure 4.9.1). From the *linearity* of the problem the response after the instant $t = \tau_n$ equals the sum

$$\begin{aligned} y(t) = f(0)a(t) &+ [f(\tau_1) - f(0)]a(t - \tau_1) + [f(\tau_2) - f(\tau_1)]a(t - \tau_2) \\ &+ \cdots + [f(\tau_n) - f(\tau_{n-1})]a(t - \tau_n). \end{aligned} \tag{4.9.22}$$

If we write $f(\tau_k) - f(\tau_{k-1}) = \Delta f_k$ and $\tau_k - \tau_{k-1} = \Delta\tau_k$, (4.9.22) becomes

$$y(t) = f(0)a(t) + \sum_{k=1}^{n} a(t - \tau_k) \frac{\Delta f_k}{\Delta \tau_k}\, \Delta\tau_k. \tag{4.9.23}$$

Finally, proceeding to the limit as the number n of jumps becomes infinite, in such a manner that all jumps and intervals between successive jumps tend to zero, this sum has the limit

$$y(t) = f(0)a(t) + \int_0^t f'(\tau)a(t-\tau)\, d\tau. \tag{4.9.24}$$

Because the total response of the system equals the weighted sum [the weights being $a(t)$] of the forcing from the initial moment up to the time t, we refer to (4.9.24) as the *superposition integral*, or *Duhamel's integral*.[26]

We can also express (4.9.24) in several different forms. Integration by parts yields

$$y(t) = f(t)a(0) + \int_0^t f(\tau)a'(t-\tau)\, d\tau \tag{4.9.25}$$

$$= \frac{d}{dt}\left[\int_0^t f(\tau)a(t-\tau)\, d\tau\right]. \tag{4.9.26}$$

• **Example 4.9.3**

Suppose that a system has the step response of $a(t) = A[1-e^{-t/T}]$, where A and T are positive constants. Let us find the response if we force this system by $f(t) = kt$, where k is a constant.

From the superposition integral (4.9.24),

$$y(t) = 0 + \int_0^t kA[1 - e^{-(t-\tau)/T}]\, d\tau \tag{4.9.27}$$

$$= kA[t - T(1 - e^{-t/T})]. \tag{4.9.28}$$

Problems

For the following nonhomogeneous differential equations, find the transfer function, impulse response, and step response. Assume that all of the necessary initial conditions are zero.

1. $y' + ky = f(t)$
2. $y'' - 2y' - 3y = f(t)$

[26] Duhamel, J.-M.-C., 1833: Mémoire sur la méthode générale relative au mouvement de la chaleur dans les corps solides plongés dans des milieux dont la température varie avec le temps, *J. École Polytech.*, **22**, 20–77.

3. $y'' + 4y' + 3y = f(t)$

4. $y'' - 2y' + 5y = f(t)$

5. $y'' - 3y' + 2y = f(t)$

6. $y'' + 4y' + 4y = f(t)$

7. $y'' - 9y = f(t)$

8. $y'' + y = f(t)$

9. $y'' - y' = f(t)$

4.10 INVERSION BY CONTOUR INTEGRATION

In Sections 4.5 and 4.6 we showed how we may use partial fractions and convolution to find the inverse of the Laplace transform $F(s)$. In many instances these methods fail simply because of the complexity of the transform to be inverted. In this section we shall show how we may invert transforms through the powerful method of contour integration. Of course, the student must be proficient in the use of complex variables.

Consider the piece-wise differentiable function $f(x)$ which vanishes for $x < 0$. We can express the function $e^{-cx}f(x)$ by the complex Fourier representation of

$$f(x)e^{-cx} = \frac{1}{2\pi}\int_{-\infty}^{\infty} e^{i\omega x}\left[\int_0^{\infty} e^{-ct}f(t)e^{-i\omega t}\,dt\right]d\omega, \qquad \textbf{(4.10.1)}$$

for any value of the real constant c, where the integral

$$I = \int_0^{\infty} e^{-ct}|f(t)|\,dt \qquad \textbf{(4.10.2)}$$

exists. By multiplying both sides of (4.10.1) by e^{cx} and bringing it inside the first integral,

$$f(x) = \frac{1}{2\pi}\int_{-\infty}^{\infty} e^{(c+\omega i)x}\left[\int_0^{\infty} f(t)e^{-(c+\omega i)t}\,dt\right]d\omega. \qquad \textbf{(4.10.3)}$$

With the substitution $z = c + \omega i$, where z is a new, complex variable of integration,

$$f(x) = \frac{1}{2\pi i}\int_{c-\infty i}^{c+\infty i} e^{zx}\left[\int_0^{\infty} f(t)e^{-zt}\,dt\right]dz. \qquad \textbf{(4.10.4)}$$

The quantity inside the square brackets is the Laplace transform $F(z)$. Therefore, we can express $f(t)$ in terms of its transform by the complex contour integral:

$$f(t) = \frac{1}{2\pi i}\int_{c-\infty i}^{c+\infty i} F(z)e^{tz}\,dz. \qquad \textbf{(4.10.5)}$$

Figure 4.10.1: An outstanding mathematician at Cambridge University at the turn of the twentieth century, Thomas John I'Anson Bromwich (1875–1929) came to Heaviside's operational calculus through his interest in divergent series. Beginning a correspondence with Heaviside, Bromwich was able to justify operational calculus through the use of contour integrals by 1915. After his premature death, individuals such as J. R. Carson and Sir H. Jeffreys brought Laplace transforms to the increasing attention of scientists and engineers. (Portrait courtesy of the Royal Society of London.)

This line integral, *Bromwich's integral*,[27] runs along the line $x = c$ parallel to the imaginary axis and c units to the right of it, the so-called *Bromwich contour*. We select the value of c sufficiently large so that the integral (4.10.2) exists; subsequent analysis shows that this occurs when c is larger than the real part of any of the singularities of $F(z)$.

[27] Bromwich, T. J. I'A., 1916: Normal coordinates in dynamical systems. *Proc. London Math. Soc., Ser. 2*, **15**, 401–448.

We must now evaluate the contour integral. Because of the power of the *residue* theorem in complex variables, the contour integral is usually transformed into a closed contour through the use of *Jordan's lemma*. See Section 3.4, Equations (3.4.12) and (3.4.13). The following examples will illustrate the proper use of (4.10.5).

• Example 4.10.1

Let us invert

$$F(s) = \frac{e^{-3s}}{s^2(s-1)}. \tag{4.10.6}$$

From Bromwich's integral,

$$f(t) = \frac{1}{2\pi i}\int_{c-\infty i}^{c+\infty i} \frac{e^{(t-3)z}}{z^2(z-1)}\,dz \tag{4.10.7}$$

$$= \frac{1}{2\pi i}\oint_C \frac{e^{(t-3)z}}{z^2(z-1)}\,dz - \frac{1}{2\pi i}\int_{C_R} \frac{e^{(t-3)z}}{z^2(z-1)}\,dz, \tag{4.10.8}$$

where C_R is a semicircle of infinite radius in either the right or left half of the z-plane and C is the closed contour that includes C_R and Bromwich's contour. See Figure 4.10.2.

Our first task is to choose an appropriate contour so that the integral along C_R vanishes. By Jordan's lemma this requires a semicircle in the right half-plane if $t-3<0$ and a semicircle in the left half-plane if $t-3>0$. Consequently, by considering these two separate cases, we have forced the second integral in (4.10.8) to zero and the inversion simply equals the closed contour.

Consider the case $t<3$ first. Because Bromwich's contour lies to the right of any singularities, there are no singularities within the closed contour and $f(t)=0$.

Consider now the case $t>3$. Within the closed contour in the left half-plane, there is a second-order pole at $z=0$ and a simple pole at $z=1$. Therefore,

$$f(t) = \text{Res}\left[\frac{e^{(t-3)z}}{z^2(z-1)};0\right] + \text{Res}\left[\frac{e^{(t-3)z}}{z^2(z-1)};1\right], \tag{4.10.9}$$

where

$$\text{Res}\left[\frac{e^{(t-3)z}}{z^2(z-1)};0\right] = \lim_{z\to 0}\frac{d}{dz}\left[z^2\frac{e^{(t-3)z}}{z^2(z-1)}\right] \tag{4.10.10}$$

$$= \lim_{z\to 0}\left[\frac{(t-3)e^{(t-3)z}}{z-1} - \frac{e^{(t-3)z}}{(z-1)^2}\right] \tag{4.10.11}$$

$$= 2-t \tag{4.10.12}$$

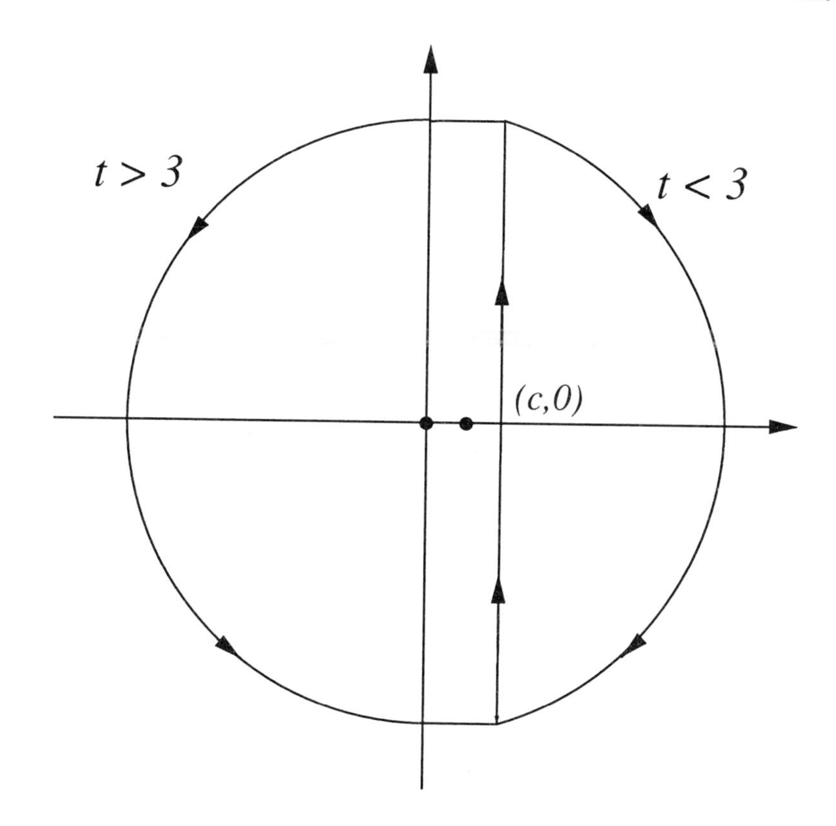

Figure 4.10.2: Contours used in the inversion of (4.10.6).

and

$$\operatorname{Res}\left[\frac{e^{(t-3)z}}{z^2(z-1)};1\right] = \lim_{z\to 1}\,(z-1)\frac{e^{(t-3)z}}{z^2(z-1)} = e^{t-3}. \tag{4.10.13}$$

Taking our earlier results into account, the inverse equals

$$f(t) = \left[e^{t-3} - (t-3) - 1\right] H(t-3) \tag{4.10.14}$$

which we would have obtained from the second shifting theorem and tables.

• **Example 4.10.2**

For our second example of the inversion of Laplace transforms by complex integration, let us find the inverse of

$$F(s) = \frac{1}{s\sinh(as)}, \tag{4.10.15}$$

where a is real. From Bromwich's integral,

$$f(t) = \frac{1}{2\pi i} \int_{c-\infty i}^{c+\infty i} \frac{e^{tz}}{z \sinh(az)}\, dz. \qquad \textbf{(4.10.16)}$$

Here c is greater than the real part of any of the singularities in (4.10.15). Using the infinite product for the hyperbolic sine,[28]

$$\frac{e^{tz}}{z \sinh(az)} = \frac{e^{tz}}{az^2[1 + a^2z^2/\pi^2][1 + a^2z^2/(4\pi^2)][1 + a^2z^2/(9\pi^2)]\cdots}. \qquad \textbf{(4.10.17)}$$

Thus, we have a second-order pole at $z = 0$ and simple poles at $z_n = \pm n\pi i/a$, where $n = 1, 2, 3, \ldots$

We may convert the line integral (4.10.16), with the Bromwich contour lying parallel and slightly to the right of the imaginary axis, into a closed contour using Jordan's lemma through the addition of an infinite semicircle joining $i\infty$ to $-i\infty$ as shown in Figure 4.10.3. We now apply the residue theorem. For the second-order pole at $z = 0$,

$$\operatorname{Res}\left[\frac{e^{tz}}{z \sinh(az)}; 0\right] = \frac{1}{1!} \lim_{z\to 0} \frac{d}{dz}\left[\frac{(z-0)^2 e^{tz}}{z \sinh(az)}\right] \qquad \textbf{(4.10.18)}$$

$$= \lim_{z\to 0} \frac{d}{dz}\left[\frac{z e^{tz}}{\sinh(az)}\right] \qquad \textbf{(4.10.19)}$$

$$= \lim_{z\to 0}\left[\frac{e^{tz}}{\sinh(az)} + \frac{zte^{tz}}{\sinh(az)} - \frac{az\cosh(az)e^{tz}}{\sinh^2(az)}\right] \qquad \textbf{(4.10.20)}$$

$$= \frac{t}{a} \qquad \textbf{(4.10.21)}$$

after using $\sinh(az) = az + O(z^3)$. For the simple poles $z_n = \pm n\pi i/a$,

$$\operatorname{Res}\left[\frac{e^{tz}}{z \sinh(az)}; z_n\right] = \lim_{z\to z_n} \frac{(z - z_n)e^{tz}}{z \sinh(az)} \qquad \textbf{(4.10.22)}$$

$$= \lim_{z\to z_n} \frac{e^{tz}}{\sinh(az) + az\cosh(az)} \qquad \textbf{(4.10.23)}$$

$$= \frac{\exp(\pm n\pi i t/a)}{(-1)^n(\pm n\pi i)}, \qquad \textbf{(4.10.24)}$$

[28] Gradshteyn, I. S. and Ryzhik, I. M., 1965: *Table of Integrals, Series and Products*, Academic Press, New York. See Section 1.431, formula 2.

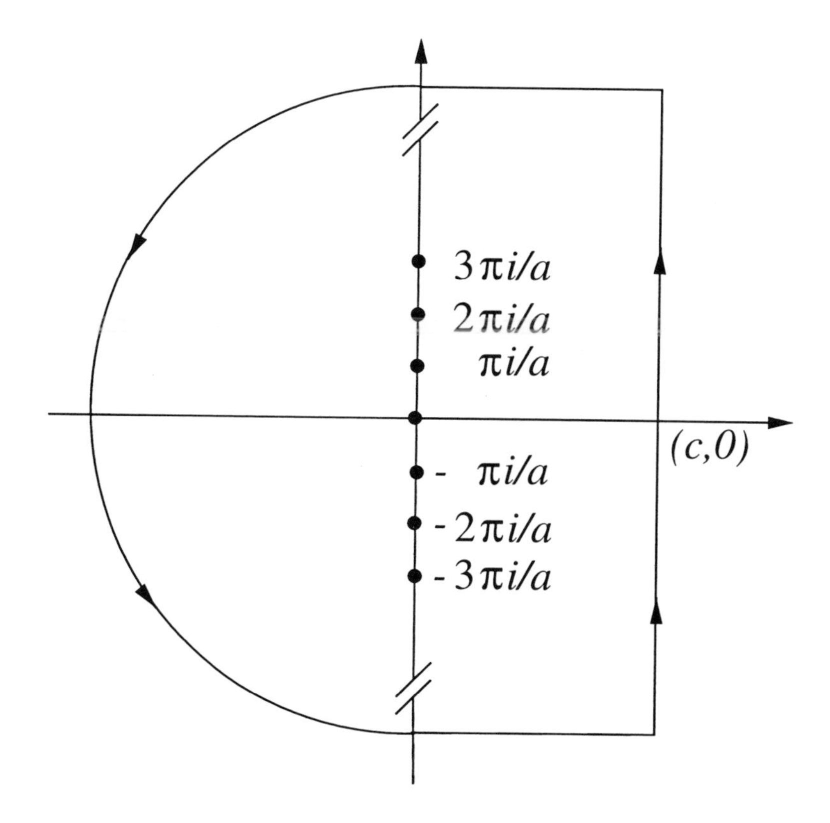

Figure 4.10.3: Contours used in the inversion of (4.10.15).

because $\cosh(\pm n\pi i) = \cos(n\pi) = (-1)^n$. Thus, summing up all of the residues gives

$$f(t) = \frac{t}{a} + \sum_{n=1}^{\infty} \frac{(-1)^n \exp(n\pi i t/a)}{n\pi i} - \sum_{n=1}^{\infty} \frac{(-1)^n \exp(-n\pi i t/a)}{n\pi i} \tag{4.10.25}$$

$$= \frac{t}{a} + \frac{2}{\pi} \sum_{n=1}^{\infty} \frac{(-1)^n}{n} \sin(n\pi t/a). \tag{4.10.26}$$

In addition to computing the inverse of Laplace transforms, Bromwich's integral places certain restrictions on $F(s)$ in order that an inverse exists. If α denotes the minimum value that c may possess, the restrictions are threefold.[29] First, $F(z)$ must be analytic in the half-plane $x \geq \alpha$, where $z = x + iy$. Second, in the same half-plane it must behave as z^{-k}, where $k > 1$. Finally, $F(x)$ must be real when $x \geq \alpha$.

[29] For the proof, see Churchill, R. V., 1972: *Operational Mathematics*, McGraw-Hill, New York, Section 67.

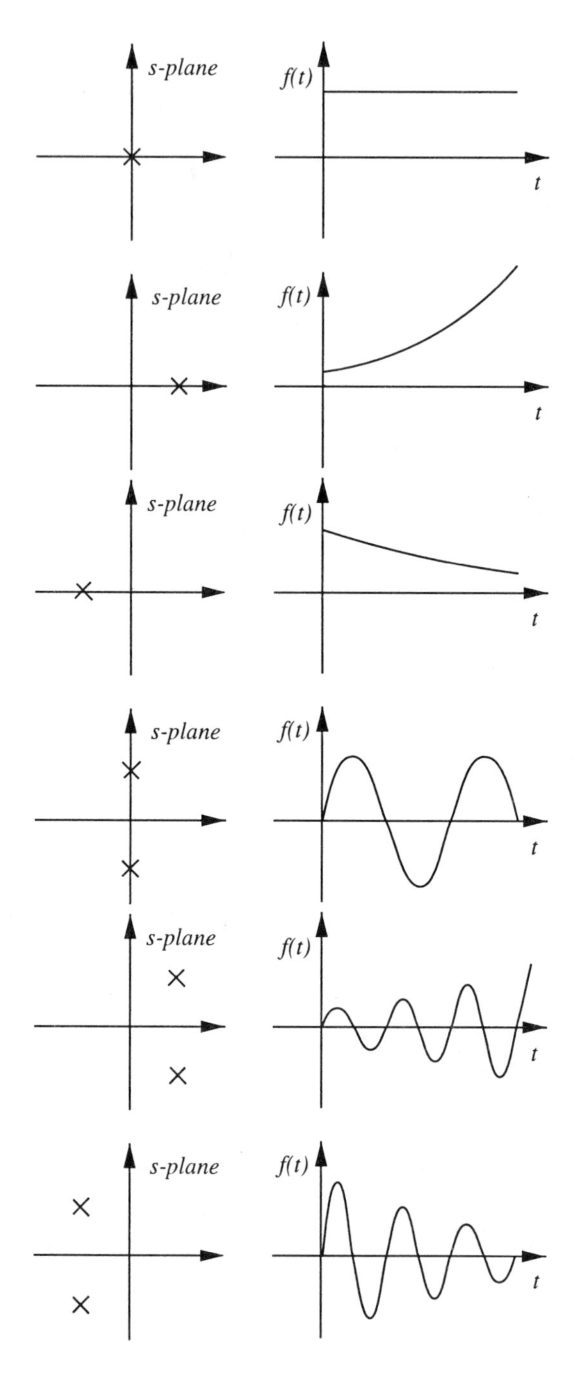

Figure 4.10.4: The correspondence between the location of the simple poles of the Laplace transform $F(s)$ and the behavior of $f(t)$.

• Example 4.10.3

Is the function $\sin(s)/(s^2+4)$ a proper Laplace transform? Although the function satisfies the first and third criteria listed in the previous paragraph on the half-plane $x > 2$, the function becomes unbounded as $y \to \pm\infty$ for any fixed $x > 2$. Thus, $\sin(s)/(s^2+4)$ cannot be a Laplace transform.

• Example 4.10.4

An additional benefit of understanding inversion by the residue method is the ability to qualitatively anticipate the inverse by knowing the location of the poles of $F(s)$. This intuition is important because many engineering analyses discuss stability and performance entirely in terms of the properties of the system's Laplace transform. In Figure 4.10.4 we have graphed the location of the poles of $F(s)$ and the corresponding $f(t)$. The student should go through the mental exercise of connecting the two pictures.

Problems

Use Bromwich's integral to invert the following Laplace transform:

1. $F(s) = \dfrac{s+1}{(s+2)^2(s+3)}$

2. $F(s) = \dfrac{1}{s^2(s+a)^2}$

3. $F(s) = \dfrac{1}{s(s-2)^3}$

4. $F(s) = \dfrac{1}{s(s+a)^2(s^2+b^2)}$

5. $F(s) = \dfrac{e^{-s}}{s^2(s+2)}$

6. $F(s) = \dfrac{1}{s(1+e^{-as})}$

7. $F(s) = \dfrac{1}{(s+b)\cosh(as)}$

8. $F(s) = \dfrac{1}{s(1-e^{-as})}$

9. Consider a function $f(t)$ which has the Laplace transform $F(z)$ which is analytic in the half-plane $\mathrm{Re}(z) > s_0$. Can we use this knowledge to find $g(t)$ whose Laplace transform $G(z)$ equals $F[\varphi(z)]$, where $\varphi(z)$ is also analytic for $\mathrm{Re}(z) > s_0$? The answer to this question leads to the Schouten[30] – Van der Pol[31] theorem.

---

[30] Schouten, J. P., 1935: A new theorem in operational calculus together with an application of it. *Physica*, **2**, 75–80.

[31] Van der Pol, B., 1934: A theorem on electrical networks with applications to filters. *Physica*, **1**, 521–530.

Step 1: Show that the following relationships hold true:

$$G(z) = F[\varphi(z)] = \int_0^\infty f(\tau) e^{-\varphi(z)\tau}\, d\tau$$

and

$$g(t) = \frac{1}{2\pi i} \int_{c-\infty i}^{c+\infty i} F[\varphi(z)] e^{tz}\, dz.$$

Step 2: Using the results from Step 1, show that

$$g(t) = \int_0^\infty f(\tau) \left[\frac{1}{2\pi i} \int_{c-\infty i}^{c+\infty i} e^{-\varphi(z)\tau} e^{tz}\, dz \right] d\tau.$$

This is the Schouten-Van der Pol theorem.

Step 3: If $G(z) = F(\sqrt{z}\,)$ show that

$$g(t) = \frac{1}{2\sqrt{\pi t^3}} \int_0^\infty \tau f(\tau) \exp\left(-\frac{\tau^2}{4t} \right) d\tau.$$

Hint: Do not evaluate the contour integral. Instead, ask yourself: What function of time has a Laplace transform that equals $e^{-\varphi(z)\tau}$, where τ is a parameter? Then use tables.

Chapter 5
The Z-Transform

Since the Second World War, the rise of digital technology has resulted in a corresponding demand for designing and understanding discrete-time (data sampled) systems. These systems are governed by *difference equations* in which members of the sequence y_n are coupled to each other.

One source of difference equations is the numerical evaluation of integrals on a digital computer. Because we can only have values at discrete time points $t_k = kT$ for $k = 0, 1, 2, \ldots$, the value of the integral $y(t) = \int_0^t f(\tau)\, d\tau$ is

$$y(kT) = \int_0^{kT} f(\tau)\, d\tau = \int_0^{(k-1)T} f(\tau)\, d\tau + \int_{(k-1)T}^{kT} f(\tau)\, d\tau \tag{5.0.1}$$

$$= y[(k-1)T] + \int_{(k-1)T}^{kT} f(\tau)\, d\tau \tag{5.0.2}$$

$$= y[(k-1)T] + Tf(kT), \tag{5.0.3}$$

because $\int_{(k-1)T}^{kT} f(\tau)\, d\tau \approx Tf(kT)$. Equation (5.0.3) is an example of a first-order difference equation because the numerical scheme couples the sequence value $y(kT)$ directly to the previous sequence value $y[(k-1)T]$. If (5.0.3) had contained $y[(k-2)T]$, then it would have been a second-order difference equation, and so forth.

Although we could use the conventional Laplace transform to solve these difference equations, the use of z-transforms can greatly facilitate the analysis, especially when we only desire responses at the sampling instants. Often the entire analysis can be done using only the transforms and the analyst does not actually find the sequence $y(kT)$.

In this chapter we shall first define the z-transform and discuss its properties. Then we will show how to find its inverse. Finally we shall use them to solve difference equations.

5.1 THE RELATIONSHIP OF THE Z-TRANSFORM TO THE LAPLACE TRANSFORM

Let $f(t)$ be a continuous function that an instrument samples every T units of time. We denote this data-sampled function by $f_S^*(t)$. See Figure 5.1.1. Taking ϵ, the duration of an individual sampling event, to be small, we may approximate the narrow-width pulse in Figure 5.1.1 by flat-topped pulses. Then $f_S^*(t)$ approximately equals

$$f_S^*(t) \approx \frac{1}{\epsilon} \sum_{n=0}^{\infty} f(nT) \left[H(t - nT + \epsilon/2) - H(t - nT - \epsilon/2) \right] \tag{5.1.1}$$

if $\epsilon \ll T$.

Clearly the presence of ϵ is troublesome in (5.1.1); it adds one more parameter to our problem. For this reason we introduce the concept of the *ideal sampler*, where the sampling time becomes infinitesimally small so that

$$f_S(t) = \lim_{\epsilon \to 0} \sum_{n=0}^{\infty} f(nT) \left[\frac{H(t - nT + \epsilon/2) - H(t - nT - \epsilon/2)}{\epsilon} \right] \tag{5.1.2}$$

$$= \sum_{n=0}^{\infty} f(nT) \delta(t - nT). \tag{5.1.3}$$

Let us now find the Laplace transform of this data-sampled function. We find from the linearity property of Laplace transforms that

$$F_S(s) = \mathcal{L}[f_S(t)] = \mathcal{L}\left[\sum_{n=0}^{\infty} f(nT) \delta(t - nT) \right] \tag{5.1.4}$$

$$= \sum_{n=0}^{\infty} f(nT) \mathcal{L}[\delta(t - nT)]. \tag{5.1.5}$$

Because $\mathcal{L}[\delta(t - nT)] = e^{-nsT}$, (5.1.5) simplifies to

$$F_S(s) = \sum_{n=0}^{\infty} f(nT) e^{-nsT}. \tag{5.1.6}$$

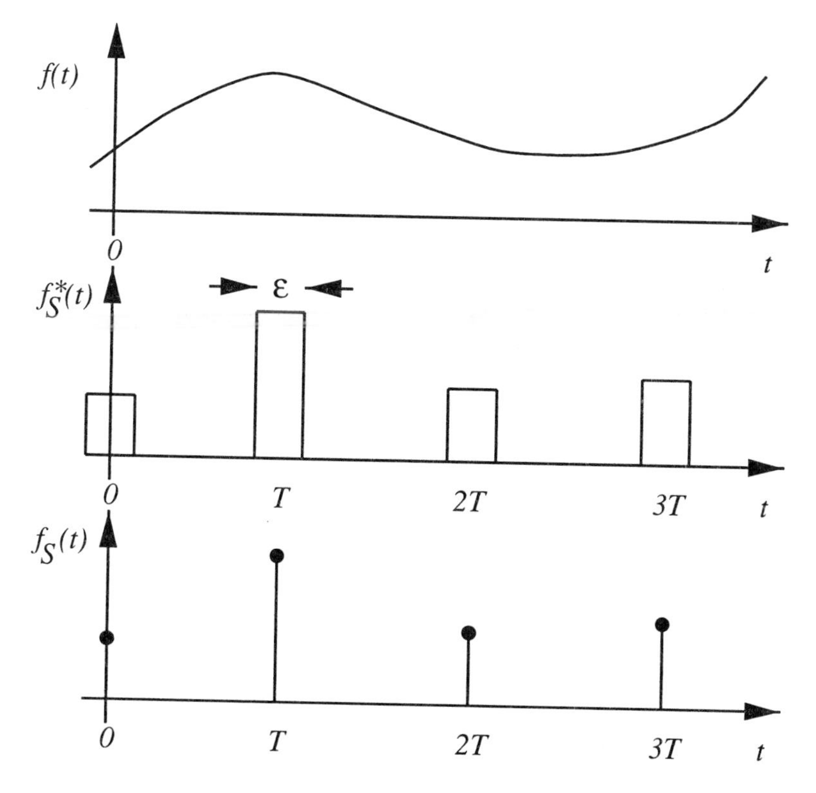

Figure 5.1.1: Schematic of how a continuous function $f(t)$ is sampled by a narrow-width pulse sampler $f_S^*(t)$ and an ideal sampler $f_S(t)$.

If we now make the substitution that $z = e^{sT}$, then $F_S(s)$ becomes

$$F(z) = \mathcal{Z}(f_n) = \sum_{n=0}^{\infty} f_n z^{-n}, \tag{5.1.7}$$

where $F(z)$ is the one-sided z-transform[1] of the sequence $f(nT)$, which we shall denote from now on by f_n. Here $\mathcal{Z}$ denotes the operation of taking the z-transform while $\mathcal{Z}^{-1}$ represents the inverse z-transformation. We will consider methods for finding the inverse z-transform in Section 5.3.

---

[1] The standard reference is Jury, E. I., 1964: *Theory and Application of the z-Transform Method*, John Wiley & Sons, New York.

Just as the Laplace transform was defined by an integration in t, the z-transform is defined by a power series (Laurent series) in z. Consequently, every z-transform has a region of convergence which must be implicitly understood if not explicitly stated. Furthermore, just as the Laplace integral diverged for certain functions, there are sequences where the associated power series will diverge and its z-transform does not exist.

Consider now the following examples of how to find the z-transform.

• **Example 5.1.1**

Given the unit sequence $f_n = 1$, $n \geq 0$, let us find $F(z)$. Substituting f_n into the definition of the z-transform leads to

$$F(z) = \sum_{n=0}^{\infty} z^{-n} = \frac{z}{z-1}, \tag{5.1.8}$$

because $\sum_{n=0}^{\infty} z^{-n}$ is a complex-valued *geometric series* with common ratio z^{-1}. This series converges if $|z^{-1}| < 1$ or $|z| > 1$, which gives the region of convergence of $F(z)$.

• **Example 5.1.2**

Let us find the z-transform of the sequence

$$f_n = e^{-anT}, \qquad n \geq 0, \tag{5.1.9}$$

for a real and a imaginary.

For a real, substitution of the sequence into the definition of the z-transform yields

$$F(z) = \sum_{n=0}^{\infty} e^{-anT} z^{-n} = \sum_{n=0}^{\infty} \left(e^{-aT} z^{-1}\right)^n. \tag{5.1.10}$$

If $u = e^{-aT} z^{-1}$, then (5.1.10) is a geometric series so that

$$F(z) = \sum_{n=0}^{\infty} u^n = \frac{1}{1-u}. \tag{5.1.11}$$

Because $|u| = e^{-aT}|z^{-1}|$, the condition for convergence is that $|z| > e^{-aT}$. Thus,

$$F(z) = \frac{z}{z - e^{-aT}}, \qquad |z| > e^{-aT}. \tag{5.1.12}$$

For imaginary a, the infinite series in (5.1.10) converges if $|z| > 1$, because $|u| = |z^{-1}|$ when a is imaginary. Thus,

$$F(z) = \frac{z}{z - e^{-aT}}, \qquad |z| > 1. \tag{5.1.13}$$

Although the z-transforms in (5.1.12) and (5.1.13) are the same in these two cases, the corresponding regions of convergence are different. If a is a complex number, then

$$F(z) = \frac{z}{z - e^{-aT}}, \qquad |z| > |e^{-aT}|. \tag{5.1.14}$$

• **Example 5.1.3**

Let us find the z-transform of the sinusoidal sequence

$$f_n = \cos(n\omega T), \qquad n \geq 0. \tag{5.1.15}$$

Substituting (5.1.15) into the definition of the z-transform results in

$$F(z) = \sum_{n=0}^{\infty} \cos(n\omega T) z^{-n}. \tag{5.1.16}$$

From Euler's formula,

$$\cos(n\omega T) = \tfrac{1}{2}(e^{in\omega T} + e^{-in\omega T}), \tag{5.1.17}$$

so that (5.1.16) becomes

$$F(z) = \frac{1}{2} \sum_{n=0}^{\infty} \left(e^{in\omega T} z^{-n} + e^{-in\omega T} z^{-n} \right) \tag{5.1.18}$$

or

$$F(z) = \tfrac{1}{2}\left[\mathcal{Z}(e^{in\omega T}) + \mathcal{Z}(e^{-in\omega T})\right]. \tag{5.1.19}$$

From (5.1.13),

$$\mathcal{Z}(e^{\pm in\omega T}) = \frac{z}{z - e^{\pm i\omega T}}, \qquad |z| > 1. \tag{5.1.20}$$

Substituting (5.1.20) into (5.1.19) and simplifying yields

$$F(z) = \frac{z[z - \cos(\omega T)]}{z^2 - 2z\cos(\omega T) + 1}, \qquad |z| > 1. \tag{5.1.21}$$

Table 5.1.1: Z-Transforms of Some Commonly Used Sequences.

	$\mathbf{f_n,\ n \geq 0}$	$\mathbf{F(z)}$	**Region of convergence**
1.	$f_0 = k = \text{const.}$ $f_n = 0,\ n \geq 1$	k	$\|z\| > 0$
2.	$f_m = k = \text{const.}$ $f_n = 0$, all other n's	kz^{-m}	$\|z\| > 0$
3.	$k = \text{constant}$	$kz/(z-1)$	$\|z\| > 1$
4.	kn	$kz/(z-1)^2$	$\|z\| > 1$
5.	kn^2	$kz(z+1)/(z-1)^3$	$\|z\| > 1$
6.	ke^{-anT}, a complex	$kz/\left(z-e^{-aT}\right)$	$\|z\| > \|e^{-aT}\|$
7.	kne^{-anT}, a complex	$\frac{kze^{-aT}}{(z-e^{-aT})^2}$	$\|z\| > \|e^{-aT}\|$
8.	$\sin(\omega_0 nT)$	$\frac{z\sin(\omega_0 T)}{z^2-2z\cos(\omega_0 T)+1}$	$\|z\| > 1$
9.	$\cos(\omega_0 nT)$	$\frac{z[z-\cos(\omega_0 T)]}{z^2-2z\cos(\omega_0 T)+1}$	$\|z\| > 1$
10.	$e^{-anT}\sin(\omega_0 nT)$	$\frac{ze^{-aT}\sin(\omega_0 T)}{z^2-2ze^{-aT}\cos(\omega_0 T)+e^{-2aT}}$	$\|z\| > e^{-aT}$
11.	$e^{-anT}\cos(\omega_0 nT)$	$\frac{ze^{-aT}[ze^{aT}-\cos(\omega_0 T)]}{z^2-2ze^{-aT}\cos(\omega_0 T)+e^{-2aT}}$	$\|z\| > e^{-aT}$
12.	α^n, α constant	$z/(z-\alpha)$	$\|z\| > \alpha$
13.	$n\alpha^n$	$\alpha z/(z-\alpha)^2$	$\|z\| > \alpha$
14.	$n^2\alpha^n$	$\alpha z(z+\alpha)/(z-\alpha)^3$	$\|z\| > \alpha$
15.	$\sinh(\omega_0 nT)$	$\frac{z\sinh(\omega_0 T)}{z^2-2z\cosh(\omega_0 T)+1}$	$\|z\| > \cosh(\omega_0 T)$
16.	$\cosh(\omega_0 nT)$	$\frac{z[z-\cosh(\omega_0 T)]}{z^2-2z\cosh(\omega_0 T)+1}$	$\|z\| > \sinh(\omega_0 T)$
17.	$a^n/n!$	$e^{a/z}$	$\|z\| > 0$
18.	$[\ln(a)]^n/n!$	$a^{1/z}$	$\|z\| > 0$

• **Example 5.1.4**

Let us find the z-transform for the sequence

$$f_n = \begin{cases} 1, & 0 \le n \le 5 \\ (\frac{1}{2})^n, & n \ge 6. \end{cases} \tag{5.1.22}$$

From the definition of the z-transform,

$$\mathcal{Z}(f_n) = F(z) = \sum_{n=0}^{5} z^{-n} + \sum_{n=6}^{\infty} \left(\frac{1}{2z}\right)^n. \tag{5.1.23}$$

Because

$$\sum_{n=0}^{N} q^n = \frac{1 - q^{N+1}}{1 - q}, \tag{5.1.24}$$

$$F(z) = \frac{1 - z^{-6}}{1 - z^{-1}} + \left(\frac{1}{2z}\right)^6 \sum_{m=0}^{\infty} \left(\frac{1}{2z}\right)^m \tag{5.1.25}$$

$$= \frac{z^6 - 1}{z^6 - z^5} + \left(\frac{1}{2z}\right)^6 \frac{1}{1 - \frac{1}{2z}} \tag{5.1.26}$$

$$= \frac{z^6 - 1}{z^6 - z^5} + \frac{1}{(2z)^6 - (2z)^5}, \tag{5.1.27}$$

if $n = m + 6$ and $|z| > 1/2$. We summarize some of the more commonly encountered sequences and their transforms in Table 5.1.1 along with their regions of convergence.

• **Example 5.1.5**

In many engineering studies, the analysis is done entirely using transforms without actually finding any inverses. Consequently, it is useful to compare and contrast how various transforms behave in very simple test problems.

Consider the simple time function $f(t) = ae^{-at}H(t)$, $a > 0$. Its Laplace and Fourier transform are identical, namely $a/(a + i\omega)$, if we set $s = i\omega$. In Figure 5.1.2 we have illustrated its behavior as a function of positive ω.

Let us now generate the sequence of observations that we would measure if we sampled $f(t)$ every T units of time apart: $f_n = ae^{-anT}$. Taking the z-transform of this sequence, it equals $az/\left(z - e^{-aT}\right)$. Recalling that $z = e^{sT} = e^{i\omega T}$, we can also plot this transform as a function of positive ω. For small ω, the transforms agree, but as ω becomes larger they diverge markedly. Why does this occur?

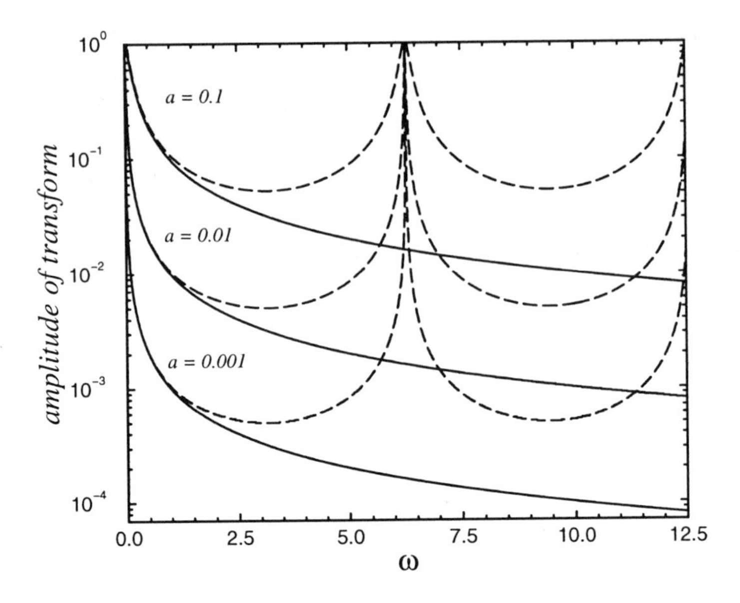

Figure 5.1.2: The amplitude of the Laplace or Fourier transform (solid line) for $ae^{-at}H(t)$ and the z-transform (dashed line) for $f_n = ae^{-anT}$ as a function of frequency ω for various positive a's and $T = 1$.

Recall that the z-transform is computed from a sequence comprised of samples from a continuous signal. One very important flaw in sampled data is the possible misrepresentation of high-frequency effects as lower-frequency phenomena. It is this *aliasing* or *folding* effect that we are observing here. Consequently, the z-transform of a sampled record can differ markedly from the corresponding Laplace or Fourier transforms of the continuous record at frequencies above one half of the sampling frequency. This also suggests that care should be exercised in interpolating between sampling instants. Indeed, in those applications where the output between sampling instants is very important, such as in a hybrid mixture of digital and analog systems, we must apply the so-called "modified z-transform".

Problems

From the fundamental definition of the z-transform, find the transform of the following sequences, where $n \geq 0$:

1. $f_n = \left(\frac{1}{2}\right)^n$

2. $f_n = e^{in\theta}$

3. $f_n = \begin{cases} 1, & 0 \le n \le 5 \\ 0, & n > 5 \end{cases}$

4. $f_n = \begin{cases} \left(\frac{1}{2}\right)^n, & n = 0, 1, \ldots, 10 \\ \left(\frac{1}{4}\right)^n, & n \ge 11 \end{cases}$

5. $f_n = \begin{cases} 0, & n = 0 \\ -1, & n = 1 \\ a^n, & n \ge 2 \end{cases}$

5.2 SOME USEFUL PROPERTIES

In principle we could construct any desired transform from the definition of the z-transform. However, there are several general theorems that are much more effective in finding new transforms.

Linearity

From the definition of the z-transform, it immediately follows that

$$\text{if} \quad h_n = c_1 f_n + c_2 g_n, \quad \text{then} \quad H(z) = c_1 F(z) + c_2 G(z), \tag{5.2.1}$$

where $F(z) = \mathcal{Z}(f_n)$, $G(z) = \mathcal{Z}(g_n)$, $H(z) = \mathcal{Z}(h_n)$, and c_1, c_2 are arbitrary constants.

Multiplication by an exponential sequence

$$\text{If} \quad g_n = e^{-anT} f_n, n \ge 0, \quad \text{then} \quad G(z) = F(ze^{aT}). \tag{5.2.2}$$

This follows from

$$G(z) = \mathcal{Z}(g_n) = \sum_{n=0}^{\infty} g_n z^{-n} = \sum_{n=0}^{\infty} e^{-anT} f_n z^{-n} \tag{5.2.3}$$

$$= \sum_{n=0}^{\infty} f_n (ze^{aT})^{-n} = F(ze^{aT}). \tag{5.2.4}$$

This is the z-transform analog to the first shifting theorem in Laplace transforms.

Table 5.2.1: Examples of Shifting Involving Sequences.

n	f_n	f_{n-2}	f_{n+2}
0	1	0	4
1	2	0	8
2	4	1	16
3	8	2	64
4	16	4	128
$\vdots$	$\vdots$	$\vdots$	$\vdots$

Shifting

The effect of shifting depends upon whether it is to the right or to the left, as Table 5.2.1 illustrates. For the sequence f_{n-2}, no values from the sequence f_n are lost; thus, we anticipate that the z-transform of f_{n-2} only involves $F(z)$. However, in forming the sequence f_{n+2}, the first two values of f_n are lost, and we anticipate that the z-transform of f_{n+2} cannot be expressed solely in terms of $F(z)$ but must include those two lost pieces of information.

Let us now confirm these conjectures by finding the z-transform of f_{n+1} which is a sequence that has been shifted one step to the left. From the definition of the z-transform, it follows that

$$\mathcal{Z}(f_{n+1}) = \sum_{n=0}^{\infty} f_{n+1} z^{-n} = z \sum_{n=0}^{\infty} f_{n+1} z^{-(n+1)} \tag{5.2.5}$$

$$= z \sum_{k=1}^{\infty} f_k z^{-k} - z f_0 + z f_0, \tag{5.2.6}$$

where we have added zero in (5.2.6). This algebraic trick allows us to collapse the first two terms on the right side of (5.2.6) to

$$\mathcal{Z}(f_{n+1}) = zF(z) - zf_0. \tag{5.2.7}$$

In a similar manner, repeated applications of (5.2.7) yield

$$\mathcal{Z}(f_{n+m}) = z^m F(z) - z^m f_0 - z^{m-1} f_1 - \ldots - z f_{m-1}, \tag{5.2.8}$$

where $m > 0$. This shifting operation transforms f_{n+m} into an algebraic expression involving m. Furthermore, we have introduced initial sequence values, just as we introduced initial conditions when we took the Laplace transform of the nth derivative of $f(t)$. We will make frequent use of this property in solving difference equations in Section 5.4.

Consider now shifting to the right by the positive integer k,

$$g_n = f_{n-k} H_{n-k}, \quad n \geq 0, \tag{5.2.9}$$

where $H_{n-k} = 0$ for $n < k$ and 1 for $n \geq k$. Then the z-transform of (5.2.9) is

$$G(z) = z^{-k} F(z), \tag{5.2.10}$$

where $G(z) = \mathcal{Z}(g_n)$ and $F(z) = \mathcal{Z}(f_n)$. This follows from

$$G(z) = \sum_{n=0}^{\infty} g_n z^{-n} = \sum_{n=0}^{\infty} f_{n-k} H_{n-k} z^{-n} \tag{5.2.11}$$

$$= z^{-k} \sum_{n=k}^{\infty} f_{n-k} z^{-(n-k)} = z^{-k} \sum_{m=0}^{\infty} f_m z^{-m} \tag{5.2.12}$$

$$= z^{-k} F(z). \tag{5.2.13}$$

This result is the z-transform analog to the second shifting theorem in Laplace transforms.

Initial-value theorem

The initial value of the sequence f_n, f_0, can be computed from $F(z)$ using the initial-value theorem:

$$f_0 = \lim_{z \to \infty} F(z). \tag{5.2.14}$$

From the definition of the z-transform,

$$F(z) = \sum_{n=0}^{\infty} f_n z^{-n} = f_0 + f_1 z^{-1} + f_2 z^{-2} + \ldots \tag{5.2.15}$$

In the limit of $z \to \infty$, we obtain the desired result.

Final-value theorem

The value of f_n, as $n \to \infty$, is given by the final-value theorem:

$$f_\infty = \lim_{z \to 1} (z-1)F(z), \tag{5.2.16}$$

where $F(z)$ is the z-transform of f_n.

We begin by noting that

$$\mathcal{Z}(f_{n+1} - f_n) = \lim_{n \to \infty} \sum_{k=0}^{n} (f_{k+1} - f_k) z^{-k}. \tag{5.2.17}$$

Using the shifting theorem on the left side of (5.2.17),

$$zF(z) - zf_0 - F(z) = \lim_{n \to \infty} \sum_{k=0}^{n} (f_{k+1} - f_k) z^{-k}. \tag{5.2.18}$$

Applying the limit as z approaches 1 to both sides of (5.2.18):

$$\lim_{z \to 1} (z-1)F(z) - f_0 = \lim_{n \to \infty} \sum_{k=0}^{n} (f_{k+1} - f_k) \tag{5.2.19}$$

$$= \lim_{n \to \infty} \big[(f_1 - f_0) + (f_2 - f_1) + \ldots + (f_n - f_{n-1}) + (f_{n+1} - f_n) + \ldots\big] \tag{5.2.20}$$

$$= \lim_{n \to \infty} (-f_0 + f_{n+1}) \tag{5.2.21}$$

$$= -f_0 + f_\infty. \tag{5.2.22}$$

Consequently,

$$f_\infty = \lim_{z \to 1} (z-1)F(z). \tag{5.2.23}$$

Note that this limit has meaning only if f_∞ exists. This occurs if $F(z)$ has no second-order or higher poles on the unit circle and no poles outside the unit circle.

Multiplication by n

Given

$$g_n = nf_n, \qquad n \ge 0, \tag{5.2.24}$$

this theorem states that

$$G(z) = -z\frac{dF(z)}{dz}, \tag{5.2.25}$$

where $G(z) = \mathcal{Z}(g_n)$ and $F(z) = \mathcal{Z}(f_n)$.

This follows from

$$G(z) = \sum_{n=0}^{\infty} g_n z^{-n} = \sum_{n=0}^{\infty} n f_n z^{-n} = z\sum_{n=0}^{\infty} n f_n z^{-n-1} = -z\frac{dF(z)}{dz}. \tag{5.2.26}$$

Periodic sequence theorem

Consider the N-periodic sequence:

$$f_n = \{\underbrace{f_0 f_1 f_2 \ldots f_{N-1}}_{\text{first period}} f_0 f_1 \ldots\} \tag{5.2.27}$$

and the related sequence:

$$x_n = \begin{cases} f_n, & 0 \le n \le N-1 \\ 0, & n \ge N. \end{cases} \tag{5.2.28}$$

This theorem allows us to find the z-transform of f_n if we can find the z-transform of x_n via the relationship

$$F(z) = \frac{X(z)}{1 - z^{-N}}, \qquad |z^N| > 1, \tag{5.2.29}$$

where $X(z) = \mathcal{Z}(x_n)$.

This follows from

$$F(z) = \sum_{n=0}^{\infty} f_n z^{-n} \tag{5.2.30}$$

$$= \sum_{n=0}^{N-1} x_n z^{-n} + \sum_{n=N}^{2N-1} x_{n-N} z^{-n} + \sum_{n=2N}^{3N-1} x_{n-2N} z^{-n} + \ldots \tag{5.2.31}$$

Application of the shifting theorem in (5.2.31) leads to

$$F(z) = X(z) + z^{-N}X(z) + z^{-2N}X(z) + \ldots \tag{5.2.32}$$

$$= X(z)\left[1 + z^{-N} + z^{-2N} + \ldots\right]. \tag{5.2.33}$$

Equation (5.2.33) contains an infinite geometric series with common ratio z^{-N}, which converges if $|z^{-N}| < 1$. Thus,

$$F(z) = \frac{X(z)}{1 - z^{-N}}, \qquad |z^N| > 1. \tag{5.2.34}$$

Convolution

Given the sequences f_n and g_n, the convolution product of these two sequences is

$$w_n = f_n * g_n = \sum_{k=0}^{n} f_k g_{n-k} = \sum_{k=0}^{n} f_{n-k} g_k. \tag{5.2.35}$$

Given $F(z)$ and $G(z)$, we then have that $W(z) = F(z)G(z)$.

This follows from

$$W(z) = \sum_{n=0}^{\infty} \left[\sum_{k=0}^{n} f_k g_{n-k} \right] z^{-n} = \sum_{n=0}^{\infty} \sum_{k=0}^{\infty} f_k g_{n-k} z^{-n}, \tag{5.2.36}$$

because $g_{n-k} = 0$ for $k > n$. Reversing the order of summation and letting $m = n - k$,

$$W(z) = \sum_{k=0}^{\infty} \sum_{m=-k}^{\infty} f_k g_m z^{-(m+k)} \tag{5.2.37}$$

$$= \left[\sum_{k=0}^{\infty} f_k z^{-k} \right] \left[\sum_{m=0}^{\infty} g_m z^{-m} \right] = F(z)G(z). \tag{5.2.38}$$

Consider now the following examples of the properties discussed in this section.

• Example 5.2.1

From

$$\mathcal{Z}(a^n) = \frac{1}{1 - az^{-1}} \tag{5.2.39}$$

for $n \geq 0$ and $|z| < a$, we have that

$$\mathcal{Z}\left(e^{inx}\right) = \frac{1}{1 - e^{ix} z^{-1}} \tag{5.2.40}$$

and

$$\mathcal{Z}\left(e^{-inx}\right) = \frac{1}{1 - e^{-ix}z^{-1}}, \tag{5.2.41}$$

if $n \geq 0$ and $|z| < 1$. Therefore, the sequence $f_n = \cos(nx)$ has the z-transform

$$F(z) = \mathcal{Z}[\cos(nx)] = \tfrac{1}{2}\mathcal{Z}\left(e^{inx}\right) + \tfrac{1}{2}\mathcal{Z}\left(e^{-inx}\right) \tag{5.2.42}$$

$$= \frac{1}{2}\frac{1}{1 - e^{ix}z^{-1}} + \frac{1}{2}\frac{1}{1 - e^{-ix}z^{-1}} = \frac{1 - \cos(x)z^{-1}}{1 - 2\cos(x)z^{-1} + z^{-2}}. \tag{5.2.43}$$

• **Example 5.2.2**

Using the z-transform,

$$\mathcal{Z}\left(a^n\right) = \frac{1}{1 - az^{-1}}, \qquad n \geq 0, \tag{5.2.44}$$

we find that

$$\mathcal{Z}\left(na^n\right) = -z\frac{d}{dz}\left[\left(1 - az^{-1}\right)^{-1}\right] \tag{5.2.45}$$

$$= (-z)(-1)\left(1 - az^{-1}\right)^{-2}(-a)(-1)z^{-2} \tag{5.2.46}$$

$$= \frac{az^{-1}}{(1 - az^{-1})^2} = \frac{az}{(z - a)^2}. \tag{5.2.47}$$

• **Example 5.2.3**

Consider $F(z) = 2az^{-1}/(1 - az^{-1})^3$, where $|a| < |z|$ and $|a| < 1$. Here we have that

$$f_0 = \lim_{z\to\infty} F(z) = \lim_{z\to\infty} \frac{2az^{-1}}{(1 - az^{-1})^3} = 0 \tag{5.2.48}$$

from the initial-value theorem. This agrees with the inverse of $X(z)$:

$$F(z) = \mathcal{Z}\left[n(n+1)a^n\right], \qquad n \geq 0. \tag{5.2.49}$$

• **Example 5.2.4**

Given the z-transform $F(z) = (1 - a)z/[(z - 1)(z - a)]$, where $|z| > 1 > a > 0$, then from the final-value theorem we have that

$$\lim_{n\to\infty} f_n = \lim_{z\to 1}(z - 1)F(z) = \lim_{z\to 1} \frac{1 - a}{1 - az^{-1}} = 1. \tag{5.2.50}$$

This is consistent with the inverse transform $f_n = 1 - a^n$ with $n \geq 0$.

- **Example 5.2.5**

Using the sequences $f_n = 1$ and $g_n = a^n$, where a is real, verify the convolution theorem.

We first compute the convolution of f_n with g_n, namely

$$w_n = f_n * g_n = \sum_{k=0}^{n} a^k = \frac{1}{1-a} - \frac{a^{n+1}}{1-a}. \tag{5.2.51}$$

Taking the z-transform of w_n,

$$W(z) = \frac{z}{(1-a)(z-1)} - \frac{az}{(1-a)(z-a)} = \frac{z^2}{(z-1)(z-a)} = F(z)G(z) \tag{5.2.52}$$

and convolution theorem holds true for this special case.

Problems

Use the properties and Table 5.1.1 to find the z-transform of the following sequences:

1. $f_n = nTe^{-anT}$

2. $f_n = \begin{cases} 0, & n = 0 \\ na^{n-1}, & n \geq 1 \end{cases}$

3. $f_n = \begin{cases} 0, & n = 0 \\ n^2 a^{n-1}, & n \geq 1 \end{cases}$

4. $f_n = a^n \cos(n)$
   [Use $\cos(n) = \frac{1}{2}(e^{in} + e^{-in})$]

5. $f_n = \cos(n-2)H_{n-2}$

6. $f_n = 3 + e^{-2nT}$

7. $f_n = \sin(n\omega_0 T + \theta)$

8. $f_n = \begin{cases} 0, & n = 0 \\ 1, & n = 1 \\ 2, & n = 2 \\ 1, & n = 3, \end{cases} \qquad f_{n+4} = f_n$

9. $f_n = (-1)^n$
   (Hint: It's periodic.)

10. Using the property stated in (5.2.24)–(5.2.25) *twice*, find the z-transform of $n^2 = n[n(1)^n]$.

11. Verify the convolution theorem using the sequences $f_n = g_n = 1$.

12. Verify the convolution theorem using the sequences $f_n = 1$ and $g_n = n$.

13. Verify the convolution theorem using the sequences $f_n = g_n = 1/(n!)$. [Hint: Use the binomial theorem with $x = 1$ to evaluate the summation.]

14. If a is a real, show that $\mathcal{Z}(a^n f_n) = F(z/a)$, where $\mathcal{Z}(f_n) = F(z)$.

5.3 INVERSE Z-TRANSFORMS

In the previous two sections we have dealt with finding the z-transform. In this section we find f_n by inverting the z-transform $F(z)$. There are four methods for finding the inverse: (1) power series, (2) recursion, (3) partial fractions, and (4) the residue method. We will discuss each technique individually. The first three apply only to those $F(z)$'s that are *rational* functions while the residue method is more general.

Power series

By means of the long-division process, we can always rewrite $F(z)$ as the Laurent expansion:

$$F(z) = a_0 + a_1 z^{-1} + a_2 z^{-2} + \dots \tag{5.3.1}$$

From the definition of the z-transform

$$F(z) = \sum_{n=0}^{\infty} f_n z^{-n} = f_0 + f_1 z^{-1} + f_2 z^{-2} + \dots, \tag{5.3.2}$$

the desired sequence f_n is given by a_n.

• Example 5.3.1

Let

$$F(z) = \frac{z+1}{2z-2} = \frac{N(z)}{D(z)}. \tag{5.3.3}$$

Using long division, $N(z)$ is divided by $D(z)$ and we obtain

$$F(z) = \tfrac{1}{2} + z^{-1} + z^{-2} + z^{-3} + z^{-4} + \dots \tag{5.3.4}$$

Therefore,

$$a_0 = \tfrac{1}{2},\ a_1 = 1,\ a_2 = 1,\ a_3 = 1,\ a_4 = 1,\ \text{etc.} \tag{5.3.5}$$

which suggests that $f_0 = \frac{1}{2}$ and $f_n = 1$ for $n \geq 1$ is the inverse of $F(z)$.

• **Example 5.3.2**

Let us find the inverse of the z-transform:

$$F(z) = \frac{2z^2 - 1.5z}{z^2 - 1.5z + 0.5}. \tag{5.3.6}$$

By the long-division process, we have that

$$\begin{array}{r|lllll}
 & 2 & +\,1.5z^{-1} & +\,1.25z^{-2} & +\,1.125z^{-3} & +\cdots \\ \hline
z^2 - 1.5z + 0.5 & 2z^2 - 1.5z & & & & \\
 & 2z^2 - 3z & +\,1 & & & \\ \cline{2-3}
 & 1.5z & -\,1 & & & \\
 & 1.5z & -\,2.25 & +\,0.75z^{-1} & & \\ \cline{2-4}
 & & 1.25 & -\,0.75z^{-1} & & \\
 & & 1.25 & -\,1.87z^{-1} & +\cdots & \\ \cline{3-5}
 & & & 1.125z^{-1} & +\cdots &
\end{array}$$

Thus, $f_0 = 2$, $f_1 = 1.5$, $f_2 = 1.25$, $f_3 = 1.125$, and so forth, or $f_n = 1 + (\frac{1}{2})^n$. In general, this technique only produces numerical values for some of the elements of the sequence. Note also that our long division must always yield the power series (5.3.1) in order for this method to be of any use.

Recursive method

An alternative to long division was suggested[2] several years ago. It obtains the inverse recursively.

We begin by assuming that the z-transform is of the form

$$F(z) = \frac{a_0 z^m + a_1 z^{m-1} + a_2 z^{m-2} + \cdots + a_{m-1} z + a_m}{b_0 z^m + b_1 z^{m-1} + b_2 z^{m-2} + \cdots + b_{m-1} z + b_m}, \tag{5.3.7}$$

where some of the coefficients a_i and b_i may be zero and $b_0 \neq 0$. Applying the final-value theorem,

$$f_0 = \lim_{z \to \infty} F(z) = a_0/b_0. \tag{5.3.8}$$

[2] Jury, E. I., 1964: *Theory and Application of the z-Transform Method,* John Wiley & Sons, New York, p. 41; Pierre, D. A., 1963: A tabular algorithm for z-transform inversion. *Control Eng.*, **10(9)**, 110–111. The present derivation is by Jenkins, L. B., 1967: A useful recursive form for obtaining inverse z-transforms. *Proc. IEEE*, **55**, 574–575. ©IEEE.

Next, we apply the final-value theorem to $z[F(z) - f_0]$ and find that

$$f_1 = \lim_{z\to\infty} z[F(z) - f_0] \tag{5.3.9}$$

$$= \lim_{z\to\infty} z \frac{(a_0 - b_0 f_0)z^m + (a_1 - b_1 f_0)z^{m-1} + \cdots + (a_m - b_m f_0)}{b_0 z^m + b_1 z^{m-1} + b_2 z^{m-2} + \cdots + b_{m-1} z + b_m} \tag{5.3.10}$$

$$= (a_1 - b_1 f_0)/b_0. \tag{5.3.11}$$

Note that the coefficient $a_0 - b_0 f_0 = 0$ from (5.3.8). Similarly,

$$f_2 = \lim_{z\to\infty} z[zF(z) - zf_0 - f_1] \tag{5.3.12}$$

$$= \lim_{z\to\infty} z \frac{(a_0 - b_0 f_0)z^{m+1} + (a_1 - b_1 f_0 - b_0 f_1)z^m + (a_2 - b_2 f_0 - b_1 f_1)z^{m-1} + \cdots - b_m f_1}{b_0 z^m + b_1 z^{m-1} + b_2 z^{m-2} + \cdots + b_{m-1} z + b_m} \tag{5.3.13}$$

$$= (a_2 - b_2 f_0 - b_1 f_1)/b_0 \tag{5.3.14}$$

because $a_0 - b_0 f_0 = a_1 - b_1 f_0 - f_1 b_0 = 0$. Continuing this process, we finally have that

$$f_n = (a_n - b_n f_0 - b_{n-1} f_1 - \cdots - b_1 f_{n-1})/b_0, \tag{5.3.15}$$

where $a_n = b_n \equiv 0$ for $n > m$.

- **Example 5.3.3**

Let us redo Example 5.3.2 using the recursive method. Comparing (5.3.7) to (5.3.6), $a_0 = 2$, $a_1 = -1.5$, $a_2 = 0$, $b_0 = 1$, $b_1 = -1.5$, $b_2 = 0.5$ and $a_n = b_n = 0$ if $n \geq 3$. From (5.3.15),

$$f_0 = a_0/b_0 = 2/1 = 2, \tag{5.3.16}$$

$$f_1 = (a_1 - b_1 f_0)/b_0 = [-1.5 - (-1.5)(2)]/1 = 1.5, \tag{5.3.17}$$

$$f_2 = (a_2 - b_2 f_0 - b_1 f_1)/b_0 \tag{5.3.18}$$

$$= [0 - (0.5)(2) - (-1.5)(1.5)]/1 = 1.25 \tag{5.3.19}$$

and

$$f_3 = (a_3 - b_3 f_0 - b_2 f_1 - b_1 f_2)/b_0 \tag{5.3.20}$$

$$= [0 - (0)(2) - (0.5)(1.5) - (-1.5)(1.25)]/1 = 1.125. \tag{5.3.21}$$

Partial fraction expansion

One of the popular methods for inverting Laplace transforms is partial fractions. A similar, but slightly different scheme works here.

• Example 5.3.4

Given $F(z) = z/\left(z^2 - 1\right)$, let us find f_n. The first step is to obtain the partial fraction expansion of $F(z)/z$. Why we want $F(z)/z$ rather than $F(z)$ will be made clear in a moment. Thus,

$$\frac{F(z)}{z} = \frac{1}{(z-1)(z+1)} = \frac{A}{z-1} + \frac{B}{z+1}, \tag{5.3.22}$$

where

$$A = (z-1)\left.\frac{F(z)}{z}\right|_{z=1} = \frac{1}{2} \tag{5.3.23}$$

and

$$B = (z+1)\left.\frac{F(z)}{z}\right|_{z=-1} = -\frac{1}{2}. \tag{5.3.24}$$

Multiplying (5.3.22) by z,

$$F(z) = \frac{1}{2}\left(\frac{z}{z-1} - \frac{z}{z+1}\right). \tag{5.3.25}$$

Next, we find the inverse z-transform of each of the terms $z/(z-1)$ and $z/(z+1)$ in Table 5.1.1. This yields

$$\mathcal{Z}^{-1}\left(\frac{z}{z-1}\right) = 1 \quad \text{and} \quad \mathcal{Z}^{-1}\left(\frac{z}{z+1}\right) = (-1)^n. \tag{5.3.26}$$

Thus, the inverse is

$$f_n = \tfrac{1}{2}\left[1 - (-1)^n\right], \; n \geq 0. \tag{5.3.27}$$

From this example it is clear that there are two steps involved: (1) obtain the partial fraction expansion of $F(z)/z$, and (2) finding the inverse z-transform by referring to Table 5.1.1.

• Example 5.3.5

Given $F(z) = 2z^2/[(z+2)(z+1)^2]$, let us find f_n. We begin by expanding $F(z)/z$ as

$$\frac{F(z)}{z} = \frac{2z}{(z+2)(z+1)^2} = \frac{A}{z+2} + \frac{B}{z+1} + \frac{C}{(z+1)^2}, \tag{5.3.28}$$

where

$$A = (z+2)\left.\frac{F(z)}{z}\right|_{z=-2} = -4, \tag{5.3.29}$$

$$B = \frac{d}{dz}\left[(z+1)^2\frac{F(z)}{z}\right]\Bigg|_{z=-1} = 4 \tag{5.3.30}$$

and

$$C = (z+1)^2\left.\frac{F(z)}{z}\right|_{z=-1} = -2 \tag{5.3.31}$$

so that

$$F(z) = \frac{4z}{z+1} - \frac{4z}{z+2} - \frac{2z}{(z+1)^2} \tag{5.3.32}$$

or

$$f_n = \mathcal{Z}^{-1}\left[\frac{4z}{z+1}\right] - \mathcal{Z}^{-1}\left[\frac{4z}{z+2}\right] - \mathcal{Z}^{-1}\left[\frac{2z}{(z+1)^2}\right]. \tag{5.3.33}$$

From Table 5.1.1,

$$\mathcal{Z}^{-1}\left(\frac{z}{z+1}\right) = (-1)^n, \tag{5.3.34}$$

$$\mathcal{Z}^{-1}\left(\frac{z}{z+2}\right) = (-2)^n \tag{5.3.35}$$

and

$$\mathcal{Z}^{-1}\left[\frac{z}{(z+1)^2}\right] = -\ \mathcal{Z}^{-1}\left[\frac{-z}{(z+1)^2}\right] = -n(-1)^n = n(-1)^{n+1}. \tag{5.3.36}$$

Applying (5.3.34)–(5.3.36) to (5.3.33),

$$f_n = 4(-1)^n - 4(-2)^n + 2n(-1)^n,\ \ n \geq 0. \tag{5.3.37}$$

• **Example 5.3.6**

Given $F(z) = (z^2+z)/(z-2)^2$, let us determine f_n. Because

$$\frac{F(z)}{z} = \frac{z+1}{(z-2)^2} = \frac{1}{z-2} + \frac{3}{(z-2)^2}, \tag{5.3.38}$$

$$f_n = \mathcal{Z}^{-1}\left[\frac{z}{z-2}\right] + \mathcal{Z}^{-1}\left[\frac{3z}{(z-2)^2}\right]. \tag{5.3.39}$$

Referring to Table 5.1.1,

$$\mathcal{Z}^{-1}\left(\frac{z}{z-2}\right) = 2^n \quad\text{and}\quad \mathcal{Z}^{-1}\left[\frac{3z}{(z-2)^2}\right] = \tfrac{3}{2}n2^n. \tag{5.3.40}$$

Substituting (5.3.40) into (5.3.39) yields

$$f_n = \left(\tfrac{3}{2}n + 1\right) 2^n, \quad n \geq 0. \tag{5.3.41}$$

Residue method

The power series, recursive, and partial fraction expansion methods are rather limited. We will now prove that f_n may be computed from the following *inverse integral formula*:

$$f_n = \frac{1}{2\pi i} \oint_C z^{n-1} F(z)\, dz, \quad n \geq 0, \tag{5.3.42}$$

where C is any simple curve, taken in the positive sense, that encloses all of the singularities of $F(z)$. It is readily shown that the power series and partial fraction methods are *special cases* of the residue method.

Proof: Starting with the definition of the z-transform

$$F(z) = \sum_{n=0}^{\infty} f_n z^{-n}, \qquad |z| > R_1, \tag{5.3.43}$$

we multiply (5.3.43) by z^{n-1} and integrating both sides around any contour C which includes all of the singularities,

$$\frac{1}{2\pi i} \oint_C z^{n-1} F(z)\, dz = \sum_{m=0}^{\infty} f_m \frac{1}{2\pi i} \oint_C z^{n-m} \frac{dz}{z}. \tag{5.3.44}$$

Let C be a circle of radius R, where $R > R_1$. Then, changing variables to $z = R\, e^{i\theta}$ and $dz = iz\, d\theta$,

$$\frac{1}{2\pi i} \oint_C z^{n-m} \frac{dz}{z} = \frac{R^{n-m}}{2\pi} \int_0^{2\pi} e^{i(n-m)\theta}\, d\theta = \begin{cases} 1, & m = n \\ 0, & \text{otherwise.} \end{cases} \tag{5.3.45}$$

Substituting (5.3.45) into (5.3.44) yields the desired result that

$$\frac{1}{2\pi i} \oint_C z^{n-1} F(z)\, dz = f_n. \tag{5.3.46}$$

□

We can easily evaluate the inversion integral (5.3.42) using Cauchy's residue theorem.

- **Example 5.3.7**

Let us find the inverse z-transform of

$$F(z) = \frac{1}{(z-1)(z-2)}. \tag{5.3.47}$$

From the inversion integral,

$$f_n = \frac{1}{2\pi i} \oint_C \frac{z^{n-1}}{(z-1)(z-2)}\, dz. \tag{5.3.48}$$

Clearly the integral has simple poles at $z = 1$ and $z = 2$. However, when $n = 0$ we also have a simple pole at $z = 0$. Thus the cases $n = 0$ and $n > 0$ must be considered separately.

Case 1: $n = 0$. The residue theorem yields

$$f_0 = \operatorname{Res}\left[\frac{1}{z(z-1)(z-2)}; 0\right] + \operatorname{Res}\left[\frac{1}{z(z-1)(z-2)}; 1\right] + \operatorname{Res}\left[\frac{1}{z(z-1)(z-2)}; 2\right]. \tag{5.3.49}$$

Evaluating these residues,

$$\operatorname{Res}\left[\frac{1}{z(z-1)(z-2)}; 0\right] = \left.\frac{1}{(z-1)(z-2)}\right|_{z=0} = \frac{1}{2}, \tag{5.3.50}$$

$$\operatorname{Res}\left[\frac{1}{z(z-1)(z-2)}; 1\right] = \left.\frac{1}{z(z-2)}\right|_{z=1} = -1 \tag{5.3.51}$$

and

$$\operatorname{Res}\left[\frac{1}{z(z-1)(z-2)}; 2\right] = \left.\frac{1}{z(z-1)}\right|_{z=2} = \frac{1}{2}. \tag{5.3.52}$$

Substituting (5.3.50)–(5.3.52) into (5.3.49) yields $f_0 = 0$.

Case 2: $n > 0$. Here we only have contributions from $z = 1$ and $z = 2$.

$$f_n = \operatorname{Res}\left[\frac{z^{n-1}}{(z-1)(z-2)}; 1\right] + \operatorname{Res}\left[\frac{z^{n-1}}{(z-1)(z-2)}; 2\right], n > 0, \tag{5.3.53}$$

where

$$\operatorname{Res}\left[\frac{z^{n-1}}{(z-1)(z-2)}; 1\right] = \left.\frac{z^{n-1}}{z-2}\right|_{z=1} = -1 \tag{5.3.54}$$

and

$$\operatorname{Res}\left[\frac{z^{n-1}}{(z-1)(z-2)};2\right] = \left.\frac{z^{n-1}}{z-1}\right|_{z=2} = 2^{n-1}, \; n>0. \tag{5.3.55}$$

Thus,

$$f_n = 2^{n-1} - 1, \quad n>0. \tag{5.3.56}$$

Combining our results,

$$f_n = \begin{cases} 0, & n=0 \\ \frac{1}{2}\left(2^n - 2\right), & n>0. \end{cases} \tag{5.3.57}$$

• Example 5.3.8

Let us use the inversion integral to find the inverse of

$$F(z) = \frac{z^2+2z}{(z-1)^2}. \tag{5.3.58}$$

The inversion theorem gives

$$f_n = \frac{1}{2\pi i}\oint_C \frac{z^{n+1}+2z^n}{(z-1)^2}\,dz = \operatorname{Res}\left[\frac{z^{n+1}+2z^n}{(z-1)^2};1\right], \tag{5.3.59}$$

where the pole at $z=1$ is second order. Consequently, the corresponding residue is

$$\operatorname{Res}\left[\frac{z^{n+1}+2z^n}{(z-1)^2};1\right] = \left.\frac{d}{dz}\left(z^{n+1}+2z^n\right)\right|_{z=1} = 3n+1. \tag{5.3.60}$$

Thus, the inverse z-transform of (5.3.58) is

$$f_n = 3n+1, \quad n \ge 0. \tag{5.3.61}$$

• Example 5.3.9

Let $F(z)$ be a z-transform whose poles lie within the unit circle $|z|=1$. Then

$$F(z) = \sum_{n=0}^{\infty} f_n z^{-n}, \qquad |z|>1 \tag{5.3.62}$$

and

$$F(z)F(z^{-1}) = \sum_{n=0}^{\infty} f_n^2 + \sum_{\substack{n=0 \\ n\neq m}}^{\infty}\sum_{m=0}^{\infty} f_m f_n z^{m-n}. \tag{5.3.63}$$

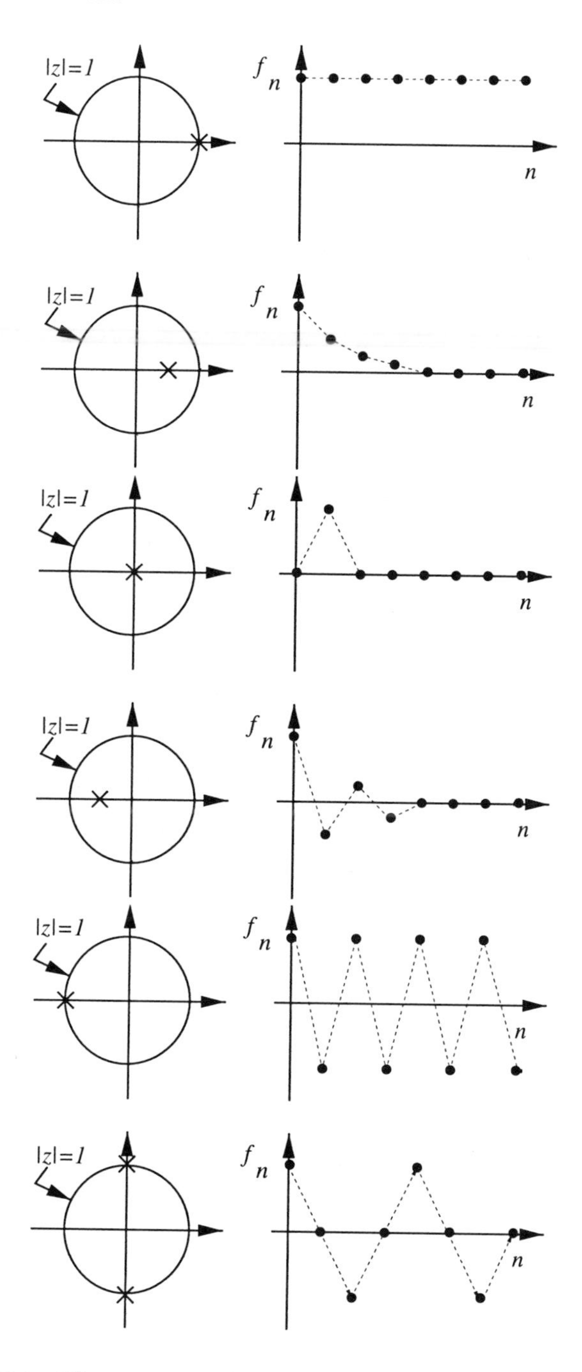

Figure 5.3.1: The correspondence between the location of the simple poles of the z-transform $F(z)$ and the behavior of f_n.

We now multiply both sides of (5.3.63) by z^{-1} and integrate around the unit circle C. Therefore,

$$\oint_{|z|=1} F(z)F(z^{-1})z^{-1}\,dz = \sum_{n=0}^{\infty} \oint_{|z|=1} f_n^2 z^{-1}\,dz + \sum_{\substack{n=0 \\ n\neq m}}^{\infty} \sum_{m=0}^{\infty} f_m f_n \oint_{|z|=1} z^{m-n-1}\,dz \quad (5.3.64)$$

after interchanging the order of integration and summation. Performing the integration,

$$\sum_{n=0}^{\infty} f_n^2 = \frac{1}{2\pi i} \oint_{|z|=1} F(z)F(z^{-1})z^{-1}\,dz, \quad (5.3.65)$$

which is *Parseval's theorem* for one-sided z-transforms. Recall that there are similar theorems for Fourier series and transforms.

• Example 5.3.10

An additional benefit of understanding inversion by the residue method is the ability to *qualitatively* anticipate the inverse by knowing the location of the poles of $F(z)$. This intuition is important because many engineering analyses discuss stability and performance entirely in terms of the properties of the system's z-transform. In Figure 5.3.1 we have graphed the location of the poles of $F(z)$ and the corresponding f_n. The student should go through the mental exercise of connecting the two pictures.

Problems

Use the power series or recursive method to compute the first few f_n's of the following z-transforms:

1. $F(z) = \dfrac{0.09z^2 + 0.9z + 0.09}{12.6z^2 - 24z + 11.4}$

2. $F(z) = \dfrac{z+1}{2z^4 - 2z^3 + 2z - 2}$

3. $F(z) = \dfrac{1.5z^2 + 1.5z}{15.25z^2 - 36.75z + 30.75}$

4. $F(z) = \dfrac{6z^2 + 6z}{19z^3 - 33z^2 + 21z - 7}$

Use partial fractions to find the inverse of the following z-transforms:

5. $F(z) = \dfrac{z(z+1)}{(z-1)(z^2 - z + 1/4)}$

6. $F(z) = \dfrac{(1 - e^{-aT})z}{(z-1)(z - e^{-aT})}$

7. $F(z) = \dfrac{z^2}{(z-1)(z-\alpha)}$

8. $F(z) = \dfrac{(2z - a - b)z}{(z-a)(z-b)}$

9. Using the property that the z-transform of $g_n = f_{n-k}H_{n-k}$ if $n \geq 0$ is $G(z) = z^{-k}F(z)$, find the inverse of

$$F(z) = \frac{z+1}{z^{10}(z-1/2)}.$$

Use the residue method to find the inverse z-transform of the following z-transforms:

10. $F(z) = \dfrac{z^2+3z}{(z-1/2)^3}$

11. $F(z) = \dfrac{z}{(z+1)^2(z-2)}$

12. $F(z) = \dfrac{z}{(z+1)^2(z-1)^2}$

13. $F(z) = e^{a/z}$

5.4 SOLUTION OF DIFFERENCE EQUATIONS

Having reached the point where we can take a z-transform and then find its inverse, we are ready to use it to solve difference equations. The procedure parallels that of solving ordinary differential equations by Laplace transforms. Essentially we reduce the difference equation to an algebraic problem. We then find the solution by inverting $Y(z)$.

• Example 5.4.1

Let us solve the second-order difference equation

$$2y_{n+2} - 3y_{n+1} + y_n = 5\ 3^n, \quad n \geq 0, \tag{5.4.1}$$

where $y_0 = 0$ and $y_1 = 1$.

Taking the z-transform of both sides of (5.4.1), we obtain

$$2\mathcal{Z}(y_{n+2}) - 3\mathcal{Z}(y_{n+1}) + \mathcal{Z}(y_n) = 5\,\mathcal{Z}(3^n). \tag{5.4.2}$$

From the shifting theorem and Table 5.1.1,

$$2z^2Y(z) - 2z^2y_0 - 2zy_1 - 3[zY(z) - zy_0] + Y(z) = \frac{5z}{z-3}. \tag{5.4.3}$$

Substituting $y_0 = 0$ and $y_1 = 1$ into (5.4.3) and simplifying yields

$$(2z-1)(z-1)Y(z) = \frac{z(2z-1)}{z-3}. \tag{5.4.4}$$

or

$$Y(z) = \frac{z}{(z-3)(z-1)}. \tag{5.4.5}$$

To obtain y_n from $Y(z)$ we can employ partial fractions or the residue method. Applying partial fractions yields

$$\frac{Y(z)}{z} = \frac{A}{z-1} + \frac{B}{z-3}, \tag{5.4.6}$$

where

$$A = (z-1)\left.\frac{Y(z)}{z}\right|_{z=1} = -\frac{1}{2} \tag{5.4.7}$$

and

$$B = (z-3)\left.\frac{Y(z)}{z}\right|_{z=3} = \frac{1}{2}. \tag{5.4.8}$$

Thus,

$$Y(z) = -\frac{1}{2}\frac{z}{z-1} + \frac{1}{2}\frac{z}{z-3} \tag{5.4.9}$$

or

$$y_n = -\frac{1}{2}\mathcal{Z}^{-1}\left(\frac{z}{z-1}\right) + \frac{1}{2}\mathcal{Z}^{-1}\left(\frac{z}{z-3}\right). \tag{5.4.10}$$

From (5.4.10) and Table 5.1.1,

$$y_n = \tfrac{1}{2}\left(3^n - 1\right), \qquad n \geq 0. \tag{5.4.11}$$

Two checks confirm that we have the *correct* solution. First, our solution must satisfy the initial values of the sequence. Computing y_0 and y_1,

$$y_0 = \tfrac{1}{2}(3^0 - 1) = \tfrac{1}{2}(1-1) = 0 \tag{5.4.12}$$

and

$$y_1 = \tfrac{1}{2}(3^1 - 1) = \tfrac{1}{2}(3-1) = 1. \tag{5.4.13}$$

Thus, our solution gives the correct initial values.

Our sequence y_n must also satisfy the difference equation. Now

$$y_{n+2} = \tfrac{1}{2}(3^{n+2} - 1) = \tfrac{1}{2}(9\ 3^n - 1) \tag{5.4.14}$$

and

$$y_{n+1} = \tfrac{1}{2}(3^{n+1} - 1) = \tfrac{1}{2}(3\ 3^n - 1). \tag{5.4.15}$$

Therefore,

$$2y_{n+2} - 3y_{n+1} + y_n = \left(9 - \tfrac{9}{2} + \tfrac{1}{2}\right)3^n - 1 + \tfrac{3}{2} - \tfrac{1}{2} = 5\ 3^n \tag{5.4.16}$$

and our solution is correct.

Finally, we note that the term $3^n/2$ is necessary to give the right side of (5.4.1); it is the particular solution. The $-1/2$ term is necessary so that the sequence satisfies the initial values; it is the complementary solution.

• **Example 5.4.2**

Let us find the y_n in the difference equation

$$y_{n+2} - 2y_{n+1} + y_n = 1, \quad n \geq 0 \tag{5.4.17}$$

with the initial conditions $y_0 = 0$ and $y_1 = 3/2$.

From (5.4.17),

$$\mathcal{Z}(y_{n+2}) - 2\mathcal{Z}(y_{n+1}) + \mathcal{Z}(y_n) = \mathcal{Z}(1). \tag{5.4.18}$$

The z-transform of the left side of (5.4.18) is obtained from the shifting theorem and Table 5.1.1 yields $\mathcal{Z}(1)$. Thus,

$$z^2Y(z) - z^2y_0 - zy_1 - 2zY(z) + 2zy_0 + Y(z) = \frac{z}{z-1}. \tag{5.4.19}$$

Substituting $y_0 = 0$ and $y_1 = 3/2$ in (5.4.19) and simplifying yields

$$Y(z) = \frac{3z^2 - z}{2(z-1)^3} \tag{5.4.20}$$

or

$$y_n = \mathcal{Z}^{-1}\left[\frac{3z^2 - z}{2(z-1)^3}\right]. \tag{5.4.21}$$

We find the inverse z-transform of (5.4.21) by the residue method or

$$y_n = \frac{1}{2\pi i}\oint_C \frac{3z^{n+1} - z^n}{2(z-1)^3}\,dz = \frac{1}{2!}\frac{d^2}{dz^2}\left[\frac{3z^{n+1}}{2} - \frac{z^n}{2}\right]\bigg|_{z=1} \tag{5.4.22}$$

$$= \tfrac{1}{2}n^2 + n. \tag{5.4.23}$$

Thus,

$$y_n = \tfrac{1}{2}n^2 + n, \quad n \geq 0. \tag{5.4.24}$$

Note that $n^2/2$ gives the particular solution to (5.4.17), while n is there so that y_n satisfies the initial conditions. This problem is particularly interesting because our constant forcing produces a response that grows as n^2, just as in the case of resonance in a time-continuous system when a finite forcing such as $\sin(\omega_0 t)$ results in a response whose amplitude grows as t^m.

• **Example 5.4.3**

Let us solve the difference equation

$$b^2 y_n + y_{n+2} = 0, \tag{5.4.25}$$

where $|b| < 1$ and the initial conditions are $y_0 = b^2$ and $y_1 = 0$.

We begin by taking the z-transform of each term in (5.4.25). This yields

$$b^2 \mathcal{Z}(y_n) + \mathcal{Z}(y_{n+2}) = 0. \tag{5.4.26}$$

From the shifting theorem, it follows that

$$b^2 Y(z) + z^2 Y(z) - z^2 y_0 - z y_1 = 0. \tag{5.4.27}$$

Substituting $y_0 = b^2$ and $y_1 = 0$ into (5.4.27),

$$b^2 Y(z) + z^2 Y(z) - b^2 z^2 = 0 \tag{5.4.28}$$

or

$$Y(z) = \frac{b^2 z^2}{z^2 + b^2}. \tag{5.4.29}$$

To find y_n we employ the residue method or

$$y_n = \frac{1}{2\pi i} \oint_C \frac{b^2 z^{n+1}}{(z - ib)(z + ib)}\, dz. \tag{5.4.30}$$

Thus,

$$y_n = \left. \frac{b^2 z^{n+1}}{z + ib} \right|_{z=ib} + \left. \frac{b^2 z^{n+1}}{z - ib} \right|_{z=-ib} = \frac{b^{n+2} i^n}{2} + \frac{b^{n+2}(-i)^n}{2} \tag{5.4.31}$$

$$= \frac{b^{n+2} e^{in\pi/2}}{2} + \frac{b^{n+2} e^{-in\pi/2}}{2} = b^{n+2} \cos\left(\frac{n\pi}{2}\right), \tag{5.4.32}$$

because $\cos(x) = \frac{1}{2}\left(e^{ix} + e^{-ix}\right)$. Consequently, we obtain the desired result that

$$y_n = b^{n+2} \cos\left(\frac{n\pi}{2}\right) \quad \text{for } n \geq 0. \tag{5.4.33}$$

• **Example 5.4.4: Compound interest**

Finite difference equations arise in finance because the increase or decrease in an account occurs in discrete steps. For example, the amount of money in a compound interest saving account after $n + 1$ conversion periods (the time period between interest payments) is

$$y_{n+1} = y_n + r y_n, \tag{5.4.34}$$

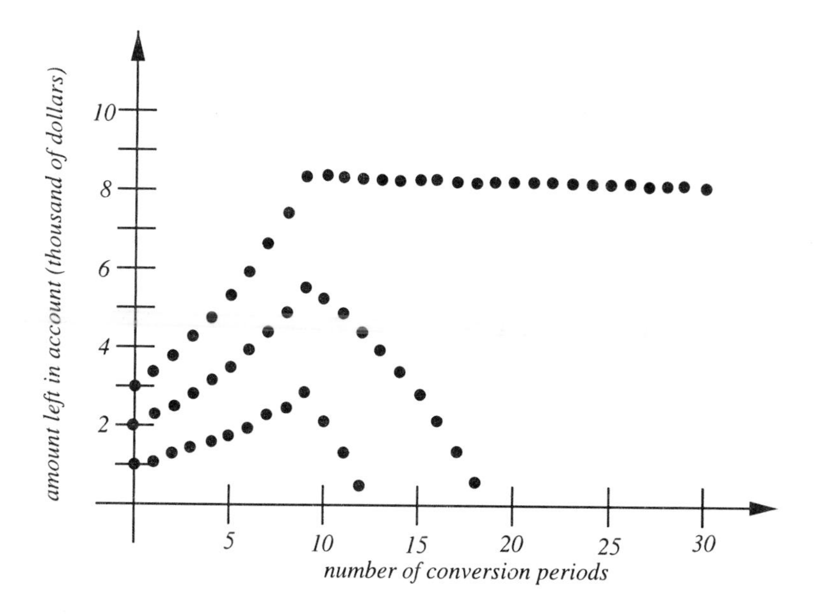

Figure 5.4.1: The amount in a saving account as a function of an annual conversion period when interest is compounded at the annual rate of 12% and a $1000 is taken from the account every period starting with period 10.

where r is the interest rate per conversion period. The second term on the right side of (5.4.34) is the amount of interest paid at the end of each period.

Let us ask a somewhat more difficult question of how much money we will have if we withdraw the amount A at the end of every period starting after the period ℓ. Now the difference equation becomes

$$y_{n+1} = y_n + ry_n - AH_{n-\ell-1}. \tag{5.4.35}$$

Taking the z-transform of (5.4.35),

$$zY(z) - zy_0 = (1+r)Y(z) - \frac{Az^{2-\ell}}{z-1} \tag{5.4.36}$$

after using (5.2.10) or

$$Y(z) = \frac{y_0 z}{z-(1+r)} - \frac{Az^{2-\ell}}{(z-1)[z-(1+r)]}. \tag{5.4.37}$$

Taking the inverse of (5.4.37),

$$y_n = y_0(1+r)^n - \frac{A}{r}\left[(1+r)^{n-\ell+1} - 1\right] H_{n-\ell}. \tag{5.4.38}$$

The first term in (5.4.38) represents the growth of money by compound interest while the second term gives the depletion of the account by withdrawals. Figure 5.4.1 gives the values of y_n for various starting amounts assuming an annual conversion period with $r = 0.12$, $\ell = 10$ years, and $A = \$1000$. It shows that if an investor places an initial amount of \$3000 in an account bearing 12% annually, after 10 years he can withdraw \$1000 annually, essentially forever. This is because the amount that he removes every year is replaced by the interest on the funds that remain in the account.

• Example 5.4.5

Let us solve the following system of difference equations:

$$x_{n+1} = 4x_n + 2y_n \tag{5.4.39}$$

and

$$y_{n+1} = 3x_n + 3y_n \tag{5.4.40}$$

with the initial values of $x_0 = 0$ and $y_0 = 5$.

Taking the z-transform of (5.4.39)–(5.4.40),

$$zX(z) - x_0 z = 4X(z) + 2Y(z) \tag{5.4.41}$$

$$zY(z) - y_0 z = 3X(z) + 3Y(z) \tag{5.4.42}$$

or

$$(z-4)X(z) - 2Y(z) = 0 \tag{5.4.43}$$

$$3X(z) - (z-3)Y(z) = -5z. \tag{5.4.44}$$

Solving for $X(z)$ and $Y(z)$,

$$X(z) = -\frac{10z}{(z-6)(z-1)} = \frac{2z}{z-1} - \frac{2z}{z-6} \tag{5.4.45}$$

and

$$Y(z) = \frac{5z(z-4)}{(z-6)(z-1)} = \frac{2z}{z-6} + \frac{3z}{z-1}. \tag{5.4.46}$$

Taking the inverse of (5.4.45)–(5.4.46) term by term,

$$x_n = 2 - 2\ 6^n \qquad \text{and} \qquad y_n = 3 + 2\ 6^n. \tag{5.4.47}$$

Problems

Solve the following difference equations using z-transforms, where $n \geq 0$.

1. $y_{n+1} - y_n = n^2, \quad y_0 = 1.$

2. $y_{n+2} - 2y_{n+1} + y_n = 0, \quad y_0 = y_1 = 1.$

3. $y_{n+2} - 2y_{n+1} + y_n = 1, \quad y_0 = y_1 = 0.$

4. $y_{n+1} + 3y_n = n, \quad y_0 = 0.$

5. $y_{n+1} - 5y_n = \cos(n\pi), \quad y_0 = 0.$

6. $y_{n+2} - 4y_n = 1, \quad y_0 = 1, y_1 = 0.$

7. $y_{n+2} - \frac{1}{4}y_n = (\frac{1}{2})^n, \quad y_0 = y_1 = 0.$

8. $y_{n+2} - 5y_{n+1} + 6y_n = 0, \quad y_0 = y_1 = 1.$

9. $y_{n+2} - 3y_{n+1} + 2y_n = 1, \quad y_0 = y_1 = 0.$

10. $y_{n+2} - 2y_{n+1} + y_n = 2, \quad y_0 = 0, \quad y_1 = 2.$

11. $x_{n+1} = 3x_n - 4y_n,\ y_{n+1} = 2x_n - 3y_n, \quad x_0 = 3,\ y_0 = 2.$

12. $x_{n+1} = 2x_n - 10y_n,\ y_{n+1} = -x_n - y_n, \quad x_0 = 3,\ y_0 = -2.$

13. $x_{n+1} = x_n - 2y_n,\ y_{n+1} = -6y_n, \quad x_0 = -1,\ y_0 = -7.$

14. $x_{n+1} = 4x_n - 5y_n,\ y_{n+1} = x_n - 2y_n, \quad x_0 = 6,\ y_0 = 2.$

5.5 STABILITY OF DISCRETE-TIME SYSTEMS

When we discussed the solution of ordinary differential equations by Laplace transforms, we introduced the concept of transfer function and impulse response. In the case of discrete-time systems, similar considerations come into play.

Consider the recursive system

$$y_n = a_1 y_{n-1} H_{n-1} + a_2 y_{n-2} H_{n-2} + x_n, \quad n \geq 0, \tag{5.5.1}$$

where H_{n-k} is the unit step function. It equals 0 for $n < k$ and 1 for $n \geq k$. Equation (5.5.1) is called a *recursive system* because future values of the sequence depend upon all of the previous values. At present, a_1 and a_2 are free parameters which we shall vary.

Using (5.2.10),

$$z^2Y(z) - a_1zY(z) - a_2Y(z) = z^2X(z) \tag{5.5.2}$$

or

$$G(z) = \frac{Y(z)}{X(z)} = \frac{z^2}{z^2 - a_1z - a_2}. \tag{5.5.3}$$

As in the case of Laplace transforms, the ratio $Y(z)/X(z)$ is the transfer function. The inverse of the transfer function gives the impulse response for our discrete-time system. This particular transfer function has two poles, namely

$$z_{1,2} = \frac{a_1}{2} \pm \sqrt{\frac{a_1^2}{4} + a_2}. \tag{5.5.4}$$

At this point, we consider three cases.

Case 1: $a_1^2/4 + a_2 < 0$. In this case z_1 and z_2 are complex conjugates. Let us write them as $z_{1,2} = re^{\pm i\omega_0 T}$. Then

$$G(z) = \frac{z^2}{(z - re^{i\omega_0 T})(z - re^{-i\omega_0 T})} = \frac{z^2}{z^2 - 2r\cos(\omega_0 T)z + r^2}, \tag{5.5.5}$$

where $r^2 = -a_2$ and $\omega_0 T = \cos^{-1}(a_1/2r)$. From the inversion integral,

$$\begin{aligned} g_n = \operatorname{Res}&\left[\frac{z^{n+1}}{z^2 - 2r\cos(\omega_0 T)z + r^2}; z_1\right] \\ &+ \operatorname{Res}\left[\frac{z^{n+1}}{z^2 - 2r\cos(\omega_0 T)z + r^2}; z_2\right], \end{aligned} \tag{5.5.6}$$

where g_n denotes the impulse response. Now

$$\operatorname{Res}\left[\frac{z^{n+1}}{z^2 - 2r\cos(\omega_0 T)z + r^2}; z_1\right] = \lim_{z\to z_1} \frac{(z - z_1)z^{n+1}}{(z - z_1)(z - z_2)} \tag{5.5.7}$$

$$= r^n \frac{\exp[i(n+1)\omega_0 T]}{e^{i\omega_0 T} - e^{-i\omega_0 T}} \tag{5.5.8}$$

$$= \frac{r^n \exp[i(n+1)\omega_0 T]}{2i\sin(\omega_0 T)}. \tag{5.5.9}$$

Similarly,

$$\operatorname{Res}\left[\frac{z^{n+1}}{z^2 - 2r\cos(\omega_0 T)z + r^2}; z_2\right] = -\frac{r^n \exp[-i(n+1)\omega_0 T]}{2i\sin(\omega_0 T)} \tag{5.5.10}$$

and

$$g_n = \frac{r^n \sin[(n+1)\omega_0 T]}{\sin(\omega_0 T)}. \tag{5.5.11}$$

A graph of $\sin[(n+1)\omega_0 T]/\sin(\omega_0 T)$ with respect to n gives a sinusoidal envelope. More importantly, if $|r| < 1$ these oscillations will vanish as $n \to \infty$ and the system is stable. On the other hand, if $|r| > 1$ the oscillation will grow without bound as $n \to \infty$ and the system is unstable.

Recall that $|r| > 1$ corresponds to poles that lie outside the unit circle while $|r| < 1$ is exactly the opposite. Our example suggests that for discrete-time systems to be stable, all of the poles of the transfer function must lie within the unit circle while an unstable system has at least one pole that lies outside of this circle.

Case 2: $a_1^2/4 + a_2 > 0$. This case leads to two real roots, z_1 and z_2. From the inversion integral, the sum of the residues gives the impulse response

$$g_n = \frac{z_1^{n+1} - z_2^{n+1}}{z_1 - z_2}. \tag{5.5.12}$$

Once again, if the poles lie within the unit circle, $|z_1| < 1$ and $|z_2| < 1$, the system is stable.

Case 3: $a_1^2/4 + a_2 = 0$. This case yields $z_1 = z_2$,

$$G(z) = \frac{z^2}{(z - a_1/2)^2} \tag{5.5.13}$$

and

$$g_n = \frac{1}{2\pi i} \oint_C \frac{z^{n+1}}{(z - a_1/2)^2} dz = \left(\frac{a_1}{2}\right)^n (n+1). \tag{5.5.14}$$

This system is obviously stable if $|a_1/2| < 1$ and the pole of the transfer function lies within the unit circle.

In summary, finding the transfer function of a discrete-time system is important in determining its stability. Because the location of the poles of $G(z)$ determines the response of the system, a stable system will have all of its poles within the unit circle. Conversely, if any of the poles of $G(z)$ lie outside of the unit circle, the system is unstable. Finally, if $\lim_{n\to\infty} g_n = c$, the system is marginally stable. For example, if $G(z)$ has simple poles, some of the poles must lie *on* the unit circle.

• Example 5.5.1

Numerical methods of integration provide some of the simplest, yet most important, difference equations in the literature. In this example,[3]

---

[3] From Salzer, J. M., 1954: Frequency analysis of digital computers operating in real time. *Proc. IRE*, **42**, 457–466. ©IRE (now IEEE).

we show how z-transforms can be used to highlight the strengths and weaknesses of such schemes.

Consider the trapezoidal integration rule in numerical analysis. The integral y_n is updated by adding the latest trapezoidal approximation of the continuous curve. Thus, the integral is computed by

$$y_n = \tfrac{1}{2}T(x_n + x_{n-1}H_{n-1}) + y_{n-1}H_{n-1}, \tag{5.5.15}$$

where T is the interval between evaluations of the integrand.

We first determine the stability of this rule because it is of little value if it is not stable. Using (5.2.10), the transfer function is

$$G(z) = \frac{Y(z)}{X(z)} = \frac{T}{2}\left(\frac{z+1}{z-1}\right). \tag{5.5.16}$$

To find the impulse response, we use the inversion integral and find that

$$g_n = \frac{T}{4\pi i}\oint_C z^{n-1}\,\frac{z+1}{z-1}\,dz. \tag{5.5.17}$$

At this point, we must consider two cases: $n = 0$ and $n > 0$. For $n = 0$,

$$g_0 = \frac{T}{2}\text{Res}\left[\frac{z+1}{z(z-1)};0\right] + \frac{T}{2}\text{Res}\left[\frac{z+1}{z(z-1)};1\right] = \frac{T}{2}. \tag{5.5.18}$$

For $n > 0$,

$$g_0 = \frac{T}{2}\text{Res}\left[\frac{z^{n-1}(z+1)}{z-1};1\right] = T. \tag{5.5.19}$$

Therefore, the impulse response for this numerical scheme is $g_0 = \frac{T}{2}$ and $g_n = T$ for $n > 0$. Note that this is a marginally stable system (the solution neither grows nor decays with n) because the pole associated with the transfer function lies *on* the unit circle.

Having discovered that the system is not unstable, let us continue and explore some of its properties. Recall now that $z = e^{sT} = e^{i\omega T}$ if $s = i\omega$. Then the transfer function becomes

$$G(\omega) = \frac{T}{2}\frac{1+e^{-i\omega T}}{1-e^{-i\omega T}} = -\frac{iT}{2}\cot\left(\frac{\omega T}{2}\right). \tag{5.5.20}$$

On the other hand, the transfer function of an ideal integrator is $1/s$ or $-i/\omega$. Thus, the trapezoidal rule has ideal phase but its shortcoming lies in its amplitude characteristic; it lies below the ideal integrator for

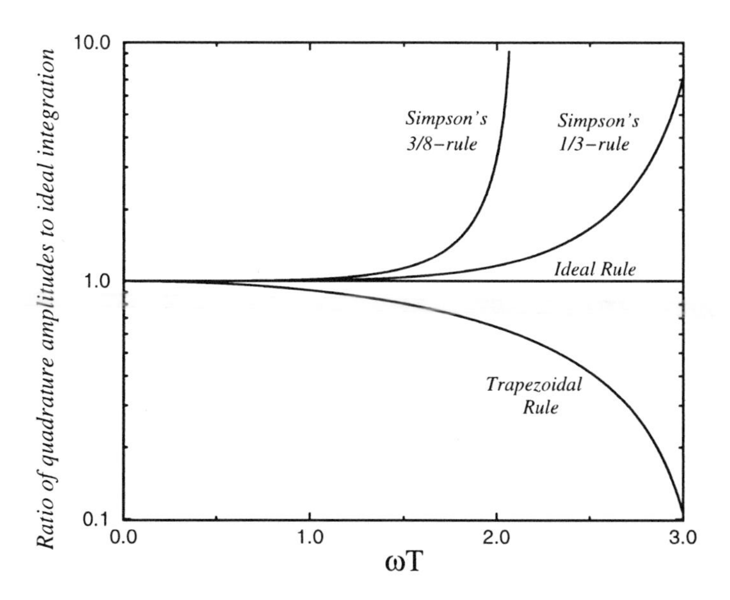

Figure 5.5.1: Comparison of various quadrature formulas by ratios of their amplitudes to that of an ideal integrator. [From Salzer, J. M., 1954: Frequency analysis of digital computers operating in real time. *Proc. IRE*, **42**, p. 463. ©IRE (now IEEE).]

$0 < \omega T < \pi$. We show this behavior, along with that for Simpson's $\frac{1}{3}$rd-rule and Simpson's $\frac{3}{8}$th-rule, in Figure 5.5.1.

Figure 5.5.1 confirms the superiority of Simpson's $\frac{1}{3}$rd rule over his $\frac{3}{8}$th rule. The figure also shows that certain schemes are better at suppressing noise at higher frequencies; an effect not generally emphasized in numerical calculus but often important in system design. For example, the trapezoidal rule is inferior to all others at low frequencies but only to Simpson's $\frac{1}{3}$rd rule at higher frequencies. Furthermore, the trapezoidal rule might actually be preferred not only because of its simplicity but also because it attenuates at higher frequencies, thereby counteracting the effect of noise.

• **Example 5.5.2**

Given the transfer function

$$G(z) = \frac{z^2}{(z-1)(z-1/2)}, \tag{5.5.21}$$

is this discrete-time system stable or marginally stable?

This transfer function has two simple poles. The pole at $z = 1/2$ gives rise to a term that varies as $(\frac{1}{2})^n$ in the impulse response while the $z = 1$ pole will give a constant. Because this constant will neither grow nor decay with n, the system is marginally stable.

Problems

For the following time-discrete systems, find the transfer function and determine whether the systems are unstable, marginally stable, or stable.

1. $y_n = y_{n-1}H_{n-1} + x_n$

2. $y_n = 2y_{n-1}H_{n-1} - y_{n-2}H_{n-2} + x_n$

3. $y_n = 3y_{n-1}H_{n-1} + x_n$

4. $y_n = \frac{1}{4}y_{n-2}H_{n-2} + x_n$

Chapter 6

The Sturm-Liouville Problem

In the next three chapters we shall be solving partial differential equations using the technique of separation of variables. This technique requires that we expand a piece-wise continuous function $f(x)$ as a linear sum of *eigenfunctions*, much as we used sines and cosines to reexpress $f(x)$ in a Fourier series. The purpose of this chapter is to explain and illustrate these eigenfunction expansions.

6.1 EIGENVALUES AND EIGENFUNCTIONS

Repeatedly, in the next three chapters on partial differential equations, we will solve the following second-order linear differential equation:

$$\frac{d}{dx}\left[p(x)\frac{dy}{dx}\right] + [q(x) + \lambda r(x)]y = 0, \quad a \leq x \leq b, \tag{6.1.1}$$

together with the boundary conditions:

$$\alpha y(a) + \beta y'(a) = 0 \qquad \text{and} \qquad \gamma y(b) + \delta y'(b) = 0. \tag{6.1.2}$$

In (6.1.1), $p(x)$, $q(x)$ and $r(x)$ are real functions of x; λ is a parameter; and $p(x)$ and $r(x)$ are functions that are continuous and positive on the interval $a \le x \le b$. Taken together, (6.1.1) and (6.1.2) constitute a regular *Sturm-Liouville problem.* This name honors the French mathematicians Sturm and Liouville[1] who first studied these equations in the 1830s. In the case when $p(x)$ or $r(x)$ vanishes at one of the endpoints of the interval $[a, b]$ or when the interval is of infinite length, the problem is a *singular Sturm-Liouville problem.*

Consider now the solutions of the Sturm-Liouville problem. Clearly there is the trivial solution $y = 0$ for all λ. However, nontrivial solutions will exist only if λ takes on specific values; these values are called *characteristic values* or *eigenvalues*. The corresponding nontrivial solutions are called the *characteristic functions* or *eigenfunctions*. In particular, we have the following theorems.

Theorem: *For a regular Sturm-Liouville problem with $p(x) > 0$, all of the eigenvalues are real if $p(x)$, $q(x)$, and $r(x)$ are real functions and the eigenfunctions are differentiable and continuous.*

Proof: Let $y(x) = u(x) + iv(x)$ be an eigenfunction corresponding to an eigenvalue $\lambda = \lambda_r + i\lambda_i$, where λ_r, λ_i are real numbers and $u(x), v(x)$ are real functions of x. Substituting into the Sturm-Liouville equation yields

$$\{p(x)[u'(x)+iv'(x)]\}'+[q(x)+(\lambda_r+i\lambda_i)r(x)][u(x)+iv(x)] = 0. \tag{6.1.3}$$

Separating the real and imaginary parts yields

$$[p(x)u'(x)]' + [q(x) + \lambda_r]u(x) - \lambda_i r(x)v(x) = 0 \tag{6.1.4}$$

and

$$[p(x)v'(x)]' + [q(x) + \lambda_r]v(x) + \lambda_i r(x)u(x) = 0. \tag{6.1.5}$$

---

[1] For the complete history as well as the relevant papers, see Lützen, J., 1984: Sturm and Liouville's work on ordinary linear differential equations. The emergence of Sturm-Liouville theory. *Arch. Hist. Exact Sci.*, **29**, 309–376.

Figure 6.1.1: By the time that Charles-François Sturm (1803–1855) met Joseph Liouville in the early 1830s, he had already gained fame for his work on the compression of fluids and his celebrated theorem on the number of real roots of a polynomial. An eminent teacher, Sturm spent most of his career teaching at various Parisian colleges. (Portrait courtesy of the Archives de l'Académie des sciences, Paris.)

If we multiply (6.1.4) by v and (6.1.5) by u and subtract the results, we find that

$$u(x)[p(x)v'(x)]' - v(x)[p(x)u'(x)]' + \lambda_i r(x)[u^2(x) + v^2(x)] = 0. \quad \textbf{(6.1.6)}$$

The derivative terms in (6.1.6) can be rewritten in such a manner that it becomes

$$\frac{d}{dx}\left\{[p(x)v'(x)]u(x) - [p(x)u'(x)]v(x)\right\} + \lambda_i r(x)[u^2(x) + v^2(x)] = 0. \quad \textbf{(6.1.7)}$$

Figure 6.1.2: Although educated as an engineer, Joseph Liouville (1809–1882) would devote his life to teaching pure and applied mathematics in the leading Parisian institutions of higher education. Today he is most famous for founding and editing for almost 40 years the *Journal de Liouville*. (Portrait courtesy of the Archives de l'Académie des sciences, Paris.)

Integrating from a to b, we find that

$$-\lambda_i \int_a^b r(x)[u^2(x) + v^2(x)]\, dx = \{p(x)[u(x)v'(x) - v(x)u'(x)]\}\big|_a^b. \tag{6.1.8}$$

From the boundary conditions (6.1.2),

$$\alpha[u(a) + iv(a)] + \beta[u'(a) + iv'(a)] = 0 \tag{6.1.9}$$

and

$$\gamma[u(b) + iv(b)] + \delta[u'(b) + iv'(b)] = 0. \tag{6.1.10}$$

Separating the real and imaginary parts yields

$$\alpha u(a) + \beta u'(a) = 0 \quad \text{and} \quad \alpha v(a) + \beta v'(a) = 0 \tag{6.1.11}$$

and

$$\gamma u(b) + \delta u'(b) = 0 \quad \text{and} \quad \gamma v(b) + \delta v'(b) = 0. \tag{6.1.12}$$

Both α and β cannot be zero; otherwise, there would be no boundary condition at $x = a$. Similar considerations hold for γ and δ. Therefore,

$$u(a)v'(a) - u'(a)v(a) = 0 \quad \text{and} \quad u(b)v'(b) - u'(b)v(b) = 0, \tag{6.1.13}$$

if we treat α, β, γ, and δ as unknowns in a system of homogeneous equations (6.1.11)–(6.1.12) and require that the corresponding determinants equal zero. Applying (6.1.13) to the right side of (6.1.8), we obtain

$$\lambda_i \int_a^b r(x)[u^2(x) + v^2(x)]\, dx = 0. \tag{6.1.14}$$

Because $r(x) > 0$, the integral is positive and $\lambda_i = 0$. Since $\lambda_i = 0$, λ is purely real. This implies that the eigenvalues are real. □

If there is only one independent eigenfunction for each eigenvalue, that eigenvalue is *simple*. When more than one eigenfunction belongs to a single eigenvalue, the problem is *degenerate*.

Theorem: *The regular Sturm-Liouville problem has infinitely many real and simple eigenvalues* λ_n, $n = 0, 1, 2, \ldots$, *which can be arranged in a monotonically increasing sequence* $\lambda_0 < \lambda_1 < \lambda_2 < \cdots$ *such that* $\lim_{n\to\infty} \lambda_n = \infty$. *Every eigenfunction* $y_n(x)$ *associated with the corresponding eigenvalue* λ_n *has exactly* n *zeros in the interval* (a, b). *For each eigenvalue there exists only one eigenfunction (up to a multiplicative constant).*

The proof is beyond the scope of this book but may be found in more advanced treatises.[2]

In the following examples we will illustrate how to find these real eigenvalues and their corresponding eigenfunctions.

---

[2] See, for example, Birkhoff, G. and Rota, G.-C., 1989: *Ordinary Differential Equations*, John Wiley & Sons, New York, chaps. 10 and 11; Sagan, H., 1961: *Boundary and Eigenvalue Problems in Mathematical Physics*, John Wiley & Sons, New York, chap. 5.

• **Example 6.1.1**

Let us find the eigenvalues and eigenfunctions of

$$y'' + \lambda y = 0 \tag{6.1.15}$$

subject to the boundary conditions

$$y(0) = 0 \qquad \text{and} \qquad y(\pi) - y'(\pi) = 0. \tag{6.1.16}$$

Our first task is to check to see whether the problem is indeed a regular Sturm-Liouville problem. A comparison between (6.1.1) and (6.1.15) shows that they are the same if $p(x) = 1$, $q(x) = 0$, and $r(x) = 1$. Similarly, the boundary conditions (6.1.16) are identical to (6.1.2) if $\alpha = \gamma = 1$, $\delta = -1$, $\beta = 0$, $a = 0$, and $b = \pi$.

Because the form of the solution to (6.1.15) depends on λ, we consider three cases: λ negative, positive, or equal to zero. The general solution of the differential equation is

$$y(x) = A\cosh(mx) + B\sinh(mx) \quad \text{if} \quad \lambda < 0, \tag{6.1.17}$$

$$y(x) = C + Dx \quad \text{if} \quad \lambda = 0 \tag{6.1.18}$$

and

$$y(x) = E\cos(kx) + F\sin(kx) \quad \text{if} \quad \lambda > 0, \tag{6.1.19}$$

where for convenience $\lambda = -m^2 < 0$ in (6.1.17) and $\lambda = k^2 > 0$ in (6.1.19). Both k and m are real and positive by these definitions.[3]

---

[3] In many differential equations courses, the solution to

$$y'' - m^2 y = 0, \qquad m > 0$$

is written

$$y(x) = c_1 e^{mx} + c_2 e^{-mx}.$$

However, we can rewrite this solution as

$$y(x) = (c_1 + c_2)\tfrac{1}{2}(e^{mx} + e^{-mx}) + (c_1 - c_2)\tfrac{1}{2}(e^{mx} - e^{-mx})$$

$$= A\cosh(mx) + B\sinh(mx),$$

where $\cosh(mx) = (e^{mx} + e^{-mx})/2$ and $\sinh(mx) = (e^{mx} - e^{-mx})/2$. The advantage of using these hyperbolic functions over exponentials is the simplification that occurs when we substitute the hyperbolic functions into the boundary conditions.

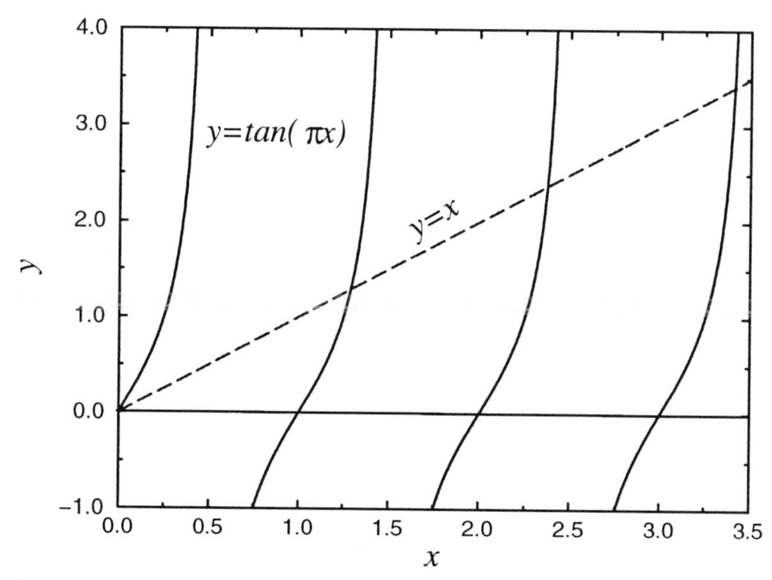

Figure 6.1.3: Graphical solution of $\tan(\pi x) = x$.

Turning to the condition that $y(0) = 0$, we find that $A = C = E = 0$. The other boundary condition $y(\pi) - y(\pi) = 0$ gives

$$B[\sinh(m\pi) - m\cosh(m\pi)] = 0, \tag{6.1.20}$$

$$D = 0 \tag{6.1.21}$$

and

$$F[\sin(k\pi) - k\cos(k\pi)] = 0. \tag{6.1.22}$$

If we graph $\sinh(m\pi) - m\cosh(m\pi)$ for all positive m, this quantity is always negative. Consequently, $B = 0$. However, in (6.1.22), a nontrivial solution (i.e., $F \neq 0$) occurs if

$$F\cos(k\pi)[\tan(k\pi) - k] = 0 \quad \text{or} \quad \tan(k\pi) = k. \tag{6.1.23}$$

In summary, we have found nontrivial solutions only when $\lambda_n = k_n^2 > 0$, where k_n is the nth root of the transcendental Equation (6.1.23). We may find the roots either graphically or through the use of a numerical algorithm. Figure 6.1.3 illustrates the graphical solution to the problem. We exclude the root $k = 0$ because λ must be greater than zero.

Let us now find the corresponding eigenfunctions. Because $A = B = C = D = E = 0$, we are left with $y(x) = F\sin(kx)$. Consequently, the eigenfunction, traditionally written without the arbitrary amplitude constant, is

$$y_n(x) = \sin(k_n x), \tag{6.1.24}$$

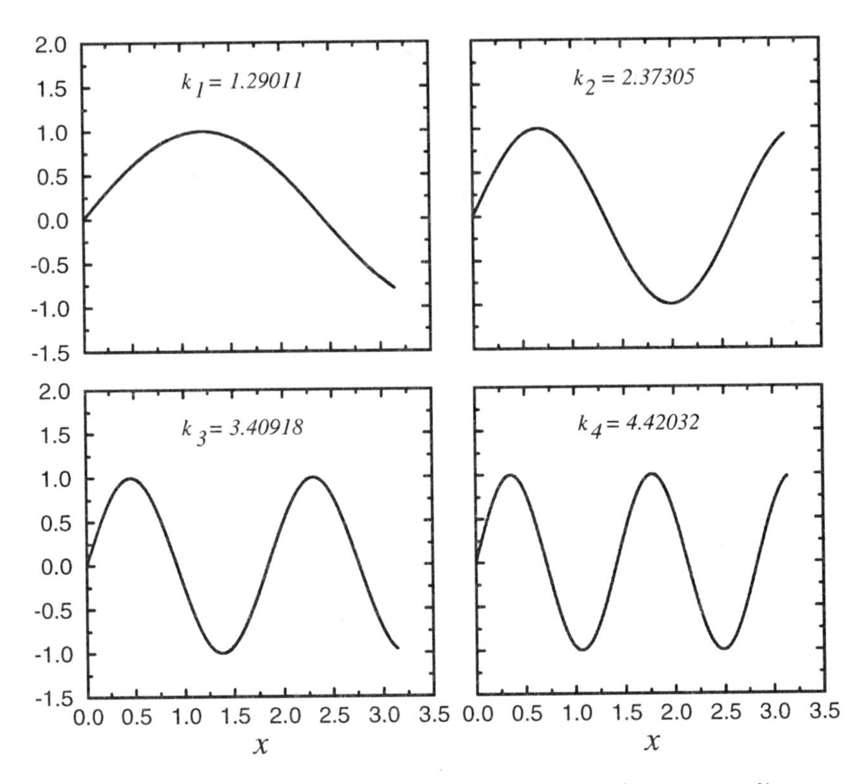

Figure 6.1.4: The first four eigenfunctions $\sin(k_n x)$ corresponding to the eigenvalue problem $\tan(k\pi) = k$.

because k must equal k_n. Figure 6.1.4 shows the first four eigenfunctions.

• **Example 6.1.2**

For our second example let us solve the Sturm-Liouville problem,

$$y'' + \lambda y = 0 \tag{6.1.25}$$

with the boundary conditions

$$y(0) - y'(0) = 0 \quad \text{and} \quad y(\pi) - y'(\pi) = 0. \tag{6.1.26}$$

Once again the three possible solutions to (6.1.25) are

$$y(x) = A\cosh(mx) + B\sinh(mx) \quad \text{if} \quad \lambda = -m^2 < 0, \tag{6.1.27}$$

$$y(x) = C + Dx \quad \text{if} \quad \lambda = 0 \tag{6.1.28}$$

and

$$y(x) = E\cos(kx) + F\sin(kx) \quad \text{if} \quad \lambda = k^2 > 0. \tag{6.1.29}$$

Let us first check and see if there are any nontrivial solutions for $\lambda < 0$. Two simultaneous equations result from the substitution of (6.1.27) into (6.1.26):

$$A - mB = 0 \tag{6.1.30}$$

$$[\cosh(m\pi) - m\sinh(m\pi)]A + [\sinh(m\pi) - m\cosh(m\pi)]B = 0. \tag{6.1.31}$$

The elimination of A between the two equations yields

$$\sinh(m\pi)(1 - m^2)B = 0. \tag{6.1.32}$$

If (6.1.27) is a nontrivial solution, then $B \neq 0$ and

$$\sinh(m\pi) = 0 \tag{6.1.33}$$

or

$$m^2 = 1. \tag{6.1.34}$$

Equation (6.1.33) cannot hold because it implies $m = \lambda = 0$ which contradicts the assumption used in deriving (6.1.27) that $\lambda < 0$. On the other hand, (6.1.34) is quite acceptable. It corresponds to the eigenvalue $\lambda = -1$ and the eigenfunction is

$$y_0 = \cosh(x) + \sinh(x) = e^x, \tag{6.1.35}$$

because it satisfies the differential equation

$$y_0'' - y_0 = 0 \tag{6.1.36}$$

and the boundary conditions

$$y_0(0) - y_0'(0) = 0 \tag{6.1.37}$$

and

$$y_0(\pi) - y_0'(\pi) = 0. \tag{6.1.38}$$

An alternative method of finding m, which is quite popular because of its use in more difficult problems, follows from viewing (6.1.30) and (6.1.31) as a system of homogeneous linear equations, where A and B are the unknowns. It is well known[4] that in order for (6.1.30)–(6.1.31)

---

[4] See Chapter 11.

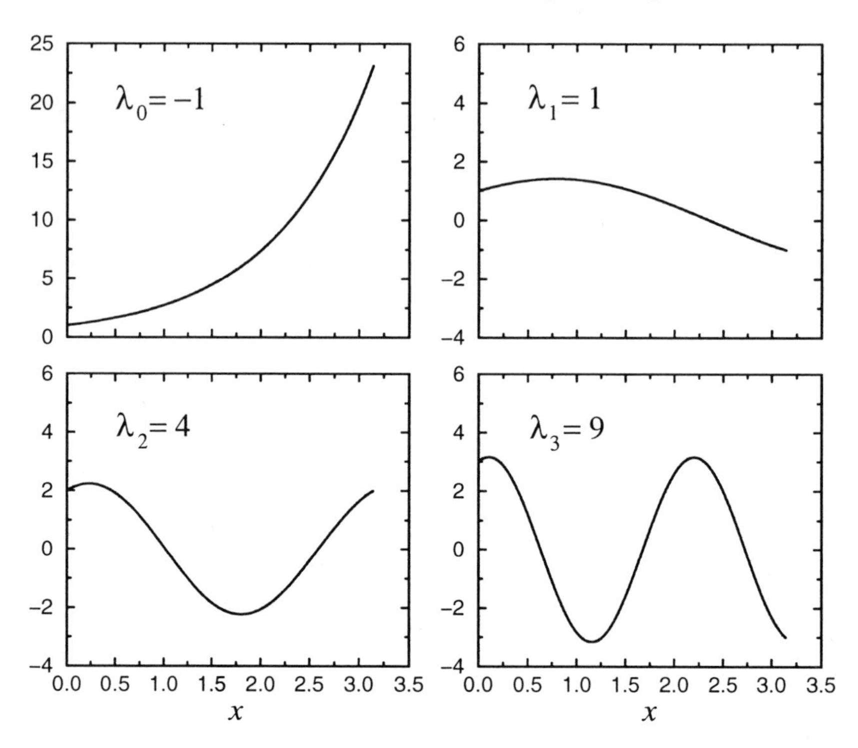

Figure 6.1.5: The first four eigenfunctions for the Sturm-Liouville problem (6.1.25)–(6.1.26).

to have a nontrivial solution (i.e., $A \neq 0$ and/or $B \neq 0$) the determinant of the coefficients must vanish:

$$\begin{vmatrix} 1 & -m \\ \cosh(m\pi) - m\sinh(m\pi) & \sinh(m\pi) - m\cosh(m\pi) \end{vmatrix} = 0. \quad \textbf{(6.1.39)}$$

Expanding the determinant,

$$\sinh(m\pi)(1 - m^2) = 0, \quad \textbf{(6.1.40)}$$

which leads directly to (6.1.33) and (6.1.34).

We consider next the case of $\lambda = 0$. Substituting (6.1.28) into (6.1.26), we find that

$$C - D = 0 \quad \textbf{(6.1.41)}$$

and

$$C + D\pi - D = 0. \quad \textbf{(6.1.42)}$$

This set of simultaneous equations yields $C = D = 0$ and we have only trivial solutions for $\lambda = 0$.

Finally, we examine the case when $\lambda > 0$. Substituting (6.1.29) into (6.1.26), we obtain

$$E - kF = 0 \tag{6.1.43}$$

and

$$[\cos(k\pi) + k\sin(k\pi)]E + [\sin(k\pi) - k\cos(k\pi)]F = 0. \tag{6.1.44}$$

The elimination of E from (6.1.43) and (6.1.44) gives

$$F(1 + k^2)\sin(k\pi) = 0. \tag{6.1.45}$$

In order that (6.1.29) be nontrivial, $F \neq 0$ and

$$k^2 = -1 \tag{6.1.46}$$

or

$$\sin(k\pi) = 0. \tag{6.1.47}$$

Condition (6.1.46) violates the assumption that k is real, which follows from the fact that $\lambda = k^2 > 0$. On the other hand, we can satisfy (6.1.47) if $k = 1, 2, 3, \ldots$; a negative k yields the same λ. Consequently we have the additional eigenvalues $\lambda_n = n^2$.

Let us now find the corresponding eigenfunctions. Because $E = kF$, $y(x) = F\sin(kx) + Fk\cos(kx)$ from (6.1.29). Thus, the eigenfunctions for $\lambda > 0$ are

$$y_n(x) = \sin(nx) + n\cos(nx). \tag{6.1.48}$$

Figure 6.1.4 illustrates some of the eigenfunctions given by (6.1.35) and (6.1.48).

• Example 6.1.3

Consider now the Sturm-Liouville problem

$$y'' + \lambda y = 0 \tag{6.1.49}$$

with

$$y(\pi) = y(-\pi) \quad \text{and} \quad y'(\pi) = y'(-\pi). \tag{6.1.50}$$

This is *not* a regular Sturm-Liouville problem because the boundary conditions are periodic and do not conform to the canonical boundary condition (6.1.2).

The general solution to (6.1.49) is

$$y(x) = A\cosh(mx) + B\sinh(mx) \quad \text{if} \quad \lambda = -m^2 < 0, \tag{6.1.51}$$

$$y(x) = C + Dx \quad \text{if} \quad \lambda = 0 \tag{6.1.52}$$

and

$$y(x) = E\cos(kx) + F\sin(kx) \quad \text{if} \quad \lambda = k^2 > 0. \tag{6.1.53}$$

Substituting these solutions into the boundary condition (6.1.50),

$$A\cosh(m\pi) + B\sinh(m\pi) = A\cosh(-m\pi) + B\sinh(-m\pi), \tag{6.1.54}$$

$$C + D\pi = C - D\pi \tag{6.1.55}$$

and

$$E\cos(k\pi) + F\sin(k\pi) = E\cos(-k\pi) + F\sin(-k\pi) \tag{6.1.56}$$

or

$$B\sinh(m\pi) = 0, \quad D = 0 \quad \text{and} \quad F\sin(k\pi) = 0, \tag{6.1.57}$$

because $\cosh(-m\pi) = \cosh(m\pi)$, $\sinh(-m\pi) = -\sinh(m\pi)$, $\cos(-k\pi) = \cos(k\pi)$, and $\sin(-k\pi) = -\sin(k\pi)$. Because m must be positive, $\sinh(m\pi)$ cannot equal zero and $B = 0$. On the other hand, if $\sin(k\pi) = 0$ or $k = n$, $n = 1, 2, 3, \ldots$, we have a nontrivial solution for positive λ and $\lambda_n = n^2$. Note that we still have A, C, E, and F as free constants.

From the boundary condition (6.1.50),

$$A\sinh(m\pi) = A\sinh(-m\pi) \tag{6.1.58}$$

and

$$-E\sin(k\pi) + F\cos(k\pi) = -E\sin(-k\pi) + F\cos(-k\pi). \tag{6.1.59}$$

The solution $y_0(x) = C$ identically satisfies the boundary condition (6.1.50) for all C. Because m and $\sinh(m\pi)$ must be positive, $A = 0$. From (6.1.57), we once again have $\sin(k\pi) = 0$ and $k = n$. Consequently, the eigenfunction solutions to (6.1.49)–(6.1.50) are

$$\lambda_0 = 0, \qquad y_0(x) = 1 \tag{6.1.60}$$

and

$$\lambda_n = n^2, \qquad y_n(x) = \begin{cases} \sin(nx) \\ \cos(nx) \end{cases} \tag{6.1.61}$$

and we have a degenerate set of eigenfunctions to the Sturm-Liouville problem (6.1.49) with the periodic boundary condition (6.1.50).

Problems

Find the eigenvalues and eigenfunctions for each of the following:

1. $y'' + \lambda y = 0, \quad y'(0) = 0, \quad y(L) = 0$

2. $y'' + \lambda y = 0, \quad y'(0) = 0, \quad y'(\pi) = 0$

3. $y'' + \lambda y = 0, \quad y(0) + y'(0) = 0, \quad y(\pi) + y'(\pi) = 0$

4. $y'' + \lambda y = 0, \quad y'(0) = 0, \quad y(\pi) - y'(\pi) = 0$

5. $y^{(iv)} + \lambda y = 0, \quad y(0) = y''(0) = 0, \quad y(L) = y''(L) = 0$

Find an equation from which you could find λ and give the form of the eigenfunction for each of the following:

6. $y'' + \lambda y = 0, \quad y(0) + y'(0) = 0, \quad y(1) = 0$

7. $y'' + \lambda y = 0, \quad y(0) = 0, \quad y(\pi) + y'(\pi) = 0$

8. $y'' + \lambda y = 0, \quad y'(0) = 0, \quad y(1) - y'(1) = 0$

9. $y'' + \lambda y = 0, \quad y(0) + y'(0) = 0, \quad y'(\pi) = 0$

10. $y'' + \lambda y = 0, \quad y(0) + y'(0) = 0, \quad y(\pi) - y'(\pi) = 0$

11. Find the eigenvalues and eigenfunctions of the Sturm-Liouville problem

$$\frac{d}{dx}\left[x\frac{dy}{dx}\right] + \frac{\lambda}{x}y = 0, \quad 1 \le x \le e$$

for each of the following boundary conditions: (a) $u(1) = u(e) = 0$, (b) $u(1) = u'(e) = 0$, and (c) $u'(1) = u'(e) = 0$.

Find the eigenvalues and eigenfunctions of the following Sturm-Liouville problems:

12.
$$x^2y'' + 2xy' + \lambda y = 0, \quad y(1) = y(e) = 0, \quad 1 \le x \le e.$$

13.
$$\frac{d}{dx}\left[x^3y'\right] + \lambda xy = 0, \quad y(1) = y(e^\pi) = 0, \quad 1 \le x \le e^\pi.$$

14.
$$\frac{d}{dx}\left[\frac{1}{x}y'\right] + \frac{\lambda}{x}y = 0, \quad y(1) = y(e) = 0, \quad 1 \le x \le e.$$

6.2 ORTHOGONALITY OF EIGENFUNCTIONS

In the previous section we saw how nontrivial solutions to the regular Sturm-Liouville problem consist of eigenvalues and eigenfunctions. The most important property of eigenfunctions is orthogonality.

Theorem: *Let the functions $p(x)$, $q(x)$, and $r(x)$ of the regular Sturm-Liouville problem (6.1.1)–(6.1.2) be real and continuous on the interval $[a, b]$. If $y_n(x)$ and $y_m(x)$ are continuously differentiable eigenfunctions corresponding to the distinct eigenvalues λ_n and λ_m, respectively, then $y_n(x)$ and $y_m(x)$ satisfy the* orthogonality *condition:*

$$\int_a^b r(x) y_n(x) y_m(x)\, dx = 0, \tag{6.2.1}$$

if $\lambda_n \neq \lambda_m$. When (6.2.1) is satisfied, the eigenfunction $y_n(x)$ and $y_m(x)$ are said to be *orthogonal* to each other with respect to the *weight function $r(x)$*. The term *orthogonality* appears to be borrowed from linear algebra where a similar relationship holds between two perpendicular or orthogonal vectors.

Proof: Let y_n and y_m denote the eigenfunctions associated with two different eigenvalues λ_n and λ_m. Then

$$\frac{d}{dx}\left[p(x)\frac{dy_n}{dx}\right] + [q(x) + \lambda_n r(x)]y_n(x) = 0, \tag{6.2.2}$$

$$\frac{d}{dx}\left[p(x)\frac{dy_m}{dx}\right] + [q(x) + \lambda_m r(x)]y_m(x) = 0 \tag{6.2.3}$$

and both solutions satisfy the boundary conditions. Let us multiply the first differential equation by y_m; the second by y_n. Next, we subtract these two equations and move the terms containing $y_n y_m$ to the right side. The resulting equation is

$$y_n \frac{d}{dx}\left[p(x)\frac{dy_m}{dx}\right] - y_m \frac{d}{dx}\left[p(x)\frac{dy_n}{dx}\right] = (\lambda_n - \lambda_m) r(x) y_n y_m. \tag{6.2.4}$$

Integrating (6.2.4) from a to b yields

$$\int_a^b \left\{ y_n \frac{d}{dx}\left[p(x)\frac{dy_m}{dx}\right] - y_m \frac{d}{dx}\left[p(x)\frac{dy_n}{dx}\right]\right\} dx$$
$$= (\lambda_n - \lambda_m)\int_a^b r(x) y_n y_m\, dx. \tag{6.2.5}$$

We may simplify the left side of (6.2.5) by integrating by parts to give

$$\int_a^b \left\{ y_n \frac{d}{dx}\left[p(x)\frac{dy_m}{dx}\right] - y_n \frac{d}{dx}\left[p(x)\frac{dy_m}{dx}\right]\right\} dx$$

$$= [p(x)y_m' y_n - p(x)y_n' y_m]_a^b - \int_a^b p(x)[y_n' y_m' - y_n' y_m']dx. \quad (6.2.6)$$

The second integral equals zero since the integrand vanishes identically. Because $y_n(x)$ and $y_m(x)$ satisfy the boundary condition at $x = a$,

$$\alpha y_n(a) + \beta y_n'(a) = 0 \quad (6.2.7)$$

and

$$\alpha y_m(a) + \beta y_m'(a) = 0. \quad (6.2.8)$$

These two equations are simultaneous equations in α and β. Hence, the determinant of the equations must be zero:

$$y_n'(a)y_m(a) - y_m'(a)y_n(a) = 0. \quad (6.2.9)$$

Similarly, at the other end,

$$y_n'(b)y_m(b) - y_m'(b)y_n(b) = 0. \quad (6.2.10)$$

Consequently, the right side of (6.2.6) vanishes and (6.2.5) reduces to (6.2.1). □

- **Example 6.2.1**

Let us verify the orthogonality condition for the eigenfunctions that we found in Example 6.1.1.

Because $r(x) = 1$, $a = 0$, $b = \pi$, and $y_n(x) = \sin(k_n x)$, we find that

$$\int_a^b r(x)y_n y_m \, dx = \int_0^\pi \sin(k_n x)\sin(k_m x)\, dx \quad (6.2.11)$$

$$= \frac{1}{2}\int_0^\pi \{\cos[(k_n - k_m)x] - \cos[(k_n + k_m)x]\, dx \quad (6.2.12)$$

$$= \left.\frac{\sin[(k_n - k_m)x]}{2(k_n - k_m)}\right|_0^\pi - \left.\frac{\sin[(k_n + k_m)x]}{2(k_n + k_m)}\right|_0^\pi \quad (6.2.13)$$

$$= \frac{\sin[(k_n - k_m)\pi]}{2(k_n - k_m)} - \frac{\sin[(k_n + k_m)\pi]}{2(k_n + k_m)} \quad (6.2.14)$$

$$= \frac{\sin(k_n\pi)\cos(k_m\pi) - \cos(k_n\pi)\sin(k_m\pi)}{2(k_n - k_m)}$$

$$- \frac{\sin(k_n\pi)\cos(k_m\pi) + \cos(k_n\pi)\sin(k_m\pi)}{2(k_n + k_m)} \tag{6.2.15}$$

$$\int_a^b r(x) y_n y_m \, dx = \frac{k_n \cos(k_n\pi)\cos(k_m\pi) - k_m \cos(k_n\pi)\cos(k_m\pi)}{2(k_n - k_m)} - \frac{k_n \cos(k_n\pi)\cos(k_m\pi) + k_m \cos(k_n\pi)\cos(k_m\pi)}{2(k_n + k_m)} \tag{6.2.16}$$

$$= \frac{(k_n - k_m)\cos(k_n\pi)\cos(k_m\pi)}{2(k_n - k_m)} - \frac{(k_n + k_m)\cos(k_n\pi)\cos(k_m\pi)}{2(k_n + k_m)} = 0. \tag{6.2.17}$$

We have used the relationships $k_n = \tan(k_n\pi)$ and $k_m = \tan(k_m\pi)$ to simplify (6.2.15). Note, however, that if $n = m$,

$$\int_0^\pi \sin(k_n x)\sin(k_n x)\, dx = \tfrac{1}{2}\int_0^\pi [1 - \cos(2k_n x)]\, dx \tag{6.2.18}$$

$$= \frac{\pi}{2} - \frac{\sin(2k_n\pi)}{4k_n} \tag{6.2.19}$$

$$= \tfrac{1}{2}[\pi - \cos^2(k_n\pi)] > 0 \tag{6.2.20}$$

because $\sin(2A) = 2\sin(A)\cos(A)$ and $k_n = \tan(k_n\pi)$. That is, any eigenfunction *cannot* be orthogonal to itself.

In closing, we note that had we defined the eigenfunction in our example as

$$y_n(x) = \frac{\sin(k_n x)}{\sqrt{[\pi - \cos^2(k_n\pi)]/2}} \tag{6.2.21}$$

rather than $y_n(x) = \sin(k_n x)$, the orthogonality condition would read

$$\int_0^\pi y_n(x) y_m(x)\, dx = \begin{cases} 0, & m \neq n \\ 1, & m = n. \end{cases} \tag{6.2.22}$$

This process of *normalizing* an eigenfunction so that the orthogonality condition becomes

$$\int_a^b r(x) y_n(x) y_m(x)\, dx = \begin{cases} 0, & m \neq n \\ 1, & m = n \end{cases} \tag{6.2.23}$$

generates *orthonormal* eigenfunctions. We will see the convenience of doing this in the next section.

Problems

1. The Sturm-Liouville problem $y'' + \lambda y = 0$, $y(0) = y(L) = 0$ has the eigenfunction solution $y_n(x) = \sin(n\pi x/L)$. By direct integration verify the orthogonality condition (6.2.1).

2. The Sturm-Liouville problem $y'' + \lambda y = 0$, $y'(0) = y'(L) = 0$ has the eigenfunction solutions $y_0(x) = 1$ and $y_n(x) = \cos(n\pi x/L)$. By direct integration verify the orthogonality condition (6.2.1).

3. The Sturm-Liouville problem $y'' + \lambda y = 0$, $y(0) = y'(L) = 0$ has the eigenfunction solution $y_n(x) = \sin[(2n-1)\pi x/(2L)]$. By direct integration verify the orthogonality condition (6.2.1).

4. The Sturm-Liouville problem $y'' + \lambda y = 0$, $y'(0) = y(L) = 0$ has the eigenfunction solution $y_n(x) = \cos[(2n-1)\pi x/(2L)]$. By direct integration verify the orthogonality condition (6.2.1).

6.3 EXPANSION IN SERIES OF EIGENFUNCTIONS

In calculus we learned that under certain conditions we could represent a function $f(x)$ by a linear and infinite sum of polynomials $(x-x_0)^n$. In this section we show that an analogous procedure exists for representing a piece-wise continuous function by a linear sum of eigenfunctions. These *eigenfunction expansions* will be used in the next three chapters to solve partial differential equations.

Let the function $f(x)$ be defined in the interval $a < x < b$. We wish to reexpress $f(x)$ in terms of the eigenfunctions $y_n(x)$ given by a regular Sturm-Liouville problem. Assuming that the function $f(x)$ can be represented by a uniformly convergent series,[5] we write

$$f(x) = \sum_{n=1}^{\infty} c_n y_n(x). \tag{6.3.1}$$

The orthogonality relation (6.2.1) gives us the method for computing the coefficients c_n. First we multiply both sides of (6.3.1) by $r(x)y_m(x)$, where m is a fixed integer, and then integrate from a to b. Because this

[5] If $S_n(x) = \sum_{k=1}^{n} u_k(x)$, $S(x) = \lim_{n\to\infty} S_n(x)$ and $0 < |S_n(x) - S(x)| < \epsilon$ for all $n > M > 0$, the series $\sum_{k=1}^{\infty} u_k(x)$ is uniformly convergent if M is dependent on ϵ alone and not x.

series is uniformly convergent and $y_n(x)$ is continuous, we can integrate the series term by term or

$$\int_a^b r(x)f(x)y_m(x)\,dx = \sum_{n=1}^{\infty} c_n \int_a^b r(x)y_n(x)y_m(x)\,dx. \qquad \textbf{(6.3.2)}$$

The orthogonality relationship states that all of the terms on the right side of (6.3.2) must disappear except the one for which $n = m$. Thus, we are left with

$$\int_a^b r(x)f(x)y_m(x)\,dx = c_m \int_a^b r(x)y_m(x)y_m(x)\,dx \qquad \textbf{(6.3.3)}$$

or

$$\boxed{c_n = \frac{\int_a^b r(x)f(x)y_n(x)\,dx}{\int_a^b r(x)y_n^2(x)\,dx},} \qquad \textbf{(6.3.4)}$$

if we replace m by n in (6.3.3).

The series (6.3.1) with the coefficients found by (6.3.4) is a *generalized Fourier series* of the function $f(x)$ with respect to the eigenfunction $y_n(x)$. It is called a generalized Fourier series because we have generalized the procedure of reexpressing a function $f(x)$ by sines and cosines into one involving solutions to regular Sturm-Liouville problems. Note that if we had used an orthonormal set of eigenfunctions, then the denominator of (6.3.4) would equal one and we reduce our work by half. The coefficients c_n are the *Fourier coefficients.*

One of the most remarkable facts about generalized Fourier series is their applicability even when the function has a finite number of bounded discontinuities in the range $[a, b]$. We may formally express this fact by the following theorem:

Theorem: *If both $f(x)$ and $f'(x)$ are piece-wise continuous in $a \le x \le b$, then $f(x)$ can be expanded in a uniformly convergent Fourier series (6.3.1), whose coefficients c_n are given by (6.3.4). It converges to $[f(x^+) + f(x^-)]/2$ at any point x in the open interval $a < x < b$.*

The proof is beyond the scope of this book but may be found in more advanced treatises.[6] If we are willing to include stronger constraints,

[6] For example, Titchmarsh, E. C., 1962: *Eigenfunction Expansions Associated with Second-Order Differential Equations. Part I*, Oxford University Press, Oxford, pp. 12–16.

we can make even stronger statements about convergence. For example,[7] if we require that $f(x)$ be a continuous function with a piece-wise continuous first derivative, then the eigenfunction expansion (6.3.1) will converge to $f(x)$ uniformly and absolutely in $[a, b]$ if $f(x)$ satisfies the same boundary conditions as does $y_n(x)$.

In the case when $f(x)$ is discontinuous, we are not merely rewriting $f(x)$ in a new form. We are actually choosing the c_n's so that the eigenfunctions fit $f(x)$ in the "least squares" sense that

$$\int_a^b r(x) \left| f(x) - \sum_{n=1}^{\infty} c_n y_n(x) \right|^2 dx = 0. \tag{6.3.5}$$

Consequently we should expect peculiar things, such as spurious oscillations, to occur in the neighborhood of the discontinuity. This is *Gibbs phenomena*,[8] the same phenomena discovered with Fourier series. See Section 2.2.

• Example 6.3.1

To illustrate the concept of an eigenfunction expansion, let us find the expansion for $f(x) = x$ over the interval $0 < x < \pi$ using the solution to the regular Sturm-Liouville problem of

$$y'' + \lambda y = 0, \qquad y(0) = y(\pi) = 0. \tag{6.3.6}$$

This problem will arise when we solve the wave or heat equation by separation of variables in the next two chapters.

Because the eigenfunctions are $y_n(x) = \sin(nx)$, $n = 1, 2, 3, \ldots$, $r(x) = 1$, $a = 0$, and $b = \pi$, (6.3.4) gives

$$c_n = \frac{\int_0^\pi x \sin(nx)\, dx}{\int_0^\pi \sin^2(nx)\, dx} = \frac{-x\cos(nx)/n + \sin(nx)/n^2 \big|_0^\pi}{x/2 - \sin(2nx)/(4n) \big|_0^\pi} \tag{6.3.7}$$

$$= -\frac{2}{n}\cos(n\pi) = \frac{2}{n}(-1)^n. \tag{6.3.8}$$

Equation (6.3.1) then gives

$$f(x) = -2 \sum_{n=1}^{\infty} \frac{(-1)^n}{n} \sin(nx). \tag{6.3.9}$$

[7] Tolstov, G. P., 1962: *Fourier Series*, Dover Publishers, Mineola, NY, p. 255.

[8] Apparently first discussed by Weyl, H., 1910: Die Gibbs'sche Erscheinung in der Theorie der Sturm-Liouvilleschen Reihen. *Rend. Circ. Mat. Palermo*, **29**, 321–323.

This particular example is in fact an example of a half-range sine expansion.

Finally we must state the values of x for which (6.3.9) is valid. At $x = \pi$ the series converges to zero while $f(\pi) = \pi$. At $x = 0$ both the series and the function converge to zero. Hence the series expansion (6.3.9) is valid for $0 \le x < \pi$.

• Example 6.3.2

For our second example let us find the expansion for $f(x) = x$ over the interval $0 \le x < \pi$ using the solution to the regular Sturm-Liouville problem of

$$y'' + \lambda y = 0, \qquad y(0) = y(\pi) - y'(\pi) = 0. \tag{6.3.10}$$

We will encounter this problem when we solve the heat equation with radiative boundary conditions by separation of variables.

Because $r(x) = 1$, $a = 0$, $b = \pi$ and the eigenfunctions are $y_n(x) = \sin(k_n x)$, where $k_n = \tan(k_n \pi)$, (6.3.4) give

$$c_n = \frac{\int_0^\pi x \sin(k_n x)\, dx}{\int_0^\pi \sin^2(k_n x)\, dx} = \frac{\int_0^\pi x \sin(k_n x)\, dx}{\frac{1}{2}\int_0^\pi [1 - \cos(2k_n x)]\, dx} \tag{6.3.11}$$

$$= \frac{2\sin(k_n x)/k_n^2 - 2x\cos(k_n x)/k_n \big|_0^\pi}{x - \sin(2k_n x)/(2k_n)\big|_0^\pi} \tag{6.3.12}$$

$$= \frac{2\sin(k_n \pi)/k_n^2 - 2\pi\cos(k_n \pi)/k_n}{\pi - \sin(2k_n \pi)/(2k_n)} \tag{6.3.13}$$

$$= \frac{2[\cos(k_n \pi) - \pi\cos(k_n \pi)]/k_n}{\pi - \cos^2(k_n \pi)}, \tag{6.3.14}$$

where we have used the property that $\sin(k_n \pi) = k_n \cos(k_n \pi)$. Equation (6.3.1) then gives

$$f(x) = 2(1 - \pi)\sum_{n=1}^{\infty} \frac{\cos(k_n \pi)}{k_n[\pi - \cos^2(k_n \pi)]} \sin(k_n x). \tag{6.3.15}$$

Problems

1. The Sturm-Liouville problem $y'' + \lambda y = 0$, $y(0) = y(L) = 0$ has the eigenfunction solution $y_n(x) = \sin(n\pi x/L)$. Find the eigenfunction expansion for $f(x) = x$ using this eigenfunction.

2. The Sturm-Liouville problem $y'' + \lambda y = 0$, $y'(0) = y'(L) = 0$ has the eigenfunction solutions $y_0(x) = 1$ and $y_n(x) = \cos(n\pi x/L)$. Find the eigenfunction expansion for $f(x) = x$ using these eigenfunctions.

3. The Sturm-Liouville problem $y'' + \lambda y = 0$, $y(0) = y'(L) = 0$ has the eigenfunction solution $y_n(x) = \sin[(2n-1)\pi x/(2L)]$. Find the eigenfunction expansion for $f(x) = x$ using this eigenfunction.

4. The Sturm-Liouville problem $y'' + \lambda y = 0$, $y'(0) = y(L) = 0$ has the eigenfunction solution $y_n(x) = \cos[(2n-1)\pi x/(2L)]$. Find the eigenfunction expansion for $f(x) = x$ using this eigenfunction.

6.4 A SINGULAR STURM-LIOUVILLE PROBLEM: LEGENDRE'S EQUATION

In the previous sections we used solutions to a regular Sturm-Liouville problem in the eigenfunction expansion of the function $f(x)$. The fundamental reason why we could form such an expansion was the orthogonality condition (6.2.1). This crucial property allowed us to solve for the Fourier coefficient c_n given by (6.3.4).

In the next few chapters, when we solve partial differential equations in cylindrical and spherical coordinates, we will find that $f(x)$ must be expanded in terms of eigenfunctions from singular Sturm-Liouville problems. Is this permissible? How do we compute the Fourier coefficients in this case? The final two sections of this chapter deal with these questions by examining the two most frequently encountered singular Sturm-Liouville problems, those involving Legendre's and Bessel's equations.

We begin by determining the orthogonality condition for singular Sturm-Liouville problems. Returning to the beginning portions of Section 6.2, we combine (6.2.5) and (6.2.6) to obtain

$$(\lambda_n - \lambda_m)\int_a^b r(x) y_n y_m \, dx = [p(b)y_m'(b)y_n(b) - p(b)y_n'(b)y_m(b) - p(a)y_m'(a)y_n(a) + p(a)y_n'(a)y_m(a)]. \tag{6.4.1}$$

From (6.4.1) the right side vanishes and we preserve orthogonality if $y_n(x)$ is finite and $p(x)y_n'(x)$ tends to zero at both endpoints. This is not the only choice but let us see where it leads.

Consider now Legendre's equation:

$$(1-x^2)\frac{d^2y}{dx^2} - 2x\frac{dy}{dx} + n(n+1)y = 0 \tag{6.4.2}$$

or

$$\frac{d}{dx}\left[(1-x^2)\frac{dy}{dx}\right] + n(n+1)y = 0, \tag{6.4.3}$$

where we set $a = -1$, $b = 1$, $\lambda = n(n+1)$, $p(x) = 1 - x^2$, $q(x) = 0$, and $r(x) = 1$. This equation arises in the solution of partial differential

Figure 6.4.1: Born into an affluent family, Adrien-Marie Legendre's (1752–1833) modest family fortune was sufficient to allow him to devote his life to research in celestial mechanics, number theory, and the theory of elliptic functions. In July 1784 he read before the *Académie des sciences* his *Recherches sur la figure des planètes*. It is in this paper that Legendre polynomials first appeared. (Portrait courtesy of the Archives de l'Académie des sciences, Paris.)

equations involving spherical geometry. Because $p(-1) = p(1) = 0$, we are faced with a singular Sturm-Liouville problem. Before we can determine if any of its solutions can be used in an eigenfunction expansion, we must find them.

Equation (6.4.2) does not have a simple general solution. [If $n = 0$, then $y(x) = 1$ is a solution.] Consequently we try to solve it with the power series:

$$y(x) = \sum_{k=0}^{\infty} A_k x^k, \tag{6.4.4}$$

$$y'(x) = \sum_{k=0}^{\infty} kA_k x^{k-1} \tag{6.4.5}$$

and

$$y''(x) = \sum_{k=0}^{\infty} k(k-1)A_k x^{k-2}. \tag{6.4.6}$$

Substituting into (6.4.2),

$$\sum_{k=0}^{\infty} k(k-1)A_k x^{k-2} + \sum_{k=0}^{\infty} [n(n+1) - 2k - k(k-1)]\, A_k x^k = 0, \tag{6.4.7}$$

which equals

$$\sum_{m=2}^{\infty} m(m-1)A_m x^{m-2} + \sum_{k=0}^{\infty} [n(n+1) - k(k+1)]\, A_k x^k = 0. \tag{6.4.8}$$

If we define $k = m + 2$ in the first summation, then

$$\sum_{k=0}^{\infty} (k+2)(k+1)A_{k+2} x^k + \sum_{k=0}^{\infty} [n(n+1) - k(k+1)]\, A_k x^k = 0. \tag{6.4.9}$$

Because (6.4.9) must be true for any x, each power of x must vanish separately. It then follows that

$$(k+2)(k+1)A_{k+2} = [k(k+1) - n(n+1)]A_k \tag{6.4.10}$$

or

$$A_{k+2} = \frac{[k(k+1) - n(n+1)]}{(k+1)(k+2)} A_k, \tag{6.4.11}$$

where $k = 0, 1, 2, \ldots$ Note that we still have the two arbitrary constants A_0 and A_1 that are necessary for the general solution of (6.4.2).

The first few terms of the solution associated with A_0 are

$$u_p(x) = 1 - \frac{n(n+1)}{2!}x^2 + \frac{n(n-2)(n+1)(n+3)}{4!}x^4 - \frac{n(n-2)(n-4)(n+1)(n+3)(n+5)}{6!}x^6 + \cdots \tag{6.4.12}$$

while the first few terms associated with the A_1 coefficient are

$$v_p(x) = x - \frac{(n-1)(n+2)}{3!}x^3 + \frac{(n-1)(n-3)(n+2)(n+4)}{5!}x^5 - \frac{(n-1)(n-3)(n-5)(n+2)(n+4)(n+6)}{7!}x^7 + \cdots \tag{6.4.13}$$

If n is an *even* positive integer (including $n = 0$), then the series (6.6.12) terminates with the term involving x^n: the solution is a polynomial of degree n. Similarly, if n is an *odd* integer, the series (6.4.13) terminates with the term involving x^n. Otherwise, for n noninteger the expressions are infinite series.

For reasons that will become apparent, we restrict ourselves to positive integers n. Actually, this includes all possible integers because the negative integer $-n-1$ has the same Legendre's equation and solution as the positive integer n. These polynomials are *Legendre polynomials*[9] and we may compute them by the power series:

$$P_n(x) = \sum_{k=0}^{m} (-1)^k \frac{(2n-2k)!}{2^n k!(n-k)!(n-2k)!} x^{n-2k}, \qquad \textbf{(6.4.14)}$$

where $m = n/2$ or $m = (n-1)/2$, depending upon which is an integer. We have chosen to use (6.4.14) over (6.4.12) or (6.4.13) because (6.4.14) has the advantage that $P_n(1) = 1$. Table 6.4.1 gives the first ten Legendre polynomials.

The other solution, the infinite series, is the Legendre function of the second kind, $Q_n(x)$. Figure 6.4.2 illustrates the first four Legendre polynomials $P_n(x)$ while Figure 6.4.3 gives the first four Legendre functions of the second kind Q_n. From this figure we see that $Q_n(x)$ becomes infinite at the points $x = \pm 1$. As shown earlier, this is important because we are only interested in solutions to Legendre's equation that are finite over the interval $[-1, 1]$. On the other hand, in problems where we exclude the points $x = \pm 1$, Legendre functions of the second kind will appear in the general solution.[10]

In the case that n is not an integer, we can construct a solution[11] that remains finite at $x = 1$ but not at $x = -1$. Furthermore, we can

---

[9] Legendre, A. M., 1785: Sur l'attraction des sphéroïdes homogénes. *Mém. math. phys. présentés à l'Acad. sci. pars divers savants*, **10**, 411–434. The best reference on Legendre polynomials is given by Hobson, E. W., 1965: *The Theory of Spherical and Ellipsoidal Harmonics*, Chelsea Publishing Co., New York.

[10] See Smythe, W. R., 1950: *Static and Dynamic Electricity*, McGraw-Hill, New York, Section 5.215 for an example.

[11] See Carrier, G. F., Krook, M., and Pearson, C. E., 1966: *Functions of the Complex Variable: Theory and Technique*, McGraw-Hill, New York, pp. 212–213.

Table 6.4.1: The First Ten Legendre Polynomials.

$$P_0(x) = 1$$

$$P_1(x) = x$$

$$P_2(x) = \tfrac{1}{2}(3x^2 - 1)$$

$$P_3(x) = \tfrac{1}{2}(5x^3 - 3x)$$

$$P_4(x) = \tfrac{1}{8}(35x^4 - 30x^2 + 3)$$

$$P_5(x) = \tfrac{1}{8}(63x^5 - 70x^3 + 15x)$$

$$P_6(x) = \tfrac{1}{16}(231x^6 - 315x^4 + 105x^2 - 5)$$

$$P_7(x) = \tfrac{1}{16}(429x^7 - 693x^5 + 315x^3 - 35x)$$

$$P_8(x) = \tfrac{1}{128}(6435x^8 - 12012x^6 + 6930x^4 - 1260x^2 + 35)$$

$$P_9(x) = \tfrac{1}{128}(12155x^9 - 25740x^7 + 18018x^5 - 4620x^3 + 315x)$$

$$P_{10}(x) = \tfrac{1}{256}(46189x^{10} - 109395x^8 + 90090x^6 - 30030x^4 + 3465x^2 - 63)$$

construct a solution which is finite at $x = -1$ but not at $x = 1$. Because our solutions must be finite at both endpoints so that we can use them in an eigenfunction expansion, we must reject these solutions from further consideration and are left only with Legendre polynomials. From now on, we will only consider the properties and uses of these polynomials.

Although we have the series (6.4.14) to compute $P_n(x)$, there are several alternative methods. We obtain the first method, known as *Rodrigues' formula*,[12] by writing (6.4.14) in the form

$$P_n(x) = \frac{1}{2^n n!}\sum_{k=0}^{n}(-1)^k \frac{n!}{k!(n-k)!}\frac{(2n-2k)!}{(n-2k)!}x^{n-2k} \tag{6.4.15}$$

$$= \frac{1}{2^n n!}\frac{d^n}{dx^n}\left[\sum_{k=0}^{n}(-1)^k \frac{n!}{k!(n-k)!}x^{2n-2k}\right]. \tag{6.4.16}$$

The last summation is the binomial expansion of $(x^2-1)^n$ so that

[12] Rodriques, O., 1816: Mémoire sur l'attraction des sphéroïdes. *Correspond. l'Ecole Polytech.*, **3**, 361–385.

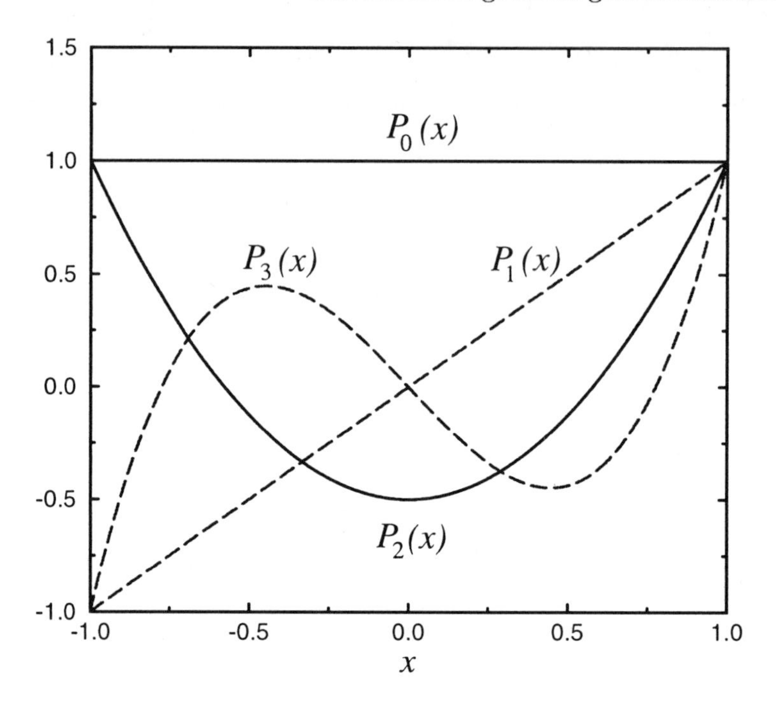

Figure 6.4.2: The first four Legendre functions of the first kind.

$$P_n(x) = \frac{1}{2^n n!} \frac{d^n}{dx^n}(x^2 - 1)^n. \qquad \textbf{(6.4.17)}$$

Another method for computing $P_n(x)$ involves the use of recurrence formulas. The first step in finding these formulas is to establish the fact that

$$(1 + h^2 - 2xh)^{-1/2} = P_0(x) + hP_1(x) + h^2P_2(x) + \cdots \qquad \textbf{(6.4.18)}$$

The function $(1 + h^2 - 2xh)^{-1/2}$ is the *generating function* for $P_n(x)$. We obtain the expansion via the formal binomial expansion

$$(1+h^2-2xh)^{-1/2} = 1 + \tfrac{1}{2}(2xh - h^2) + \tfrac{1}{2}\tfrac{3}{2}\tfrac{1}{2!}(2xh - h^2)^2 + \cdots \quad \textbf{(6.4.19)}$$

Upon expanding the terms contained in $2x - h^2$ and grouping like powers of h,

$$(1 + h^2 - 2xh)^{-1/2} = 1 + xh + (\tfrac{3}{2}x^2 - \tfrac{1}{2})h^2 + \cdots \qquad \textbf{(6.4.20)}$$

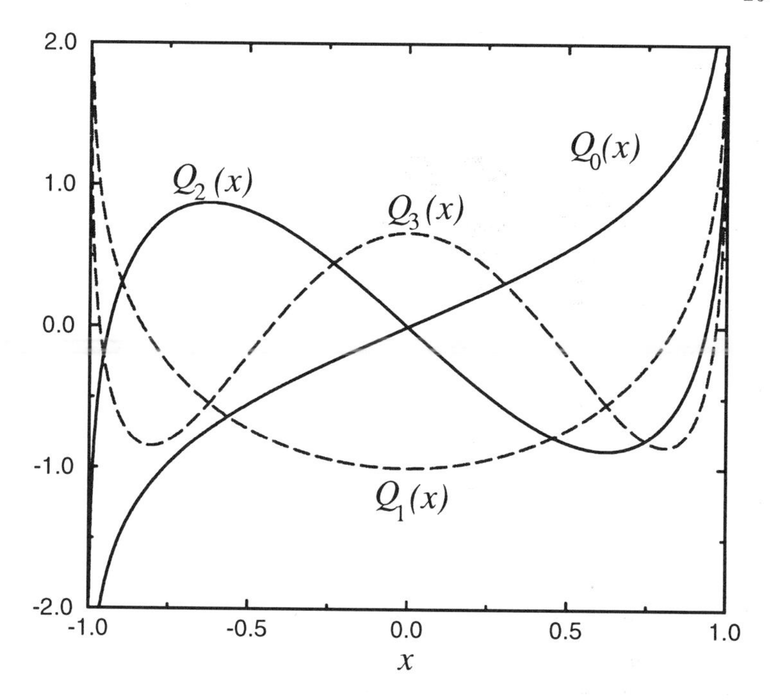

Figure 6.4.3: The first four Legendre functions of the second kind.

A direct comparison between the coefficients of each power of h and the Legendre polynomial $P_n(x)$ completes the demonstration. Note that these results hold only if $|x|$ and $|h| < 1$.

Next we define $W(x,h) = (1+h^2-2xh)^{-1/2}$. A quick check shows that $W(x,h)$ satisfies the first-order partial differential equation

$$(1-2xh+h^2)\frac{\partial W}{\partial h} + (h-x)W = 0. \qquad \textbf{(6.4.21)}$$

The substitution of (6.4.18) into (6.4.21) yields

$$(1-2xh+h^2)\sum_{n=0}^{\infty} nP_n(x)h^{n-1} + (h-x)\sum_{n=0}^{\infty} P_n(x)h^n = 0. \qquad \textbf{(6.4.22)}$$

Setting the coefficients of h^n equal to zero, we find that

$$(n+1)P_{n+1}(x) - 2nxP_n(x) + (n-1)P_{n-1}(x) + P_{n-1}(x) - xP_n(x) = 0 \qquad \textbf{(6.4.23)}$$

or

$$\boxed{(n+1)P_{n+1}(x) - (2n+1)xP_n(x) + nP_{n-1}(x) = 0} \qquad \textbf{(6.4.24)}$$

with $n = 1, 2, 3, \ldots$

Similarly, the first-order partial differential equation

$$(1 - 2xh + h^2)\frac{\partial W}{\partial x} - hW = 0 \tag{6.4.25}$$

leads to

$$(1 - 2xh + h^2)\sum_{n=0}^{\infty} P_n'(x)h^n - \sum_{n=0}^{\infty} P_n(x)h^{n+1} = 0, \tag{6.4.26}$$

which implies

$$P_{n+1}'(x) - 2xP_n'(x) + P_{n-1}'(x) - P_n(x) = 0. \tag{6.4.27}$$

Differentiating (6.4.24), we first eliminate $P_{n-1}'(x)$ and then $P_{n+1}'(x)$ from the resulting equations and (6.4.27). This gives two further recurrence relationships:

$$P_{n+1}'(x) - xP_n'(x) - (n+1)P_n(x) = 0, \quad n = 0, 1, 2, \ldots \tag{6.4.28}$$

and

$$xP_n'(x) - P_{n-1}'(x) - nP_n(x) = 0, \quad n = 1, 2, 3, \ldots \tag{6.4.29}$$

Adding (6.4.28) and (6.4.29), we obtain the more symmetric formula

$$\boxed{P_{n+1}'(x) - P_{n-1}'(x) = (2n+1)P_n(x), \qquad n = 1, 2, 3, \ldots} \tag{6.4.30}$$

Given any two of the polynomials $P_{n+1}(x)$, $P_n(x)$ and $P_{n-1}(x)$, (6.4.24) or (6.4.30) yields the third.

Having determined several methods for finding the Legendre polynomial $P_n(x)$, we now turn to the actual orthogonality condition.[13] Consider the integral

$$J = \int_{-1}^{1} \frac{dx}{\sqrt{1 + h^2 - 2xh}\,\sqrt{1 + t^2 - 2xt}}, \qquad |h|, |t| < 1 \tag{6.4.31}$$

$$= \int_{-1}^{1} [P_0(x) + hP_1(x) + \cdots + h^n P_n(x) + \cdots] \times [P_0(x) + tP_1(x) + \cdots + t^n P_n(x) + \cdots]\, dx \tag{6.4.32}$$

$$= \sum_{n=0}^{\infty}\sum_{m=0}^{\infty} h^n t^m \int_{-1}^{1} P_n(x)P_m(x)\, dx. \tag{6.4.33}$$

[13] From Symons, B., 1982: Legendre polynomials and their orthogonality. *Math. Gaz.*, **66**, 152–154 with permission.

On the other hand, if $a = (1 + h^2)/2h$ and $b = (1 + t^2)/2t$, the integral J is

$$J = \int_{-1}^{1} \frac{dx}{\sqrt{1 + h^2 - 2xh}\,\sqrt{1 + t^2 - 2xt}} \tag{6.4.34}$$

$$= \frac{1}{2\sqrt{ht}} \int_{-1}^{1} \frac{dx}{\sqrt{a - x}\,\sqrt{b - x}} = \frac{1}{\sqrt{ht}} \int_{-1}^{1} \frac{\frac{1}{2}\left(\frac{1}{\sqrt{a-x}} + \frac{1}{\sqrt{b-x}}\right)}{\sqrt{a - x} + \sqrt{b - x}}\, dx \tag{6.4.35}$$

$$= -\frac{1}{\sqrt{ht}} \ln(\sqrt{a - x} + \sqrt{b - x})\Big|_{-1}^{1} = \frac{1}{\sqrt{ht}} \ln\left(\frac{\sqrt{a + 1} + \sqrt{b + 1}}{\sqrt{a - 1} + \sqrt{b - 1}}\right). \tag{6.4.36}$$

But $a + 1 = (1 + h^2 + 2h)/2h = (1 + h)^2/2h$ and $a - 1 = (1 - h)^2/2h$. After a little algebra,

$$J = \frac{1}{\sqrt{ht}} \ln\left(\frac{1 + \sqrt{ht}}{1 - \sqrt{ht}}\right) = \frac{2}{\sqrt{ht}}\left(\sqrt{ht} + \frac{1}{3}\sqrt{(ht)^3} + \frac{1}{5}\sqrt{(ht)^5} + \cdots\right) \tag{6.4.37}$$

$$= 2\left(1 + \frac{ht}{3} + \frac{h^2t^2}{5} + \cdots + \frac{h^nt^n}{2n + 1} + \cdots\right). \tag{6.4.38}$$

As we noted earlier, the coefficients of h^nt^m in this series is $\int_{-1}^{1} P_n(x)$ $P_m(x)\, dx$. If we match the powers of h^nt^m, the orthogonality condition is

$$\int_{-1}^{1} P_n(x)P_m(x)\, dx = \begin{cases} 0, & m \neq n \\ \frac{2}{2n+1}, & m = n. \end{cases} \tag{6.4.39}$$

With the orthogonality condition (6.4.39) we are ready to show that we can represent a function $f(x)$, which is piece-wise differentiable in the interval $(-1, 1)$, by the series:

$$f(x) = \sum_{m=0}^{\infty} A_m P_m(x), \qquad -1 \leq x \leq 1. \tag{6.4.40}$$

To find A_m we multiply both sides of (6.4.40) by $P_n(x)$ and integrate from -1 to 1:

$$\int_{-1}^{1} f(x)P_n(x)\,dx = \sum_{m=0}^{\infty} A_m \int_{-1}^{1} P_n(x)P_m(x)\,dx. \qquad \textbf{(6.4.41)}$$

All of the terms on the right side vanish except for $n = m$ because of the orthogonality condition (6.4.39). Consequently, the coefficient A_n is

$$A_n \int_{-1}^{1} P_n^2(x)\,dx = \int_{-1}^{1} f(x)P_n(x)\,dx \qquad \textbf{(6.4.42)}$$

or

$$\boxed{A_n = \frac{2n+1}{2} \int_{-1}^{1} f(x)P_n(x)\,dx.} \qquad \textbf{(6.4.43)}$$

In the special case when $f(x)$ and its first n derivatives are continuous throughout the interval $(-1, 1)$, we may use Rodrigues' formula to evaluate

$$\int_{-1}^{1} f(x)P_n(x)\,dx = \frac{1}{2^n n!}\int_{-1}^{1} f(x)\frac{d^n(x^2-1)^n}{dx^n}\,dx \qquad \textbf{(6.4.44)}$$

$$= \frac{(-1)^n}{2^n n!}\int_{-1}^{1} (x^2-1)^n f^{(n)}(x)\,dx \qquad \textbf{(6.4.45)}$$

by integrating by parts n times. Consequently,

$$A_n = \frac{2n+1}{2^{n+1}n!}\int_{-1}^{1} (1-x^2)^n f^{(n)}(x)\,dx. \qquad \textbf{(6.4.46)}$$

A particularly useful result follows from (6.4.46) if $f(x)$ is a polynomial of degree k. Because all derivatives of $f(x)$ of order n vanish identically when $n > k$, $A_n = 0$ if $n > k$. Consequently, any polynomial of degree k can be expressed as a linear combination of the first $k + 1$ Legendre polynomials $[P_0(x), \ldots, P_k(x)]$. Another way of viewing this result is to recognize that any polynomial of degree k is an expansion in powers of x. When we expand in Legendre polynomials we are merely regrouping these powers of x into new groups that can be identified as $P_0(x), P_1(x), P_2(x), \ldots, P_k(x)$.

- **Example 6.4.1**

Let us use Rodrigues' formula to compute $P_2(x)$. From (6.4.17) with $n = 2$,

$$P_2(x) = \frac{1}{2^2 2!}\frac{d^2}{dx^2}[(x^2-1)^2] = \frac{1}{8}\frac{d^2}{dx^2}(x^4 - 2x^2 - 1) = \frac{1}{2}(3x^2 - 1). \tag{6.4.47}$$

- **Example 6.4.2**

Let us compute $P_3(x)$ from a recurrence relation. From (6.4.24) with $n = 2$,

$$3P_3(x) - 5xP_2(x) + 2P_1(x) = 0. \tag{6.4.48}$$

But $P_2(x) = (3x^2 - 1)/2$ and $P_1(x) = x$, so that

$$3P_3(x) = 5xP_2(x) - 2P_1(x) = 5x[(3x^2-1)/2] - 2x = \tfrac{15}{2}x^3 - \tfrac{9}{2}x \tag{6.4.49}$$

or

$$P_3(x) = (5x^3 - 3x)/2. \tag{6.4.50}$$

- **Example 6.4.3**

We want to show that

$$\int_{-1}^{1} P_n(x)\, dx = 0. \tag{6.4.51}$$

From (6.4.30),

$$(2n+1)\int_{-1}^{1} P_n(x)\, dx = \int_{-1}^{1} [P'_{n+1}(x) - P'_{n-1}(x)]\, dx \tag{6.4.52}$$

$$= P_{n+1}(x) - P_{n-1}(x)|_{-1}^{1} \tag{6.4.53}$$

$$= P_{n+1}(1) - P_{n-1}(1) - P_{n+1}(-1) + P_{n-1}(-1) = 0, \tag{6.4.54}$$

because $P_n(1) = 1$ and $P_n(-1) = (-1)^n$.

- **Example 6.4.4**

Let us express $f(x) = x^2$ in terms of Legendre polynomials. The results from (6.4.46) mean that we need only worry about $P_0(x)$, $P_1(x)$, and $P_2(x)$:

$$x^2 = A_0P_0(x) + A_1P_1(x) + A_2P_2(x). \tag{6.4.55}$$

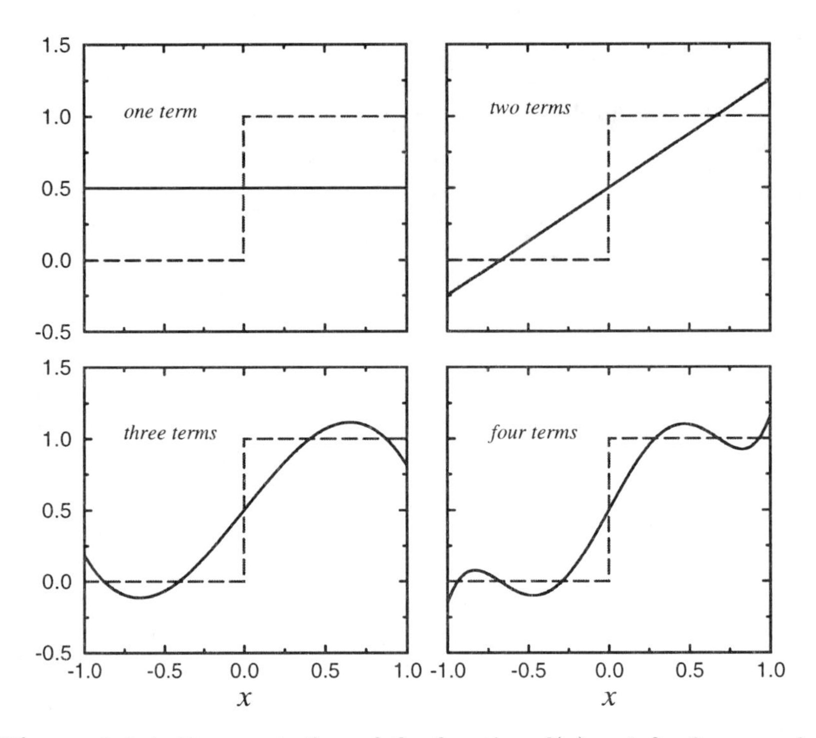

Figure 6.4.4: Representation of the function $f(x) = 1$ for $0 < x < 1$ and 0 for $-1 < x < 0$ by various partial summations of its Legendre polynomial expansion. The dashed lines denote the exact function.

Substituting for the Legendre polynomials,

$$x^2 = A_0 + A_1 x + \tfrac{1}{2} A_2 (3x^2 - 1) \tag{6.4.56}$$

and

$$A_0 = \tfrac{1}{3}, \quad A_1 = 0 \quad \text{and} \quad A_2 = \tfrac{2}{3}. \tag{6.4.57}$$

• **Example 6.4.5**

Let us find the expansion in Legendre polynomials of the function:

$$f(x) = \begin{cases} 0, & -1 < x < 0 \\ 1, & 0 < x < 1. \end{cases} \tag{6.4.58}$$

We could have done this expansion as a Fourier series but in the solution of partial differential equations on a sphere we must make the expansion in Legendre polynomials.

In this problem, we find that

$$A_n = \frac{2n+1}{2} \int_0^1 P_n(x)\, dx. \tag{6.4.59}$$

Therefore,

$$A_0 = \tfrac{1}{2}\int_0^1 1\,dx = \tfrac{1}{2}, \qquad A_1 = \tfrac{3}{2}\int_0^1 x\,dx = \tfrac{3}{4}, \tag{6.4.60}$$

$$A_2 = \tfrac{5}{2}\int_0^1 \tfrac{1}{2}(3x^2-1)\,dx = 0 \quad \text{and} \quad A_3 = \tfrac{7}{2}\int_0^1 \tfrac{1}{2}(5x^3-3x)\,dx = -\tfrac{7}{16} \tag{6.4.61}$$

so that

$$f(x) = \tfrac{1}{2}P_0(x) + \tfrac{3}{4}P_1(x) - \tfrac{7}{16}P_3(x) + \tfrac{11}{32}P_5(x) + \cdots \tag{6.4.62}$$

Figure 6.4.4 illustrates the expansion (6.4.62) where we have used only the first four terms. As we add each additional term in the orthogonal expansion, the expansion fits $f(x)$ better in the "least squares" sense of (6.3.5). The spurious oscillations arise from trying to represent a discontinuous function by four continuous, oscillatory functions. Even if we add additional terms, the spurious oscillations will persist although located nearer to the discontinuity. This is another example of *Gibbs phenomena*.[14] See Section 2.2.

• Example 6.4.6: Iterative solution of the radiative transfer equation

One of the fundamental equations of astrophysics is the integro-differential equation that describes radiative transfer (the propagation of energy by radiative, rather than conductive or convective, processes) in a gas.

Consider a gas which varies in only one spatial direction that we divide into infinitesimally thin slabs. As radiation enters a slab, it is absorbed and scattered. If we assume that all of the radiation undergoes isotropic scattering, the radiative transfer equation is

$$\mu\frac{dI}{d\tau} = I - \tfrac{1}{2}\int_{-1}^1 I\,d\mu, \tag{6.4.63}$$

where I is the intensity of the radiation, τ is the optical depth (a measure of the absorbing power of the gas and related to the distance that you have traveled within the gas), $\mu = \cos(\theta)$, and θ is the angle at which radiation enters the slab. In this example, we show how the Fourier-Legendre expansion[15]

$$I(\tau,\mu) = \sum_{n=0}^{\infty} I_n(\tau)P_n(\mu) \tag{6.4.64}$$

---

[14] Weyl, H., 1910: Die Gibbs'sche Erscheinung in der Theorie der Kugelfunktionen. *Rend. Circ. Mat. Palermo*, **29**, 308–321.

[15] Chandrasekhar, S., 1944: On the radiative equilibrium of a stellar atmosphere. *Astrophys. J.*, **99**, 180–190. Published by University of Chicago Press, ©1944.

may be used to solve (6.4.63). Here $I_n(\tau)$ is the Fourier coefficient in the Fourier-Legendre expansion involving the Legendre polynomial $P_n(\mu)$.

We begin by substituting (6.4.64) into (6.4.63),

$$\sum_{n=0}^{\infty} \frac{[(n+1)P_{n+1}(\mu) + nP_{n-1}(\mu)]}{2n+1} \frac{dI_n}{d\tau} = \sum_{n=0}^{\infty} I_n P_n(\mu) - I_0, \qquad (6.4.65)$$

where we have used (6.4.24) to eliminate $\mu P_n(\mu)$. Note that only the $I_0(\tau)$ term remains after integrating because of the orthogonality condition:

$$\int_{-1}^{1} 1 \cdot P_n(\mu)\, d\mu = \int_{-1}^{1} P_0(\mu) P_n(\mu)\, d\mu = 0, \qquad (6.4.66)$$

if $n > 0$. Equating the coefficients of the various Legendre polynomials,

$$\frac{n}{2n-1} \frac{dI_{n-1}}{d\tau} + \frac{n+1}{2n+3} \frac{dI_{n+1}}{d\tau} = I_n \qquad (6.4.67)$$

for $n = 1, 2, \ldots$ and

$$\frac{dI_1}{d\tau} = 0. \qquad (6.4.68)$$

Thus, the solution for I_1 is I_1 = constant $= 3F/4$, where F is the net integrated flux and an observable quantity.

For $n = 1$,

$$\frac{dI_0}{d\tau} + \frac{2}{5} \frac{dI_2}{d\tau} = I_1 = \frac{3F}{4}. \qquad (6.4.69)$$

Therefore,

$$I_0 + \tfrac{2}{5} I_2 = \tfrac{3}{4} F\tau + A. \qquad (6.4.70)$$

The next differential equation arises from $n = 2$ and equals

$$\frac{2}{3} \frac{dI_1}{d\tau} + \frac{3}{7} \frac{dI_3}{d\tau} = I_2. \qquad (6.4.71)$$

Because I_1 is a constant and we only retain I_0, I_1, and I_2 in the simplest approximation, we neglect $dI_3/d\tau$ and $I_2 = 0$. Thus, the simplest approximate solution is

$$I_0 = \tfrac{3}{4} F\tau + A, \quad I_1 = \tfrac{3}{4} F \quad \text{and} \quad I_2 = 0. \qquad (6.4.72)$$

To complete our approximate solution, we must evaluate A. If we are dealing with a stellar atmosphere where we assume no external radiation incident on the star, $I(0, \mu) = 0$ for $-1 \le \mu < 0$. Therefore,

$$\int_{-1}^{1} I(\tau,\mu) P_n(\mu)\, d\mu = \sum_{m=0}^{\infty} I_m(\tau) \int_{-1}^{1} P_m(\mu) P_n(\mu)\, d\mu = \frac{2}{2n+1} I_n(\tau). \qquad (6.4.73)$$

Taking the limit $\tau \to 0$ and using the boundary condition,

$$\frac{2}{2n+1} I_n(0) = \int_0^1 I(0,\mu) P_n(\mu)\, d\mu = \sum_{m=0}^{\infty} I_m(0) \int_0^1 P_n(\mu) P_m(\mu)\, d\mu. \tag{6.4.74}$$

Thus, we must satisfy, in principle, an infinite set of equations. For example, for $n = 0$, 1, and 2,

$$2I_0(0) = I_0(0) + \tfrac{1}{2} I_1(0) - \tfrac{1}{8} I_3(0) + \tfrac{1}{16} I_5(0) + \cdots \tag{6.4.75}$$

$$\tfrac{2}{3} I_1(0) = \tfrac{1}{2} I_0(0) + \tfrac{1}{3} I_1(0) + \tfrac{1}{8} I_2(0) - \tfrac{1}{48} I_4(0) + \cdots \tag{6.4.76}$$

and

$$\tfrac{2}{5} I_2(0) = \tfrac{1}{8} I_1(0) + \tfrac{1}{5} I_2(0) + \tfrac{1}{8} I_3(0) - \tfrac{5}{128} I_5(0) + \cdots. \tag{6.4.77}$$

Using $I_1(0) = 3F/4$,

$$\tfrac{1}{2} I_0(0) + \tfrac{1}{16} I_3(0) - \tfrac{1}{32} I_5(0) + \cdots = \tfrac{3}{16} F, \tag{6.4.78}$$

$$\tfrac{1}{2} I_0(0) + \tfrac{1}{8} I_2(0) - \tfrac{1}{48} I_4(0) + \cdots = \tfrac{1}{4} F \tag{6.4.79}$$

and

$$\tfrac{2}{5} I_2(0) - \tfrac{1}{4} I_3(0) + \tfrac{5}{64} I_5(0) + \cdots = \tfrac{3}{16} F. \tag{6.4.80}$$

Of the two possible Equations (6.4.78)–(6.4.79), Chandrasekhar chose (6.4.79) from physical considerations. Thus, to first approximation, the solution is

$$I(\mu, \tau) = \tfrac{3}{4} F \left(\tau + \tfrac{2}{3}\right) + \tfrac{3}{4} F \mu + \cdots. \tag{6.4.81}$$

Better approximations can be obtained by including more terms; the interested reader is referred to the original article. In the early 1950s, Wang and Guth[16] improved the procedure for finding the successive approximations and formulating the approximate boundary conditions.

Problems

Find the first three nonvanishing coefficients in the Legendre polynomial expansion for the following functions:

1. $f(x) = \begin{cases} 0, & -1 < x < 0 \\ x, & 0 < x < 1 \end{cases}$

2. $f(x) = \begin{cases} 1/(2\epsilon), & |x| < \epsilon \\ 0, & \epsilon < |x| < 1 \end{cases}$

3. $f(x) = |x|, \quad |x| < 1$

4. $f(x) = x^3, \quad |x| < 1$

5. $f(x) = \begin{cases} -1, & -1 < x < 0 \\ 1, & 0 < x < 1 \end{cases}$

6. $f(x) = \begin{cases} -1, & -1 < x < 0 \\ x, & 0 < x < 1 \end{cases}$

[16] Wang, M. C. and Guth, E., 1951: On the theory of multiple scattering, particularly of charged particles. *Phys. Rev., Ser. 2*, **84**, 1092–1111.

7. Use Rodrigues' formula to show that $P_4(x) = \frac{1}{8}(35x^4 - 30x^2 + 3)$.

8. Given $P_5(x) = \frac{63}{8}x^5 - \frac{70}{8}x^3 + \frac{15}{8}x$ and $P_4(x)$ from problem 7, use the recurrence formula for $P_{n+1}(x)$ to find $P_6(x)$.

9. Show that (a) $P_n(1) = 1$, (b) $P_n(-1) = (-1)^n$, (c) $P_{2n+1}(0) = 0$ and (d) $P_{2n}(0) = (-1)^n(2n)!/(2^{2n}n!n!)$.

10. Prove that

$$\int_x^1 P_n(t)\,dt = \frac{1}{2n+1}[P_{n-1}(x) - P_{n+1}(x)].$$

11. Given[17]

$$P_n[\cos(\theta)] = \frac{2}{\pi}\int_0^\theta \frac{\cos[(n+\frac{1}{2})x]}{\sqrt{2[\cos(x) - \cos(\theta)]}}\,dx$$
$$= \frac{2}{\pi}\int_\theta^\pi \frac{\sin[(n+\frac{1}{2})x]}{\sqrt{2[\cos(\theta) - \cos(x)]}}\,dx,$$

show that the following generalized Fourier series hold:

$$\frac{H(\theta - t)}{\sqrt{2\cos(t) - 2\cos(\theta)}} = \sum_{n=0}^\infty P_n[\cos(\theta)]\cos\left[\left(n+\tfrac{1}{2}\right)t\right], \quad 0 \le t < \theta \le \pi,$$

if we use the eigenfunction $y_n(x) = \cos\left[\left(n+\frac{1}{2}\right)x\right]$, $0 < x < \pi$, $r(x) = 1$ and $H(\ )$ is Heaviside's step function, and

$$\frac{H(t - \theta)}{\sqrt{2\cos(\theta) - 2\cos(t)}} = \sum_{n=0}^\infty P_n[\cos(\theta)]\sin\left[\left(n+\tfrac{1}{2}\right)t\right], \quad 0 \le \theta < t \le \pi,$$

if we use the eigenfunction $y_n(x) = \sin\left[\left(n+\frac{1}{2}\right)x\right]$, $0 < x < \pi$, $r(x) = 1$ and $H(\ )$ is Heaviside's step function.

12. The series given in problem 11 are also expansions in Legendre polynomials. In that light, show that

$$\int_0^t \frac{P_n[\cos(\theta)]\sin(\theta)}{\sqrt{2\cos(\theta) - 2\cos(t)}}\,d\theta = \frac{\sin\left[\left(n+\frac{1}{2}\right)t\right]}{n+\frac{1}{2}}$$

and

$$\int_t^\pi \frac{P_n[\cos(\theta)]\sin(\theta)}{\sqrt{2\cos(t) - 2\cos(\theta)}}\,d\theta = \frac{\cos\left[\left(n+\frac{1}{2}\right)t\right]}{n+\frac{1}{2}},$$

where $0 < t < \pi$.

---

[17] Hobson, E. W., 1965: *The Theory of Spherical and Ellipsoidal Harmonics*, Chelsea Publishing Co., New York, pp. 26–27.

6.5 ANOTHER SINGULAR STURM-LIOUVILLE PROBLEM: BESSEL'S EQUATION

In the previous section we discussed the solutions to Legendre's equation, especially with regard to their use in orthogonal expansions. In the section we consider another classic equation, Bessel's equation[18]

$$x^2y'' + xy' + (\mu^2x^2 - n^2)y = 0 \tag{6.5.1}$$

or

$$\frac{d}{dx}\left(x\frac{dy}{dx}\right) + \left(\mu^2x - \frac{n^2}{x}\right)y = 0. \tag{6.5.2}$$

Once again, our ultimate goal is the use of its solutions in orthogonal expansions. These orthogonal expansions, in turn, are used in the solution of partial differential equations in cylindrical coordinates.

A quick check of Bessel's equation shows that it conforms to the canonical form of the Sturm-Liouville problem: $p(x) = x$, $q(x) = -n^2/x$, $r(x) = x$, and $\lambda = \mu^2$. Restricting our attention to the interval $[0, L]$, the Sturm-Liouville problem involving (6.5.2) is singular because $p(0) = 0$. From (6.4.1) in the previous section, the eigenfunctions to a singular Sturm-Liouville problem will still be orthogonal over the interval $[0, L]$ if (1) $y(x)$ is finite and $xy'(x)$ is zero at $x = 0$, and (2) $y(x)$ satisfies the homogeneous boundary condition (6.1.2) at $x = L$. Consequently, we will only seek solutions that satisfy these conditions.

We cannot write down the solution to Bessel's equation in a simple closed form; as in the case with Legendre's equation, we must find the solution by power series. Because we intend to make the expansion about $x = 0$ and this point is a regular singular point, we must use the method of Frobenius, where n is an integer.[19] Moreover, because the quantity n^2 appears in (6.5.2), we may take n to be nonnegative without any loss of generality.

To simplify matters, we first find the solution when $\mu = 1$; the solution for $\mu \neq 1$ follows by substituting μx for x. Consequently, we seek solutions of the form

$$y(x) = \sum_{k=0}^{\infty} B_k x^{2k+s}, \tag{6.5.3}$$

---

[18] Bessel, F. W., 1824: Untersuchung des Teils der planetarischen Störungen, welcher aus der Bewegung der Sonne entsteht. *Abh. d. K. Akad. Wiss. Berlin*, 1–52. See Dutka, J., 1995: On the early history of Bessel functions. *Arch. Hist. Exact Sci.*, **49**, 105–134. The classic reference on Bessel functions is Watson, G. N., 1966: *A Treatise on the Theory of Bessel Functions,* Cambridge University Press, Cambridge.

[19] This case is much simpler than for arbitrary n. See Hildebrand, F. B., 1962: *Advanced Calculus for Applications.* Prentice-Hall, Englewood Cliffs, NJ, Section 4.8.

Figure 6.5.1: It was Friedrich William Bessel's (1784–1846) apprenticeship to the famous mercantile firm of Kulenkamp that ignited his interest in mathematics and astronomy. As the founder of the German school of practical astronomy, Bessel discovered his functions while studying the problem of planetary motion. Bessel functions arose as coefficients in one of the series that described the gravitational interaction between the sun and two other planets in elliptic orbit. (Portrait courtesy of Photo AKG, London.)

$$y'(x) = \sum_{k=0}^{\infty} (2k+s) B_k x^{2k+s-1} \tag{6.5.4}$$

and

$$y''(x) = \sum_{k=0}^{\infty} (2k+s)(2k+s-1) B_k x^{2k+s-2}, \tag{6.5.5}$$

where we formally assume that we can interchange the order of differentiation and summation. The substitution of (6.5.3)–(6.5.5) into (6.5.1) with $\mu = 1$ yields

$$\sum_{k=0}^{\infty}(2k+s)(2k+s-1)B_k x^{2k+s} + \sum_{k=0}^{\infty}(2k+s)B_k x^{2k+s} + \sum_{k=0}^{\infty} B_k x^{2k+s+2} - n^2 \sum_{k=0}^{\infty} B_k x^{2k+s} = 0 \quad (6.5.6)$$

or

$$\sum_{k=0}^{\infty}[(2k+s)^2 - n^2]B_k x^{2k} + \sum_{k=0}^{\infty} B_k x^{2k+2} = 0. \quad (6.5.7)$$

If we explicitly separate the $k = 0$ term from the other terms in the first summation in (6.5.7),

$$(s^2 - n^2)B_0 + \sum_{m=1}^{\infty}[(2m+s)^2 - n^2]B_m x^{2m} + \sum_{k=0}^{\infty} B_k x^{2k+2} = 0. \quad (6.5.8)$$

We now change the dummy integer in the first summation of (6.5.8) by letting $m = k + 1$ so that

$$(s^2 - n^2)B_0 + \sum_{k=0}^{\infty}\{[(2k+s+2)^2 - n^2]B_{k+1} + B_k\}x^{2k+2} = 0. \quad (6.5.9)$$

Because (6.5.9) must be true for all x, each power of x must vanish identically. This yields $s = \pm n$ and

$$[(2k+s+2)^2 - n^2]B_{k+1} + B_k = 0. \quad (6.5.10)$$

Since the difference of the larger indicial root from the lower root equals the integer $2n$, we are only guaranteed a power series solution of the form (6.5.3) for $s = n$. If we use this indicial root and the recurrence formula (6.5.10), this solution, known as the Bessel function of the first kind of order n and denoted by $J_n(x)$, is

$$J_n(x) = \sum_{k=0}^{\infty} \frac{(-1)^k (x/2)^{n+2k}}{k!(n+k)!}. \quad (6.5.11)$$

To find the second general solution to Bessel's equation, the one corresponding to $s = -n$, the most economical method[20] is to express it in terms of partial derivatives of $J_n(x)$ with respect to its order n:

$$Y_n(x) = \left[\frac{\partial J_\nu(x)}{\partial \nu} - (-1)^n \frac{\partial J_{-\nu}(x)}{\partial \nu}\right]_{\nu=n}. \quad (6.5.12)$$

---

[20] See Watson, G. N., 1966: *A Treatise on the Theory of Bessel Functions,* Cambridge University Press, Cambridge, Section 3.5 for the derivation.

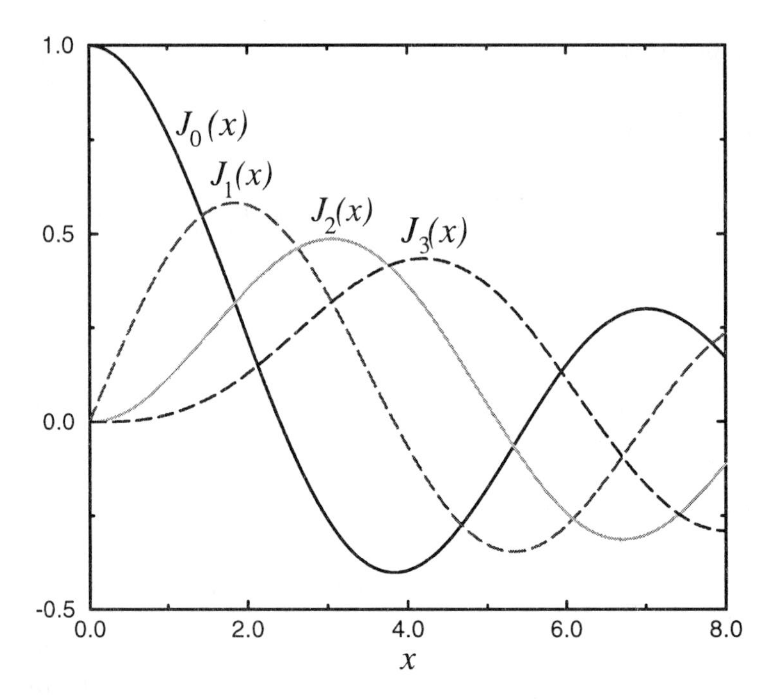

Figure 6.5.2: The first four Bessel functions of the first kind over $0 \le x \le 8$.

Upon substituting the power series representation (6.5.11) into (6.5.12),

$$Y_n(x) = \frac{2}{\pi} J_n(x) \ln(x/2) - \frac{1}{\pi} \sum_{k=0}^{n-1} \frac{(n-k-1)!}{k!} \left(\frac{x}{2}\right)^{2k-n}$$

$$- \frac{1}{\pi} \sum_{k=0}^{\infty} \frac{(-1)^k (x/2)^{n+2k}}{k!(n+k)!} [\psi(k+1) + \psi(k+n+1)], \qquad \textbf{(6.5.13)}$$

where

$$\psi(m+1) = -\gamma + 1 + \frac{1}{2} + \cdots + \frac{1}{m}, \qquad \textbf{(6.5.14)}$$

$\psi(1) = -\gamma$ and γ is Euler's constant (0.5772157). In the case $n = 0$, the first sum in (6.5.13) disappears. This function $Y_n(x)$ is Neumann's Bessel function of the second kind of order n. Consequently, the general solution to (6.5.1) is

$$y(x) = AJ_n(\mu x) + BY_n(\mu x). \qquad \textbf{(6.5.15)}$$

Figure 6.5.2 illustrates the functions $J_0(x)$, $J_1(x)$, $J_2(x)$, and $J_3(x)$ while Figure 6.5.3 gives $Y_0(x)$, $Y_1(x)$, $Y_2(x)$, and $Y_3(x)$.

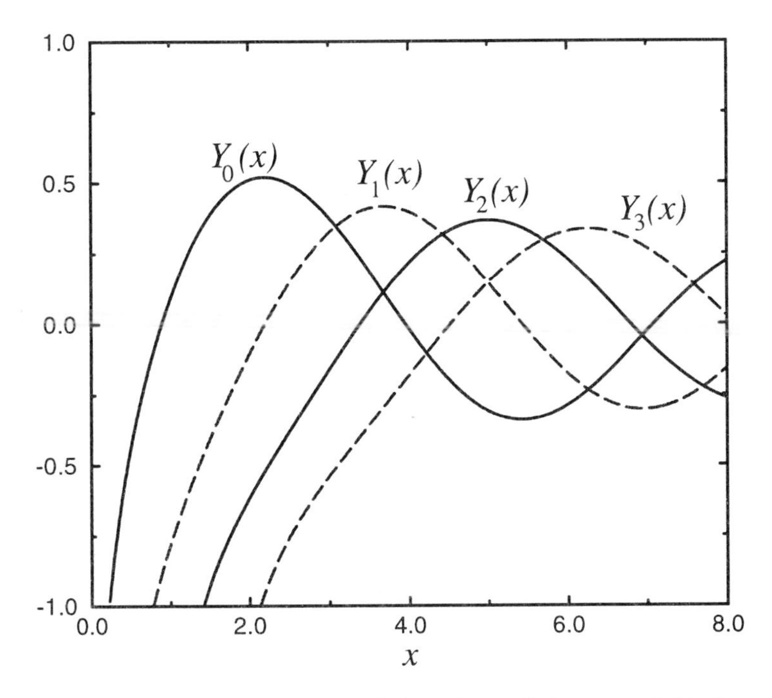

Figure 6.5.3: The first four Bessel functions of the second kind over $0 \leq x \leq 8$.

An equation which is very similar to (6.5.1) is

$$x^2 \frac{d^2y}{dx^2} + x\frac{dy}{dx} - (n^2 + x^2)y = 0. \tag{6.5.16}$$

It arises in the solution of partial differential equations in cylindrical coordinates. If we substitute $ix = t$ (where $i = \sqrt{-1}$) into (6.5.16), it becomes Bessel's equation:

$$t^2 \frac{d^2y}{dt^2} + t\frac{dy}{dt} + (t^2 - n^2)y = 0. \tag{6.5.17}$$

Consequently, we may immediately write the solution to (6.5.16) as

$$y(x) = c_1 J_n(ix) + c_2 Y_n(ix), \tag{6.5.18}$$

if n is an integer. Traditionally the solution to (6.5.16) has been written

$$y(x) = c_1 I_n(x) + c_2 K_n(x) \tag{6.5.19}$$

rather than in terms of $J_n(ix)$ and $Y_n(ix)$, where

$$I_n(x) = \sum_{k=0}^{\infty} \frac{(x/2)^{2k+n}}{k!(k+n)!} \tag{6.5.20}$$

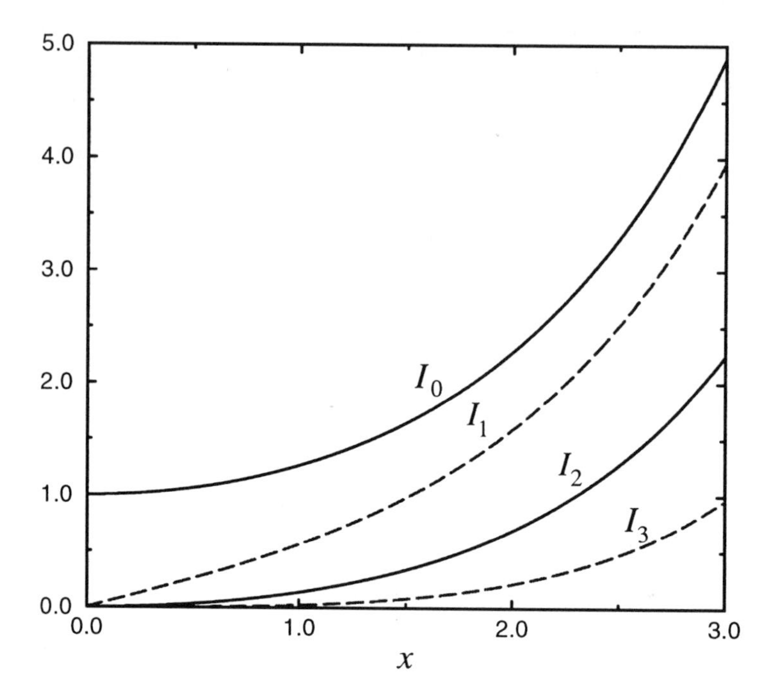

Figure 6.5.4: The first four modified Bessel functions of the first kind over $0 \le x \le 3$.

and

$$K_n(x) = \frac{\pi}{2} i^{n+1} \left[J_n(ix) + iY_n(ix) \right]. \tag{6.5.21}$$

The function $I_n(x)$ is the modified Bessel function of the first kind, of order n, while $K_n(x)$ is the modified Bessel function of the second kind, of order n. Figure 6.5.4 illustrates $I_0(x)$, $I_1(x)$, $I_2(x)$, and $I_3(x)$ while in Figure 6.5.5 $K_0(x)$, $K_1(x)$, $K_2(x)$, and $K_3(x)$ have been graphed. Note that $K_n(x)$ has no real zeros while $I_n(x)$ equals zero only at $x = 0$ for $n \ge 1$.

As our derivation suggests, modified Bessel functions are related to ordinary Bessel functions via complex variables. In particular, $J_n(iz) = i^n I_n(z)$ and $I_n(iz) = i^n J_n(z)$ for z complex.

Although we have found solutions to Bessel's equation (6.5.1), as well as (6.5.16), can we use any of them in an eigenfunction expansion? From Figures 6.5.2–6.5.5 we see that $J_n(x)$ and $I_n(x)$ remain finite at $x = 0$ while $Y_n(x)$ and $K_n(x)$ do not. Furthermore, the products $xJ_n'(x)$ and $xI_n'(x)$ tend to zero at $x = 0$. Thus, both $J_n(x)$ and $I_n(x)$ satisfy the first requirement of an eigenfunction for a Fourier-Bessel expansion.

What about the second condition that the eigenfunction must satisfy the homogeneous boundary condition (6.1.2) at $x = L$? From Figure 6.5.4 we see that $I_n(x)$ can never satisfy this condition while from Fig-

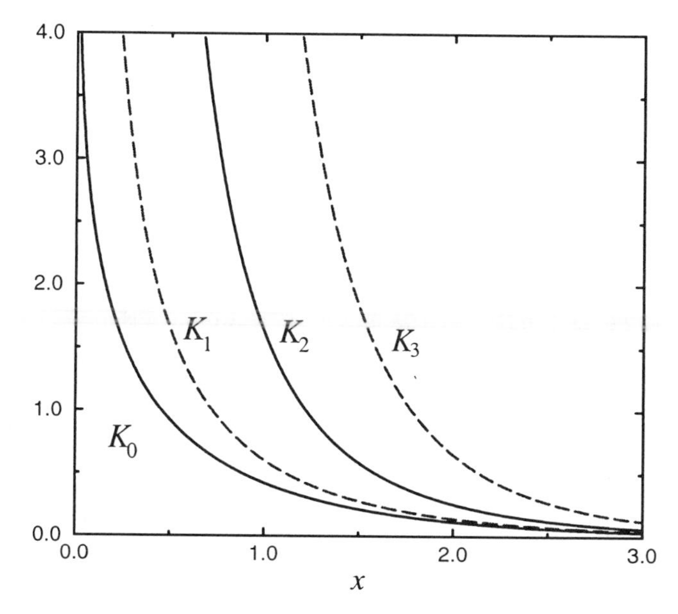

Figure 6.5.5: The first four modified Bessel functions of the second kind over $0 \leq x \leq 3$.

ure 6.5.2 $J_n(x)$ can. For that reason, we discard $I_n(x)$ from further consideration and continue our analysis only with $J_n(x)$.

Before we can derive the expressions for a Fourier-Bessel expansion, we need to find how $J_n(x)$ is related to $J_{n+1}(x)$ and $J_{n-1}(x)$. Assuming that n is a positive integer, we multiply the series (6.5.11) by x^n and then differentiate with respect to x. This gives

$$\frac{d}{dx}\left[x^n J_n(x)\right] = \sum_{k=0}^{\infty} \frac{(-1)^k (2n+2k) x^{2n+2k-1}}{2^{n+2k} k!(n+k)!} \tag{6.5.22}$$

$$= x^n \sum_{k=0}^{\infty} \frac{(-1)^k (x/2)^{n-1+2k}}{k!(n-1+k)!} \tag{6.5.23}$$

$$= x^n J_{n-1}(x) \tag{6.5.24}$$

or

$$\boxed{\frac{d}{dx}\left[x^n J_n(x)\right] = x^n J_{n-1}(x)} \tag{6.5.25}$$

for $n = 1, 2, 3, \ldots$. Similarly, multiplying (6.5.11) by x^{-n}, we find that

$$\frac{d}{dx}\left[x^{-n}J_n(x)\right] = -x^{-n}J_{n+1}(x) \tag{6.5.26}$$

for $n = 0, 1, 2, 3, \ldots$. If we now carry out the differentiation on (6.5.25) and (6.5.26) and divide by the factors $x^{\pm n}$, we have that

$$J_n'(x) + \frac{n}{x}J_n(x) = J_{n-1}(x) \tag{6.5.27}$$

and

$$J_n'(x) - \frac{n}{x}J_n(x) = -J_{n+1}(x). \tag{6.5.28}$$

Equations (6.3.27)–(6.3.28) immediately yield the *recurrence relationships*

$$J_{n-1}(x) + J_{n+1}(x) = \frac{2n}{x}J_n(x) \tag{6.5.29}$$

and

$$J_{n-1}(x) - J_{n+1}(x) = 2J_n'(x) \tag{6.5.30}$$

for $n = 1, 2, 3, \ldots$ For $n = 0$, we replace (6.5.30) by $J_0'(x) = -J_1(x)$.

Let us now construct a Fourier-Bessel series. The exact form of the expansion depends upon the boundary condition at $x = L$. There are three possible cases. One of them is the requirement that $y(L) = 0$ and results in the condition that $J_n(\mu_k L) = 0$. Another condition is $y'(L) = 0$ and gives $J_n'(\mu_k L) = 0$. Finally, if $hy(L) + y'(L) = 0$, then $hJ_n(\mu_k L) + \mu_k J_n'(\mu_k L) = 0$. In all of these cases, the eigenfunction expansion is the same, namely

$$f(x) = \sum_{k=1}^{\infty} A_k J_n(\mu_k x), \tag{6.5.31}$$

where μ_k is the kth positive solution of either $J_n(\mu_k L) = 0$, $J_n'(\mu_k L) = 0$ or $hJ_n(\mu_k L) + \mu_k J_n'(\mu_k L) = 0$.

We now need a mechanism for computing A_k. We begin by multiplying (6.5.31) by $xJ_n(\mu_m x)\,dx$ and integrate from 0 to L. This yields

$$\sum_{k=1}^{\infty} A_k \int_0^L xJ_n(\mu_k x)J(\mu_m x)\,dx = \int_0^L xf(x)J_n(\mu_m x)\,dx. \quad \textbf{(6.5.32)}$$

From the general orthogonality condition (6.2.1),

$$\int_0^L xJ_n(\mu_k x)J_n(\mu_m x)\,dx = 0 \quad \textbf{(6.5.33)}$$

if $k \neq m$. Equation (6.5.32) then simplifies to

$$A_m \int_0^L xJ_n^2(\mu_m x)\,dx = \int_0^L xf(x)J_n(\mu_m x)\,dx \quad \textbf{(6.5.34)}$$

or

$$\boxed{A_k = \frac{1}{C_k}\int_0^L xf(x)J_n(\mu_k x)\,dx,} \quad \textbf{(6.5.35)}$$

where

$$C_k = \int_0^L xJ_n^2(\mu_k x)\,dx \quad \textbf{(6.5.36)}$$

and k has replaced m in (6.5.34).

The factor C_k depends upon the nature of the boundary conditions at $x = L$. In all cases we start from Bessel's equation

$$[xJ_n'(\mu_k x)]' + \left(\mu_k^2 x - \frac{n^2}{x}\right) J_n(\mu_k x) = 0. \quad \textbf{(6.5.37)}$$

If we multiply both sides of (6.5.37) by $2xJ_n'(\mu_k x)$, the resulting equation is

$$\left(\mu_k^2 x^2 - n^2\right)\left[J_n^2(\mu_k x)\right]' = -\frac{d}{dx}\left[xJ_n'(\mu_k x)\right]^2. \quad \textbf{(6.5.38)}$$

An integration of (6.5.38) from 0 to L, followed by the subsequent use of integration by parts, results in

$$(\mu_k^2 x^2 - n^2)J_n^2(\mu_k x)\Big|_0^L - 2\mu_k^2 \int_0^L xJ_n^2(\mu_k x)\,dx = -\left[xJ_n'(\mu_k x)\right]^2\Big|_0^L. \quad \textbf{(6.5.39)}$$

Because $J_n(0) = 0$ for $n > 0$, $J_0(0) = 1$ and $xJ_n'(x) = 0$ at $x = 0$, the contribution from the lower limits vanishes. Thus,

$$C_k = \int_0^L x J_n^2(\mu_k x)\, dx \tag{6.5.40}$$

$$= \frac{1}{2\mu_k^2}\left[(\mu_k^2 L^2 - n^2) J_n^2(\mu_k L) + L^2 J_n'^{\,2}(\mu_k L)\right]. \tag{6.5.41}$$

Because

$$J_n'(\mu_k x) = \frac{n}{x} J_n(\mu_k x) - \mu_k J_{n+1}(\mu_k x) \tag{6.5.42}$$

from (6.5.28), C_k becomes

$$\boxed{C_k = \tfrac{1}{2} L^2 J_{n+1}^2(\mu_k L),} \tag{6.5.43}$$

if $J_n(\mu_k L) = 0$. Otherwise, if $J_n'(\mu_k L) = 0$, then

$$\boxed{C_k = \frac{\mu_k^2 L^2 - n^2}{2\mu_k^2} J_n^2(\mu_k L).} \tag{6.5.44}$$

Finally,

$$\boxed{C_k = \frac{\mu_k^2 L^2 - n^2 + h^2 L^2}{2\mu_k^2} J_n^2(\mu_k L),} \tag{6.5.45}$$

if $\mu_k J_n'(\mu_k L) = -h J_n(\mu_k L)$.

All of the preceding results must be slightly modified when $n = 0$ and the boundary condition is $J_0'(\mu_k L) = 0$ or $\mu_k J_1(\mu_k L) = 0$. This modification results from the additional eigenvalue $\mu_0 = 0$ being present and we must add the extra term A_0 to the expansion. For this case the series reads

$$f(x) = A_0 + \sum_{k=1}^{\infty} A_k J_0(\mu_k x), \tag{6.5.46}$$

where the equation for finding A_0 is

$$A_0 = \frac{2}{L^2}\int_0^L f(x)\,x\,dx \tag{6.5.47}$$

and (6.5.35) and (6.5.44) with $n = 0$ give the remaining coefficients.

- **Example 6.5.1**

Starting with Bessel's equation, we want to show that the solution to

$$y'' + \frac{1-2a}{x}y' + \left(b^2c^2x^{2c-2} + \frac{a^2 - n^2c^2}{x^2}\right)y = 0 \tag{6.5.48}$$

is

$$y(x) = Ax^aJ_n(bx^c) + Bx^aY_n(bx^c), \tag{6.5.49}$$

provided that $bx^c > 0$ so that $Y_n(bx^c)$ exists.

The general solution to

$$\xi^2\frac{d^2\eta}{d\xi^2} + \xi\frac{d\eta}{d\xi} + (\xi^2 - n^2)\eta = 0 \tag{6.5.50}$$

is

$$\eta = AJ_n(\xi) + BY_n(\xi). \tag{6.5.51}$$

If we now let $\eta = y(x)/x^a$ and $\xi = bx^c$, then

$$\frac{d}{d\xi} = \frac{dx}{d\xi}\frac{d}{dx} = \frac{x^{1-c}}{bc}\frac{d}{dx}, \tag{6.5.52}$$

$$\frac{d^2}{d\xi^2} = \frac{x^{2-2c}}{b^2c^2}\frac{d^2}{dx^2} - \frac{(c-1)x^{1-2c}}{b^2c^2}\frac{d}{dx}, \tag{6.5.53}$$

$$\frac{d}{dx}\left(\frac{y}{x^a}\right) = \frac{1}{x^a}\frac{dy}{dx} - \frac{a}{x^{a+1}}y \tag{6.5.54}$$

and

$$\frac{d^2}{dx^2}\left(\frac{y}{x^a}\right) = \frac{1}{x^a}\frac{d^2y}{dx^2} - \frac{2a}{x^{a+1}}\frac{dy}{dx} + \frac{a(1+a)}{x^{a+2}}y. \tag{6.5.55}$$

Substituting (6.5.52)–(6.5.55) into (6.5.51) and simplifying, yields the desired result.

- **Example 6.5.2**

We want to show that

$$x^2J_n''(x) = (n^2 - n - x^2)J_n(x) + xJ_{n+1}(x). \tag{6.5.56}$$

From (6.5.28),

$$J_n'(x) = \frac{n}{x} J_n(x) - J_{n+1}(x), \tag{6.5.57}$$

$$J_n''(x) = -\frac{n}{x^2} J_n(x) + \frac{n}{x} J_n'(x) - J_{n+1}'(x) \tag{6.5.58}$$

and

$$J_n''(x) = -\frac{n}{x^2} J_n(x) + \frac{n}{x} \left[\frac{n}{x} J_n(x) - J_{n+1}(x)\right] \\ - \left[J_n(x) - \frac{n+1}{x} J_{n+1}(x)\right] \tag{6.5.59}$$

after using (6.5.27) and (6.5.28). Simplifying,

$$J_n''(x) = \left(\frac{n^2 - n}{x^2} - 1\right) J_n(x) + \frac{J_{n+1}(x)}{x}. \tag{6.5.60}$$

After multiplying (6.5.60) by x^2, we obtain (6.5.56).

- **Example 6.5.3**

Show that

$$\int_0^a x^5 J_2(x)\, dx = a^5 J_3(a) - 2a^4 J_4(a). \tag{6.5.61}$$

We begin by integrating (6.5.61) by parts. If $u = x^2$ and $dv = x^3 J_2(x)\, dx$, then

$$\int_0^a x^5 J_2(x)\, dx = x^5 J_3(x)\Big|_0^a - 2\int_0^a x^4 J_3(x)\, dx, \tag{6.5.62}$$

because $d[x^3 J_3(x)]/dx = x^2 J_2(x)$ by (6.5.25). Finally, since $x^4 J_3(x) = d[x^4 J_4(x)]/dx$ by (6.5.25),

$$\int_0^a x^5 J_2(x)\, dx = a^5 J_3(a) - 2x^4 J_4(x)\Big|_0^a = a^5 J_3(a) - 2a^4 J_4(a). \tag{6.5.63}$$

- **Example 6.5.4**

Let us expand $f(x) = x$, $0 < x < 1$, in the series

$$f(x) = \sum_{k=1}^{\infty} A_k J_1(\mu_k x), \tag{6.5.64}$$

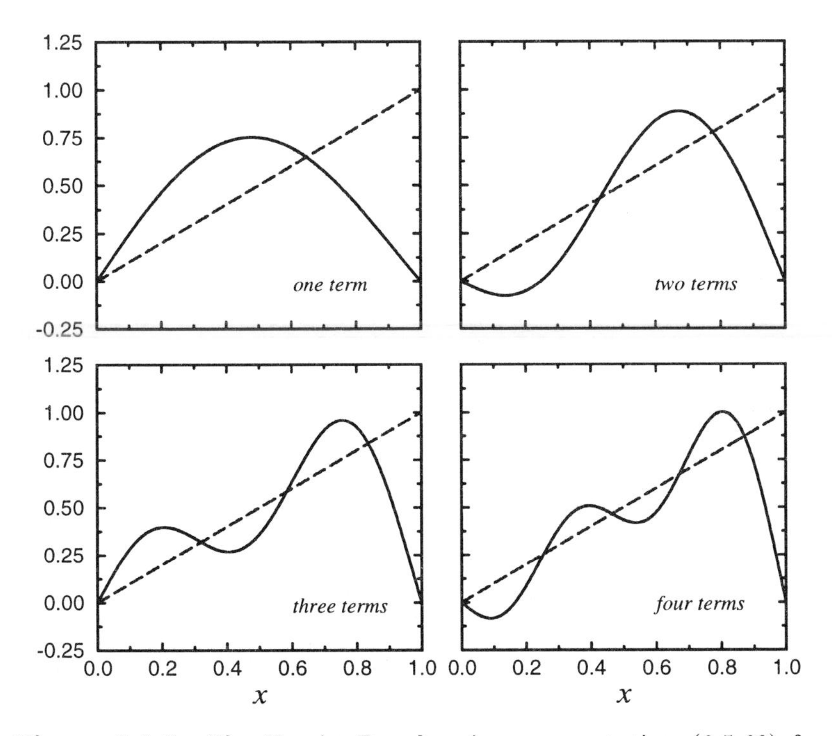

Figure 6.5.6: The Fourier-Bessel series representation (6.5.68) for $f(x) = x$, $0 < x < 1$, when we truncate the series so that it includes only the first, first two, first three, and first four terms.

where μ_k denotes the kth zero of $J_1(\mu)$. From (6.5.35) and (6.5.43),

$$A_k = \frac{2}{J_2^2(\mu_k)} \int_0^1 x^2 J_1(\mu_k x)\, dx. \tag{6.5.65}$$

However, from (6.5.25),

$$\frac{d}{dx}\left[x^2 J_2(x)\right] = x^2 J_1(x), \tag{6.5.66}$$

if $n = 2$. Therefore, (6.5.65) becomes

$$A_k = \left. \frac{2x^2 J_2(x)}{\mu_k^3 J_2^2(\mu_k)} \right|_0^{\mu_k} = \frac{2}{\mu_k J_2(\mu_k)} \tag{6.5.67}$$

and the resulting expansion is

$$x = 2 \sum_{k=1}^{\infty} \frac{J_1(\mu_k x)}{\mu_k J_2(\mu_k)}, \qquad 0 < x < 1. \tag{6.5.68}$$

Figure 6.5.6 shows the Fourier-Bessel expansion of $f(x) = x$ in truncated form when we only include one, two, three, and four terms.

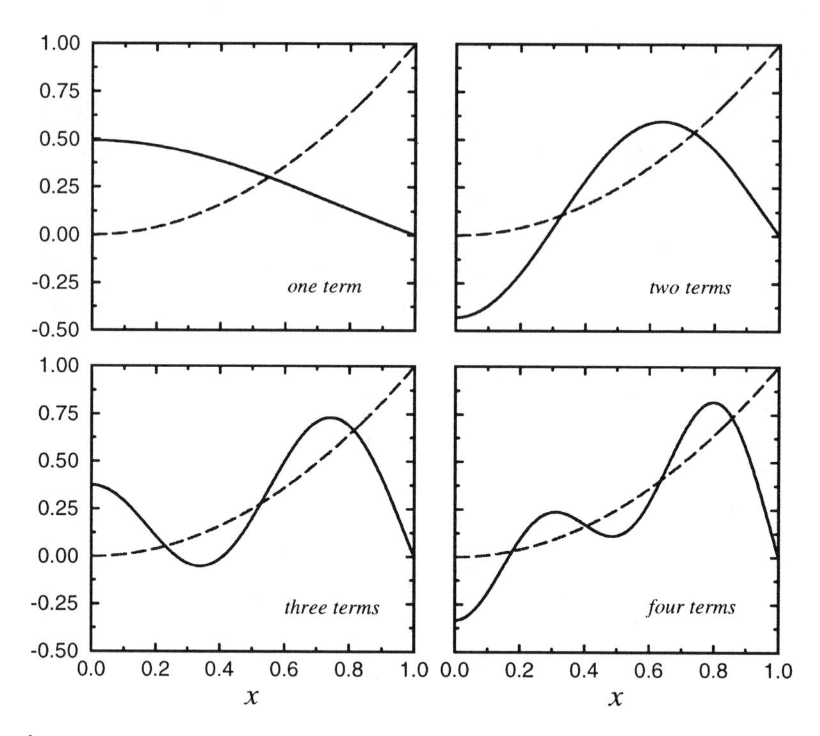

Figure 6.5.7: The Fourier-Bessel series representation (6.5.79) for $f(x) = x^2$, $0 < x < 1$, when we truncate the series so that it includes only the first, first two, first three, and first four terms.

• Example 6.5.5

Let us expand the function $f(x) = x^2$, $0 < x < 1$, in the series

$$f(x) = \sum_{k=1}^{\infty} A_k J_0(\mu_k x), \tag{6.5.69}$$

where μ_k denotes the kth positive zero of $J_0(\mu)$. From (6.5.35) and (6.5.43),

$$A_k = \frac{2}{J_1^2(\mu_k)} \int_0^1 x^3 J_0(\mu_k x)\, dx. \tag{6.5.70}$$

If we let $t = \mu_k x$, the integration (6.5.70) becomes

$$A_k = \frac{2}{\mu_k^4 J_1^2(\mu_k)} \int_0^{\mu_k} t^3 J_0(t)\, dt. \tag{6.5.71}$$

We now let $u = t^2$ and $dv = tJ_0(t)\, dt$ so that integration by parts results in

$$A_k = \frac{2}{\mu_k^4 J_1^2(\mu_k)} \left[t^3 J_1(t)\Big|_0^{\mu_k} - 2\int_0^{\mu_k} t^2 J_1(t)\, dt \right] \quad (6.5.72)$$

$$= \frac{2}{\mu_k^4 J_1^2(\mu_k)} \left[\mu_k^3 J_1(\mu_k) - 2\int_0^{\mu_k} t^2 J_1(t)\, dt \right], \quad (6.5.73)$$

because $v = tJ_1(t)$ from (6.5.25). If we integrate by parts once more, we find that

$$A_k = \frac{2}{\mu_k^4 J_1^2(\mu_k)} \left[\mu_k^3 J_1(\mu_k) - 2\mu_k^2 J_2(\mu_k) \right] \quad (6.5.74)$$

$$= \frac{2}{J_1^2(\mu_k)} \left[\frac{J_1(\mu_k)}{\mu_k} - \frac{2J_2(\mu_k)}{\mu_k^2} \right]. \quad (6.5.75)$$

However, from (6.5.29) with $n = 1$,

$$J_1(\mu_k) = \tfrac{1}{2}\mu_k \left[J_2(\mu_k) + J_0(\mu_k) \right] \quad (6.5.76)$$

or

$$J_2(\mu_k) = \frac{2J_1(\mu_k)}{\mu_k}, \quad (6.5.77)$$

because $J_0(\mu_k) = 0$. Therefore,

$$A_k = \frac{2(\mu_k^2 - 4)J_1(\mu_k)}{\mu_k^3 J_1^2(\mu_k)} \quad (6.5.78)$$

and

$$x^2 = 2\sum_{k=1}^{\infty} \frac{(\mu_k^2 - 4)J_0(\mu_k x)}{\mu_k^3 J_1(\mu_k)}, \qquad 0 < x < 1. \quad (6.5.79)$$

Figure 6.5.7 shows the representation of x^2 by the Fourier-Bessel series (6.5.79) when we truncate it so that it includes only one, two, three, or four terms. As we add each additional term in the orthogonal expansion, the expansion fits $f(x)$ better in the "least squares" sense of (6.3.5).

Problems

1. Show from the series solution that

$$\frac{d}{dx}\left[J_0(kx) \right] = -kJ_1(kx).$$

From the recurrence formulas, show these following relations:

2.

$$2J_0''(x) = J_2(x) - J_0(x)$$

3.

$$J_2(x) = J_0''(x) - J_0'(x)/x$$

4.

$$J_0'''(x) = \frac{J_0(x)}{x} + \left(\frac{2}{x^2} - 1\right) J_0'(x)$$

5.

$$\frac{J_2(x)}{J_1(x)} = \frac{1}{x} - \frac{J_0''(x)}{J_0'(x)} = \frac{2}{x} - \frac{J_0(x)}{J_1(x)} = \frac{2}{x} + \frac{J_0(x)}{J_0'(x)}$$

6.

$$J_4(x) = \left(\frac{48}{x^3} - \frac{8}{x}\right) J_1(x) - \left(\frac{24}{x^2} - 1\right) J_0(x)$$

7.

$$J_{n+2}(x) = \left[2n + 1 - \frac{2n(n^2-1)}{x^2}\right] J_n(x) + 2(n+1)J_n''(x)$$

8.

$$J_3(x) = \left(\frac{8}{x^2} - 1\right) J_1(x) - \frac{4}{x} J_0(x)$$

9.

$$4J_n''(x) = J_{n-2}(x) - 2J_n(x) + J_{n+2}(x)$$

10. Show that the maximum and minimum values of $J_n(x)$ occur when

$$x = \frac{nJ_n(x)}{J_{n+1}(x)}, \quad x = \frac{nJ_n(x)}{J_{n-1}(x)}, \quad \text{and} \quad J_{n-1}(x) = J_{n+1}(x).$$

Show that

11.

$$\frac{d}{dx}\left[x^2 J_3(2x)\right] = -xJ_3(2x) + 2x^2 J_2(2x)$$

12.

$$\frac{d}{dx}\left[xJ_0(x^2)\right] = J_0(x^2) - 2x^2 J_1(x^2)$$

13.

$$\int x^3 J_2(3x)\, dx = \tfrac{1}{3} x^3 J_3(3x) + C$$

14.

$$\int x^{-2} J_3(2x)\, dx = -\tfrac{1}{2} x^{-2} J_2(2x) + C$$

15.

$$\int x \ln(x) J_0(x)\, dx = J_0(x) + x \ln(x) J_1(x) + C$$

16.

$$\int_0^a x J_0(kx)\, dx = \frac{a^2 J_1(ka)}{ka}$$

17.

$$\int_0^1 x(1 - x^2) J_0(kx)\, dx = \frac{4}{k^3} J_1(k) - \frac{2}{k^2} J_0(k)$$

18.

$$\int_0^1 x^3 J_0(kx)\, dx = \frac{k^2 - 4}{k^3} J_1(k) + \frac{2}{k^2} J_0(k)$$

19. Show that

$$1 = 2 \sum_{k=1}^{\infty} \frac{J_0(\mu_k x)}{\mu_k J_1(\mu_k)}, \qquad 0 < x < 1,$$

where μ_k is the kth positive root of $J_0(\mu) = 0$.

20. Show that

$$\frac{1 - x^2}{8} = \sum_{k=1}^{\infty} \frac{J_0(\mu_k x)}{\mu_k^3 J_1(\mu_k)}, \qquad 0 < x < 1,$$

where μ_k is the kth positive root of $J_0(\mu) = 0$.

21. Show that

$$4x - x^3 = -16 \sum_{k=1}^{\infty} \frac{J_1(\mu_k x)}{\mu_k^3 J_0(2\mu_k)}, \qquad 0 < x < 2,$$

where μ_k is the kth positive root of $J_1(2\mu) = 0$.

22. Show that

$$x^3 = 2\sum_{k=1}^{\infty} \frac{(\mu_k^2 - 8)J_1(\mu_k x)}{\mu_k^3 J_2(\mu_k)}, \qquad 0 < x < 1,$$

where μ_k is the kth positive root of $J_1(\mu) = 0$.

23. Using the relationship[21]

$$\int_0^a J_\nu(\alpha r) J_\nu(\beta r)\, r\, dr = \frac{a\beta J_\nu(\alpha a) J_\nu'(\beta a) - a\alpha J_\nu(\beta a) J_\nu'(\alpha a)}{\alpha^2 - \beta^2},$$

show that

$$\frac{J_0(bx) - J_0(ba)}{J_0(ba)} = \frac{2b^2}{a} \sum_{k=1}^{\infty} \frac{J_0(\mu_k x)}{\mu_k(\mu_k^2 - b^2) J_1(\mu_k a)}, \qquad 0 < x < a,$$

where μ_k is the kth positive root of $J_0(\mu a) = 0$ and b is a constant.

24. Given the definite integral[22]

$$\int_0^1 \frac{x\, J_0(bx)}{\sqrt{1 - x^2}}\, dx = \frac{\sin(b)}{b}, \qquad b > 0,$$

show that

$$\frac{H(t - x)}{\sqrt{t^2 - x^2}} = 2\sum_{k=1}^{\infty} \frac{\sin(\mu_k t) J_0(\mu_k x)}{\mu_k J_1^2(\mu_k)}, \qquad 0 < x < 1, \quad 0 < t \le 1,$$

where μ_k is the kth positive root of $J_0(\mu) = 0$ and $H(\ )$ is Heaviside's step function.

25. Given the definite integral[23]

$$\int_0^a \cos(cx)\, J_0\left(b\sqrt{a^2 - x^2}\right) dx = \frac{\sin\left(a\sqrt{b^2 + c^2}\right)}{\sqrt{b^2 + c^2}}, \qquad b > 0$$

---

[21] Watson, G. N., 1966: *A Treatise on the Theory of Bessel Functions*, Cambridge University Press, Cambridge, Section 5.11, Equation (8).

[22] Gradshteyn, I. S. and Ryzhik, I. M., 1965: *Table of Integrals, Series, and Products*, Academic Press, New York. See Section 6.567, formula 1 with $\nu = 0$ and $\mu = -1/2$.

[23] *Ibid.* See Section 6.677, formula 6.

show that

$$\frac{\cosh\left(b\sqrt{t^2-x^2}\right)}{\sqrt{t^2-x^2}} H(t-x) = \frac{2}{a^2} \sum_{k=1}^{\infty} \frac{\sin\left(t\sqrt{\mu_k^2-b^2}\right) J_0(\mu_k x)}{\sqrt{\mu_k^2-b^2}\, J_1^2(\mu_k a)},$$

where $0 < x < a$, μ_k is the kth positive root of $J_0(\mu a) = 0$, $H(\ )$ is Heaviside's step function, and b is a constant.

26. Using the integral definition of the Bessel function[24] for $J_1(z)$:

$$J_1(z) = \frac{2}{\pi} \int_0^1 \frac{t \sin(zt)}{z\sqrt{1-t^2}}\, dt, \qquad z > 0,$$

show that

$$\frac{x}{t\sqrt{t^2-x^2}} H(t-x) = \sum_{k=1}^{\infty} J_1\left(\frac{n\pi t}{L}\right) \sin\left(\frac{n\pi t}{L}\right), \qquad 0 < x < L,$$

where $H(\ )$ is Heaviside's step function. [Hint: Treat this as a Fourier half-range sine expansion.]

[24] *Ibid.* See Section 3.753, formula 5.

Chapter 7
The Wave Equation

In this chapter we shall study problems associated with the equation

$$\frac{\partial^2 u}{\partial t^2} = c^2 \frac{\partial^2 u}{\partial x^2}, \tag{7.0.1}$$

where $u = u(x, t)$, x and t are the two independent variables, and c is a constant. This equation, called the *wave equation*, serves as the prototype for a wider class of *hyperbolic equations*:

$$a(x,t)\frac{\partial^2 u}{\partial x^2} + b(x,t)\frac{\partial^2 u}{\partial x \partial t} + c(x,t)\frac{\partial^2 u}{\partial t^2} = f\left(x, t, u, \frac{\partial u}{\partial x}, \frac{\partial u}{\partial t}\right), \tag{7.0.2}$$

where $b^2 > 4ac$. It arises in the study of many important physical problems involving wave propagation, such as the transverse vibrations of an elastic string and the longitudinal vibrations or torsional oscillations of a rod.

7.1 THE VIBRATING STRING

The motion of a string of length L and constant density ρ (mass per unit *length*) provides a simple example of the wave equation. See Figure 7.1.1. Assuming that the equilibrium position of the string and the

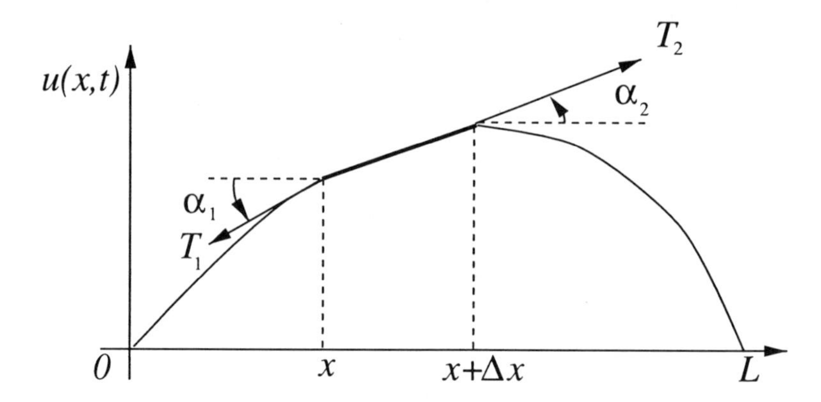

Figure 7.1.1: The vibrating string.

interval $[0, L]$ along the x-axis coincide, the equation of motion which describes the vertical displacement $u(x,t)$ of the string follows by considering a short piece whose ends are at x and $x + \Delta x$ and applying Newton's second law.

If we assume that the string is perfectly flexible and offers no resistance to bending, Figure 7.1.1 shows the forces on an element of the string. Applying Newton's second law in the x-direction, the sum of forces equals

$$-T(x)\cos(\alpha_1) + T(x + \Delta x)\cos(\alpha_2), \tag{7.1.1}$$

where $T(x)$ denotes the tensile force. If we assume that a point on the string moves only in the vertical direction, the sum of forces in (7.1.1) equals zero and the horizontal component of tension is constant:

$$-T(x)\cos(\alpha_1) + T(x + \Delta x)\cos(\alpha_2) = 0 \tag{7.1.2}$$

and

$$T(x)\cos(\alpha_1) = T(x + \Delta x)\cos(\alpha_2) = T, \text{ a constant.} \tag{7.1.3}$$

If gravity is the only external force, Newton's law in the vertical direction gives

$$-T(x)\sin(\alpha_1) + T(x + \Delta x)\sin(\alpha_2) - mg = m\frac{\partial^2 u}{\partial t^2}, \tag{7.1.4}$$

where u_{tt} is the acceleration. Because

$$T(x) = \frac{T}{\cos(\alpha_1)} \quad \text{and} \quad T(x + \Delta x) = \frac{T}{\cos(\alpha_2)}, \tag{7.1.5}$$

then

$$-T\tan(\alpha_1) + T\tan(\alpha_2) - \rho\Delta x g = \rho\Delta x\frac{\partial^2 u}{\partial t^2}. \tag{7.1.6}$$

The quantities $\tan(\alpha_1)$ and $\tan(\alpha_2)$ equal the slope of the string at x and $x + \Delta x$, respectively; that is,

$$\tan(\alpha_1) = \frac{\partial u(x,t)}{\partial x} \quad \text{and} \quad \tan(\alpha_2) = \frac{\partial u(x+\Delta x,t)}{\partial x}. \tag{7.1.7}$$

Substituting (7.1.7) into (7.1.6),

$$T\left[\frac{\partial u(x+\Delta x,t)}{\partial x} - \frac{\partial u(x,t)}{\partial x}\right] = \rho\Delta x\left(\frac{\partial^2 u}{\partial t^2} + g\right). \tag{7.1.8}$$

After dividing through by Δx, we have a difference quotient on the left:

$$\frac{T}{\Delta x}\left[\frac{\partial u(x+\Delta x,t)}{\partial x} - \frac{\partial u(x,t)}{\partial x}\right] = \rho\left(\frac{\partial^2 u}{\partial t^2} + g\right). \tag{7.1.9}$$

In the limit as $\Delta x \to 0$, this difference quotient becomes a partial derivative with respect to x, leaving Newton's second law in the form

$$T\frac{\partial^2 u}{\partial x^2} = \rho\frac{\partial^2 u}{\partial t^2} + \rho g \tag{7.1.10}$$

or

$$\frac{\partial^2 u}{\partial x^2} = \frac{1}{c^2}\frac{\partial^2 u}{\partial t^2} + \frac{g}{c^2}, \tag{7.1.11}$$

where $c^2 = T/\rho$. Because u_{tt} is generally much larger than g, we can neglect the last term, giving the equation of the vibrating string as

$$\frac{\partial^2 u}{\partial x^2} = \frac{1}{c^2}\frac{\partial^2 u}{\partial t^2}. \tag{7.1.12}$$

Equation (7.1.12) is the one-dimensional *wave equation.*

As a second example[1] we derive the threadline equation which describes how a thread composed of yard vibrates as we draw it between two eyelets spaced a distance L apart. We assume that the tension in the thread is constant, the vibrations are small, the thread is perfectly flexible, the effects of gravity and air drag are negligible and the mass of the thread per unit length is constant. Unlike the vibrating string between two fixed ends, we draw the threadline through the eyelets at a speed V so that a segment of thread experiences motion in both the x and y directions as it vibrates about its equilibrium position. The eyelets may move in the vertical direction.

---

[1] Reprinted from *J. Franklin Inst.*, **275**, Swope, R. D., and W. F. Ames, Vibrations of a moving threadline, 36–55, ©1963, with kind permission from Elsevier Science Ltd, The Boulevard, Langford Lane, Kidlington OX5 1GB, UK.

From Newton's second law:

$$\frac{d}{dt}\left(m\frac{dy}{dt}\right) = \sum \text{ forces,} \tag{7.1.13}$$

where m is the mass of the thread. But

$$\frac{dy}{dt} = \frac{\partial y}{\partial t} + \frac{dx}{dt}\frac{\partial y}{\partial x}. \tag{7.1.14}$$

Because $dx/dt = V$,

$$\frac{dy}{dt} = \frac{\partial y}{\partial t} + V\frac{\partial y}{\partial x} \tag{7.1.15}$$

and

$$\frac{d}{dt}\left(m\frac{dy}{dt}\right) = \frac{\partial}{\partial t}\left[m\left(\frac{\partial y}{\partial t} + V\frac{\partial y}{\partial x}\right)\right] + V\frac{\partial}{\partial x}\left[m\left(\frac{\partial y}{\partial t} + V\frac{\partial y}{\partial x}\right)\right]. \tag{7.1.16}$$

Because both m and V are constant, it follows that

$$\frac{d}{dt}\left(m\frac{dy}{dt}\right) = m\frac{\partial^2 y}{\partial t^2} + 2mV\frac{\partial^2 y}{\partial x \partial t} + mV^2\frac{\partial^2 y}{\partial x^2}. \tag{7.1.17}$$

The sum of the forces again equals

$$T\frac{\partial^2 y}{\partial x^2}\Delta x \tag{7.1.18}$$

so that the threadline equation is

$$T\frac{\partial^2 y}{\partial x^2}\Delta x = m\frac{\partial^2 y}{\partial t^2} + 2mV\frac{\partial^2 y}{\partial x \partial t} + mV^2\frac{\partial^2 y}{\partial x^2} \tag{7.1.19}$$

or

$$\frac{\partial^2 y}{\partial t^2} + 2V\frac{\partial^2 y}{\partial x \partial t} + \left(V^2 - \frac{gT}{\rho}\right)\frac{\partial^2 y}{\partial x^2} = 0, \tag{7.1.20}$$

where ρ is the density of the thread. Although (7.1.20) is *not* the classic wave equation given in (7.1.12), it is an example of a hyperbolic equation. As we shall see, the solutions to hyperbolic equations share the same behavior, namely, wave-like motion.

7.2 INITIAL CONDITIONS: CAUCHY PROBLEM

Any mathematical model of a physical process must include not only the governing differential equation but also any conditions that are imposed on the solution. For example, in time-dependent problems the solution must conform to the initial condition of the modeled process.

Finding those solutions that satisfy the initial conditions (initial data) is called the *Cauchy problem*.

In the case of partial differential equations with second-order derivatives in time, such as the wave equation, we correctly pose the *Cauchy boundary condition* if we specify the value of the solution $u(x, t_0) = f(t)$ *and* its time derivative $u_t(x, t_0) = g(t)$ at some initial time t_0, usually taken to be $t_0 = 0$. The functions $f(t)$ and $g(t)$ are called the *Cauchy data*. We require two conditions involving time because the differential equation has two time derivatives.

In addition to the initial conditions, we must specify boundary conditions in the spatial direction. For example, we may require that the end of the string be fixed. In the next chapter, we discuss the boundary conditions in greater depth. However, one boundary condition that is uniquely associated with the wave equation on an open domain is the *radiation condition*. It requires that the waves radiate off to infinity and remain finite as they propagate there.

In summary, the Cauchy boundary condition, along with the appropriate spatial boundary conditions, uniquely determines the solution to the wave equation; any additional information is extraneous. Having developed the differential equations and initial conditions necessary to solve the wave equation, let us now turn to the actual methods used to solve this equation.

7.3 SEPARATION OF VARIABLES

We begin by presenting the most classical method of solving the wave equation: *separation of variables*. Despite its current widespread use, its initial application to the vibrating string problem was immersed in controversy involving the application of a half-range Fourier sine series to represent the initial conditions. On one side, Daniel Bernoulli claimed (in 1775) that he could represent any general initial condition with this technique. To d'Alembert and Euler, however, the half-range Fourier sine series, with its period of $2L$, could not possibly represent any arbitrary function.[2] However, by 1807 Bernoulli was proven correct by the use of separation of variables in the heat conduction problem and it rapidly grew in acceptance.[3] In the following examples we show how to apply this method to solve the wave equation.

---

[2] See Hobson, E. W., 1957: *The Theory of Functions of a Real Variable and the Theory of Fourier's Series, Vol. 2,* Dover Publishers, Mineola, NY, Sections 312–314.

[3] Lützen, J., 1984: Sturm and Liouville's work on ordinary linear differential equations. The emergence of Sturm-Liouville theory. *Arch. Hist. Exact Sci.*, **29**, 317.

Separation of variables consists of four distinct steps. The basic idea is to convert a second-order partial differential equation into two ordinary differential equations. First, we *assume* that the solution equals the product $X(x)T(t)$. Direct substitution into the partial differential equation and boundary conditions yields two ordinary differential equations and the corresponding boundary conditions. Step two involves solving a boundary-value problem of the Sturm-Liouville type. In step three we find the corresponding time dependence. Finally we construct the complete solution as a sum of all product solutions. Upon applying the initial conditions, we have an eigenfunction expansion and must compute the Fourier coefficients. The substitution of these coefficients into the summation yields the final solution.

• Example 7.3.1

Let us solve the wave equation for the special case when we clamp the string at $x = 0$ and $x = L$. Mathematically, we find the solution to the wave equation

$$\frac{\partial^2 u}{\partial t^2} = c^2 \frac{\partial^2 u}{\partial x^2}, \qquad 0 < x < L, 0 < t \tag{7.3.1}$$

which satisfies the initial conditions

$$u(x,0) = f(x), \qquad \frac{\partial u(x,0)}{\partial t} = g(x), \quad 0 < x < L, \tag{7.3.2}$$

and the boundary conditions

$$u(0,t) = u(L,t) = 0, \quad 0 < t. \tag{7.3.3}$$

For the present, we have left the Cauchy data quite arbitrary.

We begin by assuming that the solution $u(x,t)$ equals the product $X(x)T(t)$. (Here T no longer denotes tension.) Because

$$\frac{\partial^2 u}{\partial t^2} = X(x)T''(t) \tag{7.3.4}$$

and

$$\frac{\partial^2 u}{\partial x^2} = X''(x)T(t), \tag{7.3.5}$$

the wave equation becomes

$$c^2 X''T = T''X \tag{7.3.6}$$

or

$$\frac{X''}{X} = \frac{T''}{c^2 T}. \tag{7.3.7}$$

after dividing through by $c^2X(x)T(t)$. Because the left side of (7.3.7) depends only on x and the right side depends only on t, both sides must equal a constant. We write this separation constant $-\lambda$ and separate (7.3.7) into two ordinary differential equations:

$$T'' + c^2\lambda T = 0, \qquad 0 < t \tag{7.3.8}$$

and

$$X'' + \lambda X = 0, \qquad 0 < x < L. \tag{7.3.9}$$

We now rewrite the boundary conditions in terms of $X(x)$ by noting that the boundary conditions become

$$u(0,t) = X(0)T(t) = 0 \tag{7.3.10}$$

and

$$u(L,t) = X(L)T(t) = 0 \tag{7.3.11}$$

for $0 < t$. If we were to choose $T(t) = 0$, then we would have a trivial solution for $u(x,t)$. Consequently,

$$X(0) = X(L) = 0. \tag{7.3.12}$$

This concludes the first step.

In the second step we consider three possible values for λ: $\lambda < 0$, $\lambda = 0$, and $\lambda > 0$. We begin by assuming that $\lambda = -m^2 < 0$. We have chosen $\lambda = -m^2$ so that square roots of λ will not appear later on and m is real. The general solution of (7.3.9) is

$$X(x) = A\cosh(mx) + B\sinh(mx). \tag{7.3.13}$$

Because $X(0) = 0$, $A = 0$. On the other hand, $X(L) = B\sinh(mL) = 0$. The function $\sinh(mL)$ does not equal to zero since $mL \neq 0$ (recall $m > 0$). Thus, $B = 0$ and we have trivial solutions for a positive separation constant.

If $\lambda = 0$, the general solution now becomes

$$X(x) = C + Dx. \tag{7.3.14}$$

The condition $X(0) = 0$ yields $C = 0$ while $X(L) = 0$ yields $DL = 0$ or $D = 0$. Hence, we have a trivial solution for the $\lambda = 0$ separation constant.

If $\lambda = k^2 > 0$, the general solution to (7.3.9) is

$$X(x) = E\cos(kx) + F\sin(kx). \tag{7.3.15}$$

The condition $X(0) = 0$ results in $E = 0$. On the other hand, $X(L) = F\sin(kL) = 0$. If we wish to avoid a trivial solution in this case ($F \neq 0$),

$\sin(kL) = 0$ or $k_n = n\pi/L$ and $\lambda_n = n^2\pi^2/L^2$. The x-dependence equals $X_n(x) = F_n \sin(n\pi x/L)$. We have added the n subscript to k and λ to indicate that these quantities depend on n. This concludes the second step.

Turning to (7.3.8) for the third step, the solution to the $T(t)$ equation is

$$T_n(t) = G_n \cos(k_n ct) + H_n \sin(k_n ct), \tag{7.3.16}$$

where G_n and H_n are arbitrary constants. For each $n = 1, 2, 3, \ldots$, a particular solution that satisfies the wave equation and prescribed boundary conditions is

$$u_n(x,t) = F_n \sin\left(\frac{n\pi x}{L}\right)\left[G_n \cos\left(\frac{n\pi ct}{L}\right) + H_n \sin\left(\frac{n\pi ct}{L}\right)\right] \tag{7.3.17}$$

or

$$u_n(x,t) = \sin\left(\frac{n\pi x}{L}\right)\left[A_n \cos\left(\frac{n\pi ct}{L}\right) + B_n \sin\left(\frac{n\pi ct}{L}\right)\right], \tag{7.3.18}$$

where $A_n = F_n G_n$ and $B_n = F_n H_n$. This concludes the third step.

An alternative method of finding the product solution is to treat (7.3.9) along with $X(0) = X(L) = 0$ as a Sturm-Liouville problem. Consequently, by solving the Sturm-Liouville problem and finding the corresponding eigenvalue λ_n and eigenfunction, we obtain the spatial dependence. Next we solve for $T_n(t)$. Finally we form the product solution $u_n(x,t)$ by multiplying the eigenfunction times the temporal dependence.

For any choice of A_n and B_n, (7.3.18) is a solution of the partial differential equation (7.3.1) and also satisfies the end boundary conditions (7.3.3). Therefore, any linear combination of the $u_n(x,t)$ also satisfies the partial differential equation and the boundary conditions. In making this linear combination we need no new constants because A_n and B_n are still arbitrary. We have, then,

$$u(x,t) = \sum_{n=1}^{\infty} \sin\left(\frac{n\pi x}{L}\right)\left[A_n \cos\left(\frac{n\pi ct}{L}\right) + B_n \sin\left(\frac{n\pi ct}{L}\right)\right]. \tag{7.3.19}$$

Our example of using particular solutions to build up the general solution illustrates the powerful *principle of linear superposition*, which is applicable to any *linear* system. This principle states that if u_1 and u_2 are any solutions of a linear homogeneous partial differential equation in any region, then $u = c_1 u_1 + c_2 u_2$ is also a solution of that equation in that region, where c_1 and c_2 are any constants. We can generalize this to an infinite sum. It is extremely important because it allows us to construct general solutions to partial differential equations from particular solutions to the same problem.

Our fourth and final task remains to determine A_n and B_n. At $t = 0$,

$$u(x,0) = \sum_{n=1}^{\infty} A_n \sin\left(\frac{n\pi x}{L}\right) = f(x) \tag{7.3.20}$$

and

$$u_t(x,0) = \sum_{n=1}^{\infty} \frac{n\pi c}{L} B_n \sin\left(\frac{n\pi x}{L}\right) = g(x). \tag{7.3.21}$$

Both of these series are Fourier half-range sine expansions over the interval $(0, L)$. Applying the results from Section 2.3,

$$A_n = \frac{2}{L}\int_0^L f(x)\sin\left(\frac{n\pi x}{L}\right)\,dx \tag{7.3.22}$$

and

$$\frac{n\pi c}{L} B_n = \frac{2}{L}\int_0^L g(x)\sin\left(\frac{n\pi x}{L}\right)\,dx \tag{7.3.23}$$

or

$$B_n = \frac{2}{n\pi c}\int_0^L g(x)\sin\left(\frac{n\pi x}{L}\right)\,dx. \tag{7.3.24}$$

As an example, let us take the initial conditions:

$$f(x) = \begin{cases} 0, & 0 < x \le L/4 \\ 4h\left(\frac{x}{L} - \frac{1}{4}\right), & L/4 \le x \le L/2 \\ 4h\left(\frac{3}{4} - \frac{x}{L}\right), & L/2 \le x \le 3L/4 \\ 0, & 3L/4 \le x < L \end{cases} \tag{7.3.25}$$

and

$$g(x) = 0, \qquad 0 < x < L. \tag{7.3.26}$$

In this particular example, $B_n = 0$ for all n because $g(x) = 0$. On the other hand,

$$A_n = \frac{8h}{L}\int_{L/4}^{L/2}\left(\frac{x}{L} - \frac{1}{4}\right)\sin\left(\frac{n\pi x}{L}\right)\,dx$$
$$+\frac{8h}{L}\int_{L/2}^{3L/4}\left(\frac{3}{4} - \frac{x}{L}\right)\sin\left(\frac{n\pi x}{L}\right)\,dx \tag{7.3.27}$$
$$= \frac{8h}{n^2\pi^2}\left[2\sin\left(\frac{n\pi}{2}\right) - \sin\left(\frac{3n\pi}{4}\right) - \sin\left(\frac{n\pi}{4}\right)\right] \tag{7.3.28}$$
$$= \frac{8h}{n^2\pi^2}\left[2\sin\left(\frac{n\pi}{2}\right) - 2\sin\left(\frac{n\pi}{2}\right)\cos\left(\frac{n\pi}{4}\right)\right] \tag{7.3.29}$$
$$= \frac{8h}{n^2\pi^2}\sin\left(\frac{n\pi}{2}\right)\left[1 - \cos\left(\frac{n\pi}{4}\right)\right] \tag{7.3.30}$$
$$= \frac{32h}{n^2\pi^2}\sin\left(\frac{n\pi}{2}\right)\sin^2\left(\frac{n\pi}{8}\right), \tag{7.3.31}$$

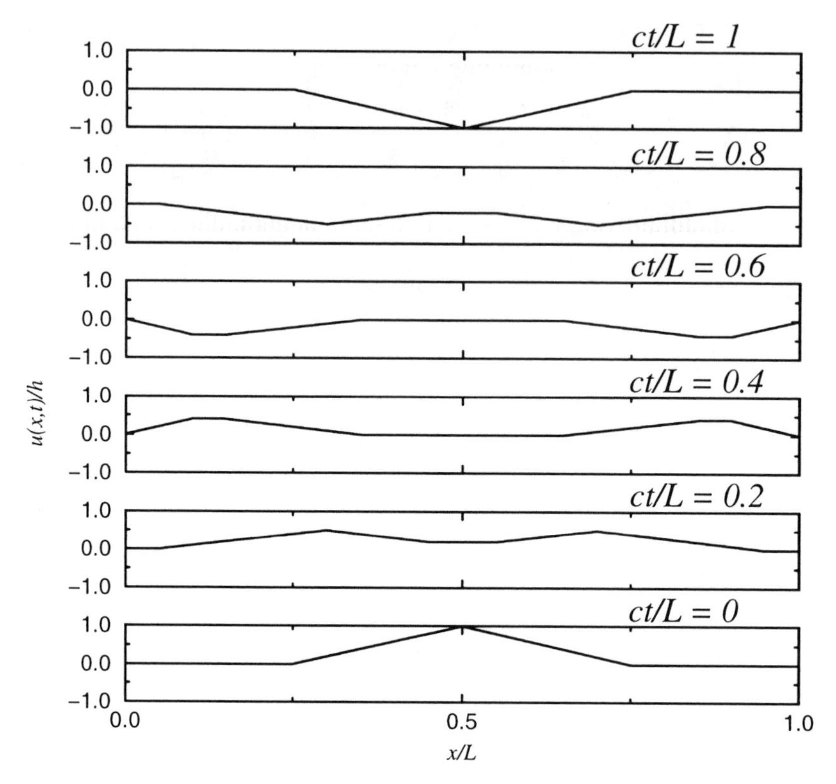

Figure 7.3.1: The vibration of a string $u(x,t)/h$ at various positions x/L at the times $ct/L = 0, 0.2, 0.4, 0.6, 0.8$, and 1. For times $1 < ct/L < 2$ the pictures appear in reverse order.

because $\sin(A)+\sin(B) = 2\sin[\frac{1}{2}(A+B)]\cos[\frac{1}{2}(A-B)]$ and $1-\cos(2A) = 2\sin^2(A)$. Therefore,

$$u(x,t) = \frac{32h}{\pi^2}\sum_{n=1}^{\infty} \sin\left(\frac{n\pi}{2}\right)\sin^2\left(\frac{n\pi}{8}\right)\frac{1}{n^2}\sin\left(\frac{n\pi x}{L}\right)\cos\left(\frac{n\pi ct}{L}\right). \tag{7.3.32}$$

Because $\sin(n\pi/2)$ vanishes for n even, so does A_n. If (7.3.32) were evaluated on a computer, considerable time and effort would be wasted. Consequently it is preferable to rewrite (7.3.32) so that we eliminate these vanishing terms. The most convenient method introduces the general expression $n = 2m-1$ for any odd integer, where $m = 1, 2, 3, \ldots$, and notes that $\sin[(2m-1)\pi/2] = (-1)^{m+1}$. Therefore, (7.3.32) becomes

$$u(x,t) = \frac{32h}{\pi^2}\sum_{m=1}^{\infty} \frac{(-1)^{m+1}}{(2m-1)^2}\sin^2\left[\frac{(2m-1)\pi}{8}\right]$$

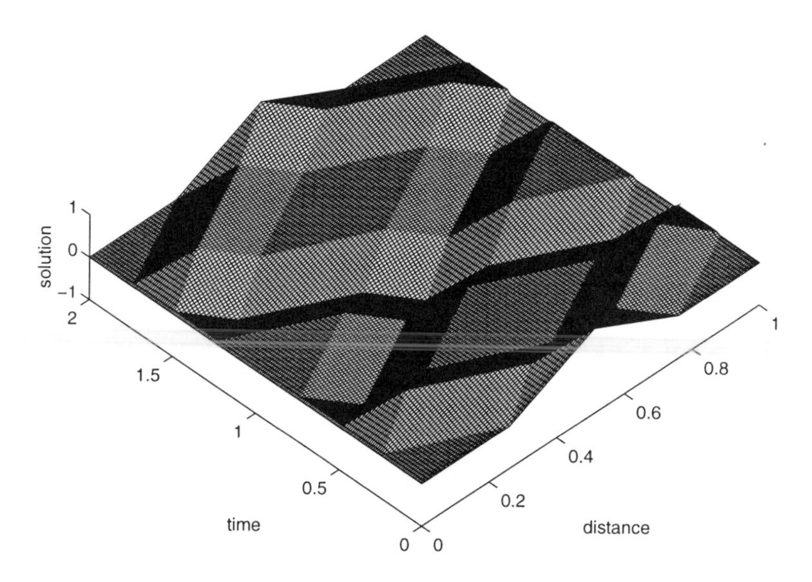

Figure 7.3.2: Two-dimensional plot of the vibration of a string $u(x,t)/h$ at various times ct/L and positions x/L.

$$\times \sin\left[\frac{(2m-1)\pi x}{L}\right]\cos\left[\frac{(2m-1)\pi ct}{L}\right]. \quad \textbf{(7.3.33)}$$

Although we have completely solved the problem, it is useful to rewrite (7.3.33) as

$$u(x,t) = \frac{1}{2}\sum_{n=1}^{\infty} A_n\left\{\sin\left[\frac{n\pi}{L}(x-ct)\right] + \sin\left[\frac{n\pi}{L}(x+ct)\right]\right\} \quad \textbf{(7.3.34)}$$

through the application of the trigonometric identity $\sin(A)\cos(B) = \frac{1}{2}\sin(A-B) + \frac{1}{2}\sin(A+B)$. From general physics we find expressions like $\sin[k_n(x-ct)]$ or $\sin(kx-\omega t)$ arising in studies of simple wave motions. The quantity $\sin(kx-\omega t)$ is the mathematical description of a propagating wave in the sense that we must move to the right at the speed c if we wish to keep in the same position relative to the nearest crest and trough. The quantities k, ω, and c are the wavenumber, frequency, and phase speed or wave-velocity, respectively. The relationship $\omega = kc$ holds between the frequency and phase speed.

It may seem paradoxical that we are talking about traveling waves in a problem dealing with waves confined on a string of length L. Actually we are dealing with standing waves because at the same time that a wave is propagating to the right its mirror image is running to the left so that there is no resultant progressive wave motion. Figures 7.3.1 and 7.3.2 illustrate our solution; Figure 7.3.1 gives various cross sections of the continuous solution plotted in Figure 7.3.2. The single large peak at $t = 0$ breaks into two smaller peaks which race towards the two ends. At each end, they reflect and turn upside down as they propagate back towards $x = L/2$ at $ct/L = 1$. This large, negative peak at $x = L/2$ again breaks apart, with the two smaller peaks propagating towards the endpoints. They reflect and again become positive peaks as they propagate back to $x = L/2$ at $ct/L = 2$. After that time, the whole process repeats itself.

An important dimension to the vibrating string problem is the fact that the wavenumber k_n is not a free parameter but has been restricted to the values of $n\pi/L$. This restriction on wavenumber is common in wave problems dealing with limited domains (for example, a building, ship, lake, or planet) and these oscillations are given the special name of *normal modes* or *natural vibrations*.

In our problem of the vibrating string, all of the components propagate with the same phase speed. That is, all of the waves, regardless of wavenumber k_n, will move the characteristic distance $c\Delta t$ or $-c\Delta t$ after the time interval Δt has elapsed. In the next example we will see that this is not always true.

• Example 7.3.2: Dispersion

In the preceding example, the solution to the vibrating string problem consisted of two simple waves, each propagating with a phase speed c to the right and left. In problems where the equations of motion are a little more complicated than (7.3.1), all of the harmonics no longer propagate with the same phase speed but at a speed that depends upon the wavenumber. In such systems the phase relation varies between the harmonics and these systems are referred to as *dispersive*.

A modification of the vibrating string problem provides a simple illustration. We now subject each element of the string to an additional applied force which is proportional to its displacement:

$$\frac{\partial^2 u}{\partial t^2} = c^2\frac{\partial^2 u}{\partial x^2} - hu, \quad 0 < x < L, 0 < t, \tag{7.3.35}$$

where $h > 0$ is constant. For example, if we embed the string in a thin sheet of rubber, then in addition to the restoring force due to tension, there will be a restoring force due to the rubber on each portion of the string. From its use in the quantum mechanics of "scalar" mesons, (7.3.35) is often referred to as the *Klein-Gordon* equation.

We shall again look for particular solutions of the form $u(x,t) = X(x)T(t)$. This time, however,

$$XT'' - c^2X''T + hXT = 0 \tag{7.3.36}$$

or

$$\frac{T''}{c^2T} + \frac{h}{c^2} = \frac{X''}{X} = -\lambda, \tag{7.3.37}$$

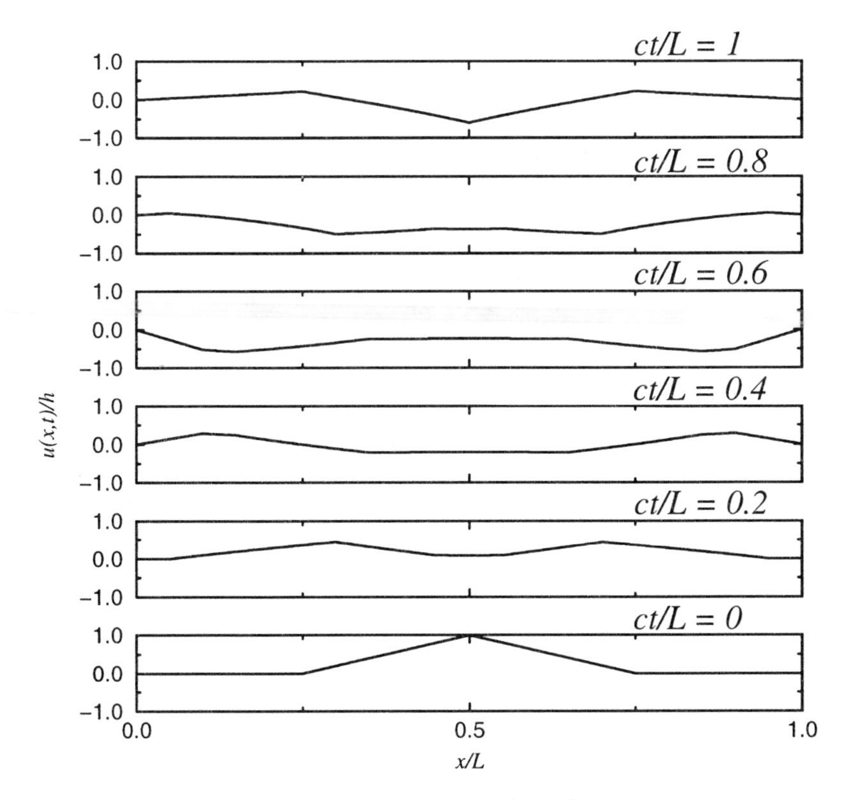

Figure 7.3.3: The vibration of a string $u(x,t)/h$ embedded in a thin sheet of rubber at various positions x/L at the times $ct/L = 0$, 0.2, 0.4, 0.6, 0.8, and 1 for $hL^2/c^2 = 10$. The same parameters were used as in Figure 7.3.1.

which leads to two ordinary differential equations

$$X'' + \lambda X = 0 \tag{7.3.38}$$

and

$$T'' + (\lambda c^2 + h)T = 0. \tag{7.3.39}$$

If we attach the string at $x = 0$ and $x = L$, the $X(x)$ solution is

$$X_n(x) = \sin\left(\frac{n\pi x}{L}\right) \tag{7.3.40}$$

with $k_n = n\pi/L$ and $\lambda_n = n^2\pi^2/L^2$. On the other hand, the $T(t)$ solution becomes

$$T_n(t) = A_n \cos\left(\sqrt{k_n^2 c^2 + h}\, t\right) + B_n \sin\left(\sqrt{k_n^2 c^2 + h}\, t\right) \tag{7.3.41}$$

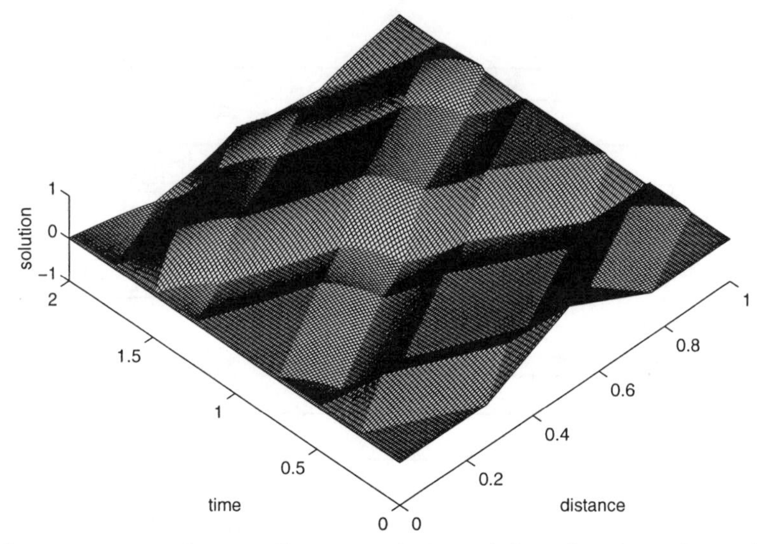

Figure 7.3.4: The two-dimensional plot of the vibration of a string $u(x,t)/h$ embedded in a thin sheet of rubber at various times ct/L and positions x/L for $hL^2/c^2 = 10$.

so that the product solution is

$$u_n(x,t) = \sin\left(\frac{n\pi x}{L}\right)\left[A_n \cos\left(\sqrt{k_n^2c^2+h}\,t\right) + B_n \sin\left(\sqrt{k_n^2c^2+h}\,t\right)\right]. \tag{7.3.42}$$

Finally, the general solution becomes

$$u(x,t) = \sum_{n=1}^{\infty} \sin\left(\frac{n\pi x}{L}\right)\left[A_n \cos\left(\sqrt{k_n^2c^2+h}\,t\right) + B_n \sin\left(\sqrt{k_n^2c^2+h}\,t\right)\right] \tag{7.3.43}$$

from the principle of linear superposition. Let us consider the case when $B_n = 0$. Then we can write (7.3.43)

$$u(x,t) = \sum_{n=1}^{\infty} \frac{A_n}{2}\left[\sin\left(k_n x + \sqrt{k_n^2c^2+h}\,t\right) + \sin\left(k_n x - \sqrt{k_n^2c^2+h}\,t\right)\right]. \tag{7.3.44}$$

Comparing our results with (7.3.34), the distance that a particular mode k_n moves during the time interval Δt depends not only upon external parameters such as h, the tension and density of the string, but also upon its wavenumber (or equivalently, wavelength). Furthermore, the frequency of a particular harmonic is larger than that when $h = 0$.

This result is not surprising, because the added stiffness of the medium should increase the natural frequencies.

The importance of dispersion lies in the fact that if the solution $u(x,t)$ is a superposition of progressive waves in the same direction, then the phase relationship between the different harmonics will change with time. Because most signals consist of an infinite series of these progressive waves, dispersion causes the signal to become garbled. We show this by comparing the solution (7.3.43) given in Figures 7.3.3 and 7.3.4 for the initial conditions (7.3.25) and (7.3.26) with $hL^2/c^2 = 10$ to the results given in Figures 7.3.1 and 7.3.2. Note how garbled the picture becomes at $ct/L = 2$ in Figure 7.3.4 compared to the nondispersive solution at the same time in Figure 7.3.2.

• Example 7.3.3: Damped wave equation

In the previous example a slight modification of the wave equation resulted in a wave solution where each Fourier harmonic propagates with its own particular phase speed. In this example we introduce a modification of the wave equation that will result not only in dispersive waves but also in the exponential decay of the amplitude as the wave propagates.

So far we have neglected the reaction of the surrounding medium (air or water, for example) on the motion of the string. For small-amplitude motions this reaction opposes the motion of each element of the string and is proportional to the element's velocity. The equation of motion, when we account for the tension and friction in the medium but not its stiffness or internal friction, is

$$\frac{\partial^2 u}{\partial t^2} + 2h\frac{\partial u}{\partial t} = c^2\frac{\partial^2 u}{\partial x^2}, \quad 0 < x < L, 0 < t. \tag{7.3.45}$$

Because (7.3.45) first arose in the mathematical description of the telegraph,[4] it is generally known as the *equation of telegraphy*. The effect of friction is, of course, to damp out the free vibration.

Let us assume a solution of the form $u(x,t) = X(x)T(t)$ and separate the variables to obtain the two ordinary differential equations:

$$X'' + \lambda X = 0 \tag{7.3.46}$$

and

$$T'' + 2hT' + \lambda c^2 T = 0 \tag{7.3.47}$$

[4] The first published solution was by Kirchhoff, G., 1857: Über die Bewegung der Electrität in Drähten. *Ann. Phys. Chem.*, **100**, 193–217. English translation: Kirchhoff, G., 1857: On the motion of electricity in wires. *Philos. Mag., Ser. 4*, **13**, 393–412.

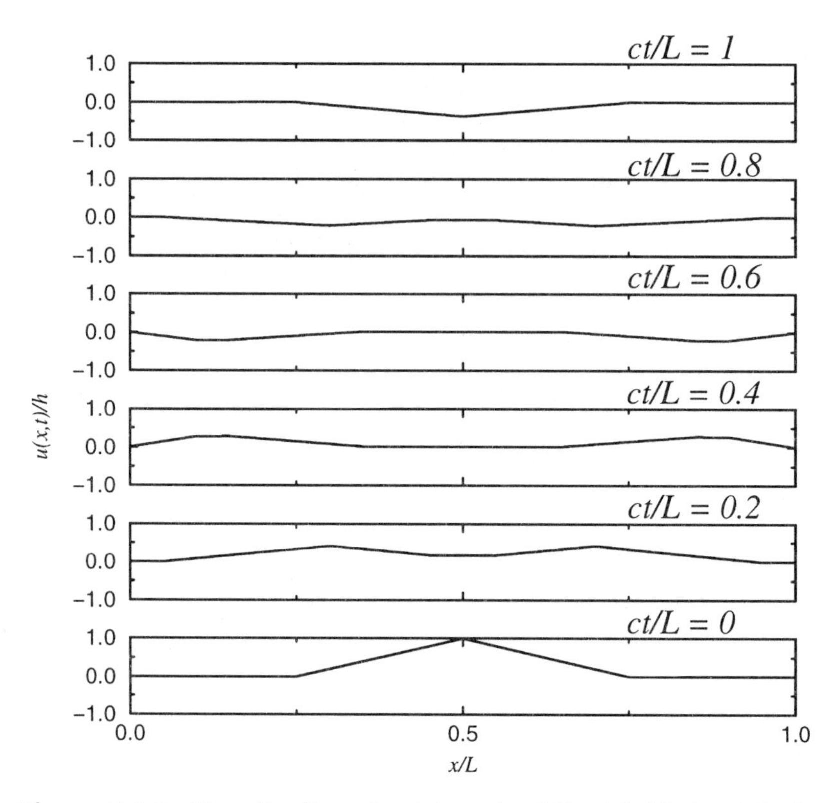

Figure 7.3.5: The vibration of a string $u(x,t)/h$ with frictional dissipation at various positions x/L at the times $ct/L = 0$, 0.2, 0.4, 0.6, 0.8, and 1 for $hL/c = 1$. The same parameters were used as in Figure 7.3.1.

with $X(0) = X(L) = 0$. Friction does not affect the shape of the normal modes; they are still

$$X_n(x) = \sin\left(\frac{n\pi x}{L}\right) \tag{7.3.48}$$

with $k_n = n\pi/L$ and $\lambda_n = n^2\pi^2/L^2$.

The solution for the $T(t)$ equation is

$$T_n(t) = e^{-ht}\left[A_n \cos\left(\sqrt{k_n^2 c^2 - h^2}\,t\right) + B_n \sin\left(\sqrt{k_n^2 c^2 - h^2}\,t\right)\right] \tag{7.3.49}$$

with the condition that $k_n c > h$. If we violate this condition, the solutions are two exponentially decaying functions in time. Because most physical problems usually fulfill this condition, we will concentrate on this solution.

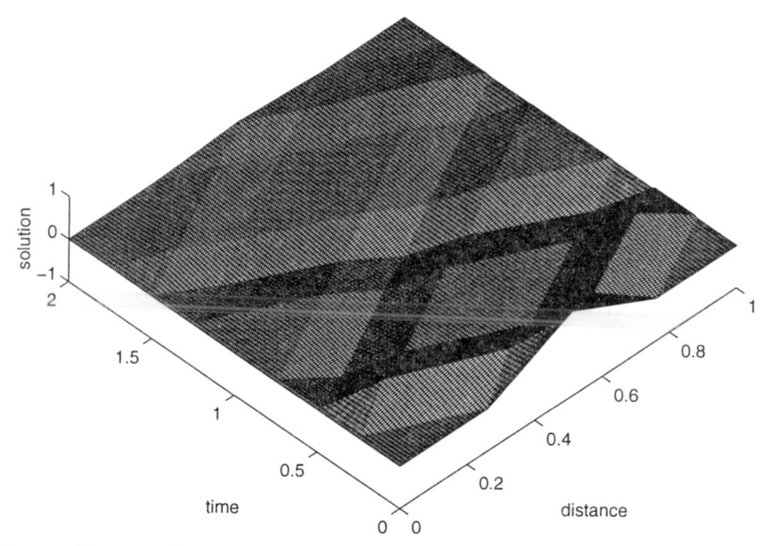

Figure 7.3.6: The vibration of a string $u(x,t)/h$ with frictional dissipation at various times ct/L and positions x/L for $hL/c = 1$.

From the principle of linear superposition, the general solution is

$$u(x,t) = e^{-ht} \sum_{n=1}^{\infty} \sin\left(\frac{n\pi x}{L}\right) \left[A_n \cos\left(\sqrt{k_n^2 c^2 - h^2}\, t\right) + B_n \sin\left(\sqrt{k_n^2 c^2 - h^2}\, t\right)\right], \quad \textbf{(7.3.50)}$$

where $\pi c > hL$. From (7.3.50) we see two important effects. First, the presence of friction slows all of the harmonics. Furthermore, friction dampens all of the harmonics. Figures 7.3.5 and 7.3.6 illustrate the solution using the initial conditions given by (7.3.25) and (7.3.26) with $hL/c = 1$. This is a rather large coefficient of friction and these figures show the rapid damping that results with a small amount of dispersion.

This damping and dispersion of waves also occurs in solutions of the equation of telegraphy where the solutions are progressive waves. Because early telegraph lines were short, time delay effects were negligible. However, when engineers laid the first transoceanic cables in the 1850s, the time delay became seconds and differences in the velocity of propagation of different frequencies, as predicted by (7.3.50), became noticeable to the operators. Table 7.3.1 gives the transmission rate for various transatlantic submarine telegraph lines. As it shows, increases in the transmission rates during the nineteenth century were due primarily to improvements in terminal technology.

When they instituted long-distance telephony just before the turn of the twentieth century, this difference in velocity between frequencies

Table 7.3.1: Technological Innovation on Transatlantic Telegraph Cables.

Year	Technological Innovation	Performance (words/min)
1857–58	Mirror galvanometer	3–7
1870	Condensers	12
1872	Siphon recorder	17
1879	Duplex	24
1894	Larger diameter cable	72–90
1915–20	Brown drum repeater and Heurtley magnifier	100
1923–28	Magnetically loaded lines	300–320
1928–32	Electronic signal shaping amplifiers and time division multiplexing	480
1950	Repeaters on the continental shelf	100–300
1956	Repeatered telephone cables	21600

From Coates, V. T. and Finn, B., 1979: *A Retrospective Technology Assessment: Submarine Telegraphy. The Transatlantic Cable of 1866*, San Francisco Press, Inc.

should have limited the circuits to a few tens of miles.[5] However, in 1899, Prof. Michael Pupin, at Columbia University, showed that by adding inductors ("loading coils") to the line at regular intervals the velocities at the different frequencies could be equalized.[6] Heaviside[7] and the French engineer Vaschy[8] made similar suggestions in the nineteenth century. Thus, adding resistance and inductance, which would seem to make things worse, actually made possible long-distance telephony. Today

[5] Rayleigh, J. W., 1884: On telephoning through a cable. *Br. Assoc. Rep.*, 632–633; Jordan, D. W., 1982: The adoption of self-induction by telephony, 1886–1889. *Ann. Sci.*, **39**, 433–461.

[6] There is considerable controversy on this subject. See Brittain, J. E., 1970: The introduction of the loading coil: George A. Campbell and Michael I. Pupin. *Tech. Culture*, **11**, 36–57.

[7] First published 3 June 1887. Reprinted in Heaviside, O., 1970: *Electrical Papers, Vol. II*, Chelsea Publishing, Bronx, NY, pp. 119–124.

[8] See Devaux-Charbonnel, X. G. F., 1917: La contribution des ingénieurs français à la téléphonie à grande distance par câbles souterrains: Vaschy et Barbarat. *Rev. Gén. Électr.*, **2**, 288–295.

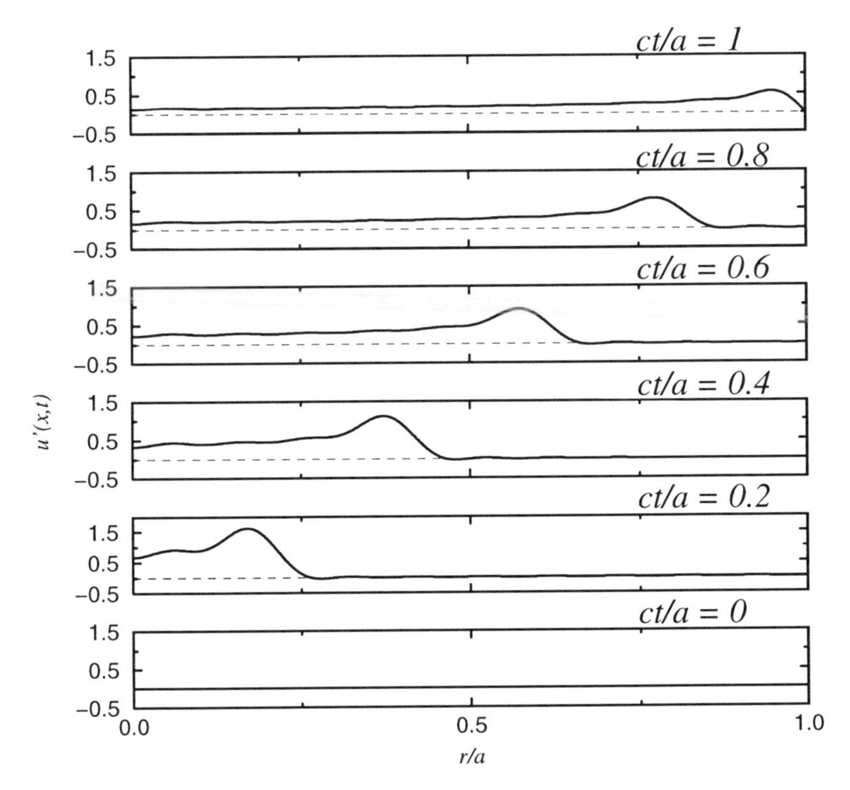

Figure 7.3.7: The axisymmetric vibrations $u'(r,t) = ca\rho u(r,t)/P$ of a circular membrane when struck by a hammer at various positions r/a at the times $ct/a = 0$, 0.2, 0.4, 0.6, 0.8, and 1 for $\epsilon = a/4$.

you can see these loading coils as you drive along the street; they are the black cylinders, approximately one between each pair of telephone poles, spliced into the telephone cable. The loading of long submarine telegraph cables had to wait for the development of permalloy and mu-metal materials of high magnetic induction.

• **Example 7.3.4: Axisymmetric vibrations of a circular membrane**

The wave equation

$$\frac{\partial^2 u}{\partial r^2} + \frac{1}{r}\frac{\partial u}{\partial r} = \frac{1}{c^2}\frac{\partial^2 u}{\partial t^2}, \quad 0 \le r < a, 0 < t \qquad (7.3.51)$$

governs axisymmetric vibrations of a circular membrane, where $u(r,t)$ is the vertical displacement of the membrane, r is the radial distance, t is time, c is the square root of the ratio of the tension of the membrane to its density, and a is the radius of the membrane. We shall solve (7.3.51)

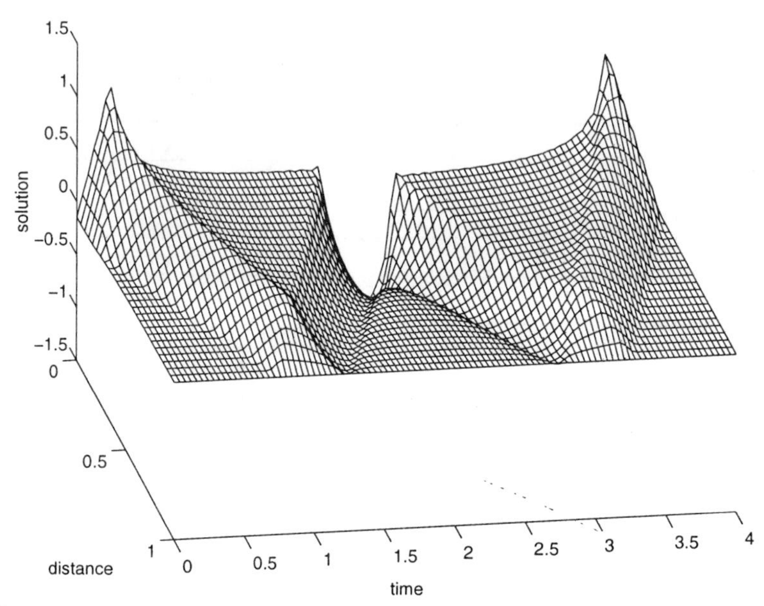

Figure 7.3.8: The axisymmetric vibrations $ca\rho u(r,t)/P$ of a circular membrane when struck by a hammer at various times ct/a and positions r/a for $\epsilon = a/4$.

when the membrane is initially at rest, $u(r,0) = 0$, and struck so that its initial velocity is

$$\frac{\partial u(r,0)}{\partial t} = \begin{cases} P/(\pi\epsilon^2\rho), & 0 \le r < \epsilon \\ 0, & \epsilon < r < a. \end{cases} \tag{7.3.52}$$

If this problem can be solved by separation of variables, then $u(r,t) = R(r)T(t)$. Following the substitution of this $u(r,t)$ into (7.3.51), separation of variables leads to

$$\frac{1}{rR}\frac{d}{dr}\left(r\frac{dR}{dr}\right) = \frac{1}{c^2T}\frac{d^2T}{dt^2} = -k^2 \tag{7.3.53}$$

or

$$\frac{1}{r}\frac{d}{dr}\left(r\frac{dR}{dr}\right) + k^2R = 0 \tag{7.3.54}$$

and

$$\frac{d^2T}{dt^2} + k^2c^2T = 0. \tag{7.3.55}$$

The separation constant $-k^2$ must be negative so that we obtain solutions that remain bounded in the region $0 \le r < a$ and can satisfy the boundary condition. This boundary condition is $u(a,t) = R(a)T(t) = 0$ or $R(a) = 0$.

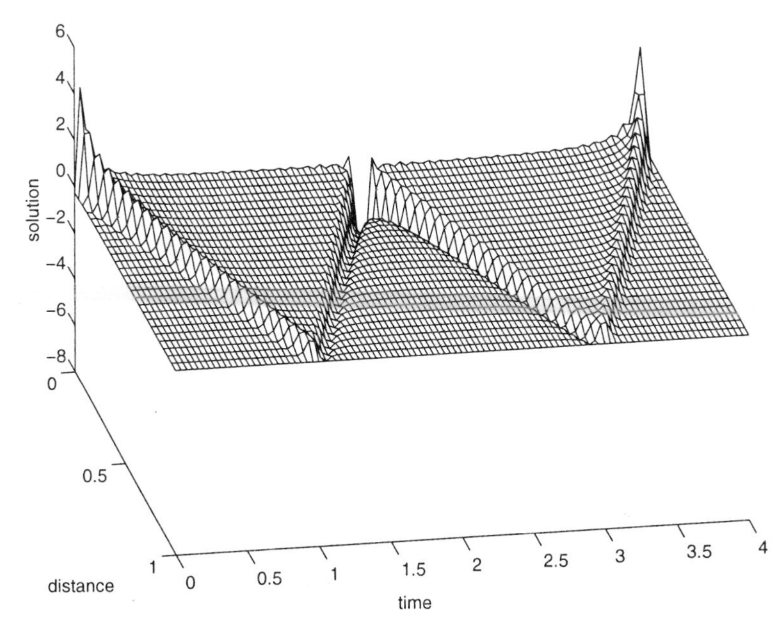

Figure 7.3.9: Same as Figure 7.3.8 except $\epsilon = a/20$.

The solutions of (7.3.54)–(7.3.55), subject to the boundary condition, are

$$R_n(r) = J_0\left(\frac{\lambda_n r}{a}\right) \tag{7.3.56}$$

and

$$T_n(t) = A_n \sin\left(\frac{\lambda_n ct}{a}\right) + B_n \cos\left(\frac{\lambda_n ct}{a}\right), \tag{7.3.57}$$

where λ_n satisfies the equation $J_0(\lambda) = 0$. Because $u(r,0) = 0$ and $T_n(0) = 0$, $B_n = 0$. Consequently, the product solution is

$$u(r,t) = \sum_{n=1}^{\infty} A_n J_0\left(\frac{\lambda_n r}{a}\right) \sin\left(\frac{\lambda_n ct}{a}\right). \tag{7.3.58}$$

To determine A_n, we use the condition

$$\frac{\partial u(r,0)}{\partial t} = \sum_{n=1}^{\infty} \frac{\lambda_n c}{a} A_n J_0\left(\frac{\lambda_n r}{a}\right) = \begin{cases} P/(\pi\epsilon^2\rho), & 0 \le r < \epsilon \\ 0, & \epsilon < r < a. \end{cases} \tag{7.3.59}$$

Equation (7.3.59) is a Fourier-Bessel expansion in the orthogonal function $J_0(\lambda_n r/a)$, where A_n equals

$$\frac{\lambda_n c}{a} A_n = \frac{2}{a^2 J_1^2(\lambda_n)} \int_0^{\epsilon} \frac{P}{\pi\epsilon^2\rho} J_0\left(\frac{\lambda_n r}{a}\right)\, dr \tag{7.3.60}$$

from (6.5.35) and (6.5.43) in Section 6.5. Carrying out the integration,

$$A_n = \frac{2PJ_1(\lambda_n\epsilon/a)}{c\pi\epsilon\rho\lambda_n^2 J_1^2(\lambda_n)} \tag{7.3.61}$$

or

$$u(r,t) = \frac{2P}{c\pi\epsilon\rho}\sum_{n=1}^{\infty}\frac{J_1(\lambda_n\epsilon/a)}{\lambda_n^2 J_1^2(\lambda_n)} J_0\left(\frac{\lambda_n r}{a}\right)\sin\left(\frac{\lambda_n ct}{a}\right). \tag{7.3.62}$$

Figures 7.3.7, 7.3.8, and 7.3.9 illustrate the solution (7.3.62) for various times and positions when $\epsilon = a/4$ and $\epsilon = a/20$. Figures 7.3.8 and 7.3.9 show that striking the membrane with a hammer generates a pulse that propagates out to the rim, reflects, inverts, and propagates back to the center. This process then repeats itself forever.

Problems

Solve the wave equation $u_{tt} = c^2 u_{xx}$, $0 < x < L$, $0 < t$ subject to the boundary conditions that $u(0,t) = u(L,t) = 0, t < 0$ and the following initial conditions for $0 < x < L$:

1. $u(x,0) = 0, \quad u_t(x,0) = 1$

2. $u(x,0) = 1, \quad u_t(x,0) = 0$

3. $u(x,0) = \begin{cases} 3hx/2L, & 0 < x < 2L/3 \\ 3h(L-x)/L, & 2L/3 < x < L, \end{cases} \quad u_t(x,0) = 0$

4. $u(x,0) = [3\sin(\pi x/L) - \sin(3\pi x/L)]/4, \quad u_t(x,0) = 0,$

5. $u(x,0) = \sin(\pi x/L), \quad u_t(x,0) = \begin{cases} 0, & 0 < x < L/4 \\ a, & L/4 < x < 3L/4 \\ 0, & 3L/4 < x < L \end{cases}$

6. $u(x,0) = 0, \quad u_t(x,0) = \begin{cases} ax/L, & 0 < x < L/2 \\ a(L-x)/L, & L/2 < x < L \end{cases}$

7. $u(x,0) = \begin{cases} x, & 0 < x < L/2 \\ L-x, & L/2 < x < L, \end{cases} \quad u_t(x,0) = 0$

8. Solve the wave equation

$$\frac{\partial^2 u}{\partial t^2} = c^2\frac{\partial^2 u}{\partial x^2}, \qquad 0 < x < \pi, 0 < t$$

subject to the boundary conditions

$$\frac{\partial u(0,t)}{\partial x} = \frac{\partial u(\pi,t)}{\partial x} = 0, \qquad 0 < t$$

and the initial conditions

$$u(x,0) = 0 \quad \text{and} \quad \frac{\partial u(x,0)}{\partial t} = 1 + \cos^3(x), \qquad 0 < x < \pi.$$

[Hint: You must include the separation constant of zero.]

9. The differential equation for the longitudinal vibrations of a rod within a viscous fluid is

$$\frac{\partial^2 u}{\partial t^2} + 2h\frac{\partial u}{\partial t} = c^2 \frac{\partial^2 u}{\partial x^2}, \qquad 0 < x < L, 0 < t,$$

where c is the velocity of sound in the rod and h is the damping coefficient. If the rod is fixed at $x = 0$ so that $u(0,t) = 0$ and allowed to freely oscillate at the other end $x = L$ so that $u_x(L,t) = 0$, find the vibrations for any location x and subsequent time t if the rod has the initial displacement of $u(x,0) = x$ and the initial velocity $u_t(x,0) = 0$ for $0 < x < L$. Assume that $h < c\pi/(2L)$. Why?

10. A closed pipe of length L contains air whose density is slightly greater than that of the outside air in the ratio of $1+s_0$ to 1. Everything being at rest, we suddenly draw aside the disk closing one end of the pipe. We want to determine what happens *inside* the pipe after we remove the disk.

As the air rushes outside, it generates sound waves within the pipe. The wave equation

$$\frac{\partial^2 u}{\partial t^2} = c^2 \frac{\partial^2 u}{\partial x^2}$$

governs these waves, where c is the speed of sound and $u(x,t)$ is the velocity potential. Without going into the fluid mechanics of the problem, the boundary conditions are

a. No flow through the closed end: $u_x(0,t) = 0$.

b. No infinite acceleration at the open end: $u_{xx}(L,t) = 0$.

c. Air is initially at rest: $u_x(x,0) = 0$.

d. Air initially has a density greater than the surrounding air by the amount s_0: $u_t(x,0) = -c^2 s_0$.

Find the velocity potential at all positions within the pipe and all subsequent times.

11. One of the classic applications of the wave equation has been the explanation of the acoustic properties of string instruments. Usually we excite a string in one of three ways: by plucking (as in the harp, zither, etc.), by striking with a hammer (piano), or by bowing (violin, violoncello, etc.). In all these case, the governing partial differential equation is

$$\frac{\partial^2 u}{\partial t^2} = c^2 \frac{\partial^2 u}{\partial x^2}$$

with the boundary conditions $u(0,t) = u(L,t) = 0$, $0 < t$. For each of following methods of exciting a string instrument, find the complete solution to the problem:

(a) Plucked string

For the initial conditions:

$$u(x,0) = \begin{cases} \beta x/a, & 0 < x < a \\ \beta(L-x)/(L-a), & a < x < L \end{cases}$$

and

$$u_t(x,0) = 0, \qquad 0 < x < L,$$

show that

$$u(x,t) = \frac{2\beta L^2}{\pi^2 a(L-a)} \sum_{n=1}^{\infty} \frac{1}{n^2} \sin\left(\frac{n\pi a}{L}\right) \sin\left(\frac{n\pi x}{L}\right) \cos\left(\frac{n\pi ct}{L}\right)$$

We note that the harmonics are absent where $\sin(n\pi a/L) = 0$. Thus, if we pluck the string at the center, all of the harmonics of even order will be absent. Furthermore, the intensity of the successive harmonics will vary as n^{-2}. The higher harmonics (overtones) are therefore relatively feeble compared to the $n = 1$ term (the fundamental).

(b) String excited by impact

The effect of the impact of a hammer depends upon the manner and duration of the contact, and is more difficult to estimate. However, as a first estimate, let

$$u(x,0) = 0, \qquad 0 < x < L$$

and

$$u_t(x,0) = \begin{cases} \mu, & a - \epsilon < x < a + \epsilon \\ 0, & \text{otherwise}, \end{cases}$$

where $\epsilon \ll 1$. Show that the solution in this case is

$$u(x,t) = \frac{4\mu L}{\pi^2 c} \sum_{n=1}^{\infty} \frac{1}{n^2} \sin\left(\frac{n\pi\epsilon}{L}\right) \sin\left(\frac{n\pi a}{L}\right) \sin\left(\frac{n\pi x}{L}\right) \sin\left(\frac{n\pi ct}{L}\right).$$

As in part (a), the nth mode is absent if the origin is at a node. The intensity of the overtones are now of the same order of magnitude; higher harmonics (overtones) are relatively more in evidence than in part (a).

(c) Bowed violin string

The theory of the vibration of a string when excited by bowing is poorly understood. The bow drags the string for a time until the string springs back. After awhile the process repeats. It can be shown[9] that the proper initial conditions are

$$u(x,0) = 0, \qquad 0 < x < L$$

and

$$u_t(x,0) = 4\beta c(L-x)/L^2, \qquad 0 < x < L,$$

where β is the maximum displacement. Show that the solution is now

$$u(x,t) = \frac{8\beta}{\pi^2} \sum_{n=1}^{\infty} \frac{1}{n^2} \sin\left(\frac{n\pi x}{L}\right) \sin\left(\frac{n\pi ct}{L}\right).$$

[9] See Lamb, H., 1960: *The Dynamical Theory of Sound.* Dover Publishers, Mineola, NY, Section 27.

7.4 D'ALEMBERT'S FORMULA

In the previous section we sought solutions to the homogeneous wave equation in the form of a product $X(x)T(t)$. For the one-dimensional wave equation there is a more general method for constructing the solution. D'Alembert[10] derived it in 1747.

Let us determine a solution to the homogeneous wave equation

$$\frac{\partial^2 u}{\partial t^2} = c^2 \frac{\partial^2 u}{\partial x^2}, \quad -\infty < x < \infty, 0 < t \tag{7.4.1}$$

which satisfies the initial conditions

$$u(x,0) = f(x) \quad \text{and} \quad \frac{\partial u(x,0)}{\partial t} = g(x), \quad -\infty < x < \infty. \tag{7.4.2}$$

We begin by introducing two new variables ξ, η defined by $\xi = x+ct$ and $\eta = x-ct$ and set $u(x,t) = w(\xi,\eta)$. The variables ξ and η are called the *characteristics* of the wave equation. Using the chain rule,

$$\frac{\partial}{\partial x} = \frac{\partial \xi}{\partial x}\frac{\partial}{\partial \xi} + \frac{\partial \eta}{\partial x}\frac{\partial}{\partial \eta} = \frac{\partial}{\partial \xi} + \frac{\partial}{\partial \eta} \tag{7.4.3}$$

$$\frac{\partial}{\partial t} = \frac{\partial \xi}{\partial t}\frac{\partial}{\partial \xi} + \frac{\partial \eta}{\partial t}\frac{\partial}{\partial \eta} = c\frac{\partial}{\partial \xi} - c\frac{\partial}{\partial \eta} \tag{7.4.4}$$

$$\frac{\partial^2}{\partial x^2} = \frac{\partial \xi}{\partial x}\frac{\partial}{\partial \xi}\left(\frac{\partial}{\partial \xi} + \frac{\partial}{\partial \eta}\right) + \frac{\partial \eta}{\partial x}\frac{\partial}{\partial \eta}\left(\frac{\partial}{\partial \xi} + \frac{\partial}{\partial \eta}\right) \tag{7.4.5}$$

$$= \frac{\partial^2}{\partial \xi^2} + 2\frac{\partial^2}{\partial \xi \partial \eta} + \frac{\partial^2}{\partial \eta^2} \tag{7.4.6}$$

and similarly

$$c^2\frac{\partial^2}{\partial t^2} = c^2\left(\frac{\partial^2}{\partial \xi^2} - 2\frac{\partial^2}{\partial \xi \partial \eta} + \frac{\partial^2}{\partial \eta^2}\right), \tag{7.4.7}$$

so that the wave equation becomes

$$\frac{\partial^2 w}{\partial \xi \partial \eta} = 0. \tag{7.4.8}$$

The general solution of (7.4.8) is

$$w(\xi,\eta) = F(\xi) + G(\eta). \tag{7.4.9}$$

[10] D'Alembert, J., 1747: Recherches sur la courbe que forme une corde tenduë mise en vibration. *Hist. Acad. R. Sci. Belles Lett., Berlin*, 214–219.

Figure 7.4.1: Although largely self-educated in mathematics, Jean Le Rond d'Alembert (1717–1783) gained equal fame as a mathematician and *philosophe* of the continental Enlightenment. By the middle of the eighteenth century, he stood with such leading European mathematicians and mathematical physicists as Clairaut, D. Bernoulli, and Euler. Today we best remember him for his work in fluid dynamics and applying partial differential equations to problems in physics. (Portrait courtesy of the Archives de l'Académie des sciences, Paris.)

Thus, the general solution of (7.4.1) is of the form

$$u(x,t) = F(x+ct) + G(x-ct), \qquad \textbf{(7.4.10)}$$

where F and G are arbitrary functions of one variable and are assumed to be twice differentiable. Setting $t = 0$ in (7.4.10) and using the initial condition that $u(x,0) = f(x)$,

$$F(x) + G(x) = f(x). \qquad \textbf{(7.4.11)}$$

The partial derivative of (7.4.10) with respect to t yields

$$\frac{\partial u(x,t)}{\partial t} = cF'(x+ct) - cG'(x-ct). \tag{7.4.12}$$

Here primes denote differentiation with respect to the argument of the function. If we set $t = 0$ in (7.4.12) and apply the initial condition that $u_t(x,0) = g(x)$,

$$cF'(x) - cG'(x) = g(x). \tag{7.4.13}$$

Integrating (7.4.13) from 0 to any point x gives

$$F(x) - G(x) = \frac{1}{c}\int_0^x g(\tau)\,d\tau + C, \tag{7.4.14}$$

where C is the constant of integration. Combining this result with (7.4.11),

$$F(x) = \frac{f(x)}{2} + \frac{1}{2c}\int_0^x g(\tau)\,d\tau + \frac{C}{2} \tag{7.4.15}$$

and

$$G(x) = \frac{g(x)}{2} - \frac{1}{2c}\int_0^x g(\tau)\,d\tau - \frac{C}{2}. \tag{7.4.16}$$

If we replace the variable x in the expression for F and G by $x + ct$ and $x - ct$, respectively, and substitute the results into (7.4.10), we finally arrive at the formula

$$u(x,t) = \frac{f(x+ct) + f(x-ct)}{2} + \frac{1}{2c}\int_{x-ct}^{x+ct} g(\tau)\,d\tau. \tag{7.4.17}$$

This is known as *d'Alembert's formula* for the solution of the wave equation (7.4.1) subject to the initial conditions (7.4.2). It gives a *representation* of the solution in terms of *known* initial conditions.

• **Example 7.4.1**

To illustrate d'Alembert's formula, let us find the solution to the wave equation (7.4.1) satisfying the initial conditions $u(x,0) = \sin(x)$ and $u_t(x,0) = 0$, $-\infty < x < \infty$. By d'Alembert's formula (7.4.17),

$$u(x,t) = \frac{\sin(x-ct) + \sin(x+ct)}{2} = \sin(x)\cos(ct). \tag{7.4.18}$$

• **Example 7.4.2**

Let us find the solution to the wave equation (7.4.1) when $u(x,0) = 0$ and $u_t(x,0) = \sin(2x)$, $-\infty < x < \infty$. By d'Alembert's formula, the solution is

$$u(x,t) = \frac{1}{2c}\int_{x-ct}^{x+ct} \sin(2\tau)\,d\tau = \frac{\sin(2x)\sin(2ct)}{2}. \tag{7.4.19}$$

In addition to providing a method of solving the wave equation, d'Alembert's solution may also be used to gain physical insight into the vibration of a string. Consider the case when we release a string with zero velocity after giving it an initial displacement of $f(x)$. According to (7.4.17), the displacement at a point x at any time t is

$$u(x,t) = \frac{f(x+ct) + f(x-ct)}{2}. \tag{7.4.20}$$

Because the function $f(x-ct)$ is the same as the function of $f(x)$ translated to the right by a distance equal to ct, $f(x-ct)$ represents a wave of form $f(x)$ traveling to the right with the velocity c, a forward wave. Similarly, we can interpret the function $f(x+ct)$ as representing a wave with the shape $f(x)$ traveling to the left with the velocity c, a backward wave. Thus, the solution (7.4.17) is a superposition of forward and backward waves traveling with the same velocity c and having the shape of the initial profile $f(x)$ with half of the amplitude. Clearly the characteristics $x+ct$ and $x-ct$ give the propagation paths along which the waveform $f(x)$ propagates.

• **Example 7.4.3**

To illustrate our physical interpretation of d'Alembert's solution, suppose that the string has an initial displacement defined by

$$f(x) = \begin{cases} a - |x|, & -a \le x \le a \\ 0, & \text{otherwise.} \end{cases} \tag{7.4.21}$$

In Figure 7.4.2(A) the forward and backward waves, indicated by the dashed line, coincide at $t = 0$. As time advances, both waves move in opposite directions. In particular, at $t = a/(2c)$, they have moved through a distance $a/2$, resulting in the displacement of the string shown in Figure 7.4.2(B). Eventually, at $t = a/c$, the forward and backward waves completely separate. Finally, Figures 4.7.2(D) and 4.7.2(E) show how the waves radiate off to infinity at the speed of c. Note that at each point the string returns to its original position of rest after the passage of each wave.

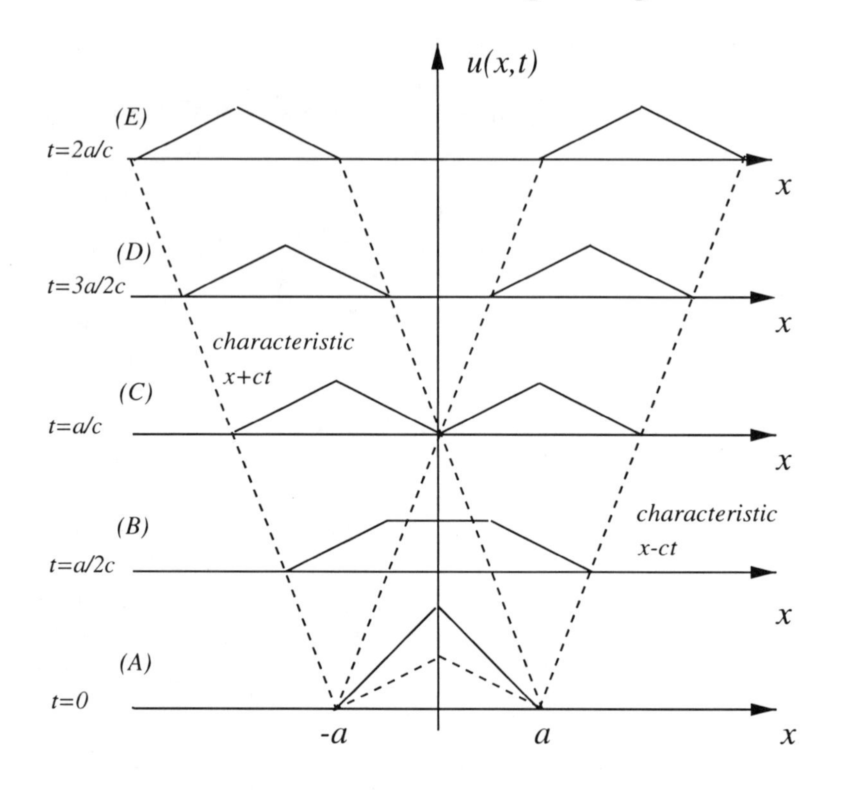

Figure 7.4.2: The propagation of waves due to an initial displacement according to d'Alembert's formula.

Consider now the opposite situation when $u(x,0) = 0$ and $u_t(x,0) = g(x)$. The displacement is

$$u(x,t) = \frac{1}{2c}\int_{x-ct}^{x+ct} g(\tau)\,d\tau. \tag{7.4.22}$$

If we introduce the function

$$\varphi(x) = \frac{1}{2c}\int_{0}^{x} g(\tau)\,d\tau, \tag{7.4.23}$$

then we can write (7.4.22) as

$$u(x,t) = \varphi(x+ct) - \varphi(x-ct), \tag{7.4.24}$$

which again shows that the solution is a superposition of a forward wave $-\varphi(x-ct)$ and a backward wave $\varphi(x+ct)$ traveling with the same velocity c. The function φ, which we compute from (7.4.23) and the initial velocity $g(x)$, determines the exact form of these waves.

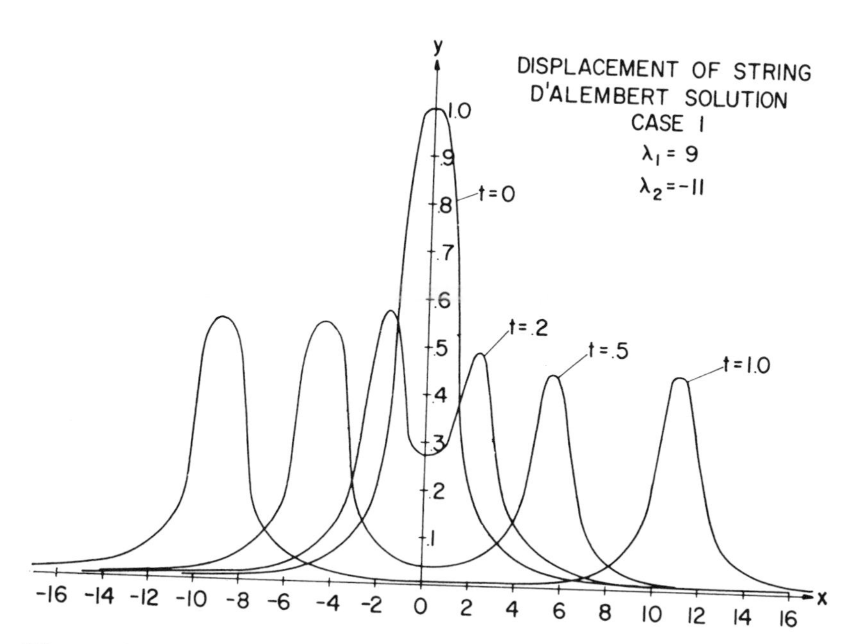

Figure 7.4.3: Displacement of an infinite, moving threadline when $c = 10$ and $V = 1$.

• **Example 7.4.4: Vibration of a moving threadline**

The characterization and analysis of the oscillations of a string or yarn have an important application in the textile industry because they describe the way that yarn winds on a bobbin[11]. As we showed in Section 7.4.1, the governing equation, the "threadline equation," is

$$\frac{\partial^2 y}{\partial t^2} + \alpha \frac{\partial^2 y}{\partial x \partial t} + \beta \frac{\partial^2 y}{\partial x^2} = 0, \qquad \textbf{(7.4.25)}$$

where $\alpha = 2V$, $\beta = V^2 - gT/\rho$, V is the windup velocity, g is the gravitational attraction, T is the tension in the yarn, and ρ is the density of the yarn. We now introduce the characteristics $\xi = x + \lambda_1 t$ and $\eta = x + \lambda_2 t$, where λ_1 and λ_2 are yet undetermined. Upon substituting ξ and η into (7.4.25),

$$(\lambda_1^2 + 2V\lambda_1 + V^2 - gT/\rho)u_{\xi\xi} + (\lambda_2^2 + 2V\lambda_2 + V^2 - gT/\rho)u_{\eta\eta} \\ + [2V^2 - 2gT/\rho + 2V(\lambda_1 + \lambda_2) + 2\lambda_1\lambda_2]u_{\xi\eta} = 0. \qquad \textbf{(7.4.26)}$$

[11] Reprinted from *J. Franklin Inst.*, **275**, Swope, R. D., and W. F. Ames, Vibrations of a moving threadline, 36–55, ©1963, with kind permission from Elsevier Science Ltd, The Boulevard, Langford Lane, Kidlington OX5 1GB, UK.

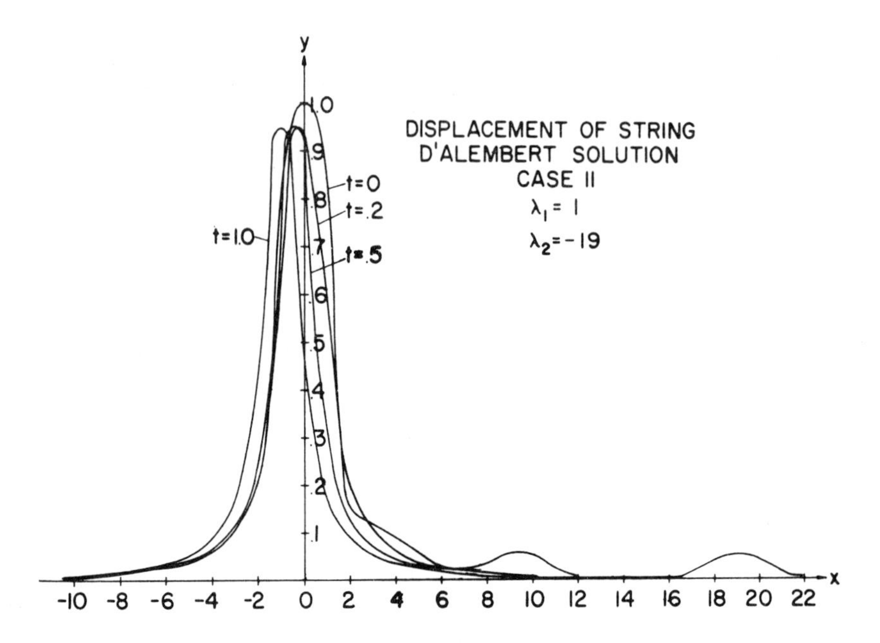

Figure 7.4.4: Displacement of an infinite, moving threadline when $c = 11$ and $V = 10$.

If we choose λ_1 and λ_2 to be roots of the equation:

$$\lambda^2 + 2V\lambda + V^2 - gT/\rho = 0, \tag{7.4.27}$$

(7.4.26) reduces to the simple form

$$u_{\xi\eta} = 0, \tag{7.4.28}$$

which has the general solution

$$u(x,t) = F(\xi) + G(\eta) = F(x + \lambda_1 t) + G(x + \lambda_2 t). \tag{7.4.29}$$

Solving (7.4.27) yields

$$\lambda_1 = c - V \qquad \text{and} \qquad \lambda_2 = -c - V, \tag{7.4.30}$$

where $c = \sqrt{gT/\rho}$. If the initial conditions are

$$u(x,0) = f(x) \qquad \text{and} \qquad u_t(x,0) = g(x), \tag{7.4.31}$$

then

$$u(x,t) = \frac{1}{2c}\left[\lambda_1 f(x + \lambda_2 t) - \lambda_2 f(x + \lambda_1 t) + \int_{x+\lambda_2 t}^{x+\lambda_1 t} g(\tau)\, d\tau\right]. \tag{7.4.32}$$

Because λ_1 does not generally equal to λ_2, the two waves that constitute the motion of the string move with different speeds and have different shapes and forms. For example, if

$$f(x) = \frac{1}{x^2+1} \quad \text{and} \quad g(x) = 0, \tag{7.4.33}$$

$$u(x,t) = \frac{1}{2c}\left\{\frac{c-V}{1+[x-(c+V)t]^2} + \frac{c+V}{1+[x-(c-V)t]^2}\right\}. \tag{7.4.34}$$

Figures 7.4.3 and 7.4.4 illustrate this solution for several different parameters.

Problems

Use d'Alembert's formula to solve the wave equation (7.4.1) for the following initial conditions defined for $|x| < \infty$.

1. $u(x,0) = 2\sin(x)\cos(x)$ $\quad u_t(x,0) = \cos(x)$
2. $u(x,0) = x\sin(x)$ $\quad u_t(x,0) = \cos(2x)$
3. $u(x,0) = 1/(x^2+1)$ $\quad u_t(x,0) = e^x$
4. $u(x,0) = e^{-x}$ $\quad u_t(x,0) = 1/(x^2+1)$
5. $u(x,0) = \cos(\pi x/2)$ $\quad u_t(x,0) = \sinh(ax)$
6. $u(x,0) = \sin(3x)$ $\quad u_t(x,0) = \sin(2x) - \sin(x)$

7.5 THE LAPLACE TRANSFORM METHOD

The solution of linear partial differential equations by Laplace transforms is the most commonly employed analytic technique after the method of separation of variables. Because the transform consists solely of an integration with respect to time, we obtain a transform which varies both in x and s, namely

$$U(x,s) = \int_0^\infty u(x,t)e^{-st}\,dt. \tag{7.5.1}$$

Partial derivatives involving time have transforms similar to those that we encountered in the case of functions of a single variable. They include

$$\mathcal{L}[u_t(x,t)] = sU(x,s) - u(x,0) \tag{7.5.2}$$

and

$$\mathcal{L}[u_{tt}(x,t)] = s^2U(x,s) - su(x,0) - u_t(x,0). \tag{7.5.3}$$

These transforms introduce the initial conditions via $u(x,0)$ and $u_t(x,0)$. On the other hand, derivatives involving x become

$$\mathcal{L}[u_x(x,t)] = \frac{d}{dx}\{\mathcal{L}[u(x,t)]\} = \frac{dU(x,s)}{dx} \tag{7.5.4}$$

and

$$\mathcal{L}[u_{xx}(x,t)] = \frac{d^2}{dx^2}\{\mathcal{L}[u(x,t)]\} = \frac{d^2U(x,s)}{dx^2}. \tag{7.5.5}$$

Because the transformation has eliminated the time variable, only $U(x,s)$ and its derivatives remain in the equation. Consequently, we have transformed the partial differential equation into a boundary-value problem for an ordinary differential equation. Because this equation is often easier to solve than a partial differential equation, the use of Laplace transforms has considerably simplified the original problem. Of course, the Laplace transforms must exist for this technique to work.

To summarize this method, we have constructed the following schematic:

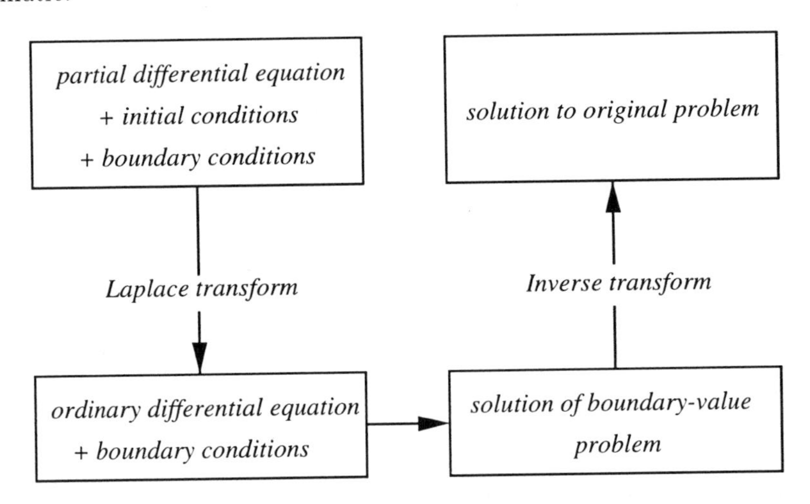

In the following examples, we will illustrate transform methods by solving the classic equation of telegraphy as it applies to a uniform transmission line. The line has a resistance R, an inductance L, a capacitance C, and a leakage conductance G per unit length. We denote the current in the direction of positive x by I; V is the voltage drop across the transmission line at the point x. The dependent variables I and V are functions of both distance x along the line and time t.

To derive the differential equations that govern the current and voltage in the line, consider the points A at x and B at $x + \Delta x$ in Figure 7.5.1. The current and voltage at A are $I(x,t)$ and $V(x,t)$; at B, $I + \frac{\partial I}{\partial x}\Delta x$ and $V + \frac{\partial V}{\partial x}\Delta x$. Therefore, the voltage drop from A to B is $-\frac{\partial V}{\partial x}\Delta x$ and the current in the line is $I + \frac{\partial I}{\partial x}\Delta x$. Neglecting terms that are proportional to $(\Delta x)^2$,

$$\left(L\frac{\partial I}{\partial t} + RI\right)\Delta x = -\frac{\partial V}{\partial x}\Delta x. \tag{7.5.6}$$

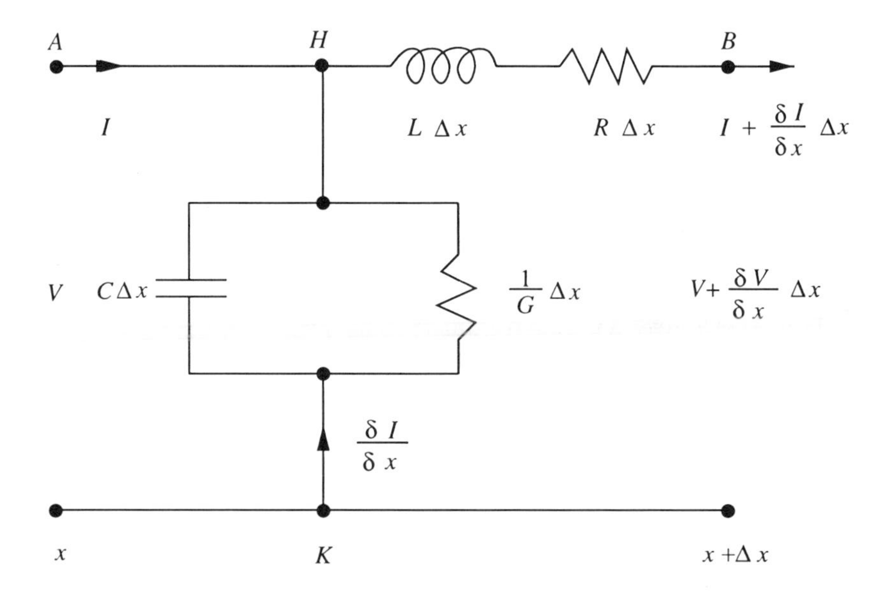

Figure 7.5.1: Schematic of an uniform transmission line.

The voltage drop over the parallel portion HK of the line is V while the current in this portion of the line is $-\frac{\partial I}{\partial x}\Delta x$. Thus,

$$\left(C\frac{\partial V}{\partial t} + GV\right)\Delta x = -\frac{\partial I}{\partial x}\Delta x. \tag{7.5.7}$$

Therefore, the differential equations for I and V are

$$L\frac{\partial I}{\partial t} + RI = -\frac{\partial V}{\partial x} \tag{7.5.8}$$

and

$$C\frac{\partial V}{\partial t} + GV = -\frac{\partial I}{\partial x}. \tag{7.5.9}$$

Turning to the initial conditions, we solve these simultaneous partial differential equations with the initial conditions:

$$I(x, 0) = I_0(x) \tag{7.5.10}$$

and

$$V(x, 0) = V_0(x) \tag{7.5.11}$$

for $0 < t$. There are also boundary conditions at the ends of the line; we will introduce them for each specific problem. For example, if the line is short-circuited at $x = a$, $V = 0$ at $x = a$; if there is an open circuit at $x = a$, $I = 0$ at $x = a$.

To solve (7.5.8)–(7.5.9) by Laplace transforms, we take the Laplace transform of both sides of these equations, which yields

$$(Ls+R)\overline{I}(x,s) = -\frac{d\overline{V}(x,s)}{dx} + LI_0(x) \tag{7.5.12}$$

and

$$(Cs+G)\overline{V}(x,s) = -\frac{d\overline{I}(x,s)}{dx} + CV_0(x). \tag{7.5.13}$$

Eliminating $\overline{I}$ gives an ordinary differential equation in $\overline{V}$:

$$\frac{d^2\overline{V}}{dx^2} - q^2\overline{V} = L\frac{dI_0(x)}{dx} - C(Ls+R)V_0(x), \tag{7.5.14}$$

where $q^2 = (Ls+R)(Cs+G)$. After finding $\overline{V}$, we may compute $\overline{I}$ from

$$\overline{I} = -\frac{1}{Ls+R}\frac{d\overline{V}}{dx} + \frac{LI_0(x)}{Ls+R}. \tag{7.5.15}$$

At this point we treat several classic cases.

• Example 7.5.1: The semi-infinite transmission line

We consider the problem of a semi-infinite line $x > 0$ with no initial current and charge. The end $x = 0$ has a constant voltage E for $0 < t$.

In this case,

$$\frac{d^2\overline{V}}{dx^2} - q^2\overline{V} = 0, \qquad x > 0. \tag{7.5.16}$$

The boundary conditions at the ends of the line are

$$V(0,t) = E, \qquad 0 < t \tag{7.5.17}$$

and $V(x,t)$ is finite as $x \to \infty$. The transform of these boundary conditions is

$$\overline{V}(0,s) = E/s \quad \text{and} \quad \lim_{x\to\infty} \overline{V}(x,s) < \infty. \tag{7.5.18}$$

The general solution of (7.5.16) is

$$\overline{V}(x,s) = Ae^{-qx} + Be^{qx}. \tag{7.5.19}$$

The requirement that $\overline{V}$ remains finite as $x \to \infty$ forces $B = 0$. The boundary condition at $x = 0$ gives $A = E/s$. Thus,

$$\overline{V}(x,s) = \frac{E}{s}\exp[-\sqrt{(Ls+R)(Cs+G)}\,x]. \tag{7.5.20}$$

We will discuss the general case later. However, for the so-called "loss-less" line, where $R = G = 0$,

$$\overline{V}(x,s) = \frac{E}{s}\exp(-sx/c), \tag{7.5.21}$$

where $c = 1/\sqrt{LC}$. Consequently,

$$V(x,t) = EH\left(t - \frac{x}{c}\right), \tag{7.5.22}$$

where $H(t)$ is Heaviside's step function. The physical interpretation of this solution is as follows: $V(x,t)$ is zero up to the time x/c at which time a wave traveling with speed c from $x = 0$ would arrive at the point x. $V(x,t)$ has the constant value E afterwards.

For the so-called "distortionless" line, $R/L = G/C = \rho$,

$$V(x,t) = Ee^{-\rho x/c}H\left(t - \frac{x}{c}\right). \tag{7.5.23}$$

In this case, the disturbance not only propagates with velocity c but also attenuates as we move along the line.

Suppose now, that instead of applying a constant voltage E at $x = 0$, we apply a time-dependent voltage, $f(t)$. The only modification is that in place of (7.5.20),

$$\overline{V}(x,s) = F(s)e^{-qx}. \tag{7.5.24}$$

In the case of the distortionless line, $q = (s+\rho)/c$, this becomes

$$\overline{V}(x,s) = F(s)e^{-(s+\rho)x/c} \tag{7.5.25}$$

and

$$V(x,t) = e^{-\rho x/c}f\left(t - \frac{x}{c}\right)H\left(t - \frac{x}{c}\right). \tag{7.5.26}$$

Thus, our solution shows that the voltage at x is zero up to the time x/c. Afterwards $V(x,t)$ follows the voltage at $x = 0$ with a time lag of x/c and decreases in magnitude by $e^{-\rho x/c}$.

• Example 7.5.2: The finite transmission line

We now discuss the problem of a finite transmission line $0 < x < l$ with zero initial current and charge. We ground the end $x = 0$ and maintain the end $x = l$ at constant voltage E for $0 < t$.

The transformed partial differential equation becomes

$$\frac{d^2\overline{V}}{dx^2} - q^2\overline{V} = 0, \qquad 0 < x < l. \tag{7.5.27}$$

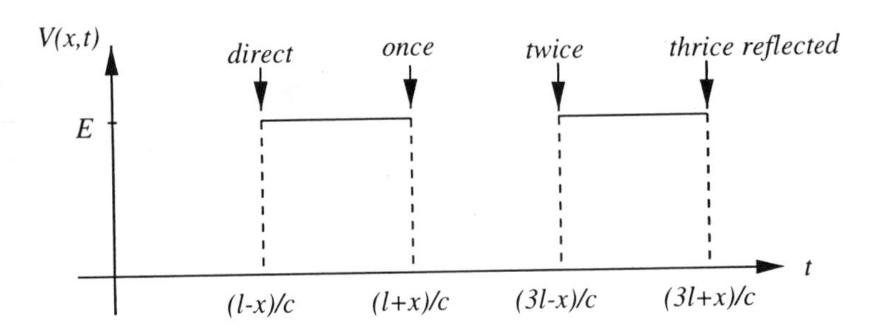

Figure 7.5.2: The voltage within a lossless, finite transmission line of length l as a function of time t.

The boundary conditions are

$$V(0,t) = 0 \quad \text{and} \quad V(l,t) = E, \qquad 0 < t. \tag{7.5.28}$$

The Laplace transform of these boundary conditions is

$$\overline{V}(0,s) = 0 \quad \text{and} \quad \overline{V}(l,s) = E/s. \tag{7.5.29}$$

The solution of (7.5.27) which satisfies the boundary conditions is

$$\overline{V}(x,s) = \frac{E \sinh(qx)}{s \sinh(ql)}. \tag{7.5.30}$$

Let us rewrite (7.5.30) in a form involving negative exponentials and expand the denominator by the binomial theorem,

$$\overline{V}(x,s) = \frac{E}{s} e^{-q(l-x)} \frac{1-\exp(-2qx)}{1-\exp(-2ql)} \tag{7.5.31}$$

$$= \frac{E}{s} e^{-q(l-x)} (1 - e^{-2qx}) \left(1 + e^{-2ql} + e^{-4ql} + \cdots\right) \tag{7.5.32}$$

$$= \frac{E}{s} \left[e^{-q(l-x)} - e^{-q(l+x)} + e^{-q(3l-x)} - e^{-q(3l+x)} + \cdots\right]. \tag{7.5.33}$$

In the special case of the lossless line where $q = s/c$,

$$\overline{V}(x,s) = \frac{E}{s} \left[e^{-s(l-x)/c} - e^{-s(l+x)/c} + e^{-s(3l-x)/c} - e^{-s(3l+x)/c} + \cdots\right] \tag{7.5.34}$$

or

$$V(x,t) = E\left[H\left(t - \frac{l-x}{c}\right) - H\left(t - \frac{l+x}{c}\right) + H\left(t - \frac{3l-x}{c}\right) - H\left(t - \frac{3l+x}{c}\right) + \cdots\right]. \tag{7.5.35}$$

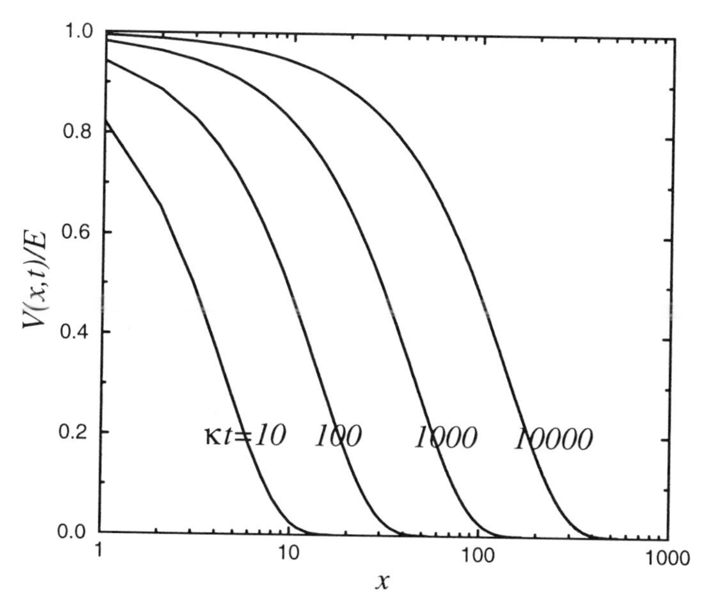

Figure 7.5.3: The voltage within a submarine cable as a function of distance for various κt's.

We illustrate (7.5.35) in Figure 7.5.2. The voltage at x is zero up to the time $(l-x)/c$, at which time a wave traveling directly from the end $x=l$ would reach the point x. The voltage then has the constant value E up to the time $(l+x)/c$, at which time a wave traveling from the end $x=l$ and reflected back from the end $x=0$ would arrive. From this time up to the time of arrival of a twice-reflected wave, it has the value zero, and so on.

• **Example 7.5.3: The semi-infinite transmission line reconsidered**

In the first example, we showed that the transform of the solution for the semi-infinite line is

$$\overline{V}(x,s) = \frac{E}{s} e^{-qx}, \tag{7.5.36}$$

where $q^2 = (Ls+R)(Cs+G)$. In the case of a lossless line ($R=G=0$), we found traveling wave solutions.

In this example, we shall examine the case of a submarine cable[12] where $L=G=0$. In this special case,

$$\overline{V}(x,s) = \frac{E}{s} e^{-x\sqrt{s/\kappa}}, \tag{7.5.37}$$

[12] First solved by Thomson, W., 1855: On the theory of the electric telegraph. *Proc. R. Soc. London*, **A7**, 382–399.

where $\kappa = 1/(RC)$. From a table of Laplace transforms,[13] we can immediately invert (7.5.37) and find that

$$V(x,t) = E \text{ erfc}\left(\frac{x}{2\sqrt{\kappa t}}\right), \tag{7.5.38}$$

where erfc is the complementary error function. Unlike the traveling wave solution, the voltage diffuses into the cable as time increases. We illustrate (7.5.38) in Figure 7.5.3.

• Example 7.5.4: A short-circuited, finite transmission line

Let us find the voltage of a lossless transmission line of length l that initially has the constant voltage E. At $t = 0$, we ground the line at $x = 0$ while we leave the end $x = l$ insulated.

The transformed partial differential equation now becomes

$$\frac{d^2\overline{V}}{dx^2} - \frac{s^2}{c^2}\overline{V} = -\frac{sE}{c^2}, \tag{7.5.39}$$

where $c = 1/\sqrt{LC}$. The boundary conditions are

$$\overline{V}(0,s) = 0 \tag{7.5.40}$$

and

$$\overline{I}(l,s) = -\frac{1}{Ls}\frac{d\overline{V}(l,s)}{dx} = 0 \tag{7.5.41}$$

from (7.5.15).

The solution to this boundary-value problem is

$$\overline{V}(x,s) = \frac{E}{s} - \frac{E\cosh[s(l-x)/c]}{s\cosh(sl/c)}. \tag{7.5.42}$$

The first term on the right side of (7.5.42) is easy to invert and equals E. The second term is much more difficult to handle. We will use Bromwich's integral.

In Section 4.10 we showed that

$$\mathcal{L}^{-1}\left\{\frac{\cosh[s(l-x)/c]}{s\cosh(sl/c)}\right\} = \frac{1}{2\pi i}\int_{c-\infty i}^{c+\infty i}\frac{\cosh[z(l-x)/c]e^{tz}}{z\cosh(zl/c)}\,dz. \tag{7.5.43}$$

---

[13] See Churchill, R. V., 1972: *Operational Mathematics*, McGraw-Hill Book, New York, Section 27.

To evaluate this integral we must first locate and then classify the singularities. Using the product formula for the hyperbolic cosine,

$$\frac{\cosh[z(l-x)/c]}{z\cosh(zl/c)} = \frac{[1+\frac{4z^2(l-x)^2}{c^2\pi^2}][1+\frac{4z^2(l-x)^2}{9c^2\pi^2}]\cdots}{z[1+\frac{4z^2l^2}{c^2\pi^2}][1+\frac{4z^2l^2}{9c^2\pi^2}]\cdots}. \tag{7.5.44}$$

This shows that we have an infinite number of simple poles located at $z=0$ and $z_n = \pm(2n-1)\pi ci/(2l)$, where $n=1,2,3,\ldots$. Therefore, Bromwich's contour can lie along, and just to the right of, the imaginary axis. By Jordan's lemma we close the contour with a semicircle of infinite radius in the left half of the complex plane. Computing the residues,

$$\operatorname{Res}\left\{\frac{\cosh[z(l-x)/c]e^{tz}}{z\cosh(zl/c)};0\right\} = \lim_{z\to 0}\frac{\cosh[z(l-x)/c]e^{tz}}{\cosh(zl/c)} = 1 \tag{7.5.45}$$

and

$$\operatorname{Res}\left\{\frac{\cosh[z(l-x)/c]e^{tz}}{z\cosh(zl/c)};z_n\right\}$$

$$= \lim_{z\to z_n}\frac{(z-z_n)\cosh[z(l-x)/c]e^{tz}}{z\cosh(zl/c)} \tag{7.5.46}$$

$$= \frac{\cosh[(2n-1)\pi(l-x)i/(2l)]\exp[\pm(2n-1)\pi cti/(2l)]}{[(2n-1)\pi i/2]\sinh[(2n-1)\pi i/2]} \tag{7.5.47}$$

$$= \frac{(-1)^{n+1}}{(2n-1)\pi/2}\cos\left[\frac{(2n-1)\pi(l-x)}{2l}\right]\exp\left[\pm\frac{(2n-1)\pi cti}{2l}\right]. \tag{7.5.48}$$

Summing the residues and using the relationship that $\cos(t) = (e^{ti} + e^{-ti})/2$,

$$V(x,t) = E - E\left\{1 - \frac{4}{\pi}\sum_{n=1}^{\infty}\frac{(-1)^{n+1}}{2n-1}\cos\left[\frac{(2n-1)\pi(l-x)}{2l}\right] \times\cos\left[\frac{(2n-1)\pi ct}{2l}\right]\right\} \tag{7.5.49}$$

$$= \frac{4E}{\pi}\sum_{n=1}^{\infty}\frac{(-1)^{n+1}}{2n-1}\cos\left[\frac{(2n-1)\pi(l-x)}{2l}\right]\cos\left[\frac{(2n-1)\pi ct}{2l}\right]. \tag{7.5.50}$$

An alternative to contour integration is to rewrite (7.5.42) as

$$\overline{V}(x,s) = \frac{E}{s}\left(1 - \frac{\exp(-sx/c)\{1+\exp[-2s(l-x)/c]\}}{1+\exp(-2sl/c)}\right) \tag{7.5.51}$$

$$= \frac{E}{s}\left[1 - e^{-sx/c} - e^{-s(2l-x)/c} + e^{-s(2l+x)/c} + \cdots\right] \tag{7.5.52}$$

so that

$$V(x,t) = E\left[1 - H\left(t - \frac{x}{c}\right) - H\left(t - \frac{2l - x}{c}\right) + H\left(t - \frac{2l + x}{c}\right) + \cdots\right]. \tag{7.5.53}$$

• Example 7.5.5: The general solution of the equation of telegraphy

In this example we solve the equation of telegraphy without any restrictions on R, C, G or L. We begin by eliminating the dependent variable $I(x,t)$ from the set of Equations (7.5.8)–(7.5.9). This yields

$$CL\frac{\partial^2 V}{\partial t^2} + (GL + RC)\frac{\partial V}{\partial t} + RG\,V = \frac{\partial^2 V}{\partial x^2}. \tag{7.5.54}$$

We next take the Laplace transform of (7.5.54) assuming that $V(x,0) = f(x)$ and $V_t(x,0) = g(x)$. The transformed version of (7.5.54) is

$$\frac{d^2\overline{V}}{dx^2} - [CLs^2 + (GL+RC)s + RG]\overline{V} = -CLg(x) - (CLs + GL + RC)f(x) \tag{7.5.55}$$

or

$$\frac{d^2\overline{V}}{dx^2} - \frac{(s+\rho)^2 - \sigma^2}{c^2}\overline{V} = -\frac{g(x)}{c^2} - \left(\frac{s}{c^2} + \frac{2\rho}{c^2}\right)f(x), \tag{7.5.56}$$

where $c^2 = 1/LC$, $\rho = c^2(RC + GL)/2$ and $\sigma = c^2(RC - GL)/2$.

We solve (7.5.56) by Fourier transforms (see Section 3.6) with the requirement that the solution dies away as $|x| \to \infty$. The most convenient way of expressing this solution is the convolution product (see Section 3.5)

$$\overline{V}(x,s) = \left[\frac{g(x)}{c} + \left(\frac{s}{c} + \frac{2\rho}{c}\right)f(x)\right] * \frac{\exp[-|x|\sqrt{(s+\rho)^2 - \sigma^2}/c]}{2\sqrt{(s+\rho)^2 - \sigma^2}}. \tag{7.5.57}$$

From a table of Laplace transforms,

$$\mathcal{L}^{-1}\left[\frac{\exp\left(-b\sqrt{s^2 - a^2}\right)}{\sqrt{s^2 - a^2}}\right] = I_0\left(a\sqrt{t^2 - b^2}\right)H(t - b), \tag{7.5.58}$$

where $b > 0$ and $I_0(\;)$ is the zeroth order modified Bessel function of the first kind. Therefore, by the first shifting theorem,

$$\mathcal{L}^{-1}\left\{\frac{\exp\left[-|x|\sqrt{(s+\rho)^2 - \sigma^2}/c\right]}{\sqrt{(s+\rho)^2 - \sigma^2}}\right\} = e^{-\rho t} I_0\left[\sigma\sqrt{t^2 - (x/c)^2}\right] H\left(t - \frac{|x|}{c}\right). \tag{7.5.59}$$

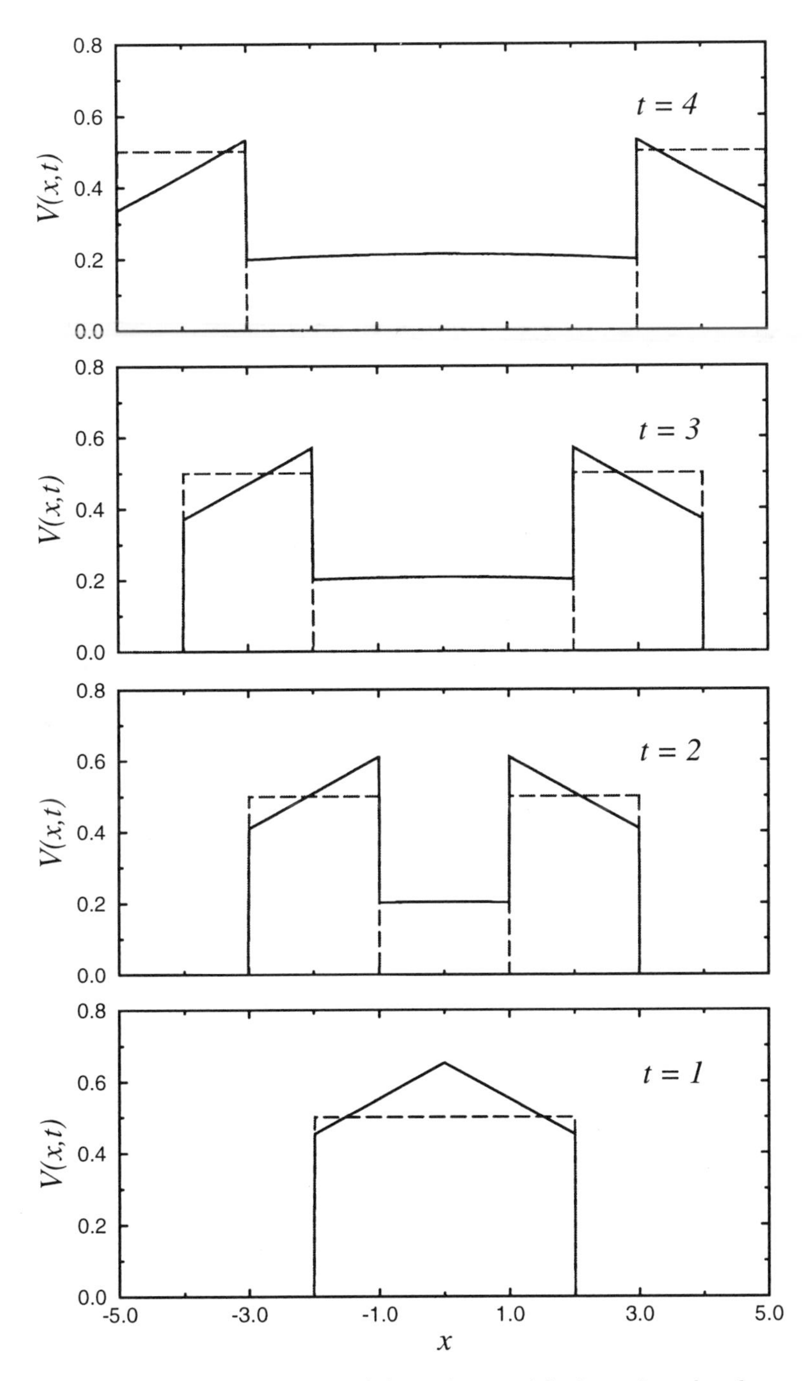

Figure 7.5.4: The evolution of the voltage with time given by the general equation of telegraphy for initial conditions and parameters stated in the text.

Using (7.5.59) to invert (7.5.57), we have that

$$\begin{aligned} V(x,t) &= \tfrac{1}{2c}e^{-\rho t}g(x) * I_0\left[\sigma\sqrt{t^2-(x/c)^2}\right]H(t-|x|/c) \\ &+ \tfrac{1}{2c}e^{-\rho t}f(x) * \frac{\partial}{\partial t}\left\{I_0[\sigma\sqrt{t^2-(x/c)^2}]\right\}H(t-|x|/c) \\ &+ \tfrac{\rho}{c}e^{-\rho t}f(x) * I_0\left[\sigma\sqrt{t^2-(x/c)^2}\right]H(t-|x|/c) \\ &+ \tfrac{1}{2}e^{-\rho t}[f(x+ct)+f(x-ct)]. \end{aligned} \tag{7.5.60}$$

The last term in (7.5.60) arises from noting that $sF(s) = \mathcal{L}[f(t)] + f(0)$. If we explicitly write out the convolution, the final form of the solution is

$$\begin{aligned} V(x,t) &= \tfrac{1}{2}e^{-\rho t}[f(x+ct)+f(x-ct)] \\ &+ \tfrac{1}{2c}e^{-\rho t}\int_{x-ct}^{x+ct}[g(\eta)+2\rho f(\eta)]I_0\left[\sigma\sqrt{c^2t^2-(x-\eta)^2}\Big/c\right]d\eta \\ &+ \tfrac{1}{2c}e^{-\rho t}\int_{x-ct}^{x+ct}f(\eta)\frac{\partial}{\partial t}\left\{I_0\left[\sigma\sqrt{c^2t^2-(x-\eta)^2}\Big/c\right]\right\}d\eta. \end{aligned} \tag{7.5.61}$$

The physical interpretation of the first line of (7.5.61) is straightforward. It represents damped progressive waves; one is propagating to the right and the other to the left. In addition to these progressive waves, there is a contribution from the integrals, even after the waves have passed. These integrals include all of the points where $f(x)$ and $g(x)$ are nonzero within a distance ct from the point in question. This effect persists through all time, although dying away, and constitutes a residue or tail. Figure 7.5.4 illustrates this for $\rho = 0.1$, $\sigma = 0.2$, and $c = 1$. We evaluated the integrals by Simpson's rule for the initial conditions $f(x) = H(x+1) - H(x-1)$ and $g(x) = 0$. We have also included the solution for the lossless case for comparison. If there was no loss, then two pulses would propagate to the left and right as shown by the dashed line. However, with resistance and leakage the waves leave a residue after their leading edge has passed.

Problems

1. Use transform methods to solve the wave equation

$$\frac{\partial^2 u}{\partial t^2} = \frac{\partial^2 u}{\partial x^2}, \qquad 0 < x < 1, 0 < t$$

for the boundary conditions

$$u(0,t) = u(1,t) = 0, \qquad 0 < t$$

and the initial conditions

$$u(x,0) = 0, \qquad \frac{\partial u(0,t)}{\partial t} = 1, \quad 0 < x < 1.$$

2. Use transform methods to solve the wave equation

$$\frac{\partial^2 u}{\partial t^2} = \frac{\partial^2 u}{\partial x^2}, \quad 0 < x < 1, 0 < t$$

for the boundary conditions

$$u(0,t) = u_x(1,t) = 0, \qquad 0 < t$$

and the initial conditions

$$u(x,0) = 0, \qquad \frac{\partial u(0,t)}{\partial t} = x, \quad 0 < x < 1.$$

3. Use transform methods to solve the wave equation

$$\frac{\partial^2 u}{\partial t^2} = \frac{\partial^2 u}{\partial x^2}, \quad 0 < x < 1, 0 < t$$

for the boundary conditions

$$u(0,t) = u(1,t) = 0, \qquad 0 < t$$

and the initial conditions

$$u(x,0) = \sin(\pi x), \qquad \frac{\partial u(x,0)}{\partial t} = -\sin(\pi x), \quad 0 < x < 1.$$

4. Use transform methods to solve the wave equation

$$\frac{\partial^2 u}{\partial t^2} = c^2 \frac{\partial^2 u}{\partial x^2}, \quad 0 < x < a, 0 < t$$

for the boundary conditions

$$u(0,t) = \sin(\omega t), \quad u(a,t) = 0, \qquad 0 < t$$

and the initial conditions

$$u(x,0) = 0, \qquad \frac{\partial u(x,0)}{\partial t} = 0, \quad 0 < x < a.$$

Assume that $\omega a/c$ is *not* an integer multiple of π. Why?

5. Use transform methods to solve the wave equation

$$\frac{\partial^2 u}{\partial t^2} - \frac{\partial^2 u}{\partial x^2} = te^{-x}, \quad 0 < x < \infty, 0 < t$$

for the boundary conditions

$$u(0,t) = 1 - e^{-t}, \quad \lim_{x\to\infty} |u(x,t)| \sim x^n, \ n \text{ finite}, \quad 0 < t$$

and the initial conditions

$$u(x,0) = 0, \quad \frac{\partial u(x,0)}{\partial t} = x, \quad 0 < x < \infty.$$

6. Use transform methods to solve the wave equation

$$\frac{\partial^2 u}{\partial t^2} - \frac{\partial^2 u}{\partial x^2} = xe^{-t}, \quad 0 < x < \infty, 0 < t$$

for the boundary conditions

$$u(0,t) = \cos(t), \quad \lim_{x\to\infty} |u(x,t)| \sim x^n, \ n \text{ finite}, \quad 0 < t$$

and the initial conditions

$$u(x,0) = 1, \quad \frac{\partial u(x,0)}{\partial t} = 0, \quad 0 < x < \infty.$$

7. Use transform methods to solve the wave equation

$$\frac{\partial^2 u}{\partial t^2} = \frac{\partial^2 u}{\partial x^2}, \quad 0 < x < L, 0 < t$$

for the boundary conditions

$$u(0,t) = 0, \quad \frac{\partial^2 u(L,t)}{\partial t^2} + \frac{k}{m}\frac{\partial u(L,t)}{\partial x} = g, \quad 0 < t$$

and the initial conditions

$$u(x,0) = 0, \quad \frac{\partial u(x,0)}{\partial t} = 0, \quad 0 < x < L,$$

where c, k, m, and g are constants.

8. Use transform methods[14] to solve the wave equation

$$\frac{\partial^2 u}{\partial t^2} = c^2 \frac{\partial}{\partial x}\left(x \frac{\partial u}{\partial x}\right), \quad 0 < x < 1, 0 < t$$

for the boundary conditions

$$\lim_{x \to \infty} |u(x,t)| < \infty, \quad u(1,t) = A\sin(\omega t), \quad 0 < t$$

and the initial conditions

$$u(x,0) = 0, \quad \frac{\partial u(x,0)}{\partial t} = 0, \quad 0 < x < 1.$$

Assume that $2\omega \neq c\beta_n$, where $J_0(\beta_n) = 0$. [Hint: The ordinary differential equation

$$\frac{d}{dx}\left(x \frac{dU}{dx}\right) - \frac{s^2}{c^2} U = 0$$

has the solution

$$U(x,s) = c_1 I_0\left(\frac{s}{c}\sqrt{x}\right) + c_2 K_0\left(\frac{s}{c}\sqrt{x}\right),$$

where $I_0(x)$ and $K_0(x)$ are modified Bessel functions of the first and second kind, respectively. Note that $J_n(iz) = i^n I_n(z)$ and $I_n(iz) = i^n J_n(z)$ for complex z.]

9. A lossless transmission line of length ℓ has a constant voltage E applied to the end $x = 0$ while we insulate the other end $[u_x(\ell,t) = 0]$. Find the voltage at any point on the line if the initial current and charge are zero.

10. Solve the equation of telegraphy without leakage

$$\frac{\partial^2 u}{\partial x^2} = CR\frac{\partial u}{\partial t} + CL\frac{\partial^2 u}{\partial t^2}, \quad 0 < x < \ell, 0 < t$$

subject to the boundary conditions

$$u(0,t) = 0 \quad \text{and} \quad u(\ell,t) = E, \quad 0 < t$$

and the initial conditions

$$u(x,0) = u_t(x,0) = 0, \quad 0 < x < \ell.$$

---

[14] Suggested by a problem solved by Brown, J., 1975: Stresses in towed cables during re-entry. *J. Spacecr. Rockets*, **12**, 524–527.

Assume that $4\pi^2 L/CR^2\ell^2 > 1$. Why?

11. The pressure and velocity oscillations from water hammer in a pipe without friction[15] are given by the equations

$$\frac{\partial p}{\partial t} = -\rho c^2 \frac{\partial u}{\partial x}$$

and

$$\frac{\partial u}{\partial t} = -\frac{1}{\rho}\frac{\partial p}{\partial x},$$

where $p(x,t)$ denotes the pressure perturbation, $u(x,t)$ is the velocity perturbation, c is the speed of sound in water, and ρ is the density of water. These two first-order partial differential equations may be combined to yield

$$\frac{\partial^2 p}{\partial t^2} = c^2 \frac{\partial^2 p}{\partial x^2}.$$

Find the solution to this partial differential equation if $p(0,t) = p_0$ and $u(L,t) = 0$ and the initial conditions are $p(x,0) = p_0$, $p_t(x,0) = 0$ and $u(x,t) = u_0$.

12. Use Laplace transforms to solve the wave equation[16]

$$\frac{\partial^2 (ru)}{\partial t^2} = c^2 \frac{\partial^2 (ru)}{\partial r^2}, \qquad a < r < \infty,\ 0 < t$$

subject to the boundary conditions that

$$-\rho c^2 \left(\frac{\partial^2 u}{\partial r^2} + \frac{2}{3r}\frac{\partial u}{\partial r} \right)\bigg|_{r=a} = p_0 e^{-\alpha t} H(t) \text{ and } \lim_{r\to\infty} |u(r,t)| < \infty, \quad 0 < t,$$

where $\alpha > 0$, and the initial conditions that

$$u(r,0) = u_t(r,0) = 0, \qquad a < r < \infty.$$

13. Consider a vertical rod or column of length L that is supported at both ends. The elastic waves that arise when the support at the bottom is suddenly removed are governed by the wave equation[17]

$$\frac{\partial^2 u}{\partial t^2} = c^2 \frac{\partial^2 u}{\partial x^2} + g, \qquad 0 < x < L,\ 0 < t,$$

---

[15] See Rich, G. R., 1945: Water-hammer analysis by the Laplace-Mellin transformation. *Trans. ASME*, **67**, 361–376.

[16] Originally solved using Fourier transforms by Sharpe, J. A., 1942: The production of elastic waves by explosion pressures. I. Theory and empirical field observations. *Geophysics*, **7**, 144–154.

[17] Abstracted with permission from Hall, L. H., 1953: Longitudinal vibrations of a vertical column by the method of Laplace transform. *Am. J. Phys.*, **21**, 287–292. ©1953 American Association of Physics Teachers.

where g denotes the gravitational acceleration, $c^2 = E/\rho$, E is Young's modulus and ρ is the mass density. Find the wave solution if the boundary conditions are

$$\frac{\partial u(0,t)}{\partial x} = \frac{\partial u(L,t)}{\partial x} = 0, \qquad 0 < t$$

and the initial conditions are

$$u(x,0) = -\frac{gx^2}{2c^2}, \qquad \frac{\partial u(x,0)}{\partial t} = 0, \qquad 0 < x < L.$$

14. Solve the telegraph-like equation[18]

$$\frac{\partial^2 u}{\partial t^2} + k\frac{\partial u}{\partial t} = c^2\left(\frac{\partial^2 u}{\partial x^2} + \alpha\frac{\partial u}{\partial x}\right), \qquad 0 < x < \infty,\ 0 < t$$

subject to the boundary conditions

$$\frac{\partial u(0,t)}{\partial x} = -u_0\delta(t), \qquad \lim_{x\to\infty} |u(x,t)| < \infty, \qquad 0 < x < \infty$$

and the initial conditions

$$u(x,0) = u_0, \quad u_t(x,0) = 0, \quad 0 < t$$

with $\alpha c > k$.

Step 1: Take the Laplace transform of the partial differential equation and boundary conditions and show that

$$\frac{d^2U(x,s)}{dx^2} + \alpha\frac{dU(x,s)}{dx} - \left(\frac{s^2+ks}{c^2}\right)U(x,s) = -\left(\frac{s+k}{c^2}\right)u_0$$

with $U'(0,s) = -u_0$ and $\lim_{x\to\infty} |U(x,s)| < \infty$.

Step 2: Show that the solution to the previous step is

$$U(x,s) = \frac{u_0}{s} + u_0 e^{-\alpha x/2}\,\frac{\exp\left[-x\sqrt{\left(s+\frac{k}{2}\right)^2 + a^2/c}\right]}{\frac{\alpha}{2} + \sqrt{(s+\frac{k}{2})^2 + a^2/c}},$$

[18] From Abbott, M. R., 1959: The downstream effect of closing a barrier across an estuary with particular reference to the Thames. *Proc. R. Soc. London*, **A251**, 426–439 with permission.

where $4a^2 = \alpha^2 c^2 - k^2 > 0$.

Step 3: Using the first and second shifting theorems and the property that

$$F\left(\sqrt{s^2+a^2}\right) = \mathcal{L}\left[f(t) - a\int_0^t \frac{J_1\left(a\sqrt{t^2-\tau^2}\right)}{\sqrt{t^2-\tau^2}}\tau f(\tau)\,d\tau\right],$$

show that

$$u(x,t) = u_0 + u_0 c e^{-kt/2} H(t - x/c)$$
$$\times\left[e^{-\alpha ct/2} - a\int_{x/c}^t \frac{J_1\left(a\sqrt{t^2-\tau^2}\right)}{\sqrt{t^2-\tau^2}}\tau e^{-\alpha c\tau/2}d\tau\right].$$

15. As an electric locomotive travels down a track at the speed V, the pantograph (the metallic framework that connects the overhead power lines to the locomotive) pushes up the line with a force P. Let us find the behavior[19] of the overhead wire as a pantograph passes between two supports of the electrical cable that are located a distance L apart. We model this system as a vibrating string with a point load:

$$\frac{\partial^2 u}{\partial t^2} = c^2\frac{\partial^2 u}{\partial x^2} + \frac{P}{\rho V}\delta\left(t - \frac{x}{V}\right), \qquad 0 < x < L,\ 0 < t.$$

Let us assume that the wire is initially at rest $[u(x,0) = u_t(x,0) = 0$ for $0 < x < L]$ and fixed at both ends $[u(0,t) = u(L,t) = 0$ for $0 < t]$.

Step 1: Take the Laplace transform of the partial differential equation and show that

$$s^2 U(x,s) = c^2\frac{d^2U(x,s)}{dx^2} + \frac{P}{\rho V}e^{-xs/V}.$$

Step 2: Solve the ordinary differential equation in Step 1 as a Fourier half-range sine series:

$$U(x,s) = \sum_{n=1}^{\infty} B_n(s)\sin\left(\frac{n\pi x}{L}\right),$$

[19] From Oda, O. and Ooura, Y., 1976: Vibrations of catenary overhead wire. *Q. Rep., (Tokyo) Railway Tech. Res. Inst.*, **17**, 134–135 with permission.

where

$$B_n(s) = \frac{2P\beta_n}{\rho L(\beta_n^2 - \alpha_n^2)}\left[\frac{1}{s^2+\alpha_n^2} - \frac{1}{s^2+\beta_n^2}\right]\left[1-(-1)^n e^{-Ls/V}\right],$$

$\alpha_n = n\pi c/L$ and $\beta_n = n\pi V/L$. This solution satisfies the boundary conditions.

Step 3: By inverting the solution in Step 2, show that

$$\begin{aligned}
u(x,t) &= \frac{2P}{\rho L}\sum_{n=1}^{\infty}\left[\frac{\sin(\beta_n t)}{\alpha_n^2-\beta_n^2} - \frac{V}{c}\frac{\sin(\alpha_n t)}{\alpha_n^2-\beta_n^2}\right]\sin\left(\frac{n\pi x}{L}\right) \\
&\quad - \frac{2P}{\rho L}H\left(t-\frac{L}{V}\right)\sum_{n=1}^{\infty}(-1)^n\sin\left(\frac{n\pi x}{L}\right) \\
&\qquad \times\left\{\frac{\sin[\beta_n(t-L/V)]}{\alpha_n^2-\beta_n^2} - \frac{V}{c}\frac{\sin[\alpha_n(t-L/V)]}{\alpha_n^2-\beta_n^2}\right\} \\
&= \frac{2P}{\rho L}\sum_{n=1}^{\infty}\left[\frac{\sin(\beta_n t)}{\alpha_n^2-\beta_n^2} - \frac{V}{c}\frac{\sin(\alpha_n t)}{\alpha_n^2-\beta_n^2}\right]\sin\left(\frac{n\pi x}{L}\right) \\
&\quad - \frac{2P}{\rho L}H\left(t-\frac{L}{V}\right)\sum_{n=1}^{\infty}\sin\left(\frac{n\pi x}{L}\right) \\
&\qquad \times\left\{\frac{\sin(\beta_n t)}{\alpha_n^2-\beta_n^2} - \frac{V}{c}(-1)^n\frac{\sin[\alpha_n(t-L/V)]}{\alpha_n^2-\beta_n^2}\right\}.
\end{aligned}$$

The first term in both summations represents the static uplift on the line; this term disappears after the pantograph has passed. The second term in both summations represents the vibrations excited by the traveling force. Even after the pantograph passes, they will continue to exist.

16. Solve the wave equation

$$\frac{1}{c^2}\frac{\partial^2 u}{\partial t^2} - \frac{\partial^2 u}{\partial r^2} - \frac{1}{r}\frac{\partial u}{\partial r} + \frac{u}{r^2} = \frac{\delta(r-\alpha)}{\alpha^2}, \qquad 0 \le r < a,\ 0 < t,$$

where $0 < \alpha < a$, subject to the boundary conditions

$$\lim_{r\to 0}|u(r,t)| < \infty \quad\text{and}\quad \frac{\partial u(a,t)}{\partial r} + \frac{h}{a}u(a,t) = 0, \quad 0 < t$$

and the initial conditions

$$u(r,0) = u_t(r,0) = 0, \qquad 0 \le r < a.$$

Step 1: Take the Laplace transform of the partial differential equation and show that

$$\frac{d^2U(r,s)}{dr^2}+\frac{1}{r}\frac{dU(r,s)}{dr}-\left(\frac{s^2}{c^2}+\frac{1}{r^2}\right)U(r,s)=-\frac{\delta(r-\alpha)}{s\alpha^2},\qquad 0\le r<a$$

with

$$\lim_{r\to 0}|U(r,s)|<\infty \quad\text{and}\quad \frac{dU(a,s)}{dr}+\frac{h}{a}U(a,s)=0.$$

Step 2: Show that the Dirac delta function can be reexpressed as the Fourier-Bessel series:

$$\delta(r-\alpha)=\frac{2\alpha}{a^2}\sum_{n=1}^{\infty}\frac{\beta_n^2\,J_1(\beta_n\alpha/a)}{(\beta_n^2+h^2-1)J_1^2(\beta_n)}J_1(\beta_n r/a),\qquad 0\le r<a,$$

where β_n is the nth root of $\beta J_1'(\beta)+h\,J_1(\beta)=\beta J_0(\beta)+(h-1)J_1(\beta)=0$ and $J_0(\;)$, $J_1(\;)$ are the zeroth and first-order Bessel functions of the first kind, respectively.

Step 3: Show that solution to the ordinary differential equation in Step 1 is

$$U(r,s)=\frac{2}{\alpha}\sum_{n=1}^{\infty}\frac{J_1(\beta_n\alpha/a)J_1(\beta_n r/a)}{(\beta_n^2+h^2-1)\,J_1^2(\beta_n)}\left[\frac{1}{s}-\frac{s}{s^2+c^2\beta_n^2/a^2}\right].$$

Note that this solution satisfies the boundary conditions.

Step 4: Taking the inverse of the Laplace transform in Step 3, show that the solution to the partial differential equation is

$$u(r,t)=\frac{2}{\alpha}\sum_{n=1}^{\infty}\frac{J_1(\beta_n\alpha/a)J_1(\beta_n r/a)}{(\beta_n^2+h^2-1)\,J_1^2(\beta_n)}\left[1-\cos\left(\frac{c\beta_n t}{a}\right)\right].$$

17. A powerful method for solving certain partial differential equations is the joint application of Laplace and Fourier transforms. To illustrate this *joint transform method*, let us find the Green's function for the Klein-Gordon equation

$$\frac{\partial^2 u}{\partial x^2}-\frac{1}{c^2}\frac{\partial^2 u}{\partial t^2}-\beta^2 u=-\delta(x)\delta(t),\qquad -\infty<x<\infty, 0<t$$

subject to the boundary condition

$$\lim_{x\to\pm\infty}|u(x,t)|<\infty,\quad 0<t$$

and the initial conditions

$$u(x,0) = u_t(x,0) = 0, \qquad -\infty < x < \infty.$$

Step 1: Take the Laplace transform of the partial differential equation and show that

$$\frac{d^2U(x,s)}{dx^2} - \left(\frac{s^2}{c^2} + \beta^2\right) U(x,s) = -\delta(x), \qquad -\infty < x < \infty$$

with the boundary condition

$$\lim_{x \to \pm\infty} |U(x,s)| < \infty.$$

Step 2: Using Fourier transforms, show that the solution to the ordinary differential equation in Step 1 is

$$U(x,s) = \frac{\exp\left(-|x|\sqrt{s^2/c^2 + \beta^2}\right)}{2\sqrt{s^2/c^2 + \beta^2}}.$$

You may need to review Section 3.6.

Step 3: Using tables, show that the Green's function is

$$u(x,t) = \frac{c}{2} J_0\left(\beta\sqrt{c^2t^2 - x^2}\right) H(ct - x),$$

where $J_0(\,)$ is the zeroth order Bessel function of the first kind.

7.6 NUMERICAL SOLUTION OF THE WAVE EQUATION

Despite the powerful techniques shown in the previous sections for solving the wave equation, often these analytic techniques fail and we must resort to numerical techniques. In counterpoint to the continuous solutions, finite difference methods, a type of numerical solution technique, give discrete numerical values at a specific location (x_m, t_n), called a *grid point*. These numerical values represent a numerical approximation of the continuous solution over the region $(x_m - \Delta x/2, x_m + \Delta x/2)$ and $(t_n - \Delta t/2, t_n + \Delta t/2)$, where Δx and Δt are the distance and time intervals between grid points, respectively. Clearly, in the limit of $\Delta x, \Delta t \to 0$, we recover the continuous solution. However, practical considerations such as computer memory or execution time often require that Δx and Δt, although small, are not negligibly small.

The first task in the numerical solution of a partial differential equation is the replacement of its continuous derivatives with finite differences. The most popular approach employs Taylor expansions. If we focus on the x-derivative, then the value of the solution at $u[(m+1)\Delta x, n\Delta t]$ in terms of the solution at $(m\Delta x, n\Delta t)$ is

$$u[(m+1)\Delta x, n\Delta t] = u(x_m, t_n) + \frac{\Delta x}{1!}\frac{\partial u(x_m, t_n)}{\partial x} + \frac{(\Delta x)^2}{2!}\frac{\partial^2 u(x_m, t_n)}{\partial x^2} + \frac{(\Delta x)^3}{3!}\frac{\partial^3 u(x_m, t_n)}{\partial x^3} + \frac{(\Delta x)^4}{4!}\frac{\partial^4 u(x_m, t_n)}{\partial x^4} + \cdots \tag{7.6.1}$$

$$= u(x_m, t_n) + \Delta x \frac{\partial u(x_m, t_n)}{\partial x} + O[(\Delta x)^2], \tag{7.6.2}$$

where $O[(\Delta x)^2]$ gives a measure of the magnitude of neglected terms.[20]

From (7.6.2), one possible approximation for u_x is

$$\frac{\partial u(x_m, t_n)}{\partial x} = \frac{u_{m+1}^n - u_m^n}{\Delta x} + O(\Delta x), \tag{7.6.3}$$

where we have used the standard notation that $u_m^n = u(x_m, t_n)$. This is an example of a *one-sided finite difference* approximation of the partial derivative u_x. The error in using this approximation behaves as Δx.

Another possible approximation for the derivative arises from using $u(m\Delta x, n\Delta t)$ and $u[(m-1)\Delta x, n\Delta t]$. From the Taylor expansion:

$$u[(m-1)\Delta x, n\Delta t] = u(x_m, t_n) - \frac{\Delta x}{1!}\frac{\partial u(x_m, t_n)}{\partial x} + \frac{(\Delta x)^2}{2!}\frac{\partial^2 u(x_m, t_n)}{\partial x^2}$$

---

[20] The symbol O is a mathematical notation indicating relative magnitude of terms, namely that $f(\epsilon) = O(\epsilon^n)$ provided $\lim_{\epsilon \to 0} |f(\epsilon)/\epsilon^n| < \infty$. For example, as $\epsilon \to 0$, $\sin(\epsilon) = O(\epsilon)$, $\sin(\epsilon^2) = O(\epsilon^2)$, and $\cos(\epsilon) = O(1)$.

$$- \frac{(\Delta x)^3}{3!} \frac{\partial^3 u(x_m, t_n)}{\partial x^3} + \frac{(\Delta x)^4}{4!} \frac{\partial^4 u(x_m, t_n)}{\partial x^4} - \cdots, \tag{7.6.4}$$

we can also obtain the one-sided difference formula

$$\frac{u(x_m, t_n)}{\partial x} = \frac{u_m^n - u_{m-1}^n}{\Delta x} + O(\Delta x). \tag{7.6.5}$$

A third possibility arises from subtracting (7.6.4) from (7.6.1):

$$u_{m+1}^n - u_{m-1}^n = 2\Delta x \frac{\partial u(x_m, t_n)}{\partial x} + O[(\Delta x)^3] \tag{7.6.6}$$

or

$$\frac{\partial u(x_m, t_n)}{\partial x} = \frac{u_{m+1}^n - u_{m-1}^n}{2\Delta x} + O[(\Delta x)^2]. \tag{7.6.7}$$

Thus, the choice of the finite differencing scheme can produce profound differences in the accuracy of the results. In the present case, *centered finite differences* can yield results that are markedly better than using one-sided differences.

To solve the wave equation, we need to approximate u_{xx}. If we add (7.6.1) and (7.6.4),

$$u_{m+1}^n + u_{m-1}^n = 2u_m^n + \frac{\partial^2 u(x_m, t_n)}{\partial x^2}(\Delta x)^2 + O[(\Delta x)^4] \tag{7.6.8}$$

or

$$\frac{\partial^2 u(x_m, t_n)}{\partial x^2} = \frac{u_{m+1}^n - 2u_m^n + u_{m-1}^n}{(\Delta x)^2} + O[(\Delta x)^2]. \tag{7.6.9}$$

Similar considerations hold for the time derivative. Thus, by neglecting errors of $O[(\Delta x)^2]$ and $O[(\Delta t)^2]$, we may approximate the wave equation by

$$\frac{u_m^{n+1} - 2u_m^n + u_m^{n-1}}{(\Delta t)^2} = c^2 \frac{u_{m+1}^n - 2u_m^n + u_{m-1}^n}{(\Delta x)^2}. \tag{7.6.10}$$

Because the wave equation represents evolutionary change of some quantity, (7.6.10) is generally used as a predictive equation where we forecast u_m^{n+1} by

$$u_m^{n+1} = 2u_m^n - u_m^{n-1} + \left(\frac{c\Delta t}{\Delta x}\right)^2 \left(u_{m+1}^n - 2u_m^n + u_{m-1}^n\right). \tag{7.6.11}$$

Figure 7.6.1 illustrates this numerical scheme.

The greatest challenge in using (7.6.11) occurs with the very first prediction. When $n = 0$, clearly u_{m+1}^0, u_m^0 and u_{m-1}^0 are specified from the initial condition $u(m\Delta x, 0) = f(x_m)$. But what about u_m^{-1}? Recall

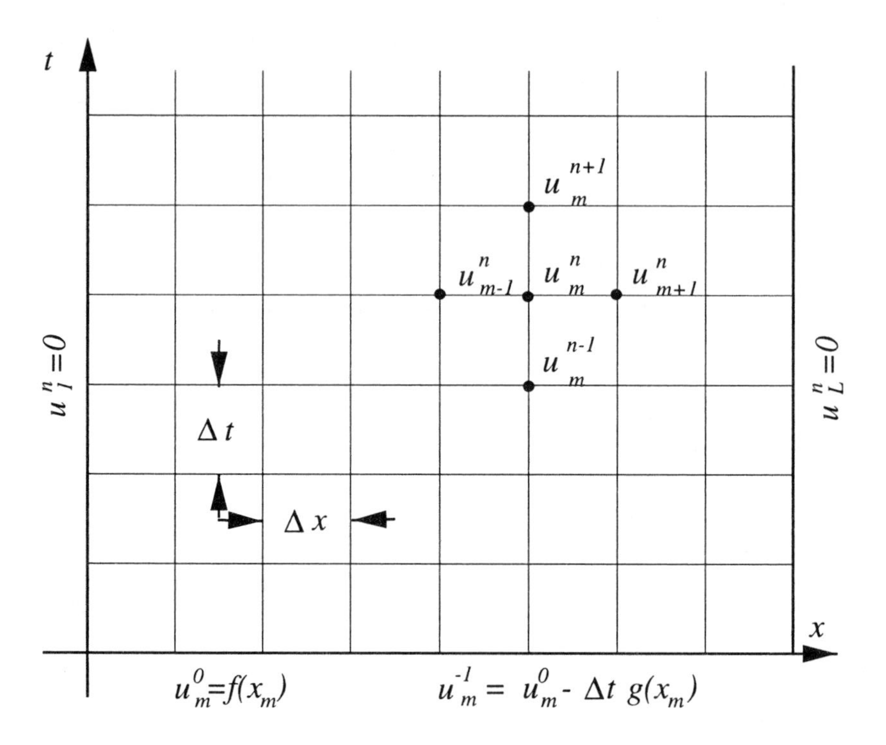

Figure 7.6.1: Schematic of the numerical solution of the wave equation with fixed end points.

that we still have $u_t(x,0) = g(x)$. If we use the backward difference formula (7.6.5),

$$\frac{u_m^0 - u_m^{-1}}{\Delta t} = g(x_m). \tag{7.6.12}$$

Solving for u_m^{-1},

$$u_m^{-1} = u_m^0 - \Delta t g(x_m). \tag{7.6.13}$$

One of the disadvantages of using the backward finite-difference formula is the larger error associated with this term compared to those associated with the finite-differenced form of the wave equation. In the case of the barotropic vorticity equation, a partial differential equation with wave-like solutions, this inconsistency eventually leads to a separation of solution between adjacent time levels.[21] This difficulty is avoided by stopping after a certain number of time steps, averaging the solution, and starting again.

[21] Gates, W. L., 1959: On the truncation error, stability, and convergence of difference solutions of the barotropic vorticity equation. *J. Meteorol.*, **16**, 556–568. See Section 4.

A better solution for computing that first time step employs the centered difference form

$$\frac{u_m^1 - u_m^{-1}}{2\Delta t} = g(x_m) \tag{7.6.14}$$

along with the wave equation

$$\frac{u_m^1 - 2u_m^0 + u_m^{-1}}{(\Delta t)^2} = c^2 \frac{u_{m+1}^0 - 2u_m^0 + u_{m-1}^0}{(\Delta x)^2} \tag{7.6.15}$$

so that

$$u_m^1 = \left(\frac{c\Delta t}{\Delta x}\right)^2 \frac{f(x_{m+1}) + f(x_{m-1})}{2} + \left[1 - \left(\frac{c\Delta t}{\Delta x}\right)^2\right] f(x_m) + \Delta t g(x_m). \tag{7.6.16}$$

Although it appears that we are ready to start calculating, we need to check whether our numerical scheme possesses three properties: convergence, stability, and consistency. By *consistency* we mean that the difference equations approach the differential equation as $\Delta x, \Delta t \to 0$. To prove consistency, we first write u_{m+1}^n, u_{m-1}^n, u_m^{n-1}, and u_m^{n+1} in terms of $u(x,t)$ and its derivatives evaluated at (x_m, t_n). From Taylor expansions,

$$u_{m+1}^n = u_m^n + \Delta x \left.\frac{\partial u}{\partial x}\right|_n^m + \tfrac{1}{2}(\Delta x)^2 \left.\frac{\partial^2 u}{\partial x^2}\right|_n^m + \tfrac{1}{6}(\Delta x)^3 \left.\frac{\partial^3 u}{\partial x^3}\right|_n^m + \cdots, \tag{7.6.17}$$

$$u_{m-1}^n = u_m^n - \Delta x \left.\frac{\partial u}{\partial x}\right|_n^m + \tfrac{1}{2}(\Delta x)^2 \left.\frac{\partial^2 u}{\partial x^2}\right|_n^m - \tfrac{1}{6}(\Delta x)^3 \left.\frac{\partial^3 u}{\partial x^3}\right|_n^m + \cdots, \tag{7.6.18}$$

$$u_m^{n+1} = u_m^n + \Delta t \left.\frac{\partial u}{\partial t}\right|_n^m + \tfrac{1}{2}(\Delta t)^2 \left.\frac{\partial^2 u}{\partial t^2}\right|_n^m + \tfrac{1}{6}(\Delta t)^3 \left.\frac{\partial^3 u}{\partial t^3}\right|_n^m + \cdots \tag{7.6.19}$$

and

$$u_m^{n-1} = u_m^n - \Delta t \left.\frac{\partial u}{\partial t}\right|_n^m + \tfrac{1}{2}(\Delta t)^2 \left.\frac{\partial^2 u}{\partial t^2}\right|_n^m - \tfrac{1}{6}(\Delta t)^3 \left.\frac{\partial^3 u}{\partial t^3}\right|_n^m + \cdots \tag{7.6.20}$$

Substituting (7.6.17)–(7.6.20) into (7.6.10), we obtain

$$\begin{aligned} \frac{u_m^{n+1} - 2u_m^n + u_m^{n-1}}{(\Delta t)^2} &- c^2 \frac{u_{m+1}^n - 2u_m^n + u_{m-1}^n}{(\Delta x)^2} \\ &= \left.\left(\frac{\partial^2 u}{\partial t^2} - c^2 \frac{\partial^2 u}{\partial x^2}\right)\right|_n^m \\ &+ \tfrac{1}{12}(\Delta t)^2 \left.\frac{\partial^4 u}{\partial t^4}\right|_n^m - \tfrac{1}{12}(c\Delta x)^2 \left.\frac{\partial^4 u}{\partial x^4}\right|_n^m + \cdots \end{aligned} \tag{7.6.21}$$

The first term on the right side of (7.6.21) vanishes because $u(x,t)$ satisfies the wave equation. As $\Delta x \to 0$, $\Delta t \to 0$, the remaining terms on the right side of (7.6.21) tend to zero and (7.6.10) is a consistent finite difference approximation of the wave equation.

Stability is another question. Under certain conditions the small errors inherent in fixed precision arithmetic (round off) can grow for certain choices of Δx and Δt. During the 1920s the mathematicians Courant, Friedrichs, and Lewy[22] found that if $c\Delta t/\Delta x > 1$, then our scheme is unstable. This CFL criteria has its origin in the fact that if $c\Delta t > \Delta x$, then we are asking signals in the numerical scheme to travel faster than their real-world counterparts and this unrealistic expectation leads to instability!

One method of determining *stability*, commonly called the von Neumann method,[23] involves examining solutions to (7.6.11) that have the form

$$u_m^n = e^{im\theta} e^{in\lambda}, \tag{7.6.22}$$

where θ is an arbitrary real number and λ is a complex number that has yet to be determined. Our choice of (7.6.22) is motivated by the fact that the initial condition u_m^0 can be represented by a Fourier series where a typical term behaves as $e^{im\theta}$.

If we substitute (7.6.22) into (7.6.10) and divide out the common factor $e^{im\theta}e^{in\lambda}$, we have that

$$\frac{e^{i\lambda} - 2 + e^{-i\lambda}}{(\Delta t)^2} = c^2 \frac{e^{i\theta} - 2 + e^{-i\theta}}{(\Delta x)^2} \tag{7.6.23}$$

or

$$\sin^2\left(\frac{\lambda}{2}\right) = \left(\frac{c\Delta t}{\Delta x}\right)^2 \sin^2\left(\frac{\theta}{2}\right). \tag{7.6.24}$$

The behavior of u_m^n is determined by the values of λ given by (7.6.24). If $c\Delta t/\Delta x \le 1$, then λ is real and u_m^n is bounded for all θ as $n \to \infty$. If $c\Delta t/\Delta x > 1$, then it is possible to find a value of θ such that the right side of (7.6.24) exceeds unity and the corresponding λ's occur as complex conjugate pairs. The λ with the negative imaginary part produces a solution with exponential growth because $n = t_n/\Delta t \to \infty$ as $\Delta t \to 0$ for a fixed t_n and $c\Delta t/\Delta x$. Thus, the value of u_m^n becomes infinitely large, even though the initial data may be arbitrarily small.

---

[22] Courant, R., Friedrichs, K. O., and Lewy, H., 1928: Über die partiellen Differenzengleichungen der mathematischen Physik. *Math. Annalen*, **100**, 32–74. Translated into English in *IBM J. Res. Dev.*, **11**, 215–234.

[23] After its inventor, J. von Neumann. See O'Brien, G. G., Hyman, M. A., and Kaplan, S., 1950: A study of the numerical solution of partial differential equations. *J. Math. Phys.*, **29**, 223–251.

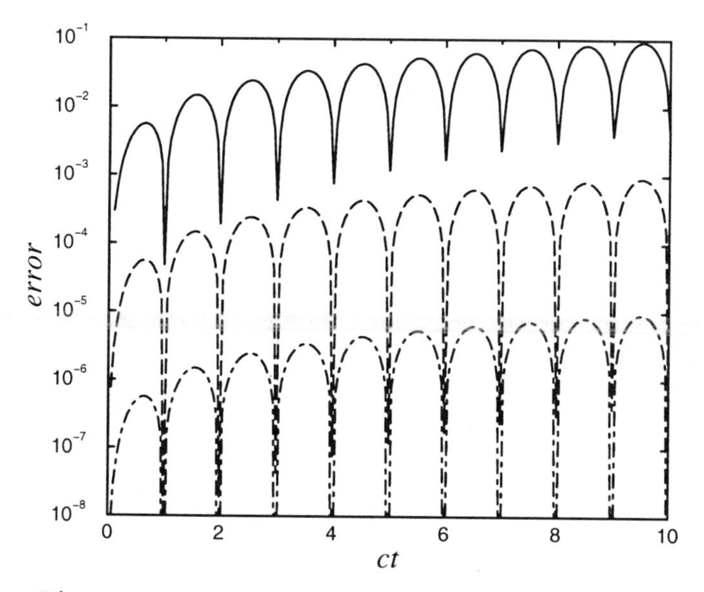

Figure 7.6.2: The growth of error $||e_n||$ as a function of ct for various resolutions. For the top line, $\Delta x = 0.1$; for the middle line, $\Delta x = 0.01$; and for the bottom line, $\Delta x = 0.001$.

Finally, we must check for convergence. A numerical scheme is *convergent* if the numerical solution approaches the continuous solution as $\Delta x, \Delta t \to 0$. The general procedure for proving convergence involves the evolution of the error term e_m^n which gives the difference between the true solution $u(x_m, t_n)$ and the finite difference solution u_m^n. From (7.6.21),

$$e_m^{n+1} = \left(\frac{c\Delta t}{\Delta x}\right)^2 (e_{m+1}^n + e_{m-1}^n) + 2\left[1 - \left(\frac{c\Delta t}{\Delta x}\right)^2\right] e_m^n - e_m^{n-1} + O[(\Delta t)^4] + O[(\Delta x)^2(\Delta t)^2]. \tag{7.6.25}$$

Let us apply (7.6.25) to work backwards from the point (x_m, t_n) by changing n to $n-1$. The nonvanishing terms in e_m^n reduce to a sum of $n+1$ values on the line $n = 1$ plus $\frac{1}{2}(n+1)n$ terms of the form $A(\Delta x)^4$. If we define the max norm $||e_n|| = \max_m |e_m^n|$, then

$$||e_n|| \le nB(\Delta x)^3 + \tfrac{1}{2}(n+1)nA(\Delta x)^4. \tag{7.6.26}$$

Because $n\Delta x \le ct_n$, (7.6.26) simplifies to

$$||e_n|| \le ct_n B(\Delta x)^2 + \tfrac{1}{2}c^2 t_n^2 A(\Delta x)^2. \tag{7.6.27}$$

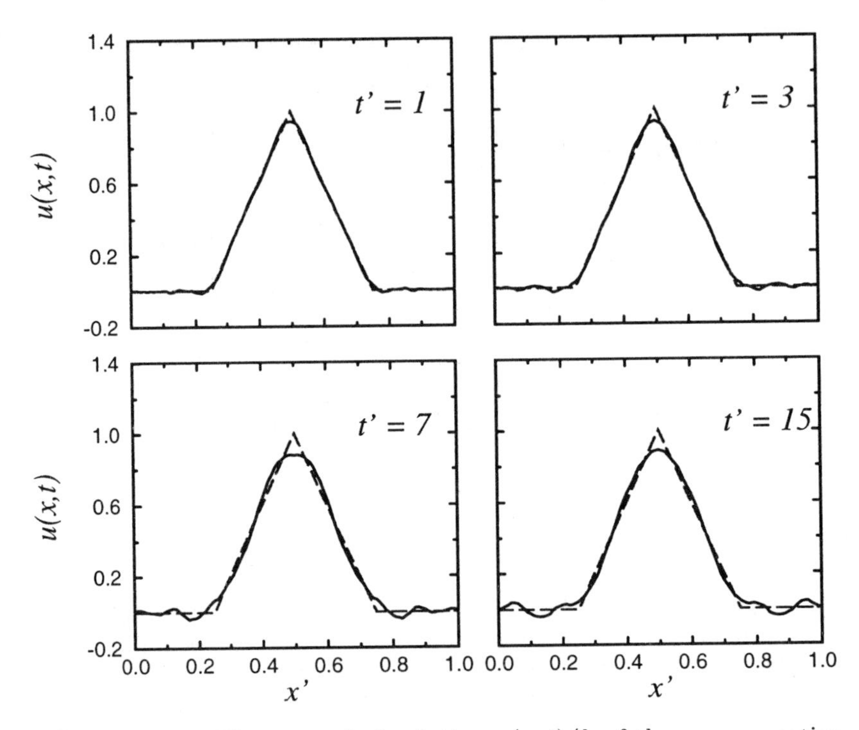

Figure 7.6.3: The numerical solution $u(x,t)/h$ of the wave equation with $c\Delta t/\Delta x = \frac{1}{2}$ using (7.6.11) at various positions $x' = x/L$ and times $t' = ct/L$. We have plotted the exact solution as a dashed line.

Thus, the error tends to zero as $\Delta x \to 0$, verifying convergence. We have illustrated (7.6.27) by using the finite difference equation (7.6.11) to compute $||e_n||$ during a numerical experiment that used $c\Delta t/\Delta x = 0.5$, $f(x) = \sin(\pi x)$ and $g(x) = 0$. Note how each increase of resolution by 10 results in a drop in the error by 100.

In the following examples we apply our scheme to solve a few simple initial and boundary conditions:

• Example 7.6.1

For our first example, we resolve (7.3.1) – (7.3.3) and (7.3.25) – (7.3.26) numerically using (7.6.11) with $c\Delta t/\Delta x = 1/2$ and $\Delta x = 0.01$. Figure 7.6.3 shows the resulting numerical solution at the nondimensional times $ct/L = 1, 3, 7$, and 15. We also included the exact solution as a dashed line.

Overall, the numerical solution approximates the exact or analytic solution well. However, we note small-scale noise in the numerical solution. Why does this occur? Recall that the exact solution could be written as an infinite sum of sines in the x dimension. Each successive

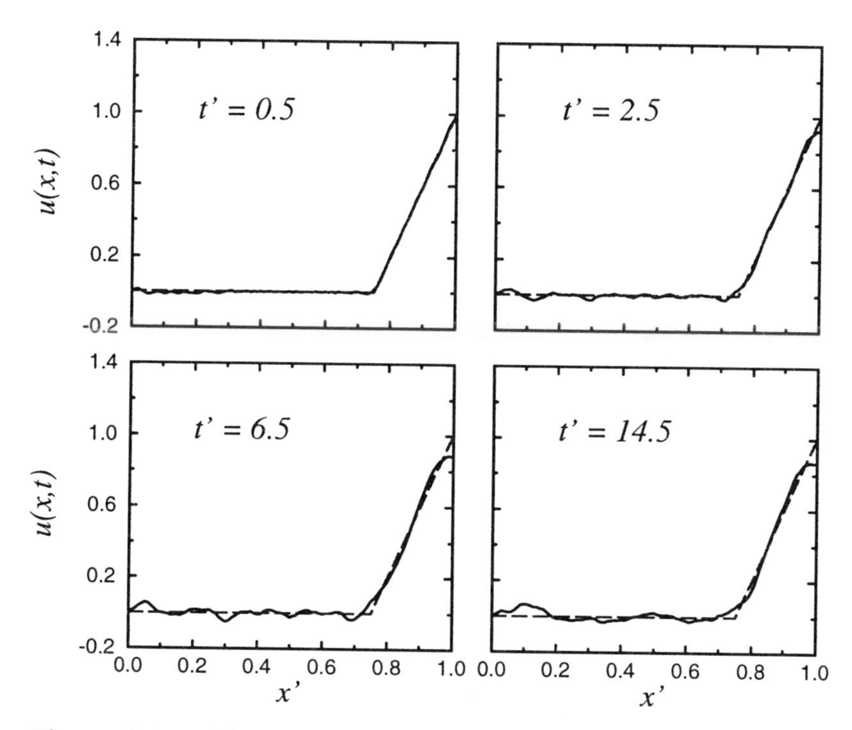

Figure 7.6.4: The numerical solution $u(x,t)/h$ of the wave equation when the right end moves freely with $c\Delta t/\Delta x = \frac{1}{2}$ using (7.6.11) and (7.6.30) at various positions $x' = x/L$ and times $t' = ct/L$. We have plotted the exact solution as a dashed line.

harmonic adds a contribution from waves of shorter and shorter wavelength. In the case of the numerical solution, the longer-wavelength harmonics are well represented by the numerical scheme because there are many grid points available to resolve a given wavelength. As the wavelengths become shorter, the higher harmonics are poorly resolved by the numerical scheme, move at incorrect phase speeds, and their misplacement (dispersion) creates the small-scale noise that you observe rather than giving the sharp angular features of the exact solution. The only method for avoiding this problem is to devise schemes that resolve the smaller-scale waves better.

- **Example 7.6.2**

Let us redo Example 7.6.1 except that we will introduce the boundary condition that $u_x(L,t) = 0$. This corresponds to a string where we fix the left end and allow the right end to freely move up and down. This requires a new difference condition along the right boundary. If we employ centered differencing,

$$\frac{u_{L+1}^n - u_{L-1}^n}{2\Delta x} = 0 \tag{7.6.28}$$

and

$$u_L^{n+1} = 2u_L^n - u_L^{n-1} + \left(\frac{c\Delta t}{\Delta x}\right)^2 \left(u_{L+1}^n - 2u_L^n + u_{L-1}^n\right). \tag{7.6.29}$$

Eliminating u_{L+1}^n between (7.6.28)–(7.6.29),

$$u_L^{n+1} = 2u_L^n - u_L^{n-1} + \left(\frac{c\Delta t}{\Delta x}\right)^2 \left(2u_{L-1}^n - 2u_L^n\right). \tag{7.6.30}$$

Figure 7.6.4 is the same as Figure 7.6.3 except for the new boundary condition. In this case the exact solution is

$$\begin{aligned} u(x,t) = {} & \frac{32h}{\pi^2} \sum_{n=1}^{\infty} \frac{1}{(2n-1)^2} \\ & \times \left\{ 2\sin\left[\frac{(2n-1)\pi}{4}\right] - \sin\left[\frac{3(2n-1)\pi}{8}\right] - \sin\left[\frac{(2n-1)\pi}{8}\right] \right\} \\ & \times \sin\left[\frac{(2n-1)\pi x}{2L}\right] \cos\left[\frac{(2n-1)\pi ct}{2L}\right]. \end{aligned} \tag{7.6.31}$$

We have highlighted those times when the solution has its maximum amplitude at the free right end. The results are consistent with those presented in Example 7.6.1, especially the small-scale noise due the dispersion. Overall, however, the numerical solution does approximate the exact solution well.

Project: Numerical Solution of First-Order Hyperbolic Equations

The equation $u_t + u_x = 0$ is the simplest possible hyperbolic partial differential equation. Indeed the classic wave equation can be written as a system of these equations: $u_t + cv_x = 0$ and $v_t + cu_x = 0$. In this project you will examine several numerical schemes for solving such a partial differential equation.

Step 1: One of the simplest numerical schemes is the forward-in-time, centered-in-space of

$$\frac{u_m^{n+1} - u_m^n}{\Delta t} + \frac{u_{m+1}^n - u_{m-1}^n}{2\Delta x} = 0.$$

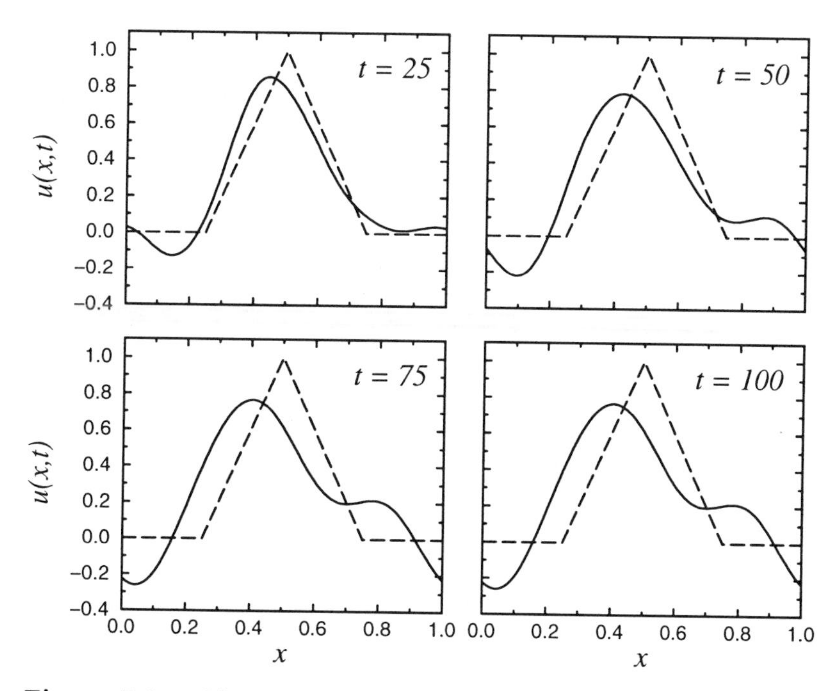

Figure 7.6.5: The numerical solution $u(x, t)$ of the first-order hyperbolic partial differential equation $u_t + u_x = 0$ using the Lax-Wendroff formula. The initial conditions are given by (7.3.25) with $h = 1$ and $\Delta t/\Delta x = \frac{1}{2}$. We have plotted the exact solution as a dashed line.

Use von Neumann's stability analysis to show that this scheme is *always* unstable.

Step 2: The most widely used method for numerically integrating first-order hyperbolic equations is the *Lax-Wendroff* method:

$$u_m^{n+1} = u_m^n - \frac{\Delta t}{2\Delta x}\left(u_{m+1}^n - u_{m-1}^n\right) + \frac{(\Delta t)^2}{2(\Delta x)^2}\left(u_{m+1}^n - 2u_m^n + u_{m-1}^n\right).$$

This methods introduces errors of $O[(\Delta t)^2]$ and $O[(\Delta x)^2]$. Show that this scheme is stable if it satisfies the CFL criteria of $\Delta t/\Delta x \leq 1$.

Using the initial condition given by (7.3.25), write code that uses this scheme to numerically integrate $u_t + u_x = 0$. Plot the results over the interval $0 < x < 1$ given the *periodic* boundary conditions of $u(0, t) = u(1, t)$ for the temporal interval $0 < t \leq 100$. Discuss the strengths and weaknesses of the scheme with respect to dissipation or damping of the numerical solution and preserving the phase of the solution. Most numerical methods books will discuss this.[24]

[24] For example, Lapidus, L. and Pinder, G. F., 1982: *Numerical Solu-*

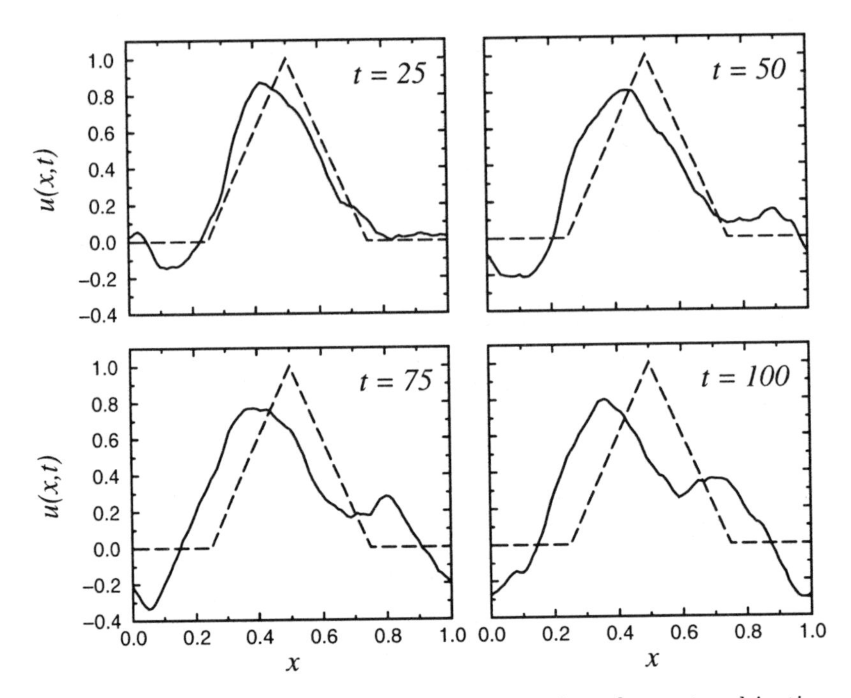

Figure 7.6.6: Same as Figure 7.6.5 except that the centered-in-time, centered-in-space scheme was used.

Step 3: Another simple scheme is the centered-in-time, centered-in-space of

$$\frac{u_m^{n+1} - u_m^{n-1}}{2\Delta t} + \frac{u_{m+1}^n - u_{m-1}^n}{2\Delta x} = 0.$$

This methods introduces errors of $O[(\Delta t)^2]$ and $O[(\Delta x)^2]$. Show that this scheme is stable if it satisfies the CFL criteria of $\Delta t/\Delta x \leq 1$.

Using the initial condition given by (7.3.25), write code that uses this scheme to numerically integrate $u_t + u_x = 0$ over the interval $0 < x < 1$ given the *periodic* boundary conditions of $u(0,t) = u(1,t)$. Plot the results over the spatial interval for the temporal interval $0 < t \leq 100$. One of the difficulties is taking the first time step. Use the scheme in Step 1 to take this first time step. Discuss the strengths and weaknesses of the scheme with respect to dissipation or damping of the numerical solution and preserving the phase of the solution.

---

tion of Partial Differential Equations in Science and Engineering, John Wiley & Sons, New York.

Chapter 8

The Heat Equation

In this chapter we deal with the linear parabolic differential equation

$$\frac{\partial u}{\partial t} = a^2 \frac{\partial^2 u}{\partial x^2} \tag{8.0.1}$$

in the two independent variables x and t. This equation, known as the one-dimensional heat equation, serves as the prototype for a wider class of *parabolic equations*:

$$a(x,t)\frac{\partial^2 u}{\partial x^2} + b(x,t)\frac{\partial^2 u}{\partial x \partial t} + c(x,t)\frac{\partial^2 u}{\partial t^2} = f\left(x,t,u,\frac{\partial u}{\partial x},\frac{\partial u}{\partial t}\right), \tag{8.0.2}$$

where $b^2 = 4ac$. It arises in the study of heat conduction in solids as well as in a variety of diffusive phenomena. The heat equation is similar to the wave equation in that it is also an equation of evolution. However, the heat equation is not "conservative" because if we reverse the sign of t, we obtain a different solution. This reflects the presence of entropy which must always increase during heat conduction.

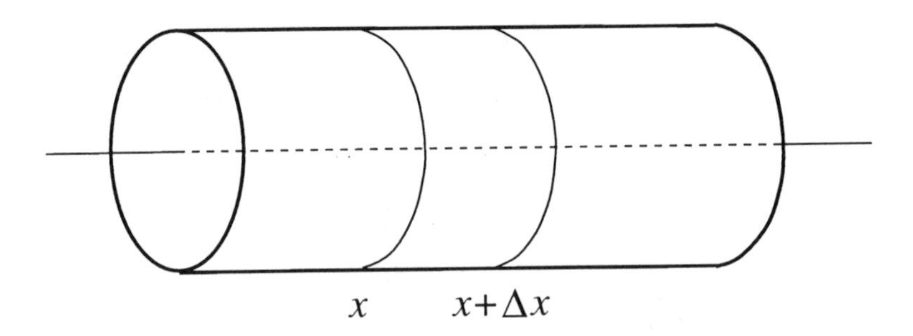

Figure 8.1.1: Heat conduction in a thin bar.

8.1 DERIVATION OF THE HEAT EQUATION

To derive the heat equation, consider a heat-conducting homogeneous rod, extending from $x = 0$ to $x = L$ along the x-axis (see Figure 8.1.1). The rod has uniform cross section A and constant density ρ, is insulated laterally so that heat flows only in the x-direction and is sufficiently thin so that the temperature at all points on a cross section is constant. Let $u(x,t)$ denote the temperature of the cross section at the point x at any instant of time t, and let c denote the specific heat of the rod (the amount of heat required to raise the temperature of a unit mass of the rod by a degree). In the segment of the rod between the cross section at x and the cross section at $x + \Delta x$, the amount of heat is

$$Q(t) = \int_x^{x+\Delta x} c\rho A u(s,t)\,ds. \tag{8.1.1}$$

On the other hand, the rate at which heat flows into the segment across the cross section at x is proportional to the cross section and the gradient of the temperature at the cross section (Fourier's law of heat conduction):

$$-\kappa A \frac{\partial u(x,t)}{\partial x}, \tag{8.1.2}$$

where κ denotes the thermal conductivity of the rod. The sign in (8.1.2) indicates that heat flows in the direction of decreasing temperature. Similarly, the rate at which heat flows out of the segment through the cross section at $x + \Delta x$ equals

$$-\kappa A \frac{\partial u(x+\Delta x,t)}{\partial x}. \tag{8.1.3}$$

The difference between the amount of heat that flows in through the cross section at x and the amount of heat that flows out through the cross section at $x + \Delta x$ must equal the change in the heat content of the segment $x \leq s \leq x+\Delta x$. Hence, by subtracting (8.1.3) from (8.1.2) and equating the result to the time derivative of (8.1.1),

$$\frac{\partial Q}{\partial t} = \int_x^{x+\Delta x} c\rho A \frac{\partial u(s,t)}{\partial t}\, ds = \kappa A \left[\frac{\partial u(x+\Delta x,t)}{\partial x} - \frac{\partial u(x,t)}{\partial x} \right]. \tag{8.1.4}$$

Assuming that the integrand in (8.1.4) is a continuous function of s, then by the mean value theorem for integrals,

$$\int_x^{x+\Delta x} \frac{\partial u(s,t)}{\partial t}\, ds = \frac{\partial u(\xi,t)}{\partial t} \Delta x, \qquad x < \xi < x+\Delta x, \tag{8.1.5}$$

so that (8.1.4) becomes

$$c\rho\Delta x \frac{\partial u(\xi,t)}{\partial t} = \kappa \left[\frac{\partial u(x+\Delta x,t)}{\partial x} - \frac{\partial u(x,t)}{\partial x} \right]. \tag{8.1.6}$$

Dividing both sides of (8.1.6) by $c\rho\Delta x$ and taking the limit as $\Delta x \to 0$,

$$\frac{\partial u(x,t)}{\partial t} = a^2 \frac{\partial^2 u(x,t)}{\partial x^2} \tag{8.1.7}$$

with $a^2 = \kappa/(c\rho)$. Equation (8.1.7) is called the one-dimensional *heat equation*. The constant a^2 is called the *diffusivity* within the solid.

If an external source supplies heat to the rod at a rate $f(x,t)$ per unit volume per unit time, we must add the term $\int_x^{x+\Delta x} f(s,t)\, ds$ to the time derivative term of (8.1.4). Thus, in the limit $\Delta x \to 0$,

$$\frac{\partial u(x,t)}{\partial t} - a^2 \frac{\partial^2 u(x,t)}{\partial x^2} = F(x,t), \tag{8.1.8}$$

where $F(x,t) = f(x,t)/(c\rho)$ is the source density. This equation is called the *nonhomogeneous heat equation*.

8.2 INITIAL AND BOUNDARY CONDITIONS

In the case of heat conduction in a thin rod, the temperature function $u(x,t)$ must satisfy not only the heat equation (8.1.7) but also how the two ends of the rod exchange heat energy with the surrounding medium. If (1) there is no heat source, (2) the function $f(x)$, $0 < x < L$ describes the temperature in the rod at $t = 0$, and (3) we maintain both ends at zero temperature for all time, then the partial differential equation

$$\frac{\partial u(x,t)}{\partial t} = a^2 \frac{\partial^2 u(x,t)}{\partial x^2}, \qquad 0 < x < L,\ 0 < t \tag{8.2.1}$$

describes the temperature distribution $u(x,t)$ in the rod at any later time $0 < t$ subject to the condition

$$u(x,0) = f(x), \qquad 0 < x < L \tag{8.2.2}$$

and

$$u(0,t) = u(L,t) = 0, \qquad 0 < t. \tag{8.2.3}$$

Equations (8.2.1)–(8.2.3) describe the *initial-boundary value problem* for this particular heat conduction problem; (8.2.3) is the boundary condition while (8.2.2) gives the initial condition. Note that in the case of the heat equation, the problem only demands the initial value of $u(x,t)$ and not $u_t(x,0)$, as with the wave equation.

Historically most linear boundary conditions have been classified in one of three ways. The condition (8.2.3) is an example of a *Dirichlet problem*[1] or *condition of the first kind.* This type of boundary condition gives the value of the solution (which is not necessarily equal to zero) along a boundary.

The next simplest condition involves derivatives. If we insulate both ends of the rod so that no heat flows from the ends, then according to (8.1.2) the boundary condition assumes the form

$$\frac{\partial u(0,t)}{\partial x} = \frac{\partial u(L,t)}{\partial x} = 0, \qquad 0 < t. \tag{8.2.4}$$

This is an example of a *Neumann problem*[2] or *condition of the second kind.* This type of boundary condition specifies the value of the normal derivative (which may not be equal to zero) of the solution along the boundary.

Finally, if there is radiation of heat from the ends of the rod into the surrounding medium, we shall show that the boundary condition is of the form

$$\frac{\partial u(0,t)}{\partial x} - hu(0,t) = \text{a constant} \tag{8.2.5}$$

and

$$\frac{\partial u(L,t)}{\partial x} + hu(L,t) = \text{another constant} \tag{8.2.6}$$

---

[1] Dirichlet, P. G. L., 1850: Über einen neuen Ausdruck zur Bestimmung der Dichtigkeit einer unendlich dünnen Kugelschale, wenn der Werth des Potentials derselben in jedem Punkte ihrer Oberfläche gegeben ist. *Abh. Königlich. Preuss. Akad. Wiss.*, 99–116.

[2] Neumann, C. G., 1877: *Untersuchungen über das Logarithmische und Newton'sche Potential.* Leibzig.

for $0 < t$, where h is a positive constant. This is an example of a *condition of the third kind* or *Robin problem*[3] and is a linear combination of Dirichlet and Neumann conditions.

8.3 SEPARATION OF VARIABLES

As with the wave equation, the most popular and widely used technique for solving the heat equation is separation of variables. Its success depends on our ability to express the solution $u(x,t)$ as the product $X(x)T(t)$. If we cannot achieve this separation, then the technique must be abandon for others. In the following examples we show how to apply this technique even if it takes a little work to get it right.

• Example 8.3.1

Let us find the solution to the homogeneous heat equation

$$\frac{\partial u}{\partial t} = a^2 \frac{\partial^2 u}{\partial x^2}, \quad 0 < x < L, 0 < t \tag{8.3.1}$$

which satisfies the initial condition

$$u(x,0) = f(x), \quad 0 < x < L \tag{8.3.2}$$

and the boundary conditions

$$u(0,t) = u(L,t) = 0, \quad 0 < t. \tag{8.3.3}$$

This system of equations models heat conduction in a thin metallic bar where both ends are held at the constant temperature of zero and the bar initially has the temperature $f(x)$.

We shall solve this problem by the method of separation of variables. Accordingly, we seek particular solutions of (8.3.1) of the form

$$u(x,t) = X(x)T(t), \tag{8.3.4}$$

which satisfy the boundary conditions (8.3.3). Because

$$\frac{\partial u}{\partial t} = X(x)T'(t) \tag{8.3.5}$$

and

$$\frac{\partial^2 u}{\partial x^2} = X''(x)T(t), \tag{8.3.6}$$

[3] Robin, G., 1886: Sur la distribution de l'électricité à la surface des conducteurs fermés et des conducteurs ouverts. *Ann. Sci. l'Ecole Norm. Sup., Ser. 3*, **3**, S1–S58.

(8.3.1) becomes

$$T'(t)X(x) = a^2X''(x)T(t). \tag{8.3.7}$$

Dividing both sides of (8.3.7) by $a^2X(x)T(t)$ gives

$$\frac{T'}{a^2T} = \frac{X''}{X} = -\lambda, \tag{8.3.8}$$

where $-\lambda$ is the separation constant. Equation (8.3.8) immediately yields two ordinary differential equations:

$$X'' + \lambda X = 0 \tag{8.3.9}$$

and

$$T' + a^2\lambda T = 0 \tag{8.3.10}$$

for the functions $X(x)$ and $T(t)$, respectively.

We now rewrite the boundary conditions in terms of $X(x)$ by noting that the boundary conditions are $u(0,t) = X(0)T(t) = 0$ and $u(L,t) = X(L)T(t) = 0$ for $0 < t$. If we were to choose $T(t) = 0$, then we would have a trivial solution for $u(x,t)$. Consequently, $X(0) = X(L) = 0$.

There are three possible cases: $\lambda = -m^2$, $\lambda = 0$, and $\lambda = k^2$. If $\lambda = -m^2 < 0$, then we must solve the boundary-value problem:

$$X'' - m^2X = 0, \qquad X(0) = X(L) = 0. \tag{8.3.11}$$

The general solution to (8.3.11) is

$$X(x) = A\cosh(mx) + B\sinh(mx). \tag{8.3.12}$$

Because $X(0) = 0$, it follows that $A = 0$. The condition $X(L) = 0$ yields $B\sinh(mL) = 0$. Since $\sinh(mL) \neq 0$, $B = 0$ and we have a trivial solution for $\lambda < 0$.

If $\lambda = 0$, the corresponding boundary-value problem is

$$X''(x) = 0, \qquad X(0) = X(L) = 0. \tag{8.3.13}$$

The general solution is

$$X(x) = C + Dx. \tag{8.3.14}$$

From $X(0) = 0$, we have that $C = 0$. From $X(L) = 0$, $DL = 0$ or $D = 0$. Again, we obtain a trivial solution.

Finally, we assume that $\lambda = k^2 > 0$. The corresponding boundary-value problem is

$$X'' + k^2X = 0, \qquad X(0) = X(L) = 0. \tag{8.3.15}$$

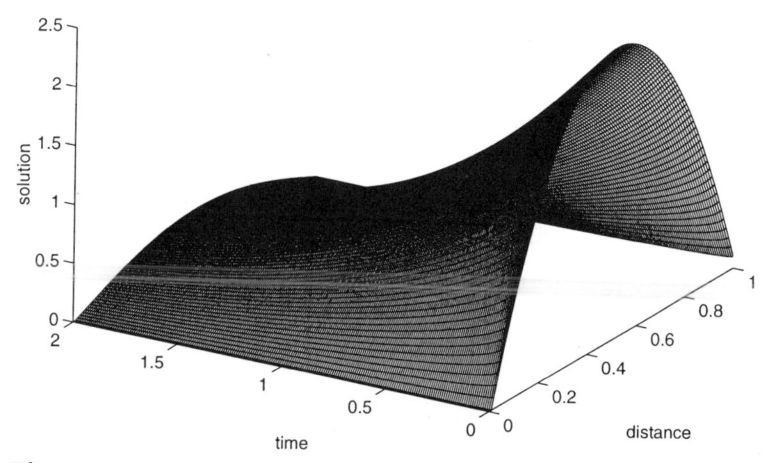

Figure 8.3.1: The temperature $u(x,t)$ within a thin bar as a function of position x/π and time a^2t when we maintain both ends at zero and the initial temperature equals $x(\pi - x)$.

The general solution to (8.3.15) is

$$X(x) = E\cos(kx) + F\sin(kx). \tag{8.3.16}$$

Because $X(0) = 0$, it follows that $E = 0$; from $X(L) = 0$, we obtain $F\sin(kL) = 0$. To have a nontrivial solution, $F \neq 0$ and $\sin(kL) = 0$. This implies that $k_nL = n\pi$, where $n = 1, 2, 3, \ldots$. In summary, the x-dependence of the solution is

$$X_n(x) = F_n \sin\left(\frac{n\pi x}{L}\right), \tag{8.3.17}$$

where $\lambda_n = n^2\pi^2/L^2$.

Turning to the time dependence, we use $\lambda_n = n^2\pi^2/L^2$ in (8.3.10):

$$T_n' + \frac{a^2n^2\pi^2}{L^2}T_n = 0. \tag{8.3.18}$$

The corresponding general solution is

$$T_n(t) = G_n \exp\left(-\frac{a^2n^2\pi^2}{L^2}t\right). \tag{8.3.19}$$

Thus, the functions

$$u_n(x,t) = B_n \sin\left(\frac{n\pi x}{L}\right) \exp\left(-\frac{a^2n^2\pi^2}{L^2}t\right), n = 1, 2, 3, \ldots, \tag{8.3.20}$$

where $B_n = F_nG_n$, are particular solutions of (8.3.1) and satisfy the homogeneous boundary conditions (8.3.3).

As we noted in the case of wave equation, we can solve the x-dependence equation as a regular Sturm-Liouville problem. After finding the eigenvalue λ_n and eigenfunction, we solve for $T_n(t)$. The product solution $u_n(x,t)$ equals the product of the eigenfunction and $T_n(t)$.

Having found particular solutions to our problem, the most general solution equals a linear sum of these particular solutions:

$$u(x,t) = \sum_{n=1}^{\infty} B_n \sin\left(\frac{n\pi x}{L}\right) \exp\left(-\frac{a^2 n^2 \pi^2}{L^2} t\right). \tag{8.3.21}$$

The coefficient B_n is chosen so that (8.3.21) yields the initial condition (8.3.2) if $t = 0$. Thus, setting $t = 0$ in (8.3.21), we see from (8.3.2) that the coefficients B_n must satisfy the relationship

$$f(x) = \sum_{n=1}^{\infty} B_n \sin\left(\frac{n\pi x}{L}\right), \qquad 0 < x < L. \tag{8.3.22}$$

This is precisely a Fourier half-range sine series for $f(x)$ on the interval $(0, L)$. Therefore, the formula

$$B_n = \frac{2}{L}\int_0^L f(x)\sin\left(\frac{n\pi x}{L}\right) dx, \qquad n = 1, 2, 3, \ldots \tag{8.3.23}$$

gives the coefficients B_n. For example, if $L = \pi$ and $u(x,0) = x(\pi - x)$, then

$$B_n = \frac{2}{\pi}\int_0^{\pi} x(\pi - x)\sin(nx)\, dx \tag{8.3.24}$$

$$= 2\int_0^{\pi} x\sin(nx)\, dx - \frac{2}{\pi}\int_0^{\pi} x^2\sin(nx)\, dx \tag{8.3.25}$$

$$= 4\frac{1-(-1)^n}{n^3\pi}. \tag{8.3.26}$$

Hence,

$$u(x,t) = \frac{8}{\pi}\sum_{n=1}^{\infty} \frac{\sin[(2n-1)x]}{(2n-1)^3} e^{-(2n-1)^2 a^2 t}. \tag{8.3.27}$$

Figure 8.3.1 illustrates (8.3.27) for various times. Note that both ends of the bar satisfy the boundary conditions, namely that the temperature equals zero. As time increases, heat flows out from the center of the bar to both ends where it is removed. This process is reflected in the collapse of the original parabolic shape of the temperature profile towards zero as time increases.

- **Example 8.3.2**

As a second example, let us solve the heat equation

$$\frac{\partial u}{\partial t} = a^2 \frac{\partial^2 u}{\partial x^2}, \qquad 0 < x < L, 0 < t \tag{8.3.28}$$

which satisfies the initial condition

$$u(x, 0) = x, \quad 0 < x < L \tag{8.3.29}$$

and the boundary conditions

$$\frac{\partial u(0,t)}{\partial x} = u(L,t) = 0, \quad 0 < t. \tag{8.3.30}$$

The condition $u_x(0,t) = 0$ expresses mathematically the constraint that no heat flows through the left boundary (insulated end condition).

Once again, we employ separation of variables; as in the previous example, the positive and zero separation constants yield trivial solutions. For a negative separation constant, however,

$$X'' + k^2 X = 0 \tag{8.3.31}$$

with

$$X'(0) = X(L) = 0, \tag{8.3.32}$$

because $u_x(0,t) = X'(0)T(t) = 0$ and $u(L,t) = X(L)T(t) = 0$. This regular Sturm-Liouville problem has the solution

$$X_n(x) = \cos\left[\frac{(2n-1)\pi x}{2L}\right], \qquad n = 1, 2, 3, \ldots \tag{8.3.33}$$

The temporal solution then becomes

$$T_n(t) = B_n \exp\left[-\frac{a^2(2n-1)^2\pi^2 t}{4L^2}\right]. \tag{8.3.34}$$

Consequently, a linear superposition of the particular solutions gives the total solution which equals

$$u(x,t) = \sum_{n=1}^{\infty} B_n \cos\left[\frac{(2n-1)\pi x}{2L}\right] \exp\left[-\frac{a^2(2n-1)^2\pi^2}{4L^2} t\right]. \tag{8.3.35}$$

Our final task remains to find the B_n's. Evaluating (8.3.35) at $t = 0$,

$$u(x,0) = x = \sum_{n=1}^{\infty} B_n \cos\left[\frac{(2n-1)\pi x}{2L}\right], \qquad 0 < x < L. \tag{8.3.36}$$

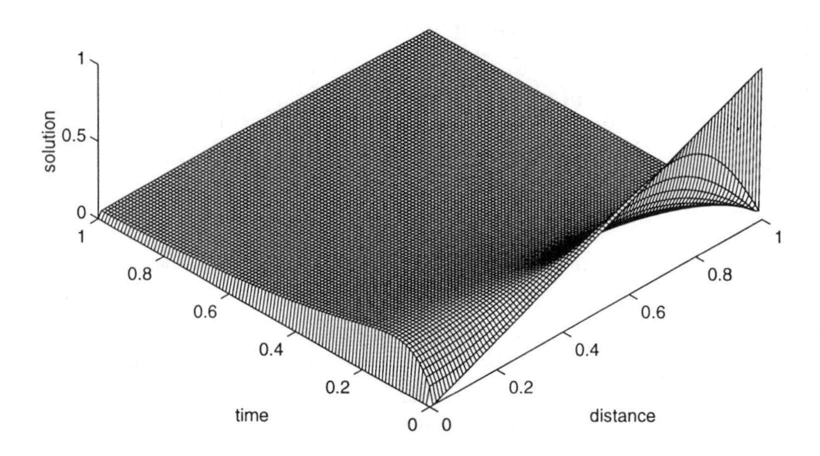

Figure 8.3.2: The temperature $u(x,t)/L$ within a thin bar as a function of position x/L and time a^2t/L^2 when we insulate the left end and hold the right end at the temperature of zero. The initial temperature equals x.

Equation (8.3.36) is not a half-range cosine expansion; it is an expansion in the orthogonal functions $\cos[(2n-1)\pi x/(2L)]$ corresponding to the regular Sturm-Liouville problem (8.3.31)–(8.3.32). Consequently, B_n is given by (6.3.4) with $r(x) = 1$ as

$$B_n = \frac{\int_0^L x\cos[(2n-1)\pi x/(2L)]\,dx}{\int_0^L \cos^2[(2n-1)\pi x/(2L)]\,dx} \tag{8.3.37}$$

$$= \frac{\frac{4L^2}{(2n-1)^2\pi^2}\cos\left[\frac{(2n-1)\pi x}{2L}\right]\Big|_0^L + \frac{2Lx}{(2n-1)\pi}\sin\left[\frac{(2n-1)\pi x}{2L}\right]\Big|_0^L}{\frac{x}{2}\Big|_0^L + \frac{L}{2(2n-1)\pi}\sin\left[\frac{(2n-1)\pi x}{L}\right]\Big|_0^L} \tag{8.3.38}$$

$$= \frac{8L}{(2n-1)^2\pi^2}\left\{\cos\left[\frac{(2n-1)\pi}{2}\right]-1\right\} + \frac{4L}{(2n-1)\pi}\sin\left[\frac{(2n-1)\pi}{2}\right] \tag{8.3.39}$$

$$= -\frac{8L}{(2n-1)^2\pi^2} - \frac{4L(-1)^n}{(2n-1)\pi}, \tag{8.3.40}$$

as $\cos[(2n-1)\pi/2] = 0$ and $\sin[(2n-1)\pi/2] = (-1)^{n+1}$. Consequently, the final solution is

$$u(x,t) = -\frac{4L}{\pi}\sum_{n=1}^{\infty}\left[\frac{2}{(2n-1)^2\pi} + \frac{(-1)^n}{2n-1}\right]\cos\left[\frac{(2n-1)\pi x}{2L}\right] \times \exp\left[-\frac{(2n-1)^2\pi^2a^2t}{4L^2}\right]. \tag{8.3.41}$$

Figure 8.3.2 illustrates the evolution of the temperature field with time. Initially, heat near the center of the bar flows towards the cooler, insulated end, resulting in an increase of temperature there. On the right side, heat flows out of the bar because the temperature is maintained at zero at $x = L$. Eventually the heat that has accumulated at the left end flows rightward because of the continual heat loss on the right end. In the limit of $t \to \infty$, all of the heat has left the bar.

• **Example 8.3.3**

A slight variation on Example 8.3.1 is

$$\frac{\partial u}{\partial t} = a^2 \frac{\partial^2 u}{\partial x^2}, \qquad 0 < x < L, 0 < t, \tag{8.3.42}$$

where

$$u(x,0) = u(0,t) = 0 \quad \text{and} \quad u(L,t) = \theta. \tag{8.3.43}$$

We begin by blindly employing the technique of separation of variables. Once again, we obtain the ordinary differential equation (8.3.9) and (8.3.10). The initial and boundary conditions become, however,

$$X(0) = T(0) = 0 \tag{8.3.44}$$

and

$$X(L)T(t) = \theta. \tag{8.3.45}$$

Although (8.3.44) is acceptable, (8.3.45) gives us an impossible condition because $T(t)$ cannot be constant. If it were, it would have to equal to zero by (8.3.44).

To find a way around this difficulty, suppose we wanted the solution to our problem at a time long after $t = 0$. From experience we know that heat conduction with time-independent boundary conditions eventually results in an evolution from the initial condition to some time-independent (steady-state) equilibrium. If we denote this steady-state solution by $w(x)$, it must satisfy the heat equation

$$a^2 w''(x) = 0 \tag{8.3.46}$$

and the boundary conditions

$$w(0) = 0 \quad \text{and} \quad w(L) = \theta. \tag{8.3.47}$$

We can integrate (8.3.46) immediately to give

$$w(x) = A + Bx \tag{8.3.48}$$

and the boundary condition (8.3.47) results in

$$w(x) = \frac{\theta x}{L}. \tag{8.3.49}$$

Clearly (8.3.49) cannot hope to satisfy the initial conditions; that was never expected of it. However, if we add a time-varying (transient) solution $v(x,t)$ to $w(x)$ so that

$$u(x,t) = w(x) + v(x,t), \tag{8.3.50}$$

we could satisfy the initial condition if

$$v(x,0) = u(x,0) - w(x) \tag{8.3.51}$$

and $v(x,t)$ tends to zero as $t \to \infty$. Furthermore, because $w''(x) = w(0) = 0$ and $w(L) = \theta$,

$$\frac{\partial v}{\partial t} = a^2 \frac{\partial^2 v}{\partial x^2}, \qquad 0 < x < L, 0 < t \tag{8.3.52}$$

with the boundary conditions

$$v(0,t) = 0 \quad \text{and} \quad v(L,t) = 0, \quad 0 < t. \tag{8.3.53}$$

We can solve (8.3.51), (8.3.52), and (8.3.53) by separation of variables; we did it in Example 8.3.1. However, in place of $f(x)$ we now have $u(x,0) - w(x)$ or $-w(x)$ because $u(x,0) = 0$. Therefore, the solution $v(x,t)$ is

$$v(x,t) = \sum_{n=1}^{\infty} B_n \sin\left(\frac{n\pi x}{L}\right) \exp\left(-\frac{a^2 n^2 \pi^2}{L^2} t\right) \tag{8.3.54}$$

with

$$B_n = \frac{2}{L}\int_0^L -w(x)\sin\left(\frac{n\pi x}{L}\right) dx \tag{8.3.55}$$

$$= \frac{2}{L}\int_0^L -\frac{\theta x}{L}\sin\left(\frac{n\pi x}{L}\right) dx \tag{8.3.56}$$

$$= -\frac{2\theta}{L^2}\left[\frac{L^2}{n^2\pi^2}\sin\left(\frac{n\pi x}{L}\right) - \frac{xL}{n\pi}\cos\left(\frac{n\pi x}{L}\right)\right]_0^L \tag{8.3.57}$$

$$= (-1)^n \frac{2\theta}{n\pi}. \tag{8.3.58}$$

Thus, the entire solution is

$$u(x,t) = \frac{\theta x}{L} + \frac{2\theta}{\pi}\sum_{n=1}^{\infty} \frac{(-1)^n}{n} \sin\left(\frac{n\pi x}{L}\right) \exp\left(-\frac{a^2 n^2 \pi^2}{L^2} t\right). \tag{8.3.59}$$

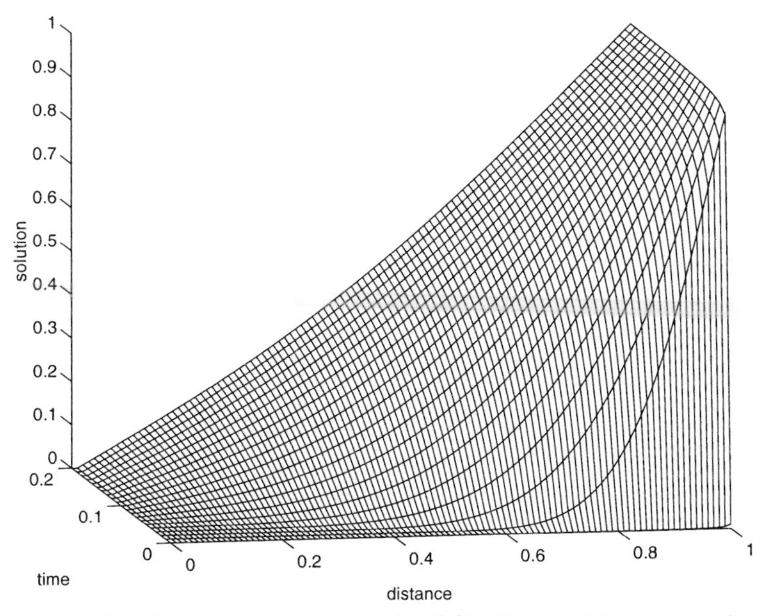

Figure 8.3.3: The temperature $u(x,t)/\theta$ within a thin bar as a function of position x/L and time a^2t/L^2 with the left end held at a temperature of zero and right end held at a temperature θ while the initial temperature of the bar is zero.

The quantity a^2t/L^2 is the *Fourier number*.

Figure 8.3.3 illustrates our solution. Clearly it satisfies the boundary conditions. Initially, heat flows rapidly from right to left. As time increases, the rate of heat transfer decreases until the final equilibrium (steady-state) is established and no more heat flows.

• Example 8.3.4

Let us find the solution to the heat equation

$$\frac{\partial u}{\partial t} = a^2 \frac{\partial^2 u}{\partial x^2}, \qquad 0 < x < L, 0 < t \tag{8.3.60}$$

subject to the Neumann boundary conditions

$$\frac{\partial u(0,t)}{\partial x} = \frac{\partial u(L,t)}{\partial x} = 0, \qquad 0 < t \tag{8.3.61}$$

and the initial condition that

$$u(x,0) = x, \qquad 0 < x < L. \tag{8.3.62}$$

We have now insulated *both* ends of the bar.

Assuming that $u(x,t) = X(x)T(t)$,

$$\frac{T'}{a^2T} = \frac{X''}{X} = -k^2, \tag{8.3.63}$$

where we have presently assumed that the separation constant is negative. The Neumann conditions give $u_x(0,t) = X'(0)T(t) = 0$ and $u_x(L,t) = X'(L)T(t) = 0$ so that $X'(0) = X'(L) = 0$.

The Sturm-Liouville problem

$$X'' + k^2X = 0 \tag{8.3.64}$$

and

$$X'(0) = X'(L) = 0 \tag{8.3.65}$$

gives the x-dependence. The eigenfunction solution is

$$X_n(x) = \cos\left(\frac{n\pi x}{L}\right), \tag{8.3.66}$$

where $k_n = n\pi/L$ and $n = 1, 2, 3, \ldots$

The corresponding temporal part equals the solution of

$$T_n' + a^2k_n^2T_n = T_n' + \frac{a^2n^2\pi^2}{L^2}T_n = 0, \tag{8.3.67}$$

which is

$$T_n(t) = A_n \exp\left(-\frac{a^2n^2\pi^2}{L^2}t\right). \tag{8.3.68}$$

Thus, the product solution given by a negative separation constant is

$$u_n(x,t) = X_n(x)T_n(t) = A_n \cos\left(\frac{n\pi x}{L}\right) \exp\left(-\frac{a^2n^2\pi^2}{L^2}t\right). \tag{8.3.69}$$

Unlike our previous problems, there is a nontrivial solution for a separation constant that equals zero. In this instance, the x-dependence equals

$$X(x) = Ax + B. \tag{8.3.70}$$

The boundary conditions $X'(0) = X'(L) = 0$ force A to be zero but B is completely free. Consequently, the eigenfunction in this particular case is

$$X_0(x) = 1. \tag{8.3.71}$$

Because $T_0'(t) = 0$ in this case, the temporal part equals a constant which we shall take to be $A_0/2$. Therefore, the product solution corresponding to the zero separation constant is

$$u_0(x,t) = X_0(x)T_0(t) = A_0/2. \tag{8.3.72}$$

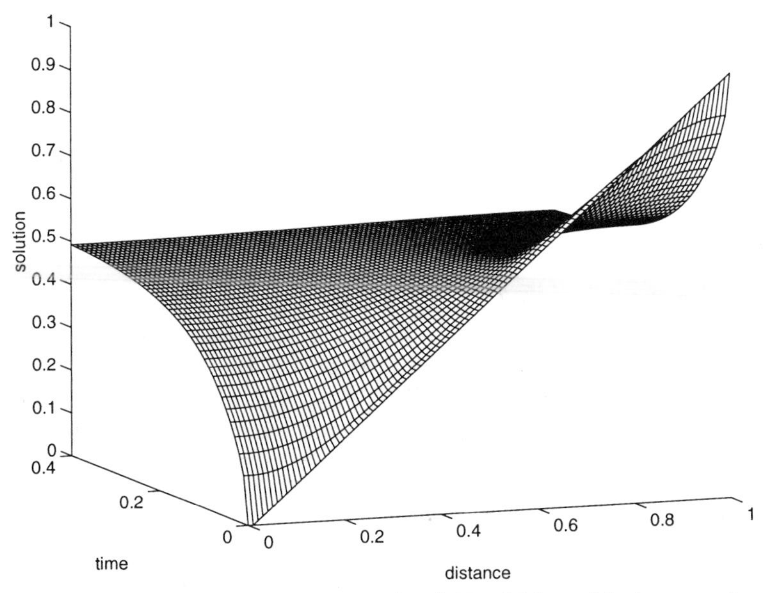

Figure 8.3.4: The temperature $u(x,t)/L$ within a thin bar as a function of position x/L and time a^2t/L^2 when we insulate both ends. The initial temperature of the bar is x.

The most general solution to our problem equals the sum of all of the possible solutions:

$$u(x,t) = \frac{A_0}{2} + \sum_{n=1}^{\infty} A_n \cos\left(\frac{n\pi x}{L}\right) \exp\left(-\frac{a^2n^2\pi^2}{L^2}t\right). \qquad \textbf{(8.3.73)}$$

Upon substituting $t = 0$ into (8.3.73), we can determine A_n because

$$u(x,0) = x = \frac{A_0}{2} + \sum_{n=1}^{\infty} A_n \cos\left(\frac{n\pi x}{L}\right) \qquad \textbf{(8.3.74)}$$

is merely a half-range Fourier cosine expansion of the function x over the interval $(0, L)$. From (2.1.23)–(2.1.24),

$$A_0 = \frac{2}{L}\int_0^L x\,dx = L \qquad \textbf{(8.3.75)}$$

and

$$A_n = \frac{2}{L}\int_0^L x\cos\left(\frac{n\pi x}{L}\right)dx \qquad \textbf{(8.3.76)}$$

$$= \frac{2}{L}\left[\frac{L^2}{n^2\pi^2}\cos\left(\frac{n\pi x}{L}\right) + \frac{xL}{n\pi}\sin\left(\frac{n\pi x}{L}\right)\right]_0^L \qquad \textbf{(8.3.77)}$$

$$= \frac{2L}{n^2\pi^2}\left[(-1)^n - 1\right]. \qquad \textbf{(8.3.78)}$$

The final solution is

$$u(x,t) = \frac{L}{2} - \frac{4L}{\pi^2} \sum_{m=1}^{\infty} \frac{1}{(2m-1)^2} \cos\left[\frac{(2m-1)\pi x}{L}\right] \times \exp\left[-\frac{a^2(2m-1)^2\pi^2}{L^2} t\right], \tag{8.3.79}$$

because all of the even harmonics vanish and we may rewrite the odd harmonics using $n = 2m - 1$, where $m = 1, 2, 3, 4, \ldots$.

Figure 8.3.4 illustrates (8.3.79) for various positions and times. The physical interpretation is quite simple. Since heat cannot flow in or out of the rod because of the insulation, it can only redistribute itself. Thus, heat flows from the warm right end to the cooler left end. Eventually the temperature achieves steady-state when the temperature is uniform throughout the bar.

• Example 8.3.5

So far we have dealt with problems where the temperature or flux of heat has been specified at the ends of the rod. In many physical applications, one or both of the ends may radiate to free space at temperature u_0. According to Stefan's law, the amount of heat radiated from a given area dA in a given time interval dt is

$$\sigma(u^4 - u_0^4)\, dA\, dt, \tag{8.3.80}$$

where σ is called the Stefan-Boltzmann constant. On the other hand, the amount of heat that reaches the surface from the interior of the body, assuming that we are at the right end of the bar, equals

$$-\kappa \frac{\partial u}{\partial x}\, dA\, dt, \tag{8.3.81}$$

where κ is the thermal conductivity. Because these quantities must be equal,

$$-\kappa \frac{\partial u}{\partial x} = \sigma(u^4 - u_0^4) = \sigma(u - u_0)(u^3 + u^2 u_0 + u u_0^2 + u_0^3). \tag{8.3.82}$$

If u and u_0 are nearly equal, we may approximate the second bracketed term on the right side of (8.3.82) as $4u_0^3$. We write this approximate form of (8.3.82) as

$$-\frac{\partial u}{\partial x} = h(u - u_0), \tag{8.3.83}$$

where h, the *surface conductance* or the *coefficient of surface heat transfer*, equals $4\sigma u_0^3/\kappa$. Equation (8.3.83) is a "radiation" boundary condition. Sometimes someone will refer to it as "Newton's law" because

(8.3.83) is mathematically identical to Newton's law of cooling of a body by forced convection.

Let us now solve the problem of a rod that we initially heat to the uniform temperature of 100. We then allow it to cool by maintaining the temperature at zero at $x = 0$ and radiatively cooling to the surrounding air at the temperature of zero[4] at $x = L$. We may restate the problem as

$$\frac{\partial u}{\partial t} = a^2 \frac{\partial^2 u}{\partial x^2}, \quad 0 < x < L, 0 < t \tag{8.3.84}$$

with

$$u(x,0) = 100, \quad 0 < x < L \tag{8.3.85}$$

$$u(0,t) = 0, \quad 0 < t \tag{8.3.86}$$

and

$$\frac{\partial u(L,t)}{\partial x} + hu(L,t) = 0, \quad 0 < t. \tag{8.3.87}$$

Once again, we assume a product solution $u(x,t) = X(x)T(t)$ with a negative separation constant so that

$$\frac{X''}{X} = \frac{T'}{a^2 T} = -k^2. \tag{8.3.88}$$

We obtain for the x-dependence that

$$X'' + k^2 X = 0 \tag{8.3.89}$$

but the boundary conditions are now

$$X(0) = 0 \quad \text{and} \quad X'(L) + hX(L) = 0. \tag{8.3.90}$$

The most general solution of (8.3.89) is

$$X(x) = A\cos(kx) + B\sin(kx). \tag{8.3.91}$$

However, $A = 0$ because $X(0) = 0$. On the other hand,

$$k\cos(kL) + h\sin(kL) = kL\cos(kL) + hL\sin(kL) = 0, \tag{8.3.92}$$

if $B \neq 0$. The nondimensional number hL is the *Biot number* and is completely dependent upon the physical characteristics of the rod.

---

[4] Although this would appear to make $h = 0$, we have merely chosen a temperature scale so that the air temperature is zero and the absolute temperature used in Stefan's law is nonzero.

Table 8.3.1: The First Ten Roots of (8.3.93) and C_n for $hL = 1$.

n	α_n	Approximate α_n	C_n
1	2.0288	2.2074	118.9193
2	4.9132	4.9246	31.3402
3	7.9787	7.9813	27.7554
4	11.0856	11.0865	16.2878
5	14.2075	14.2079	14.9923
6	17.3364	17.3366	10.8359
7	23.6044	23.6043	8.0989
8	26.7410	26.7409	7.7483
9	29.8786	29.8776	6.4625
10	33.0170	33.0170	6.2351

In Chapter 6 we saw how to find the roots of the transcendental equation

$$\alpha + hL\tan(\alpha) = 0, \tag{8.3.93}$$

where $\alpha = kL$. Consequently, if α_n is the nth root of (8.3.93), then the eigenfunction is

$$X_n(x) = \sin(\alpha_n x/L). \tag{8.3.94}$$

In Table 8.3.1, we list the first ten roots of (8.3.93) for $hL = 1$.

In general, we must solve (8.3.93) either numerically or graphically. If α is large, however, we can find approximate values by noting that

$$\cot(\alpha) = -hL/\alpha \approx 0 \tag{8.3.95}$$

or

$$\alpha_n = (2n-1)\pi/2, \tag{8.3.96}$$

where $n = 1, 2, 3, \ldots$ We may obtain a better approximation by setting

$$\alpha_n = (2n-1)\pi/2 - \epsilon_n, \tag{8.3.97}$$

where $\epsilon_n \ll 1$. Substituting into (8.3.95),

$$[(2n-1)\pi/2 - \epsilon_n]\cot[(2n-1)\pi/2 - \epsilon_n] + hL = 0. \tag{8.3.98}$$

We can simplify (8.3.98) to

$$\epsilon_n^2 + (2n-1)\pi\epsilon_n/2 + hL = 0 \tag{8.3.99}$$

because $\cot[(2n-1)\pi/2-\theta] = \tan(\theta)$ and $\tan(\theta) \approx \theta$ for $\theta \ll 1$. Solving for ϵ_n,

$$\epsilon_n \approx -\frac{2hL}{(2n-1)\pi} \tag{8.3.100}$$

and

$$\alpha_n \approx \frac{(2n-1)\pi}{2} + \frac{2hL}{(2n-1)\pi}. \tag{8.3.101}$$

In Table 8.3.1 we compare the approximate roots given by (8.3.101) with the actual roots.

The temporal part equals

$$T_n(t) = C_n \exp\left(-k_n^2 a^2 t\right) = C_n \exp\left(-\frac{\alpha_n^2 a^2 t}{L^2}\right). \tag{8.3.102}$$

Consequently, the general solution is

$$u(x,t) = \sum_{n=1}^{\infty} C_n \sin\left(\frac{\alpha_n x}{L}\right) \exp\left(-\frac{\alpha_n^2 a^2 t}{L^2}\right), \tag{8.3.103}$$

where α_n is the nth root of (8.3.93).

To determine C_n, we use the initial condition (8.3.85) and find that

$$100 = \sum_{n=1}^{\infty} C_n \sin\left(\frac{\alpha_n x}{L}\right). \tag{8.3.104}$$

Equation (8.3.104) is an eigenfunction expansion of 100 employing the eigenfunctions from the Sturm-Liouville problem

$$X'' + k^2 X = 0 \tag{8.3.105}$$

and

$$X(0) = X'(L) + hX(L) = 0. \tag{8.3.106}$$

Thus, the coefficient C_n is given by (6.3.4) or

$$C_n = \frac{\int_0^L 100 \sin(\alpha_n x/L)\, dx}{\int_0^L \sin^2(\alpha_n x/L)\, dx}, \tag{8.3.107}$$

as $r(x) = 1$. Performing the integrations,

$$C_n = \frac{100L[1-\cos(\alpha_n)]/\alpha_n}{\frac{1}{2}[L - L\sin(2\alpha_n)/(2\alpha_n)]} = \frac{200[1-\cos(\alpha_n)]}{\alpha_n[1+\cos^2(\alpha_n)/(hL)]}, \tag{8.3.108}$$

because $\sin(2\alpha_n) = 2\cos(\alpha_n)\sin(\alpha_n)$ and $\alpha_n = -hL\tan(\alpha_n)$. The final solution is

$$u(x,t) = \sum_{n=1}^{\infty} \frac{200[1-\cos(\alpha_n)]}{\alpha_n[1+\cos^2(\alpha_n)/(hL)]} \sin\left(\frac{\alpha_n x}{L}\right) \exp\left(-\frac{\alpha_n^2 a^2 t}{L^2}\right). \tag{8.3.109}$$

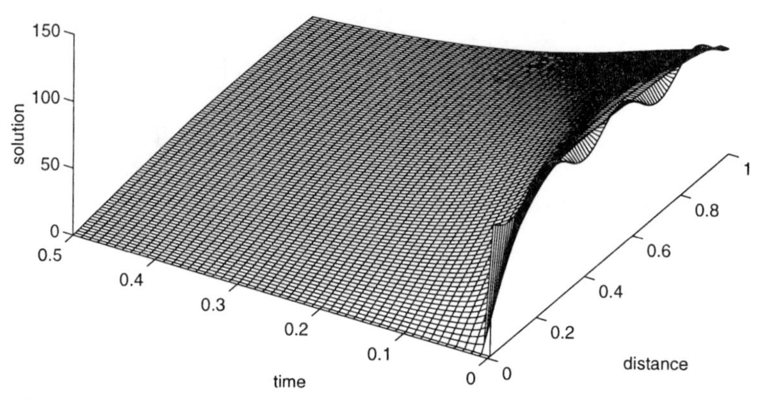

Figure 8.3.5: Tbe temperature $u(x,t)$ within a thin bar as a function of position x/L and time a^2t/L^2 when we allow the bar to radiatively cool at $x = L$ while the temperature is zero at $x = 0$. Initially the temperature was 100.

Figure 8.3.5 illustrates this solution for $hL = 1$ at various times and positions. It is similar to Example 8.3.1 in that the heat lost to the environment occurs either because the temperature at an end is zero or because it radiates heat to space which has the temperature of zero. The oscillations in the initial temperature distribution arise from Gibbs phenomena. We are using eigenfunctions that satisfy the boundary conditions (8.3.90) to fit a curve that equals 100 for all x.

• Example 8.3.6: Refrigeration of apples

Some decades ago, shiploads of apples, going from Australia to England, deteriorated from a disease called "brown heart," which occurred under insufficient cooling conditions. Apples, when placed on shipboard, are usually warm and must be cooled to be carried in cold storage. They also generate heat by their respiration. It was suspected that this heat generation effectively counteracted the refrigeration of the apples, resulting in the "brown heart."

This was the problem which induced Awberry[5] to study the heat distribution within a sphere in which heat is being generated. Awberry first assumed that the apples are initially at a uniform temperature. We can take this temperature to be zero by the appropriate choice of temperature scale. At time $t = 0$, the skins of the apples assume the temperature θ immediately when we introduce them into the hold.

---

[5] Awberry, J. H., 1927: The flow of heat in a body generating heat. *Philos. Mag., Ser. 7*, **4**, 629–638.

Because of the spherical geometry, the nonhomogeneous heat equation becomes

$$\frac{1}{a^2}\frac{\partial u}{\partial t} = \frac{1}{r^2}\frac{\partial}{\partial r}\left(r^2\frac{\partial u}{\partial r}\right) + \frac{G}{\kappa}, \quad 0 \le r < b, 0 < t, \qquad \textbf{(8.3.110)}$$

where a^2 is the thermal diffusivity, b is the radius of the apple, κ is the thermal conductivity, and G is the heating rate (per unit time per unit volume).

If we try to use separation of variables on (8.3.110), we find that it does not work because of the G/κ term. To circumvent this difficulty, we ask the simpler question of what happens after a very long time. We anticipate that a balance will eventually be established where conduction transports the heat produced within the apple to the surface of the apple where the surroundings absorb it. Consequently, just as we introduced a steady-state solution in Example 8.3.3, we again anticipate a steady-state solution $w(r)$ where the heat conduction removes the heat generated within the apples. The ordinary differential equation

$$\frac{1}{r^2}\frac{d}{dr}\left(r^2\frac{dw}{dr}\right) = -\frac{G}{\kappa} \qquad \textbf{(8.3.111)}$$

gives the steady-state. Furthermore, just as we introduced a transient solution which allowed our solution to satisfy the initial condition, we must also have one here and the governing equation is

$$\frac{\partial v}{\partial t} = \frac{a^2}{r^2}\frac{\partial}{\partial r}\left(r^2\frac{\partial v}{\partial r}\right). \qquad \textbf{(8.3.112)}$$

Solving (8.3.111) first,

$$w(r) = C + \frac{D}{r} - \frac{Gr^2}{6\kappa}. \qquad \textbf{(8.3.113)}$$

The constant D equals zero because the solution must be finite at $r = 0$. Because the steady-state solution must satisfy the boundary condition $w(b) = \theta$,

$$C = \theta + \frac{Gb^2}{6\kappa}. \qquad \textbf{(8.3.114)}$$

Turning to the transient problem, we introduce a new dependent variable $y(r,t) = rv(r,t)$. This new dependent variable allows us to replace (8.3.112) with

$$\frac{\partial y}{\partial t} = a^2\frac{\partial^2 y}{\partial r^2}, \qquad \textbf{(8.3.115)}$$

which we can solve. If we assume that $y(r,t) = R(r)T(t)$ and we only have a negative separation constant, the $R(r)$ equation becomes

$$\frac{d^2R}{dr^2} + k^2R = 0, \tag{8.3.116}$$

which has the solution

$$R(r) = A\cos(kr) + B\sin(kr). \tag{8.3.117}$$

The constant A equals zero because the solution (8.3.117) must vanish at $r = 0$ in order that $v(0,t)$ remains finite. However, because $\theta = w(b) + v(b,t)$ for all time and $v(b,t) = R(b)T(t)/b = 0$, then $R(b) = 0$. Consequently, $k_n = n\pi/b$ and

$$v_n(r,t) = \frac{B_n}{r}\sin\left(\frac{n\pi r}{b}\right)\exp\left(-\frac{n^2\pi^2a^2t}{b^2}\right). \tag{8.3.118}$$

Superposition gives the total solution which equals

$$u(r,t) = \theta + \frac{G}{6\kappa}(b^2 - r^2) + \sum_{n=1}^{\infty}\frac{B_n}{r}\sin\left(\frac{n\pi r}{b}\right)\exp\left(-\frac{n^2\pi^2a^2t}{b^2}\right). \tag{8.3.119}$$

Finally, we determine the B_n's by the initial condition that $u(r,0) = 0$. Therefore,

$$B_n = -\frac{2}{b}\int_0^b r\left[\theta + \frac{G}{6\kappa}(b^2 - r^2)\right]\sin\left(\frac{n\pi r}{b}\right)dr \tag{8.3.120}$$

$$= \frac{2\theta b}{n\pi}(-1)^n + \frac{2G}{\kappa}\left(\frac{b}{n\pi}\right)^3(-1)^n. \tag{8.3.121}$$

The final solution is

$$u(r,t) = \theta + \frac{2\theta b}{r\pi}\sum_{n=1}^{\infty}\frac{(-1)^n}{n}\sin\left(\frac{n\pi r}{b}\right)\exp\left(-\frac{n^2\pi^2a^2t}{b^2}\right)$$
$$+ \frac{G}{6\kappa}(b^2 - r^2) + \frac{2Gb^3}{r\kappa\pi^3}\sum_{n=1}^{\infty}\frac{(-1)^n}{n^3}\sin\left(\frac{n\pi r}{b}\right)\exp\left(-\frac{n^2\pi^2a^2t}{b^2}\right). \tag{8.3.122}$$

The first line of (8.3.122) gives the temperature distribution due to the imposition of the temperature θ on the surface of the apple while the second line gives the rise in the temperature due to the interior heating.

Returning to our original problem of whether the interior heating is strong enough to counteract the cooling by refrigeration, we merely use the second line of (8.3.122) to find how much the temperature deviates from what we normally expect. Because the highest temperature exists at the center of each apple, its value there is the only one of interest in this problem. Assuming $b = 4$ cm as the radius of the apple, $a^2G/\kappa = 1.33 \times 10^{-5}$ °C/s and $a^2 = 1.55 \times 10^{-3}$ cm^2/s, the temperature effect of the heat generation is very small, only 0.0232 °C when, after about 2 hours, the temperatures within the apples reach equilibrium. Thus, we must conclude that heat generation within the apples is not the cause of brown heart.

We now know that brown heart results from an excessive concentration of carbon dioxide and a deficient amount of oxygen in the storage hold.[6] Presumably this atmosphere affects the metabolic activities that are occurring in the apple[7] and leads to low-temperature breakdown.

• Example 8.3.7

In this example we illustrate how separation of variables may be employed in solving the axisymmetric heat equation in an infinitely long cylinder. In circular coordinates the heat equation is

$$\frac{\partial u}{\partial t} = a^2\left(\frac{\partial^2 u}{\partial r^2} + \frac{1}{r}\frac{\partial u}{\partial r}\right), \qquad 0 \le r < b, 0 < t, \tag{8.3.123}$$

where r denotes the radial distance and a^2 denotes the thermal diffusivity. Let us assume that we have heated this cylinder of radius b to the uniform temperature T_0 and then allowed it to cool by having its surface held at the temperature of zero starting from the time $t = 0$.

We begin by assuming that the solution is of the form $u(r,t) = R(r)T(t)$ so that

$$\frac{1}{R}\left(\frac{d^2R}{dr^2} + \frac{1}{r}\frac{dR}{dr}\right) = \frac{1}{a^2T}\frac{dT}{dt} = -\frac{k^2}{b^2}. \tag{8.3.124}$$

The only values of the separation constant that yield nontrivial solutions are negative. The nontrivial solutions are $R(r) = J_0(kr/b)$, where J_0 is the Bessel function of the first kind and zeroth order. A separation constant of zero gives $R(r) = \ln(r)$ which becomes infinite at the

[6] Thornton, N. C., 1931: The effect of carbon dioxide on fruits and vegetables in storage. *Contrib. Boyce Thompson Inst.*, **3**, 219–244.

[7] Fidler, J. C. and North, C. J., 1968: The effect of conditions of storage on the respiration of apples. IV. Changes in concentration of possible substrates of respiration, as related to production of carbon dioxide and uptake of oxygen by apples at low temperatures. *J. Hortic. Sci.*, **43**, 429–439.

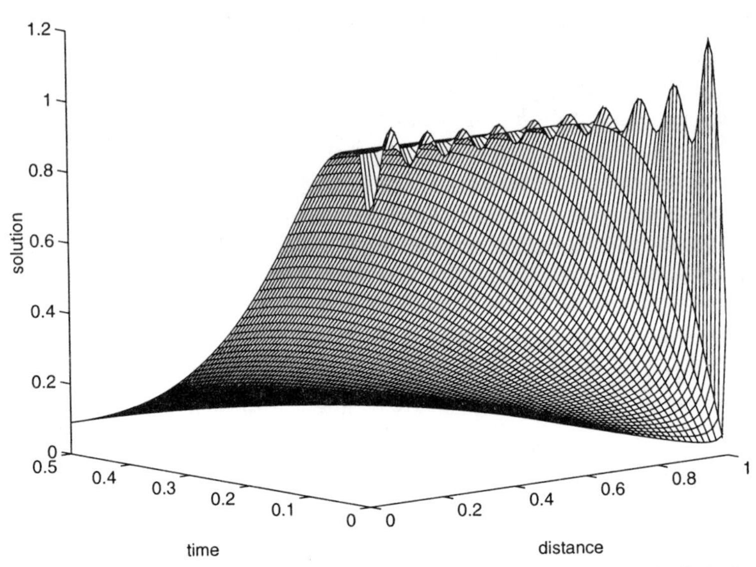

Figure 8.3.6: The temperature $u(r,t)/T_0$ within an infinitely long cylinder at various positions r/b and times a^2t/b^2 that we initially heated to the uniform temperature T_0 and then allowed to cool by forcing its surface to equal zero.

origin. Positive separation constants yield the modified Bessel function $I_0(kr/b)$. Although this function is finite at the origin, it cannot satisfy the boundary condition that $u(b,t) = R(b)T(t) = 0$ or $R(b) = 0$.

The boundary condition that $R(b) = 0$ requires that $J_0(k) = 0$. This transcendental equation yields an infinite number of k_n's. For each of these k_n's, the temporal part of the solution satisfies the differential equation

$$\frac{dT_n}{dt} + \frac{k_n^2 a^2}{b^2} T_n = 0, \tag{8.3.125}$$

which has the solution

$$T_n(t) = A_n \exp\left(-\frac{k_n^2 a^2}{b^2} t\right). \tag{8.3.126}$$

Consequently, the product solutions are

$$u_n(r,t) = A_n J_0\left(k_n \frac{r}{b}\right) \exp\left(-\frac{k_n^2 a^2}{b^2} t\right). \tag{8.3.127}$$

The total solution is a linear superposition of all of the particular solutions or

$$u(r,t) = \sum_{n=1}^{\infty} A_n J_0\left(k_n \frac{r}{b}\right) \exp\left(-\frac{k_n^2 a^2}{b^2} t\right). \tag{8.3.128}$$

Our final task remains to determine A_n. From the initial condition that $u(r,0) = T_0$,

$$u(r,0) = T_0 = \sum_{n=1}^{\infty} A_n J_0\left(k_n \frac{r}{b}\right). \tag{8.3.129}$$

From (6.5.35) and (6.5.43),

$$A_n = \frac{2T_0}{J_1^2(k_n)b^2} \int_0^b r J_0\left(k_n \frac{r}{b}\right) dr \tag{8.3.130}$$

$$= \frac{2T_0}{k_n^2 J_1^2(k_n)} \left(\frac{k_n r}{b}\right) J_1\left(k_n \frac{r}{b}\right)\Big|_0^b = \frac{2T_0}{k_n J_1(k_n)} \tag{8.3.131}$$

from (6.5.25). Thus, the final solution is

$$u(r,t) = 2T_0 \sum_{n=1}^{\infty} \frac{1}{k_n J_1(k_n)} J_0\left(k_n \frac{r}{b}\right) \exp\left(-\frac{k_n^2 a^2}{b^2} t\right). \tag{8.3.132}$$

Figure 8.3.6 illustrates the solution (8.3.132) for various Fourier numbers a^2t/b^2. It is similar to Example 8.3.1 except that we are in cylindrical coordinates. Heat flows from the interior and is removed at the cylinder's surface where the temperature equals zero. The initial oscillations of the solution result from Gibbs phenomena because we have a jump in the temperature field at $r = b$.

- **Example 8.3.8**

In this example we find the evolution of the temperature field within a cylinder of radius b as it radiatively cools from an initial uniform temperature T_0. The heat equation is

$$\frac{\partial u}{\partial t} = a^2 \left(\frac{\partial^2 u}{\partial r^2} + \frac{1}{r}\frac{\partial u}{\partial r}\right), \quad 0 \le r < b, 0 < t, \tag{8.3.133}$$

which we shall solve by separation of variables $u(r,t) = R(r)T(t)$. Therefore,

$$\frac{1}{R}\left(\frac{d^2R}{dr^2} + \frac{1}{r}\frac{dR}{dr}\right) = \frac{1}{a^2T}\frac{dT}{dt} = -\frac{k^2}{b^2}, \tag{8.3.134}$$

because only a negative separation constant yields a $R(r)$ which is finite at the origin and satisfies the boundary condition. This solution is $R(r) = J_0(kr/b)$, where J_0 is the Bessel function of the first kind and zeroth order.

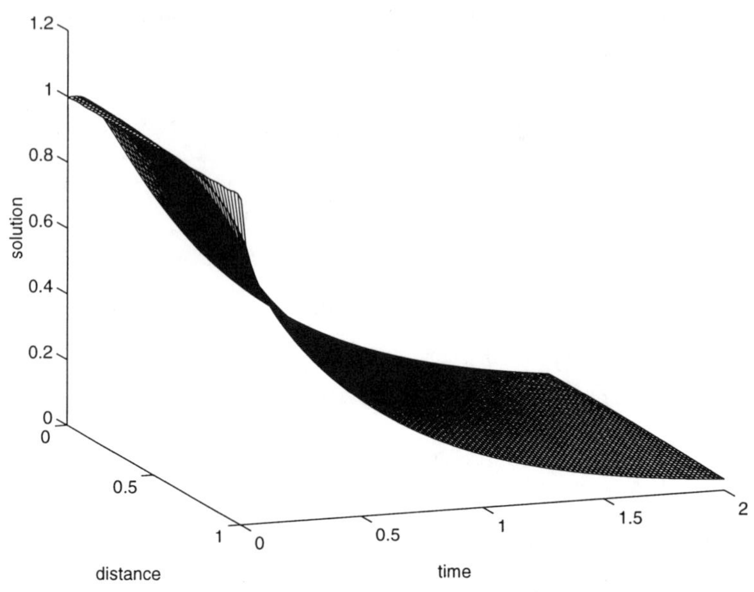

Figure 8.3.7: The temperature $u(r,t)/T_0$ within an infinitely long cylinder at various positions r/b and times a^2t/b^2 that we initially heated to the temperature T_0 and then allowed to radiatively cool with $hb = 1$.

The radiative boundary condition may be expressed as

$$\frac{\partial u(b,t)}{\partial r} + hu(b,t) = T(t)\left[\frac{dR(b)}{dr} + hR(b)\right] = 0. \tag{8.3.135}$$

Because $T(t) \neq 0$,

$$kJ_0'(k) + hbJ_0(k) = -kJ_1(k) + hbJ_0(k) = 0, \tag{8.3.136}$$

where the product hb is the Biot number. The solution of the transcendental equation (8.3.136) yields an infinite number of distinct k_n's. For each of these k_n's, the temporal part equals the solution of

$$\frac{dT_n}{dt} + \frac{k_n^2 a^2}{b^2} T_n = 0, \tag{8.3.137}$$

or

$$T_n(t) = A_n \exp\left(-\frac{k_n^2 a^2}{b^2} t\right). \tag{8.3.138}$$

The product solution is, therefore,

$$u_n(r,t) = A_n J_0\left(k_n \frac{r}{b}\right) \exp\left(-\frac{k_n^2 a^2}{b^2} t\right) \tag{8.3.139}$$

and the most general solution is a sum of these product solutions

$$u(r,t) = \sum_{n=1}^{\infty} A_n J_0\left(k_n \frac{r}{b}\right) \exp\left(-\frac{k_n^2 a^2}{b^2} t\right). \tag{8.3.140}$$

Finally, we must determine A_n. From the initial condition that $u(r,0) = T_0$,

$$u(r,0) = T_0 = \sum_{n=1}^{\infty} A_n J_0\left(k_n \frac{r}{b}\right), \tag{8.3.141}$$

where

$$A_n = \frac{2k_n^2 T_0}{b^2[k_n^2 + b^2h^2]J_0^2(k_n)} \int_0^b r J_0\left(k_n \frac{r}{b}\right)\, dr \tag{8.3.142}$$

$$= \frac{2k_n^2 T_0}{[k_n^2 + b^2h^2]J_0^2(k_n)} \left(\frac{k_n r}{b}\right) J_1\left(k_n \frac{r}{b}\right)\Big|_0^b \tag{8.3.143}$$

$$= \frac{2k_n T_0 J_1(k_n)}{[k_n^2 + b^2h^2]J_0^2(k_n)} = \frac{2k_n T_0 J_1(k_n)}{k_n^2 J_0^2(k_n) + b^2h^2J_0^2(k_n)} \tag{8.3.144}$$

$$= \frac{2k_n T_0 J_1(k_n)}{k_n^2 J_0^2(k_n) + k_n^2 J_1^2(k_n)} = \frac{2T_0 J_1(k_n)}{k_n[J_0^2(k_n) + J_1^2(k_n)]}, \tag{8.3.145}$$

which follows from (6.5.25), (6.5.35), (6.5.45), and (8.3.136). Consequently, the final solution is

$$u(r,t) = 2T_0 \sum_{n=1}^{\infty} \frac{J_1(k_n)}{k_n[J_0^2(k_n) + J_1^2(k_n)]} J_0\left(k_n \frac{r}{b}\right) \exp\left(-\frac{k_n^2 a^2}{b^2} t\right). \tag{8.3.146}$$

Figure 8.3.7 illustrates the solution (8.3.146) for various Fourier numbers a^2t/b^2 with $hb = 1$. It is similar to Example 8.3.5 except that we are in cylindrical coordinates. Heat flows from the interior and is removed at the cylinder's surface where it radiates to space at the temperature zero. Note that we do *not* suffer from Gibbs phenomena in this case because there is no initial jump in the temperature distribution.

• Example 8.3.9: Temperature within an electrical cable

In the design of cable installations we need the temperature reached within an electrical cable as a function of current and other parameters. To this end,[8] let us solve the nonhomogeneous heat equation in cylindrical coordinates with a radiation boundary condition.

The derivation of the heat equation follows from the conservation of energy:

heat generated = heat dissipated + heat stored

---

[8] Iskenderian, H. P. and Horvath, W. J., 1946: Determination of the temperature rise and the maximum safe current through multiconductor electric cables. *J. Appl. Phys.*, **17**, 255–262.

or

$$I^2RN\,dt = -\kappa\left[2\pi r\left.\frac{\partial u}{\partial r}\right|_r - 2\pi(r+\Delta r)\left.\frac{\partial u}{\partial r}\right|_{r+\Delta r}\right]dt + 2\pi r\Delta r c\rho\,du, \tag{8.3.147}$$

where I is the current through each wire, R is the resistance of each conductor, N is the number of conductors in the shell between radii r and $r+\Delta r = 2\pi m r\Delta r/(\pi b^2)$, b is the radius of the cable, m is the total number of conductors in the cable, κ is the thermal conductivity, ρ is the density, c is the average specific heat, and u is the temperature. In the limit of $\Delta r \to 0$, (8.3.147) becomes

$$\frac{\partial u}{\partial t} = A + a^2\frac{1}{r}\frac{\partial}{\partial r}\left(r\frac{\partial u}{\partial r}\right), \quad 0 \le r < b, 0 < t, \tag{8.3.148}$$

where $A = I^2Rm/(\pi b^2 c\rho)$ and $a^2 = \kappa/(\rho c)$.

Equation (8.3.148) is the nonhomogeneous heat equation for an infinitely long, axisymmetric cylinder. From Example 8.3.3, we know that we must write the temperature as the sum of a steady-state and transient solution: $u(r,t) = w(r) + v(r,t)$. The steady-state solution $w(r)$ satisfies

$$\frac{1}{r}\frac{d}{dr}\left(r\frac{dw}{dr}\right) = -\frac{A}{a^2} \tag{8.3.149}$$

or

$$w(r) = T_c - \frac{Ar^2}{4a^2}, \tag{8.3.150}$$

where T_c is the (yet unknown) temperature in the center of the cable.

The transient solution $v(r,t)$ is govern by

$$\frac{\partial v}{\partial t} = a^2\frac{1}{r}\frac{\partial}{\partial r}\left(r\frac{\partial v}{\partial r}\right), \quad 0 \le r < b, 0 < t \tag{8.3.151}$$

with the initial condition that $u(r,0) = T_c - Ar^2/(4a^2) + v(0,t) = 0$. At the surface $r = b$ heat radiates to free space so that the boundary condition is $u_r = -hu$, where h is the surface conductance. Because the steady-state temperature must be true when all transient effects die away, it must satisfy this radiation boundary condition regardless of the transient solution. This requires that

$$T_c = \frac{A}{a^2}\left(\frac{b^2}{4} + \frac{b}{2h}\right). \tag{8.3.152}$$

Therefore, $v(r,t)$ must satisfy $v_r(b,t) = -hv(b,t)$ at $r = b$.

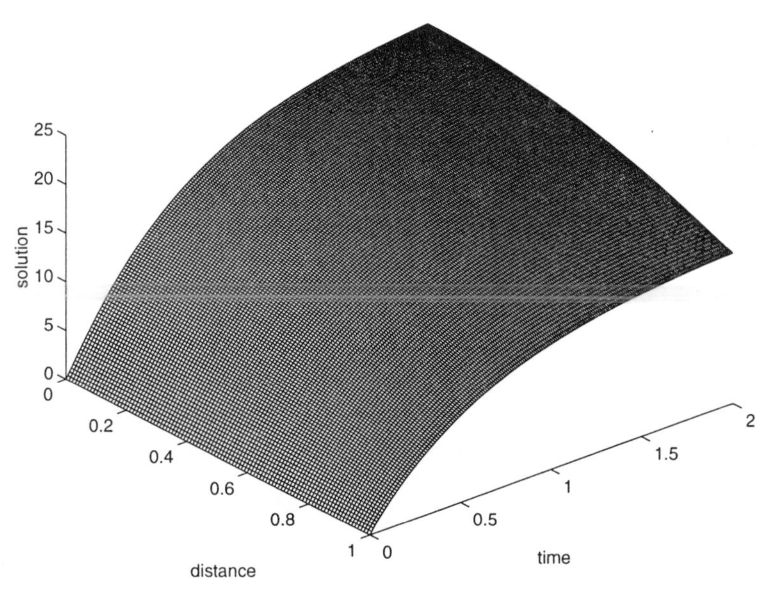

Figure 8.3.8: The temperature field (in degrees Celsius) within an electric copper cable containing 37 wires and a current of 22 amperes at various positions r/b and times a^2t/b^2. Initially the temperature was zero and then we allow the cable to cool radiatively as it is heated. The parameters are $hb = 1$ and the radius of the cable $b = 4$ cm.

We find the transient solution $v(r,t)$ by separation of variables $v(r,t) = R(r)T(t)$. Substituting into (8.3.151),

$$\frac{1}{rR}\frac{d}{dr}\left(r\frac{dR}{dr}\right) = \frac{1}{a^2T}\frac{dT}{dt} = -k^2 \tag{8.3.153}$$

or

$$\frac{d}{dr}\left(r\frac{dR}{dr}\right) + k^2rR = 0 \tag{8.3.154}$$

and

$$\frac{dT}{dt} + k^2a^2T = 0, \tag{8.3.155}$$

with $R'(b) = -hR(b)$. The only solution of (8.3.154) which remains finite at $r = 0$ and satisfies the boundary condition is $R(r) = J_0(kr)$, where J_0 is the zero-order Bessel function of the first kind. Substituting $J_0(kr)$ into the boundary condition, the transcendental equation is

$$kbJ_1(kb) - hbJ_0(kb) = 0. \tag{8.3.156}$$

For a given value of h and b, (8.3.156) yields an infinite number of unique zeros k_n.

The corresponding temporal solution to the problem is

$$T_n(t) = A_n \exp(-a^2 k_n^2 t), \tag{8.3.157}$$

so that the sum of the product solutions is

$$v(r,t) = \sum_{n=1}^{\infty} A_n J_0(k_n r) \exp(-a^2 k_n^2 t). \tag{8.3.158}$$

Our final task remains to compute A_n. By evaluating (8.3.158) at $t = 0$,

$$v(r,0) = \frac{Ar^2}{4a^2} - T_c = \sum_{n=1}^{\infty} A_n J_0(k_n r), \tag{8.3.159}$$

which is a Fourier-Bessel series in $J_0(k_n r)$. In Section 6.5 we showed that the coefficient of a Fourier-Bessel series with the orthogonal function $J_0(k_n r)$ and the boundary condition (8.3.156) equals

$$A_n = \frac{2k_n^2}{(k_n^2 b^2 + h^2 b^2) J_0^2(k_n b)} \int_0^b r \left(\frac{Ar^2}{4a^2} - T_c \right) J_0(k_n r)\, dr \tag{8.3.160}$$

from (6.5.35) and (6.5.45). Carrying out the indicated integrations,

$$A_n = \frac{2}{(k_n^2 + h^2) J_0^2(k_n b)} \times \left[\left(\frac{Ak_n b}{4a^2} - \frac{A}{k_n b a^2} - \frac{T_c k_n}{b} \right) J_1(k_n b) + \frac{A}{2a^2} J_0(k_n b) \right]. \tag{8.3.161}$$

We obtained (8.3.161) by using (6.5.25) and integrating by parts in a similar manner as was done in Example 6.5.5.

To illustrate this solution, let us compute it for the typical parameters $b = 4$ cm, $hb = 1$, $a^2 = 1.14$ cm^2/s, $A = 2.2747$ °C/s, and $T_c = 23.94$°C. The value of A corresponds to 37 wires of #6 AWG copper wire within a cable carrying a current of 22 amp.

Figure 8.3.8 illustrates the solution as a function of radius at various times. From an initial temperature of zero, the temperature rises due to the constant electrical heating. After a short period of time, it reaches its steady-state distribution given by (8.3.150). The cable is coolest at the surface where heat is radiating away. Heat flows from the interior to replace the heat lost by radiation.

Problems

For problems 1–5, solve the heat equation $u_t = a^2 u_{xx}$, $0 < x < \pi$, $0 < t$ subject to the boundary conditions that $u(0,t) = u(\pi,t) = 0$, $0 < t$ and the following initial conditions for $0 < x < \pi$:

1. $u(x,0) = A$, a constant

2. $u(x,0) = \sin^3(x) = [3\sin(x) - \sin(3x)]/4$

3. $u(x,0) = x$

4. $u(x,0) = \pi - x$

5. $u(x,0) = \begin{cases} x, & 0 < x < \pi/2 \\ \pi - x, & \pi/2 < x < \pi \end{cases}$

For problems 6–10, solve the heat equation $u_t = a^2 u_{xx}$, $0 < x < \pi$, $0 < t$ subject to the boundary conditions that $u_x(0,t) = u_x(\pi,t) = 0$, $0 < t$ and the following initial conditions for $0 < x < \pi$:

6. $u(x,0) = 1$

7. $u(x,0) = x$

8. $u(x,0) = \cos^2(x) = [1 + \cos(2x)]/2$

9. $u(x,0) = \pi - x$

10. $u(x,0) = \begin{cases} T_0, & 0 < x < \pi/2 \\ T_1, & \pi/2 < x < \pi \end{cases}$

For problems 11–17, solve the heat equation $u_t = a^2 u_{xx}$, $0 < x < \pi$, $0 < t$ subject to the following boundary conditions and initial condition:

11. $u_x(0,t) = u(\pi,t) = 0$, $0 < t$; $u(x,0) = x^2 - \pi^2$, $0 < x < \pi$

12. $u(0,t) = u(\pi,t) = T_0$, $0 < t$; $u(x,0) = T_1 \neq T_0$, $0 < x < \pi$

13. $u(0,t) = 0$, $u_x(\pi,t) = 0$, $0 < t$; $u(x,0) = 1$, $0 < x < \pi$

14. $u(0,t) = 0$, $u_x(\pi,t) = 0$, $0 < t$; $u(x,0) = x$, $0 < x < \pi$

15. $u(0,t) = 0$, $u_x(\pi,t) = 0$, $0 < t$; $u(x,0) = \pi - x$, $0 < x < \pi$

16. $u(0,t) = T_0$, $u_x(\pi,t) = 0$, $0 < t$; $u(x,0) = T_1 \neq T_0$, $0 < x < \pi$

17. $u(0,t) = 0, u(\pi,t) = T_0$, $0 < t$; $u(x,0) = T_0$, $0 < x < \pi$

18. It is well known that a room with masonry walls is often very difficult to heat. Consider a wall of thickness L, conductivity κ, and diffusivity a^2 which we heat at a constant rate H. The temperature of the outside (out-of-doors) face of the wall remains constant at T_0 and the entire wall initially has the uniform temperature T_0. Let us find the temperature of the inside face as a function of time.[9]

We begin by solving the heat conduction problem

$$\frac{\partial u}{\partial t} = a^2 \frac{\partial^2 u}{\partial x^2}, \qquad 0 < x < L, 0 < t$$

subject to the boundary conditions that

$$\frac{\partial u(0,t)}{\partial x} = -\frac{H}{\kappa} \qquad \text{and} \qquad u(L,t) = T_0$$

and the initial condition that $u(x,0) = T_0$. Show that the temperature field equals

$$u(x,t) = T_0 + \frac{HL}{\kappa}\left\{1 - \frac{x}{L} - \frac{8}{\pi^2}\sum_{n=1}^{\infty}\frac{1}{(2n-1)^2}\cos\left[\frac{(2n-1)\pi x}{2L}\right] \times \exp\left[-\frac{(2n-1)^2\pi^2a^2t}{4L^2}\right]\right\}.$$

Therefore, the rise of temperature at the interior wall $x = 0$ is

$$\frac{HL}{\kappa}\left\{1 - \frac{8}{\pi^2}\sum_{n=1}^{\infty}\frac{1}{(2n-1)^2}\exp\left[-\frac{(2n-1)^2\pi^2a^2t}{4L^2}\right]\right\}$$

or

$$\frac{8HL}{\kappa\pi^2}\sum_{n=1}^{\infty}\frac{1}{(2n-1)^2}\left\{1 - \exp\left[-\frac{(2n-1)^2\pi^2a^2t}{4L^2}\right]\right\}.$$

For $a^2t/L^2 \leq 1$ this last expression can be approximated[10] by $4Hat^{1/2}/\pi^{1/2}\kappa$. We thus see that the temperature will initially rise as the square

---

[9] Reproduced with acknowledgement to Taylor and Francis, Publishers, from Dufton, A. F., 1927: The warming of walls. *Philos. Mag., Ser. 7*, **4**, 888–889.

[10] Let us define the function

$$f(t) = \sum_{n=1}^{\infty}\frac{1 - \exp[-(2n-1)^2\pi^2a^2t/L^2]}{(2n-1)^2}.$$

root of time and diffusivity and inversely with conductivity. For an average rock $\kappa = 0.0042$ g/cm-s and $a^2 = 0.0118$ cm^2/s while for wood (Spruce) $\kappa = 0.0003$ g/cm-s and $a^2 = 0.0024$ cm^2/s.

The same set of equations applies to heat transfer within a transistor operating at low frequencies.[11] At the junction ($x = 0$) heat is produced at the rate of H and flows to the transistor's supports ($x = \pm L$) where it is removed. The supports are maintained at the temperature T_0 which is also the initial temperature of the transistor.

19. The linearized Boussinesq equation[12]

$$\frac{\partial u}{\partial t} = \frac{\partial^2 u}{\partial x^2}, \qquad 0 < x < L, 0 < t$$

governs the height of the water table $u(x,t)$ above some reference point, where a^2 is the product of the storage coefficient times the hydraulic coefficient divided by the aquifer thickness. A typical value of a^2 is 10 m^2/min. Consider the problem of a strip of land of width L that separates two reservoirs of depth h_1. Initially the height of the water table would be h_1. Suddenly we lower the reservoir on the right $x = L$ to a depth h_2 [$u(0,t) = h_1$, $u(L,t) = h_2$, and $u(x,0) = h_1$]. Find the

---

Then

$$f'(t) = \frac{a^2\pi^2}{L^2} \sum_{n=1}^{\infty} \exp[-(2n-1)^2\pi^2 a^2 t/L^2].$$

Consider now the integral

$$\int_0^\infty \exp\left(-\frac{a^2\pi^2 t}{L^2} x^2\right)\, dx = \frac{L}{2a\sqrt{\pi t}}.$$

If we approximate this integral by using the trapezoidal rule with $\Delta x = 2$, then

$$\int_0^\infty \exp\left(-\frac{a^2\pi^2 t}{L^2} x^2\right)\, dx \approx 2 \sum_{n=1}^{\infty} \exp[-(2n-1)^2\pi^2 a^2 t/L^2]$$

and $f'(t) \approx a\pi^{3/2}/(4Lt^{1/2})$. Integrating and using $f(0) = 0$, we finally have $f(t) \approx a\pi^{3/2}t^{1/2}/(2L)$. The smaller a^2t/L^2 is, the smaller the error will be. For example, if $t = L^2/a^2$, then the error is 2.4 %.

[11] Mortenson, K. E., 1957: Transistor junction temperature as a function of time. *Proc. IRE*, **45**, 504–513. Eq. (2a) should read $T_x = -F/k$.

[12] See, for example, Van Schilfgaarde, J., 1970: Theory of flow to drains in *Advances in Hydroscience*, Academic Press, New York, pp. 81–85.

height of the water table at any position x within the aquifer and any time $t > 0$.

20. The equation (see Problem 19)

$$\frac{\partial u}{\partial t} = \frac{\partial^2 u}{\partial x^2}, \qquad 0 < x < L, 0 < t$$

governs the height of the water table $u(x,t)$. Consider the problem[13] of a piece of land that suddenly has two drains placed at the points $x = 0$ and $x = L$ so that $u(0,t) = u(L,t) = 0$. If the water table initially has the profile:

$$u(x,0) = 8H(L^3x - 3L^2x^2 + 4Lx^3 - 2x^4)/L^4,$$

find the height of the water table at any point within the aquifer and any time $t > 0$.

21. We want to find the rise of the water table of an aquifer which we sandwich between a canal and impervious rocks if we suddenly raise the water level in the canal h_0 units above its initial elevation and then maintain the canal at this level. The linearized Boussinesq equation (see Problem 19)

$$\frac{\partial u}{\partial t} = \frac{\partial^2 u}{\partial x^2}, \qquad 0 < x < L, 0 < t$$

governs the level of the water table with the boundary conditions $u(0,t) = h_0$ and $u_x(L,t) = 0$ and the initial condition $u(x,0) = 0$. Find the height of the water table at any point in the aquifer and any time $t > 0$.

22. Solve the nonhomogeneous heat equation

$$\frac{\partial u}{\partial t} - a^2\frac{\partial^2 u}{\partial x^2} = e^{-x}, \qquad 0 < x < \pi, 0 < t$$

subject to the boundary conditions $u(0,t) = u_x(\pi,t) = 0$, $0 < t$, and the initial condition $u(x,0) = f(x)$, $0 < x < \pi$.

23. Solve the nonhomogeneous heat equation

$$\frac{\partial u}{\partial t} - \frac{\partial^2 u}{\partial x^2} = -1, \qquad 0 < x < 1, 0 < t$$

[13] For a similar problem, see Dumm, L. D., 1954: New formula for determining depth and spacing of subsurface drains in irrigated lands. *Agric. Eng.*, **35**, 726–730.

subject to the boundary conditions $u_x(0,t) = u_x(1,t) = 0$, $0 < t$, and the initial condition $u(x,0) = \frac{1}{2}(1-x^2)$, $0 < x < 1$. [Hint: Note that any function of time satisfies the boundary conditions.]

24. Solve the nonhomogeneous heat equation

$$\frac{\partial u}{\partial t} - a^2 \frac{\partial^2 u}{\partial x^2} = A\cos(\omega t), \qquad 0 < x < \pi, 0 < t$$

subject to the boundary conditions $u_x(0,t) = u_x(\pi,t) = 0$, $0 < t$, and the initial condition $u(x,0) = f(x)$, $0 < x < \pi$. [Hint: Note that any function of time satisfies the boundary conditions.]

25. Solve the nonhomogeneous heat equation

$$\frac{\partial u}{\partial t} - \frac{\partial^2 u}{\partial x^2} = \begin{cases} x, & 0 < x \leq \pi/2 \\ \pi - x, & \pi/2 \leq x < \pi, \end{cases} \qquad 0 < x < \pi, 0 < t$$

subject to the boundary conditions $u(0,t) = u(\pi,t) = 0$, $0 < t$, and the initial condition $u(x,0) = 0$, $0 < x < \pi$. [Hint: Represent the forcing function as a half-range Fourier sine expansion over the interval $(0,\pi)$.]

26. A uniform, conducting rod of length L and thermometric diffusivity a^2 is initially at temperature zero. We supply heat uniformly throughout the rod so that the heat conduction equation is

$$a^2 \frac{\partial^2 u}{\partial x^2} = \frac{\partial u}{\partial t} - P, \qquad 0 < x < L, 0 < t,$$

where P is the rate at which the temperature would rise if there was no conduction. If we maintain the ends of the rod at the temperature of zero, find the temperature at any position and subsequent time.

27. Solve the nonhomogeneous heat equation

$$\frac{\partial u}{\partial t} = a^2 \frac{\partial^2 u}{\partial x^2} + \frac{A_0}{c\rho} \qquad 0 < x < L, 0 < t,$$

where $a^2 = \kappa/c\rho$ with the boundary conditions that

$$\frac{\partial u(0,t)}{\partial x} = 0 \quad \text{and} \quad \kappa \frac{\partial u(L,t)}{\partial x} + hu(L,t) = 0, \qquad 0 < t$$

and the initial condition that $u(x,0) = 0$, $0 < x < L$.

28. Find the solution of

$$\frac{\partial u}{\partial t} = \frac{\partial^2 u}{\partial x^2} - u, \qquad 0 < x < L, 0 < t$$

with the boundary conditions $u(0,t) = 1$ and $u(L,t) = 0$, $0 < t$, and the initial condition $u(x,0) = 0$, $0 < x < L$.

29. Solve the heat equation in spherical coordinates

$$\frac{\partial u}{\partial t} = \frac{a^2}{r^2}\frac{\partial}{\partial r}\left(r^2\frac{\partial u}{\partial r}\right), \qquad 0 \le r < 1, 0 < t$$

subject to the boundary conditions $\lim_{r\to 0}|u(r,t)| < \infty$ and $u(1,t) = 0$, $0 < t$, and the initial condition $u(r,0) = 1$, $0 \le r < 1$.

30. Solve the heat equation in cylindrical coordinates

$$\frac{\partial u}{\partial t} = \frac{a^2}{r}\frac{\partial}{\partial r}\left(r\frac{\partial u}{\partial r}\right), \qquad 0 \le r < b, 0 < t$$

subject to the boundary conditions $\lim_{r\to 0}|u(r,t)| < \infty$ and $u(b,t) = \theta$, $0 < t$, and the initial condition $u(r,0) = 1$, $0 \le r < b$.

31. The equation[14]

$$\frac{\partial u}{\partial t} = \frac{G}{\rho} + \nu\left(\frac{\partial^2 u}{\partial r^2} + \frac{1}{r}\frac{\partial u}{\partial r}\right), \qquad 0 \le r < b, 0 < t$$

governs the velocity $u(r,t)$ of an incompressible fluid of density ρ and kinematic viscosity ν flowing in a long circular pipe of radius b with an imposed, constant pressure gradient $-G$. If the fluid is initially at rest $u(r,0) = 0$, $0 \le r < b$, and there is no slip at the wall $u(b,t) = 0$, $0 < t$, find the velocity at any subsequent time and position.

32. Solve the heat equation in cylindrical coordinates

$$\frac{\partial u}{\partial t} = \frac{a^2}{r}\frac{\partial}{\partial r}\left(r\frac{\partial u}{\partial r}\right), \qquad 0 \le r < b, 0 < t$$

subject to the boundary conditions $\lim_{r\to 0}|u(r,t)| < \infty$ and $u_r(b,t) = -hu(b,t)$, $0 < t$, and the initial condition $u(r,0) = b^2 - r^2$, $0 \le r < b$.

---

[14] From Szymanski, P., 1932: Quelques solutions exactes des équations de l'hydrodynamique du fluide visqueux dans le cas d'un tube cylindrique. *J. math. pures appl., Ser. 9*, **11**, 67–107. ©Gauthier-Villars

33. In their study of heat conduction within a thermocouple through which a steady current flows, Reich and Madigan[15] solved the following nonhomogeneous heat conduction problem:

$$\frac{\partial u}{\partial t} - a^2 \frac{\partial^2 u}{\partial x^2} = J - P\,\delta(x-b), \qquad 0 < x < L, 0 < t, 0 < b < L,$$

where J represents the Joule heating generated by the steady current and the P term represents the heat loss from Peltier cooling.[16] Find $u(x,t)$ if both ends are kept at zero $[u(0,t) = u(L,t) = 0]$ and initially the temperature is zero $[u(x,0) = 0]$. The interesting aspect of this problem is the presence of the delta function.

Step 1: Assuming that $u(x,t)$ equals the sum of a steady-state solution $w(x)$ and a transient solution $v(x,t)$, show that the steady-state solution is governed by

$$a^2 \frac{d^2 w}{dx^2} = P\,\delta(x-b) - J, \qquad w(0) = w(L) = 0.$$

Step 2: Show that the steady-state solution is

$$w(x) = \begin{cases} Jx(L-x)/2a^2 + Ax, & 0 < x < b \\ Jx(L-x)/2a^2 + B(L-x), & b < x < L. \end{cases}$$

Step 3: The temperature must be continuous at $x = b$; otherwise, we would have infinite heat conduction there. Use this condition to show that $Ab = B(L-b)$.

Step 4: To find a second relationship between A and B, integrate the steady-state differential equation across the interface at $x = b$ and show that

$$\lim_{\epsilon \to 0} a^2 \left. \frac{dw}{dx} \right|_{b-\epsilon}^{b+\epsilon} = P.$$

Step 5: Using the result from Step 4, show that $A + B = -P/a^2$ and

$$w(x) = \begin{cases} Jx(L-x)/2a^2 - Px(L-b)/a^2 L, & 0 < x < b \\ Jx(L-x)/2a^2 - Pb(L-x)/a^2 L, & b < x < L. \end{cases}$$

---

[15] Reich, A. D. and Madigan, J. R., 1961: Transient response of a thermocouple circuit under steady currents. *J. Appl. Phys.*, **32**, 294–301.

[16] In 1834 Jean Charles Athanase Peltier (1785–1845) discovered that there is a heating or cooling effect, quite apart from ordinary resistance heating, whenever an electric current flows through the junction between two different metals.

Step 6: Reexpress $w(x)$ as a half-range Fourier sine expansion and show that

$$w(x) = \frac{4JL^2}{a^2\pi^3}\sum_{m=1}^{\infty}\frac{\sin[(2m-1)\pi x/L]}{(2m-1)^3} - \frac{2LP}{a^2\pi^2}\sum_{n=1}^{\infty}\frac{\sin(n\pi b/L)\sin(n\pi x/L)}{n^2}.$$

Step 7: Use separation of variables to find the transient solution by solving

$$\frac{\partial v}{\partial t} = a^2\frac{\partial^2 v}{\partial x^2}, \qquad 0 < x < L, 0 < t$$

subject to the boundary conditions $v(0,t) = v(L,t) = 0$, $0 < t$, and the initial condition $v(x,0) = -w(x), 0 < x < L$.

Step 8: Add the steady-state and transient solutions together and show that

$$u(x,t) = \frac{4JL^2}{a^2\pi^3}\sum_{m=1}^{\infty}\frac{\sin[(2m-1)\pi x/L]}{(2m-1)^3}\left[1 - e^{-a^2(2m-1)^2\pi^2 t/L^2}\right] - \frac{2LP}{a^2\pi^2}\sum_{n=1}^{\infty}\frac{\sin(n\pi b/L)\sin(n\pi x/L)}{n^2}\left[1 - e^{-a^2n^2\pi^2 t/L^2}\right].$$

8.4 THE LAPLACE TRANSFORM METHOD

In the previous chapter we showed that we may solve the wave equation by the method of Laplace transforms. This is also true for the heat equation. Once again, we take the Laplace transform with respect to time. From the definition of Laplace transforms,

$$\mathcal{L}[u(x,t)] = U(x,s), \tag{8.4.1}$$

$$\mathcal{L}[u_t(x,t)] = sU(x,s) - u(x,0) \tag{8.4.2}$$

and

$$\mathcal{L}[u_{xx}(x,t)] = \frac{d^2U(x,s)}{dx^2}. \tag{8.4.3}$$

We next solve the resulting ordinary differential equation, known as the *auxiliary equation*, along with the corresponding Laplace transformed boundary conditions. The initial condition gives us the value of $u(x,0)$. The final step is the inversion of the Laplace transform $U(x,s)$. We typically use the inversion integral.

• **Example 8.4.1**

To illustrate these concepts, we solve a heat conduction problem[17] in a plane slab of thickness $2L$. Initially the slab has a constant temperature of unity. For $0 < t$ we allow both faces of the slab to radiatively cool in a medium which has a temperature of zero.

If $u(x,t)$ denotes the temperature, a^2 is the thermal diffusivity, h is the relative emissivity, t is the time, and x is the distance perpendicular to the face of the slab and measured from the middle of the slab, then the governing equation is

$$\frac{\partial u}{\partial t} = a^2 \frac{\partial^2 u}{\partial x^2}, \qquad -L < x < L, 0 < t \tag{8.4.4}$$

with the initial condition

$$u(x,0) = 1, \qquad -L < x < L \tag{8.4.5}$$

and boundary conditions

$$\frac{\partial u(L,t)}{\partial x} + hu(L,t) = 0 \quad \text{and} \quad \frac{\partial u(-L,t)}{\partial x} + hu(-L,t) = 0, \quad 0 < t. \tag{8.4.6}$$

Taking the Laplace transform of (8.4.4) and substituting the initial condition,

$$a^2 \frac{d^2 U(x,s)}{dx^2} - sU(x,s) = -1. \tag{8.4.7}$$

If we write $s = a^2 q^2$, (8.4.7) becomes

$$\frac{d^2 U(x,s)}{dx^2} - q^2 U(x,s) = -\frac{1}{a^2}. \tag{8.4.8}$$

From the boundary conditions $U(x,s)$ is an even function in x and we may conveniently write the solution as

$$U(x,s) = \frac{1}{s} + A\cosh(qx). \tag{8.4.9}$$

From (8.4.6),

$$qA\sinh(qL) + \frac{h}{s} + hA\cosh(qL) = 0 \tag{8.4.10}$$

and

$$U(x,s) = \frac{1}{s} - \frac{h\cosh(qx)}{s[q\sinh(qL) + h\cosh(qL)]}. \tag{8.4.11}$$

---

[17] Goldstein, S., 1932: The application of Heaviside's operational method to the solution of a problem in heat conduction. *Zeit. Angew. Math. Mech.*, **12**, 234–243.

The inverse of $U(x,s)$ consists of two terms. The inverse of the first term is simply unity. We will invert the second term by contour integration.

We begin by examining the nature and location of the singularities in the second term. Using the product formulas for the hyperbolic cosine and sine functions, the second term equals

$$\frac{h\left(1+\frac{4q^2x^2}{\pi^2}\right)\left(1+\frac{4q^2x^2}{9\pi^2}\right)\cdots}{s\left[q^2L\left(1+\frac{q^2L^2}{\pi^2}\right)\left(1+\frac{q^2L^2}{4\pi^2}\right)\cdots+h\left(1+\frac{4q^2L^2}{\pi^2}\right)\left(1+\frac{4q^2L^2}{9\pi^2}\right)\cdots\right]}. \tag{8.4.12}$$

Because $q^2 = s/a^2$, (8.4.12) shows that we do not have any $\sqrt{s}$ in the transform and we need not concern ourselves with branch points and cuts. Furthermore, we have only simple poles: one located at $s = 0$ and the others where

$$q\sinh(qL) + h\cosh(qL) = 0. \tag{8.4.13}$$

If we set $q = i\lambda$, (8.4.13) becomes

$$h\cos(\lambda L) - \lambda\sin(\lambda L) = 0 \tag{8.4.14}$$

or

$$\lambda L\tan(\lambda L) = hL. \tag{8.4.15}$$

From Bromwich's integral,

$$\mathcal{L}^{-1}\left\{\frac{h\cosh(qx)}{s[q\sinh(qL)+h\cosh(qL)]}\right\} = \frac{1}{2\pi i}\oint_C \frac{h\cosh(qx)e^{tz}}{z[q\sinh(qL)+h\cosh(qL)]}\,dz, \tag{8.4.16}$$

where $q = z^{1/2}/a$ and the closed contour C consists of Bromwich's contour plus a semicircle of infinite radius in the left half of the z-plane. The residue at $z = 0$ is 1 while at $z_n = -a^2\lambda_n^2$,

$$\operatorname{Res}\left\{\frac{h\cosh(qx)e^{tz}}{z[q\sinh(qL)+h\cosh(qL)]}; z_n\right\}$$

$$= \lim_{z\to z_n}\frac{h(z+a^2\lambda_n^2)\cosh(qx)e^{tz}}{z[q\sinh(qL)+h\cosh(qL)]} \tag{8.4.17}$$

$$= \lim_{z\to z_n}\frac{h\cosh(qx)e^{tz}}{z[(1+hL)\sinh(qL)+h\cosh(qL)]/(2a^2q)} \tag{8.4.18}$$

$$= \frac{2ha^2\lambda_n i\cosh(i\lambda_n x)\exp(-\lambda_n^2a^2t)}{(-a^2\lambda_n^2)[(1+hL)i\sin(\lambda_n L)+i\lambda_n L\cos(\lambda_n L)]} \tag{8.4.19}$$

$$= -\frac{2h\cos(\lambda_n x)\exp(-a^2\lambda_n^2t)}{\lambda_n[(1+hL)\sin(\lambda_n L)+\lambda_n L\cos(\lambda_n L)]}. \tag{8.4.20}$$

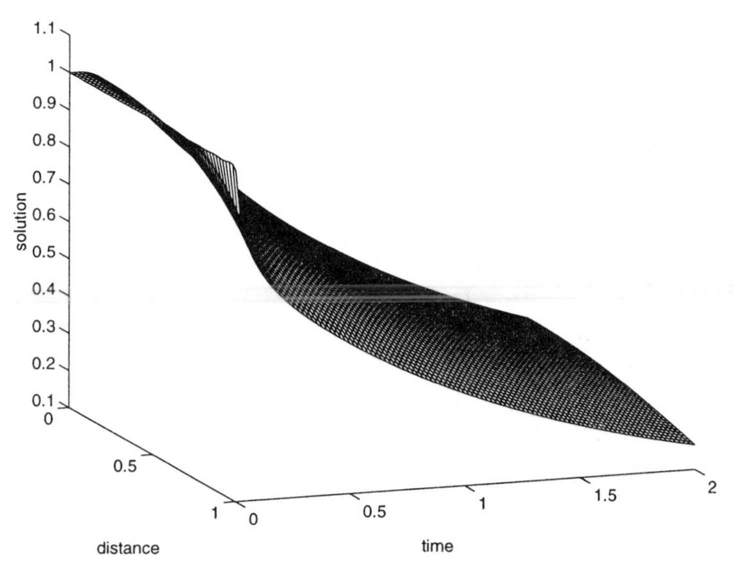

Figure 8.4.1: The temperature within the portion of a slab $0 < x/L < 1$ at various times a^2t/L^2 if the faces of the slab radiate to free space at temperature zero and the slab initially has the temperature 1. The parameter $hL = 1$.

Therefore, the inversion of $U(x,s)$ is

$$u(x,t) = 1 - \left\{1 - 2h \sum_{n=1}^{\infty} \frac{\cos(\lambda_n x)\exp(-a^2\lambda_n^2 t)}{\lambda_n[(1+hL)\sin(\lambda_n L) + \lambda_n L\cos(\lambda_n L)]}\right\} \tag{8.4.21}$$

or

$$u(x,t) = 2h \sum_{n=1}^{\infty} \frac{\cos(\lambda_n x)\exp(-a^2\lambda_n^2 t)}{\lambda_n[(1+hL)\sin(\lambda_n L) + \lambda_n L\cos(\lambda_n L)]}. \tag{8.4.22}$$

We can further simplify (8.4.22) by using $h/\lambda_n = \tan(\lambda_n L)$ and $hL = \lambda_n L \tan(\lambda_n L)$. Substituting these relationships into (8.4.22) and simplifying,

$$u(x,t) = 2 \sum_{n=1}^{\infty} \frac{\sin(\lambda_n L)\cos(\lambda_n x)\exp(-a^2\lambda_n^2 t)}{\lambda_n L + \sin(\lambda_n L)\cos(\lambda_n L)}. \tag{8.4.23}$$

• **Example 8.4.2: Heat dissipation in disc brakes**

Disc brakes consist of two blocks of frictional material known as pads which press against each side of a rotating annulus, usually made of a ferrous material. In this problem we determine the transient temperatures reached in a disc brake during a single brake application.[18] If we ignore the errors introduced by replacing the cylindrical portion of the drum by a rectangular plate, we can model our disc brakes as a one-dimensional solid which friction heats at both ends. Assuming symmetry about $x = 0$, the boundary condition there is $u_x(0,t) = 0$. To model the heat flux from the pads, we assume a uniform disc deceleration that generates heat from the frictional surfaces at the rate $N(1 - Mt)$, where M and N are experimentally determined constants.

If $u(x,t)$, κ and a^2 denote the temperature, thermal conductivity, and diffusivity of the rotating annulus, respectively, then the heat equation is

$$\frac{\partial u}{\partial t} = a^2 \frac{\partial^2 u}{\partial x^2}, \qquad 0 < x < L, 0 < t \qquad \textbf{(8.4.24)}$$

with the boundary conditions

$$\frac{\partial u(0,t)}{\partial x} = 0 \quad \text{and} \quad \kappa \frac{\partial u(L,t)}{\partial x} = N(1 - Mt), \qquad 0 < t. \qquad \textbf{(8.4.25)}$$

The boundary condition at $x = L$ gives the frictional heating of the disc pads.

Introducing the Laplace transform of $u(x,t)$, defined as

$$U(x,s) = \int_0^\infty u(x,t) e^{-st}\, dt, \qquad \textbf{(8.4.26)}$$

the equation to be solved becomes

$$\frac{d^2 U}{dx^2} - \frac{s}{a^2} U = 0, \qquad \textbf{(8.4.27)}$$

subject to the boundary conditions that

$$\frac{dU(0,s)}{dx} = 0 \quad \text{and} \quad \frac{dU(L,s)}{dx} = \frac{N}{\kappa}\left(\frac{1}{s} - \frac{M}{s^2}\right). \qquad \textbf{8.4.28)}$$

The solution of (8.4.27) is

$$U(x,s) = A\cosh(qx) + B\sinh(qx), \qquad \textbf{(8.4.29)}$$

[18] From Newcomb, T. P., 1958: The flow of heat in a parallel-faced infinite solid. *Br. J. Appl. Phys.*, **9**, 370–372. See also Newcomb, T. P., 1958/59: Transient temperatures in brake drums and linings. *Proc. Inst. Mech. Eng., Auto. Div.*, 227–237; Newcomb, T. P., 1959: Transient temperatures attained in disk brakes. *Br. J. Appl. Phys.*, **10**, 339–340.

where $q = s^{1/2}/a$. Using the boundary conditions, the solution becomes

$$U(x,s) = \frac{N}{\kappa}\left(\frac{1}{s} - \frac{M}{s^2}\right)\frac{\cosh(qx)}{q\sinh(qL)}. \tag{8.4.30}$$

It now remains to invert the transform (8.4.30). We will invert $\cosh(qx)/[sq\sinh(qL)]$; the inversion of the second term follows by analog.

Our first concern is the presence of $s^{1/2}$ because this is a multivalued function. However, when we replace the hyperbolic cosine and sine functions with their Taylor expansions, $\cosh(qx)/[sq\sinh(qL)]$ contains only powers of s and is, in fact, a single-valued function.

From Bromwich's integral,

$$\mathcal{L}^{-1}\left[\frac{\cosh(qx)}{sq\sinh(qL)}\right] = \frac{1}{2\pi i}\int_{c-\infty i}^{c+\infty i}\frac{\cosh(qx)e^{tz}}{zq\sinh(qL)}\,dz, \tag{8.4.31}$$

where $q = z^{1/2}/a$. Just as in the previous example, we replace the hyperbolic cosine and sine with their product expansion and find that $z = 0$ is a second-order pole. The remaining poles are located where $z_n^{1/2}L/a = n\pi i$ or $z_n = -n^2\pi^2a^2/L^2$, where $n = 1, 2, 3, \ldots$. We have chosen the positive sign because $z^{1/2}$ must be single-valued; if we had chosen the negative sign the answer would have been the same. Our expansion also shows that the poles are simple.

Having classified the poles, we now close Bromwich's contour, which lies slightly to the right of the imaginary axis, with an infinite semicircle in the left half-plane, and use the residue theorem. The values of the residues are

$$\operatorname{Res}\left[\frac{\cosh(qx)e^{tz}}{zq\sinh(qL)};0\right]$$

$$= \frac{1}{1!}\lim_{z\to 0}\frac{d}{dz}\left\{\frac{(z-0)^2\cosh(qx)e^{tz}}{zq\sinh(qL)}\right\} \tag{8.4.32}$$

$$= \lim_{z\to 0}\frac{d}{dz}\left\{\frac{z\cosh(qx)e^{tz}}{q\sinh(qL)}\right\} \tag{8.4.33}$$

$$= \frac{a^2}{L}\lim_{z\to 0}\frac{d}{dz}\left\{\frac{z\left[1+\frac{zx^2}{2!a^2}+\cdots\right]\left[1+tz+\frac{t^2z^2}{2!}+\cdots\right]}{z+\frac{L^2z^2}{3!a^2}+\cdots}\right\} \tag{8.4.34}$$

$$= \frac{a^2}{L}\lim_{z\to 0}\frac{d}{dz}\left\{1+tz+\frac{zx^2}{2a^2}-\frac{zL^2}{3!a^2}+\cdots\right\} \tag{8.4.35}$$

$$= \frac{a^2}{L}\left\{t+\frac{x^2}{2a^2}-\frac{L^2}{6a^2}\right\} \tag{8.4.36}$$

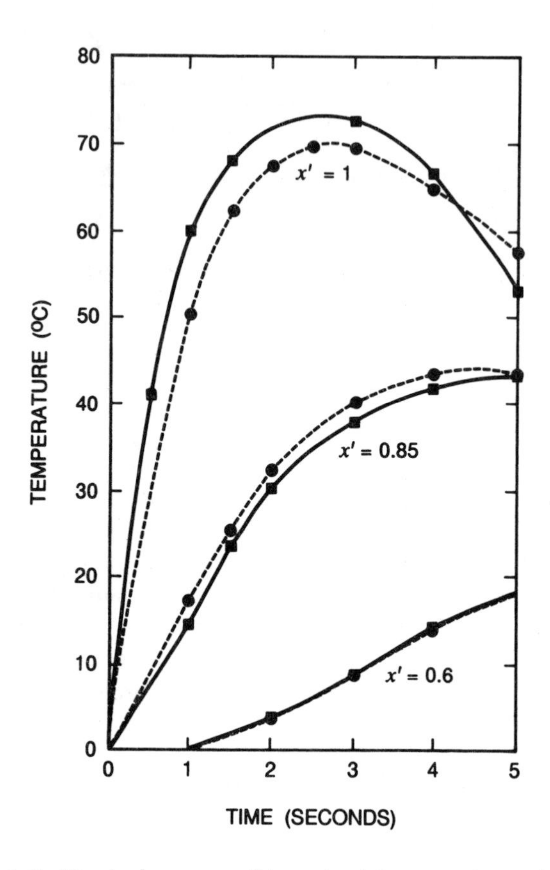

Figure 8.4.2: Typical curves of transient temperature at different locations in a brake lining. Circles denote computed values while squares are experimental measurements. (From Newcomb, T. P., 1958: The flow of heat in a parallel-faced infinite solid. *Br. J. Appl. Phys.*, **9**, 372 with permission.)

and

$$\operatorname{Res}\left[\frac{\cosh(qx)e^{tz}}{zq\sinh(qL)}; z_n\right] = \left[\lim_{z\to z_n} \frac{\cosh(qx)}{zq}e^{tz}\right]\left[\lim_{z\to z_n} \frac{z-z_n}{\sinh(qL)}\right] \quad (8.4.37)$$

$$= \lim_{z\to z_n} \frac{\cosh(qx)e^{tz}}{zq\cosh(qL)L/(2a^2q)} \quad (8.4.38)$$

$$= \frac{\cosh(n\pi xi/L)\exp(-n^2\pi^2a^2t/L^2)}{(-n^2\pi^2a^2/L^2)\cosh(n\pi i)L/(2a^2)} \quad (8.4.39)$$

$$= -\frac{2L(-1)^n}{n^2\pi^2}\cos(n\pi x/L)e^{-n^2\pi^2a^2t/L^2}. \quad (8.4.40)$$

When we sum all of the residues from both inversions, the solution is

$$
\begin{aligned}
u(x,t) = {} & \frac{a^2 N}{\kappa L}\left\{t + \frac{x^2}{2a^2} - \frac{L^2}{6a^2}\right\} \\
& - \frac{2LN}{\kappa\pi^2}\sum_{n=1}^{\infty}\frac{(-1)^n}{n^2}\cos(n\pi x/L)e^{-n^2\pi^2a^2t/L^2} \\
& - \frac{a^2NM}{\kappa L}\left\{\frac{t^2}{2} + \frac{tx^2}{2a^2} - \frac{tL^2}{6a^2} + \frac{x^4}{24a^4} - \frac{x^2L^2}{12a^4} + \frac{7L^4}{360a^4}\right\} \\
& - \frac{2L^3NM}{a^2\kappa\pi^4}\sum_{n=1}^{\infty}\frac{(-1)^n}{n^4}\cos(n\pi x/L)e^{-n^2\pi^2a^2t/L^2}. \qquad (8.4.41)
\end{aligned}
$$

Figure 8.4.2 shows the temperature in the brake lining at various places within the lining $[x' = x/L]$ if $a^2 = 3.3 \times 10^{-3}$ cm^2/sec, $\kappa = 1.8 \times 10^{-3}$ cal/(cm sec°C), $L = 0.48$ cm and $N = 1.96$ cal/(cm^2 sec). Initially the frictional heating results in an increase in the disc brake's temperature. As time increases, the heating rate decreases and radiative cooling becomes sufficiently large that the temperature begins to fall.

- **Example 8.4.3**

In the previous example we showed that Laplace transforms are particularly useful when the boundary conditions are time dependent. Consider now the case when one of the boundaries is moving.

We wish to solve the heat equation

$$\frac{\partial u}{\partial t} = a^2\frac{\partial^2 u}{\partial x^2}, \qquad \beta t < x < \infty, 0 < t \qquad (8.4.42)$$

subject to the boundary conditions

$$u(x,t)\big|_{x=\beta t} = f(t) \quad \text{and} \quad \lim_{x\to\infty}|u(x,t)| < \infty, \qquad 0 < t \qquad (8.4.43)$$

and the initial condition

$$u(x,0) = 0, \qquad 0 < x < \infty. \qquad (8.4.44)$$

This type of problems arises in combustion problems where the boundary moves due to the burning of the fuel.

We begin by introducing the coordinate $\eta = x - \beta t$. Then the problem can be reformulated as

$$\frac{\partial u}{\partial t} - \beta\frac{\partial u}{\partial \eta} = a^2\frac{\partial^2 u}{\partial \eta^2}, \qquad 0 < \eta < \infty, 0 < t \qquad (8.4.45)$$

subject to the boundary conditions

$$u(0,t) = f(t) \quad \text{and} \quad \lim_{\eta\to\infty} |u(\eta,t)| < \infty, \qquad 0 < t \tag{8.4.46}$$

and the initial condition

$$u(\eta,0) = 0, \qquad 0 < \eta < \infty. \tag{8.4.47}$$

Taking the Laplace transform of (8.4.45), we have that

$$\frac{d^2U(\eta,s)}{d\eta^2} + \frac{\beta}{a^2}\frac{dU(\eta,s)}{d\eta} - \frac{s}{a^2}U(\eta,s) = 0 \tag{8.4.48}$$

with

$$U(0,s) = F(s) \quad \text{and} \quad \lim_{\eta\to\infty} |U(\eta,s)| < \infty. \tag{8.4.49}$$

The solution to (8.4.48)–(8.4.49) is

$$U(\eta,s) = F(s)\exp\left(-\frac{\beta\eta}{2a^2} - \frac{\eta}{a}\sqrt{s + \frac{\beta^2}{4a^2}}\right). \tag{8.4.50}$$

Because

$$\mathcal{L}\left[\Phi(\eta,t)\right] = \exp\left(-\frac{\eta}{a}\sqrt{s + \frac{\beta^2}{4a^2}}\right), \tag{8.4.51}$$

where

$$\Phi(\eta,t) = \tfrac{1}{2}\left[e^{-\beta\eta/2a^2}\operatorname{erfc}\left(\frac{\eta}{2a\sqrt{t}} - \frac{\beta\sqrt{t}}{2a}\right) + e^{\beta\eta/2a^2}\operatorname{erfc}\left(\frac{\eta}{2a\sqrt{t}} + \frac{\beta\sqrt{t}}{2a}\right)\right] \tag{8.4.52}$$

and

$$\operatorname{erfc}(x) = 1 - \frac{2}{\sqrt{\pi}}\int_0^x e^{-\eta^2}\,d\eta, \tag{8.4.53}$$

we have by the convolution theorem that

$$u(\eta,t) = e^{-\beta\eta/2a^2}\int_0^t f(t-\tau)\Phi(\eta,\tau)\,d\tau \tag{8.4.54}$$

or

$$u(x,t) = e^{-\beta(x-\beta t)/2a^2}\int_0^t f(t-\tau)\Phi(x-\beta\tau,\tau)\,d\tau. \tag{8.4.55}$$

Problems

1. Solve

$$\frac{\partial u}{\partial t} = \frac{\partial^2 u}{\partial x^2} - a^2(u - T_0), \qquad 0 < x < 1, 0 < t$$

subject to the boundary conditions

$$\frac{\partial u(0,t)}{\partial x} = \frac{\partial u(1,t)}{\partial x} = 0, \qquad 0 < t$$

and the initial condition

$$u(x,0) = 0, \qquad 0 < x < 1.$$

2. Solve

$$\frac{\partial u}{\partial t} = \frac{\partial^2 u}{\partial x^2}, \qquad 0 < x < 1, 0 < t$$

subject to the boundary conditions

$$\frac{\partial u(0,t)}{\partial x} = 0, \quad u(1,t) = t, \quad 0 < t$$

and the initial condition

$$u(x,0) = 0, \qquad 0 < x < 1.$$

3. Solve

$$\frac{\partial u}{\partial t} = \frac{\partial^2 u}{\partial x^2}, \qquad 0 < x < 1, 0 < t$$

subject to the boundary conditions

$$u(0,t) = 0, \quad u(1,t) = 1, \quad 0 < t$$

and the initial condition

$$u(x,0) = 0, \qquad 0 < x < 1.$$

4. Solve

$$\frac{\partial u}{\partial t} = \frac{\partial^2 u}{\partial x^2}, \qquad -\tfrac{1}{2} < x < \tfrac{1}{2}, 0 < t$$

subject to the boundary conditions

$$u_x\left(-\tfrac{1}{2}, t\right) = 0, \quad u_x\left(\tfrac{1}{2}, t\right) = \delta(t), \qquad 0 < t$$

and the initial condition

$$u(x,0) = 0, \qquad -\tfrac{1}{2} < x < \tfrac{1}{2}.$$

5. Solve

$$\frac{\partial u}{\partial t} - \frac{\partial^2 u}{\partial x^2} = 1, \qquad 0 < x < 1, 0 < t$$

subject to the boundary conditions

$$u(0,t) = u(1,t) = 0, \qquad 0 < t$$

and the initial condition

$$u(x,0) = 0, \qquad 0 < x < 1.$$

6. Solve[19]

$$\frac{\partial u}{\partial t} = a^2 \frac{\partial^2 u}{\partial x^2}, \qquad 0 < x < \infty, 0 < t$$

subject to the boundary conditions

$$u(0,t) = 1, \quad \lim_{x\to\infty} |u(x,t)| < \infty, \quad 0 < t$$

and the initial condition

$$u(x,0) = 0, \qquad 0 < x < \infty.$$

[Hint: Use tables to invert the Laplace transform.]

7. Solve

$$\frac{\partial u}{\partial t} = \frac{\partial^2 u}{\partial x^2}, \qquad 0 < x < \infty, 0 < t$$

subject to the boundary conditions

$$\frac{\partial u(0,t)}{\partial x} = 1, \quad \lim_{x\to\infty} |u(x,t)| < \infty, \quad 0 < t$$

and the initial condition

$$u(x,0) = 0, \qquad 0 < x < \infty.$$

---

[19] If $u(x,t)$ denotes the Eulerian velocity of a viscous fluid in the half space $x > 0$ and parallel to the wall located at $x = 0$, then this problem was first solved by Stokes, G. G., 1850: On the effect of the internal friction of fluids on the motions of pendulums. *Proc. Cambridge Philos. Soc.*, **9**, **Part II**, [8]–[106].

[Hint: Use tables to invert the Laplace transform.]

8. Solve

$$\frac{\partial u}{\partial t} = \frac{\partial^2 u}{\partial x^2}, \qquad 0 < x < \infty, 0 < t$$

subject to the boundary conditions

$$u(0,t) = 1, \quad \lim_{x\to\infty} |u(x,t)| < \infty, \quad 0 < t$$

and the initial condition

$$u(x,0) = e^{-x}, \qquad 0 < x < \infty.$$

[Hint: Use tables to invert the Laplace transform.]

9. Solve

$$\frac{\partial u}{\partial t} = a^2 \left[\frac{\partial^2 u}{\partial x^2} + (1+\delta)\frac{\partial u}{\partial x} + \delta u \right], \qquad 0 < x < \infty, 0 < t,$$

where δ is a constant, subject to the boundary conditions

$$u(0,t) = u_0, \quad \lim_{x\to\infty} |u(x,t)| < \infty, \quad 0 < t$$

and the initial condition

$$u(x,0) = 0, \qquad 0 < x < \infty.$$

Note that

$$\mathcal{L}^{-1}\left[\frac{1}{s}\exp\left(-2\alpha\sqrt{s+\beta^2}\,\right)\right] = \tfrac{1}{2}e^{2\alpha\beta}\operatorname{erfc}\left(\frac{\alpha}{\sqrt{t}} + \beta\sqrt{t}\right) + \tfrac{1}{2}e^{-2\alpha\beta}\operatorname{erfc}\left(\frac{\alpha}{\sqrt{t}} - \beta\sqrt{t}\right),$$

where erfc is the complementary error function.

10. Solve

$$\frac{\partial u}{\partial t} = a^2\frac{\partial^2 u}{\partial x^2} + Ae^{-kx}, \qquad 0 < x < \infty, 0 < t$$

subject to the boundary conditions

$$\frac{\partial u(0,t)}{\partial x} = 0, \quad \lim_{x\to\infty} u(x,t) = u_0, \quad 0 < t$$

and the initial condition

$$u(x,0) = u_0, \qquad 0 < x < \infty.$$

11. Solve

$$\frac{\partial u}{\partial t} = \frac{\partial^2 u}{\partial x^2} - P, \qquad 0 < x < L, 0 < t$$

subject to the boundary conditions

$$u(0,t) = t, \quad u(L,t) = 0, \quad 0 < t$$

and the initial condition

$$u(x,0) = 0, \qquad 0 < x < L.$$

12. An electric fuse protects electrical devices by using resistance heating to melt an enclosed wire when excessive current passes through it. A knowledge of the distribution of temperature along the wire is important in the design of the fuse. If the temperature rises to the melting point only over a small interval of the element, the melt will produce a small gap, resulting in an unnecessary prolongation of the fault and a considerable release of energy. Therefore, the desirable temperature distribution should melt most of the wire. For this reason, Guile and Carne[20] solved the heat conduction equation

$$\frac{\partial u}{\partial t} = a^2 \frac{\partial^2 u}{\partial x^2} + q(1 + \alpha u), \qquad -L < x < L, 0 < t$$

to understand the temperature structure within the fuse just before meltdown. The second term on the right side of the heat conduction equation gives the resistance heating which is assumed to vary linearly with temperature. If the terminals at $x = \pm L$ remain at a constant temperature, which we can take to be zero, the boundary conditions are

$$u(-L,t) = u(L,t) = 0, \qquad 0 < t.$$

The initial condition is

$$u(x,0) = 0, \qquad -L < x < L.$$

---

[20] From Guile, A. E. and Carne, E. B., 1954: An analysis of an analogue solution applied to the heat conduction problem in a cartridge fuse. *AIEE Trans., Part I*, **72**, 861–868. ©AIEE (now IEEE).

Find the temperature field as a function of the parameters a, q, and α.

13. Solve[21]

$$\frac{\partial u}{\partial t} = \frac{\partial^2 u}{\partial r^2} + \frac{2}{r}\frac{\partial u}{\partial r}, \qquad 0 \le r < 1, 0 < t$$

subject to the boundary conditions

$$\lim_{r\to 0} |u(r,t)| < \infty, \qquad \frac{\partial u(1,t)}{\partial r} = 1, \quad 0 < t$$

and the initial condition

$$u(r,0) = 0, \qquad 0 \le r < 1.$$

[Hint: Use the new dependent variable $v(r,t) = ru(r,t)$.]

14. Consider[22] a viscous fluid located between two fixed walls $x = \pm L$. At $x = 0$ we introduce a thin, infinitely long rigid barrier of mass m per unit area and let it fall under the force of gravity which points in the direction of positive x. We wish to find the velocity of the fluid $u(x,t)$. The fluid is governed by the partial differential equation

$$\frac{\partial u}{\partial t} = \nu\frac{\partial^2 u}{\partial x^2}, \qquad 0 < x < L, 0 < t$$

subject to the boundary conditions

$$u(L,t) = 0 \quad \text{and} \quad \frac{\partial u(0,t)}{\partial t} - \frac{2\mu}{m}\frac{\partial u(0,t)}{\partial x} = g, \qquad 0 < t$$

and the initial condition

$$u(x,0) = 0, \qquad 0 < x < L.$$

15. Consider[23] a viscous fluid located between two fixed walls $x = \pm L$. At $x = 0$ we introduce a thin, infinitely long rigid barrier of mass m per

[21] From Reismann, H., 1962: Temperature distribution in a spinning sphere during atmospheric entry. *J. Aerosp. Sci.*, **29**, 151–159 with permission.

[22] Reproduced with acknowledgement to Taylor and Francis, Publishers, from Havelock, T. H., 1921: The solution of an integral equation occurring in certain problems of viscous fluid motion. *Philos. Mag., Ser. 6*, **42**, 620–628.

[23] Reproduced with acknowledgement to Taylor and Francis, Publishers, from Havelock, T. H., 1921: On the decay of oscillation of a solid body in a viscous fluid. *Philos. Mag., Ser. 6*, **42**, 628–634.

unit area. The barrier is acted upon an elastic force in such a manner that it would vibrate with a frequency ω if the liquid were absent. We wish to find the barrier's deviation from equilibrium, $y(t)$. The fluid is governed by the partial differential equation

$$\frac{\partial u}{\partial t} = \nu \frac{\partial^2 u}{\partial x^2}, \quad 0 < x < L, 0 < t.$$

The boundary conditions are

$$u(L,t) = m\frac{d^2 y}{dt^2} - 2\mu \frac{\partial u(0,t)}{\partial x} + m\omega^2 y = 0 \text{ and } \frac{dy}{dt} = u(0,t), \quad 0 < t$$

and the initial conditions are

$$u(x,0) = 0, \; 0 < x < L \qquad \text{and} \qquad y(0) = A, \; y'(0) = 0.$$

16. Solve

$$\frac{\partial u}{\partial t} = \frac{\partial^2 u}{\partial x^2}, \qquad 0 \le x < 1, 0 < t$$

subject to the boundary conditions

$$u(0,t) = 0, \qquad 3a\left[\frac{\partial u(1,t)}{\partial x} - u(1,t)\right] + \frac{\partial u(1,t)}{\partial t} = \delta(t), \qquad 0 < t$$

and the initial condition

$$u(x,0) = 0, \qquad 0 \le x < 1.$$

17. Solve[24] the partial differential equation

$$\frac{\partial u}{\partial t} + V\frac{\partial u}{\partial x} = \frac{\partial^2 u}{\partial x^2}, \quad 0 < x < 1, 0 < t,$$

where V is a constant, subject to the boundary conditions

$$u(0,t) = 1 \qquad \text{and} \qquad u_x(1,t) = 0, \qquad 0 < t$$

---

[24] Reprinted from *Solar Energy*, **56**, Yoo, H., and E.-T. Pak, Analytical solutions to a one-dimensional finite-domain model for stratified thermal storage tanks, 315–322, ©1996, with kind permission from Elsevier Science Ltd, The Boulevard, Langford Lane, Kidlington OX5 1GB, UK.

and the initial condition

$$u(x,0) = 0, \qquad 0 < x < 1.$$

18. Solve

$$\frac{1}{r}\frac{\partial}{\partial r}\left(r\frac{\partial u}{\partial r}\right) - \frac{\partial u}{\partial t} = \delta(t), \qquad 0 \le r < a, 0 < t$$

subject to the boundary conditions

$$\lim_{r\to 0} |u(r,t)| < \infty, \qquad u(a,t) = 0, \qquad 0 < t$$

and the initial condition

$$u(r,0) = 0, \qquad 0 \le r < a.$$

Note that $J_n(iz) = i^n I_n(z)$ and $I_n(iz) = i^n J_n(z)$ for all complex z.

19. Solve

$$\frac{\partial u}{\partial t} = \frac{1}{r}\frac{\partial}{\partial r}\left(r\frac{\partial u}{\partial r}\right) + H(t), \qquad 0 \le r < a, 0 < t$$

subject to the boundary conditions

$$\lim_{r\to 0} |u(r,t)| < \infty, \qquad u(a,t) = 0, \qquad 0 < t$$

and the initial condition

$$u(r,0) = 0, \qquad 0 \le r < a.$$

Note that $J_n(iz) = i^n I_n(z)$ and $I_n(iz) = i^n J_n(z)$ for all complex z.

20. Solve

$$\frac{\partial u}{\partial t} = \frac{1}{r}\frac{\partial}{\partial r}\left(r\frac{\partial u}{\partial r}\right), \qquad 0 \le r < a, 0 < t$$

subject to the boundary conditions

$$\lim_{r\to 0} |u(r,t)| < \infty, \qquad u(a,t) = e^{-t/\tau_0}, \qquad 0 < t$$

and the initial condition

$$u(r,0) = 1, \qquad 0 \le r < a.$$

Note that $J_n(iz) = i^n I_n(z)$ and $I_n(iz) = i^n J_n(z)$ for all complex z.

21. Solve the nonhomogeneous heat equation for the spherical shell[25]

$$\frac{\partial u}{\partial t} = a^2\left(\frac{\partial^2 u}{\partial r^2} + \frac{2}{r}\frac{\partial u}{\partial r} + \frac{A}{r^4}\right), \qquad \alpha < r < \beta, 0 < t$$

subject to the boundary conditions

$$\frac{\partial u(\alpha,t)}{\partial r} = u(\beta,t) = 0, \qquad 0 < t$$

and the initial condition

$$u(r,0) = 0, \qquad \alpha < r < \beta.$$

Step 1: By introducing $v(r,t) = r\,u(r,t)$, show that the problem simplifies to

$$\frac{\partial v}{\partial t} = a^2\left(\frac{\partial^2 v}{\partial r^2} + \frac{A}{r^3}\right), \qquad \alpha < r < \beta, 0 < t$$

subject to the boundary conditions

$$\frac{\partial v(\alpha,t)}{\partial r} - \frac{v(\alpha,t)}{\alpha} = v(\beta,t) = 0, \qquad 0 < t$$

and the initial condition

$$v(r,0) = 0, \qquad \alpha < r < \beta.$$

Step 2: Using Laplace transforms and variation of parameters, show that the Laplace transform of $u(r,t)$ is

$$U(r,s) = \frac{A}{srq}\left\{\frac{\sinh[q(\beta-r)]}{\alpha q\cosh(q\ell) + \sinh(q\ell)}\int_0^{\ell}\frac{\alpha q\cosh(q\eta) + \sinh(q\eta)}{(\alpha+\eta)^3}\,d\eta - \int_0^{\beta-r}\frac{\sinh(q\eta)}{(r+\eta)^3}\,d\eta\right\},$$

where $q = \sqrt{s}/a$ and $\ell = \beta - \alpha$.

Step 3: Take the inverse of $U(r,s)$ and show that

$$u(r,t) = A\left\{\left(\frac{1}{r} - \frac{1}{\beta}\right)\left[\frac{1}{\alpha} - \frac{1}{2}\left(\frac{1}{r} + \frac{1}{\beta}\right)\right] - \frac{2\alpha^2}{r\ell^2}\sum_{n=0}^{\infty}\frac{\sin[\gamma_n(\beta-r)]\exp(-a^2\gamma_n^2 t)}{\sin^2(\gamma_n\ell)(\beta+\alpha^2\ell\gamma_n^2)}\int_0^1\frac{\sin(\gamma_n\ell\eta)}{(\delta-\eta)^3}\,d\eta\right\},$$

where γ_n is the nth root of $\alpha\gamma + \tan(\ell\gamma) = 0$ and $\delta = 1 + \alpha/\ell$.

---

[25] Abstracted with permission from Malkovich, R. Sh., 1977: Heating of a spherical shell by a radial current. *Sov. Phys. Tech. Phys.*, **22**, 636. ©1977 American Institute of Physics.

8.5 THE FOURIER TRANSFORM METHOD

We now consider the problem of one-dimensional heat flow in a rod of infinite length with insulated sides. Although there are no boundary conditions because the slab is of infinite dimension, we do require that the solution remains bounded as we go to either positive or negative infinity. The initial temperature within the rod is $u(x,0) = f(x)$.

Employing the product solution technique of Section 8.3, $u(x,t) = X(x)T(t)$ with

$$T' + a^2\lambda T = 0 \tag{8.5.1}$$

and

$$X'' + \lambda X = 0. \tag{8.5.2}$$

Solutions to (8.5.1)–(8.5.2) which remain finite over the entire x-domain are

$$X(x) = E\cos(kx) + F\sin(kx) \tag{8.5.3}$$

and

$$T(t) = C\exp(-k^2a^2t). \tag{8.5.4}$$

Because we do not have any boundary conditions, we must include *all* possible values of k. Thus, when we sum all of the product solutions according to the principle of linear superposition, we obtain the integral

$$u(x,t) = \int_0^\infty [A(k)\cos(kx) + B(k)\sin(kx)]\exp(-k^2a^2t)\,dk. \tag{8.5.5}$$

We can satisfy the initial condition by choosing

$$A(k) = \frac{1}{\pi}\int_{-\infty}^\infty f(x)\cos(kx)\,dx \tag{8.5.6}$$

and

$$B(k) = \frac{1}{\pi}\int_{-\infty}^\infty f(x)\sin(kx)\,dx, \tag{8.5.7}$$

because the initial condition has the form of a Fourier integral

$$f(x) = \int_0^\infty [A(k)\cos(kx) + B(k)\sin(kx)]\,dk, \tag{8.5.8}$$

when $t = 0$.

Several important results follow by rewriting (8.5.8) as

$$u(x,t) = \frac{1}{\pi}\int_0^\infty \left[\int_{-\infty}^\infty f(\xi)\cos(k\xi)\cos(kx)\,d\xi + \int_{-\infty}^\infty f(\xi)\sin(k\xi)\sin(kx)\,d\xi\right]\exp(-k^2a^2t)\,dk. \tag{8.5.9}$$

Combining terms,

$$u(x,t) = \frac{1}{\pi}\int_0^\infty \left\{\int_{-\infty}^\infty f(\xi)[\cos(k\xi)\cos(kx) + \sin(k\xi)\sin(kx)]\,d\xi\right\} e^{-k^2a^2t}\,dk \tag{8.5.10}$$

$$= \frac{1}{\pi}\int_0^\infty \left[\int_{-\infty}^\infty f(\xi)\cos[k(\xi - x)]\,d\xi\right] e^{-k^2a^2t}\,dk. \tag{8.5.11}$$

Reversing the order of integration,

$$u(x,t) = \frac{1}{\pi}\int_{-\infty}^\infty f(\xi)\left[\int_0^\infty \cos[k(\xi - x)]\exp(-k^2a^2t)\,dk\right] d\xi. \tag{8.5.12}$$

The inner integral is called the *source function*. We may compute its value through an integration on the complex plane; it equals

$$\int_0^\infty \cos[k(\xi - x)]\exp(-k^2a^2t)\,dk = \left(\frac{\pi}{4a^2t}\right)^{1/2} \exp\left[-\frac{(\xi - x)^2}{4a^2t}\right], \tag{8.5.13}$$

if $0 < t$. This gives the final form for the temperature distribution:

$$u(x,t) = \frac{1}{\sqrt{4a^2\pi t}}\int_{-\infty}^\infty f(\xi)\exp\left[-\frac{(\xi - x)^2}{4a^2t}\right] d\xi. \tag{8.5.14}$$

• Example 8.5.1

Let us find the temperature field if the initial distribution is

$$u(x,0) = \begin{cases} T_0, & x > 0 \\ -T_0, & x < 0. \end{cases} \tag{8.5.15}$$

Then

$$u(x,t) = -\frac{T_0}{\sqrt{4a^2\pi t}}\int_{-\infty}^0 \exp\left[-\frac{(\xi - x)^2}{4a^2t}\right] d\xi + \frac{T_0}{\sqrt{4a^2\pi t}}\int_0^\infty \exp\left[-\frac{(\xi - x)^2}{4a^2t}\right] d\xi \tag{8.5.16}$$

$$= \frac{T_0}{\sqrt{\pi}}\left[\int_{-x/2a\sqrt{t}}^\infty e^{-\tau^2}\,d\tau - \int_{x/2a\sqrt{t}}^\infty e^{-\tau^2}\,d\tau\right] \tag{8.5.17}$$

$$= \frac{T_0}{\sqrt{\pi}}\int_{-x/2a\sqrt{t}}^{x/2a\sqrt{t}} e^{-\tau^2}\,d\tau = \frac{2T_0}{\sqrt{\pi}}\int_0^{x/2a\sqrt{t}} e^{-\tau^2}\,d\tau \tag{8.5.18}$$

$$= T_0\,\mathrm{erf}\left(\frac{x}{2a\sqrt{t}}\right), \tag{8.5.19}$$

where erf is the error function.

• Example 8.5.2: Kelvin's estimate of the age of the earth

In the middle of the nineteenth century Lord Kelvin[26] estimated the age of the earth using the observed vertical temperature gradient at the earth's surface. He hypothesized that the earth was initially formed at a uniform high temperature T_0 and that its surface was subsequently maintained at the lower temperature of T_S. Assuming that most of the heat conduction occurred near the earth's surface, he reasoned that he could neglect the curvature of the earth, consider the earth's surface planar, and employ our one-dimensional heat conduction model in the vertical direction to compute the observed heat flux.

Following Kelvin, we model the earth's surface as a flat plane with an infinitely deep earth below ($z > 0$). Initially the earth has the temperature T_0. Suddenly we drop the temperature at the surface to T_S. We wish to find the heat flux across the boundary at $z = 0$ from the earth into an infinitely deep atmosphere.

The first step is to redefine our temperature scale $v(z,t) = u(z,t) + T_S$, where $v(z,t)$ is the observed temperature so that $u(0,t) = 0$ at the surface. Next, in order to use (8.5.14), we must define our initial state for $z < 0$. To maintain the temperature $u(0,t) = 0$, $f(z)$ must be an odd function or

$$f(z) = \begin{cases} T_0 - T_S, & z > 0 \\ T_S - T_0, & z < 0. \end{cases} \tag{8.5.20}$$

From (8.5.14)

$$u(z,t) = -\frac{T_0 - T_S}{\sqrt{4a^2\pi t}} \int_{-\infty}^{0} \exp\left[-\frac{(\xi - z)^2}{4a^2t}\right] d\xi + \frac{T_0 - T_S}{\sqrt{4a^2\pi t}} \int_{0}^{\infty} \exp\left[-\frac{(\xi - z)^2}{4a^2t}\right] d\xi \tag{8.5.21}$$

$$= (T_0 - T_S)\operatorname{erf}\left(\frac{z}{2a\sqrt{t}}\right), \tag{8.5.22}$$

following the work in the previous example.

The heat flux q at the surface $z = 0$ is obtained by differentiating (8.5.22) according to Fourier's law and evaluating the result at $z = 0$:

$$q = -\kappa \left.\frac{\partial v}{\partial z}\right|_{z=0} = \frac{\kappa(T_S - T_0)}{a\sqrt{\pi t}}. \tag{8.5.23}$$

---

[26] Thomson, W., 1863: On the secular cooling of the earth. *Philos. Mag., Ser. 4*, **25**, 157–170.

The surface heat flux is infinite at $t = 0$ because of the sudden application of the temperature T_S at $t = 0$. After that time, the heat flux decreases with time. Consequently, the time t at which we have the temperature gradient $\partial v(0,t)/\partial z$ is

$$t = \frac{(T_0 - T_S)^2}{\pi a^2 [\partial v(0,t)/\partial z]^2}. \tag{8.5.24}$$

For the present near-surface thermal gradient of 25 K/km, $T_0 - T_S =$ 2000 K and $a^2 = 1$ mm^2/s, the age of the earth from (8.5.24) is 65 million years.

Although Kelvin realized that this was a very rough estimate, his calculation showed that the earth had a finite age. This was a direct frontal assault on the contemporary geological principle of *uniformitarianism* that the earth's surface and upper crust had remained unchanged in temperature and other physical quantities for millions and millions of years. This debate would rage throughout the latter half of the nineteenth century and feature such luminaries as Kelvin, Charles Darwin, Thomas Huxley, and Oliver Heaviside.[27] Eventually Kelvin's arguments would prevail and uniformitarianism would fade into history.

Today, Kelvin's estimate is of academic interest because of the discovery of radioactivity at the turn of the twentieth century. The radioactivity was assumed to be uniformly distributed around the globe and restricted to the upper few tens of kilometers of the crust. Then geologists would use observed heat fluxes to discover the distribution of radioactivity within the solid earth.[28] Today we know that the interior of the earth is quite dynamic; the oceans and continents are mobile and interconnected according to the theory of plate tectonics. However, geophysicists still use measured surface heat fluxes to infer the interior[29] of the earth.

Problems

For problems 1–4, find the solution of the heat equation

$$\frac{\partial u}{\partial t} = a^2 \frac{\partial^2 u}{\partial x^2}, \qquad -\infty < x < \infty, 0 < t$$

---

[27] See Burchfield, J. D., 1975: *Lord Kelvin and the Age of the Earth*, Science History Publ., 260 pp.

[28] See Slichter, L. B., 1941: Cooling of the earth. *Bull. Geol. Soc. Am.*, **52**, 561–600.

[29] Sclater, J. G., Jaupart, C., and Galson, D., 1980: The heat flow through oceanic and continental crust and the heat loss of the earth. *Rev. Geophys. Space Phys.*, **18**, 269–311.

subject to the stated initial conditions.

1.

$$u(x,0) = \begin{cases} 1, & |x| < b \\ 0, & |x| > b \end{cases}$$

Lovering[30] has applied this solution to problems involving the cooling of lava.

2.

$$u(x,0) = e^{-b|x|}$$

3.

$$u(x,0) = \begin{cases} 0, & -\infty < x < 0 \\ T_0, & 0 < x < b \\ 0, & b < x < \infty \end{cases}$$

4.

$$u(x,0) = \delta(x)$$

5. Solve the spherically symmetric equation of diffusion,[31]

$$\frac{\partial u}{\partial t} = a^2 \left(\frac{\partial^2 u}{\partial r^2} + \frac{2}{r} \frac{\partial u}{\partial r} \right), \qquad 0 \le r < \infty, 0 < t$$

with $u(r,0) = u_0(r)$.

Step 1: Assuming $v(r,t) = r\,u(r,t)$, show that the problem can be recast as

$$\frac{\partial v}{\partial t} = a^2 \frac{\partial^2 v}{\partial r^2} \qquad 0 \le r < \infty, 0 < t$$

with $v(r,0) = r\,u_0(r)$.

Step 2: Using (8.5.14), show that the general solution is

$$u(r,t) = \frac{1}{2ar\sqrt{\pi t}} \int_0^\infty u_0(\rho) \left\{ \exp\left[- \left(\frac{r-\rho}{2a\sqrt{t}} \right)^2 \right] - \exp\left[- \left(\frac{r+\rho}{2a\sqrt{t}} \right)^2 \right] \right\} \rho \, d\rho.$$

---

[30] Lovering, T. S., 1935: Theory of heat conduction applied to geological problems. *Bull. Geol. Soc. Am.*, **46**, 69–94.

[31] From Shklovskii, I. S. and Kurt, V. G., 1960: Determination of atmospheric density at a height of 430 km by means of the diffusion of sodium vapors. *ARS J.*, **30**, 662–667 with permission.

Hint: What is the constraint on (8.5.14) so that the solution remains radially symmetric.

Step 3: For the initial concentration of

$$u_0(r) = \begin{cases} N_0, & 0 \le r < r_0 \\ 0, & r > r_0, \end{cases}$$

show that

$$u(r,t) = \tfrac{1}{2}N_0\left[\operatorname{erf}\left(\frac{r_0 - r}{2a\sqrt{t}}\right) + \operatorname{erf}\left(\frac{r_0 + r}{2a\sqrt{t}}\right) + \frac{2a\sqrt{t}}{r\sqrt{\pi}}\left\{\exp\left[-\left(\frac{r_0 + r}{2a\sqrt{t}}\right)^2\right] - \exp\left[-\left(\frac{r_0 - r}{2a\sqrt{t}}\right)^2\right]\right\}\right],$$

where erf is the error function.

8.6 THE SUPERPOSITION INTEGRAL

In our study of Laplace transforms, we showed that we may construct solutions to ordinary differential equations with a general forcing $f(t)$ by first finding the solution to a similar problem where the forcing equals Heaviside's step function. Then we can write the general solution in terms of a superposition integral according to Duhamel's theorem. In this section we show that similar considerations hold in solving the heat equation with time-dependent boundary conditions or forcings.

Let us solve the heat condition problem

$$\frac{\partial u}{\partial t} = a^2 \frac{\partial^2 u}{\partial x^2}, \qquad 0 < x < L, 0 < t \tag{8.6.1}$$

with the boundary conditions

$$u(0,t) = 0, \quad u(L,t) = f(t), \quad 0 < t \tag{8.6.2}$$

and the initial condition

$$u(x,0) = 0, \quad 0 < x < L. \tag{8.6.3}$$

The solution of (8.6.1)–(8.6.3) is difficult because of the time-dependent boundary condition. Instead of solving this system directly, let us solve the easier problem

$$\frac{\partial A}{\partial t} = a^2 \frac{\partial^2 A}{\partial x^2}, \quad 0 < x < L, 0 < t \tag{8.6.4}$$

with the boundary conditions

$$A(0,t) = 0, \quad A(L,t) = 1, \quad 0 < t \tag{8.6.5}$$

and the initial condition

$$A(x,0) = 0, \quad 0 < x < L. \tag{8.6.6}$$

Separation of variables yields the solution

$$A(x,t) = \frac{x}{L} + \frac{2}{\pi}\sum_{n=1}^{\infty}\frac{(-1)^n}{n}\sin\left(\frac{n\pi x}{L}\right)\exp\left(-\frac{a^2n^2\pi^2t}{L^2}\right). \tag{8.6.7}$$

Consider the following case. Suppose that we maintain the temperature at zero at the end $x = L$ until $t = \tau_1$ and then raise it to the value of unity. The resulting temperature distribution equals zero everywhere when $t < \tau_1$ and equals $A(x, t - \tau_1)$ for $t > \tau_1$. We have merely shifted our time axis so that the initial condition occurs at $t = \tau_1$.

Consider an analogous, but more complicated, situation of the temperature at the end position $x = L$ held at $f(0)$ from $t = 0$ to $t = \tau_1$ at which time we abruptly change it by the amount $f(\tau_1) - f(0)$ to the value $f(\tau_1)$. This temperature remains until $t = \tau_2$ when we again abruptly change it by an amount $f(\tau_2) - f(\tau_1)$. We can imagine this process continuing up to the instant $t = \tau_n$. Because of linear superposition, the temperature distribution at any given time equals the sum of these temperature increments:

$$\begin{aligned} u(x,t) = {} & f(0)A(x,t) + [f(\tau_1) - f(0)]A(x,t-\tau_1) \\ & + [f(\tau_2) - f(\tau_1)]A(x,t-\tau_2) + \cdots \\ & + [f(\tau_n) - f(\tau_{n-1})]A(x,t-\tau_n), \end{aligned} \tag{8.6.8}$$

where τ_n is the time of the most recent temperature change. If we write

$$\Delta f_k = f(\tau_k) - f(\tau_{k-1}) \quad \text{and} \quad \Delta\tau_k = \tau_k - \tau_{k-1}, \tag{8.6.9}$$

(8.6.8) becomes

$$u(x,t) = f(0)A(x,t) + \sum_{k=1}^{n} A(x,t-\tau_k)\frac{\Delta f_k}{\Delta\tau_k}\Delta\tau_k. \tag{8.6.10}$$

Consequently, in the limit of $\Delta\tau_k \to 0$, (8.6.10) becomes

$$u(x,t) = f(0)A(x,t) + \int_0^t A(x,t-\tau)f'(\tau)\,d\tau, \tag{8.6.11}$$

assuming that $f(t)$ is differentiable. Equation (8.6.11) is the *superposition integral.* We can obtain an alternative form by integration by parts:

$$u(x,t) = f(t)A(x,0) - \int_0^t f(\tau)\frac{\partial A(x,t-\tau)}{\partial \tau}\,d\tau \tag{8.6.12}$$

or

$$u(x,t) = f(t)A(x,0) + \int_0^t f(\tau)\frac{\partial A(x,t-\tau)}{\partial t}\,d\tau, \tag{8.6.13}$$

because

$$\frac{\partial A(x,t-\tau)}{\partial \tau} = -\frac{\partial A(x,t-\tau)}{\partial t}. \tag{8.6.14}$$

To illustrate the superposition integral, suppose $f(t) = t$. Then, by (8.6.11),

$$u(x,t) = \int_0^t \left\{\frac{x}{L} + \frac{2}{\pi}\sum_{n=1}^{\infty}\frac{(-1)^n}{n}\sin\left(\frac{n\pi x}{L}\right)\exp\left[-\frac{a^2n^2\pi^2}{L^2}(t-\tau)\right]\right\}d\tau \tag{8.6.15}$$

$$= \frac{xt}{L} - \frac{2L^2}{a^2\pi^3}\sum_{n=1}^{\infty}\frac{(-1)^n}{n^3}\sin\left(\frac{n\pi x}{L}\right)\left[1 - \exp\left(-\frac{a^2n^2\pi^2 t}{L^2}\right)\right]. \tag{8.6.16}$$

• Example 8.6.1: Temperature oscillations in a wall heated by an alternating current

In addition to finding solutions to heat conduction problems with time-dependent boundary conditions, we may also apply the superposition integral to the nonhomogeneous heat equation when the source is time dependent. Jeglic[32] used this technique in obtaining the temperature distribution within a slab heated by alternating electric current. If we assume that the flat plate has a surface area A and depth L, then the heat equation for the plate when electrically heated by an alternating current of frequency ω is

$$\frac{\partial u}{\partial t} - a^2\frac{\partial^2 u}{\partial x^2} = \frac{2q}{\rho C_p AL}\sin^2\omega t, \quad 0 < x < L, 0 < t, \tag{8.6.17}$$

---

[32] Jeglic, F. A., 1962: An analytical determination of temperature oscillations in a wall heated by alternating current. *NASA Tech. Note No. D-1286.*

where q is the average heat rate caused by the current, ρ is the density, C_p is the specific heat at constant pressure, and a^2 is the diffusivity of the slab. We will assume that we have insulated the inner wall so that

$$\frac{\partial u(0,t)}{\partial x} = 0, \qquad 0 < t, \tag{8.6.18}$$

while we allow the outer wall to radiatively cool to free space at the temperature of zero

$$\kappa \frac{\partial u(L,t)}{\partial x} + hu(L,t) = 0, \qquad 0 < t, \tag{8.6.19}$$

where κ is the thermal conductivity and h is the heat transfer coefficient. The slab is initially at the temperature of zero

$$u(x,0) = 0, \qquad 0 < x < L. \tag{8.6.20}$$

To solve the heat equation, we first solve the simpler problem of

$$\frac{\partial A}{\partial t} - a^2 \frac{\partial^2 A}{\partial x^2} = 1, \quad 0 < x < L, 0 < t \tag{8.6.21}$$

with the boundary conditions

$$\frac{\partial A(0,t)}{\partial x} = 0 \quad \text{and} \quad \kappa \frac{\partial A(L,t)}{\partial x} + hA(L,t) = 0, \quad 0 < t \tag{8.6.22}$$

and the initial condition

$$A(x,0) = 0, \qquad 0 < x < L. \tag{8.6.23}$$

The solution $A(x,t)$ is the *indicial admittance* because it is the response of a system to forcing by the step function $H(t)$.

We will solve (8.6.21)–(8.6.23) by separation of variables. We begin by assuming that $A(x,t)$ consists of a steady-state solution $w(x)$ plus a transient solution $v(x,t)$, where

$$a^2 w''(x) = -1, \quad w'(0) = 0, \quad \kappa w'(L) + hw(L) = 0, \tag{8.6.24}$$

$$\frac{\partial v}{\partial t} = a^2 \frac{\partial^2 v}{\partial x^2}, \quad \frac{\partial v(0,t)}{\partial x} = 0, \quad \kappa \frac{\partial v(L,t)}{\partial x} + hv(L,t) = 0 \tag{8.6.25}$$

and

$$v(x,0) = -w(x). \tag{8.6.26}$$

Solving (8.6.24),

$$w(x) = \frac{L^2 - x^2}{2a^2} + \frac{\kappa L}{ha^2}. \tag{8.6.27}$$

Table 8.6.1: The First Six Roots of the Equation $k_n \tan(k_n) = h^*$.

h^*	k_1	k_2	k_3	k_4	k_5	k_6
0.001	0.03162	3.14191	6.28334	9.42488	12.56645	15.70803
0.002	0.04471	3.14223	6.28350	9.42499	12.56653	15.70809
0.005	0.07065	3.14318	6.28398	9.42531	12.56677	15.70828
0.010	0.09830	3.14477	6.28478	9.42584	12.56717	15.70860
0.020	0.14095	3.14795	6.28637	9.42690	12.56796	15.70924
0.050	0.22176	3.15743	6.29113	9.43008	12.57035	15.71115
0.100	0.31105	3.17310	6.29906	9.43538	12.57432	15.71433
0.200	0.43284	3.20393	6.31485	9.44595	12.58226	15.72068
0.500	0.65327	3.29231	6.36162	9.47748	12.60601	15.73972
1.000	0.86033	3.42562	6.43730	9.52933	12.64529	15.77128
2.000	1.07687	3.64360	6.57833	9.62956	12.72230	15.83361
5.000	1.31384	4.03357	6.90960	9.89275	12.93522	16.01066
10.000	1.42887	4.30580	7.22811	10.20026	13.21418	16.25336
20.000	1.49613	4.49148	7.49541	10.51167	13.54198	16.58640
∞	1.57080	4.71239	7.85399	10.99557	14.13717	17.27876

Turning to the transient solution $v(x,t)$, we use separation of variables and find that

$$v(x,t) = \sum_{n=1}^{\infty} C_n \cos\left(\frac{k_n x}{L}\right) \exp\left(-\frac{a^2 k_n^2 t}{L^2}\right), \tag{8.6.28}$$

where k_n is the nth root of the transcendental equation:

$$k_n \tan(k_n) = hL/\kappa = h^*. \tag{8.6.29}$$

Table 8.6.1 gives the first six roots for various values of hL/κ.

Our final task is to compute C_n. After substituting $t = 0$ into (8.6.28), we are left with a orthogonal expansion of $-w(x)$ using the eigenfunctions $\cos(k_n x/L)$. From (6.3.4),

$$C_n = \frac{\int_0^L -w(x)\cos(k_n x/L)\,dx}{\int_0^L \cos^2(k_n x/L)\,dx} = \frac{-L^3 \sin(k_n)/(a^2 k_n^3)}{L[k_n + \sin(2k_n)/2]/(2k_n)} \tag{8.6.30}$$

$$= -\frac{2L^2 \sin(k_n)}{a^2 k_n^2 [k_n + \sin(2k_n)/2]}. \tag{8.6.31}$$

Combining (8.6.28) and (8.6.31),

$$v(x,t) = -\frac{2L^2}{a^2} \sum_{n=1}^{\infty} \frac{\sin(k_n)}{k_n^2 [k_n + \sin(2k_n)/2]} \cos\left(\frac{k_n x}{L}\right) \exp\left(-\frac{a^2 k_n^2 t}{L^2}\right). \tag{8.6.32}$$

Consequently, $A(x,t)$ equals

$$A(x,t) = \frac{L^2 - x^2}{2a^2} + \frac{\kappa L}{ha^2} - \frac{2L^2}{a^2} \sum_{n=1}^{\infty} \frac{\sin(k_n)}{k_n^2[k_n + \sin(2k_n)/2]} \cos\left(\frac{k_n x}{L}\right) \exp\left(-\frac{a^2 k_n^2 t}{L^2}\right). \tag{8.6.33}$$

We now wish to use the solution (8.6.33) to find the temperature distribution within the slab when it is heated by a time-dependent source $f(t)$. As in the case of time-dependent boundary conditions, we imagine that we can break the process into an infinite number of small changes to the heating which occur at the times $t = \tau_1$, $t = \tau_2$, etc. Consequently, the temperature distribution at the time t following the change at $t = \tau_n$ and before the change at $t = \tau_{n+1}$ is

$$u(x,t) = f(0)A(x,t) + \sum_{k=1}^{n} A(x, t - \tau_k) \frac{\Delta f_k}{\Delta \tau_k} \Delta \tau_k, \tag{8.6.34}$$

where

$$\Delta f_k = f(\tau_k) - f(\tau_{k-1}) \qquad \text{and} \qquad \Delta \tau_k = \tau_k - \tau_{k-1}. \tag{8.6.35}$$

In the limit of $\Delta \tau_k \to 0$,

$$u(x,t) = f(0)A(x,t) + \int_0^t A(x, t-\tau) f'(\tau)\, d\tau \tag{8.6.36}$$

$$= f(t)A(x,0) + \int_0^t f(\tau) \frac{\partial A(x, t-\tau)}{\partial \tau}\, d\tau. \tag{8.6.37}$$

In our present problem,

$$f(t) = \frac{2q}{\rho C_p AL} \sin^2(\omega t), \qquad f'(t) = \frac{2q\omega}{\rho C_p AL} \sin(2\omega t). \tag{8.6.38}$$

Therefore,

$$u(x,t) = \frac{2q\omega}{\rho C_p AL} \int_0^t \sin(2\omega\tau) \left\{ \frac{L^2 - x^2}{2a^2} + \frac{\kappa L}{ha^2} - \frac{2L^2}{a^2} \sum_{n=1}^{\infty} \frac{\sin(k_n)}{k_n^2[k_n + \sin(2k_n)/2]} \cos\left(\frac{k_n x}{L}\right) \times \exp\left[-\frac{a^2 k_n^2 (t-\tau)}{L^2}\right] \right\} d\tau \tag{8.6.39}$$

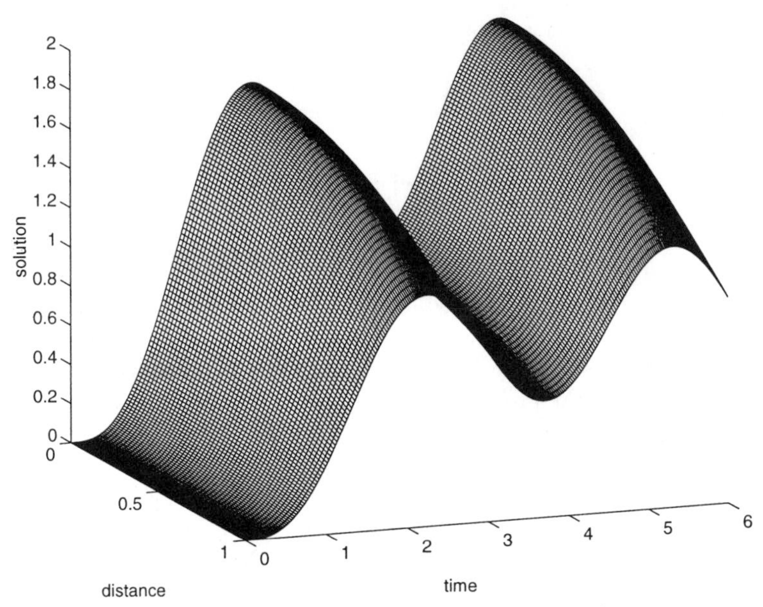

Figure 8.6.1: The nondimensional temperature $a^2 A\rho C_p u(x,t)/qL$ within a slab that we heat by alternating electric current as a function of position x/L and time a^2t/L^2 when we insulate the $x=0$ end and let the $x=L$ end radiate to free space at temperature zero. The initial temperature is zero, $hL/\kappa = 1$, and $a^2/(L^2\omega)=1$.

$$\begin{aligned} u(x,t) &= -\frac{q}{\rho C_p AL}\left(\frac{L^2-x^2}{2a^2}+\frac{\kappa L}{ha^2}\right)\cos(2\omega\tau)\Big|_0^t \\ &\quad -\frac{4L^2 q\omega}{a^2\rho C_p AL}\sum_{n=1}^{\infty}\frac{\sin(k_n)\exp(-a^2k_n^2t/L^2)}{k_n^2[k_n+\sin(2k_n)/2]}\cos\left(\frac{k_n x}{L}\right) \\ &\quad \times\int_0^t \sin(2\omega\tau)\exp\left(\frac{a^2k_n^2\tau}{L^2}\right)\,d\tau \end{aligned} \tag{8.6.40}$$

$$\begin{aligned} &= \frac{qL}{a^2A\rho C_p}\left\{\left[\frac{L^2-x^2}{2L^2}+\frac{\kappa}{hL}\right][1-\cos(2\omega t)]\right. \\ &\quad -\sum_{n=1}^{\infty}\frac{4\sin(k_n)\cos(k_nx/L)}{k_n^2[k_n+\sin(2k_n)/2][4+a^4k_n^4/(L^4\omega^2)]} \\ &\quad \left.\times\left[\frac{a^2k_n^2}{\omega L^2}\sin(2\omega t)-2\cos(2\omega t)+2\exp\left(-\frac{a^2k_n^2t}{L^2}\right)\right]\right\}. \end{aligned} \tag{8.6.41}$$

Figure 8.6.1 illustrates (8.6.41) for $hL/\kappa = 1$ and $a^2/(L^2\omega)=1$. The oscillating solution, reflecting the periodic heating by the alternating current, rapidly reaches equilibrium. Because heat is radiated to space at $x=L$, the temperature is maximum at $x=0$ at any given instant as heat flows from $x=0$ to $x=L$.

Problems

1. Solve the heat equation[33]

$$\frac{\partial u}{\partial t} = a^2 \frac{\partial^2 u}{\partial x^2}, \quad 0 < x < L, 0 < t$$

subject to the boundary conditions $u(0,t) = u(L,t) = f(t)$, $0 < t$ and the initial condition $u(x,0) = 0$, $0 < x < L$.

Step 1: First solve the heat conduction problem

$$\frac{\partial A}{\partial t} = a^2 \frac{\partial^2 A}{\partial x^2}, \quad 0 < x < L, 0 < t$$

subject to the boundary conditions $A(0,t) = A(L,t) = 1$, $0 < t$ and the initial condition $A(x,0) = 0$, $0 < x < L$. Show that

$$A(x,t) = 1 - \frac{4}{\pi} \sum_{n=1}^{\infty} \frac{\sin[(2n-1)\pi x/L]}{2n-1} e^{-a^2(2n-1)^2\pi^2 t/L^2}.$$

Step 2: Use Duhamel's theorem and show that

$$u(x,t) = \frac{4\pi a^2}{L^2} \sum_{n=1}^{\infty} (2n-1) \sin\left[\frac{(2n-1)\pi x}{L}\right] e^{-a^2(2n-1)^2\pi^2 t/L^2} \times \int_0^t f(\tau) e^{a^2(2n-1)^2\pi^2\tau/L^2}\, d\tau.$$

2. A thermometer measures temperature by the thermal expansion of a liquid (usually mercury or alcohol) stored in a bulb into a glass stem containing an empty cylindrical channel. Under normal conditions, temperature changes occur sufficiently slow so that the temperature within the liquid is uniform. However, for rapid temperature changes (such as those that would occur during the rapid ascension of an airplane or meteorological balloon), significant errors could occur. In such situations the recorded temperature would lag behind the actual temperature because of the time needed for the heat to conduct in or out of the bulb.

---

[33] From Tao, L. N., 1960: Magnetohydrodynamic effects on the formation of Couette flow. *J. Aerosp. Sci.*, **27**, 334–338 with permission.

During his investigation of this question, McLeod[34] solved

$$\frac{\partial u}{\partial t} = a^2 \frac{1}{r}\frac{\partial}{\partial r}\left(r\frac{\partial u}{\partial r}\right), \qquad 0 \le r < b, 0 < t$$

subject to the boundary conditions $\lim_{r\to 0} |u(r,t)| < \infty$ and $u(b,t) = \varphi(t)$, $0 < t$ and the initial condition $u(r,0) = 0$, $0 \le r < b$. The analysis was as follows:

Step 1: First solve the heat conduction problem

$$\frac{\partial A}{\partial t} = a^2 \frac{1}{r}\frac{\partial}{\partial r}\left(r\frac{\partial A}{\partial r}\right), \qquad 0 \le r < b, 0 < t$$

subject to the boundary conditions $\lim_{r\to 0} |A(r,t)| < \infty$ and $A(b,t) = 1$, $0 < t$ and the initial condition $A(r,0) = 0$, $0 \le r < b$. Show that

$$A(r,t) = 1 - 2\sum_{n=1}^{\infty} \frac{J_0(k_n r/b)}{k_n\, J_1(k_n)} e^{-a^2k_n^2t/b^2},$$

where $J_0(k_n) = 0$.

Step 2: Use Duhamel's theorem and show that

$$u(r,t) = \frac{2a^2}{b^2}\sum_{n=1}^{\infty} \frac{k_n\, J_0(k_n r/b)}{J_1(k_n)} \int_0^t \varphi(\tau) e^{-a^2k_n^2(t-\tau)/b^2}\, d\tau.$$

Step 3: If $\varphi(t) = Gt$, show that

$$u(r,t) = 2G\sum_{n=1}^{\infty} \frac{J_0(k_n r/b)}{k_n\, J_1(k_n)} \left[t + \frac{b^2}{a^2k_n^2}\left(e^{-a^2k_n^2t/b^2} - 1\right)\right].$$

McLeod found that for a mercury thermometer of 10-cm length a lag of 0.01°C would occur for a warming rate of 0.032°C s^{-1} (a warming gradient of 1.9°C per thousand feet and a descent of one thousand feet per minute). Although this is a very small number, when he included

---

[34] Reproduced with acknowledgement to Taylor and Francis, Publishers, from McLeod, A. R., 1919: On the lags of thermometers with spherical and cylindrical bulbs in a medium whose temperature is changing at a constant rate. *Philos. Mag., Ser. 6*, **37**, 134–144. See also Bromwich, T. J. I'A., 1919: Examples of operational methods in mathematical physics. *Philos. Mag., Ser. 6*, **37**, 407–419; McLeod, A. R., 1922: On the lags of thermometers. *Philos. Mag., Ser. 6*, **43**, 49–70.

the surface conductance of the glass tube, the lag increased to 0.85°C. Similar problems plague bimetal thermometers[35] and thermistors[36] used in radiosondes (meteorological sounding balloons).

3. A classic problem[37] in fluid mechanics is the motion of a semi-infinite viscous fluid that results from the sudden movement of the adjacent wall starting at $t = 0$. Initially the fluid is at rest. If we denote the velocity of the fluid parallel to the wall by $u(x,t)$, the governing equation is

$$\frac{\partial u}{\partial t} = \nu \frac{\partial^2 u}{\partial x^2}, \qquad 0 < x < \infty, 0 < t$$

with the boundary conditions

$$u(0,t) = V(t), \qquad \lim_{x\to\infty} u(x,t) \to 0, \qquad 0 < t$$

and the initial condition $u(x,0) = 0$, $0 < x < \infty$.

Step 1: Find the step response by solving

$$\frac{\partial A}{\partial t} = \nu \frac{\partial^2 A}{\partial x^2}, \quad 0 < x < \infty, 0 < t$$

subject to the boundary conditions

$$A(0,t) = 1 \quad \text{and} \quad \lim_{x\to\infty} A(x,t) \to 0, \qquad 0 < t$$

and the initial condition $A(x,0) = 0$, $0 < x < \infty$. Show that

$$A(x,t) = \operatorname{erfc}\left(\frac{x}{2\sqrt{\nu t}}\right) = \frac{2}{\sqrt{\pi}} \int_{x/2\sqrt{\nu t}}^{\infty} e^{-\eta^2}\, d\eta,$$

where erf is the error function. Hint: Use Laplace transforms.

Step 2: Use Duhamel's theorem and show that the solution is

$$\begin{aligned} u(x,t) &= \int_0^t V(t-\tau) \frac{x \exp(-x^2/4\nu\tau)}{2\sqrt{\pi\nu\tau^3}}\, d\tau \\ &= \frac{2}{\pi} \int_{x/\sqrt{4\nu t}}^{\infty} V\left(t - \frac{x^2}{4\nu\eta^2}\right) e^{-\eta^2}\, d\eta. \end{aligned}$$

---

[35] Mitra, H. and Datta, M. B., 1954: Lag coefficient of bimetal thermometer of chronometric radiosonde. *Indian J. Meteorol. Geophys.*, **5**, 257–261.

[36] Badgley, F. I., 1957: Response of radiosonde thermistors. *Rev. Sci. Instrum.*, **28**, 1079–1084.

[37] This problem was first posed and partially solved by Stokes, G. G., 1850: On the effect of the internal friction of fluids on the motions of pendulums. *Proc. Cambridge Philos. Soc.*, **9, Part II**, [8]–[106].

8.7 NUMERICAL SOLUTION OF THE HEAT EQUATION

In the previous chapter we showed how we may use finite difference techniques to solve the wave equation. In this section we show that similar considerations hold for the heat equation.

Starting with the heat equation

$$\frac{\partial u}{\partial t} = a^2 \frac{\partial^2 u}{\partial x^2}, \tag{8.7.1}$$

we must first replace the exact derivatives with finite differences. Drawing upon our work in Section 7.6,

$$\frac{\partial u(x_m, t_n)}{\partial t} = \frac{u_m^{n+1} - u_m^n}{\Delta t} + O(\Delta t) \tag{8.7.2}$$

and

$$\frac{\partial^2 u(x_m, t_n)}{\partial x^2} = \frac{u_{m+1}^n - 2u_m^n + u_{m-1}^n}{(\Delta x)^2} + O[(\Delta x)^2], \tag{8.7.3}$$

where the notation u_m^n denotes $u(x_m, t_n)$. Figure 8.7.1 illustrates our numerical scheme when we hold both ends at the temperature of zero. Substituting (8.7.2)–(8.7.3) into (8.7.1) and rearranging,

$$u_m^{n+1} = u_m^n + \frac{a^2 \Delta t}{(\Delta x)^2} \left(u_{m+1}^n - 2u_m^n + u_{m-1}^n \right). \tag{8.7.4}$$

The numerical integration begins with $n = 0$ and the value of u_{m+1}^0, u_m^0 and u_{m-1}^0 are given by $f(m\Delta x)$.

Once again we must check the *convergence*, *stability* and *consistency* of our scheme. We begin by writing u_{m+1}^n, u_{m-1}^n and u_m^{n+1} in terms of the exact solution u and its derivatives evaluated at the point $x_m = m\Delta x$ and $t_n = n\Delta t$. By Taylor's expansion,

$$u_{m+1}^n = u_m^n + \Delta x \left.\frac{\partial u}{\partial x}\right|_n^m + \tfrac{1}{2}(\Delta x)^2 \left.\frac{\partial^2 u}{\partial x^2}\right|_n^m + \tfrac{1}{6}(\Delta x)^3 \left.\frac{\partial^3 u}{\partial x^3}\right|_n^m + \cdots, \tag{8.7.5}$$

$$u_{m-1}^n = u_m^n - \Delta x \left.\frac{\partial u}{\partial x}\right|_n^m + \tfrac{1}{2}(\Delta x)^2 \left.\frac{\partial^2 u}{\partial x^2}\right|_n^m - \tfrac{1}{6}(\Delta x)^3 \left.\frac{\partial^3 u}{\partial x^3}\right|_n^m + \cdots \tag{8.7.6}$$

and

$$u_m^{n+1} = u_m^n + \Delta t \left.\frac{\partial u}{\partial t}\right|_n^m + \tfrac{1}{2}(\Delta t)^2 \left.\frac{\partial^2 u}{\partial t^2}\right|_n^m + \tfrac{1}{6}(\Delta t)^3 \left.\frac{\partial^3 u}{\partial t^3}\right|_n^m + \cdots \tag{8.7.7}$$

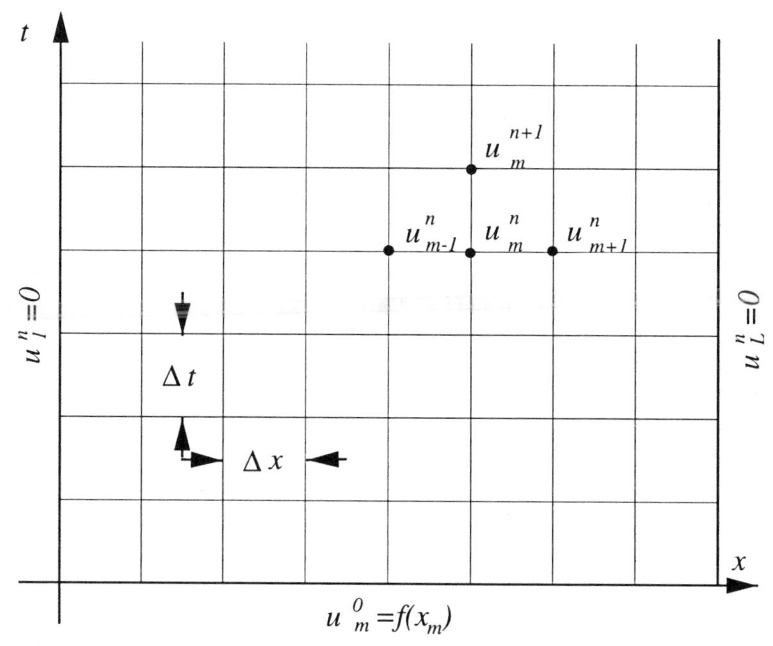

Figure 8.7.1: Schematic of the numerical solution of the heat equation when we hold both ends at a temperature of zero.

Substituting into (8.7.4), we obtain

$$\frac{u_m^{n+1}-u_m^n}{\Delta t} - a^2\frac{u_{m+1}^n - 2u_m^n + u_{m-1}^n}{(\Delta x)^2}$$
$$= \left(\frac{\partial u}{\partial t} - a^2\frac{\partial^2 u}{\partial x^2}\right)\Bigg|_n^m + \tfrac{1}{2}\Delta t\frac{\partial^2 u}{\partial t^2}\Bigg|_n^m - \tfrac{1}{12}(a\Delta x)^2\frac{\partial^4 u}{\partial x^4}\Bigg|_n^m + \cdots \tag{8.7.8}$$

The first term on the right side of (8.7.8) vanishes because $u(x,t)$ satisfies the heat equation. Thus, in the limit of $\Delta x \to 0$, $\Delta t \to 0$, the right side of (8.7.8) vanishes and the scheme is *consistent*.

To determine the *stability* of the explicit scheme, we again use the Fourier method. Assuming a solution of the form:

$$u_n^m = e^{im\theta}e^{in\lambda}, \tag{8.7.9}$$

we substitute (8.7.9) into (8.7.4) and find that

$$\frac{e^{i\lambda}-1}{\Delta t} = a^2\frac{e^{i\theta}-2+e^{-i\theta}}{(\Delta x)^2} \tag{8.7.10}$$

or

$$e^{i\lambda} = 1 - 4\frac{a^2\Delta t}{(\Delta x)^2}\sin^2\left(\frac{\theta}{2}\right). \tag{8.7.11}$$

The quantity $e^{i\lambda}$ will grow exponentially unless

$$-1 \leq 1 - 4\frac{a^2\Delta t}{(\Delta x)^2}\sin^2\left(\frac{\theta}{2}\right) < 1. \tag{8.7.12}$$

The right inequality is trivially satisfied if $a^2\Delta t/(\Delta x)^2 > 0$ while the left inequality yields

$$\frac{a^2\Delta t}{(\Delta x)^2} \leq \frac{1}{2\sin^2(\theta/2)}, \tag{8.7.13}$$

leading to the stability condition $0 < a^2\Delta t/(\Delta x)^2 \leq \frac{1}{2}$. This is a rather restrictive condition because doubling the resolution (halfing Δx) requires that we reduce the time step by a quarter. Thus, for many calculations the required time step may be unacceptably small. For this reason, many use an implicit form of the finite differencing (Crank-Nicolson implicit method[38]):

$$\frac{u_m^{n+1} - u_m^n}{\Delta t} = \frac{a^2}{2}\left[\frac{u_{m+1}^n - 2u_m^n + u_{m-1}^n}{(\Delta x)^2} + \frac{u_{m+1}^{n+1} - 2u_m^{n+1} + u_{m-1}^{n+1}}{(\Delta x)^2}\right], \tag{8.7.14}$$

although it requires the solution of a simultaneous set of linear equations. However, there are several efficient methods for their solution.

Finally we must check and see if our explicit scheme *converges* to the true solution. If we let e_m^n denote the difference between the exact and our finite differenced solution to the heat equation, we can use (8.7.8) to derive the equation governing e_m^n and find that

$$e_m^{n+1} = e_m^n + \frac{a^2\Delta t}{(\Delta x)^2}\left(e_{m+1}^n - 2e_m^n + e_{m-1}^n\right) + O[(\Delta t)^2 + \Delta t(\Delta x)^2], \tag{8.7.15}$$

for $m = 1, 2, \ldots, M$. Assuming that $a^2\Delta t/(\Delta x)^2 \leq \frac{1}{2}$, then

$$\begin{aligned} \left|e_m^{n+1}\right| &\leq \frac{a^2\Delta t}{(\Delta x)^2}\left|e_{m-1}^n\right| + \left[1 - 2\frac{a^2\Delta t}{(\Delta x)^2}\right]\left|e_m^n\right| + \frac{a^2\Delta t}{(\Delta x)^2}\left|e_{m+1}^n\right| \\ &\quad + A[(\Delta t)^2 + \Delta t(\Delta x)^2] & (8.7.16) \\ &\leq ||e_n|| + A[(\Delta t)^2 + \Delta t(\Delta x)^2], & (8.7.17) \end{aligned}$$

where $||e_n|| = \max_{m=0,1,\ldots,M} |e_m^n|$. Consequently,

$$||e_{n+1}|| \leq ||e_n|| + A[(\Delta t)^2 + \Delta t(\Delta x)^2]. \tag{8.7.18}$$

---

[38] Crank, J. and Nicoison, P., 1947: A practical method for numerical evaluation of solutions of partial differential equations of the heat-conduction type. *Proc. Cambridge. Philos. Soc.*, **43**, 50–67.

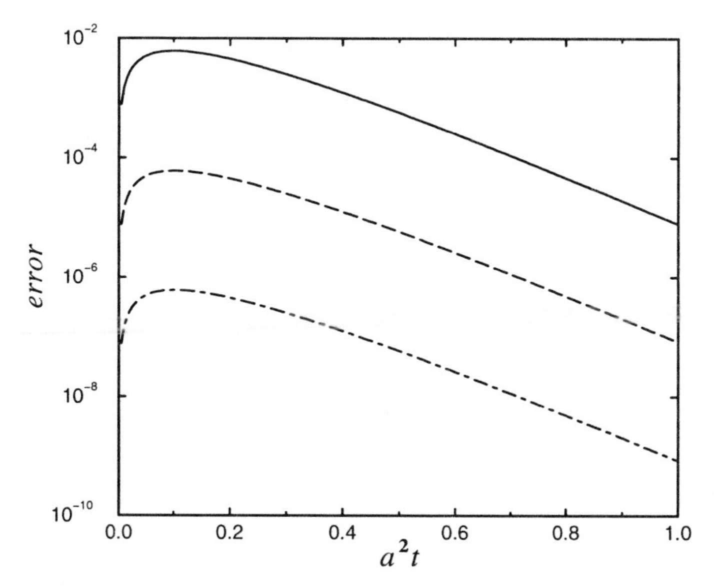

Figure 8.7.2: The growth of error $||e_n||$ as a function of a^2t for various resolutions. For the top line, $\Delta x = 0.1$; for the middle line, $\Delta x = 0.01$; and for the bottom line, $\Delta x = 0.001$.

Because $||e_0|| = 0$ and $n\Delta t \le t_n$, we find that

$$||e_{n+1}|| \le An[(\Delta t)^2 + \Delta t(\Delta x)^2] \le At_n[\Delta t + (\Delta x)^2]. \qquad \textbf{(8.7.19)}$$

As $\Delta x \to 0$, $\Delta t \to 0$, the errors tend to zero and we have convergence. We have illustrated (8.7.19) in Figure 8.7.2 by using the finite difference equation (8.7.4) to compute $||e_n||$ during a numerical experiment that used $a^2\Delta t/(\Delta x)^2 = 0.5$ and $f(x) = \sin(\pi x)$. Note how each increase of resolution by 10 results in a drop in the error by 100.

The following examples illustrate the use of numerical methods.

• **Example 8.7.1**

For our first example, we redo Example 8.3.1 with $a^2\Delta t/(\Delta x)^2 = 0.499$ and 0.501. As Figure 8.7.3 shows, the solution with $a^2\Delta t/(\Delta x)^2 < 1/2$ performs well while small-scale, growing disturbances occur for $a^2\Delta t/(\Delta x)^2 > 1/2$. This is best seen at $t' = 0.2$. It should be noted that for the reasonable $\Delta x = L/100$, it takes approximately *20,000* time steps before we reach $a^2t/L^2 = 1$.

• **Example 8.7.2**

In this example, we redo the previous example with an insulated end at $x = L$. Using the centered differencing formula,

$$u_{L+1}^n - u_{L-1}^n = 0, \qquad \textbf{(8.7.20)}$$

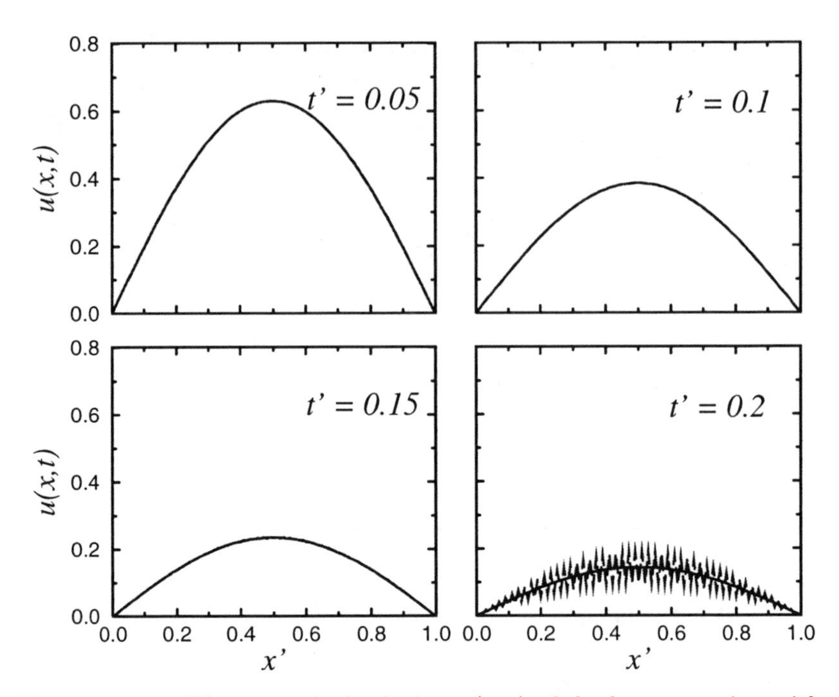

Figure 8.7.3: The numerical solution $u(x,t)$ of the heat equation with $a^2\Delta t/(\Delta x)^2 = 0.499$ (solid line) and 0.501 (jagged line) at various positions $x' = x/L$ and times $t' = a^2t/L^2$ using (8.7.4). The initial temperature $u(x,0)$ equals $4x'(1-x')$ and we hold both ends at a temperature of zero.

because $u_x(0,t) = 0$. Also, at $i = L$,

$$u_L^{n+1} = u_L^n + \frac{a^2\Delta t}{(\Delta x)^2}\left(u_{L+1}^n - 2u_L^n + u_{L-1}^n\right). \tag{8.7.21}$$

Eliminating u_{L+1}^n between the two equations,

$$u_L^{n+1} = u_L^n + \frac{a^2\Delta t}{(\Delta x)^2}\left(2u_{L-1}^n - 2u_L^n\right). \tag{8.7.22}$$

Figure 8.7.4 illustrates our numerical solution at various positions and times.

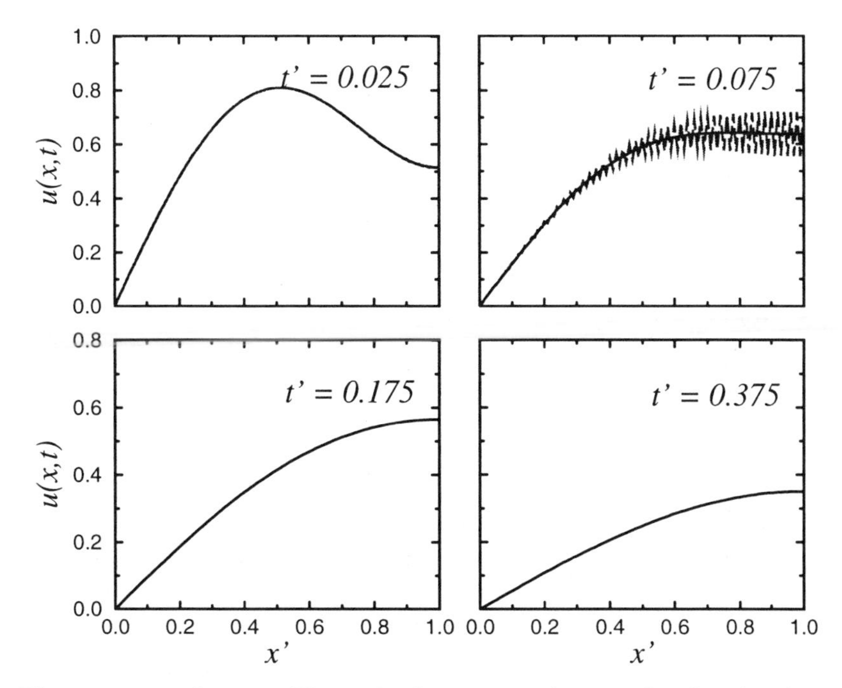

Figure 8.7.4: Same as Figure 8.7.3 except we have an insulated end at $x = L$. We have not plotted the jagged line in the bottom two frames because the solution has grown very large.

Project: Implicit Numerical Integration of the Heat Equation

The difficulty in using explicit time differencing to solve the heat equation is the very small time step that must be taken at moderate spatial resolutions to ensure stability. This small time step translates into an unacceptably long execution time. In this project you will investigate the Crank-Nicolson implicit scheme which allows for a much more reasonable time step.

Step 1: Develop code to use the Crank-Nicolson equation (8.7.14) to numerically integrate the heat equation. To do this, you will need a tridiagonal solver to find u_m^{n+1}. This is explained at the end of Section 11.1. However, many numerical methods books[39] actually have code already developed for your use. You might as well use this code.

Step 2: Test out your code by solving the heat equation given the initial condition $u(x,0) = \sin(\pi x)$ and the boundary conditions $u(0,t) =$

---

[39] For example, Press, W. H., Flannery, B. P., Teukolsky, S. A., and Vetterling, W. T., 1986: *Numerical Recipes: The Art of Scientific Computing*, Cambridge University Press, Cambridge, Section 2.6.

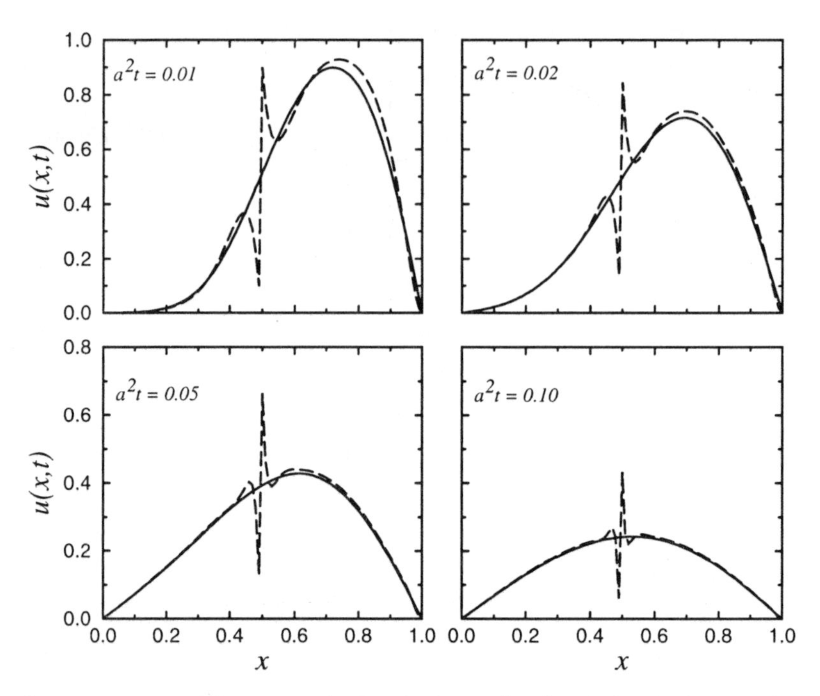

Figure 8.7.5: The numerical solution $u(x,t)$ of the heat equation $u_t = a^2 u_{xx}$ using the Crank-Nicolson method. The solid line gives the numerical solution with $a^2\Delta t = 0.0005$ while the dashed line gives the solution for $a^2\Delta t = 0.005$. Both ends are held at zero with an initial condition of $u(x,0) = 0$ for $0 \le x < \frac{1}{2}$ and $u(x,0) = 1$ for $\frac{1}{2} < x \le 1$.

$u(1,t) = 0$. Find the solution for various Δt's with $\Delta x = 0.01$. Compare this numerical solution against the exact solution which you can find. How does the error (between the numerical and exact solutions) change with Δt? For small Δt, the errors should be small. If not, then you have a mistake in your code.

Step 3: Once you have confidence in your code, discuss the behavior of the scheme for various values of Δx and Δt for the initial condition $u(x,0) = 0$ for $0 \le x < \frac{1}{2}$ and $u(x,0) = 1$ for $\frac{1}{2} < x \le 1$ with the boundary conditions $u(0,t) = u(1,t) = 0$. Although you can take quite a large Δt, what happens? Did a similar problem arise in Step 2? Explain your results.

Chapter 9

Laplace's Equation

In the previous chapter we solved the one-dimensional heat equation. Quite often we found that the transient solution died away, leaving a steady state. The partial differential equation that describes the steady state for two-dimensional heat conduction is Laplace's equation:

$$\frac{\partial^2 u}{\partial x^2} + \frac{\partial^2 u}{\partial y^2} = 0. \tag{9.0.1}$$

In general, this equation governs physical processes where *equilibrium* has been reached. It also serves as the prototype for a wider class of *elliptic equations*:

$$a(x,t)\frac{\partial^2 u}{\partial x^2} + b(x,t)\frac{\partial^2 u}{\partial x \partial t} + c(x,t)\frac{\partial^2 u}{\partial t^2} = f\left(x, t, u, \frac{\partial u}{\partial x}, \frac{\partial u}{\partial t}\right), \tag{9.0.2}$$

where $b^2 < 4ac$. Unlike the heat and wave equations, there are no initial conditions and the boundary conditions completely specify the solution. In this chapter we present some of the common techniques for solving this equation.

9.1 DERIVATION OF LAPLACE'S EQUATION

Let us imagine a thin, flat plate of heat-conducting material between two sheets of insulation. A sufficient time has passed so that the temperature depends only on the spatial coordinates x and y. We now apply the law of conservation of energy (in rate form) to a small rectangle with sides Δx and Δy.

Let $q_x(x, y)$ and $q_y(x, y)$ denote the heat flow rates in the x- and y-direction, respectively. Conservation of energy requires that the heat flow into the slab must equal the heat flow out of the slab if there is no storage or generation of heat. Now

$$\text{rate in} = q_x(x, y + \Delta y/2)\Delta y + q_y(x + \Delta x/2, y)\Delta x \tag{9.1.1}$$

and

$$\text{rate out} = q_x(x+\Delta x, y+\Delta y/2)\Delta y + q_y(x+\Delta x/2, y+\Delta y)\Delta x. \tag{9.1.2}$$

If the plate has unit thickness,

$$\begin{aligned}&[q_x(x, y + \Delta y/2) - q_x(x + \Delta x, y + \Delta y/2)]\Delta y \\ &\quad + [q_y(x + \Delta x/2, y) - q_y(x + \Delta x/2, y + \Delta y)]\Delta x = 0.\end{aligned} \tag{9.1.3}$$

Upon dividing through by $\Delta x\Delta y$, we obtain two differences quotients on the left side of (9.1.3). In the limit as Δx and Δy tend to zero, they become partial derivatives, giving

$$\frac{\partial q_x}{\partial x} + \frac{\partial q_y}{\partial y} = 0 \tag{9.1.4}$$

for any point (x, y).

We now employ Fourier's law to eliminate the rates q_x and q_y, yielding

$$\frac{\partial}{\partial x}\left(a^2\frac{\partial u}{\partial x}\right) + \frac{\partial}{\partial y}\left(a^2\frac{\partial u}{\partial y}\right) = 0, \tag{9.1.5}$$

if we have an isotropic (same in all directions) material. Finally, if a^2 is constant, (9.1.5) reduces to

$$\frac{\partial^2 u}{\partial x^2} + \frac{\partial^2 u}{\partial y^2} = 0, \tag{9.1.6}$$

which is the two-dimensional, steady-state heat equation (i.e., $u_t \approx 0$ as $t \to \infty$).

Solutions of Laplace's equation (called harmonic functions) differ fundamentally from those encountered with the heat and wave equations. These latter two equations describe the evolution of some phenomena. Laplace's equation, on the other hand, describes things at

Figure 9.1.1: Today we best remember Pierre-Simon Laplace (1749–1827) for his work in celestial mechanics and probability. In his five volumes *Traité de Mécanique céleste* (1799–1825), he accounted for the theoretical orbits of the planets and their satellites. Laplace's equation arose during this study of gravitational attraction. (Portrait courtesy of the Archives de l'Académie des sciences, Paris.)

equilibrium. Consequently, any change in the boundary conditions will affect to some degree the *entire* domain because a change to any one point will cause its neighbors to change in order to reestablish the equilibrium. Those points will, in turn, affect others. Because all of these points are in equilibrium, this modification must occur instantaneously.

Further insight follows from the *maximum principle*. If Laplace's equation governs a region, then its solution cannot have a relative maximum or minimum *inside* the region unless the solution is constant.[1] If

[1] See Courant, R. and Hilbert, D., 1962: *Methods of Mathematical Physics, Vol. II: Partial Differential Equations*, Interscience, New York,

we think of the solution as a steady-state temperature distribution, this principle is clearly true because at any one point the temperature cannot be greater than at all other nearby points. If that were so, heat would flow away from the hot point to cooler points nearby, thus eliminating the hot spot when equilibrium was once again restored.

It is often useful to consider the two-dimensional Laplace's equation in other coordinate systems. In polar coordinates, where $x = r\cos(\theta)$, $y = r\sin(\theta)$, and $z = z$, Laplace's equation becomes

$$\frac{\partial^2 u}{\partial r^2} + \frac{1}{r}\frac{\partial u}{\partial r} + \frac{\partial^2 u}{\partial z^2} = 0, \tag{9.1.7}$$

if the problem possesses axisymmetry. On the other hand, if the solution is independent of z, Laplace's equation becomes

$$\frac{\partial^2 u}{\partial r^2} + \frac{1}{r}\frac{\partial u}{\partial r} + \frac{1}{r^2}\frac{\partial^2 u}{\partial \theta^2} = 0. \tag{9.1.8}$$

In spherical coordinates, $x = r\cos(\varphi)\sin(\theta)$, $y = r\sin(\varphi)\sin(\theta)$, and $z = r\cos(\theta)$, where $r^2 = x^2 + y^2 + z^2$, θ is the angle measured *down* to the point from the z-axis (colatitude) and φ is the angle made between the x-axis and the projection of the point on the xy plane. In the case of axisymmetry (no φ dependence), Laplace's equation becomes

$$\frac{\partial}{\partial r}\left(r^2\frac{\partial u}{\partial r}\right) + \frac{1}{\sin(\theta)}\frac{\partial}{\partial \theta}\left[\sin(\theta)\frac{\partial u}{\partial \theta}\right] = 0. \tag{9.1.9}$$

9.2 BOUNDARY CONDITIONS

Because Laplace's equation involves time-independent phenomena, we must only specify boundary conditions. As we discussed in Section 8.2, we may classify these boundary conditions as follows:

1. Dirichlet condition: u given
2. Neumann condition: $\dfrac{\partial u}{\partial n}$ given, where n is the unit normal direction
3. Robin condition: $u + \alpha\dfrac{\partial u}{\partial n}$ given

along any section of the boundary. In the case of Laplace's equation, if all of the boundaries have Neumann conditions, then the solution is

326–331 for the proof.

not unique. This follows from the fact that if $u(x, y)$ is a solution, so is $u(x, y) + c$, where c is any constant.

Finally we note that we must specify the boundary conditions along each side of the boundary. These sides may be at infinity as in problems with semi-infinite domains. We must specify values along the entire boundary because we could not have an equilibrium solution if any portion of the domain was undetermined.

9.3 SEPARATION OF VARIABLES

As in the case of the heat and wave equations, separation of variables is the most popular technique for solving Laplace's equation. Although the same general procedure carries over from the previous two chapters, the following examples fill out the details.

• Example 9.3.1: Groundwater flow in a valley

Over a century ago, a French hydraulic engineer named Henri-Philibert-Gaspard Darcy (1803–1858) published the results of a laboratory experiment on the flow of water through sand. He showed that the *apparent* fluid velocity $\mathbf{q}$ relative to the sand grains is directly proportional to the gradient of the hydraulic potential $-k\nabla\varphi$, where the hydraulic potential φ equals the sum of the elevation of the point of measurement plus the pressure potential $(p/\rho g)$. In the case of steady flow, the combination of Darcy's law with conservation of mass $\nabla \cdot \mathbf{q} = 0$ yields Laplace's equation $\nabla^2\varphi = 0$ if the aquifer is isotropic (same in all directions) and homogeneous.

To illustrate how separation of variables may be used to solve Laplace's equation, we shall determine the hydraulic potential within a small drainage basin that lies in a shallow valley. See Figure 9.3.1. Following Tóth,[2] the governing equation is the two-dimensional Laplace equation

$$\frac{\partial^2 u}{\partial x^2} + \frac{\partial^2 u}{\partial y^2} = 0, \qquad 0 < x < L, 0 < y < z_0 \tag{9.3.1}$$

along with the boundary conditions

$$u(x, z_0) = gz_0 + gcx, \tag{9.3.2}$$

$$u_x(0, y) = u_x(L, y) = 0 \qquad \text{and} \qquad u_y(x, 0) = 0, \tag{9.3.3}$$

where $u(x, y)$ is the hydraulic potential, g is the acceleration due to gravity, and c gives the slope of the topography. The conditions $u_x(L, y) = 0$

[2] Tóth, J., *J. Geophys. Res.*, **67**, 4375–4387, 1962, copyright by the American Geophysical Union.

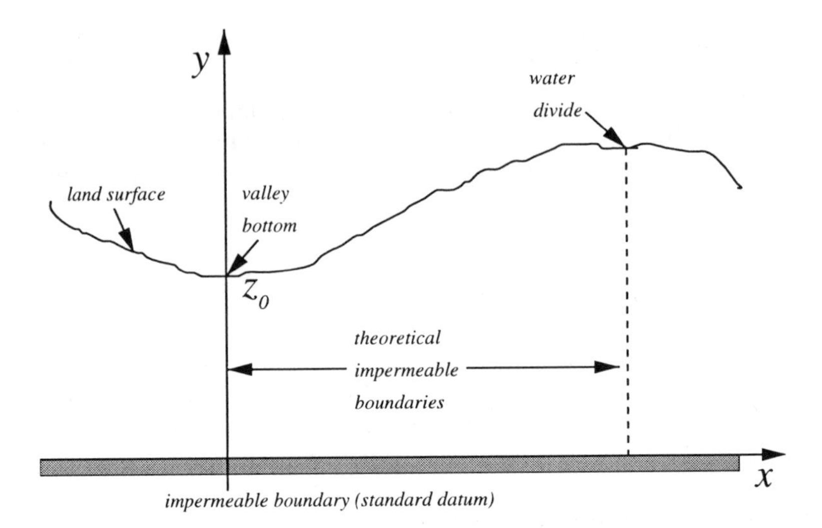

Figure 9.3.1: Cross section of a valley.

and $u_y(x,0) = 0$ specify a no-flow condition through the bottom and sides of the aquifer. The condition $u_x(0,y) = 0$ ensures symmetry about the $x = 0$ line. Equation (9.3.2) gives the fluid potential at the water table, where z_0 is the elevation of the water table above the standard datum. The term gcx in (9.3.2) expresses the increase of the potential from the valley bottom toward the water divide. On average it closely follows the topography.

Following the pattern set in the previous two chapters, we assume that $u(x,y) = X(x)Y(y)$. Then (9.3.1) becomes

$$X''Y + XY'' = 0. \tag{9.3.4}$$

Separating the variables yields

$$\frac{X''}{X} = -\frac{Y''}{Y}. \tag{9.3.5}$$

Both sides of (9.3.5) must be constant, but the sign of that constant is not obvious. From previous experience we anticipate that the ordinary differential equation in the x-direction will lead to a Sturm-Liouville problem because it possesses homogeneous boundary conditions. Proceeding along this line of reasoning, we consider three separation constants.

Trying a positive constant (say, m^2), (9.3.5) separates into the two ordinary differential equations

$$X'' - m^2X = 0 \qquad \text{and} \qquad Y'' + m^2Y = 0, \tag{9.3.6}$$

which have the solutions

$$X(x) = A\cosh(mx) + B\sinh(mx) \tag{9.3.7}$$

and

$$Y(y) = C\cos(my) + D\sin(my). \tag{9.3.8}$$

Because the boundary conditions (9.3.3) imply $X'(0) = X'(L) = 0$, both A and B must be zero, leading to the trivial solution $u(x,y) = 0$.

When the separation constant equals zero, we find a nontrivial solution given by the eigenfunction $X_0(x) = 1$ and $Y_0(y) = \frac{1}{2}A_0 + B_0 y$. However, because $Y_0'(0) = 0$ from (9.3.3), $B_0 = 0$. Thus, the particular solution for a zero separation constant is $u_0(x,y) = A_0/2$.

Finally, taking both sides of (9.3.5) equal to $-k^2$,

$$X'' + k^2 X = 0 \qquad \text{and} \qquad Y'' - k^2 Y = 0. \tag{9.3.9}$$

The first of these equations, along with the boundary conditions $X'(0) = X'(L) = 0$, gives the eigenfunction $X_n(x) = \cos(k_n x)$, with $k_n = n\pi/L$, $n = 1, 2, 3, \ldots$. The function $Y_n(y)$ for the same separation constant is

$$Y_n(y) = A_n \cosh(k_n y) + B_n \sinh(k_n y). \tag{9.3.10}$$

We must take $B_n = 0$ because $Y_n'(0) = 0$.

We now have the product solution $X_n(x)Y_n(y)$, which satisfies Laplace's equation and all of the boundary conditions except (9.3.2). By the principle of superposition, the general solution is

$$u(x,y) = \frac{A_0}{2} + \sum_{n=1}^{\infty} A_n \cos\left(\frac{n\pi x}{L}\right) \cosh\left(\frac{n\pi y}{L}\right). \tag{9.3.11}$$

Applying (9.3.2), we find that

$$u(x, z_0) = gz_0 + gcx = \frac{A_0}{2} + \sum_{n=1}^{\infty} A_n \cos\left(\frac{n\pi x}{L}\right) \cosh\left(\frac{n\pi z_0}{L}\right), \tag{9.3.12}$$

which we recognize as a Fourier half-range cosine series such that

$$A_0 = \frac{2}{L}\int_0^L (gz_0 + gcx)\, dx \tag{9.3.13}$$

and

$$\cosh\left(\frac{n\pi z_0}{L}\right) A_n = \frac{2}{L}\int_0^L (gz_0 + gcx)\cos\left(\frac{n\pi x}{L}\right) dx. \tag{9.3.14}$$

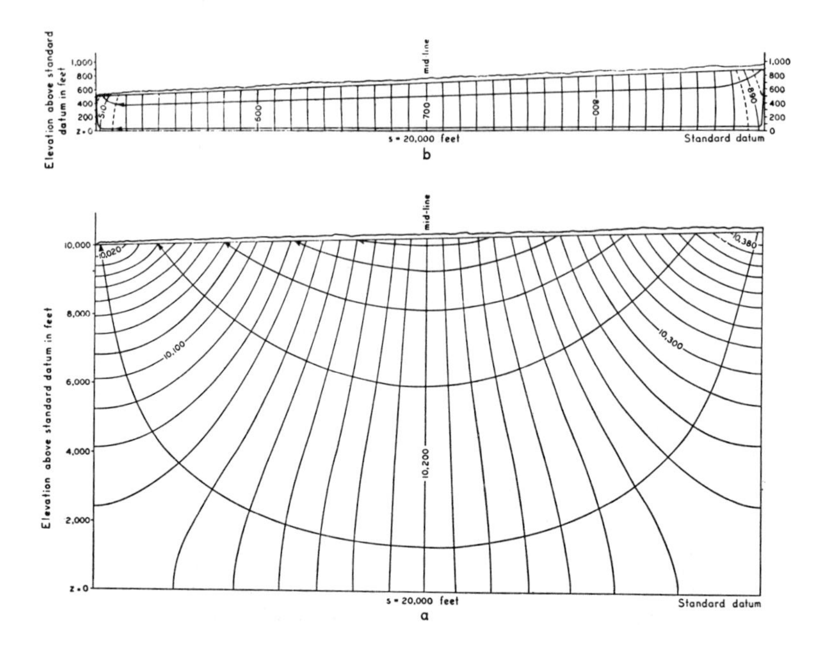

Figure 9.3.2: Two-dimensional potential distribution and flow patterns for different depths of the horizontally impermeable boundary.

Performing the integrations,

$$A_0 = 2gz_0 + gcL \tag{9.3.15}$$

and

$$A_n = -\frac{2gcL[1-(-1)^n]}{n^2\pi^2 \cosh(n\pi z_0/L)}. \tag{9.3.16}$$

Finally, the final solution is

$$u(x,y) = gz_0 + \frac{gcL}{2} - \frac{4gcL}{\pi^2}\sum_{m=1}^{\infty} \frac{\cos[(2m-1)\pi x/L]\cosh[(2m-1)\pi y/L]}{(2m-1)^2 \cosh[(2m-1)\pi z_0/L]}. \tag{9.3.17}$$

Figure 9.3.2 presents two graphs by Tóth for two different aquifers. We see that the solution satisfies the boundary condition at the bottom and side boundaries. Water flows from the elevated land (on the right) into the valley (on the left), from regions of high to low hydraulic potential.

• Example 9.3.2

In the previous example, we had the advantage of homogeneous boundary conditions along $x = 0$ and $x = L$. In a different hydraulic problem, Kirkham[3] solved the more difficult problem of

$$\frac{\partial^2 u}{\partial x^2} + \frac{\partial^2 u}{\partial y^2} = 0 \qquad 0 < x < L, 0 < y < h \tag{9.3.18}$$

subject to the Dirichlet boundary conditions

$$u(x,0) = Rx, \qquad u(x,h) = RL, \qquad u(L,y) = RL \tag{9.3.19}$$

and

$$u(0,y) = \begin{cases} 0, & 0 < y < a \\ \frac{RL}{b-a}(y-a), & a < y < b \\ RL, & b < y < h. \end{cases} \tag{9.3.20}$$

This problem arises in finding the steady flow within an aquifer resulting from the introduction of water at the top due to a steady rainfall and its removal along the sides by drains. The parameter L equals half of the distance between the drains, h is the depth of the aquifer, and R is the rate of rainfall.

The point of this example is *We need homogeneous boundary conditions along either the x or y boundaries for separation of variables to work.* We achieve this by breaking the original problem into two parts, namely

$$u(x,y) = v(x,y) + w(x,y) + RL, \tag{9.3.21}$$

where

$$\frac{\partial^2 v}{\partial x^2} + \frac{\partial^2 v}{\partial y^2} = 0 \qquad 0 < x < L, 0 < y < h \tag{9.3.22}$$

with

$$v(0,y) = v(L,y) = 0, \qquad v(x,h) = 0 \tag{9.3.23}$$

and

$$v(x,0) = R(x-L); \tag{9.3.24}$$

$$\frac{\partial^2 w}{\partial x^2} + \frac{\partial^2 w}{\partial y^2} = 0, \qquad 0 < x < L, 0 < y < h \tag{9.3.25}$$

with

$$w(x,0) = w(x,h) = 0, \qquad w(L,y) = 0 \tag{9.3.26}$$

and

$$w(0,y) = \begin{cases} -RL, & 0 < y < a \\ \frac{RL}{b-a}(y-a) - RL, & a < y < b \\ 0, & b < y < h. \end{cases} \tag{9.3.27}$$

---

[3] Kirkham, D., *Trans. Am. Geophys. Union*, **39**, 892–908, 1958, copyright by the American Geophysical Union.

Employing the same technique as in Example 9.3.1, we find that

$$v(x,y) = \sum_{n=1}^{\infty} A_n \sin\left(\frac{n\pi x}{L}\right) \frac{\sinh[n\pi(h-y)/L]}{\sinh(n\pi h/L)}, \tag{9.3.28}$$

where

$$A_n = \frac{2}{L}\int_0^L R(x-L)\sin\left(\frac{n\pi x}{L}\right)dx = -\frac{2RL}{n\pi}. \tag{9.3.29}$$

Similarly, the solution to $w(x,y)$ is found to be

$$w(x,y) = \sum_{n=1}^{\infty} B_n \sin\left(\frac{n\pi y}{h}\right) \frac{\sinh[n\pi(L-x)/h]}{\sinh(n\pi L/h)}, \tag{9.3.30}$$

where

$$B_n = \frac{2}{h}\left[-RL\int_0^a \sin\left(\frac{n\pi y}{h}\right)dy + RL\int_a^b \left(\frac{y-a}{b-a}-1\right)\sin\left(\frac{n\pi y}{h}\right)dy\right] \tag{9.3.31}$$

$$= \frac{2RL}{\pi}\left\{\frac{h}{(b-a)n^2\pi}\left[\sin\left(\frac{n\pi b}{h}\right) - \sin\left(\frac{n\pi a}{h}\right)\right] - \frac{1}{n}\right\}. \tag{9.3.32}$$

The final answer consists of substituting (9.3.28) and (9.3.30) into (9.3. 21).

• **Example 9.3.3**

The *electrostatic potential* is defined as the amount of work which must be done against electric forces to bring a unit charge from a reference point to a given point. It is readily shown[4] that the electrostatic potential is described by Laplace's equation if there is no charge within the domain. Let us find the electrostatic potential $u(r,z)$ inside a closed cylinder of length L and radius a. The base and lateral surfaces have the potential 0 while the upper surface has the potential V.

Because the potential varies in only r and z, Laplace's equation in cylindrical coordinates reduces to

$$\frac{1}{r}\frac{\partial}{\partial r}\left(r\frac{\partial u}{\partial r}\right) + \frac{\partial^2 u}{\partial z^2} = 0, \qquad 0 \le r < a, 0 < z < L \tag{9.3.33}$$

---

[4] For static fields, $\nabla \times \mathbf{E} = \mathbf{0}$, where $\mathbf{E}$ is the electric force. From Section 10.4, we can introduce a potential φ such that $\mathbf{E} = \nabla\varphi$. From Gauss' law, $\nabla \cdot \mathbf{E} = \nabla^2\varphi = 0$.

subject to the boundary conditions

$$u(a, z) = u(r, 0) = 0 \quad \text{and} \quad u(r, L) = V. \tag{9.3.34}$$

To solve this problem by separation of variables, let $u(r, z) = R(r)Z(z)$ and

$$\frac{1}{rR}\frac{d}{dr}\left(r\frac{dR}{dr}\right) = -\frac{1}{Z}\frac{d^2Z}{dz^2} = -\frac{k^2}{a^2}. \tag{9.3.35}$$

Only a negative separation constant yields nontrivial solutions in the radial direction. In that case, we have that

$$\frac{1}{r}\frac{d}{dr}\left(r\frac{dR}{dr}\right) + \frac{k^2}{a^2}R = 0. \tag{9.3.36}$$

The solutions of (9.3.36) are the Bessel functions $J_0(kr/a)$ and $Y_0(kr/a)$. Because $Y_0(kr/a)$ becomes infinite at $r = 0$, the only permissible solution is $J_0(kr/a)$. The condition that $u(a, z) = R(a)Z(z) = 0$ forces us to choose k's such that $J_0(k) = 0$. Therefore, the solution in the radial direction is $J_0(k_n r/a)$, where k_n is the nth root of $J_0(k) = 0$.

In the z direction,

$$\frac{d^2Z_n}{dz^2} + \frac{k_n^2}{a^2}Z_n = 0. \tag{9.3.37}$$

The general solution to (9.3.37) is

$$Z_n(z) = A_n \sinh\left(\frac{k_n z}{a}\right) + B_n \cosh\left(\frac{k_n z}{a}\right). \tag{9.3.38}$$

Because $u(r, 0) = R(r)Z(0) = 0$ and $\cosh(0) = 1$, B_n must equal zero. Therefore, the general product solution is

$$u(r, z) = \sum_{n=1}^{\infty} A_n J_0\left(\frac{k_n r}{a}\right) \sinh\left(\frac{k_n z}{a}\right). \tag{9.3.39}$$

The condition that $u(r, L) = V$ determines the arbitrary constant A_n. Along $z = L$,

$$u(r, L) = V = \sum_{n=1}^{\infty} A_n J_0\left(\frac{k_n r}{a}\right) \sinh\left(\frac{k_n L}{a}\right), \tag{9.3.40}$$

where

$$\sinh\left(\frac{k_n L}{a}\right) A_n = \frac{2V}{a^2 J_1^2(k_n)} \int_0^L r\, J_0\left(\frac{k_n r}{a}\right) dr \tag{9.3.41}$$

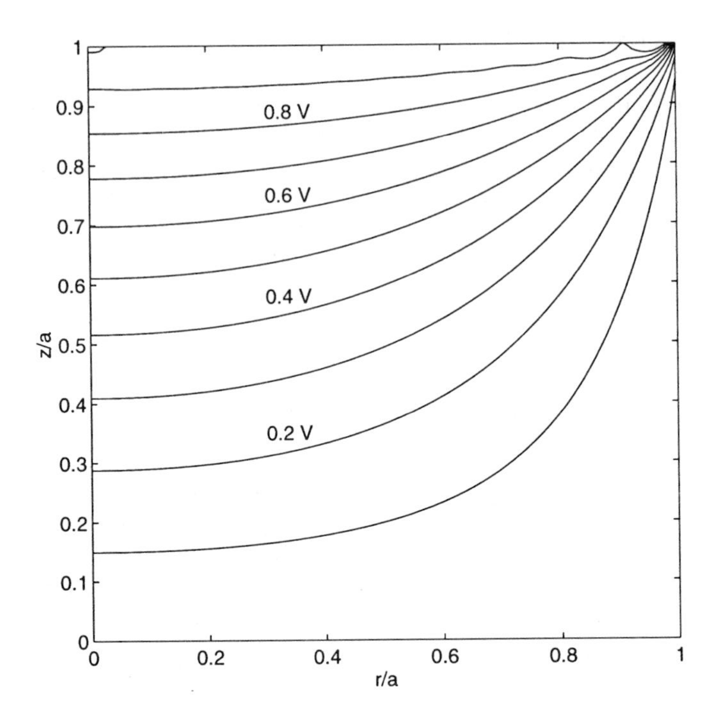

Figure 9.3.3: The steady-state potential within a cylinder of equal radius and height a when the top has the potential V while the lateral side and bottom are at potential 0.

from (6.5.35) and (6.5.43). Thus,

$$\sinh\left(\frac{k_n L}{a}\right) A_n = \frac{2V}{k_n^2 J_1^2(k_n)} \left(\frac{k_n r}{a}\right) J_1\left(\frac{k_n r}{a}\right)\bigg|_0^a = \frac{2V}{k_n J_1(k_n)}. \tag{9.3.42}$$

The solution is then

$$u(r,z) = 2V \sum_{n=1}^{\infty} \frac{J_0(k_n r/a)}{k_n J_1(k_n)} \frac{\sinh(k_n z/a)}{\sinh(k_n L/a)}. \tag{9.3.43}$$

Figure 9.3.3 illustrates (9.3.43) for the case when $L = a$ where we have included the first 20 terms of the series. Of particular interest is the convergence of the isolines in the upper right corner. At that point, the solution must jump from 0 along the line $r = a$ to V along the line $z = a$. For that reason our solution suffers from Gibbs phenomena near the top boundary. Outside of that region the electrostatic potential varies smoothly.

• **Example 9.3.4**

Let us now consider a similar, but slightly different, version of example 9.3.3, where the ends are held at zero potential while the lateral side has the value V. Once again, the governing equation is (9.3.33) with the boundary conditions

$$u(r,0) = u(r,L) = 0 \quad \text{and} \quad u(a,z) = V. \tag{9.3.44}$$

Separation of variables yields

$$\frac{1}{rR}\frac{d}{dr}\left(r\frac{dR}{dr}\right) = -\frac{1}{Z}\frac{d^2Z}{dz^2} = \frac{k^2}{L^2} \tag{9.3.45}$$

with $Z(0) = Z(L) = 0$. We have chosen a positive separation constant because a negative constant would give hyperbolic functions in z which cannot satisfy the boundary conditions. A separation constant of zero would give a straight line for $Z(z)$. Applying the boundary conditions gives a trivial solution. Consequently, the only solution in the z direction which satisfies the boundary conditions is $Z_n(z) = \sin(n\pi z/L)$.

In the radial direction, the differential equation is

$$\frac{1}{r}\frac{d}{dr}\left(r\frac{dR_n}{dr}\right) - \frac{n^2\pi^2}{L^2}R_n = 0. \tag{9.3.46}$$

As we showed in Section 6.5, the general solution is

$$R_n(r) = A_n I_0\left(\frac{n\pi r}{L}\right) + B_n K_0\left(\frac{n\pi r}{L}\right), \tag{9.3.47}$$

where I_0 and K_0 are modified Bessel functions of the first and second kind, respectively, of order zero. Because $K_0(x)$ behaves as $-\ln(x)$ as $x \to 0$, we must discard it and our solution in the radial direction becomes $R_n(r) = I_0(n\pi r/L)$. Hence, the product solution is

$$u_n(r,z) = A_n I_0\left(\frac{n\pi r}{L}\right)\sin\left(\frac{n\pi z}{L}\right) \tag{9.3.48}$$

and the general solution is a sum of these particular solutions, namely

$$u(r,z) = \sum_{n=1}^{\infty} A_n I_0\left(\frac{n\pi r}{L}\right)\sin\left(\frac{n\pi z}{L}\right). \tag{9.3.49}$$

Finally, we use the boundary conditions that $u(a,z) = V$ to compute A_n. This condition gives

$$u(a,z) = V = \sum_{n=1}^{\infty} A_n I_0\left(\frac{n\pi a}{L}\right)\sin\left(\frac{n\pi z}{L}\right) \tag{9.3.50}$$

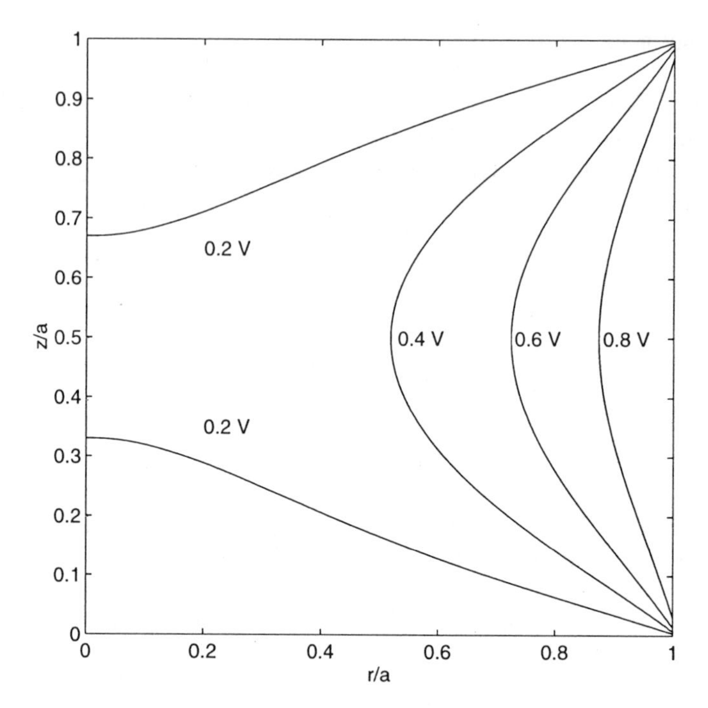

Figure 9.3.4: Potential within a conducting cylinder when the top and bottom have a potential 0 while the lateral side have a potential V.

so that

$$I_0\left(\frac{n\pi a}{L}\right) A_n = \frac{2}{L}\int_0^L V \sin\left(\frac{n\pi z}{L}\right)\, dz = \frac{2V[1-(-1)^n]}{n\pi}. \qquad \textbf{(9.3.51)}$$

Therefore, the final answer is

$$u(r,z) = \frac{4V}{\pi}\sum_{m=1}^{\infty} \frac{I_0[(2m-1)\pi r/L]\sin[(2m-1)\pi z/L]}{(2m-1)I_0[(2m-1)\pi a/L]}. \qquad \textbf{(9.3.52)}$$

Figure 9.3.4 illustrates the solution (9.3.52) for the case when $L = a$. Once again, there is a convergence of equipotentials at the corners along the right side. If we had plotted more contours, we would have observed Gibbs phenomena in the solution along the top and bottom of the cylinder.

• **Example 9.3.5**

Let us find the potential at any point P within a conducting sphere of radius a. At the surface, the potential is held at V_0 in the hemisphere $0 < \theta < \pi/2$ and $-V_0$ for $\pi/2 < \theta < \pi$.

Laplace's equation in spherical coordinates is

$$\frac{\partial}{\partial r}\left(r^2\frac{\partial u}{\partial r}\right)+\frac{1}{\sin(\theta)}\frac{\partial}{\partial\theta}\left[\sin(\theta)\frac{\partial u}{\partial\theta}\right]=0, \qquad 0\le r<a, 0<\theta<\pi. \tag{9.3.53}$$

To solve (9.3.53) we use the separation of variables $u(r,\theta)=R(r)\Theta(\theta)$. Substituting into (9.3.53), we have that

$$\frac{1}{R}\frac{d}{dr}\left(r^2\frac{dR}{dr}\right)=-\frac{1}{\sin(\theta)\Theta}\frac{d}{d\theta}\left[\sin(\theta)\frac{d\Theta}{d\theta}\right]=k^2 \tag{9.3.54}$$

or

$$r^2R''+2rR'-k^2R=0 \tag{9.3.55}$$

and

$$\frac{1}{\sin(\theta)}\frac{d}{d\theta}\left[\sin(\theta)\frac{d\Theta}{d\theta}\right]+k^2\Theta=0. \tag{9.3.56}$$

A common substitution replaces θ with $\mu=\cos(\theta)$. Then, as θ varies from 0 to π, μ varies from 1 to -1. With this substitution (9.3.56) becomes

$$\frac{d}{d\mu}\left[(1-\mu^2)\frac{d\Theta}{d\mu}\right]+k^2\Theta=0. \tag{9.3.57}$$

This is Legendre's equation which we examined in Section 6.4. Consequently, because the solution must remain finite at the poles, $k^2=n(n+1)$ and

$$\Theta_n(\theta)=P_n(\mu)=P_n[\cos(\theta)], \tag{9.3.58}$$

where $n=0,1,2,3,\ldots$

Turning to (9.3.55), this equation is the equidimensional or Euler-Cauchy linear differential equation. One method of solving this equation consists of introducing a new independent variable s so that $r=e^s$ or $s=\ln(r)$. Because

$$\frac{d}{dr}=\frac{ds}{dr}\frac{d}{ds}=e^{-s}\frac{d}{ds}, \tag{9.3.59}$$

it follows that

$$\frac{d^2}{dr^2}=\frac{d}{dr}\left(e^{-s}\frac{d}{ds}\right)=e^{-s}\frac{d}{ds}\left(e^{-s}\frac{d}{ds}\right)=e^{-2s}\left(\frac{d^2}{ds^2}-\frac{d}{ds}\right). \tag{9.3.60}$$

Substituting into (9.3.55),

$$\frac{d^2R_n}{ds^2}+\frac{dR_n}{ds}-n(n+1)R_n=0. \tag{9.3.61}$$

Equation (9.3.61) is a second-order, constant coefficient ordinary differential equation which has the solution

$$R_n(s) = C_n e^{ns} + D_n e^{-(n+1)s} \tag{9.3.62}$$

$$R_n(r) = C_n \exp[n\ln(r)] + D_n \exp[-(n+1)\ln(r)] \tag{9.3.63}$$

$$= C_n \exp[\ln(r^n)] + D_n \exp[\ln(r^{-1-n})] \tag{9.3.64}$$

$$= C_n r^n + D_n r^{-1-n}. \tag{9.3.65}$$

A more convenient form of the solution is

$$R_n(r) = A_n \left(\frac{r}{a}\right)^n + B_n \left(\frac{r}{a}\right)^{-1-n}, \tag{9.3.66}$$

where $A_n = a^n C_n$ and $B_n = D_n/a^{n+1}$. We introduced the constant a, the radius of the sphere, to simplify future calculations.

Using the results from (9.3.58) and (9.3.66), the solution to Laplace's equation in axisymmetric problems is

$$u(r,\theta) = \sum_{n=0}^{\infty}\left[A_n \left(\frac{r}{a}\right)^n + B_n \left(\frac{r}{a}\right)^{-1-n}\right] P_n[\cos(\theta)]. \tag{9.3.67}$$

In our particular problem we must take $B_n = 0$ because the solution becomes infinite at $r = 0$ otherwise. If the problem had involved the domain $a < r < \infty$, then $A_n = 0$ because the potential must remain finite as $r \to \infty$.

Finally, we must evaluate A_n. Finding the potential at the surface,

$$u(a,\mu) = \sum_{n=0}^{\infty} A_n P_n(\mu) = \begin{cases} V_0, & 0 < \mu \le 1 \\ -V_0, & -1 \le \mu < 0. \end{cases} \tag{9.3.68}$$

Upon examing (9.3.68), it is merely an expansion in Legendre polynomials of the function

$$f(\mu) = \begin{cases} V_0, & 0 < \mu \le 1 \\ -V_0, & -1 \le \mu < 0. \end{cases} \tag{9.3.69}$$

Consequently, from (9.3.69),

$$A_n = \frac{2n+1}{2}\int_{-1}^{1} f(\mu) P_n(\mu)\, d\mu. \tag{9.3.70}$$

Because $f(\mu)$ is an odd function, $A_n = 0$ if n is even. When n is odd, however,

$$A_n = (2n+1)\int_0^1 V_0 P_n(\mu)\, d\mu. \tag{9.3.71}$$

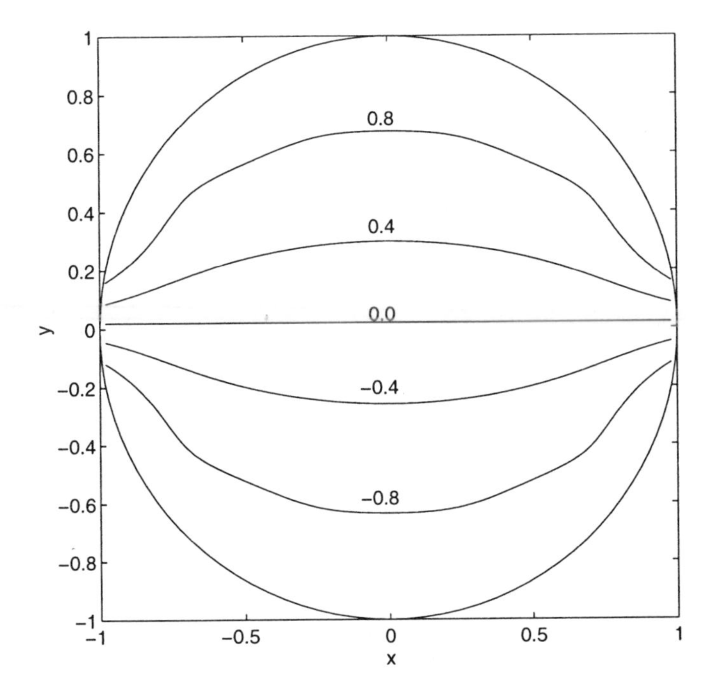

Figure 9.3.5: Electrostatic potential within a conducting sphere when the upper hemispheric surface has the potential 1 and the lower surface has the potential -1.

We can further simplify (9.3.71) by using the relationship that

$$\int_x^1 P_n(t)\,dt = \frac{1}{2n+1}\left[P_{n-1}(x) - P_{n+1}(x)\right], \tag{9.3.72}$$

where $n \geq 1$. In our problem, then,

$$A_n = \begin{cases} V_0[P_{n-1}(0) - P_{n+1}(0)], & n \text{ odd} \\ 0, & n \text{ even}. \end{cases} \tag{9.3.73}$$

This first few terms are $A_1 = 3V_0/2$, $A_3 = -7V_0/8$, and $A_5 = 11V_0/16$.

Figure 9.3.5 illustrates our solution. Here we have the convergence of the equipotentials along the equator and at the surface. The slow rate at which the coefficients are approaching zero suggests that the solution will suffer from Gibbs phenomena along the surface.

• Example 9.3.6

We now find the steady-state temperature field within a metallic sphere of radius a, which we place in direct sunlight and allow to radiatively cool. This classic problem, first solved by Rayleigh,[5] requires the use of spherical coordinates with its origin at the center of sphere and its z-axis pointing toward the sun. With this choice for the coordinate system, the incident sunlight is

$$D(\theta) = \begin{cases} D(0)\cos(\theta), & 0 \le \theta \le \pi/2 \\ 0, & \pi/2 \le \theta \le \pi. \end{cases} \tag{9.3.74}$$

If the heat dissipation takes place at the surface $r = a$ according to Newton's law of cooling and the temperature of the surrounding medium is zero, the solar heat absorbed by the surface dA must balance the Newtonian cooling at the surface plus the energy absorbed into the sphere's interior. This physical relationship is

$$(1-\rho)D(\theta)\,dA = \epsilon u(a,\theta)\,dA + \kappa \frac{\partial u(a,\theta)}{\partial r}\,dA, \tag{9.3.75}$$

where ρ is the reflectance of the surface (the albedo), ϵ is the surface conductance or coefficient of surface heat transfer, and κ is the thermal conductivity. Simplifying (9.3.75), we have that

$$\frac{\partial u(a,\theta)}{\partial r} = \frac{1-\rho}{\kappa} D(\theta) - \frac{\epsilon}{\kappa} u(a,\theta) \tag{9.3.76}$$

for $r = a$.

If the sphere has reached thermal equilibrium, Laplace's equation describes the temperature field within the sphere. In the previous example, we showed that the solution to Laplace's equation in axisymmetric problems is

$$u(r,\theta) = \sum_{n=0}^{\infty} \left[A_n \left(\frac{r}{a}\right)^n + B_n \left(\frac{r}{a}\right)^{-1-n} \right] P_n[\cos(\theta)]. \tag{9.3.77}$$

In this problem, $B_n = 0$ because the solution would become infinite at $r = 0$ otherwise. Therefore,

$$u(r,\theta) = \sum_{n=0}^{\infty} A_n \left(\frac{r}{a}\right)^n P_n[\cos(\theta)]. \tag{9.3.78}$$

Differentiation gives

$$\frac{\partial u}{\partial r} = \sum_{n=0}^{\infty} A_n \frac{nr^{n-1}}{a^n} P_n[\cos(\theta)]. \tag{9.3.79}$$

---

[5] Rayleigh, J. W., 1870: On the values of the integral $\int_0^1 Q_n Q_{n'}\,d\mu$, Q_n, $Q_{n'}$ being Laplace's coefficients of the orders n, n', with application to the theory of radiation. *Philos. Trans. R. Soc., London*, **160**, 579–590.

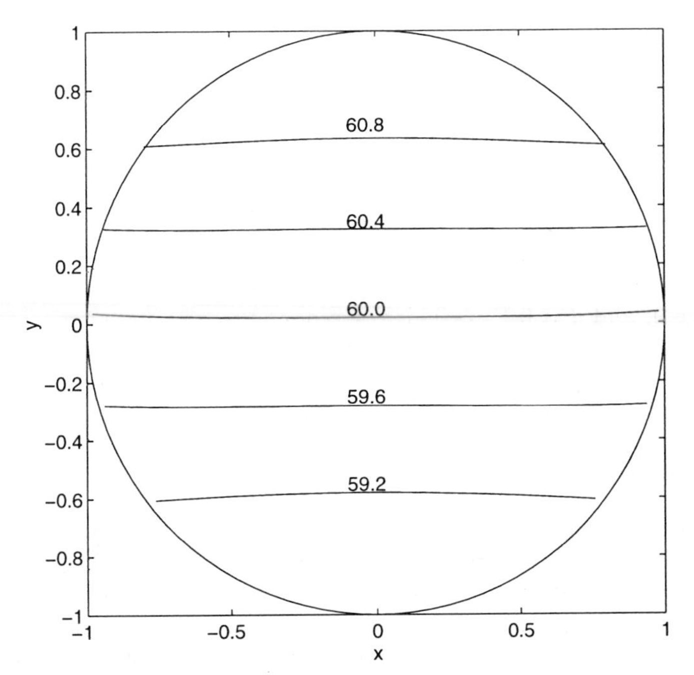

Figure 9.3.6: The difference (in °C) between the temperature field within a blackened iron surface of radius 0.1 m and the surrounding medium when we heat the surface by sunlight and allow it to radiatively cool.

Substituting into the boundary conditions leads to

$$\sum_{n=0}^{\infty} A_n \left(\frac{n}{a} + \frac{\epsilon}{\kappa}\right) P_n[\cos(\theta)] = \left(\frac{1-\rho}{\kappa}\right) D(\theta) \tag{9.3.80}$$

or

$$D(\mu) = \sum_{n=0}^{\infty} \left[\frac{n\kappa + \epsilon a}{a(1-\rho)}\right] A_n P_n(\mu) = \sum_{n=0}^{\infty} C_n P_n(\mu), \tag{9.3.81}$$

where

$$C_n = \left[\frac{n\kappa + \epsilon a}{a(1-\rho)}\right] A_n \quad \text{and} \quad \mu = \cos(\theta). \tag{9.3.82}$$

We determine the coefficients by

$$C_n = \frac{2n+1}{2} \int_{-1}^{1} D(\mu) P_n(\mu)\, d\mu = \frac{2n+1}{2} D(0) \int_{-1}^{1} \mu P_n(\mu)\, d\mu. \tag{9.3.83}$$

Evaluation of the first few coefficients gives

$$A_0 = \frac{(1-\rho)D(0)}{4\epsilon}, \quad A_1 = \frac{a(1-\rho)D(0)}{2(\kappa+\epsilon a)}, \quad A_2 = \frac{5a(1-\rho)D(0)}{16(2\kappa+\epsilon a)}, \tag{9.3.84}$$

$$A_3 = 0 \quad \text{and} \quad A_4 = -\frac{3a(1-\rho)D(0)}{32(4\kappa+\epsilon a)}. \tag{9.3.85}$$

Figure 9.3.6 illustrates the temperature field within the interior of the sphere with $D(0) = 1200$ W/m^2, $\kappa = 45$ W/m K, $\epsilon = 5$ W/m^2 K, $\rho = 0$, and $a = 0.1$ m. This corresponds to a cast iron sphere with blackened surface in sunlight. The temperature is quite warm with the highest temperature located at the position where the solar radiation is largest; the coolest temperatures are located in the shadow region.

- **Example 9.3.7**

In this example we will find the potential at any point P which results from a point charge $+q$ placed at $z = a$ on the z-axis when we introduce a conducting, grounded sphere at $z = 0$. See Figure 9.3.7. From the principle of linear superposition, the total potential $u(r,\theta)$ equals the sum of the potential from the point charge and the potential $v(r,\theta)$ due to the induced charge on the sphere

$$u(r,\theta) = \frac{q}{s} + v(r,\theta). \tag{9.3.86}$$

In common with the first term q/s, $v(r,\theta)$ must be a solution of Laplace's equation. In Example 9.3.5 we showed that the general solution to Laplace's equation in axisymmetric problems is

$$v(r,\theta) = \sum_{n=0}^{\infty} \left[A_n \left(\frac{r}{r_0}\right)^n + B_n \left(\frac{r}{r_0}\right)^{-1-n} \right] P_n[\cos(\theta)]. \tag{9.3.87}$$

Because the solutions must be valid *anywhere* outside of the sphere, $A_n = 0$; otherwise, the solution would not remain finite as $r \to \infty$. Hence,

$$v(r,\theta) = \sum_{n=0}^{\infty} B_n \left(\frac{r}{r_0}\right)^{-1-n} P_n[\cos(\theta)]. \tag{9.3.88}$$

We determine the coefficient B_n by the condition that $u(r_0,\theta) = 0$ or

$$\left.\frac{q}{s}\right|_{\text{on sphere}} + \sum_{n=0}^{\infty} B_n P_n[\cos(\theta)] = 0. \tag{9.3.89}$$

We need to expand the first term on the left side of (9.3.89) in terms of Legendre polynomials. From the law of cosines,

$$s = \sqrt{r^2 + a^2 - 2ar\cos(\theta)}. \tag{9.3.90}$$

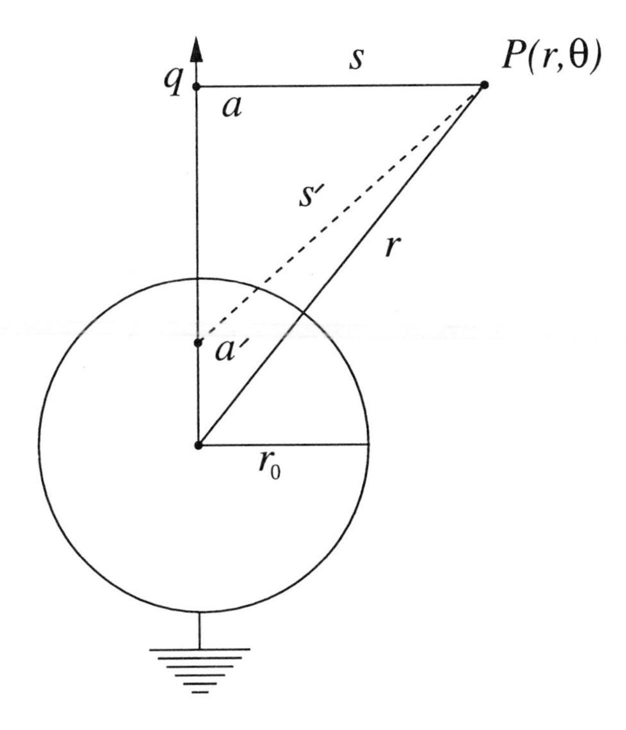

Figure 9.3.7: Point charge $+q$ in the presence of a grounded conducting sphere.

Consequently, if $a > r$, then

$$\frac{1}{s} = \frac{1}{a}\left[1 - 2\cos(\theta)\frac{r}{a} + \left(\frac{r}{a}\right)^2\right]^{-1/2} \tag{9.3.91}$$

In Section 6.4, we showed that

$$(1 - 2xz + z^2)^{-1/2} = \sum_{n=0}^{\infty} P_n(x) z^n. \tag{9.3.92}$$

Therefore,

$$\frac{1}{s} = \frac{1}{a}\sum_{n=0}^{\infty} P_n[\cos(\theta)]\left(\frac{r}{a}\right)^n. \tag{9.3.93}$$

From (9.3.89),

$$\sum_{n=0}^{\infty}\left[\frac{q}{a}\left(\frac{r_0}{a}\right)^n + B_n\right] P_n[\cos(\theta)] = 0. \tag{9.3.94}$$

We can only satisfy (9.3.94) if the square-bracketed term vanishes identically so that

$$B_n = -\frac{q}{a}\left(\frac{r_0}{a}\right)^n \tag{9.3.95}$$

On substituting (9.3.95) back into (9.3.88),

$$v(r,\theta) = -\frac{qr_0}{ra}\sum_{n=0}^{\infty}\left(\frac{r_0^2}{ar}\right)^n P_n[\cos(\theta)]. \tag{9.3.96}$$

The physical interpretation of (9.3.96) is as follows. Consider a point, such as a' (see Figure 9.3.7) on the z-axis. If $r > a'$, the expression of $1/s'$ is

$$\frac{1}{s'} = \frac{1}{r}\sum_{n=0}^{\infty} P_n[\cos(\theta)]\left(\frac{a'}{r}\right)^n, \qquad r > a'. \tag{9.3.97}$$

Using (9.3.97), we can rewrite (9.3.96) as

$$v(r,\theta) = -\frac{qr_0}{as'}, \tag{9.3.98}$$

if we set $a' = r_0^2/a$. Our final result is then

$$u(r,\theta) = \frac{q}{s} - \frac{q'}{s'}, \tag{9.3.99}$$

provided that q' equals $r_0 q/a$. In other words, when we place a grounded conducting sphere near a point charge $+q$, it changes the potential in the same manner as would a point charge of the opposite sign and magnitude $q' = r_0 q/a$, placed at the point $a' = r_0^2/a$. The charge q' is the *image* of q.

Figure 9.3.8 illustrates the solution (9.3.96). Because the charge is located above the sphere for any fixed r, the electrostatic potential is largest at the point $\theta = 0$ and weakest at $\theta = \pi$.

• Example 9.3.8: Poisson's integral formula

In this example we find the solution to Laplace's equation within a unit disc. The problem may be posed as

$$\frac{\partial^2 u}{\partial r^2} + \frac{1}{r}\frac{\partial u}{\partial r} + \frac{1}{r^2}\frac{\partial^2 u}{\partial \varphi^2} = 0, \quad 0 \le r < 1, 0 \le \varphi \le 2\pi \tag{9.3.100}$$

with the boundary condition $u(1,\varphi) = f(\varphi)$.

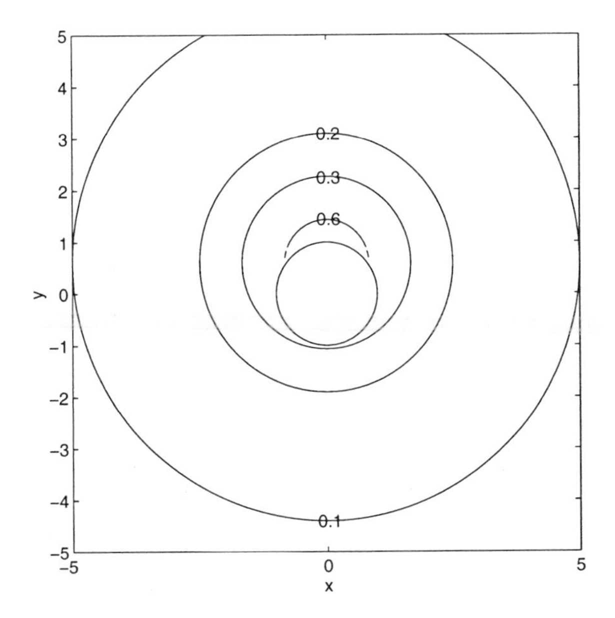

Figure 9.3.8: Electrostatic potential outside of a grounded conducting sphere in the presence of a point charge located at $a/r_0 = 2$. Contours are in units of q/r_0.

We begin by assuming the separable solution $u(r,\varphi) = R(r)\Phi(\varphi)$ so that

$$\frac{r^2R'' + rR'}{R} = -\frac{\Phi''}{\Phi} = k^2. \tag{9.3.101}$$

The solution to $\Phi'' + k^2\Phi = 0$ is

$$\Phi(\varphi) = A\cos(k\varphi) + B\sin(k\varphi). \tag{9.3.102}$$

The solution to $R(r)$ is

$$R(r) = Cr^k + Dr^{-k}. \tag{9.3.103}$$

Because the solution must be bounded for all r and periodic is φ, we must take $D = 0$ and $k = n$, where $n = 0, 1, 2, 3, \ldots$. Then, the most general solution is

$$u(r,\varphi) = \tfrac{1}{2}a_0 + \sum_{n=1}^{\infty}\left[a_n\cos(n\varphi) + b_n\sin(n\varphi)\right]r^n, \tag{9.3.104}$$

where a_n and b_n are chosen to satisfy

$$u(1,\varphi) = f(\varphi) = \tfrac{1}{2}a_0 + \sum_{n=1}^{\infty} a_n\cos(n\varphi) + b_n\sin(n\varphi). \tag{9.3.105}$$

Because

$$a_n = \frac{1}{\pi}\int_{-\pi}^{\pi} f(\varphi)\cos(n\varphi)\,d\varphi, \quad b_n = \frac{1}{\pi}\int_{-\pi}^{\pi} f(\varphi)\sin(n\varphi)\,d\varphi, \tag{9.3.106}$$

we may write $u(r, \varphi)$ as

$$u(r,\varphi) = \frac{1}{\pi}\int_{-\pi}^{\pi} f(\varphi)\left\{\tfrac{1}{2} + \sum_{n=1}^{\infty} r^n \cos[n(\theta-\varphi)]\right\} d\theta. \tag{9.3.107}$$

If we let $\alpha = \theta - \varphi$ and $z = r[\cos(\alpha) + i\sin(\alpha)]$, then

$$\sum_{n=0}^{\infty} r^n \cos(n\alpha) = \mathrm{Re}\left(\sum_{n=0}^{\infty} z^n\right) = \mathrm{Re}\left(\frac{1}{1-z}\right) \tag{9.3.108}$$

$$= \mathrm{Re}\left[\frac{1}{1 - r\cos(\alpha) - ir\sin(\alpha)}\right] \tag{9.3.109}$$

$$= \mathrm{Re}\left[\frac{1 - r\cos(\alpha) + ir\sin(\alpha)}{1 - 2r\cos(\alpha) + r^2}\right] \tag{9.3.110}$$

for all r such that $|r| < 1$. Consequently,

$$\sum_{n=0}^{\infty} r^n \cos(n\alpha) = \frac{1 - r\cos(\alpha)}{1 - 2r\cos(\alpha) + r^2} \tag{9.3.111}$$

$$\frac{1}{2} + \sum_{n=1}^{\infty} r^n \cos(n\alpha) = \frac{1 - r\cos(\alpha)}{1 - 2r\cos(\alpha) + r^2} - \frac{1}{2} \tag{9.3.112}$$

$$= \frac{1}{2}\frac{1 - r^2}{1 - 2r\cos(\alpha) + r^2}. \tag{9.3.113}$$

Substituting (9.3.113) into (9.3.107), we finally have that

$$u(r,\varphi) = \frac{1}{2\pi}\int_{-\pi}^{\pi} f(\varphi)\frac{1 - r^2}{1 - 2r\cos(\theta-\varphi) + r^2}\,d\theta. \tag{9.3.114}$$

This solution to Laplace's equation within the unit circle is referred to as *Poisson's integral formula.*[6]

[6] Poisson, S. D., 1820: Mémoire sur la manière d'exprimer les fonctions par des séries de quantités périodiques, et sur l'usage de cette transformation dans la résolution de différens problèmes. *J. École Polytech.*, **18**, 417–489.

Problems

Solve Laplace's equation over the rectangular region $0 < x < a$, $0 < y < b$ with the following boundary conditions:

1. $u(x,0) = u(x,b) = u(a,y) = 0, u(0,y) = 1$

2. $u(x,0) = u(0,y) = u(a,y) = 0, u(x,b) = x$

3. $u(x,0) = u(0,y) = u(a,y) = 0, u(x,b) = x - a$

4. $u(x,0) = u(0,y) = u(a,y) = 0,$
$$u(x,b) = \begin{cases} 2x/a, & 0 < x < a/2 \\ 2(a-x)/a, & a/2 < x < a \end{cases}$$

5. $u_x(0,y) = u(a,y) = u(x,0) = 0, u(x,b) = 1$

6. $u_y(x,0) = u(x,b) = u(a,y) = 0, u(0,y) = 1$

7. $u_y(x,0) = u_y(x,b) = 0, u(0,y) = u(a,y) = 1$

8. $u_x(a,y) = u_y(x,b) = 0, u(0,y) = u(x,0) = 1$

9. $u_y(x,0) = u(x,b) = 0, u(0,y) = u(a,y) = 1$

10. $u(a,y) = u(x,b) = 0, u(0,y) = u(x,0) = 1$

11. $u_x(0,y) = 0, u(a,y) = u(x,0) = u(x,b) = 1$

12. $u_x(0,y) = u_x(a,y) = 0, u(x,b) = u_1,$
$$u(x,0) = \begin{cases} f(x), & 0 < x < \alpha \\ 0, & \alpha < x < a \end{cases}$$

13. Variations in the earth's surface temperature can arise as a result of topographic undulations and the altitude dependence of the atmospheric temperature. These variations, in turn, affect the temperature within the solid earth. To show this, solve Laplace's equation with the surface boundary condition that

$$u(x,0) = T_0 + \Delta T \cos(2\pi x/\lambda),$$

where λ is the wavelength of the spatial temperature variation. What must be the condition on $u(x,y)$ as we go towards the center of the earth (i.e., $y \to \infty$)?

14. Tóth[7] generalized his earlier analysis of groundwater in an aquifer when the water table follows the topography. Find the groundwater potential if it varies as

$$u(x, z_0) = g[z_0 + cx + a\sin(bx)]$$

at the surface $y = z_0$ while $u_x(0, y) = u_x(L, y) = u_y(x, 0) = 0$, where g is the acceleration due to gravity. Assume that $bL \neq n\pi$, where $n = 1, 2, 3, \ldots$.

15. Solve

$$\frac{\partial^2 u}{\partial r^2} + \frac{1}{r}\frac{\partial u}{\partial r} + \frac{\partial^2 u}{\partial z^2} = 0, \quad 0 \le r < a, -L < z < L$$

with

$$u(a, z) = 0 \qquad \text{and} \qquad \frac{\partial u(r, -L)}{\partial z} = \frac{\partial u(r, L)}{\partial z} = 1.$$

16. Solve

$$\frac{\partial^2 u}{\partial r^2} + \frac{1}{r}\frac{\partial u}{\partial r} + \frac{\partial^2 u}{\partial z^2} = 0, \quad 0 \le r < a, 0 < z < h$$

with

$$\frac{\partial u(a, z)}{\partial r} = u(r, h) = 0$$

and

$$\frac{\partial u(r, 0)}{\partial z} = \begin{cases} 1, & 0 \le r < r_0 \\ 0, & r_0 < r < a. \end{cases}$$

17. Solve

$$\frac{\partial^2 u}{\partial r^2} + \frac{1}{r}\frac{\partial u}{\partial r} + \frac{\partial^2 u}{\partial z^2} = 0, \quad 0 \le r < 1, 0 < z < d$$

with

$$\frac{\partial u(1, z)}{\partial r} = \frac{\partial u(r, 0)}{\partial z} = 0$$

and

$$u(r, d) = \begin{cases} -1, & 0 \le r < a, \quad b < r < 1 \\ 1/(b^2 - a^2) - 1, & a < r < b. \end{cases}$$

---

[7] Tóth, J., *J. Geophys. Res.*, **68**, 4795–4812, 1963, copyright by the American Geophysical Union.

18. Solve

$$\frac{\partial^2 u}{\partial r^2} + \frac{1}{r}\frac{\partial u}{\partial r} - \frac{u}{r^2} + \frac{\partial^2 u}{\partial z^2} = 0, \quad 0 \le r < a, 0 < z < h$$

with

$$u(r,0) = u(a,z) = 0 \qquad \text{and} \qquad \frac{\partial u(r,h)}{\partial z} = Ar.$$

19. Solve

$$\frac{\partial^2 u}{\partial r^2} + \frac{1}{r}\frac{\partial u}{\partial r} - \frac{u}{r^2} + \frac{\partial^2 u}{\partial z^2} = 0, \quad 0 \le r < a, 0 < z < 1$$

with

$$u(r,0) = u(r,1) = 0 \qquad \text{and} \qquad u(a,z) = z.$$

20. Solve

$$\frac{\partial^2 u}{\partial r^2} + \frac{1}{r}\frac{\partial u}{\partial r} - \frac{u}{r^2} + \frac{\partial^2 u}{\partial z^2} = 0, \quad 0 \le r < a, 0 < z < h$$

with

$$\frac{\partial u(a,z)}{\partial r} = u(r,0) = 0 \qquad \text{and} \qquad \frac{\partial u(r,h)}{\partial z} = r.$$

21. Solve[8]

$$\frac{\partial^2 u}{\partial r^2} + \frac{1}{r}\frac{\partial u}{\partial r} + \frac{\partial^2 u}{\partial z^2} - \frac{\partial u}{\partial z} = 0, \quad 0 \le r < 1, 0 < z < \infty$$

with the boundary conditions

$$\lim_{r\to 0} |u(r,z)| < \infty, \qquad \frac{\partial u(1,z)}{\partial r} = -Bu(1,z), \qquad z > 0,$$

and

$$u(r,0) = 1, \qquad \lim_{z\to\infty} |u(r,z)| < \infty, \qquad 0 \le r < 1,$$

where B is a constant.

---

[8] Reprinted from *Int. J. Heat Mass Transfer*, **19**, Kern, J., and J. O. Hansen, Transient heat conduction in cylindrical systems with an axially moving boundary, 707–714, ©1976, with kind permission from Elsevier Science Ltd., The Boulevard, Langford Lane, Kidlington OX5 1GB, UK.

22. Find the steady-state temperature within a sphere of radius a if the temperature along its surface is maintained at the temperature $u(a,\theta) = 100[\cos(\theta) - \cos^5(\theta)]$.

23. Find the steady-state temperature within a sphere if the upper half of the exterior surface at radius a is maintained at the temperature 100 while the lower half is maintained at the temperature 0.

24. The surface of a sphere of radius a has a temperature of zero everywhere except in a spherical cap at the north pole (defined by the cone $\theta = \alpha$) where it equals T_0. Find the steady-state temperature within the sphere.

25. Using the relationship

$$\int_0^{2\pi} \frac{d\varphi}{1 - b\cos(\varphi)} = \frac{2\pi}{\sqrt{1-b^2}}, \qquad |b| < 1$$

and Poisson's integral formula, find the solution to Laplace's equation within a unit disc if $u(1,\varphi) = f(\varphi) = T_0$, a constant.

9.4 THE SOLUTION OF LAPLACE'S EQUATION ON THE UPPER HALF-PLANE

In this section we shall use Fourier integrals and convolution to find the solution of Laplace's equation on the upper half-plane $y > 0$. We require that the solution remains bounded over the entire domain and specify it along the x-axis, $u(x,0) = f(x)$. Under these conditions, we can take the Fourier transform of Laplace's equation and find that

$$\int_{-\infty}^{\infty} \frac{\partial^2 u}{\partial x^2} e^{-i\omega x}\, dx + \int_{-\infty}^{\infty} \frac{\partial^2 u}{\partial y^2} e^{-i\omega x}\, dx = 0. \tag{9.4.1}$$

If everything is sufficiently differentiable, we may successively integrate by parts the first integral in (9.4.1) which yields

$$\int_{-\infty}^{\infty} \frac{\partial^2 u}{\partial x^2} e^{-i\omega x}\, dx = \frac{\partial u}{\partial x} e^{-i\omega x}\Big|_{-\infty}^{\infty} + i\omega \int_{-\infty}^{\infty} \frac{\partial u}{\partial x} e^{-i\omega x}\, dx \tag{9.4.2}$$

$$= i\omega\, u(x,y) e^{-i\omega x}\Big|_{-\infty}^{\infty} - \omega^2 \int_{-\infty}^{\infty} u(x,y) e^{-i\omega x}\, dx \tag{9.4.3}$$

$$= -\omega^2 \mathcal{U}(\omega, y), \tag{9.4.4}$$

where

$$\mathcal{U}(\omega, y) = \int_{-\infty}^{\infty} u(x,y) e^{-i\omega x}\, dx. \tag{9.4.5}$$

The second integral becomes

$$\int_{-\infty}^{\infty} \frac{\partial^2 u}{\partial y^2} e^{-i\omega x}\, dx = \frac{d^2}{dy^2}\left[\int_{-\infty}^{\infty} u(x,y)e^{-i\omega x}\, dx\right] = \frac{d^2\mathcal{U}(\omega,y)}{dy^2} \quad (9.4.6)$$

along with the boundary condition that

$$F(\omega) = \mathcal{U}(\omega,0) = \int_{-\infty}^{\infty} f(x)e^{-i\omega x}\, dx. \quad (9.4.7)$$

Consequently we have reduced Laplace's equation, a partial differential equation, to an ordinary differential equation in y, where ω is merely a parameter:

$$\frac{d^2\mathcal{U}(\omega,y)}{dy^2} - \omega^2\mathcal{U}(\omega,y) = 0, \quad (9.4.8)$$

with the boundary condition $\mathcal{U}(\omega,0) = F(\omega)$. The solution to (9.4.8) is

$$\mathcal{U}(\omega,y) = A(\omega)e^{|\omega|y} + B(\omega)e^{-|\omega|y}, \quad y \geq 0. \quad (9.4.9)$$

We must discard the $e^{|\omega|y}$ term because it becomes unbounded as we go to infinity along the y-axis. The boundary condition results in $B(\omega) = F(\omega)$. Consequently,

$$\mathcal{U}(\omega,y) = F(\omega)e^{-|\omega|y}. \quad (9.4.10)$$

The inverse of the Fourier transform $e^{-|\omega|y}$ equals

$$\frac{1}{2\pi}\int_{-\infty}^{\infty} e^{-|\omega|y}e^{i\omega x}\, d\omega = \frac{1}{2\pi}\int_{-\infty}^{0} e^{\omega y}e^{i\omega x}\, d\omega + \frac{1}{2\pi}\int_{0}^{\infty} e^{-\omega y}e^{i\omega x}\, d\omega \quad (9.4.11)$$

$$= \frac{1}{2\pi}\int_{0}^{\infty} e^{-\omega y}e^{-i\omega x}\, d\omega + \frac{1}{2\pi}\int_{0}^{\infty} e^{-\omega y}e^{i\omega x}\, d\omega \quad (9.4.12)$$

$$= \frac{1}{\pi}\int_{0}^{\infty} e^{-\omega y}\cos(\omega x)\, d\omega \quad (9.4.13)$$

$$= \frac{1}{\pi}\left\{\frac{\exp(-\omega y)}{x^2+y^2}\left[-y\cos(\omega x) + x\sin(\omega x)\right]\right\}\Bigg|_{0}^{\infty} \quad (9.4.14)$$

$$= \frac{1}{\pi}\frac{y}{x^2+y^2}. \quad (9.4.15)$$

Furthermore, because (9.4.10) is a convolution of two Fourier transforms, its inverse is

$$u(x,y) = \frac{1}{\pi}\int_{-\infty}^{\infty} \frac{yf(t)}{(x-t)^2+y^2}\, dt. \quad (9.4.16)$$

Equation (9.4.16) is *Poisson's integral formula*[9] for the half-plane $y > 0$ or *Schwarz' integral formula.*[10]

• **Example 9.4.1**

As an example, let $u(x,0) = 1$ if $|x| < 1$ and $u(x,0) = 0$ otherwise. Then,

$$u(x,y) = \frac{1}{\pi}\int_{-1}^{1} \frac{y}{(x-t)^2 + y^2}\,dt \tag{9.4.17}$$

$$= \frac{1}{\pi}\left[\tan^{-1}\left(\frac{1-x}{y}\right) + \tan^{-1}\left(\frac{1+x}{y}\right)\right]. \tag{9.4.18}$$

Problems

Find the solution to Laplace's equation in the upper half-plane for the following boundary conditions:

1.

$$u(x,0) = \begin{cases} 1, & 0 < x < 1 \\ 0, & \text{otherwise} \end{cases}$$

2.

$$u(x,0) = \begin{cases} 1, & x > 0 \\ -1, & x < 0 \end{cases}$$

3.

$$u(x,0) = \begin{cases} T_0, & x < 0 \\ 0, & x > 0 \end{cases}$$

4.

$$u(x,0) = \begin{cases} 2T_0, & x < -1 \\ T_0, & -1 < x < 1 \\ 0, & x > 1 \end{cases}$$

---

[9] Poisson, S. D., 1823: Suite du mémoire sur les intégrales définies et sur la sommation des séries. *J. École Polytech.*, **19**, 404–509. See pg. 462.

[10] Schwarz, H. A., 1870: Über die Integration der partiellen Differentialgleichung $\partial^2 u/\partial x^2 + \partial^2 u/\partial y^2 = 0$ für die Fläche eines Kreises, *Vierteljahrsschr. Naturforsch. Ges. Zürich*, **15**, 113–128.

5.

$$u(x,0) = \begin{cases} T_0, & -1 < x < 0 \\ T_0 + (T_1 - T_0)x, & 0 < x < 1 \\ 0, & \text{otherwise} \end{cases}$$

6.

$$u(x,0) = \begin{cases} T_0, & x < a_1 \\ T_1, & a_1 < x < a_2 \\ T_2, & a_2 < x < a_3 \\ \vdots & \vdots \\ T_n, & a_n < x \end{cases}$$

9.5 POISSON'S EQUATION ON A RECTANGLE

Poisson's equation[11] is Laplace's equation with a source term:

$$\frac{\partial^2 u}{\partial x^2} + \frac{\partial^2 u}{\partial y^2} = f(x,y). \tag{9.5.1}$$

It arises in such diverse areas as groundwater flow, electromagnetism, and potential theory. Let us solve it if $u(0,y) = u(a,y) = u(x,0) = u(x,b) = 0$.

We begin by solving a similar partial differential equation:

$$\frac{\partial^2 u}{\partial x^2} + \frac{\partial^2 u}{\partial y^2} = \lambda u, \qquad 0 < x < a, 0 < y < b \tag{9.5.2}$$

by separation of variables. If $u(x,y) = X(x)Y(y)$, then

$$\frac{X''}{X} + \frac{Y''}{Y} = \lambda. \tag{9.5.3}$$

Because we must satisfy the boundary conditions that $X(0) = X(a) = Y(0) = Y(b) = 0$, we have the following eigenfunction solutions:

$$X_n(x) = \sin\left(\frac{n\pi x}{a}\right), \qquad Y_m(x) = \sin\left(\frac{m\pi y}{b}\right) \tag{9.5.4}$$

with $\lambda_{nm} = -n^2\pi^2/a^2 - m^2\pi^2/b^2$; otherwise, we would only have trivial solutions. The corresponding particular solutions are

$$u_{nm} = A_{nm} \sin\left(\frac{n\pi x}{a}\right) \sin\left(\frac{m\pi y}{b}\right), \tag{9.5.5}$$

[11] Poisson, S. D., 1813: Remarques sur une équation qui se présente dans la théorie des attractions des sphéroïdes. *Nouv. Bull. Soc. Philomath. Paris*, **3**, 388–392.

Figure 9.5.1: Siméon-Denis Poisson (1781–1840) was a product as well as a member of the French scientific establishment of his day. Educated at the École Polytechnique, he devoted his life to teaching, both in the classroom and with administrative duties, and to scientific research. Poisson's equation dates from 1813 when Poisson sought to extend Laplace's work on gravitational attraction. (Portrait courtesy of the Archives de l'Académie des sciences, Paris.)

where $n = 1, 2, 3, \ldots$ and $m = 1, 2, 3, \ldots$.

For a fixed y, we can expand $f(x, y)$ in the half-range Fourier sine series:

$$f(x, y) = \sum_{n=1}^{\infty} A_n(y) \sin\left(\frac{n\pi x}{a}\right), \tag{9.5.6}$$

where

$$A_n(y) = \frac{2}{a} \int_0^a f(x, y) \sin\left(\frac{n\pi x}{a}\right) dx. \tag{9.5.7}$$

However, we can also expand $A_n(y)$ in a half-range Fourier sine series:

$$A_n(y) = \sum_{m=1}^{\infty} a_{nm} \sin\left(\frac{m\pi y}{b}\right), \tag{9.5.8}$$

where

$$a_{nm} = \frac{2}{b}\int_0^b A_n(y) \sin\left(\frac{m\pi y}{b}\right) dy \tag{9.5.9}$$

$$= \frac{4}{ab}\int_0^b \int_0^a f(x,y) \sin\left(\frac{n\pi x}{a}\right) \sin\left(\frac{m\pi y}{b}\right) dx\, dy \tag{9.5.10}$$

and

$$f(x,y) = \sum_{n=1}^{\infty}\sum_{m=1}^{\infty} a_{nm} \sin\left(\frac{n\pi x}{a}\right) \sin\left(\frac{m\pi y}{b}\right). \tag{9.5.11}$$

In other words, we have reexpressed $f(x,y)$ in terms of a *double Fourier series*.

Because (9.5.2) must hold for each particular solution,

$$\frac{\partial^2 u_{nm}}{\partial x^2} + \frac{\partial^2 u_{nm}}{\partial y^2} = \lambda_{nm} u_{nm} = a_{nm} \sin\left(\frac{n\pi x}{a}\right) \sin\left(\frac{m\pi y}{b}\right), \tag{9.5.12}$$

if we now associate (9.5.1) with (9.5.2). Therefore, the solution to Poisson's equation on a rectangle where the boundaries are held at zero is the double Fourier series:

$$u(x,y) = -\sum_{n=1}^{\infty}\sum_{m=1}^{\infty} \frac{a_{nm}}{n^2\pi^2/a^2 + m^2\pi^2/b^2} \sin\left(\frac{n\pi x}{a}\right) \sin\left(\frac{m\pi y}{b}\right). \tag{9.5.13}$$

Problems

1. The equation

$$\frac{\partial^2 u}{\partial x^2} + \frac{\partial^2 u}{\partial y^2} = -\frac{R}{T}, \qquad -a < x < a, -b < y < b$$

describes the hydraulic potential (elevation of the water table) $u(x,y)$ within a rectangular island on which a recharging well is located at $(0,0)$. Here R is the rate of recharging and T is the product of the hydraulic conductivity and aquifer thickness. If the water table is at sea level around the island so that $u(-a,y) = u(a,y) = u(x,-b) = u(x,b) = 0$, find $u(x,y)$ everywhere in the island. [Hint: Use symmetry and redo the above analysis with the boundary conditions: $u_x(0, y) = u(a, y) = u_y(x, 0) = u(x, b) = 0$.]

2. Let us apply the same approach that we used to find the solution of Poisson's equation on a rectangle to solve the axisymmetric Poisson equation inside a circular cylinder

$$\frac{1}{r}\frac{\partial}{\partial r}\left(r\frac{\partial u}{\partial r}\right) + \frac{\partial^2 u}{\partial z^2} = f(r, z), \qquad 0 \le r < a, -b < z < b$$

subject to the boundary conditions

$$\lim_{r \to 0} |u(r, z)| < \infty, \qquad u(a, z) = 0, \qquad -b < z < b$$

and

$$u(r, -b) = u(r, b) = 0, \qquad 0 \le r < a.$$

Step 1. Replace the original problem with

$$\frac{1}{r}\frac{\partial}{\partial r}\left(r\frac{\partial u}{\partial r}\right) + \frac{\partial^2 u}{\partial z^2} = \lambda u, \qquad 0 \le r < a, -b < z < b$$

subject to the same boundary conditions. Use separation of variables to show that the solution to this new problem is

$$u_{nm}(r, z) = A_{nm} J_0\left(k_n \frac{r}{a}\right) \cos\left[\frac{\left(m + \frac{1}{2}\right)\pi z}{b}\right],$$

where k_n is the nth zero of $J_0(k) = 0$, $n = 1, 2, 3, \ldots$ and $m = 0, 1, 2, \ldots$.

Step 2. Show that $f(r, z)$ can be expressed as

$$f(r, z) = \sum_{n=1}^{\infty}\sum_{m=0}^{\infty} a_{nm} J_0\left(k_n \frac{r}{a}\right) \cos\left[\frac{\left(m + \frac{1}{2}\right)\pi z}{b}\right],$$

where

$$a_{nm} = \frac{2}{a^2 b J_1^2(k_n)} \int_{-b}^{b}\int_0^a f(r, z) J_0\left(k_n \frac{r}{a}\right) \cos\left[\frac{\left(m + \frac{1}{2}\right)\pi z}{b}\right] r\, dr\, dz.$$

Step 3. Show that the general solution is

$$u(r, z) = -\sum_{n=1}^{\infty}\sum_{m=0}^{\infty} a_{nm} \frac{J_0(k_n r/a) \cos\left[\left(m + \frac{1}{2}\right)\pi z/b\right]}{(k_n/a)^2 + \left[\left(m + \frac{1}{2}\right)\pi/b\right]^2}.$$

9.6 THE LAPLACE TRANSFORM METHOD

Laplace transforms are useful in solving Laplace's or Poisson's equation over a semi-infinite strip. The following problem illustrates this technique.

Let us solve Poisson's equation within an semi-infinite circular cylinder

$$\frac{1}{r}\frac{\partial}{\partial r}\left(r\frac{\partial u}{\partial r}\right)+\frac{\partial^2 u}{\partial z^2}=\frac{2}{b}n(z)\delta(r-b) \quad 0\le r<a, 0<z<\infty \tag{9.6.1}$$

subject to the boundary conditions

$$u(r,0)=0 \quad \text{and} \quad \lim_{z\to\infty}|u(r,z)|<\infty, \quad 0\le r<a \tag{9.6.2}$$

and

$$u(a,z)=0, \qquad 0<z<\infty, \tag{9.6.3}$$

where $0<b<a$. This problem gives the electrostatic potential within a semi-infinite cylinder of radius a that is grounded and has the charge density of $n(z)$ within an infinitesimally thin shell located at $r=b$.

Because the domain is semi-infinite in the z direction, we introduce the Laplace transform

$$U(r,s)=\int_0^\infty u(r,z)\,e^{-sz}\,dz. \tag{9.6.4}$$

Thus, taking the Laplace transform of (9.6.1), we have that

$$\frac{1}{r}\frac{d}{dr}\left[r\frac{dU(r,s)}{dr}\right]+s^2U(r,s)-su(r,0)-u_z(r,0)=\frac{2}{b}N(s)\delta(r-b). \tag{9.6.5}$$

Although $u(r,0)=0$, $u_z(r,0)$ is unknown and we denote its value by $f(r)$. Therefore, (9.6.5) becomes

$$\frac{1}{r}\frac{d}{dr}\left[r\frac{dU(r,s)}{dr}\right]+s^2U(r,s)=f(r)+\frac{2}{b}N(s)\delta(r-b), \quad 0\le r<a \tag{9.6.6}$$

with $\lim_{r\to 0}|U(r,s)|<\infty$ and $U(a,s)=0$.

To solve (9.6.6) we first assume that we can rewrite $f(r)$ as the Fourier-Bessel series:

$$f(r)=\sum_{n=1}^{\infty}A_nJ_0(k_nr/a), \tag{9.6.7}$$

where k_n is the nth root of the $J_0(k) = 0$ and

$$A_n = \frac{2}{a^2 J_1^2(k_n)} \int_0^a f(r)\, J_0(k_n r/a)\, r\, dr. \tag{9.6.8}$$

Similarly, the expansion for the delta function is

$$\delta(r-b) = \frac{2b}{a^2} \sum_{n=1}^{\infty} \frac{J_0(k_n b/a) J_0(k_n r/a)}{J_1^2(k_n)}, \tag{9.6.9}$$

because

$$\int_0^a \delta(r-b) J_0(k_n r/a)\, r\, dr = b\, J_0(k_n b/a). \tag{9.6.10}$$

Why we have chosen this particular expansion will become apparent shortly.

Thus, (9.6.6) may be rewritten as

$$\frac{1}{r}\frac{d}{dr}\left[r\frac{dU(r,s)}{dr}\right] + s^2 U(r,s) = \frac{2}{a^2} \sum_{n=1}^{\infty} \frac{2N(s)J_0(k_n b/a) + a_k}{J_1^2(k_n)} J_0(k_n r/a), \tag{9.6.11}$$

where $a_k = \int_0^a f(r)\, J_0(k_n r/a)\, r\, dr$.

The form of the right side of (9.6.11) suggests that we seek solutions of the form

$$U(r,s) = \sum_{n=1}^{\infty} B_n J_0(k_n r/a), \qquad 0 \le r < a. \tag{9.6.12}$$

We now understand why we rewrote the right side of (9.6.6) as a Fourier-Bessel series; the solution $U(r,s)$ automatically satisfies the boundary condition $U(a,s) = 0$. Substituting (9.6.12) into (9.6.11), we find that

$$U(r,s) = \frac{2}{a^2} \sum_{n=1}^{\infty} \frac{2N(s)J_0(k_n b/a) + a_k}{(s^2 - k_n^2/a^2) J_1^2(k_n)} J_0(k_n r/a), \qquad 0 \le r < a. \tag{9.6.13}$$

We have not yet determined a_k. Note, however, that in order for the inverse of (9.6.13) *not* to grow as $e^{k_n z/a}$, the numerator must vanish when $s = k_n/a$. Thus, $a_k = -2N(k_n/a)J_0(k_n b/a)$ and

$$U(r,s) = \frac{4}{a^2} \sum_{n=1}^{\infty} \frac{[N(s) - N(k_n/a)]J_0(k_n b/a)}{(s^2 - k_n^2/a^2) J_1^2(k_n)} J_0(k_n r/a), \quad 0 \le r < a. \tag{9.6.14}$$

The inverse of $U(r, s)$ then follows directly from simple inversions, the convolution theorem, and the definition of the Laplace transform. The final solution is

$$
\begin{aligned}
u(r, z) = \frac{2}{a} \sum_{n=1}^{\infty} \frac{J_0(k_n b/a) J_0(k_n r/a)}{k_n \, J_1^2(k_n)} \\
\times \left[\int_0^z n(\tau) e^{k_n(z-\tau)/a} \, d\tau - \int_0^z n(\tau) e^{-k_n(z-\tau)/a} \, d\tau \right. \\
\left. - \int_0^\infty n(\tau) e^{-k_n \tau/a} e^{k_n z/a} \, d\tau + \int_0^\infty n(\tau) e^{-k_n \tau/a} e^{-k_n z/a} \, d\tau \right]
\end{aligned} \tag{9.6.15}
$$

$$
\begin{aligned}
= \frac{2}{a} \sum_{n=1}^{\infty} \frac{J_0(k_n b/a) J_0(k_n r/a)}{k_n \, J_1^2(k_n)} \\
\times \left[\int_0^\infty n(\tau) e^{-k_n(z+\tau)/a} \, d\tau - \int_0^z n(\tau) e^{-k_n(z-\tau)/a} \, d\tau \right. \\
\left. - \int_z^\infty n(\tau) e^{-k_n(\tau-z)/a} \, d\tau \right].
\end{aligned} \tag{9.6.16}
$$

Problems

1. Use Laplace transforms to solve

$$\frac{\partial^2 u}{\partial x^2} + \frac{\partial^2 u}{\partial y^2} = 0, \qquad 0 < x < \infty, 0 < y < a$$

subject to the boundary conditions

$$u(0, y) = 1, \qquad \lim_{x \to \infty} |u(x, y)| < \infty, \qquad 0 < y < a$$

and

$$u(x, 0) = u(x, a) = 0, \qquad 0 < x < \infty.$$

2. Use Laplace transforms to solve

$$\frac{1}{r} \frac{\partial}{\partial r} \left(r \frac{\partial u}{\partial r} \right) + \frac{\partial^2 u}{\partial z^2} = 0, \quad 0 \le r < a, 0 < z < \infty$$

subject to the boundary conditions

$$u(r, 0) = 1, \qquad \lim_{z \to \infty} |u(r, z)| < \infty, \qquad 0 < r < a$$

and

$$\lim_{r\to 0} |u(r,z)| < \infty \quad \text{and} \quad u(a,z) = 0, \qquad 0 < z < \infty.$$

9.7 NUMERICAL SOLUTION OF LAPLACE'S EQUATION

As in the case of the heat and wave equations, numerical methods can be used to solve elliptic partial differential equations when analytic techniques fail or are too cumbersome. They are also employed when the domain differs from simple geometries.

The numerical analysis of an elliptic partial differential equation begins by replacing the continuous partial derivatives by finite-difference formulas. Employing centered differencing,

$$\frac{\partial^2 u}{\partial x^2} = \frac{u_{m+1,n} - 2u_{m,n} + u_{m-1,n}}{(\Delta x)^2} + O[(\Delta x)^2] \tag{9.7.1}$$

and

$$\frac{\partial^2 u}{\partial y^2} = \frac{u_{m,n+1} - 2u_{m,n} + u_{m,n-1}}{(\Delta y)^2} + O[(\Delta y)^2], \tag{9.7.2}$$

where $u_{m,n}$ denote the solution value at the grid point m, n. If $\Delta x = \Delta y$, Laplace's equation becomes the difference equation

$$u_{m+1,n} + u_{m-1,n} + u_{m,n+1} + u_{m,n-1} - 4u_{m,n} = 0. \tag{9.7.3}$$

Thus, we must now solve a set of simultaneous linear equations that yield the value of the solution at each grid point.

The solution of (9.7.3) is best done using techniques developed by algebraist. Later on, in Chapter 11, we will show that a very popular method for directly solving systems of linear equations is Gaussian elimination. However, for many grids at a reasonable resolution, the number of equations are generally in the tens of thousands. Because most of the coefficients in the equations are zero, Gaussian elimination is unsuitable, both from the point of view of computational expense and accuracy. For this reason alternative methods have been developed that generally use successive corrections or iterations. The most common of these point iterative methods are the Jacobi method, unextrapolated Liebmann or Gauss-Seidel method, and extrapolated Liebmann or successive over-relaxation (SOR). None of these approaches is completely satisfactory because of questions involving convergence and efficiency. Because of its simplicity we will focus on the Gauss-Seidel method.

We may illustrate the Gauss-Seidel method by considering the system:

$$10x + y + z = 39 \tag{9.7.4}$$

$$2x + 10y + z = 51 \tag{9.7.5}$$

$$2x + 2y + 10z = 64. \tag{9.7.6}$$

An important aspect of this system is the dominance of the coefficient of x in the first equation of the set and that the coefficients of y and z are dominant in the second and third equations, respectively.

The Gauss-Seidel method may be outlined as follow:

- Assign an initial value for each unknown variable. If possible, make a good first guess. If not, any arbitrarily selected values may be chosen. The initial value will not affect the convergence but will affect the number of iterations until convergence.

- Starting with (9.7.4), solve that equation for a new value of the unknown which has the largest coefficient in that equation, using the assumed values for the other unknowns.

- Go to (9.7.5) and employ the same technique used in the previous step to compute the unknown that has the largest coefficient in that equation. Where possible, use the latest values.

- Proceed to the remaining equations, always solving for the unknown having the largest coefficient in the particular equation and always using the *most recently* calculated values for the other unknowns in the equation. When the last equation (9.7.6) has been solved, you have completed a single iteration.

- Iterate until the value of each unknown does not change within a predetermined value.

Usually a compromise must be struck between the accuracy of the solution and the desired rate of convergence. The more accurate the solution is, the longer it will take for the solution to converge.

To illustrate this method, let us solve our system (9.7.4)–(9.7.6) with the initial guess $x = y = z = 0$. The first iteration yields $x = 3.9$, $y = 4.32$, and $z = 4.756$. The second iteration yields $x = 2.9924$, $y = 4.02592$, and $z = 4.996336$. As can be readily seen, the solution is converging to the correction solution of $x = 3$, $y = 4$, and $z = 5$.

Applying these techniques to (9.7.3),

$$u_{m,n}^{k+1} = \tfrac{1}{4}\left(u_{m+1,n}^{k} + u_{m-1,n}^{k+1} + u_{m,n+1}^{k} + u_{m,n-1}^{k+1}\right), \tag{9.7.7}$$

where we have assumed that the calculations occur in order of increasing m and n.

• Example 9.7.1

To illustrate the numerical solution of Laplace's equation, let us redo Example 9.3.1 with the boundary condition along $y = H$ simplified to $u(x, H) = 1 + x/L$.

We begin by finite-differencing the boundary conditions. The condition $u_x(0, y) = u_x(L, y) = 0$ leads to $u_{1,n} = u_{-1,n}$ and $u_{L+1,n} = u_{L-1,n}$ if we employ centered differences at $m = 0$ and $m = L$. Substituting these values in (9.7.7), we have the following equations for the left and right boundaries:

$$u_{0,n}^{k+1} = \tfrac{1}{4}\left(2u_{1,n}^{k} + u_{0,n+1}^{k} + u_{0,n-1}^{k+1}\right) \tag{9.7.8}$$

and

$$u_{L,n}^{k+1} = \tfrac{1}{4}\left(2u_{L-1,n}^{k+1} + u_{L,n+1}^{k} + u_{L,n-1}^{k+1}\right). \tag{9.7.9}$$

On the other hand, $u_y(x, 0) = 0$ yields $u_{m,1} = u_{m,-1}$ and

$$u_{m,0}^{k+1} = \tfrac{1}{4}\left(u_{m+1,0}^{k} + u_{m-1,0}^{k+1} + 2u_{m,1}^{k}\right). \tag{9.7.10}$$

At the bottom corners, (9.7.8)–(9.7.10) simplify to

$$u_{0,0}^{k+1} = \tfrac{1}{2}\left(u_{1,0}^{k} + u_{0,1}^{k}\right) \tag{9.7.11}$$

and

$$u_{L,0}^{k+1} = \tfrac{1}{2}\left(u_{L-1,0}^{k+1} + u_{L,1}^{k}\right). \tag{9.7.12}$$

These equations along with (9.7.7) were solved using the Gauss-Seidel method. The initial guess everywhere except along the top boundary was zero. In Figure 9.7.1 we illustrate the numerical solution after 100 and 300 iterations where we have taken 101 grid points in the x and y directions.

Project: Successive Over-Relaxation

The fundamental difficulty with relaxation methods used in solving Laplace's equation is the rate of convergence. Assuming $\Delta x = \Delta y$, the most popular method for accelerating convergence of these techniques is *successive over-relaxation*:

$$u_{m,n}^{k+1} = u_{m,n}^{k} + \omega R_{m,n},$$

where

$$R_{m,n} = \tfrac{1}{4}\left(u_{m+1,n}^{k} + u_{m-1,n}^{k+1} + u_{m,n+1}^{k} + u_{m,n-1}^{k+1}\right).$$

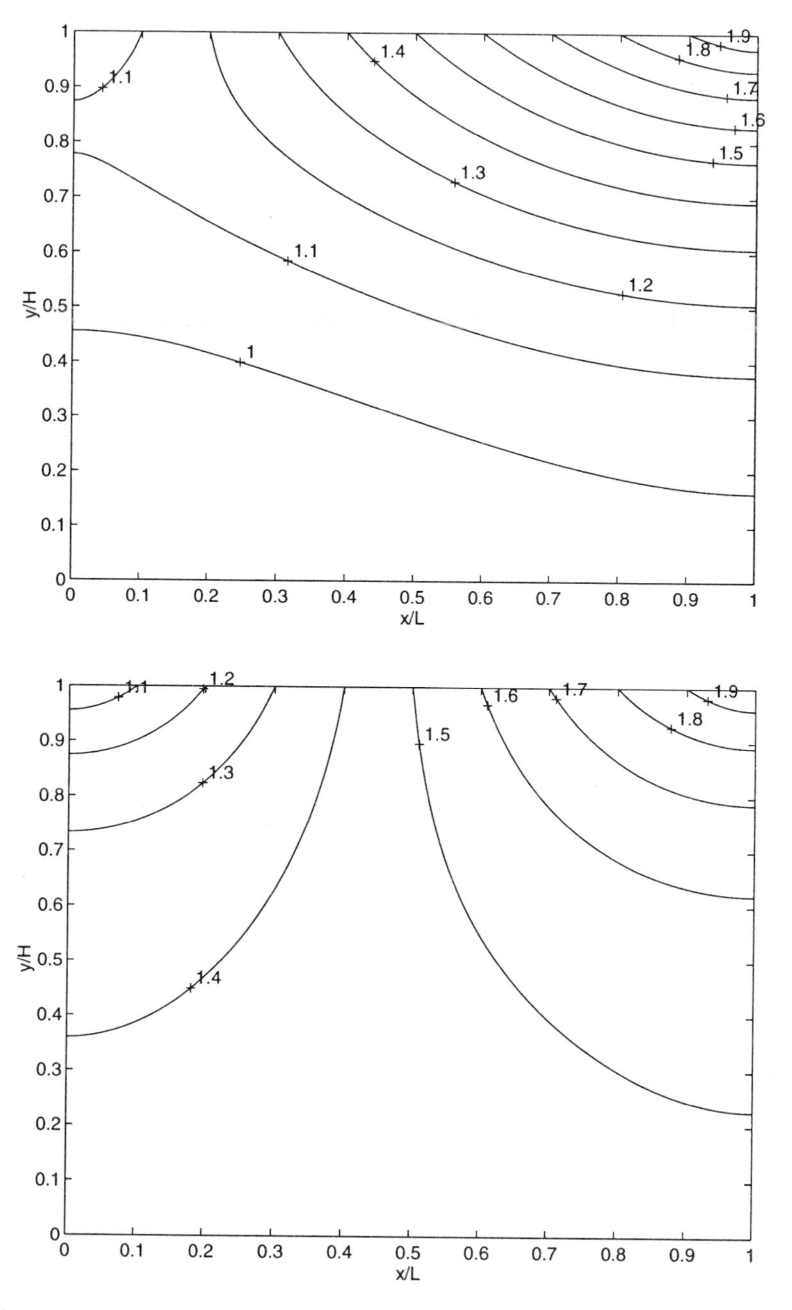

Figure 9.7.1: The solution to Laplace's equation by the Gauss-Seidel method after 100 (top) and 300 (bottom) iterations. The boundary conditions are $u_x(0, y) = u_x(L, y) = u_y(x, 0) = 0$ and $u(x, H) = 1+x/L$.

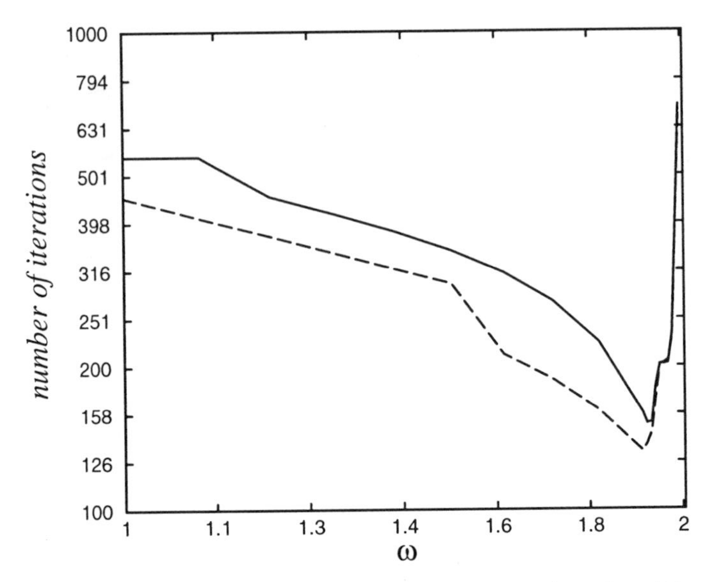

Figure 9.7.2: The number of iterations required so that $|R_{m,n}| \leq 10^{-3}$ as a function of ω during the iterative solution of the problem posed in the project. We used $\Delta x = \Delta y = 0.01$ and $L = z_0 = 1$. The iteration count for the boundary conditions stated in Step 1 are given by the solid line while the iteration count for the boundary conditions given in Step 2 are shown by the dotted line. The initial guess equaled zero.

Most numerical methods dealing with partial differential equations will discuss the theoretical reasons behind this technique;[12] the optimum value always lies between one and two.

Step 1: Solve Laplace's equation numerically $0 \leq x \leq L$, $0 \leq y \leq z_0$ with the following boundary conditions:

$$u(x, 0) = 0, u(x, z_0) = 1 + x/L, u(0, y) = y/z_0, \text{ and } u(L, y) = 2y/z_0.$$

Count the number of iterations until $|R_{m,n}| \leq 10^{-3}$ for *all* m and n. Plot this number of iterations as a function of ω. How does the curve change with resolution Δx?

Step 2: Redo Step 1 with the exception of $u(0, y) = u(L, y) = 0$. How has the convergence rate changed? Can you explain why? How sensitive are your results to the first guess?

[12] For example, Young, D. M., 1971: *Iterative Solution of Large Linear Systems*, Academic Press, New York.

Chapter 10
Vector Calculus

Physicists invented vectors and vector operations to facilitate their mathematical expression of such diverse topics as mechanics and electromagnetism. In this chapter we focus on multivariable differentiations and integrations of vector fields, such as the velocity of a fluid, where the vector field is solely a function of its position.

10.1 REVIEW

The physical sciences and engineering abound with vectors and scalars. *Scalars* are physical quantities which only possess magnitude. Examples include mass, temperature, density, and pressure. *Vectors* are physical quantities that possess both magnitude and direction. Examples include velocity, acceleration, and force. We shall denote vectors by boldface letters.

Two vectors are equal if they have the same magnitude and direction. From the limitless number of possible vectors, two special cases are the *zero vector* $\mathbf{0}$ which has no magnitude and unspecified direction and the *unit vector* which has unit magnitude.

The most convenient method for expressing a vector analytically is in terms of its components. A vector $\mathbf{a}$ in three-dimensional real space is any order triplet of real numbers (*components*) a_1, a_2, and a_3 such that $\mathbf{a} = a_1\mathbf{i} + a_2\mathbf{j} + a_3\mathbf{k}$, where $a_1\mathbf{i}$, $a_2\mathbf{j}$, and $a_3\mathbf{k}$ are vectors which lie along the coordinate axes and have their origin at a common initial point. The *magnitude*, *length*, or *norm* of a vector $\mathbf{a}$, $|\mathbf{a}|$, equals $\sqrt{a_1^2 + a_2^2 + a_3^2}$. A particularly important vector is the *position vector*, defined by $\mathbf{r} = x\mathbf{i} + y\mathbf{j} + z\mathbf{k}$.

As in the case of scalars, certain arithmetic rules hold. Addition and subtraction are very similar to their scalar counterparts:

$$\mathbf{a} + \mathbf{b} = (a_1 + b_1)\mathbf{i} + (a_2 + b_2)\mathbf{j} + (a_3 + b_3)\mathbf{k} \tag{10.1.1}$$

and

$$\mathbf{a} - \mathbf{b} = (a_1 - b_1)\mathbf{i} + (a_2 - b_2)\mathbf{j} + (a_3 - b_3)\mathbf{k}. \tag{10.1.2}$$

In contrast to its scalar counterpart, there are two types of multiplication. The *dot product* is defined as

$$\mathbf{a} \cdot \mathbf{b} = |\mathbf{a}||\mathbf{b}|\cos(\theta) = a_1 b_1 + a_2 b_2 + a_3 b_3, \tag{10.1.3}$$

where θ is the angle between the vector such that $0 \le \theta \le \pi$. The dot product yields a scalar answer. A particularly important case is $\mathbf{a} \cdot \mathbf{b} = 0$ with $|\mathbf{a}| \neq 0$ and $|\mathbf{b}| \neq 0$. In this case the vectors are orthogonal (perpendicular) to each other.

The other form of multiplication is the *cross product* which is defined by $\mathbf{a} \times \mathbf{b} = |\mathbf{a}||\mathbf{b}|\sin(\theta)\mathbf{n}$, where θ is the angle between the vectors such that $0 \le \theta \le \pi$ and $\mathbf{n}$ is a unit vector perpendicular to the plane of $\mathbf{a}$ and $\mathbf{b}$ with the direction given by the right-hand rule. A convenient method for computing the cross product from the scalar components of $\mathbf{a}$ and $\mathbf{b}$ is

$$\mathbf{a} \times \mathbf{b} = \begin{vmatrix} \mathbf{i} & \mathbf{j} & \mathbf{k} \\ a_1 & a_2 & a_3 \\ b_1 & b_2 & b_3 \end{vmatrix} = \begin{vmatrix} a_2 & a_3 \\ b_2 & b_3 \end{vmatrix} \mathbf{i} - \begin{vmatrix} a_1 & a_3 \\ b_1 & b_3 \end{vmatrix} \mathbf{j} + \begin{vmatrix} a_1 & a_2 \\ b_1 & b_2 \end{vmatrix} \mathbf{k}. \tag{10.1.4}$$

Two nonzero vectors $\mathbf{a}$ and $\mathbf{b}$ are *parallel* if and only if $\mathbf{a} \times \mathbf{b} = \mathbf{0}$.

Most of the vectors that we will use are vector-valued functions. These functions are vectors that vary either with a single parametric variable t or multiple variables, say x, y, and z.

The most commonly encountered example of a vector-valued function which varies with a single independent variable involves the trajectory of particles. If a *space curve* is parameterized by the equations $x = f(t)$, $y = g(t)$, and $z = h(t)$ with $a \le t \le b$, the position vector $\mathbf{r}(t) = f(t)\mathbf{i} + g(t)\mathbf{j} + h(t)\mathbf{k}$ gives the location of a point P as it moves from its initial position to its final position. Furthermore, because the increment quotient $\Delta\mathbf{r}/\Delta t$ is in the direction of a secant line, then the limit of this quotient as $\Delta t \to 0$, $\mathbf{r}'(t)$, gives the tangent to the curve at P.

• Example 10.1.1: Foucault pendulum

One of the great experiments of mid-nineteenth century physics was the demonstration by J. B. L. Foucault (1819–1868) in 1851 of the earth's rotation by designing a (spherical) pendulum, supported by a long wire, that essentially swings in an nonaccelerating coordinate system. This problem demonstrates many of the fundamental concepts of vector calculus.

The total force[1] acting on the bob of the pendulum is $\mathbf{F} = \mathbf{T} + m\mathbf{G}$, where $\mathbf{T}$ is the tension in the pendulum and $\mathbf{G}$ is the gravitational attraction per unit mass. Using Newton's second law,

$$\left.\frac{d^2\mathbf{r}}{dt^2}\right|_{\text{inertial}} = \frac{\mathbf{T}}{m} + \mathbf{G}, \tag{10.1.5}$$

where $\mathbf{r}$ is the position vector from a fixed point in an inertial coordinate system to the bob. This system is inconvenient because we live in a rotating coordinate system. Employing the conventional geographic coordinate system,[2] (10.1.5) becomes

$$\frac{d^2\mathbf{r}}{dt^2} + 2\boldsymbol{\Omega} \times \frac{d\mathbf{r}}{dt} + \boldsymbol{\Omega} \times (\boldsymbol{\Omega} \times \mathbf{r}) = \frac{\mathbf{T}}{m} + \mathbf{G}, \tag{10.1.6}$$

where $\boldsymbol{\Omega}$ is the angular rotation vector of the earth and $\mathbf{r}$ now denotes a position vector in the rotating reference system with its origin at the center of the earth and terminal point at the bob. If we define the gravity vector $\mathbf{g} = \mathbf{G} - \boldsymbol{\Omega} \times (\boldsymbol{\Omega} \times \mathbf{r})$, then the dynamical equation is

$$\frac{d^2\mathbf{r}}{dt^2} + 2\boldsymbol{\Omega} \times \frac{d\mathbf{r}}{dt} = \frac{\mathbf{T}}{m} + \mathbf{g}, \tag{10.1.7}$$

where the second term on the left side of (10.1.7) is called the *Coriolis force*.

Because the equation is *linear*, let us break the position vector $\mathbf{r}$ into two separate vectors: $\mathbf{r}_0$ and $\mathbf{r}_1$, where $\mathbf{r} = \mathbf{r}_0 + \mathbf{r}_1$. The vector $\mathbf{r}_0$ extends from the center of the earth to the pendulum's point of support and $\mathbf{r}_1$ extends from the support point to the bob. Because $\mathbf{r}_0$ is a constant in the geographic system,

$$\frac{d^2\mathbf{r}_1}{dt^2} + 2\boldsymbol{\Omega} \times \frac{d\mathbf{r}_1}{dt} = \frac{\mathbf{T}}{m} + \mathbf{g}. \tag{10.1.8}$$

---

[1] From Broxmeyer, C., 1960: Foucault pendulum effect in a Schuler-tuned system. *J. Aerosp. Sci.*, **27**, 343–347 with permission.

[2] For the derivation, see Marion, J. B., 1965: *Classical Dynamics of Particles and Systems,* Academic Press, New York, Sections 12.2–12.3.

If the length of the pendulum is L, then for small oscillations $\mathbf{r}_1 \approx x\mathbf{i} + y\mathbf{j} + L\mathbf{k}$ and the equations of motion are

$$\frac{d^2x}{dt^2} + 2\Omega\sin(\lambda)\frac{dy}{dt} = \frac{T_x}{m}, \tag{10.1.9}$$

$$\frac{d^2y}{dt^2} - 2\Omega\sin(\lambda)\frac{dx}{dt} = \frac{T_y}{m} \tag{10.1.10}$$

and

$$2\Omega\cos(\lambda)\frac{dy}{dt} - g = \frac{T_z}{m}, \tag{10.1.11}$$

where λ denotes the latitude of the point and Ω is the rotation rate of the earth. The relationships between the components of tension are $T_x = xT_z/L$ and $T_y = yT_z/L$. From (10.1.11),

$$\frac{T_z}{m} + g = 2\Omega\cos(\lambda)\frac{dy}{dt} \approx 0. \tag{10.1.12}$$

Substituting the definitions of T_x, T_y and (10.1.12) into (10.1.9) and (10.1.10),

$$\frac{d^2x}{dt^2} + \frac{g}{L}x + 2\Omega\sin(\lambda)\frac{dy}{dt} = 0 \tag{10.1.13}$$

and

$$\frac{d^2y}{dt^2} + \frac{g}{L}y - 2\Omega\sin(\lambda)\frac{dx}{dt} = 0. \tag{10.1.14}$$

The approximate solution to these coupled differential equations is

$$x(t) = A_0\cos[\Omega\sin(\lambda)t]\sin\left(\sqrt{g/L}\,t\right) \tag{10.1.15}$$

and

$$y(t) = A_0\sin[\Omega\sin(\lambda)t]\sin\left(\sqrt{g/L}\,t\right) \tag{10.1.16}$$

if $\Omega^2 \ll g/L$. Thus, we have a pendulum that swings with an angular frequency $\sqrt{g/L}$. However, depending upon the *latitude* λ, the direction in which the pendulum swings changes counterclockwise with time, completing a full cycle in $2\pi/[\Omega\sin(\lambda)]$. This result is most clearly seen when $\lambda = \pi/2$ and we are at the North Pole. There the earth is turning underneath the pendulum. If initially we set the pendulum swinging along the 0° longitude, the pendulum will shift with time to longitudes east of the Greenwich median. Eventually, after 24 hours, the process will repeat itself.

Consider now vector-valued functions that vary with several variables. A *vector function of position* assigns a vector value for every value

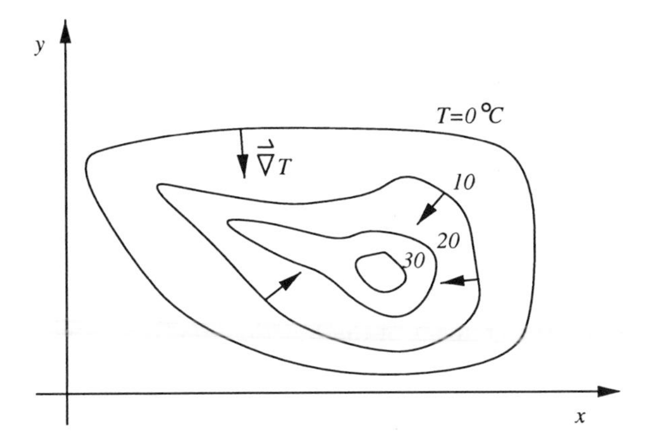

Figure 10.1.1: A graphical example of the gradient: A vector that is perpendicular to the isotherms $T(x, y) =$ constant and points in the direction of most rapidly increasing temperatures.

of x, y, and z within some domain. Examples include the velocity field of a fluid at a given instant:

$$\mathbf{v} = u(x, y, z)\mathbf{i} + v(x, y, z)\mathbf{j} + w(x, y, z)\mathbf{k}. \tag{10.1.17}$$

Another example arises in electromagnetism where electric and magnetic fields often vary as a function of the space coordinates. For us, however, probably the most useful example involves the vector differential operator, *del* or *nabla*,

$$\nabla = \frac{\partial}{\partial x}\mathbf{i} + \frac{\partial}{\partial y}\mathbf{j} + \frac{\partial}{\partial z}\mathbf{k} \tag{10.1.18}$$

which we apply to the multivariable differentiable scalar function $F(x, y, z)$ to give the *gradient* ∇F.

An important geometric interpretation of the gradient – one which we shall use frequently – is the fact that ∇f is perpendicular (normal) to the level surface at a given point P. To prove this, let the equation $F(x, y, z) = c$ describe a three-dimensional surface. If the differentiable functions $x = f(t)$, $y = g(t)$, and $z = h(t)$ are the parametric equations of a curve on the surface, then the derivative of $F[f(t), g(t), h(t)] = c$ is

$$\frac{\partial F}{\partial x}\frac{dx}{dt} + \frac{\partial F}{\partial y}\frac{dy}{dt} + \frac{\partial F}{\partial z}\frac{dz}{dt} = 0 \tag{10.1.19}$$

or

$$\nabla F \cdot \mathbf{r}' = 0. \tag{10.1.20}$$

When $\mathbf{r}' \neq \mathbf{0}$, the vector ∇F is orthogonal to the tangent vector. Because our argument holds for any differentiable curve that passes through the arbitrary point (x, y, z), then ∇F is normal to the level surface at that point.

Figure 10.1.1 gives a common application of the gradient. Consider a two-dimensional temperature field $T(x, y)$. The level curves $T(x, y) =$ constant are lines that connect points where the temperature is the same (isotherms). The gradient in this case ∇T is a vector that is perpendicular or normal to these isotherms and points in the direction of most rapidly increasing temperature.

• **Example 10.1.2**

Let us find the gradient of the function $f(x, y, z) = x^2 z^2 \sin(4y)$. Using the definition of gradient,

$$\nabla f = \frac{\partial[x^2 z^2 \sin(4y)]}{\partial x}\mathbf{i} + \frac{\partial[x^2 z^2 \sin(4y)]}{\partial y}\mathbf{j} + \frac{\partial[x^2 z^2 \sin(4y)]}{\partial z}\mathbf{k} \quad (10.1.21)$$

$$= 2xz^2 \sin(4y)\mathbf{i} + 4x^2 z^2 \cos(4y)\mathbf{j} + 2x^2 z \sin(4y)\mathbf{k}. \quad (10.1.22)$$

• **Example 10.1.3**

Let us find the unit normal to the unit sphere at any arbitrary point (x, y, z).

The surface of a unit sphere is defined by the equation $f(x, y, z) = x^2 + y^2 + z^2 = 1$. Therefore, the normal is given by the gradient

$$\mathbf{N} = \nabla f = 2x\mathbf{i} + 2y\mathbf{j} + 2z\mathbf{k} \quad (10.1.23)$$

and the unit normal

$$\mathbf{n} = \frac{\nabla f}{|\nabla f|} = \frac{2x\mathbf{i} + 2y\mathbf{j} + 2z\mathbf{k}}{\sqrt{4x^2 + 4y^2 + 4z^2}} = x\mathbf{i} + y\mathbf{j} + z\mathbf{k}, \quad (10.1.24)$$

because $x^2 + y^2 + z^2 = 1$.

A popular method for visualizing a vector field $\mathbf{F}$ is to draw space curves which are tangent to the vector field at each x, y, z. In fluid mechanics these lines are called *streamlines* while in physics they are generally called *lines of force* or *flux lines* for an electric, magnetic, or gravitational field. For a fluid with a velocity field that does not vary with time, the streamlines give the paths along which small parcels of the fluid move.

To find the streamlines of a given vector field $\mathbf{F}$ with components $P(x, y, z)$, $Q(x, y, z)$, and $R(x, y, z)$, we assume that we can parameterize the streamlines in the form $\mathbf{r}(t) = x(t)\mathbf{i} + y(t)\mathbf{j} + z(t)\mathbf{k}$. Then the tangent line is $\mathbf{r}'(t) = x'(t)\mathbf{i} + y'(t)\mathbf{j} + z'(t)\mathbf{k}$. Because the streamline must be parallel to the vector field at any t, $\mathbf{r}'(t) = \lambda\mathbf{F}$ or

$$\frac{dx}{dt} = \lambda P(x, y, z), \quad \frac{dy}{dt} = \lambda Q(x, y, z) \quad \text{and} \quad \frac{dz}{dt} = \lambda R(x, y, z) \quad (10.1.25)$$

or

$$\frac{dx}{P(x,y,z)} = \frac{dy}{Q(x,y,z)} = \frac{dz}{R(x,y,z)}. \tag{10.1.26}$$

The solution of this system of differential equations yields the streamlines.

• Example 10.1.4

Let us find the streamlines for the vector field $\mathbf{F} = \sec(x)\mathbf{i} - \cot(y)\mathbf{j} + \mathbf{k}$ that passes through the point $(\pi/4, \pi, 1)$. In this particular example, $\mathbf{F}$ represents a measured or computed fluid's velocity at a particular instant.

From (10.1.26),

$$\frac{dx}{\sec(x)} = -\frac{dy}{\cot(y)} = \frac{dz}{1}. \tag{10.1.27}$$

This yields two differential equations:

$$\cos(x)\,dx = -\frac{\sin(y)}{\cos(y)}\,dy \quad \text{and} \quad dz = -\frac{\sin(y)}{\cos(y)}\,dy. \tag{10.1.28}$$

Integrating these equations yields

$$\sin(x) = \ln|\cos(y)| + c_1 \quad \text{and} \quad z = \ln|\cos(y)| + c_2. \tag{10.1.29}$$

Substituting for the given point, we finally have that

$$\sin(x) = \ln|\cos(y)| + \sqrt{2}/2 \quad \text{and} \quad z = \ln|\cos(y)| + 1. \tag{10.1.30}$$

• Example 10.1.5

Let us find the streamlines for the vector field $\mathbf{F} = \sin(z)\mathbf{j} + e^y\mathbf{k}$ that passes through the point $(2, 0, 0)$.

From (10.1.26),

$$\frac{dx}{0} = \frac{dy}{\sin(z)} = \frac{dz}{e^y}. \tag{10.1.31}$$

This yields two differential equations:

$$dx = 0 \qquad \text{and} \qquad \sin(z)\,dz = e^y\,dy. \tag{10.1.32}$$

Integrating these equations gives

$$x = c_1 \qquad \text{and} \qquad e^y = -\cos(z) + c_2. \tag{10.1.33}$$

Substituting for the given point, we finally have that

$$x = 2 \qquad \text{and} \qquad e^y = 2 - \cos(z). \tag{10.1.34}$$

Note that (10.1.34) only applies for a certain strip in the yz-plane.

Problems

Given the following vectors $\mathbf{a}$ and $\mathbf{b}$, verify that $\mathbf{a} \cdot (\mathbf{a} \times \mathbf{b}) = 0$ and $\mathbf{b} \cdot (\mathbf{a} \times \mathbf{b}) = 0$:

1. $\mathbf{a} = 4\mathbf{i} - 2\mathbf{j} + 5\mathbf{k}, \quad \mathbf{b} = 3\mathbf{i} + \mathbf{j} - \mathbf{k}$

2. $\mathbf{a} = \mathbf{i} - 3\mathbf{j} + \mathbf{k}, \quad \mathbf{b} = 2\mathbf{i} + 4\mathbf{k}$

3. $\mathbf{a} = \mathbf{i} + \mathbf{j} + \mathbf{k}, \quad \mathbf{b} = -5\mathbf{i} + 2\mathbf{j} + 3\mathbf{k}$

4. $\mathbf{a} = 8\mathbf{i} + \mathbf{j} - 6\mathbf{k}, \quad \mathbf{b} = \mathbf{i} - 2\mathbf{j} + 10\mathbf{k}$

5. $\mathbf{a} = 2\mathbf{i} + 7\mathbf{j} - 4\mathbf{k}, \quad \mathbf{b} = \mathbf{i} + \mathbf{j} - \mathbf{k}$.

6. Prove $\mathbf{a} \times (\mathbf{b} \times \mathbf{c}) = (\mathbf{a} \cdot \mathbf{c})\mathbf{b} - (\mathbf{a} \cdot \mathbf{b})\mathbf{c}$.

7. Prove $\mathbf{a} \times (\mathbf{b} \times \mathbf{c}) + \mathbf{b} \times (\mathbf{c} \times \mathbf{a}) + \mathbf{c} \times (\mathbf{a} \times \mathbf{b}) = \mathbf{0}$.

Find the gradient of the following functions:

8. $f(x, y, z) = xy^2/z^3$

9. $f(x, y, z) = xy\cos(yz)$

10. $f(x, y, z) = \ln(x^2 + y^2 + z^2)$

11. $f(x, y, z) = x^2y^2(2z + 1)^2$

12. $f(x, y, z) = 2x - y^2 + z^2$.

Sketch the following surfaces. For each of these surfaces, find a mathematical expression for the unit normal and then sketch it.

13. $z = 3$

14. $x^2 + y^2 = 4$

15. $z = x^2 + y^2$

16. $z = \sqrt{x^2 + y^2}$

17. $z = y$

18. $x + y + z = 1$

19. $z = x^2$.

Find the streamlines for the following vector fields that pass through the specified point:

20. $\mathbf{F} = \mathbf{i} + \mathbf{j} + \mathbf{k}$; $(0, 1, 1)$

21. $\mathbf{F} = 2\mathbf{i} - y^2\mathbf{j} + z\mathbf{k}$; $(1, 1, 1)$

22. $\mathbf{F} = 3x^2\mathbf{i} - y^2\mathbf{j} + z^2\mathbf{k}$; $(2, 1, 3)$

23. $\mathbf{F} = x^2\mathbf{i} + y^2\mathbf{j} - z^3\mathbf{k}$; $(1, 1, 1)$

24. $\mathbf{F} = (1/x)\mathbf{i} + e^y\mathbf{j} - \mathbf{k}$; $(2, 0, 4)$

25. Solve the differential equations (10.1.13)–(10.1.14) with the initial conditions $x(0) = y(0) = y'(0) = 0$ and $x'(0) = A_0\sqrt{g/L}$ assuming that $\Omega^2 \ll g/L$.

26. If a fluid is bounded by a fixed surface $f(x, y, z) = c$, show that the fluid must satisfy the boundary condition $\mathbf{v} \cdot \nabla f = 0$, where $\mathbf{v}$ is the velocity of the fluid.

27. A sphere of radius a is moving in a fluid with the constant velocity $\mathbf{u}$. Show that the fluid satisfies the boundary condition $(\mathbf{v} - \mathbf{u}) \cdot (\mathbf{r} - \mathbf{u}t) = 0$ at the surface of the sphere, if the center of the sphere coincides with the origin at $t = 0$ and $\mathbf{v}$ denotes the velocity of the fluid.

10.2 DIVERGENCE AND CURL

Consider a vector field $\mathbf{v}$ defined in some region of three-dimensional space. The function $\mathbf{v}(\mathbf{r})$ can be resolved into components along the $\mathbf{i}$, $\mathbf{j}$, and $\mathbf{k}$ directions or

$$\mathbf{v}(\mathbf{r}) = u(x, y, z)\mathbf{i} + v(x, y, z)\mathbf{j} + w(x, y, z)\mathbf{k}. \tag{10.2.1}$$

If $\mathbf{v}$ is a fluid's velocity field, then we can compute the flow rate through a small (differential) rectangular box defined by increments $(\Delta x, \Delta y, \Delta z)$ centered at the point (x, y, z). See Figure 10.2.1. The flow out from the box through the face with the outwardly pointing normal $\mathbf{n} = -\mathbf{j}$ is

$$\mathbf{v} \cdot (-\mathbf{j}) = -v(x, y - \Delta y/2, z)\Delta x \Delta z \tag{10.2.2}$$

and the flow through the face with the outwardly pointing normal $\mathbf{n} = \mathbf{j}$ is

$$\mathbf{v} \cdot \mathbf{j} = v(x, y + \Delta y/2, z)\Delta x \Delta z. \tag{10.2.3}$$

The net flow through the two faces is

$$[v(x, y + \Delta y/2, z) - v(x, y - \Delta y/2, z)]\Delta x \Delta z \approx v_y(x, y, z)\Delta x \Delta y \Delta z. \tag{10.2.4}$$

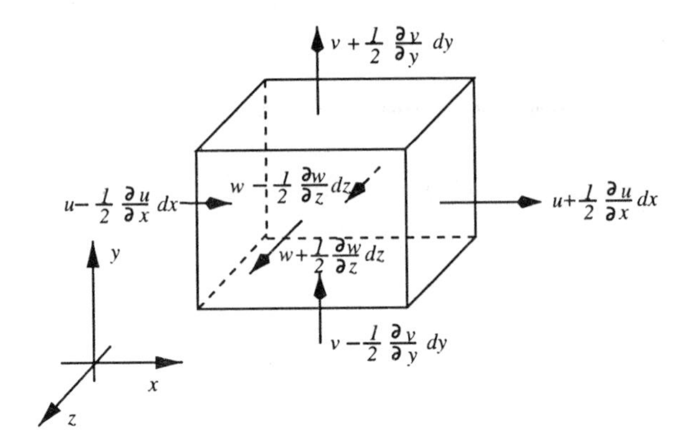

Figure 10.2.1: Divergence of a vector function $\mathbf{v}(x, y, z)$.

A similar analysis of the other faces and combination of the results give the approximate total flow from the box as

$$[u_x(x, y, z) + v_y(x, y, z) + w_z(x, y, z)]\Delta x \Delta y \Delta z. \tag{10.2.5}$$

Dividing by the volume $\Delta x \Delta y \Delta z$ and taking the limit as the dimensions of the box tend to zero yield $u_x + v_y + w_z$ as the flow out from (x, y, z) per unit volume per unit time. This scalar quantity is called the *divergence* of the vector $\mathbf{v}$:

$$\text{div}(\mathbf{v}) = \nabla \cdot \mathbf{v} = \left(\frac{\partial}{\partial x}\mathbf{i} + \frac{\partial}{\partial y}\mathbf{j} + \frac{\partial}{\partial z}\mathbf{k}\right) \cdot (u\mathbf{i} + v\mathbf{j} + w\mathbf{k}) = u_x + v_y + w_z. \tag{10.2.6}$$

Thus, if the divergence is positive, either the fluid is expanding and its density at the point is falling with time, or the point is a *source* at which fluid is entering the field. When the divergence is negative, either the fluid is contracting and its density is rising at the point, or the point is a negative source or *sink* at which fluid is leaving the field.

If the divergence of a vector field is zero everywhere within a domain, then the flux entering any element of space exactly equals that leaving it and the vector field is called *nondivergent* or *solenoidal* (from a Greek word meaning a tube). For a fluid, if there are no sources or sinks, then its density cannot change.

Some useful properties of the divergence operator are

$$\nabla \cdot (\mathbf{F} + \mathbf{G}) = \nabla \cdot \mathbf{F} + \nabla \cdot \mathbf{G}, \tag{10.2.7}$$

$$\nabla \cdot (\varphi \mathbf{F}) = \varphi \nabla \cdot \mathbf{F} + \mathbf{F} \cdot \nabla \varphi \tag{10.2.8}$$

and

$$\nabla^2 \varphi = \nabla \cdot \nabla \varphi = \varphi_{xx} + \varphi_{yy} + \varphi_{zz}. \tag{10.2.9}$$

The expression (10.2.9) is very important in physics and is given the special name of the *Laplacian*.[3]

[3] Some mathematicians write Δ instead of ∇^2.

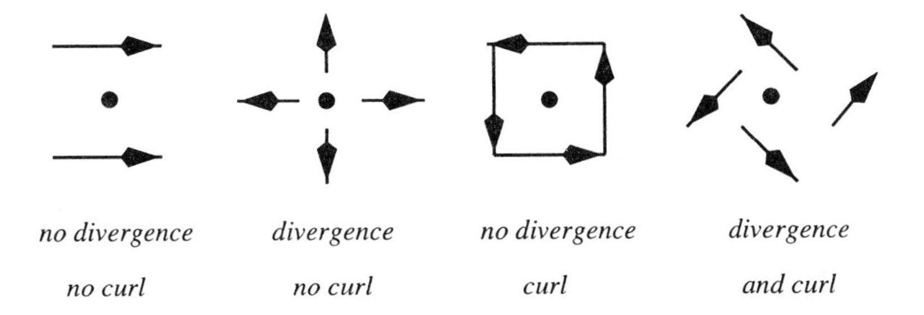

Figure 10.2.2: Examples of vector fields with and without divergence and curl.

• **Example 10.2.1**

If $\mathbf{F} = x^2z\mathbf{i} - 2y^3z^2\mathbf{j} + xy^2z\mathbf{k}$, compute the divergence of $\mathbf{F}$.

$$\nabla \cdot \mathbf{F} = \frac{\partial}{\partial x}(x^2z) + \frac{\partial}{\partial y}(-2y^3z^2) + \frac{\partial}{\partial z}(xy^2z) \tag{10.2.10}$$

$$= 2xz - 6y^2z^2 + xy^2. \tag{10.2.11}$$

• **Example 10.2.2**

If $\mathbf{r} = x\mathbf{i} + y\mathbf{j} + z\mathbf{k}$, show that $\mathbf{r}/|\mathbf{r}|^3$ is nondivergent.

$$\nabla \cdot \left(\frac{\mathbf{r}}{|\mathbf{r}|^3}\right) = \frac{\partial}{\partial x}\left[\frac{x}{(x^2+y^2+z^2)^{3/2}}\right] + \frac{\partial}{\partial y}\left[\frac{y}{(x^2+y^2+z^2)^{3/2}}\right]$$

$$+ \frac{\partial}{\partial z}\left[\frac{z}{(x^2+y^2+z^2)^{3/2}}\right] \tag{10.2.12}$$

$$= \frac{3}{(x^2+y^2+z^2)^{3/2}} - \frac{3x^2+3y^2+3z^2}{(x^2+y^2+z^2)^{5/2}} = 0. \tag{10.2.13}$$

Another important vector function involving the vector field $\mathbf{v}$ is the curl of $\mathbf{v}$, written curl($\mathbf{v}$) or rot($\mathbf{v}$) in some older textbooks. In fluid flow problems it is proportional to the instantaneous angular velocity of a fluid element. In rectangular coordinates,

$$\text{curl}(\mathbf{v}) = \nabla \times \mathbf{v} = (w_y - v_z)\mathbf{i} + (u_z - w_x)\mathbf{j} + (v_x - u_y)\mathbf{k}, \tag{10.2.14}$$

where $\mathbf{v} = u\mathbf{i} + v\mathbf{j} + w\mathbf{k}$ as before. However, it is best remembered in the mnemonic form:

$$\nabla \times \mathbf{F} = \begin{vmatrix} \mathbf{i} & \mathbf{j} & \mathbf{k} \\ \frac{\partial}{\partial x} & \frac{\partial}{\partial y} & \frac{\partial}{\partial z} \\ u & v & w \end{vmatrix} = (w_y - v_z)\mathbf{i} + (u_z - w_x)\mathbf{j} + (v_x - u_y)\mathbf{k}. \tag{10.2.15}$$

If the curl of a vector field is zero everywhere within a region, then the field is *irrotational*.

Figure 10.2.2 illustrates graphically some vector fields that do and do not possess divergence and curl. Let the vectors that are illustrated represent the motion of fluid particles. In the case of divergence only, fluid is streaming from the point, at which the density is falling. Alternatively the point could be a source. In the case where there is only curl, the fluid rotates about the point and the fluid is incompressible. Finally, the point that possesses both divergence and curl is a compressible fluid with rotation.

Some useful computational formulas exist for both the divergence and curl operations:

$$\nabla \times (\mathbf{F} + \mathbf{G}) = \nabla \times \mathbf{F} + \nabla \times \mathbf{G}, \tag{10.2.16}$$

$$\nabla \times \nabla \varphi = \mathbf{0}, \tag{10.2.17}$$

$$\nabla \cdot \nabla \times \mathbf{F} = 0, \tag{10.2.18}$$

$$\nabla \times (\varphi \mathbf{F}) = \varphi \nabla \times \mathbf{F} + \nabla \varphi \times \mathbf{F}, \tag{10.2.19}$$

$$\nabla(\mathbf{F} \cdot \mathbf{G}) = (\mathbf{F} \cdot \nabla)\mathbf{G} + (\mathbf{G} \cdot \nabla)\mathbf{F} + \mathbf{F} \times (\nabla \times \mathbf{G}) + \mathbf{G} \times (\nabla \times \mathbf{F}), \tag{10.2.20}$$

$$\nabla \times (\mathbf{F} \times \mathbf{G}) = (\mathbf{G} \cdot \nabla)\mathbf{F} - (\mathbf{F} \cdot \nabla)\mathbf{G} + \mathbf{F}(\nabla \cdot \mathbf{G}) - \mathbf{G}(\nabla \cdot \mathbf{F}), \tag{10.2.21}$$

$$\nabla \times (\nabla \times \mathbf{F}) = \nabla(\nabla \cdot \mathbf{F}) - (\nabla \cdot \nabla)\mathbf{F} \tag{10.2.22}$$

and

$$\nabla \cdot (\mathbf{F} \times \mathbf{G}) = \mathbf{G} \cdot \nabla \times \mathbf{F} - \mathbf{F} \cdot \nabla \times \mathbf{G}. \tag{10.2.23}$$

In this book the operation $\nabla \mathbf{F}$ is undefined.

• **Example 10.2.3**

If $\mathbf{F} = xz^3\mathbf{i} - 2x^2yz\mathbf{j} + 2yz^4\mathbf{k}$, compute the curl of $\mathbf{F}$ and verify that $\nabla \cdot \nabla \times \mathbf{F} = 0$.

From the definition of curl,

$$\nabla \times \mathbf{F} = \begin{vmatrix} \mathbf{i} & \mathbf{j} & \mathbf{k} \\ \frac{\partial}{\partial x} & \frac{\partial}{\partial y} & \frac{\partial}{\partial z} \\ xz^3 & -2x^2yz & 2yz^4 \end{vmatrix} \tag{10.2.24}$$

$$= \left[\frac{\partial}{\partial y}\left(2yz^4\right) - \frac{\partial}{\partial z}\left(-2x^2yz\right)\right]\mathbf{i} - \left[\frac{\partial}{\partial x}\left(2yz^4\right) - \frac{\partial}{\partial z}\left(xz^3\right)\right]\mathbf{j}$$
$$+ \left[\frac{\partial}{\partial x}\left(-2x^2yz\right) - \frac{\partial}{\partial y}\left(xz^3\right)\right]\mathbf{k} \tag{10.2.25}$$

$$= (2z^4 + 2x^2y)\mathbf{i} - (0 - 3xz^2)\mathbf{j} + (-4xyz - 0)\mathbf{k} \tag{10.2.26}$$

$$= (2z^4 + 2x^2y)\mathbf{i} + 3xz^2\mathbf{j} - 4xyz\mathbf{k}. \tag{10.2.27}$$

From the definition of divergence and (10.2.27),

$$\nabla \cdot \nabla \times \mathbf{F} = \frac{\partial}{\partial x}(2z^4 + 2x^2y) + \frac{\partial}{\partial y}(3xz^2) + \frac{\partial}{\partial z}(-4xyz) = 4xy + 0 - 4xy = 0. \tag{10.2.28}$$

• **Example 10.2.4: Potential flow theory**

One of the topics in most elementary fluid mechanics courses is the study of irrotational and nondivergent fluid flows. Because the fluid is irrotational, the velocity vector field $\mathbf{v}$ satisfies $\nabla \times \mathbf{v} = \mathbf{0}$. From (10.2.17) we can introduce a potential φ such that $\mathbf{v} = \nabla\varphi$. Because the flow field is nondivergent, $\nabla \cdot \mathbf{v} = \nabla^2\varphi = 0$. Thus, the fluid flow can be completely described in terms of solutions to Laplace's equation. This area of fluid mechanics is called *potential flow theory*.

Problems

Compute $\nabla \cdot \mathbf{F}, \nabla \times \mathbf{F}, \nabla \cdot (\nabla \times \mathbf{F})$ and $\nabla(\nabla \cdot \mathbf{F})$ for the following vector fields:

1. $\mathbf{F} = x^2 z\mathbf{i} + yz^2\mathbf{j} + xy^2\mathbf{k}$

2. $\mathbf{F} = 4x^2y^2\mathbf{i} + (2x + 2yz)\mathbf{j} + (3z + y^2)\mathbf{k}$

3. $\mathbf{F} = (x - y)^2\mathbf{i} + e^{-xy}\mathbf{j} + xze^{2y}\mathbf{k}$

4. $\mathbf{F} = 3xy\mathbf{i} + 2xz^2\mathbf{j} + y^3\mathbf{k}$

5. $\mathbf{F} = 5yz\mathbf{i} + x^2z\mathbf{j} + 3x^3\mathbf{k}$

6. $\mathbf{F} = y^3\mathbf{i} + (x^3y^2 - xy)\mathbf{j} - (x^3yz - xz)\mathbf{k}$

7. $\mathbf{F} = xe^{-y}\mathbf{i} + yz^2\mathbf{j} + 3e^{-z}\mathbf{k}$

8. $\mathbf{F} = y\ln(x)\mathbf{i} + (2 - 3yz)\mathbf{j} + xyz^3\mathbf{k}$

9. $\mathbf{F} = xyz\mathbf{i} + x^3yze^z\mathbf{j} + xye^z\mathbf{k}$

10. $\mathbf{F} = (xy^3 - z^4)\mathbf{i} + 4x^4y^2z\mathbf{j} - y^4z^5\mathbf{k}$.

11. $\mathbf{F} = xy^2\mathbf{i} + xyz^2\mathbf{j} + xy\cos(z)\mathbf{k}$

12. $\mathbf{F} = xy^2\mathbf{i} + xyz^2\mathbf{j} + xy\sin(z)\mathbf{k}$

13. $\mathbf{F} = xy^2\mathbf{i} + xyz\mathbf{j} + xy\cos(z)\mathbf{k}$.

14. (a) Assuming continuity of all partial derivatives, show that

$$\nabla \times (\nabla \times \mathbf{F}) = \nabla(\nabla \cdot \mathbf{F}) - \nabla^2\mathbf{F}.$$

(b) Using $\mathbf{F} = 3xy\mathbf{i} + 4yz\mathbf{j} + 2xz\mathbf{k}$, verify the results in part (a).

15. If $\mathbf{E} = \mathbf{E}(x, y, z, t)$ and $\mathbf{B} = \mathbf{B}(x, y, z, t)$ represent the electric and magnetic fields in a vacuum, Maxwell's field equations are

$$\nabla \cdot \mathbf{E} = 0 \qquad \nabla \times \mathbf{E} = -\frac{1}{c}\frac{\partial \mathbf{B}}{\partial t}$$

$$\nabla \cdot \mathbf{B} = 0 \qquad \nabla \times \mathbf{B} = \frac{1}{c}\frac{\partial \mathbf{E}}{\partial t},$$

where c is the speed of light. Using the results from Problem 14, show that $\mathbf{E}$ and $\mathbf{B}$ satisfy

$$\nabla^2 \mathbf{E} = \frac{1}{c^2}\frac{\partial^2 \mathbf{E}}{\partial t^2} \quad \text{and} \quad \nabla^2 \mathbf{B} = \frac{1}{c^2}\frac{\partial^2 \mathbf{B}}{\partial t^2}.$$

16. If f and g are continuously differentiable scalar fields, show that $\nabla f \times \nabla g$ is solenoidal. Hint: Show that $\nabla f \times \nabla g = \nabla \times (f \nabla g)$.

17. An inviscid (frictionless) fluid in equilibrium obeys the relationship $\nabla p = \rho \mathbf{F}$, where ρ denotes the density of the fluid, p denotes the pressure, and $\mathbf{F}$ denotes the body forces (such as gravity). Show that $\mathbf{F} \cdot \nabla \times \mathbf{F} = 0$.

10.3 LINE INTEGRALS

Line integrals are ubiquitous in physics. In mechanics they are used to compute work. In electricity and magnetism, they provide simple methods for computing the electric and magnetic fields for simple geometries.

The line integral most frequently encountered is an *oriented* one in which the path C is directed and the integrand is the dot product between the vector function $\mathbf{F}(\mathbf{r})$ and the tangent of the path $d\mathbf{r}$. It is usually written in the economical form

$$\int_C \mathbf{F} \cdot d\mathbf{r} = \int_C P(x, y, z)\, dx + Q(x, y, z)\, dy + R(x, y, z)\, dz, \qquad \textbf{(10.3.1)}$$

where $\mathbf{F} = P(x, y, z)\mathbf{i} + Q(x, y, z)\mathbf{j} + R(x, y, z)\mathbf{k}$. If the starting and terminal points are the same so that the contour is closed, then this *closed contour integral* will be denoted by $\oint_C$. In the following examples we show how to evaluate the line integrals along various types of curves.

• Example 10.3.1

If $\mathbf{F} = (3x^2 + 6y)\mathbf{i} - 14yz\mathbf{j} + 20xz^2\mathbf{k}$, let us evaluate the line integral $\int_C \mathbf{F} \cdot d\mathbf{r}$ along the parametric curves $x(t) = t$, $y(t) = t^2$, and $z(t) = t^3$ from the point $(0, 0, 0)$ to $(1, 1, 1)$. See Figure 10.3.1.

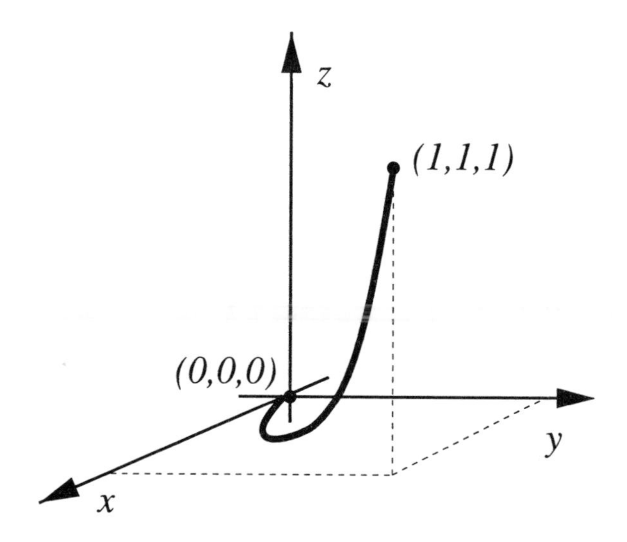

Figure 10.3.1: Diagram for the line integration in Example 10.3.1.

We begin by finding the values of t which give the corresponding end points. A quick check shows that $t = 0$ gives $(0,0,0)$ while $t = 1$ yields $(1,1,1)$. It should be noted that the same value of t must give the correct coordinates in each direction. Failure to do so suggests an error in the parameterization. Therefore,

$$\int_C \mathbf{F} \cdot d\mathbf{r} = \int_0^1 (3t^2 + 6t^2)\, dt - 14t^2(t^3)\, d(t^2) + 20t(t^3)^2 d(t^3) \quad \mathbf{(10.3.2)}$$

$$= \int_0^1 9t^2\, dt - 28t^6\, dt + 60t^9\, dt \quad \mathbf{(10.3.3)}$$

$$= \left(3t^3 - 4t^7 + 6t^{10}\right)\Big|_0^1 = 5. \quad \mathbf{(10.3.4)}$$

• **Example 10.3.2**

Let us redo the previous example with a contour that consists of three "dog legs", namely straight lines from $(0,0,0)$ to $(1,0,0)$, from $(1,0,0)$ to $(1,1,0)$, and from $(1,1,0)$ to $(1,1,1)$. See Figure 10.3.2.

In this particular problem we break the integration down into three distinct integrals:

$$\int_C \mathbf{F} \cdot d\mathbf{r} = \int_{C_1} \mathbf{F} \cdot d\mathbf{r} + \int_{C_2} \mathbf{F} \cdot d\mathbf{r} + \int_{C_3} \mathbf{F} \cdot d\mathbf{r}. \quad \mathbf{(10.3.5)}$$

For C_1, $y = z = dy = dz = 0$ and

$$\int_{C_1} \mathbf{F} \cdot d\mathbf{r} = \int_0^1 (3x^2 + 6 \cdot 0)\, dx - 14 \cdot 0 \cdot 0 \cdot 0 + 20x \cdot 0^2 \cdot 0 = \int_0^1 3x^2\, dx = 1. \quad \mathbf{(10.3.6)}$$

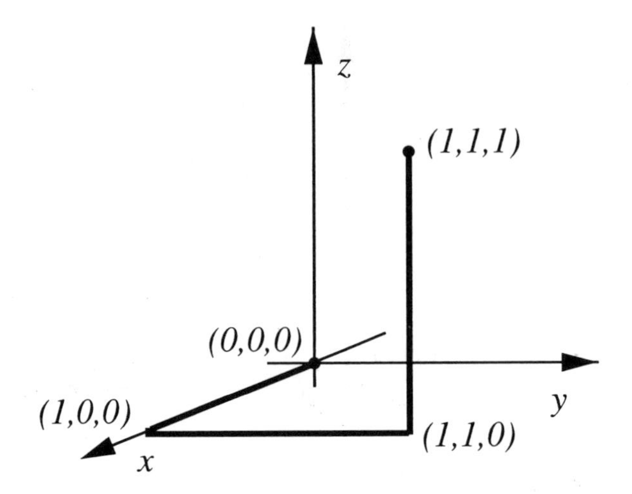

Figure 10.3.2: Diagram for the line integration in Example 10.3.2.

For C_2, $x = 1$ and $z = dx = dz = 0$ so that

$$\int_{C_2} \mathbf{F} \cdot d\mathbf{r} = \int_0^1 (3 \cdot 1^2 + 6y) \cdot 0 - 14y \cdot 0 \cdot dy + 20 \cdot 1 \cdot 0^2 \cdot 0 = 0. \quad \textbf{(10.3.7)}$$

For C_3, $x = y = 1$ and $dx = dy = 0$ so that

$$\int_{C_3} \mathbf{F} \cdot d\mathbf{r} = \int_0^1 (3 \cdot 1^2 + 6 \cdot 1) \cdot 0 - 14 \cdot 1 \cdot z \cdot 0 + 20 \cdot 1 \cdot z^2 \, dz = \int_0^1 20z^2 \, dz = \tfrac{20}{3}. \quad \textbf{(10.3.8)}$$

Therefore,

$$\int_C \mathbf{F} \cdot d\mathbf{r} = \tfrac{23}{3}. \quad \textbf{(10.3.9)}$$

• Example 10.3.3

For our third calculation, we redo the first example where the contour is a straight line. The parameterization in this case is $x = y = z = t$ with $0 \le t \le 1$. See Figure 10.3.3. Then,

$$\int_C \mathbf{F} \cdot d\mathbf{r} = \int_0^1 (3t^2 + 6t) \, dt - 14(t)(t) \, dt + 20t(t)^2 \, dt \quad \textbf{(10.3.10)}$$

$$= \int_0^1 (3t^2 + 6t - 14t^2 + 20t^3) \, dt = \tfrac{13}{3}. \quad \textbf{(10.3.11)}$$

An interesting aspect of these three examples is that, although we used a common vector field and moved from $(0, 0, 0)$ to $(1, 1, 1)$ in each

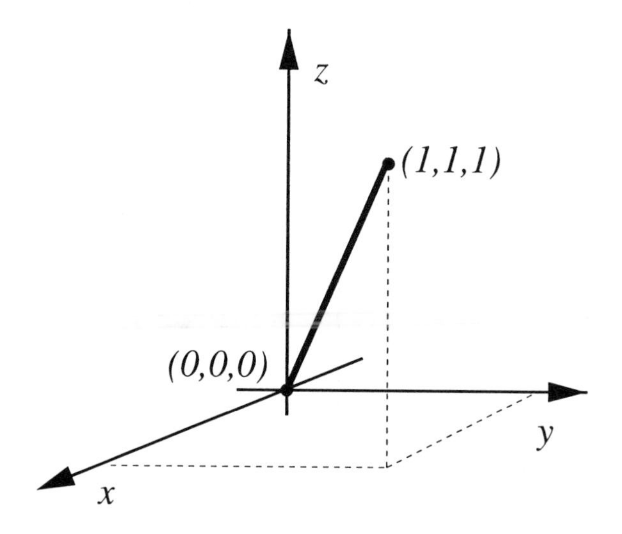

Figure 10.3.3: Diagram for the line integration in Example 10.3.3.

case, we obtained a different answer in each case. Thus, for this vector field, the line integral is *path dependent.* This is generally true. In the next section we will meet *conservative vector fields* where the results will be path independent.

• **Example 10.3.4**

If $\mathbf{F} = (x^2+y^2)\mathbf{i} - 2xy\mathbf{j} + x\mathbf{k}$, let us evaluate $\int_C \mathbf{F}\cdot d\mathbf{r}$ if the contour is that portion of the circle $x^2 + y^2 = a^2$ from the point $(a, 0, 3)$ to $(-a, 0, 3)$. See Figure 10.3.4.

The parametric equations for this example are $x = a\cos(\theta)$, $y = a\sin(\theta)$, $z = 3$ with $0 \le \theta \le \pi$. Therefore,

$$\begin{aligned}\int_C \mathbf{F}\cdot d\mathbf{r} = \int_0^\pi &[a^2\cos^2(\theta) + a^2\sin^2(\theta)][-a\sin(\theta)\,d\theta] \\ &- 2a^2\cos(\theta)\sin(\theta)[a\cos(\theta)\,d\theta] + a\cos(\theta)\cdot 0 \qquad (10.3.12)\\ &= -a^3\int_0^\pi \sin(\theta)\,d\theta - 2a^3\int_0^\pi \cos^2(\theta)\sin(\theta)\,d\theta \qquad (10.3.13)\\ &= a^3\cos(\theta)\Big|_0^\pi + \tfrac{2}{3}a^3\cos^3(\theta)\Big|_0^\pi \qquad (10.3.14)\\ &= -2a^3 - \tfrac{4}{3}a^3 = -\tfrac{10}{3}a^3. \qquad (10.3.15)\end{aligned}$$

• **Example 10.3.5: Circulation**

Let $\mathbf{v}(x, y, z)$ denote the velocity at the point (x, y, z) in a moving fluid. If it varies with time, this is the velocity at a particular instant

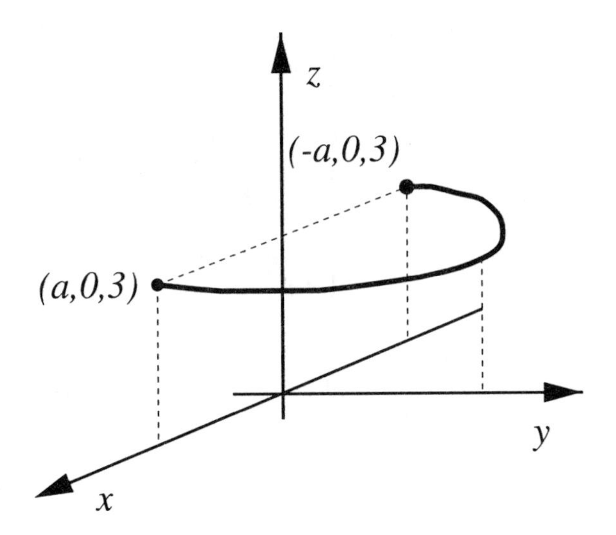

Figure 10.3.4: Diagram for the line integration in Example 10.3.4.

of time. The integral $\oint_C \mathbf{v} \cdot d\mathbf{r}$ around a closed path C is called the *circulation* around that path. The average component of velocity along the path is

$$\overline{v}_s = \frac{\oint_C v_s\, ds}{s} = \frac{\oint_C \mathbf{v} \cdot d\mathbf{r}}{s}, \tag{10.3.16}$$

where s is the total length of the path. The circulation is thus $\oint_C \mathbf{v} \cdot d\mathbf{r} = \overline{u}_s s$, the product of the length of the path and the average velocity along the path. When the circulation is positive, the flow is more in the direction of integration than opposite to it. Circulation is thus an indication and to some extent a measure of motion around the path.

Problems

Evaluate $\int_C \mathbf{F} \cdot d\mathbf{r}$ for the following vector fields and curves:

1. $\mathbf{F} = y\sin(\pi z)\mathbf{i} + x^2e^y\mathbf{j} + 3xz\mathbf{k}$ and C is the curve $x = t$, $y = t^2$ and $z = t^3$ from $(0,0,0)$ to $(1,1,1)$.

2. $\mathbf{F} = y\mathbf{i} + z\mathbf{j} + x\mathbf{k}$ and C consists of the line segments $(0,0,0)$ to $(2,3,0)$ and from $(2,3,0)$ to $(2,3,4)$.

3. $\mathbf{F} = e^x\mathbf{i} + xe^{xy}\mathbf{j} + xye^{xyz}\mathbf{k}$ and C is the curve $x = t$, $y = t^2$ and $z = t^3$ with $0 \le t \le 2$.

4. $\mathbf{F} = yz\mathbf{i} + xz\mathbf{j} + xy\mathbf{k}$ and C is the curve $x = t^3$, $y = t^2$ and $z = t$ with $1 \le t \le 2$.

5. $\mathbf{F} = y\mathbf{i} - x\mathbf{j} + 3xy\mathbf{k}$ and C consists of the semicircle $x^2 + y^2 = 4$, $z = 0$, $y > 0$ and the line segment from $(-2, 0, 0)$ to $(2, 0, 0)$.

6. $\mathbf{F} = (x + 2y)\mathbf{i} + (6y - 2x)\mathbf{j}$ and C consists of the sides of the triangle with vertices at $(0, 0, 0)$, $(1, 1, 1)$ and $(1, 1, 0)$. Proceed from $(0, 0, 0)$ to $(1, 1, 1)$ to $(1, 1, 0)$ and back to $(0, 0, 0)$.

7. $\mathbf{F} = 2xz\mathbf{i} + 4y^2\mathbf{j} + x^2\mathbf{k}$ and C is taken counterclockwise around the ellipse $x^2/4 + y^2/9 = 1$, $z = 1$.

8. $\mathbf{F} = 2x\mathbf{i} + y\mathbf{j} + z\mathbf{k}$ and C is the contour $x = t$, $y = \sin(t)$ and $z = \cos(t) + \sin(t)$ with $0 \le t \le 2\pi$.

9. $\mathbf{F} = (2y^2 + z)\mathbf{i} + 4xy\mathbf{j} + x\mathbf{k}$ and C is the spiral $x = \cos(t)$, $y = \sin(t)$ and $z = t$ with $0 \le t \le 2\pi$ between the points $(1, 0, 0)$ and $(1, 0, 2\pi)$.

10. $\mathbf{F} = x^2\mathbf{i} + y^2\mathbf{j} + (z^2 + 2xy)\mathbf{k}$ and C consists of the edges of the triangle with vertices at $(0, 0, 0)$, $(1, 1, 0)$, and $(0, 1, 0)$. Proceed from $(0, 0, 0)$ to $(1, 1, 0)$ to $(0, 1, 0)$ and back to $(0, 0, 0)$.

10.4 THE POTENTIAL FUNCTION

In Section 10.2 we showed that the curl operation applied to a gradient produces the zero vector: $\nabla \times \nabla\varphi = \mathbf{0}$. Consequently, if we have a vector field $\mathbf{F}$ such that $\nabla \times \mathbf{F} \equiv \mathbf{0}$ everywhere, then that vector field is called a *conservative* field and we may compute a potential φ such that $\mathbf{F} = \nabla\varphi$.

• Example 10.4.1

Let us show that the vector field $\mathbf{F} = ye^{xy}\cos(z)\mathbf{i} + xe^{xy}\cos(z)\mathbf{j} - e^{xy}\sin(z)\mathbf{k}$ is conservative and then find the corresponding potential function.

To show that the field is conservative, we compute the curl of $\mathbf{F}$ or

$$\nabla \times \mathbf{F} = \begin{vmatrix} \mathbf{i} & \mathbf{j} & \mathbf{k} \\ \frac{\partial}{\partial x} & \frac{\partial}{\partial y} & \frac{\partial}{\partial z} \\ ye^{xy}\cos(z) & xe^{xy}\cos(z) & -e^{xy}\sin(z) \end{vmatrix} = \mathbf{0}. \tag{10.4.1}$$

To find the potential we must solve three partial differential equations:

$$\varphi_x = ye^{xy}\cos(z) = \mathbf{F}\cdot\mathbf{i}, \tag{10.4.2}$$

$$\varphi_y = xe^{xy}\cos(z) = \mathbf{F}\cdot\mathbf{j} \tag{10.4.3}$$

and

$$\varphi_z = -e^{xy}\sin(z) = \mathbf{F}\cdot\mathbf{k}. \tag{10.4.4}$$

We begin by integrating any one of these three equations. Choosing (10.4.2),

$$\varphi(x, y, z) = e^{xy} \cos(z) + f(y, z). \tag{10.4.5}$$

To find $f(y, z)$ we differentiate (10.4.5) with respect to y and find that

$$\varphi_y = xe^{xy} \cos(z) + f_y(y, z) = xe^{xy} \cos(z) \tag{10.4.6}$$

from (10.4.3). Thus, $f_y = 0$ and $f(y, z)$ can only be a function of z, say $g(z)$. Then,

$$\varphi(x, y, z) = e^{xy} \cos(z) + g(z). \tag{10.4.7}$$

Finally,

$$\varphi_z = -e^{xy} \sin(z) + g'(z) = -e^{xy} \sin(z) \tag{10.4.8}$$

from (10.4.4) and $g'(z) = 0$. Therefore, the potential is

$$\varphi(x, y, z) = e^{xy} \cos(z) + \text{constant}. \tag{10.4.9}$$

Potentials can be very useful in computing line integrals because

$$\int_C \mathbf{F} \cdot d\mathbf{r} = \int_C \varphi_x \, dx + \varphi_y \, dy + \varphi_z \, dz = \int_C d\varphi = \varphi(B) - \varphi(A), \tag{10.4.10}$$

where the point B is the terminal point of the integration while the point A is the starting point. Thus, any path integration between any two points is *path independent.*

Finally, if we close the path so that A and B coincide, then

$$\oint_C \mathbf{F} \cdot d\mathbf{r} = 0. \tag{10.4.11}$$

It should be noted that the converse is *not* true. Just because $\oint_C \mathbf{F} \cdot d\mathbf{r} = 0$, we do not necessarily have a conservative field $\mathbf{F}$.

In summary then, an irrotational vector in a given region has three fundamental properties: (1) its integral around every simply connected circuit is zero, (2) its curl equals zero, (3) it is the gradient of a scalar function. For continuously differentiable vectors these properties are equivalent. For vectors which are only piece-wise differentiable, this is not true. Generally the first property is the most fundamental and taken as the definition of irrotationality.

- **Example 10.4.2**

Using the potential found in Example 10.4.1, let us find the value of the line integral $\int_C \mathbf{F} \cdot d\mathbf{r}$ from the point $(0, 0, 0)$ to $(-1, 2, \pi)$.

From (10.4.9),

$$\int_C \mathbf{F} \cdot d\mathbf{r} = \left[e^{xy} \cos(z) + \text{constant} \right] \Big|_{(0,0,0)}^{(-1,2,\pi)} = -1 - e^{-2}. \quad \textbf{(10.4.12)}$$

Problems

Verify that the following vector fields are conservative and then find the corresponding potential:

1. $\mathbf{F} = 2xy\mathbf{i} + (x^2 + 2yz)\mathbf{j} + (y^2 + 4)\mathbf{k}$

2. $\mathbf{F} = (2x + 2ze^{2x})\mathbf{i} + (2y - 1)\mathbf{j} + e^{2x}\mathbf{k}$

3. $\mathbf{F} = yz\mathbf{i} + xz\mathbf{j} + xy\mathbf{k}$

4. $\mathbf{F} = 2x\mathbf{i} + 3y^2\mathbf{j} + 4z^3\mathbf{k}$

5. $\mathbf{F} = [2x\sin(y) + e^{3z}]\mathbf{i} + x^2\cos(y)\mathbf{j} + (3xe^{3z} + 4)\mathbf{k}$

6. $\mathbf{F} = (2x + 5)\mathbf{i} + 3y^2\mathbf{j} + (1/z)\mathbf{k}$

7. $\mathbf{F} = e^{2z}\mathbf{i} + 3y^2\mathbf{j} + 2xe^{2z}\mathbf{k}$

8. $\mathbf{F} = y\mathbf{i} + (x + z)\mathbf{j} + y\mathbf{k}$

9. $\mathbf{F} = (z + y)\mathbf{i} + x\mathbf{j} + x\mathbf{k}$.

10.5 SURFACE INTEGRALS

Surface integrals appear in such diverse fields as electromagnetism and fluid mechanics. For example, if we were oceanographers we might be interested in the rate of volume of seawater through an instrument which has the curved surface S. The volume rate equals $\iint_S \mathbf{v} \cdot \mathbf{n}\, d\sigma$, where $\mathbf{v}$ is the velocity and $\mathbf{n}\, d\sigma$ is an infinitesimally small element on the surface of the instrument. The surface element $\mathbf{n}\, d\sigma$ must have an orientation (given by $\mathbf{n}$) because it makes a considerable difference whether the flow is directly through the surface or at right angles. More generally, if the surface encloses a three-dimensional volume, then we have a *closed surface integral.*

To illustrate the concept of computing a surface integral, we will do three examples with simple geometries. Later we will show how to use surface coordinates to do more complicated geometries.

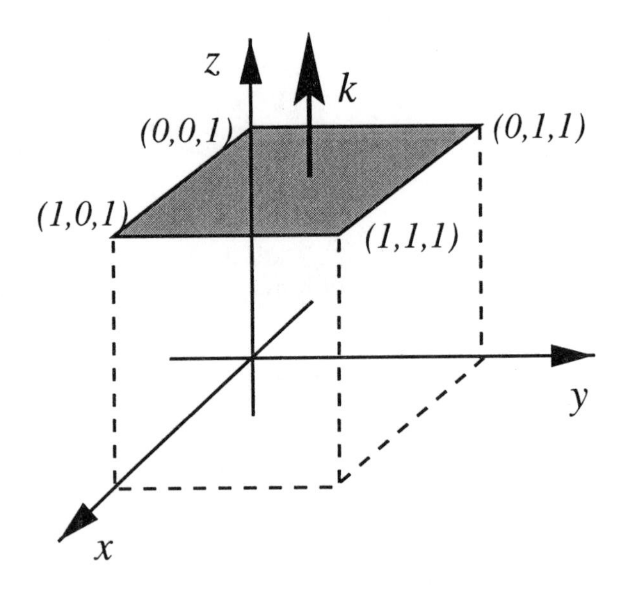

Figure 10.5.1: Diagram for the surface integration in Example 10.5.1.

• **Example 10.5.1**

Let us find the flux out the top of a unit cube if the vector field is $\mathbf{F} = x\mathbf{i} + y\mathbf{j} + z\mathbf{k}$. See Figure 10.5.1.

The top of a unit cube consists of the surface $z = 1$ with $0 \leq x \leq 1$ and $0 \leq y \leq 1$. By inspection the unit normal to this surface is $\mathbf{n} = \mathbf{k}$ or $\mathbf{n} = -\mathbf{k}$. Because we are interested in the flux *out* of the unit cube, $\mathbf{n} = \mathbf{k}$, and

$$\iint_S \mathbf{F} \cdot \mathbf{n}\, d\sigma = \int_0^1 \int_0^1 (x\mathbf{i} + y\mathbf{j} + \mathbf{k}) \cdot \mathbf{k}\, dx\, dy = 1 \qquad \textbf{(10.5.1)}$$

because $z = 1$.

• **Example 10.5.2**

Let us find the flux out of that portion of the cylinder $y^2 + z^2 = 4$ in the first octant bounded by $x = 0$, $x = 3$, $y = 0$, and $z = 0$. The vector field is $\mathbf{F} = x\mathbf{i} + 2z\mathbf{j} + y\mathbf{k}$. See Figure 10.5.2.

Because we are dealing with a cylinder, cylindrical coordinates are appropriate. Let $y = 2\cos(\theta)$, $z = 2\sin(\theta)$, and $x = x$ with $0 \leq \theta \leq \pi/2$. To find $\mathbf{n}$, we use the gradient in conjunction with the definition of the surface of the cylinder $f(x, y, z) = y^2 + z^2 = 4$. Then

$$\mathbf{n} = \frac{\nabla f}{|\nabla f|} = \frac{2y\mathbf{j} + 2z\mathbf{k}}{\sqrt{4y^2 + 4z^2}} = \frac{y}{2}\mathbf{j} + \frac{z}{2}\mathbf{k} \qquad \textbf{(10.5.2)}$$

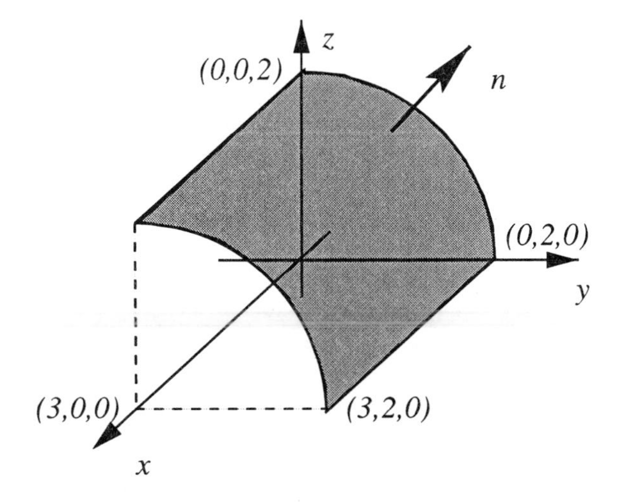

Figure 10.5.2: Diagram for the surface integration in Example 10.5.2.

because $y^2 + z^2 = 4$ along the surface. Because we want the flux *out* of the surface, then $\mathbf{n} = y\mathbf{j}/2 + z\mathbf{k}/2$ whereas the flux *into* the surface would require $\mathbf{n} = -y\mathbf{j}/2 - z\mathbf{k}/2$. Therefore,

$$\mathbf{F} \cdot \mathbf{n} = (x\mathbf{i} + 2z\mathbf{j} + y\mathbf{k}) \cdot \left(\frac{y}{2}\mathbf{j} + \frac{z}{2}\mathbf{k}\right) = \frac{3yz}{2} = 6\cos(\theta)\sin(\theta). \quad \mathbf{(10.5.3)}$$

What is $d\sigma$? Our infinitesimal surface area has a side in the x direction of length dx and a side in the θ direction of length $2\,d\theta$ because the radius equals 2. Therefore, $d\sigma = 2\,dx\,d\theta$.

Bringing all of these elements together,

$$\iint_S \mathbf{F} \cdot \mathbf{n}\,d\sigma = \int_0^3 \int_0^{\pi/2} 12\cos(\theta)\sin(\theta)\,d\theta\,dx \quad \mathbf{(10.5.4)}$$

$$= 6\int_0^3 \left[\sin^2(\theta)\Big|_0^{\pi/2}\right] dx = 6\int_0^3 dx = 18. \quad \mathbf{(10.5.5)}$$

As counterpoint to this example, let us find the flux out of the pie-shaped surface at $x = 3$. In this case, $y = r\cos(\theta)$ and $z = r\sin(\theta)$ and

$$\iint_S \mathbf{F} \cdot \mathbf{n}\,d\sigma = \int_0^{\pi/2} \int_0^2 [3\mathbf{i} + 2r\sin(\theta)\mathbf{j} + r\cos(\theta)\mathbf{k}] \cdot \mathbf{i}\,r\,dr\,d\theta \quad \mathbf{(10.5.6)}$$

$$= 3\int_0^{\pi/2} \int_0^2 r\,dr\,d\theta = 3\pi. \quad \mathbf{(10.5.7)}$$

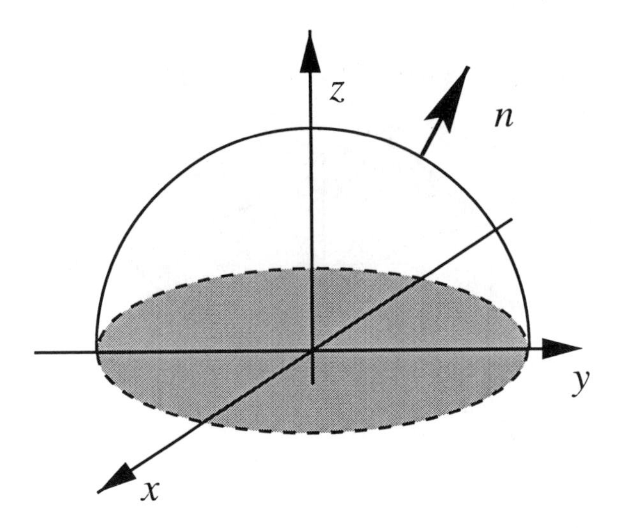

Figure 10.5.3: Diagram for the surface integration in Example 10.5.3.

• Example 10.5.3

Let us find the flux of the vector field $\mathbf{F} = y^2\mathbf{i} + x^2\mathbf{j} + 5z\mathbf{k}$ out of the hemispheric surface $x^2 + y^2 + z^2 = a^2$, $z > 0$. See Figure 10.5.3.

We begin by finding the outwardly pointing normal. Because the surface is defined by $f(x, y, z) = x^2 + y^2 + z^2 = a^2$,

$$\mathbf{n} = \frac{\nabla f}{|\nabla f|} = \frac{2x\mathbf{i} + 2y\mathbf{j} + 2z\mathbf{k}}{\sqrt{4x^2 + 4y^2 + 4z^2}} = \frac{x}{a}\mathbf{i} + \frac{y}{a}\mathbf{j} + \frac{z}{a}\mathbf{k} \tag{10.5.8}$$

because $x^2 + y^2 + z^2 = a^2$. This is also the outwardly pointing normal because $\mathbf{n} = \mathbf{r}/a$, where $\mathbf{r}$ is the radial vector.

Using spherical coordinates, $x = a\cos(\varphi)\sin(\theta)$, $y = a\sin(\varphi)\sin(\theta)$, and $z = a\cos(\theta)$, where φ is the angle made by the projection of the point onto the equatorial plane, measured from the x-axis, and θ is the colatitude or "cone angle" measured from the z-axis. To compute $d\sigma$, the infinitesimal length in the θ direction is $a\,d\theta$ while in the φ direction it is $a\sin(\theta)\,d\varphi$, where the $\sin(\theta)$ factor takes into account the convergence of the meridians. Therefore, $d\sigma = a^2\sin(\theta)\,d\theta\,d\varphi$ and

$$\iint_S \mathbf{F}\cdot\mathbf{n}\,d\sigma = \int_0^{2\pi}\int_0^{\pi/2} (y^2\mathbf{i} + x^2\mathbf{j} + 5z\mathbf{k}) \cdot \left(\frac{x}{a}\mathbf{i} + \frac{y}{a}\mathbf{j} + \frac{z}{a}\mathbf{k}\right) a^2\sin(\theta)\,d\theta\,d\varphi \tag{10.5.9}$$

$$= \int_0^{2\pi}\int_0^{\pi/2} \left(\frac{xy^2}{a} + \frac{x^2y}{a} + \frac{5z^2}{a}\right) a^2\sin(\theta)\,d\theta\,d\varphi \tag{10.5.10}$$

$$\iint_S \mathbf{F}\cdot\mathbf{n}\,d\sigma = \int_0^{\pi/2}\int_0^{2\pi} \big[a^4\cos(\varphi)\sin^2(\varphi)\sin^4(\theta) + a^4\cos^2(\varphi)\sin(\varphi)\sin^4(\theta) + 5a^3\cos^2(\theta)\sin(\theta)\big]\,d\varphi\,d\theta \tag{10.5.11}$$

$$= \int_0^{\pi/2}\left[\frac{a^4}{3}\sin^3(\varphi)\Big|_0^{2\pi}\sin^4(\theta) - \frac{a^4}{3}\cos^3(\varphi)\Big|_0^{2\pi}\sin^4(\theta) + 5a^3\cos^2(\theta)\sin(\theta)\varphi\Big|_0^{2\pi}\right]d\theta \tag{10.5.12}$$

$$= 10\pi a^3\int_0^{\pi/2}\cos^2(\theta)\sin(\theta)\,d\theta \tag{10.5.13}$$

$$= -\frac{10\pi a^3}{3}\cos^3(\theta)\Big|_0^{\pi/2} = \frac{10\pi a^3}{3}. \tag{10.5.14}$$

Although these techniques apply for simple geometries such as a cylinder or sphere, we would like a *general* method for treating any arbitrary surface. We begin by noting that a surface is an aggregate of points whose coordinates are functions of two variables. For example, in the previous example, the surface was described by the coordinates φ and θ. Let us denote these surface coordinates in general by u and v. Consequently, on any surface we can reexpress x, y, and z in terms of u and v: $x = x(u,v)$, $y = y(u,v)$, and $z = z(u,v)$.

Next, we must find an infinitesimal element of area. The position vector to the surface is $\mathbf{r} = x(u,v)\mathbf{i} + y(u,v)\mathbf{j} + z(u,v)\mathbf{k}$. Therefore, the tangent vectors along v = constant, $\mathbf{r}_u$, and along u = constant, $\mathbf{r}_v$, equal

$$\mathbf{r}_u = x_u\mathbf{i} + y_u\mathbf{j} + z_u\mathbf{k} \tag{10.5.15}$$

and

$$\mathbf{r}_v = x_v\mathbf{i} + y_v\mathbf{j} + z_v\mathbf{k}. \tag{10.5.16}$$

Consequently, the sides of the infinitesimal area is $\mathbf{r}_u\,du$ and $\mathbf{r}_v\,dv$. Therefore, the vectorial area of the parallelogram that these vectors form is

$$\mathbf{n}\,d\sigma = \mathbf{r}_u \times \mathbf{r}_v\,du\,dv \tag{10.5.17}$$

and is called the *vector element of area* on the surface. Thus, we may convert $\mathbf{F}\cdot\mathbf{n}\,d\sigma$ into an expression involving only u and v and then evaluate the surface integral by integrating over the appropriate domain in the uv-plane. Of course, we are in trouble if $\mathbf{r}_u\times\mathbf{r}_v = \mathbf{0}$. Therefore, we only treat regular points where $\mathbf{r}_u\times\mathbf{r}_v \neq \mathbf{0}$. In the next few examples, we show how to use these surface coordinates to evaluate surface integrals.

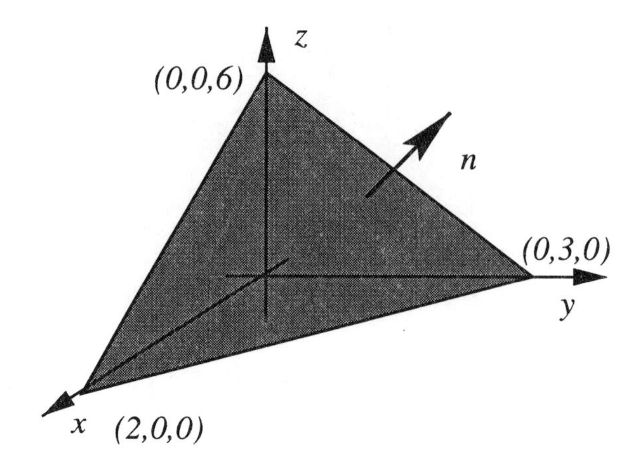

Figure 10.5.5: Diagram for the surface integration in Example 10.5.4.

• **Example 10.5.4**

Let us find the flux of the vector field $\mathbf{F} = x\mathbf{i} + y\mathbf{j} + z\mathbf{k}$ through the top of the plane $3x + 2y + z = 6$ which lies in the first octant. See Figure 10.5.5.

Our parametric equations are $x = u$, $y = v$, and $z = 6 - 3u - 2v$. Therefore,

$$\mathbf{r} = u\mathbf{i} + v\mathbf{j} + (6 - 3u - 2v)\mathbf{k} \tag{10.5.18}$$

so that

$$\mathbf{r}_u = \mathbf{i} - 3\mathbf{k}, \quad \mathbf{r}_v = \mathbf{j} - 2\mathbf{k} \tag{10.5.19}$$

and

$$\mathbf{r}_u \times \mathbf{r}_v = 3\mathbf{i} + 2\mathbf{j} + \mathbf{k}. \tag{10.5.20}$$

Bring all of these elements together,

$$\iint_S \mathbf{F} \cdot \mathbf{n}\, d\sigma = \int_0^2 \int_0^{3-3u/2} (3u + 2v + 6 - 3u - 2v)\, dv\, du \tag{10.5.21}$$

$$= 6 \int_0^2 \int_0^{3-3u/2} dv\, du = 6 \int_0^2 (3 - 3u/2)\, du \tag{10.5.22}$$

$$= 6 \left(3u - \tfrac{3}{4}u^2\right)\Big|_0^2 = 18. \tag{10.5.23}$$

To set up the limits of integration, we note that the area in u, v space corresponds to the xy-plane. On the xy-plane, $z = 0$ and $3u + 2v = 6$, along with boundaries $u = v = 0$.

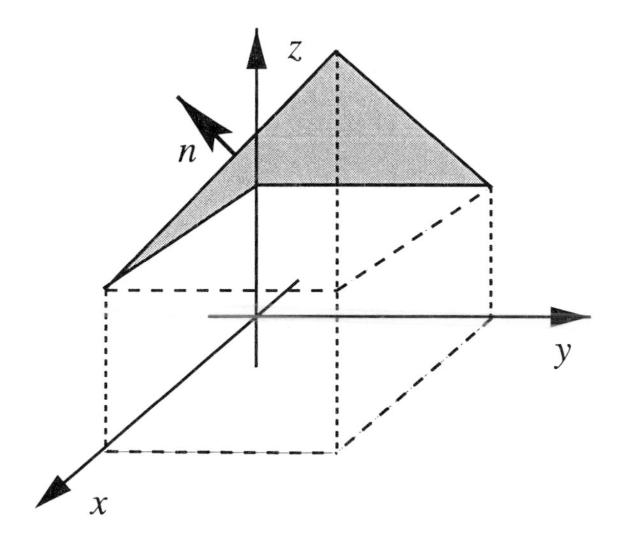

Figure 10.5.6: Diagram for the surface integration in Example 10.5.5.

• **Example 10.5.5**

Let us find the flux of the vector field $\mathbf{F} = x\mathbf{i} + y\mathbf{j} + z\mathbf{k}$ through the top of the surface $z = xy + 1$ which covers the square $0 \le x \le 1$, $0 \le y \le 1$ in the xy-plane. See Figure 10.5.6.

Our parametric equations are $x = u$, $y = v$, and $z = uv + 1$ with $0 \le u \le 1$ and $0 \le v \le 1$. Therefore,

$$\mathbf{r} = u\mathbf{i} + v\mathbf{j} + (uv + 1)\mathbf{k} \tag{10.5.24}$$

so that

$$\mathbf{r}_u = \mathbf{i} + v\mathbf{k}, \qquad \mathbf{r}_v = \mathbf{j} + u\mathbf{k} \tag{10.5.25}$$

and

$$\mathbf{r}_u \times \mathbf{r}_v = -v\mathbf{i} - u\mathbf{j} + \mathbf{k}. \tag{10.5.26}$$

Bring all of these elements together,

$$\iint_S \mathbf{F} \cdot \mathbf{n}\, d\sigma = \int_0^1 \int_0^1 [u\mathbf{i} + v\mathbf{j} + (uv + 1)\mathbf{k}] \cdot (-v\mathbf{i} - u\mathbf{j} + \mathbf{k})\, du\, dv \tag{10.5.27}$$

$$= \int_0^1 \int_0^1 (1 - uv)\, du\, dv = \int_0^1 \left(u - \tfrac{1}{2}u^2 v\right)\Big|_0^1 dv \tag{10.5.28}$$

$$= \int_0^1 \left(1 - \tfrac{1}{2}v\right) dv = \left(v - \tfrac{1}{4}v^2\right)\Big|_0^1 = \tfrac{3}{4}. \tag{10.5.29}$$

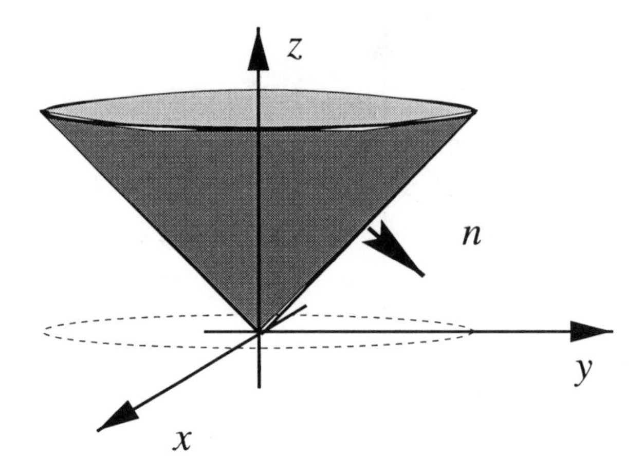

Figure 10.5.7: Diagram for the surface integration in Example 10.5.6.

• **Example 10.5.6**

Let us find the flux of the vector field $\mathbf{F} = 4xz\mathbf{i}+xyz^2\mathbf{j}+3z\mathbf{k}$ through the exterior surface of the cone $z^2 = x^2 + y^2$ above the xy-plane and below $z = 4$. See Figure 10.5.7.

A natural choice for the surface coordinates is polar coordinates r and θ. Because $x = r\cos(\theta)$ and $y = r\sin(\theta)$, $z = r$. Then

$$\mathbf{r} = r\cos(\theta)\mathbf{i} + r\sin(\theta)\mathbf{j} + r\mathbf{k} \tag{10.5.30}$$

with $0 \le r \le 4$ and $0 \le \theta \le 2\pi$ so that

$$\mathbf{r}_r = \cos(\theta)\mathbf{i} + \sin(\theta)\mathbf{j} + \mathbf{k}, \mathbf{r}_\theta = -r\sin(\theta)\mathbf{i} + r\cos(\theta)\mathbf{j} \tag{10.5.31}$$

and

$$\mathbf{r}_r \times \mathbf{r}_\theta = -r\cos(\theta)\mathbf{i} - r\sin(\theta)\mathbf{j} + r\mathbf{k}. \tag{10.5.32}$$

This is the unit area *inside* the cone. Because we want the exterior surface, we must take the negative of (10.5.32). Bring all of these elements together,

$$\iint_S \mathbf{F}\cdot\mathbf{n}\,d\sigma = \int_0^4 \int_0^{2\pi} \{[4r\cos(\theta)]r[r\cos(\theta)] + [r^2\sin(\theta)\cos(\theta)]r^2[r\sin(\theta)] - 3r^2\}\,d\theta\,dr \tag{10.5.33}$$

$$= \int_0^4 \left\{2r^3\left[\theta + \tfrac{1}{2}\sin(2\theta)\right]\Big|_0^{2\pi} + r^5\tfrac{1}{3}\sin^3(\theta)\Big|_0^{2\pi} - 3r^2\theta\Big|_0^{2\pi}\right\} dr \tag{10.5.34}$$

$$= \int_0^4 \left(4\pi r^3 - 6\pi r^2\right)\,dr = \left(\pi r^4 - 2\pi r^3\right)\Big|_0^4 = 128\pi. \tag{10.5.35}$$

Problems

Compute the surface integral $\iint_S \mathbf{F} \cdot \mathbf{n}\, d\sigma$ for the following vector fields and surfaces:

1. $\mathbf{F} = x\mathbf{i} - z\mathbf{j} + y\mathbf{k}$ and the surface is the top portion of the $z = 1$ plane where $0 \le x \le 1$ and $0 \le y \le 1$.

2. $\mathbf{F} = x\mathbf{i} + y\mathbf{j} + xz\mathbf{k}$ and the surface is the top side of the cylinder $x^2 + y^2 = 9$, $z = 0$, and $z = 1$.

3. $\mathbf{F} = xy\mathbf{i} + z\mathbf{j} + xz\mathbf{k}$ and the surface consists of both exterior *ends* of the cylinder defined by $x^2 + y^2 = 4$, $z = 0$, and $z = 2$.

4. $\mathbf{F} = x\mathbf{i} + z\mathbf{j} + y\mathbf{k}$ and the surface is the lateral and exterior side of the cylinder defined by $x^2 + y^2 = 4$, $z = -3$, and $z = 3$.

5. $\mathbf{F} = xy\mathbf{i} + z^2\mathbf{j} + y\mathbf{k}$ and the surface is the curved exterior side of the cylinder $y^2 + z^2 = 9$ in the first octant bounded by $x = 0$, $x = 1$, $y = 0$, and $z = 0$.

6. $\mathbf{F} = y\mathbf{j} + z^2\mathbf{k}$ and the surface is the exterior of the semicircular cylinder $y^2 + z^2 = 4$, $z \ge 0$ cut by the planes $x = 0$ and $x = 1$.

7. $\mathbf{F} = z\mathbf{i} + x\mathbf{j} + y\mathbf{k}$ and the surface is the curved exterior side of the cylinder $x^2 + y^2 = 4$ in the first octant cut by the planes $z = 1$ and $z = 2$.

8. $\mathbf{F} = x^2\mathbf{i} - z^2\mathbf{j} + yz\mathbf{k}$ and the surface is the exterior of the hemispheric surface of $x^2 + y^2 + z^2 = 16$ above the plane $z = 2$.

9. $\mathbf{F} = y\mathbf{i} + x\mathbf{j} + y\mathbf{k}$ and the surface is the top of the surface $z = x + 1$ where $-1 \le x \le 1$ and $-1 \le y \le 1$.

10. $\mathbf{F} = z\mathbf{i} + x\mathbf{j} - 3z\mathbf{k}$ and the surface is the top of the plane $x + y + z = 2a$ that lies above the square $0 \le x \le a$, $0 \le y \le a$ in the xy-plane.

11. $\mathbf{F} = (y^2 + z^2)\mathbf{i} + (x^2 + z^2)\mathbf{j} + (x^2 + y^2)\mathbf{k}$ and the surface is the top of the surface $z = 1 - x^2$ with $-1 \le x \le 1$ and $-2 \le y \le 2$.

12. $\mathbf{F} = y^2\mathbf{i} + xz\mathbf{j} - \mathbf{k}$ and the surface is the cone $z = \sqrt{x^2 + y^2}$, $0 \le z \le 1$ with the normal pointing away from the z-axis.

13. $\mathbf{F} = y^2\mathbf{i} + x^2\mathbf{j} + 5z\mathbf{k}$ and the surface is the top of the plane $z = y + 1$ where $-1 \le x \le 1$ and $-1 \le y \le 1$.

14. $\mathbf{F} = -y\mathbf{i} + x\mathbf{j} + z\mathbf{k}$ and the surface is the exterior or bottom of the paraboloid $z = x^2 + y^2$ where $0 \le z \le 1$.

15. $\mathbf{F} = -y\mathbf{i}+x\mathbf{j}+6z^2\mathbf{k}$ and the surface is the exterior of the paraboloids $z = 4 - x^2 - y^2$ and $z = x^2 + y^2$.

10.6 GREEN'S LEMMA

Consider a rectangle in the xy-plane which is bounded by the lines $x = a$, $x = b$, $y = c$, and $y = d$. We assume that the boundary of the rectangle is a piece-wise smooth curve which we denote by C. If we have a continuously differentiable vector function $\mathbf{F} = P(x,y)\mathbf{i} + Q(x,y)\mathbf{j}$ at each point of enclosed region R, then

$$\iint_R \frac{\partial Q}{\partial x}\,dA = \int_c^d \left[\int_a^b \frac{\partial Q}{\partial x}\,dx\right] dy \tag{10.6.1}$$

$$= \int_c^d Q(b,y)\,dy - \int_c^d Q(a,y)\,dy \tag{10.6.2}$$

$$= \oint_C Q(x,y)\,dy, \tag{10.6.3}$$

where the last integral is a closed line integral counterclockwise around the rectangle because the horizontal sides vanish since $dy = 0$. By similar arguments,

$$\iint_R \frac{\partial P}{\partial y}\,dA = -\oint_C P(x,y)\,dx \tag{10.6.4}$$

so that

$$\iint_R \left(\frac{\partial Q}{\partial x} - \frac{\partial P}{\partial y}\right) dA = \oint_C P(x,y)\,dx + Q(x,y)\,dy. \tag{10.6.5}$$

This result, often known as *Green's lemma*, may be expressed in vector form as

$$\boxed{\oint_C \mathbf{F}\cdot d\mathbf{r} = \iint_R \nabla\times\mathbf{F}\cdot\mathbf{k}\,dA.} \tag{10.6.6}$$

Although this proof was for a rectangular area, it can be generalized to *any* simply closed region on the xy-plane as follows. Consider

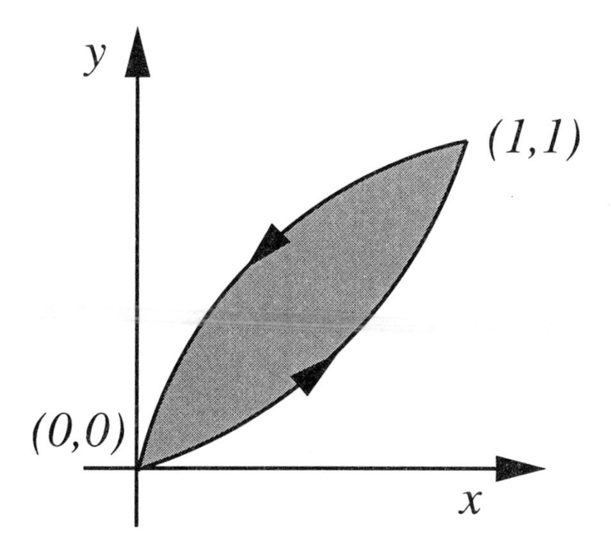

Figure 10.6.1: Diagram for the verification of Green's lemma in Example 10.6.1.

an area which is surrounded by simply closed curves. Within the closed contour we can divide the area into an infinite number of infinitesimally small rectangles and apply (10.6.6) to each rectangle. When we sum up all of these rectangles, we find $\iint_R \nabla \times \mathbf{F} \cdot \mathbf{k}\, dA$, where the integration is over the entire surface area. On the other hand, away from the boundary, the line integral along any one edge of a rectangle cancels the line integral along the same edge in a contiguous rectangle. Thus, the only nonvanishing contribution from the line integrals arises from the outside boundary of the domain $\oint_C \mathbf{F} \cdot d\mathbf{r}$.

- **Example 10.6.1**

Let us *verify* Green's lemma using the vector field $\mathbf{F} = (3x^2 - 8y^2)\mathbf{i} + (4y - 6xy)\mathbf{j}$ and the enclosed area lies between the curves $y = \sqrt{x}$ and $y = x^2$. The two curves intersect at $x = 0$ and $x = 1$. See Figure 10.6.1.

We begin with the line integral:

$$\oint_C \mathbf{F} \cdot d\mathbf{r} = \int_0^1 (3x^2 - 8x^4)\, dx + (4x^2 - 6x^3)(2x\, dx)$$

$$+ \int_1^0 (3x^2 - 8x)\, dx + (4x^{1/2} - 6x^{3/2})(\tfrac{1}{2}x^{-1/2}\, dx) \quad \textbf{(10.6.7)}$$

$$= \int_0^1 (-20x^4 + 8x^3 + 11x - 2)\, dx = \tfrac{3}{2}. \quad \textbf{(10.6.8)}$$

In (10.6.7) we used $y = x^2$ in the first integral and $y = \sqrt{x}$ in our return integration. For the areal integration,

$$\iint_R \nabla \times \mathbf{F} \cdot \mathbf{k}\, dA = \int_0^1 \int_{x^2}^{\sqrt{x}} 10y\, dy\, dx = \int_0^1 5y^2 \Big|_{x^2}^{\sqrt{x}} dx \quad \textbf{(10.6.9)}$$

$$= 5 \int_0^1 (x - x^4)\, dx = \tfrac{3}{2} \quad \textbf{(10.6.10)}$$

and Green's lemma is verified in this particular case.

• Example 10.6.2

Let us redo Example 10.6.1 except that the closed contour is the triangular region defined by the lines $x = 0$, $y = 0$, and $x + y = 1$.

The line integral is

$$\oint_C \mathbf{F} \cdot d\mathbf{r} = \int_0^1 (3x^2 - 8 \cdot 0^2) dx + (4 \cdot 0 - 6x \cdot 0) \cdot 0$$

$$+ \int_0^1 [3(1-y)^2 - 8y^2](-dy) + [4y - 6(1-y)y]\, dy$$

$$+ \int_1^0 (3 \cdot 0^2 - 8y^2) \cdot 0 + (4y - 6 \cdot 0 \cdot y)\, dy \quad \textbf{(10.6.11)}$$

$$= \int_0^1 3x^2\, dx - \int_0^1 4y\, dy + \int_0^1 (-3 + 4y + 11y^2)\, dy \quad \textbf{(10.6.12)}$$

$$= x^3\Big|_0^1 - 2y^2\Big|_0^1 + \left(-3y + 2y^2 + \tfrac{11}{3}y^3\right)\Big|_0^1 = \tfrac{5}{3}. \quad \textbf{(10.6.13)}$$

On the other hand, the areal integration is

$$\iint_R \nabla \times \mathbf{F} \cdot \mathbf{k}\, dA = \int_0^1 \int_0^{1-x} 10y\, dy\, dx = \int_0^1 5y^2\Big|_0^{1-x} dx \quad \textbf{(10.6.14)}$$

$$= 5 \int_0^1 (1-x)^2\, dx = -\tfrac{5}{3}(1-x)^3\Big|_0^1 = \tfrac{5}{3} \quad \textbf{(10.6.15)}$$

and Green's lemma is verified in this particular case.

• Example 10.6.3

Let us verify Green's lemma using the vector field $\mathbf{F} = (3x + 4y)\mathbf{i} + (2x - 3y)\mathbf{j}$ and the closed contour is a circle of radius two centered at the origin of the xy-plane. See Figure 10.6.2.

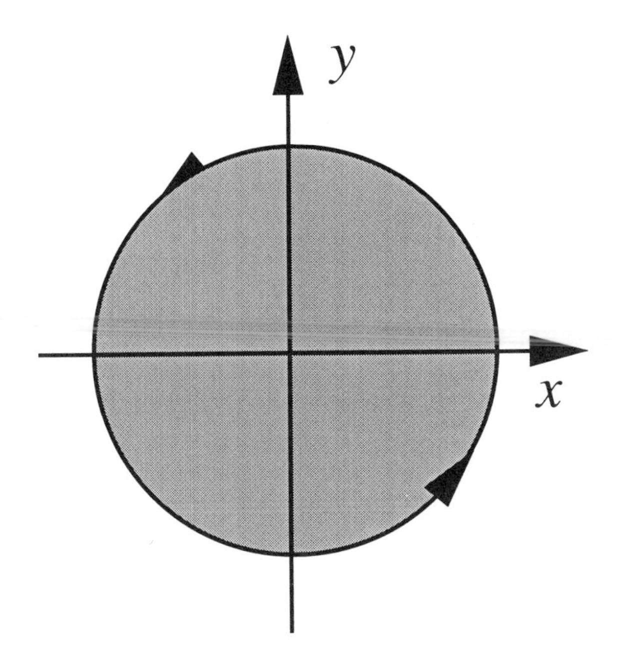

Figure 10.6.2: Diagram for the verification of Green's lemma in Example 10.6.3.

Beginning with the line integration,

$$\oint_C \mathbf{F}\cdot d\mathbf{r} = \int_0^{2\pi} [6\cos(\theta)+8\sin(\theta)][-2\sin(\theta)\,d\theta] + [4\cos(\theta)-6\sin(\theta)][2\cos(\theta)\,d\theta] \tag{10.6.16}$$

$$= \int_0^{2\pi} [-24\cos(\theta)\sin(\theta) - 16\sin^2(\theta) + 8\cos^2(\theta)]\,d\theta \tag{10.6.17}$$

$$= 12\,\cos^2(\theta)\big|_0^{2\pi} - 8\left[\theta - \tfrac{1}{2}\sin(2\theta)\right]\big|_0^{2\pi} + 4\left[\theta + \tfrac{1}{2}\sin(2\theta)\right]\big|_0^{2\pi} \tag{10.6.18}$$

$$= -8\pi. \tag{10.6.19}$$

For the areal integration,

$$\iint_R \nabla\times\mathbf{F}\cdot\mathbf{k}\,dA = \int_0^2\int_0^{2\pi} -2\,r\,d\theta\,dr = -8\pi \tag{10.6.20}$$

and Green's lemma is verified in the special case.

Problems

Verify Green's lemma for the following two-dimensional vector fields and contours:

1. $\mathbf{F} = (x^2 + 4y)\mathbf{i} + (y - x)\mathbf{j}$ and the contour is the square bounded by the lines $x = 0$, $y = 0$, $x = 1$, and $y = 1$.

2. $\mathbf{F} = (x - y)\mathbf{i} + xy\mathbf{j}$ and the contour is the square bounded by the lines $x = 0$, $y = 0$, $x = 1$, and $y = 1$.

3. $\mathbf{F} = -y^2\mathbf{i} + x^2\mathbf{j}$ and the contour is the triangle bounded by the lines $x = 1$, $y = 0$, and $y = x$.

4. $\mathbf{F} = (xy - x^2)\mathbf{i} + x^2y\mathbf{j}$ and the contour is the triangle bounded by the line $y = 0$, $x = 1$, and $y = x$.

5. $\mathbf{F} = \sin(y)\mathbf{i} + x\cos(y)\mathbf{j}$ and the contour is the triangle bounded by $x + y = 1$, $y - x = 1$, and $y = 0$.

6. $\mathbf{F} = y^2\mathbf{i} + x^2\mathbf{j}$ and the contour is the same contour used in problem 4.

7. $\mathbf{F} = -y^2\mathbf{i} + x^2\mathbf{j}$ and the contour is the circle $x^2 + y^2 = 4$.

8. $\mathbf{F} = -x^2\mathbf{i} + xy^2\mathbf{j}$ and the contour is the closed circle of radius a.

9. $\mathbf{F} = (6y + x)\mathbf{i} + (y + 2x)\mathbf{j}$ and the contour is the circle $(x - 1)^2 + (y - 2)^2 = 4$.

10. $\mathbf{F} = (x + y)\mathbf{i} + (2x^2 - y^2)\mathbf{j}$ and the contour is the boundary of the region determined by the graphs of $y = x^2$ and $y = 4$.

11. $\mathbf{F} = 3y\mathbf{i} + 2x\mathbf{j}$ and the contour is the boundary of the region determined by the graphs of $y = 0$ and $y = \sin(x)$ with $0 \le x \le \pi$.

12. $\mathbf{F} = -16y\mathbf{i} + (4e^y + 3x^2)\mathbf{j}$ and the contour is the pie wedge defined by the lines $y = x$, $y = -x$, $x^2 + y^2 = 4$, and $y > 0$.

10.7 STOKES' THEOREM

In Section 10.2 we introduced the vector quantity $\nabla \times \mathbf{v}$ which gives a measure of the rotation of a parcel of fluid lying within the velocity field $\mathbf{v}$. In this section we show how the curl may be used to simplify the calculation of certain closed line integrals.

This relationship between a closed line integral and a surface integral involving the curl is

Stokes' Theorem: *The circulation of* $\mathbf{F} = P\mathbf{i} + Q\mathbf{j} + R\mathbf{k}$ *around the closed boundary* C *of an oriented surface* S *in the direction counterclockwise with respect to the surface's unit normal vector* $\mathbf{n}$ *equals the integral of* $\nabla \times \mathbf{F} \cdot \mathbf{n}$ *over* S *or*

$$\oint_C \mathbf{F} \cdot d\mathbf{r} = \iint_S \nabla \times \mathbf{F} \cdot \mathbf{n}\, d\sigma. \tag{10.7.1}$$

Stokes' theorem requires that all of the functions and derivatives be continuous.

The proof of Stokes' theorem is as follows: Consider a finite surface S whose boundary is the loop C. We divide this surface into a number of small elements $\mathbf{n}\, d\sigma$ and compute the *circulation* $d\Gamma = \oint_L \mathbf{F} \cdot d\mathbf{r}$ around each element. When we add all of the circulations together, the contribution from an integration along a boundary line between two adjoining elements cancels out because the boundary is transversed once in each direction. For this reason, the only contributions that survive are those parts where the element boundaries form part of C. Thus, the sum of all circulations equals $\oint_C \mathbf{F} \cdot d\mathbf{r}$, the circulation around the edge of the whole surface.

Next, let us compute the circulation another way. We begin by finding the Taylor expansion for $P(x, y, z)$ about the arbitrary point (x_0, y_0, z_0):

$$\begin{aligned} P(x, y, z) = P(x_0, y_0, z_0) &+ (x - x_0)\frac{\partial P(x_0, y_0, z_0)}{\partial x} \\ &+ (y - y_0)\frac{\partial P(x_0, y_0, z_0)}{\partial y} + (z - z_0)\frac{\partial P(x_0, y_0, z_0)}{\partial z} + \cdots \end{aligned} \tag{10.7.2}$$

Figure 10.7.1: Sir George Gabriel Stokes (1819–1903) was Lucasian Professor of Mathematics at Cambridge University from 1849 until his death. Having learned of an integral theorem from his friend Lord Kelvin, Stokes included it a few years later among his questions on an examination that he wrote for the Smith prize. It is this integral theorem that we now call Stokes' theorem. (Portrait courtesy of the Royal Society of London.)

with similar expansions for $Q(x,y,z)$ and $R(x,y,z)$. Then

$$\begin{aligned} d\Gamma = \oint_L \mathbf{F}\cdot d\mathbf{r} = P(x_0,y_0,z_0)\oint_L dx + \frac{\partial P(x_0,y_0,z_0)}{\partial x}\oint_L (x-x_0)\,dx \\ + \frac{\partial P(x_0,y_0,z_0)}{\partial y}\oint_L (y-y_0)\,dy + \cdots \\ + \frac{\partial Q(x_0,y_0,z_0)}{\partial x}\oint_L (x-x_0)\,dy + \cdots, \end{aligned} \tag{10.7.3}$$

where L denotes some small loop located in the surface S. Note that integrals such as $\oint_L dx$ and $\oint_L (x-x_0)dx$ will vanish.

If we now require that the loop integrals be in the *clockwise* or *positive* sense so that we preserve the right-hand screw convention, then

$$\mathbf{n}\cdot\mathbf{k}\,\delta\sigma = \oint_L (x-x_0)\,dy = -\oint_L (y-y_0)\,dx, \tag{10.7.4}$$

$$\mathbf{n}\cdot\mathbf{j}\,\delta\sigma = \oint_L (z-z_0)\,dx = -\oint_L (x-x_0)\,dz, \tag{10.7.5}$$

$$\mathbf{n}\cdot\mathbf{i}\,\delta\sigma = \oint_L (y-y_0)\,dz = -\oint_L (z-z_0)\,dy \tag{10.7.6}$$

and

$$d\Gamma = \left(\frac{\partial R}{\partial y} - \frac{\partial Q}{\partial z}\right)\mathbf{i}\cdot\mathbf{n}\,\delta\sigma + \left(\frac{\partial P}{\partial z} - \frac{\partial R}{\partial x}\right)\mathbf{j}\cdot\mathbf{n}\,\delta\sigma$$
$$+ \left(\frac{\partial Q}{\partial x} - \frac{\partial P}{\partial y}\right)\mathbf{k}\cdot\mathbf{n}\,\delta\sigma = \nabla\times\mathbf{F}\cdot\mathbf{n}\,\delta\sigma \tag{10.7.7}$$

Therefore, the sum of all circulations in the limit when all elements are made infinitesimally small becomes the surface integral $\iint_S \nabla\times\mathbf{F}\cdot\mathbf{n}\,d\sigma$ and Stokes' theorem is proven. □

In the following examples we first apply Stokes' theorem to a few simple geometries. We then show how to apply this theorem to more complicated geometries.[4]

• Example 10.7.1

Let us verify Stokes' theorem using the vector field $\mathbf{F} = x^2\mathbf{i} + 2x\mathbf{j} + z^2\mathbf{k}$ and the closed curve is a square with vertices at $(0,0,3)$, $(1,0,3)$, $(1,1,3)$ and $(0,1,3)$. See Figure 10.7.2.

We begin with the line integral:

$$\oint_C \mathbf{F}\cdot d\mathbf{r} = \int_{C_1} \mathbf{F}\cdot d\mathbf{r} + \int_{C_2} \mathbf{F}\cdot d\mathbf{r} + \int_{C_3} \mathbf{F}\cdot d\mathbf{r} + \int_{C_4} \mathbf{F}\cdot d\mathbf{r}, \tag{10.7.8}$$

where C_1, C_2, C_3, and C_4 represent the four sides of the square. Along C_1, x varies while $y=0$ and $z=3$. Therefore,

$$\int_{C_1} \mathbf{F}\cdot d\mathbf{r} = \int_0^1 x^2\,dx + 2x\cdot 0 + 9\cdot 0 = \tfrac{1}{3}, \tag{10.7.9}$$

because $dy = dz = 0$ and $z = 3$. Along C_2, y varies with $x = 1$ and $z = 3$. Therefore,

$$\int_{C_2} \mathbf{F}\cdot d\mathbf{r} = \int_0^1 1^2\cdot 0 + 2\cdot 1\cdot dy + 9\cdot 0 = 2. \tag{10.7.10}$$

[4] Thus, different Stokes for different folks.

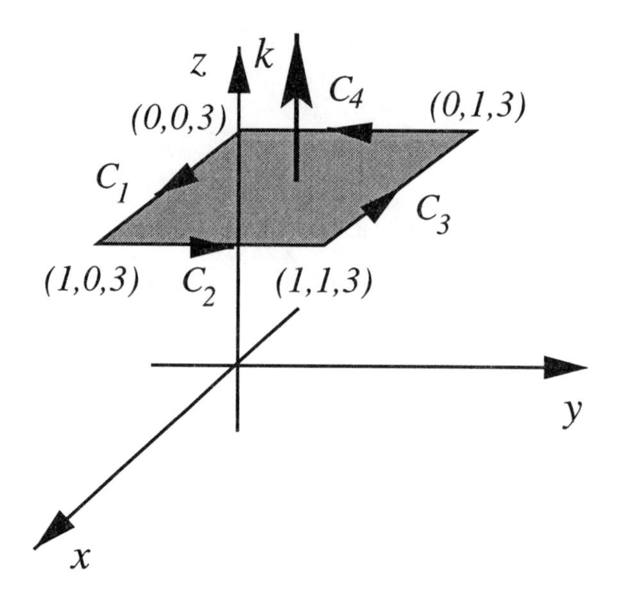

Figure 10.7.2: Diagram for the verification of Stokes' theorem in Example 10.7.1.

Along C_3, x again varies with $y = 1$ and $z = 3$, and so,

$$\int_{C_3} \mathbf{F} \cdot d\mathbf{r} = \int_1^0 x^2\, dx + 2x \cdot 0 + 9 \cdot 0 = -\tfrac{1}{3}. \tag{10.7.11}$$

Note how the limits run from 1 to 0 because x is decreasing. Finally, for C_4, y again varies with $x = 0$ and $z = 3$. Hence,

$$\int_{C_4} \mathbf{F} \cdot d\mathbf{r} = \int_1^0 0^2 \cdot 0 + 2 \cdot 0 \cdot dy + 9 \cdot 0 = 0. \tag{10.7.12}$$

Hence,

$$\oint_C \mathbf{F} \cdot d\mathbf{r} = 2. \tag{10.7.13}$$

Turning to the other side of the equation,

$$\nabla \times \mathbf{F} = \begin{vmatrix} \mathbf{i} & \mathbf{j} & \mathbf{k} \\ \frac{\partial}{\partial x} & \frac{\partial}{\partial y} & \frac{\partial}{\partial z} \\ x^2 & 2x & z^2 \end{vmatrix} = 2\mathbf{k}. \tag{10.7.14}$$

Our line integral has been such that the normal vector must be $\mathbf{n} = \mathbf{k}$. Therefore,

$$\iint_S \nabla \times \mathbf{F} \cdot \mathbf{n}\, d\sigma = \int_0^1 \int_0^1 2\mathbf{k} \cdot \mathbf{k}\, dx\, dy = 2 \tag{10.7.15}$$

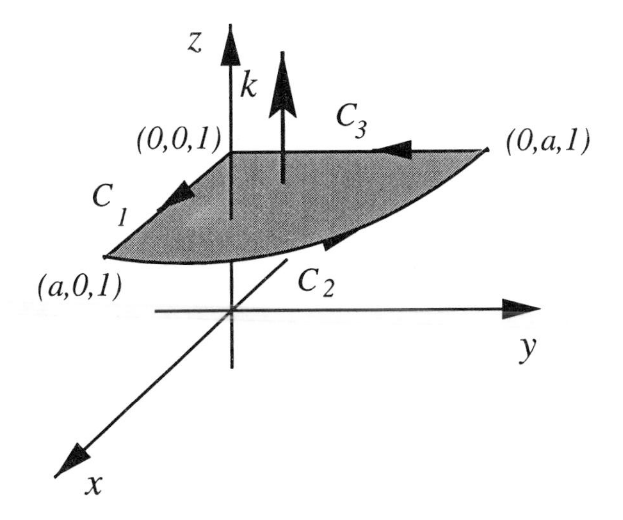

Figure 10.7.3: Diagram for the verification of Stokes' theorem in Example 10.7.2.

and Stokes' theorem is verified for this special case.

• **Example 10.7.2**

Let us verify Stokes' theorem using the vector field $\mathbf{F} = (x^2 - y)\mathbf{i} + 4z\mathbf{j} + x^2\mathbf{k}$, where the closed contour consists of the x and y coordinate axes and that portion of the circle $x^2 + y^2 = a^2$ that lies in the first quadrant with $z = 1$. See Figure 10.7.3.

The line integral consists of three parts:

$$\oint_C \mathbf{F} \cdot d\mathbf{r} = \int_{C_1} \mathbf{F} \cdot d\mathbf{r} + \int_{C_2} \mathbf{F} \cdot d\mathbf{r} + \int_{C_3} \mathbf{F} \cdot d\mathbf{r}. \tag{10.7.16}$$

Along C_1, x varies while $y = 0$ and $z = 1$. Therefore,

$$\int_{C_1} \mathbf{F} \cdot d\mathbf{r} = \int_0^a (x^2 - 0)\,dx + 4 \cdot 1 \cdot 0 + x^2 \cdot 0 = \frac{a^3}{3}. \tag{10.7.17}$$

Along the circle C_2, we use polar coordinates with $x = a\cos(t)$, $y = a\sin(t)$ and $z = 1$. Therefore,

$$\int_{C_2} \mathbf{F} \cdot d\mathbf{r} = \int_0^{\pi/2} [a^2\cos^2(t) - a\sin(t)][-a\sin(t)\,dt] + 4 \cdot 1 \cdot a\cos(t)\,dt + a^2\cos^2(t) \cdot 0 \tag{10.7.18}$$

$$\int_{C_2} \mathbf{F} \cdot d\mathbf{r} = \int_0^{\pi/2} -a^3 \cos^2(t) \sin(t)\, dt + a^2 \sin^2(t)\, dt + 4a \cos(t)\, dt \tag{10.7.19}$$

$$= \frac{a^3}{3} \cos^3(t) \Big|_0^{\pi/2} + \frac{a^2}{2} \left[t - \frac{1}{2} \sin(2t) \right] \Big|_0^{\pi/2} + 4a \sin(t)|_0^{\pi/2} \tag{10.7.20}$$

$$= -\frac{a^3}{3} + \frac{a^2 \pi}{4} + 4a, \tag{10.7.21}$$

because $dx = -a \sin(t)\, dt$ and $dy = a \cos(t)\, dt$. Finally, along C_3, y varies with $x = 0$ and $z = 1$. Therefore,

$$\int_{C_3} \mathbf{F} \cdot d\mathbf{r} = \int_a^0 (0^2 - y) \cdot 0 + 4 \cdot 1 \cdot dy + 0^2 \cdot 0 = -4a. \tag{10.7.22}$$

so that

$$\oint_C \mathbf{F} \cdot d\mathbf{r} = \frac{a^2 \pi}{4}. \tag{10.7.23}$$

Turning to the other side of the equation,

$$\nabla \times \mathbf{F} = \begin{vmatrix} \mathbf{i} & \mathbf{j} & \mathbf{k} \\ \frac{\partial}{\partial x} & \frac{\partial}{\partial y} & \frac{\partial}{\partial z} \\ x^2 - y & 4z & x^2 \end{vmatrix} = -4\mathbf{i} - 2x\mathbf{j} + \mathbf{k}. \tag{10.7.24}$$

From the path of our line integral, our unit normal vector must be $\mathbf{n} = \mathbf{k}$. Then,

$$\iint_S \nabla \times \mathbf{F} \cdot \mathbf{n}\, d\sigma = \int_0^a \int_0^{\pi/2} [-4\mathbf{i} - 2r \cos(\theta)\mathbf{j} + \mathbf{k}] \cdot \mathbf{k}\, r\, d\theta\, dr = \frac{\pi a^2}{4} \tag{10.7.25}$$

and Stokes' theorem is verified for this case.

• Example 10.7.3

Let us verify Stokes' theorem using the vector field $\mathbf{F} = 2yz\mathbf{i} - (x + 3y - 2)\mathbf{j} + (x^2 + z)\mathbf{k}$, where the closed triangular region is that portion of the plane $x + y + z = 1$ that lies in the first octant.

As shown in Figure 10.7.4, the closed line integration consists of three line integrals:

$$\oint_C \mathbf{F} \cdot d\mathbf{r} = \int_{C_1} \mathbf{F} \cdot d\mathbf{r} + \int_{C_2} \mathbf{F} \cdot d\mathbf{r} + \int_{C_3} \mathbf{F} \cdot d\mathbf{r}. \tag{10.7.26}$$

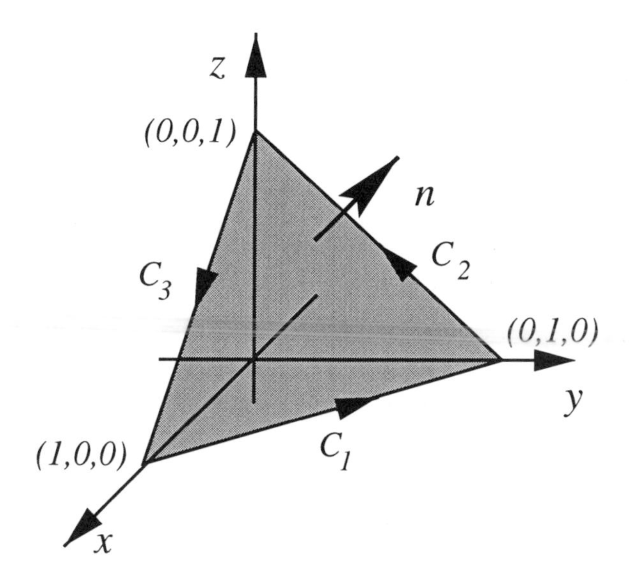

Figure 10.7.4: Diagram for the verification of Stokes' theorem in Example 10.7.3.

Along C_1, $z = 0$ and $y = 1 - x$. Therefore, using x as the independent variable,

$$\int_{C_1} \mathbf{F} \cdot d\mathbf{r} = \int_1^0 2(1-x) \cdot 0 \cdot dx - (x + 3 - 3x - 2)(-dx) + (x^2 + 0) \cdot 0$$
$$= -x^2\Big|_1^0 + x\Big|_1^0 = 0. \qquad \mathbf{(10.7.27)}$$

Along C_2, $x = 0$ and $y = 1 - z$. Thus,

$$\int_{C_2} \mathbf{F} \cdot d\mathbf{r} = \int_0^1 2(1-z)z \cdot 0 - (0 + 3 - 3z - 2)(-dz) + (0^2 + z)\, dz$$
$$= -\tfrac{3}{2}z^2 + z + \tfrac{1}{2}z^2\Big|_0^1 = 0. \qquad \mathbf{(10.7.28)}$$

Finally, along C_3, $y = 0$ and $z = 1 - x$. Hence,

$$\int_{C_3} \mathbf{F} \cdot d\mathbf{r} = \int_0^1 2 \cdot 0 \cdot (1-x)\, dx - (x + 0 - 2) \cdot 0 + (x^2 + 1 - x)(-dx)$$
$$= -\tfrac{1}{3}x^3 - x + \tfrac{1}{2}x^2\Big|_0^1 = -\tfrac{5}{6}. \qquad \mathbf{(10.7.29)}$$

Thus,

$$\oint_C \mathbf{F} \cdot d\mathbf{r} = -\tfrac{5}{6}. \qquad \mathbf{(10.7.30)}$$

On the other hand,

$$\nabla \times \mathbf{F} = \begin{vmatrix} \mathbf{i} & \mathbf{j} & \mathbf{k} \\ \frac{\partial}{\partial x} & \frac{\partial}{\partial y} & \frac{\partial}{\partial z} \\ 2yz & -x - 3y + 2 & x^2 + z \end{vmatrix} = (-2x + 2y)\mathbf{j} + (-1 - 2z)\mathbf{k}. \tag{10.7.31}$$

To find $\mathbf{n}\, d\sigma$, we use the general coordinate system $x = u$, $y = v$, and $z = 1 - u - v$. Therefore, $\mathbf{r} = u\mathbf{i} + v\mathbf{j} + (1 - u - v)\mathbf{k}$ and

$$\mathbf{r}_u \times \mathbf{r}_v = \begin{vmatrix} \mathbf{i} & \mathbf{j} & \mathbf{k} \\ 1 & 0 & -1 \\ 0 & 1 & -1 \end{vmatrix} = \mathbf{i} + \mathbf{j} + \mathbf{k}. \tag{10.7.32}$$

Thus,

$$\iint_S \nabla \times \mathbf{F} \cdot \mathbf{n}\, d\sigma = \int_0^1 \int_0^{1-u} [(-2u + 2v)\mathbf{j} + (-1 - 2 + 2u + 2v)\mathbf{k}] \cdot [\mathbf{i} + \mathbf{j} + \mathbf{k}]\, dv\, du \tag{10.7.33}$$

$$= \int_0^1 \int_0^{1-u} (4v - 3)\, dv\, du \tag{10.7.34}$$

$$= \int_0^1 [2(1 - u)^2 - 3(1 - u)]\, du \tag{10.7.35}$$

$$= \int_0^1 (-1 - u + 2u^2)\, du = -\tfrac{5}{6} \tag{10.7.36}$$

and Stokes' theorem is verified for this case.

Problems

Verify Stokes' theorem using the following vector fields and surfaces:

1. $\mathbf{F} = 5y\mathbf{i} - 5x\mathbf{j} + 3z\mathbf{k}$ and the surface S is that portion of the plane $z = 1$ with the square at the vertices $(0, 0, 1)$, $(1, 0, 1)$, $(1, 1, 1)$, and $(0, 1, 1)$.

2. $\mathbf{F} = x^2\mathbf{i} + y^2\mathbf{j} + z^2\mathbf{k}$ and the surface S is the rectangular portion of the plane $z = 2$ defined by the corners $(0, 0, 2)$, $(2, 0, 2)$, $(2, 1, 2)$, and $(0, 1, 2)$.

3. $\mathbf{F} = z\mathbf{i} + x\mathbf{j} + y\mathbf{k}$ and the surface S is the triangular portion of the plane $z = 1$ defined by the vertices $(0, 0, 1)$, $(2, 0, 1)$, and $(0, 2, 1)$.

4. $\mathbf{F} = 2z\mathbf{i} - 3x\mathbf{j} + 4y\mathbf{k}$ and the surface S is that portion of the plane $z = 5$ within the cylinder $x^2 + y^2 = 4$.

5. $\mathbf{F} = z\mathbf{i} + x\mathbf{j} + y\mathbf{k}$ and the surface S is that portion of the plane $z = 3$ bounded by the lines $y = 0$, $x = 0$, and $x^2 + y^2 = 4$.

6. $\mathbf{F} = (2z + x)\mathbf{i} + (y - z)\mathbf{j} + (x + y)\mathbf{k}$ and the surface S is the interior of the triangularly shaped plane with vertices at $(1, 0, 0)$, $(0, 1, 0)$, and $(0, 0, 1)$.

7. $\mathbf{F} = z\mathbf{i} + x\mathbf{j} + y\mathbf{k}$ and the surface S is that portion of the plane $2x + y + 2z = 6$ in the first octant.

8. $\mathbf{F} = x\mathbf{i} + xz\mathbf{j} + y\mathbf{k}$ and the surface S is that portion of the paraboloid $z = 9 - x^2 - y^2$ within the cylinder $x^2 + y^2 = 4$.

10.8 DIVERGENCE THEOREM

Although Stokes' theorem is useful in computing closed line integrals, it is usually very difficult to go the other way and convert a surface integral into a closed line integral because the integrand must have a very special form, namely $\nabla \times \mathbf{F} \cdot \mathbf{n}$. In this section we introduce a theorem that allows with equal facility the conversion of a closed surface integral into a volume integral and *vice versa*. Furthermore, if we can convert a given surface integral into a closed one by the introduction of a simple surface (for example, closing a hemispheric surface by adding an equatorial plate), it may be easier to use the divergence theorem and subtract off the contribution from the new surface integral rather than do the original problem.

This relationship between a closed surface integral and a volume integral involving the divergence operator is

The Divergence or Gauss' Theorem: *Let V be a closed and bounded region in three dimensional space with a piece-wise smooth boundary S that is oriented outward. Let $\mathbf{F} = P(x, y, z)\mathbf{i} + Q(x, y, z)\mathbf{j} + R(x, y, z)\mathbf{k}$ be a vector field for which P, Q, and R are continuous and have continuous first partial derivatives in a region of three dimensional space containing V. Then*

$$\oiint_S \mathbf{F} \cdot \mathbf{n}\, d\sigma = \iiint_V \nabla \cdot \mathbf{F}\, dV. \tag{10.8.1}$$

Figure 10.8.1: Carl Friedrich Gauss (1777–1855), the prince of mathematicians, must be on the list of the greatest mathematicians who ever lived. Gauss, a child prodigy, is almost as well known for what he did not publish during his lifetime as for what he did. This is true of Gauss' divergence theorem which he proved while working on the theory of gravitation. It was only when his notebooks were published in 1898 that his precedence over the published work of Ostrogradsky (1801–1862) was established. (Portrait courtesy of Photo AKG, London.)

Here, the circle on the double integral signs denotes a closed surface integral.

A nonrigorous proof of Gauss' theorem is as follows. Imagine that our volume V is broken down into small elements $d\tau$ of volume of any shape so long as they include all of the original volume. In general, the surfaces of these elements are composed of common interfaces between adjoining elements. However, for the elements at the periphery of V, part of their surface will be part of the surface S that encloses V. Now $d\Phi = \nabla \cdot \mathbf{F}\, d\tau$ is the net flux of the vector $\mathbf{F}$ out from the element $d\tau$.

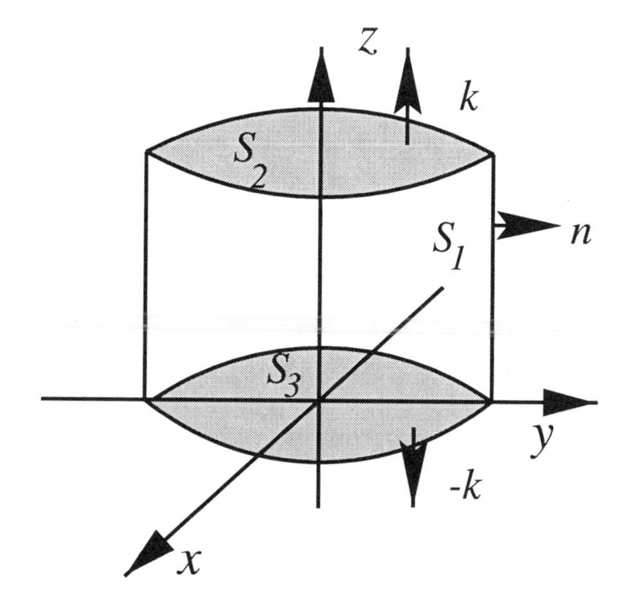

Figure 10.8.2: Diagram for the verification of the divergence theorem in Example 10.8.1.

At the common interface between elements, the flux *out* of one element equals the flux *into* its neighbor. Therefore, the sum of all such terms yields

$$\Phi = \iiint_V \nabla \cdot \mathbf{F}\, d\tau \tag{10.8.2}$$

and all the contributions from these common interfaces cancel; only the contribution from the parts on the outer surface S will be left. These contributions, when added together, give $\oiint_S \mathbf{F} \cdot \mathbf{n}\, d\sigma$ over S and the proof is completed. □

• **Example 10.8.1**

Let us verify the divergence theorem using the vector field $\mathbf{F} = 4x\mathbf{i} - 2y^2\mathbf{j} + z^2\mathbf{k}$ and the enclosed surface is the cylinder $x^2 + y^2 = 4$, $z = 0$, and $z = 3$. See Figure 10.8.2.

We begin by computing the volume integration. Because

$$\nabla \cdot \mathbf{F} = \frac{\partial(4x)}{\partial x} + \frac{\partial(-2y^2)}{\partial y} + \frac{\partial(z^2)}{\partial z} = 4 - 4y + 2z, \tag{10.8.3}$$

$$\iiint_V \nabla \cdot \mathbf{F}\, dV = \iiint_V (4 - 4y + 2z)\, dV \tag{10.8.4}$$

$$\iiint_V \nabla \cdot \mathbf{F}\, dV = \int_0^3 \int_0^2 \int_0^{2\pi} [4 - 4r\sin(\theta) + 2z]\, d\theta\, r\, dr\, dz \tag{10.8.5}$$

$$= \int_0^3 \int_0^2 \left[4\theta\Big|_0^{2\pi} + 4r\cos(\theta)\Big|_0^{2\pi} + 2z\theta\Big|_0^{2\pi} \right] r\, dr\, dz \tag{10.8.6}$$

$$= \int_0^3 \int_0^2 (8\pi + 4\pi z)\, r\, dr\, dz \tag{10.8.7}$$

$$= \int_0^3 4\pi(2+z) \tfrac{1}{2} r^2 \Big|_0^2\, dz \tag{10.8.8}$$

$$= 4\pi \int_0^3 2(2+z)\, dz = 8\pi(2z + \tfrac{1}{2}z^2)\Big|_0^3 = 84\pi. \tag{10.8.9}$$

Turning to the surface integration, we have three surfaces:

$$\oiint_S \mathbf{F}\cdot\mathbf{n}\, d\sigma = \iint_{S_1} \mathbf{F}\cdot\mathbf{n}\, d\sigma + \iint_{S_2} \mathbf{F}\cdot\mathbf{n}\, d\sigma + \iint_{S_3} \mathbf{F}\cdot\mathbf{n}\, d\sigma. \tag{10.8.10}$$

The first integral is over the exterior to the cylinder. Because the surface is defined by $f(x, y, z) = x^2 + y^2 = 4$,

$$\mathbf{n} = \frac{\nabla f}{|\nabla f|} = \frac{2x\mathbf{i} + 2y\mathbf{j}}{\sqrt{4x^2 + 4y^2}} = \frac{x}{2}\mathbf{i} + \frac{y}{2}\mathbf{j}. \tag{10.8.11}$$

Therefore,

$$\iint_{S_1} \mathbf{F}\cdot\mathbf{n}\, d\sigma = \iint_{S_1} (2x^2 - y^3)\, d\sigma \tag{10.8.12}$$

$$= \int_0^3 \int_0^{2\pi} \left\{ 2[2\cos(\theta)]^2 - [2\sin(\theta)]^3 \right\} 2\, d\theta\, dz \tag{10.8.13}$$

$$= 8 \int_0^3 \int_0^{2\pi} \left\{ \tfrac{1}{2}[1 + \cos(2\theta)] - \sin(\theta) + \cos^2(\theta)\sin(\theta) \right\} 2\, d\theta\, dz \tag{10.8.14}$$

$$= 16 \int_0^3 \left[\tfrac{1}{2}\theta + \tfrac{1}{4}\sin(2\theta) + \cos(\theta) - \tfrac{1}{3}\cos^3(\theta) \right] \Bigg|_0^{2\pi} dz \tag{10.8.15}$$

$$= 16\pi \int_0^3 dz = 48\pi, \tag{10.8.16}$$

because $x = 2\cos(\theta)$, $y = 2\sin(\theta)$ and $d\sigma = 2\, d\theta\, dz$ in cylindrical coordinates.

Along the top of the cylinder, $z = 3$, the outward pointing normal is $\mathbf{n} = \mathbf{k}$ and $d\sigma = r\, dr\, d\theta$. Then,

$$\iint_{S_2} \mathbf{F}\cdot\mathbf{n}\, d\sigma = \iint_{S_2} z^2\, d\sigma = \int_0^{2\pi} \int_0^2 9\, r\, dr\, d\theta = 2\pi \times 9 \times 2 = 36\pi. \tag{10.8.17}$$

However, along the bottom of the cylinder, $z = 0$, the outward pointing normal is $\mathbf{n} = -\mathbf{k}$ and $d\sigma = r\,dr\,d\theta$. Then,

$$\iint_{S_3} \mathbf{F} \cdot \mathbf{n}\, d\sigma = \iint_{S_3} z^2\, d\sigma = \int_0^{2\pi} \int_0^2 0\, r\, dr\, d\theta = 0. \tag{10.8.18}$$

Consequently, the flux out the entire cylinder is

$$\oiint_S \mathbf{F} \cdot \mathbf{n}\, d\sigma = 48\pi + 36\pi + 0 = 84\pi \tag{10.8.19}$$

and the divergence theorem is verified for this special case.

• Example 10.8.2

Let us verify the divergence theorem given the vector field $\mathbf{F} = 3x^2y^2\mathbf{i} + y\mathbf{j} - 6xy^2z\mathbf{k}$ and the volume is the region bounded by the paraboloid $z = x^2 + y^2$ and the plane $z = 2y$. See Figure 10.8.3.

Computing the divergence,

$$\nabla \cdot \mathbf{F} = \frac{\partial(3x^2y^2)}{\partial x} + \frac{\partial(y)}{\partial y} + \frac{\partial(-6xy^2z)}{\partial z} = 6xy^2 + 1 - 6xy^2 = 1. \tag{10.8.20}$$

Then,

$$\iiint_V \nabla \cdot \mathbf{F}\, dV = \iiint_V dV \tag{10.8.21}$$

$$= \int_0^{\pi} \int_0^{2\sin(\theta)} \int_{r^2}^{2r\sin(\theta)} dz\, r\, dr\, d\theta \tag{10.8.22}$$

$$= \int_0^{\pi} \int_0^{2\sin(\theta)} [2r\sin(\theta) - r^2]\, r\, dr\, d\theta \tag{10.8.23}$$

$$= \int_0^{\pi} \left[\tfrac{2}{3}r^3 \Big|_0^{2\sin(\theta)} \sin(\theta) - \tfrac{1}{4}r^4 \Big|_0^{2\sin(\theta)} \right] d\theta \tag{10.8.24}$$

$$= \int_0^{\pi} \left[\tfrac{16}{3}\sin^4(\theta) - 4\sin^4(\theta) \right] d\theta \tag{10.8.25}$$

$$= \int_0^{\pi} \tfrac{4}{3}\sin^4(\theta)\, d\theta \tag{10.8.26}$$

$$= \tfrac{1}{3} \int_0^{\pi} [1 - 2\cos(2\theta) + \cos^2(2\theta)]\, d\theta \tag{10.8.27}$$

$$= \tfrac{1}{3} \left[\theta \Big|_0^{\pi} - \sin(2\theta) \Big|_0^{\pi} + \tfrac{1}{2}\theta \Big|_0^{\pi} + \tfrac{1}{8}\sin(4\theta) \Big|_0^{\pi} \right] = \tfrac{\pi}{2}. \tag{10.8.28}$$

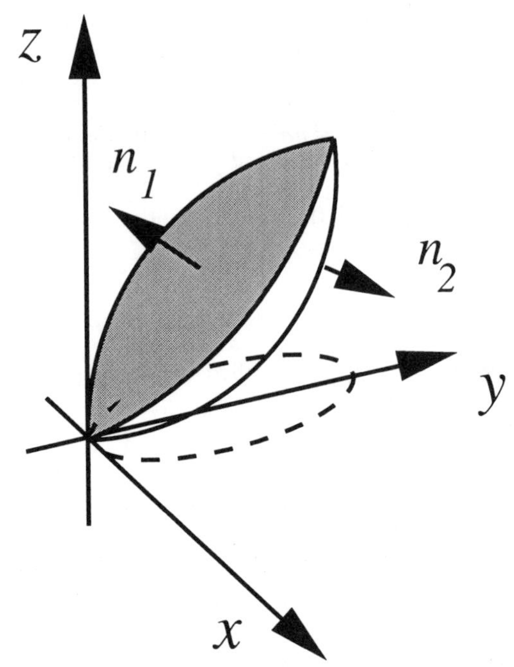

Figure 10.8.3: Diagram for the verification of the divergence theorem in Example 10.8.2. The curve $r = 2\sin(\theta)$ is denoted by a dashed line.

The limits in the radial direction are given by the intersection of the paraboloid and plane: $r^2 = 2r\sin(\theta)$ or $r = 2\sin(\theta)$ and y is greater than zero.

Turning to the surface integration, we have two surfaces:

$$\oiint_S \mathbf{F} \cdot \mathbf{n}\, d\sigma = \iint_{S_1} \mathbf{F} \cdot \mathbf{n}\, d\sigma + \iint_{S_2} \mathbf{F} \cdot \mathbf{n}\, d\sigma, \tag{10.8.29}$$

where S_1 is the plane $z = 2y$ and S_2 is the paraboloid. For either surface, polar coordinates are best so that $x = r\cos(\theta)$, $y = r\sin(\theta)$. For the integration over the plane, $z = 2r\sin(\theta)$. Therefore,

$$\mathbf{r} = r\cos(\theta)\mathbf{i} + r\sin(\theta)\mathbf{j} + 2r\sin(\theta)\mathbf{k} \tag{10.8.30}$$

so that

$$\mathbf{r}_r = \cos(\theta)\mathbf{i} + \sin(\theta)\mathbf{j} + 2\sin(\theta)\mathbf{k} \tag{10.8.31}$$

and

$$\mathbf{r}_\theta = -r\sin(\theta)\mathbf{i} + r\cos(\theta)\mathbf{j} + 2r\cos(\theta)\mathbf{k}. \tag{10.8.32}$$

Then,

$$\mathbf{r}_r \times \mathbf{r}_\theta = \begin{vmatrix} \mathbf{i} & \mathbf{j} & \mathbf{k} \\ \cos(\theta) & \sin(\theta) & 2\sin(\theta) \\ -r\sin(\theta) & r\cos(\theta) & 2r\cos(\theta) \end{vmatrix} = -2r\mathbf{j} + r\mathbf{k}. \tag{10.8.33}$$

This is an outwardly pointing normal so that we can immediately set up the surface integral:

$$\iint_{S_1} \mathbf{F}\cdot\mathbf{n}\,d\sigma = \int_0^\pi \int_0^{2\sin(\theta)} \{3r^4\cos^2(\theta)\sin^2(\theta)\mathbf{i} + r\sin(\theta)\mathbf{j} - 6[2r\sin(\theta)][r\cos(\theta)][r^2\sin^2(\theta)]\mathbf{k}\}\cdot(-2r\mathbf{j}+r\mathbf{k})\,dr\,d\theta \tag{10.8.34}$$

$$= \int_0^\pi \int_0^{2\sin(\theta)} \left[-2r^2\sin(\theta) - 12r^5\sin^3(\theta)\cos(\theta)\right] dr\,d\theta \tag{10.8.35}$$

$$= \int_0^\pi \left[-\tfrac{2}{3}r^3\Big|_0^{2\sin(\theta)}\sin(\theta) - 2r^6\Big|_0^{2\sin(\theta)}\sin^3(\theta)\cos(\theta)\right] d\theta \tag{10.8.36}$$

$$= \int_0^\pi \left[-\tfrac{16}{3}\sin^4(\theta) - 128\sin^9(\theta)\cos(\theta)\right] d\theta \tag{10.8.37}$$

$$= -\tfrac{4}{3}\left[\theta\Big|_0^\pi - \sin(2\theta)\Big|_0^\pi + \tfrac{1}{2}\theta\Big|_0^\pi + \tfrac{1}{8}\sin(4\theta)\Big|_0^\pi\right] - \tfrac{64}{5}\sin^{10}(\theta)\Big|_0^\pi \tag{10.8.38}$$

$$= -2\pi. \tag{10.8.39}$$

For the surface of the paraboloid,

$$\mathbf{r} = r\cos(\theta)\mathbf{i} + r\sin(\theta)\mathbf{j} + r^2\mathbf{k} \tag{10.8.40}$$

so that

$$\mathbf{r}_r = \cos(\theta)\mathbf{i} + \sin(\theta)\mathbf{j} + 2r\mathbf{k} \tag{10.8.41}$$

and

$$\mathbf{r}_\theta = -r\sin(\theta)\mathbf{i} + r\cos(\theta)\mathbf{j}. \tag{10.8.42}$$

Then,

$$\mathbf{r}_r \times \mathbf{r}_\theta = \begin{vmatrix} \mathbf{i} & \mathbf{j} & \mathbf{k} \\ \cos(\theta) & \sin(\theta) & 2r \\ -r\sin(\theta) & r\cos(\theta) & 0 \end{vmatrix} \tag{10.8.43}$$

$$= -2r^2\cos(\theta)\mathbf{i} - 2r^2\sin(\theta)\mathbf{j} + r\mathbf{k}. \tag{10.8.44}$$

This is an inwardly pointing normal so that we must take the negative of it before we do the surface integral. Then,

$$\iint_{S_2} \mathbf{F}\cdot\mathbf{n}\,d\sigma = \int_0^\pi \int_0^{2\sin(\theta)} \{3r^4\cos^2(\theta)\sin^2(\theta)\mathbf{i} + r\sin(\theta)\mathbf{j} - 6r^2[r\cos(\theta)][r^2\sin^2(\theta)]\mathbf{k}\}$$

$$\cdot \left[2r^2\cos(\theta)\mathbf{i} + 2r^2\sin(\theta)\mathbf{j} - r\mathbf{k}\right] dr\, d\theta \tag{10.8.45}$$

$$\iint_{S_2} \mathbf{F}\cdot\mathbf{n}\, d\sigma = \int_0^{\pi}\int_0^{2\sin(\theta)} \left[6r^6\cos^3(\theta)\sin^2(\theta) + 2r^3\sin^2(\theta) + 6r^6\cos(\theta)\sin^2(\theta)\right] dr\, d\theta \tag{10.8.46}$$

$$= \int_0^{\pi}\left[\tfrac{6}{7}r^7\Big|_0^{2\sin(\theta)}\cos^3(\theta)\sin^2(\theta) + \tfrac{1}{2}r^4\Big|_0^{2\sin(\theta)}\sin^2(\theta) + \tfrac{6}{7}r^7\Big|_0^{2\sin(\theta)}\cos(\theta)\sin^2(\theta)\right] d\theta \tag{10.8.47}$$

$$= \int_0^{\pi}\left\{\tfrac{768}{7}\sin^9(\theta)[1-\sin^2(\theta)]\cos(\theta) + 8\sin^6(\theta) + \tfrac{768}{7}\sin^9(\theta)\cos(\theta)\right\} d\theta \tag{10.8.48}$$

$$= \tfrac{1536}{70}\sin^{10}(\theta)\Big|_0^{\pi} - \tfrac{64}{7}\sin^{12}(\theta)\Big|_0^{\pi} + \int_0^{\pi}[1-\cos(2\theta)]^3\, d\theta \tag{10.8.49}$$

$$= \int_0^{\pi}\left\{1 - 3\cos(2\theta) + 3\cos^2(2\theta) - \cos(2\theta)[1-\sin^2(2\theta)]\right\} d\theta \tag{10.8.50}$$

$$= \theta\Big|_0^{\pi} - \tfrac{3}{2}\sin(2\theta)\Big|_0^{\pi} + \tfrac{3}{2}[\theta + \tfrac{1}{4}\sin(4\theta)]\Big|_0^{\pi} - \tfrac{1}{2}\sin(2\theta)\Big|_0^{\pi} + \tfrac{1}{3}\sin^3(2\theta)\Big|_0^{\pi} \tag{10.8.51}$$

$$= \pi + \tfrac{3}{2}\pi = \tfrac{5}{2}\pi. \tag{10.8.52}$$

Consequently,

$$\oiint_S \mathbf{F}\cdot\mathbf{n}\, d\sigma = -2\pi + \tfrac{5}{2}\pi = \tfrac{1}{2}\pi \tag{10.8.53}$$

and the divergence theorem is verified for this special case.

• Example 10.8.3: Archimedes' Principle

Consider a solid[5] of volume V and surface S that is immersed in a vessel filled with a fluid of density ρ. The pressure field p in the fluid is a function of the distance from the liquid/air interface and equals

$$p = p_0 - \rho g z, \tag{10.8.54}$$

[5] Adapted from Altintas, A., 1990: Archimedes' principle as an application of the divergence theorem. *IEEE Trans. Educ.*, **33**, 222. ©IEEE.

where g is the gravitational acceleration, z is the vertical distance measured from the interface (increasing in the $\mathbf{k}$ direction), and p_0 is the constant pressure along the liquid/air interface.

If we define $\mathbf{F} = -p\mathbf{k}$, then $\mathbf{F} \cdot \mathbf{n}\, d\sigma$ is the vertical component of the force on the surface due to the pressure and $\oiint_S \mathbf{F} \cdot \mathbf{n}\, d\sigma$ is the total lift. Using the divergence theorem and noting that $\nabla \cdot \mathbf{F} = \rho g$, the total lift also equals

$$\iiint_V \nabla \cdot \mathbf{F}\, dV = \rho g \iiint_V dV = \rho g V, \tag{10.8.55}$$

which is the weight of the displaced liquid. This is *Archimedes' principle*: the buoyant force on a solid immersed in a fluid of constant density equals the weight of the fluid displaced.

• Example 10.8.4: Conservation of charge

Let a charge of density ρ flow with an average velocity $\mathbf{v}$. Then the charge crossing the element $d\mathbf{S}$ per unit time is $\rho\mathbf{v} \cdot d\mathbf{S} = \mathbf{J} \cdot d\mathbf{S}$, where $\mathbf{J}$ is defined as the conduction current vector or current density vector. The current across any surface drawn in the medium is $\oiint_S \mathbf{J} \cdot d\mathbf{S}$.

The total charge inside the closed surface is $\iiint_V \rho\, dV$. If there are no sources or sinks inside the surface, the rate at which the charge is decreasing is $-\iiint_V \rho_t\, dV$. Because this change is due to the outward flow of charge,

$$-\iiint_V \frac{\partial \rho}{\partial t}\, dV = \oiint_S \mathbf{J} \cdot d\mathbf{S}. \tag{10.8.56}$$

Applying the divergence theorem,

$$\iiint_V \left(\frac{\partial \rho}{\partial t} + \nabla \cdot \mathbf{J} \right) dV = 0. \tag{10.8.57}$$

Because the result holds true for any arbitrary volume, the integrand must vanish identically and we have the equation of continuity or the *equation of conservation of charge*:

$$\frac{\partial \rho}{\partial t} + \nabla \cdot \mathbf{J} = 0. \tag{10.8.58}$$

Problems

Verify the divergence theorem using the following vector fields and volumes:

1. $\mathbf{F} = x^2\mathbf{i} + y^2\mathbf{j} + z^2\mathbf{k}$ and the volume V is the cube cut from the first octant by the planes $x = 1$, $y = 1$, and $z = 1$.

2. $\mathbf{F} = xy\mathbf{i} + yz\mathbf{j} + xz\mathbf{k}$ and the volume V is the cube bounded by $0 \le x \le 1$, $0 \le y \le 1$, and $0 \le z \le 1$.

3. $\mathbf{F} = (y-x)\mathbf{i} + (z-y)\mathbf{j} + (y-x)\mathbf{k}$ and the volume V is the cube bounded by $-1 \le x \le 1$, $-1 \le y \le 1$, and $-1 \le z \le 1$.

4. $\mathbf{F} = x^2\mathbf{i} + y\mathbf{j} + z\mathbf{k}$ and the volume V is the cylinder defined by the surfaces $x^2 + y^2 = 1$, $z = 0$, and $z = 1$.

5. $\mathbf{F} = x^2\mathbf{i} + y^2\mathbf{j} + z^2\mathbf{k}$ and the volume V is the cylinder defined by the surfaces $x^2 + y^2 = 4$, $z = 0$, and $z = 1$.

6. $\mathbf{F} = y^2\mathbf{i} + xz^3\mathbf{j} + (z-1)^2\mathbf{k}$ and the volume V is the cylinder bounded by the surface $x^2 + y^2 = 4$ and the planes $z = 1$ and $z = 5$.

7. $\mathbf{F} = 6xy\mathbf{i} + 4yz\mathbf{j} + xe^{-y}\mathbf{k}$ and the volume V is that region created by the plane $x + y + z = 1$ and the three coordinate planes.

8. $\mathbf{F} = y\mathbf{i} + xy\mathbf{j} - z\mathbf{k}$ and the volume V is that solid created by the paraboloid $z = x^2 + y^2$ and plane $z = 1$.

Chapter 11
Linear Algebra

Linear algebra involves the systematic solving of linear algebraic or differential equations. These equations arise in a wide variety of situations. They usually involve some system, either electrical, mechanical, or even human, where two or more components are interacting with each other. In this chapter we present efficient techniques for expressing these systems and their solution.

11.1 FUNDAMENTALS OF LINEAR ALGEBRA

In this chapter we shall study the solution of m simultaneous linear equations in n unknowns $x_1, x_2, x_3, \ldots, x_n$ of the form:

$$\begin{aligned}
a_{11}x_1 + a_{12}x_2 + \cdots + a_{1n}x_n &= b_1 \\
a_{21}x_1 + a_{22}x_2 + \cdots + a_{2n}x_n &= b_2 \\
&\vdots \\
&\vdots \\
a_{m1}x_1 + a_{m2}x_2 + \cdots + a_{mn}x_n &= b_m,
\end{aligned} \tag{11.1.1}$$

where the a's and b's are known real or complex numbers. *Matrix algebra* allows us to solve these systems. First, succinct notation is introduced so that we can replace (11.1.1) with rather simple expressions. Then a set of rules is used to manipulate these simple expressions. In this section we focus on developing these simple expressions.

The fundamental quantity in linear algebra is the *matrix*. A matrix is an ordered rectangular array of numbers or mathematical expressions. We shall use upper case letters to denote them. The $m \times n$ matrix

$$A = \begin{pmatrix} a_{11} & a_{12} & a_{13} & \cdot & \cdot & \cdot & a_{1n} \\ a_{21} & a_{22} & a_{23} & \cdot & \cdot & \cdot & a_{2n} \\ \cdot & \cdot & \cdot & \cdot & \cdot & \cdot & \cdot \\ \cdot & \cdot & \cdot & \cdot & a_{ij} & \cdot & \cdot \\ \cdot & \cdot & \cdot & \cdot & \cdot & \cdot & \cdot \\ a_{m1} & a_{m2} & a_{m3} & \cdot & \cdot & \cdot & a_{mn} \end{pmatrix} \tag{11.1.2}$$

has m *rows* and n *columns*. The *order* (or size) of a matrix is determined by the number of rows and columns; (11.1.2) is of order m by n. If $m = n$, the matrix is a *square* matrix; otherwise, A is *rectangular*. The numbers or expressions in the array a_{ij} are the *elements* of A and may be either real or complex. When all of the elements are real, A is a *real matrix*. If some or all of the elements are complex, then A is a *complex matrix*. For a square matrix, the diagonal from the top left corner to the bottom right corner is the *principal diagonal*.

From the limitless number of possible matrices, certain ones appear with sufficient regularity that they are given special names. A *zero* matrix (sometimes called a *null* matrix) has all of its elements equal to zero. It fulfills the role in matrix algebra that is analogous to that of zero in scalar algebra. The *unit* or *identity* matrix is a $n \times n$ matrix having 1's along its principal diagonal and zero everywhere else. The unit matrix serves essentially the same purpose in matrix algebra as does the number one in scalar algebra. A *symmetric* matrix is one where $a_{ij} = a_{ji}$ for all i and j.

• Example 11.1.1

Examples of zero, identity, and symmetric matrices are

$$O = \begin{pmatrix} 0 & 0 & 0 \\ 0 & 0 & 0 \\ 0 & 0 & 0 \end{pmatrix}, I = \begin{pmatrix} 1 & 0 \\ 0 & 1 \end{pmatrix}, \text{ and } A = \begin{pmatrix} 3 & 2 & 4 \\ 2 & 1 & 0 \\ 4 & 0 & 5 \end{pmatrix}, \tag{11.1.3}$$

respectively.

A special class of matrices are *column vectors* and *row vectors*:

$$\mathbf{x} = \begin{pmatrix} x_1 \\ x_2 \\ \vdots \\ x_m \end{pmatrix}, \qquad \mathbf{y} = \begin{pmatrix} y_1 & y_2 & \cdots & y_n \end{pmatrix}. \tag{11.1.4}$$

We denote row and column vectors by lower case, boldface letters. The length or *norm* of the vector $\mathbf{x}$ of n elements is

$$||\mathbf{x}|| = \left(\sum_{k=1}^{n} x_k^2 \right)^{1/2}. \tag{11.1.5}$$

Two matrices A and B are equal if and only if $a_{ij} = b_{ij}$ for all possible i and j and they have the same dimensions.

Having defined a matrix, let us explore some of its arithmetic properties. For two matrices A and B with the same dimensions (conformable for addition), the matrix $C = A + B$ contains the elements $c_{ij} = a_{ij} + b_{ij}$. Similarly, $C = A - B$ contains the elements $c_{ij} = a_{ij} - b_{ij}$. Because the order of addition does not matter, addition is *commutative*: $A + B = B + A$.

Consider now a scalar constant k. The product kA is formed by multiplying every element of A by k. Thus the matrix kA has elements ka_{ij}.

So far the rules for matrix arithmetic have conformed to their scalar counterparts. However, there are several possible ways of multiplying two matrices together. For example, we might simply multiply together the corresponding elements from each matrix. As we will see, the multiplication rule is designed to facilitate the solution of linear equations.

We begin by requiring that the dimensions of A be $m \times n$ while for B they are $n \times p$. That is, the number of columns in A must equal the number of rows in B. The matrices A and B are then said to be *conformable* for multiplication. If this is true, then $C = AB$ will be a matrix $m \times p$, where its elements equal

$$c_{ij} = \sum_{k=1}^{n} a_{ik} b_{kj}. \tag{11.1.6}$$

The right side of (11.1.6) is referred to as an *inner product* of the ith row of A and the jth column of B. Although (11.1.6) is the method used with a computer, an easier method for human computation is as a running sum of the products given by successive elements of the ith row of A and the corresponding elements of the jth column of B.

The product AA is usually written A^2; the product AAA, A^3, and so forth.

• **Example 11.1.2**

If

$$A = \begin{pmatrix} -1 & 4 \\ 2 & -3 \end{pmatrix} \quad \text{and} \quad B = \begin{pmatrix} 1 & 2 \\ 3 & 4 \end{pmatrix}, \qquad (11.1.7)$$

then

$$AB = \begin{pmatrix} [(-1)(1) + (4)(3)] & [(-1)(2) + (4)(4)] \\ [(2)(1) + (-3)(3)] & [(2)(2) + (-3)(4)] \end{pmatrix} \qquad (11.1.8)$$

$$= \begin{pmatrix} 11 & 14 \\ -7 & -8 \end{pmatrix}. \qquad (11.1.9)$$

Matrix multiplication is associative and distributive with respect to addition:

$$(kA)B = k(AB) = A(kB), \qquad (11.1.10)$$

$$A(BC) = (AB)C, \qquad (11.1.11)$$

$$(A + B)C = AC + BC \qquad (11.1.12)$$

and

$$C(A + B) = CA + CB. \qquad (11.1.13)$$

On the other hand, matrix multiplication is *not commutative*. In general, $AB \neq BA$.

• **Example 11.1.3**

Does $AB = BA$ if

$$A = \begin{pmatrix} 1 & 0 \\ 0 & 0 \end{pmatrix} \quad \text{and} \quad B = \begin{pmatrix} 1 & 1 \\ 1 & 0 \end{pmatrix}? \qquad (11.1.14)$$

Because

$$AB = \begin{pmatrix} 1 & 0 \\ 0 & 0 \end{pmatrix} \begin{pmatrix} 1 & 1 \\ 1 & 0 \end{pmatrix} = \begin{pmatrix} 1 & 1 \\ 0 & 0 \end{pmatrix} \qquad (11.1.15)$$

and

$$BA = \begin{pmatrix} 1 & 1 \\ 1 & 0 \end{pmatrix} \begin{pmatrix} 1 & 0 \\ 0 & 0 \end{pmatrix} = \begin{pmatrix} 1 & 0 \\ 1 & 0 \end{pmatrix}, \qquad (11.1.16)$$

$$AB \neq BA. \qquad (11.1.17)$$

- **Example 11.1.4**

Given

$$A = \begin{pmatrix} 1 & 1 \\ 3 & 3 \end{pmatrix} \quad \text{and} \quad B = \begin{pmatrix} -1 & 1 \\ 1 & -1 \end{pmatrix}, \tag{11.1.18}$$

find the product AB.

Performing the calculation, we find that

$$AB = \begin{pmatrix} 1 & 1 \\ 3 & 3 \end{pmatrix} \begin{pmatrix} -1 & 1 \\ 1 & -1 \end{pmatrix} = \begin{pmatrix} 0 & 0 \\ 0 & 0 \end{pmatrix}. \tag{11.1.19}$$

The point here is that just because $AB = 0$, this does *not* imply that either A or B equals the zero matrix.

We cannot properly speak of division when we are dealing with matrices. Nevertheless, a matrix A is said to be *nonsingular* or *invertible* if there exists a matrix B such that $AB = BA = I$. This matrix B is the multiplicative inverse of A or simply the *inverse* of A, written A^{-1}. A $n \times n$ matrix is *singular* if it does not have a multiplicative inverse.

- **Example 11.1.5**

If

$$A = \begin{pmatrix} 1 & 0 & 1 \\ 3 & 3 & 4 \\ 2 & 2 & 3 \end{pmatrix}, \tag{11.1.20}$$

let us verify that its inverse is

$$A^{-1} = \begin{pmatrix} 1 & 2 & -3 \\ -1 & 1 & -1 \\ 0 & -2 & 3 \end{pmatrix}. \tag{11.1.21}$$

We perform the check by finding AA^{-1} or $A^{-1}A$,

$$AA^{-1} = \begin{pmatrix} 1 & 0 & 1 \\ 3 & 3 & 4 \\ 2 & 2 & 3 \end{pmatrix} \begin{pmatrix} 1 & 2 & -3 \\ -1 & 1 & -1 \\ 0 & -2 & 3 \end{pmatrix} = \begin{pmatrix} 1 & 0 & 0 \\ 0 & 1 & 0 \\ 0 & 0 & 1 \end{pmatrix}. \tag{11.1.22}$$

In a later section we will show how to compute the inverse, given A.

Another matrix operation is transposition. The *transpose* of a matrix A with dimensions $m \times n$ is another matrix, written A^T, where we have interchanged the rows and columns from A. Clearly, $(A^T)^T = A$ as well as $(A + B)^T = A^T + B^T$ and $(kA)^T = kA^T$. If A and B are conformable for multiplication, then $(AB)^T = B^T A^T$. Note the reversal of order between the two sides. To prove this last result, we first show that the results are true for two 3×3 matrices A and B and then generalize to larger matrices.

Having introduced some of the basic concepts of linear algebra, we are ready to rewrite (11.1.1) in a canonical form so that we may present techniques for its solution. We begin by writing (11.1.1) as a single column vector:

$$\begin{pmatrix} a_{11}x_1 & + & a_{12}x_2 & + & \cdots & + & a_{1n}x_n \\ a_{21}x_1 & + & a_{22}x_2 & + & \cdots & + & a_{2n}x_n \\ \vdots & & \vdots & & \vdots & & \vdots \\ \vdots & & \vdots & & \vdots & & \vdots \\ a_{m1}x_1 & + & a_{m2}x_2 & + & \cdots & + & a_{mn}x_n \end{pmatrix} = \begin{pmatrix} b_1 \\ b_2 \\ \vdots \\ \vdots \\ b_m \end{pmatrix}. \tag{11.1.23}$$

On the left side of (11.1.23) we can use the multiplication rule to write

$$\begin{pmatrix} a_{11} & a_{12} & \cdots & a_{1n} \\ a_{21} & a_{22} & \cdots & a_{2n} \\ \vdots & \vdots & \vdots & \vdots \\ \vdots & \vdots & \vdots & \vdots \\ a_{m1} & a_{m2} & \cdots & a_{mn} \end{pmatrix} \begin{pmatrix} x_1 \\ x_2 \\ \vdots \\ \vdots \\ x_n \end{pmatrix} = \begin{pmatrix} b_1 \\ b_2 \\ \vdots \\ \vdots \\ b_m \end{pmatrix} \tag{11.1.24}$$

or

$$A\mathbf{x} = \mathbf{b}, \tag{11.1.25}$$

where $\mathbf{x}$ is the solution vector. If $\mathbf{b} = \mathbf{0}$, we have a *homogeneous* set of equations; otherwise, we have a *nonhomogeneous* set. In the next few sections, we will give a number of methods for finding $\mathbf{x}$.

• Example 11.1.6: Solution of a tridiagonal system

A common problem in linear algebra involves solving systems such as

$$b_1y_1 + c_1y_2 = d_1 \tag{11.1.26}$$

$$a_2y_1 + b_2y_2 + c_2y_3 = d_2 \tag{11.1.27}$$

$$\vdots$$

$$a_{N-1}y_{N-2} + b_{N-1}y_{N-1} + c_{N-1}y_N = d_{N-1} \tag{11.1.28}$$

$$b_Ny_{N-1} + c_Ny_N = d_N. \tag{11.1.29}$$

Such systems arise in the numerical solution of ordinary and partial differential equations.

We begin our analysis by rewriting (11.1.26)–(11.1.29) in the matrix notation:

$$\begin{pmatrix} b_1 & c_1 & 0 & \cdots & 0 & 0 & 0 \\ a_2 & b_2 & c_2 & \cdots & 0 & 0 & 0 \\ 0 & a_3 & b_3 & \cdots & 0 & 0 & 0 \\ \vdots & \vdots & \vdots & \vdots & \vdots & \vdots & \vdots \\ 0 & 0 & 0 & \cdots & a_{N-1} & b_{N-1} & c_{N-1} \\ 0 & 0 & 0 & \cdots & 0 & a_N & b_N \end{pmatrix} \begin{pmatrix} y_1 \\ y_2 \\ y_3 \\ \vdots \\ y_{N-1} \\ y_N \end{pmatrix} = \begin{pmatrix} d_1 \\ d_2 \\ d_3 \\ \vdots \\ d_{N-1} \\ d_N \end{pmatrix}. \quad \mathbf{(11.1.30)}$$

The matrix in (11.1.30) is an example of a *banded matrix*: a matrix where all of the elements in each row are zero except for the diagonal element and a limited number on either side of it. In our particular case, we have a *tridiagonal* matrix in which only the diagonal element and the elements immediately to its left and right in each row are nonzero.

Consider the nth equation. We can eliminate a_n by multiplying the $(n-1)$th equation by a_n/b_{n-1} and subtracting this new equation from the nth equation. The values of b_n and d_n become

$$b'_n = b_n - a_n c_{n-1}/b_{n-1} \quad \mathbf{(11.1.31)}$$

and

$$d'_n = d_n - a_n d_{n-1}/b_{n-1} \quad \mathbf{(11.1.32)}$$

for $n = 2, 3, \ldots, N$. The coefficient c_n is unaffected. Because elements a_1 and c_N are never involved, their values can be anything or they can be left undefined. The new system of equations may be written

$$\begin{pmatrix} b'_1 & c_1 & 0 & \cdots & 0 & 0 & 0 \\ 0 & b'_2 & c_2 & \cdots & 0 & 0 & 0 \\ 0 & 0 & b'_3 & \cdots & 0 & 0 & 0 \\ \vdots & \vdots & \vdots & \vdots & \vdots & \vdots & \vdots \\ 0 & 0 & 0 & \cdots & 0 & b'_{N-1} & c_{N-1} \\ 0 & 0 & 0 & \cdots & 0 & 0 & b'_N \end{pmatrix} \begin{pmatrix} y_1 \\ y_2 \\ y_3 \\ \vdots \\ y_{N-1} \\ y_N \end{pmatrix} = \begin{pmatrix} d'_1 \\ d'_2 \\ d'_3 \\ \vdots \\ d'_{N-1} \\ d'_N \end{pmatrix}. \quad \mathbf{(11.1.33)}$$

The matrix in (11.1.33) is in *upper triangular* form because all of the elements below the principal diagonal are zero. This is particularly useful because y_n may be computed by *back substitution*. That is, we first compute y_N. Next, we calculate y_{N-1} in terms of y_N. The solution y_{N-2} may then be computed in terms of y_N and y_{N-1}. We continue this process until we find y_1 in terms of $y_N, y_{N-1}, \ldots, y_2$. In the present case, we have the rather simple:

$$y_N = d'_N/b'_N \quad \mathbf{(11.1.34)}$$

and

$$y_n = (d'_n - c_n d'_{n+1})/b'_n \quad \mathbf{(11.1.35)}$$

for $n = N-1, N-2, \ldots, 2, 1$.

As we shall show shortly, this is an example of solving a system of linear equations by Gaussian elimination. For a tridiagonal case, we have the advantage that the solution can be expressed in terms of a recurrence relationship, a very convenient feature from a computational point of view. This algorithm is very robust, being stable[1] as long as $|a_i + c_i| < |b_i|$. By stability, we mean that if we change $\mathbf{b}$ by $\Delta\mathbf{b}$ so that $\mathbf{x}$ changes by $\Delta\mathbf{x}$, then $||\Delta\mathbf{x}|| < M\epsilon$, where $\epsilon \geq ||\Delta\mathbf{b}||$, $0 < M < \infty$, for *any* N.

Problems

Given $A = \begin{pmatrix} 3 & 4 \\ 1 & 2 \end{pmatrix}$ and $B = \begin{pmatrix} 1 & 1 \\ 2 & 2 \end{pmatrix}$, find

1. $A + B$, $B + A$
2. $A - B$, $B - A$
3. $3A - 2B$, $3(2A - B)$
4. A^T,B^T,$(B^T)^T$
5. $(A + B)^T$, $A^T + B^T$
6. $B + B^T$, $B - B^T$
7. AB,A^TB,BA,B^TA
8. A^2, B^2
9. BB^T, B^TB
10. $A^2 - 3A + I$
11. $A^3 + 2A$
12. $A^4 - 4A^2 + 2I$

Can multiplication occur between the following matrices? If so, compute it.

13. $\begin{pmatrix} 3 & 5 & 1 \\ -2 & 1 & 2 \end{pmatrix} \begin{pmatrix} 2 & 1 \\ 4 & 1 \\ 1 & 3 \end{pmatrix}$

14. $\begin{pmatrix} -2 & 4 \\ -4 & 6 \\ -6 & 1 \end{pmatrix} \begin{pmatrix} 1 & 2 & 3 \end{pmatrix}$

15. $\begin{pmatrix} 1 & 4 & 2 \\ 0 & 0 & 4 \\ 0 & 1 & 2 \end{pmatrix} \begin{pmatrix} 3 & 2 \\ 1 & 1 \\ 2 & 1 \end{pmatrix}$

16. $\begin{pmatrix} 4 & 6 \\ 1 & 2 \end{pmatrix} \begin{pmatrix} 1 & 3 & 6 \\ 1 & 2 & 5 \end{pmatrix}$

17. $\begin{pmatrix} 6 & 4 & 2 \\ 1 & 2 & 3 \end{pmatrix} \begin{pmatrix} 3 & 1 & 4 \\ 2 & 0 & 6 \end{pmatrix}$

If $A = \begin{pmatrix} 1 & 1 \\ 1 & 2 \\ 3 & 1 \end{pmatrix}$ verify that

18. $7A = 4A + 3A$,
19. $10A = 5(2A)$,
20. $(A^T)^T = A$.

---

[1] Torii, T., 1966: Inversion of tridiagonal matrices and the stability of tridiagonal systems of linear systems. *Tech. Rep. Osaka Univ.*, **16**, 403–414.

If $A = \begin{pmatrix} 2 & 1 \\ 3 & 1 \end{pmatrix}$, $B = \begin{pmatrix} 1 & -2 \\ 4 & 0 \end{pmatrix}$, and $C = \begin{pmatrix} 1 & 1 \\ 1 & 1 \end{pmatrix}$, verify that

21. $(A + B) + C = A + (B + C)$, 22. $(AB)C = A(BC)$,
23. $A(B + C) = AB + AC$, 24. $(A + B)C = AC + BC$.

Verify that the following A^{-1} are indeed the inverse of A:

25. $A = \begin{pmatrix} 3 & -1 \\ -5 & 2 \end{pmatrix}$ $\qquad A^{-1} = \begin{pmatrix} 2 & 1 \\ 5 & 3 \end{pmatrix}$

26. $A = \begin{pmatrix} 0 & 1 & 0 \\ 1 & 0 & 0 \\ 0 & 0 & 1 \end{pmatrix}$ $\qquad A^{-1} = \begin{pmatrix} 0 & 1 & 0 \\ 1 & 0 & 0 \\ 0 & 0 & 1 \end{pmatrix}$

Write the following linear systems of equations in matrix form: $A\mathbf{x} = \mathbf{b}$.

27.

$$x_1 - 2x_2 = 5$$
$$3x_1 + x_2 = 1$$

28.

$$2x_1 + x_2 + 4x_3 = 2$$
$$4x_1 + 2x_2 + 5x_3 = 6$$
$$6x_1 - 3x_2 + 5x_3 = 2$$

29.

$$x_2 + 2x_3 + 3x_4 = 2$$
$$3x_1 - 4x_3 - 4x_4 = 5$$
$$x_1 + x_2 + x_3 + x_4 = -3$$
$$2x_1 - 3x_2 + x_3 - 3x_4 = 7.$$

11.2 DETERMINANTS

Determinants appear naturally during the solution of simultaneous equations. Consider, for example, two simultaneous equations with two unknowns x_1 and x_2,

$$a_{11}x_1 + a_{12}x_2 = b_1 \tag{11.2.1}$$

and

$$a_{21}x_1 + a_{22}x_2 = b_2. \tag{11.2.2}$$

The solution to these equations for the value of x_1 and x_2 is

$$x_1 = \frac{b_1a_{22} - a_{12}b_2}{a_{11}a_{22} - a_{12}a_{21}} \tag{11.2.3}$$

and

$$x_2 = \frac{b_2 a_{11} - a_{21} b_1}{a_{11} a_{22} - a_{12} a_{21}}. \qquad \textbf{(11.2.4)}$$

Note that the denominator of (11.2.3) and (11.2.4) are the same. This term, which will always appear in the solution of 2×2 systems, is formally given the name of *determinant* and written

$$\det(A) = \begin{vmatrix} a_{11} & a_{12} \\ a_{21} & a_{22} \end{vmatrix} = a_{11}a_{22} - a_{12}a_{21}. \qquad \textbf{(11.2.5)}$$

Although determinants have their origin in the solution of systems of equations, any square array of numbers or expressions possesses a unique determinant, independent of whether it is involved in a system of equations or not. This determinant is evaluated (or expanded) according to a formal rule known as *Laplace's expansion of cofactors.*[2] The process revolves around expanding the determinant using any arbitrary column *or* row of A. If the ith row or jth column is chosen, the determinant is given by

$$\det(A) = a_{i1}A_{i1} + a_{i2}A_{i2} + \cdots + a_{in}A_{in} \qquad \textbf{(11.2.6)}$$

$$= a_{1j}A_{1j} + a_{2j}A_{2j} + \cdots + a_{nj}A_{nj}, \qquad \textbf{(11.2.7)}$$

where A_{ij}, the *cofactor* of a_{ij}, equals $(-1)^{i+j}M_{ij}$. The minor M_{ij} is the determinant of the $(n-1) \times (n-1)$ submatrix obtained by deleting row i, column j of A. This rule, of course, was chosen so that determinants are still useful in solving systems of equations.

• **Example 11.2.1**

Let us evaluate

$$\begin{vmatrix} 2 & -1 & 2 \\ 1 & 3 & 2 \\ 5 & 1 & 6 \end{vmatrix}$$

by an expansion in cofactors.

Using the first column,

$$\begin{vmatrix} 2 & -1 & 2 \\ 1 & 3 & 2 \\ 5 & 1 & 6 \end{vmatrix} = 2(-1)^2 \begin{vmatrix} 3 & 2 \\ 1 & 6 \end{vmatrix} + 1(-1)^3 \begin{vmatrix} -1 & 2 \\ 1 & 6 \end{vmatrix} + 5(-1)^4 \begin{vmatrix} -1 & 2 \\ 3 & 2 \end{vmatrix} \qquad \textbf{(11.2.8)}$$

$$= 2(16) - 1(-8) + 5(-8) = 0. \qquad \textbf{(11.2.9)}$$

---

[2] Laplace, P. S., 1772: Recherches sur le calcul intégral et sur le système du monde. *Hist. Acad. R. Sci., IIe Partie*, 267–376. *Œuvres*, **8**, pp. 369–501. See Muir, T., 1960: *The Theory of Determinants in the Historical Order of Development, Vol. I, Part 1, General Determinants Up to 1841*, Dover Publishers, Mineola, NY, pp. 24–33.

The greatest source of error is forgetting to take the factor $(-1)^{i+j}$ into account during the expansion.

Although Laplace's expansion does provide a method for calculating $\det(A)$, the number of calculations equals $(n!)$. Consequently, for hand calculations, an obvious strategy is to select the column or row that has the greatest number of zeros. An even better strategy would be to manipulate a determinant with the goal of introducing zeros into a particular column or row. In the remaining portion of section, we show some operations that may be performed on a determinant to introduce the desired zeros. Most of the properties follow from the expansion of determinants by cofactors.

• *Rule 1* : For every square matrix A, $\det(A^T) = \det(A)$.

The proof is left as an exercise.

• *Rule 2* : If any two rows or columns of A are identical, $\det(A) = 0$.

To see that this is true, consider the following 3×3 matrix:

$$\begin{vmatrix} b_1 & b_1 & c_1 \\ b_2 & b_2 & c_2 \\ b_3 & b_3 & c_3 \end{vmatrix} = c_1(b_2b_3 - b_3b_2) - c_2(b_1b_3 - b_3b_1) + c_3(b_1b_2 - b_2b_1) = 0. \tag{11.2.10}$$

• *Rule 3* : The determinant of a triangular matrix is equal to the product of its diagonal elements.

If A is lower triangular, successive expansions by elements in the first column give

$$\det(A) = \begin{vmatrix} a_{11} & 0 & \cdots & 0 \\ a_{21} & a_{22} & \cdots & 0 \\ \vdots & \vdots & \vdots & \vdots \\ a_{n1} & a_{n2} & \cdots & a_{nn} \end{vmatrix} = a_{11} \begin{vmatrix} a_{22} & \cdots & 0 \\ \vdots & \vdots & \vdots \\ a_{n2} & \cdots & a_{nn} \end{vmatrix} \tag{11.2.11}$$

$$= \cdots = a_{11}a_{22}\cdots a_{nn}. \tag{11.2.12}$$

If A is upper triangular, successive expansions by elements of the first row proves the property.

• *Rule 4* : If a square matrix A has either a row or a column of all zeros, then $\det(A) = 0$.

The proof is left as an exercise.

• *Rule 5* : If each element in one row (column) of a determinant is multiplied by a number c, the value of the determinant is multiplied by c.

Suppose $|B|$ has been obtained from $|A|$ by multiplying row i (column j) of $|A|$ by c. Upon expanding $|B|$ in terms of row i (column j) each term in the expansion contains c as a factor. Factor out the common c, the result is just c times the expansion $|A|$ by the same row (column).

• *Rule 6* : If each element of a row (or a column) of a determinant can be expressed as a binomial, the determinant can be written as the sum of two determinants.

To understand this property, consider the following 3×3 determinant:

$$\begin{vmatrix} a_1 + d_1 & b_1 & c_1 \\ a_2 + d_2 & b_2 & c_2 \\ a_3 + d_3 & b_3 & c_3 \end{vmatrix} = \begin{vmatrix} a_1 & b_1 & c_1 \\ a_2 & b_2 & c_2 \\ a_3 & b_3 & c_3 \end{vmatrix} + \begin{vmatrix} d_1 & b_1 & c_1 \\ d_2 & b_2 & c_2 \\ d_3 & b_3 & c_3 \end{vmatrix}. \qquad \mathbf{(11.2.13)}$$

The proof follows by expanding the determinant by the row (or column) that contains the binomials.

• *Rule 7* : If B is a matrix obtained by interchanging any two rows (columns) of a square matrix A, then $\det(B) = -\det(A)$.

The proof is by induction. It is easily shown for any 2×2 matrix. Assume that this rule holds of any $(n-1) \times (n-1)$ matrix. If A is $n \times n$, then let B be a matrix formed by interchanging rows i and j. Expanding $|B|$ and $|A|$ by a different row, say k, we have that

$$|B| = \sum_{s=1}^{n} (-1)^{k+s} b_{ks} M_{ks} \quad \text{and} \quad |A| = \sum_{s=1}^{n} (-1)^{k+s} a_{ks} N_{ks}, \quad \mathbf{(11.2.14)}$$

where M_{ks} and N_{ks} are the minors formed by deleting row k, column s from $|B|$ and $|A|$, respectively. For $s = 1, 2, \ldots, n$, we obtain N_{ks} and M_{ks} by interchanging rows i and j. By the induction hypothesis and recalling that N_{ks} and M_{ks} are $(n-1) \times (n-1)$ determinants, $N_{ks} = -M_{ks}$ for $s = 1, 2, \ldots, n$. Hence, $|B| = -|A|$. Similar arguments hold if two columns are interchanged.

• *Rule 8* : If one row (column) of a square matrix A equals to a number c times some other row (column), then $\det(A) = 0$.

Suppose one row of a square matrix A is equal to c times some other row. If $c = 0$, then $|A| = 0$. If $c \neq 0$, then $|A| = c|B|$, where $|B| = 0$ because $|B|$ has two identical rows. A similar argument holds for two columns.

• $\boxed{\textit{Rule 9}}$: The value of det(A) is unchanged if any arbitrary multiple of any line (row or column) is added to any other line.

To see that this is true, consider the simple example:

$$\begin{vmatrix} a_1 & b_1 & c_1 \\ a_2 & b_2 & c_2 \\ a_3 & b_3 & c_3 \end{vmatrix} + \begin{vmatrix} cb_1 & b_1 & c_1 \\ cb_2 & b_2 & c_2 \\ cb_3 & b_3 & c_3 \end{vmatrix} = \begin{vmatrix} a_1 + cb_1 & b_1 & c_1 \\ a_2 + cb_2 & b_2 & c_2 \\ a_3 + cb_3 & b_3 & c_3 \end{vmatrix}, \tag{11.2.15}$$

where $c \neq 0$. The first determinant on the left side is our original determinant. In the second determinant, we can again expand the first column and find that

$$\begin{vmatrix} cb_1 & b_1 & c_1 \\ cb_2 & b_2 & c_2 \\ cb_3 & b_3 & c_3 \end{vmatrix} = c \begin{vmatrix} b_1 & b_1 & c_1 \\ b_2 & b_2 & c_2 \\ b_3 & b_3 & c_3 \end{vmatrix} = 0. \tag{11.2.16}$$

• **Example 11.2.2**

Let us evaluate

$$\begin{vmatrix} 1 & 2 & 3 & 4 \\ -1 & 1 & 2 & 3 \\ 1 & -1 & 1 & 2 \\ -1 & 1 & -1 & 5 \end{vmatrix}$$

using a combination of the properties stated above and expansion by cofactors.

By adding or subtracting the first row to the other rows, we have that

$$\begin{vmatrix} 1 & 2 & 3 & 4 \\ -1 & 1 & 2 & 3 \\ 1 & -1 & 1 & 2 \\ -1 & 1 & -1 & 5 \end{vmatrix} = \begin{vmatrix} 1 & 2 & 3 & 4 \\ 0 & 3 & 5 & 7 \\ 0 & -3 & -2 & -2 \\ 0 & 3 & 2 & 9 \end{vmatrix} \tag{11.2.17}$$

$$= \begin{vmatrix} 3 & 5 & 7 \\ -3 & -2 & -2 \\ 3 & 2 & 9 \end{vmatrix} = \begin{vmatrix} 3 & 5 & 7 \\ 0 & 3 & 5 \\ 0 & -3 & 2 \end{vmatrix} \tag{11.2.18}$$

$$= 3 \begin{vmatrix} 3 & 5 \\ -3 & 2 \end{vmatrix} = 3 \begin{vmatrix} 3 & 5 \\ 0 & 7 \end{vmatrix} = 63. \tag{11.2.19}$$

Problems

Evaluate the following determinants:

1. $\begin{vmatrix} 3 & 5 \\ -2 & -1 \end{vmatrix}$

2. $\begin{vmatrix} 5 & -1 \\ -8 & 4 \end{vmatrix}$

3. $\begin{vmatrix} 3 & 1 & 2 \\ 2 & 4 & 5 \\ 1 & 4 & 5 \end{vmatrix}$

4. $\begin{vmatrix} 4 & 3 & 0 \\ 3 & 2 & 2 \\ 5 & -2 & -4 \end{vmatrix}$

5. $\begin{vmatrix} 1 & 3 & 2 \\ 4 & 1 & 1 \\ 2 & 1 & 3 \end{vmatrix}$

6. $\begin{vmatrix} 2 & -1 & 2 \\ 1 & 3 & 3 \\ 5 & 1 & 6 \end{vmatrix}$

7. $\begin{vmatrix} 2 & 0 & 0 & 1 \\ 0 & 1 & 0 & 0 \\ 1 & 6 & 1 & 0 \\ 1 & 1 & -2 & 3 \end{vmatrix}$

8. $\begin{vmatrix} 2 & 1 & 2 & 1 \\ 3 & 0 & 2 & 2 \\ -1 & 2 & -1 & 1 \\ -3 & 2 & 3 & 1 \end{vmatrix}$

9. Using the properties of determinants, show that

$$\begin{vmatrix} 1 & 1 & 1 & 1 \\ a & b & c & d \\ a^2 & b^2 & c^2 & d^2 \\ a^3 & b^3 & c^3 & d^3 \end{vmatrix} = (b-a)(c-a)(d-a)(c-b)(d-b)(d-c).$$

This determinant is called *Vandermonde's determinant.*

10. Show that

$$\begin{vmatrix} a & b+c & 1 \\ b & a+c & 1 \\ c & a+b & 1 \end{vmatrix} = 0.$$

11. Show that if all of the elements of a row or column are zero, then $\det(A) = 0$.

12. Prove that $\det(A^T) = \det(A)$.

11.3 CRAMER'S RULE

One of the most popular methods for solving simple systems of linear equations is Cramer's rule.[3] It is very useful for 2×2 systems, acceptable for 3×3 systems, and of doubtful use for 4×4 or larger systems.

Let us have n equations with n unknowns, $A\mathbf{x} = \mathbf{b}$. Cramer's rule states that

$$x_1 = \frac{\det(A_1)}{\det(A)}, \quad x_2 = \frac{\det(A_2)}{\det(A)}, \quad \ldots, \quad x_n = \frac{\det(A_n)}{\det(A)}, \tag{11.3.1}$$

where A_i is a matrix obtained from A by replacing the ith column with $\mathbf{b}$ and $n = 1, 2, 3, \ldots$. Obviously, $\det(A) \neq 0$ if Cramer's rule is to work.

To prove Cramer's rule, consider

$$x_1 \det(A) = \begin{vmatrix} a_{11}x_1 & a_{12} & a_{13} & \cdots & a_{1n} \\ a_{21}x_1 & a_{22} & a_{23} & \cdots & a_{2n} \\ a_{31}x_1 & a_{32} & a_{33} & \cdots & a_{3n} \\ \vdots & \vdots & \vdots & \vdots & \vdots \\ a_{n1}x_1 & a_{n2} & a_{n3} & \cdots & a_{nn} \end{vmatrix} \tag{11.3.2}$$

by Rule 5 from the previous section. By adding x_2 times the second column to the first column,

$$x_1 \det(A) = \begin{vmatrix} a_{11}x_1 + a_{12}x_2 & a_{12} & a_{13} & \cdots & a_{1n} \\ a_{21}x_1 + a_{22}x_2 & a_{22} & a_{23} & \cdots & a_{2n} \\ a_{31}x_1 + a_{32}x_2 & a_{32} & a_{33} & \cdots & a_{3n} \\ \vdots & \vdots & \vdots & \vdots & \vdots \\ a_{n1}x_1 + a_{n2}x_2 & a_{n2} & a_{n3} & \cdots & a_{nn} \end{vmatrix}. \tag{11.3.3}$$

Multiplying each of the columns by the corresponding x_i and adding it to the first column yields,

$$x_1 \det(A) = \begin{vmatrix} a_{11}x_1 + a_{12}x_2 + \cdots + a_{1n}x_n & a_{12} & a_{13} & \cdots & a_{1n} \\ a_{21}x_1 + a_{22}x_2 + \cdots + a_{2n}x_n & a_{22} & a_{23} & \cdots & a_{2n} \\ a_{31}x_1 + a_{32}x_2 + \cdots + a_{3n}x_n & a_{32} & a_{33} & \cdots & a_{3n} \\ \vdots & \vdots & \vdots & \vdots & \vdots \\ a_{n1}x_1 + a_{n2}x_2 + \cdots + a_{nn}x_n & a_{n2} & a_{n3} & \cdots & a_{nn} \end{vmatrix}. \tag{11.3.4}$$

[3] Cramer, G., 1750: *Introduction à l'analyse des lignes courbes algébriques*, Geneva, p. 657.

The first column of (11.3.4) equals $A\mathbf{x}$ and we replace it with $\mathbf{b}$. Thus,

$$x_1 \det(A) = \begin{vmatrix} b_1 & a_{12} & a_{13} & \cdots & a_{1n} \\ b_2 & a_{22} & a_{23} & \cdots & a_{2n} \\ b_3 & a_{32} & a_{33} & \cdots & a_{3n} \\ \vdots & \vdots & \vdots & \vdots & \vdots \\ b_n & a_{n2} & a_{n3} & \cdots & a_{nn} \end{vmatrix} = \det(A_1) \tag{11.3.5}$$

or

$$x_1 = \frac{\det(A_1)}{\det(A)} \tag{11.3.6}$$

provided $\det(A) \neq 0$. To complete the proof we do exactly the same procedure to the jth column. □

• Example 11.3.1

Let us solve the following system of equations by Cramer's rule:

$$2x_1 + x_2 + 2x_3 = -1, \tag{11.3.7}$$

$$x_1 + x_3 = -1 \tag{11.3.8}$$

and

$$-x_1 + 3x_2 - 2x_3 = 7. \tag{11.3.9}$$

From the matrix form of the equations,

$$\begin{pmatrix} 2 & 1 & 2 \\ 1 & 0 & 1 \\ -1 & 3 & -2 \end{pmatrix} \begin{pmatrix} x_1 \\ x_2 \\ x_3 \end{pmatrix} = \begin{pmatrix} -1 \\ -1 \\ 7 \end{pmatrix}, \tag{11.3.10}$$

we have that

$$\det(A) = \begin{vmatrix} 2 & 1 & 2 \\ 1 & 0 & 1 \\ -1 & 3 & -2 \end{vmatrix} = 1, \tag{11.3.11}$$

$$\det(A_1) = \begin{vmatrix} -1 & 1 & 2 \\ -1 & 0 & 1 \\ 7 & 3 & -2 \end{vmatrix} = 2, \tag{11.3.12}$$

$$\det(A_2) = \begin{vmatrix} 2 & -1 & 2 \\ 1 & -1 & 1 \\ -1 & 7 & -2 \end{vmatrix} = 1 \tag{11.3.13}$$

and

$$\det(A_3) = \begin{vmatrix} 2 & 1 & -1 \\ 1 & 0 & -1 \\ -1 & 3 & 7 \end{vmatrix} = -3. \tag{11.3.14}$$

Finally,

$$x_1 = \frac{2}{1} = 2, \quad x_2 = \frac{1}{1} = 1 \quad \text{and} \quad x_3 = \frac{-3}{1} = -3. \qquad \mathbf{(11.3.15)}$$

Problems

Solve the following systems of equations by Cramer's rule:

1. $x_1 + 2x_2 = 3, \quad 3x_1 + x_2 = 6$
2. $2x_1 + x_2 = -3, \; x_1 - x_2 = 1$
3. $x_1 + 2x_2 - 2x_3 = 4, \quad 2x_1 + x_2 + x_3 = -2, \; -x_1 + x_2 - x_3 = 2$
4. $2x_1 + 3x_2 - x_3 = -1, \; -x_1 - 2x_2 + x_3 = 5, \; 3x_1 - x_2 = -2.$

11.4 ROW ECHELON FORM AND GAUSSIAN ELIMINATION

So far, we have assumed that every system of equations has a unique solution. This is not necessary true as the following examples show.

• Example 11.4.1

Consider the system

$$x_1 + x_2 = 2 \qquad \mathbf{(11.4.1)}$$

and

$$2x_1 + 2x_2 = -1. \qquad \mathbf{(11.4.2)}$$

This system is inconsistent because the second equation does not follow after multiplying the first by 2. Geometrically (11.4.1) and (11.4.2) are parallel lines; they never intersect to give a unique x_1 and x_2.

• Example 11.4.2

Even if a system is consistent, it still may not have a unique solution. For example, the system

$$x_1 + x_2 = 2 \qquad \mathbf{(11.4.3)}$$

and

$$2x_1 + 2x_2 = 4 \qquad \mathbf{(11.4.4)}$$

is consistent, the second equation formed by multiplying the first by 2. However, there are an infinite number of solutions.

Our examples suggest the following:

Theorem: *A system of m linear equation in n unknowns may: (1) have no solution, in which case it is called an* inconsistent *system, or (2) have exactly one solution (called a unique solution), or (3) have an infinite number of solutions. In the latter two cases, the system is said to be* consistent.

Before we can prove this theorem at the end of this section, we need to introduce some new concepts.

The first one is equivalent systems. Two systems of equations involving the same variables are *equivalent* if they have the same solution set. Of course, the only reason for introducing equivalent systems is the possibility of transforming one system of linear systems into another which is easier to solve. But what operations are permissible? Also what is the ultimate goal of our transformation?

From a complete study of possible operations, there are only three operations for transforming one system of linear equations into another. These three *elementary row operations* are

(1) interchanging any two rows in the matrix,

(2) multiplying any row by a nonzero scalar, and

(3) adding any arbitrary multiple of any row to any other row.

Armed with our elementary row operations, let us now solve the following set of linear equations:

$$x_1 - 3x_2 + 7x_3 = 2, \tag{11.4.5}$$

$$2x_1 + 4x_2 - 3x_3 = -1 \tag{11.4.6}$$

and

$$-x_1 + 13x_2 - 21x_3 = 2. \tag{11.4.7}$$

We begin by writing (11.4.5)–(11.4.7) in matrix notation:

$$\begin{pmatrix} 1 & -3 & 7 \\ 2 & 4 & -3 \\ -1 & 13 & -21 \end{pmatrix} \begin{pmatrix} x_1 \\ x_2 \\ x_3 \end{pmatrix} = \begin{pmatrix} 2 \\ -1 \\ 2 \end{pmatrix}. \tag{11.4.8}$$

The matrix in (11.4.8) is called the *coefficient matrix* of the system.

We now introduce the concept of the *augmented matrix*: a matrix B composed of A plus the column vector $\mathbf{b}$ or

$$B = \left(\begin{array}{ccc|c} 1 & -3 & 7 & 2 \\ 2 & 4 & -3 & -1 \\ -1 & 13 & -21 & 2 \end{array}\right). \tag{11.4.9}$$

We can solve our original system by performing elementary row operations on the augmented matrix. Because the x_i's function essentially as placeholders, we can omit them until the end of the computation.

Returning to the problem, the first row may be used to eliminate the elements in the first column of the remaining rows. For this reason the first row is called the *pivotal* row and the element a_{11} is the *pivot.* By using the third elementary row operation twice (to eliminate the 2 and -1 in the first column), we finally have the equivalent system

$$B = \left(\begin{array}{ccc|c} 1 & -3 & 7 & 2 \\ 0 & 10 & -17 & -5 \\ 0 & 10 & -14 & 4 \end{array}\right). \qquad \textbf{(11.4.10)}$$

At this point we choose the second row as our new pivotal row and again apply the third row operation to eliminate the last element in the second column. This yields

$$B = \left(\begin{array}{ccc|c} 1 & -3 & 7 & 2 \\ 0 & 10 & -17 & -5 \\ 0 & 0 & 3 & 9 \end{array}\right). \qquad \textbf{(11.4.11)}$$

Thus, elementary row operations have transformed (11.4.5)–(11.4.7) into the triangular system:

$$x_1 - 3x_2 + 7x_3 = 2, \qquad \textbf{(11.4.12)}$$

$$10x_2 - 17x_3 = -5, \qquad \textbf{(11.4.13)}$$

$$3x_3 = 9, \qquad \textbf{(11.4.14)}$$

which is *equivalent* to the original system. The final solution is obtained by *back substitution*, solving from (11.4.14) back to (11.4.12). In the present case, $x_3 = 3$. Then, $10x_2 = 17(3) - 5$ or $x_2 = 4.6$. Finally, $x_1 = 3x_2 - 7x_3 + 2 = -5.2$.

In general, if an $n \times n$ linear system can be reduced to triangular form, then it will have a unique solution that we can obtain by performing back substitution. This reduction involves $n - 1$ steps. In the first step, a pivot element, and thus the pivotal row, is chosen from the nonzero entries in the first column of the matrix. We interchange rows (if necessary) so that the pivotal row is the first row. Multiples of the pivotal row are then subtracted from each of the remaining $n - 1$ rows so that there are 0's in the $(2,1), ..., (n,1)$ positions. In the second step, a pivot element is chosen from the nonzero entries in column 2, rows 2 through n, of the matrix. The row containing the pivot is then interchanged with the second row (if necessary) of the matrix and is used as the pivotal row. Multiples of the pivotal row are then subtracted from the remaining $n - 2$ rows, eliminating all entries below the diagonal in the second column. The same procedure is repeated for columns 3 through $n-1$. Note that in the second step, row 1 and column 1 remain unchanged, in the third step the first two rows and first two columns remain unchanged, and so on.

If elimination is carried out as described, we will arrive at an equivalent upper triangular system after $n-1$ steps. However, the procedure will fail if, at any step, all possible choices for a pivot element equal zero. Let us now examine such cases.

Consider now the system

$$x_1 + 2x_2 + x_3 = -1, \tag{11.4.15}$$

$$2x_1 + 4x_2 + 2x_3 = -2, \tag{11.4.16}$$

$$x_1 + 4x_2 + x_3 = 2. \tag{11.4.17}$$

Its augmented matrix is

$$B = \left(\begin{array}{ccc|c} 1 & 2 & 1 & -1 \\ 2 & 4 & 2 & -2 \\ 1 & 4 & 2 & 2 \end{array}\right). \tag{11.4.18}$$

Choosing the first row as our pivotal row, we find that

$$B = \left(\begin{array}{ccc|c} 1 & 2 & 1 & -1 \\ 0 & 0 & 0 & 0 \\ 0 & 2 & 1 & 3 \end{array}\right) \tag{11.4.19}$$

or

$$B = \left(\begin{array}{ccc|c} 1 & 2 & 1 & -1 \\ 0 & 2 & 1 & 3 \\ 0 & 0 & 0 & 0 \end{array}\right). \tag{11.4.20}$$

The difficulty here is the presence of the zeros in the third row. Clearly any finite numbers will satisfy the equation $0x_1 + 0x_2 + 0x_3 = 0$ and we have an infinite number of solutions. Closer examination of the original system shows a underdetermined system; (11.4.15) and (11.4.16) differ by a factor of 2. An important aspect of this problem is the fact that the final augmented matrix is of the form of a staircase or *echelon form* rather than of triangular form.

Let us modify (11.4.15)–(11.4.17) to read

$$x_1 + 2x_2 + x_3 = -1, \tag{11.4.21}$$

$$2x_1 + 4x_2 + 2x_3 = 3, \tag{11.4.22}$$

$$x_1 + 4x_2 + x_3 = 2, \tag{11.4.23}$$

then the final augmented matrix is

$$B = \left(\begin{array}{ccc|c} 1 & 2 & 1 & -1 \\ 0 & 2 & 1 & 3 \\ 0 & 0 & 0 & 5 \end{array}\right). \tag{11.4.24}$$

We again have a problem with the third row because $0x_1+0x_2+0x_3 = 5$, which is impossible. There is no solution in this case and we have an *overdetermined system*. Note, once again, that our augmented matrix has a row echelon form rather than a triangular form.

In summary, to include all possible situations in our procedure, we must rewrite the augmented matrix in row echelon form. *Row echelon form* consists of:

(1) The first nonzero entry in each row is 1.

(2) If row k does not consist entirely of zeros, the number of leading zero entries in row $k+1$ is greater than the number of leading zero entries in row k.

(3) If there are rows whose entries are all zero, they are below the rows having nonzero entries.

The number of nonzero rows in the row echelon form of a matrix is known as its *rank*. *Gaussian elimination* is the process of using elementary row operations to transform a linear system into one whose augmented matrix is in row echelon form.

• Example 11.4.3

Each of the following matrices is *not* of row echelon form because they violate one of the conditions for row echelon form:

$$\begin{pmatrix} 2 & 2 & 3 \\ 0 & 2 & 1 \\ 0 & 0 & 4 \end{pmatrix}, \begin{pmatrix} 0 & 0 & 0 \\ 0 & 2 & 0 \end{pmatrix}, \begin{pmatrix} 0 & 1 \\ 1 & 0 \end{pmatrix}. \tag{11.4.25}$$

• Example 11.4.4

The following matrices are in row echelon form:

$$\begin{pmatrix} 1 & 2 & 3 \\ 0 & 1 & 1 \\ 0 & 0 & 1 \end{pmatrix}, \begin{pmatrix} 1 & 4 & 6 \\ 0 & 0 & 1 \\ 0 & 0 & 0 \end{pmatrix}, \begin{pmatrix} 1 & 3 & 4 & 0 \\ 0 & 0 & 1 & 3 \\ 0 & 0 & 0 & 0 \end{pmatrix}. \tag{11.4.26}$$

• **Example 11.4.5**

Gaussian elimination may also be used to solve the general problem $AX = B$. One of the most common applications is in finding the inverse. For example, let us find the inverse of the matrix

$$A = \begin{pmatrix} 4 & -2 & 2 \\ -2 & -4 & 4 \\ -4 & 2 & 8 \end{pmatrix} \tag{11.4.27}$$

by Gaussian elimination.

Because the inverse is defined by $AA^{-1} = I$, our augmented matrix is

$$\left(\begin{array}{ccc|ccc} 4 & -2 & 2 & 1 & 0 & 0 \\ -2 & -4 & 4 & 0 & 1 & 0 \\ -4 & 2 & 8 & 0 & 0 & 1 \end{array}\right). \tag{11.4.28}$$

Then, by elementary row operations,

$$\left(\begin{array}{ccc|ccc} 4 & -2 & 2 & 1 & 0 & 0 \\ -2 & -4 & 4 & 0 & 1 & 0 \\ -4 & 2 & 8 & 0 & 0 & 1 \end{array}\right) = \left(\begin{array}{ccc|ccc} -2 & -4 & 4 & 0 & 1 & 0 \\ 4 & -2 & 2 & 1 & 0 & 0 \\ -4 & 2 & 8 & 0 & 0 & 1 \end{array}\right) \tag{11.4.29}$$

$$= \left(\begin{array}{ccc|ccc} -2 & -4 & 4 & 0 & 1 & 0 \\ 4 & -2 & 2 & 1 & 0 & 0 \\ 0 & 0 & 10 & 1 & 0 & 1 \end{array}\right) \tag{11.4.30}$$

$$= \left(\begin{array}{ccc|ccc} -2 & -4 & 4 & 0 & 1 & 0 \\ 0 & -10 & 10 & 1 & 2 & 0 \\ 0 & 0 & 10 & 1 & 0 & 1 \end{array}\right) \tag{11.4.31}$$

$$= \left(\begin{array}{ccc|ccc} -2 & -4 & 4 & 0 & 1 & 0 \\ 0 & -10 & 0 & 0 & 2 & -1 \\ 0 & 0 & 10 & 1 & 0 & 1 \end{array}\right) \tag{11.4.32}$$

$$= \left(\begin{array}{ccc|ccc} -2 & -4 & 0 & -2/5 & 1 & -2/5 \\ 0 & -10 & 0 & 0 & 2 & -1 \\ 0 & 0 & 10 & 1 & 0 & 1 \end{array}\right) \tag{11.4.33}$$

$$= \left(\begin{array}{ccc|ccc} -2 & 0 & 0 & -2/5 & 1/5 & 0 \\ 0 & -10 & 0 & 0 & 2 & -1 \\ 0 & 0 & 10 & 1 & 0 & 1 \end{array}\right) \tag{11.4.34}$$

$$= \left(\begin{array}{ccc|ccc} 1 & 0 & 0 & 1/5 & -1/10 & 0 \\ 0 & 1 & 0 & 0 & -1/5 & 1/10 \\ 0 & 0 & 1 & 1/10 & 0 & 1/10 \end{array}\right). \tag{11.4.35}$$

Thus, the right half of the augmented matrix yields the inverse and it equals

$$A^{-1} = \begin{pmatrix} 1/5 & -1/10 & 0 \\ 0 & -1/5 & 1/10 \\ 1/10 & 0 & 1/10 \end{pmatrix}. \tag{11.4.36}$$

Of course, we can always check our answer by multiplying A^{-1} by A.

Gaussian elimination may be used with overdetermined systems. *Overdetermined systems* are linear systems where there are more equations than unknowns ($m > n$). These systems are usually (but not always) inconsistent.

• **Example 11.4.6**

Consider the linear system

$$x_1 + x_2 = 1, \tag{11.4.37}$$

$$-x_1 + 2x_2 = -2, \tag{11.4.38}$$

$$x_1 - x_2 = 4. \tag{11.4.39}$$

After several row operations, the augmented matrix

$$\left(\begin{array}{cc|c} 1 & 1 & 1 \\ -1 & 2 & -2 \\ 1 & -1 & 4 \end{array}\right) \tag{11.4.40}$$

becomes

$$\left(\begin{array}{cc|c} 1 & 1 & 1 \\ 0 & 1 & 2 \\ 0 & 0 & -7 \end{array}\right). \tag{11.4.41}$$

From the last row of the augmented matrix (11.4.41) we see that the system is inconsistent. However, if we change the system to

$$x_1 + x_2 = 1, \tag{11.4.42}$$

$$-x_1 + 2x_2 = 5, \tag{11.4.43}$$

$$x_1 = -1, \tag{11.4.44}$$

the final form of the augmented matrix is

$$\left(\begin{array}{cc|c} 1 & 1 & 1 \\ 0 & 1 & 2 \\ 0 & 0 & 0 \end{array}\right). \tag{11.4.45}$$

which has the unique solution $x_1 = -1$ and $x_2 = 2$.

Finally, by introducing the set:

$$x_1 + x_2 = 1, \tag{11.4.46}$$

$$2x_1 + 2x_2 = 2, \tag{11.4.47}$$

$$3x_1 + 3x_3 = 3, \tag{11.4.48}$$

the final form of the augmented matrix is

$$\left(\begin{array}{cc|c} 1 & 1 & 1 \\ 0 & 0 & 0 \\ 0 & 0 & 0 \end{array}\right). \tag{11.4.49}$$

There are an infinite number of solutions: $x_1 = 1 - \alpha$ and $x_2 = \alpha$.

Gaussian elimination can also be employed with underdetermined systems. An *underdetermined linear system* is one where there are fewer equations than unknowns ($m < n$). These systems usually have an infinite number of solutions although they can be inconsistent.

• **Example 11.4.7**

Consider the underdetermined system:

$$2x_1 + 2x_2 + x_3 = -1, \tag{11.4.50}$$

$$4x_1 + 4x_2 + 2x_3 = 3. \tag{11.4.51}$$

Its augmented matrix may be transformed into the form:

$$\left(\begin{array}{ccc|c} 2 & 2 & 1 & -1 \\ 0 & 0 & 0 & 4 \end{array}\right). \tag{11.4.52}$$

Clearly this case corresponds to an inconsistent set of equations. On the other hand, if (11.4.51) is changed to

$$4x_1 + 4x_2 + 2x_3 = -2, \tag{11.4.53}$$

then the final form of the augmented matrix is

$$\left(\begin{array}{ccc|c} 2 & 2 & 1 & -1 \\ 0 & 0 & 0 & 0 \end{array}\right) \tag{11.4.54}$$

and we have an infinite number of solutions, namely $x_3 = \alpha$, $x_2 = \beta$, and $2x_1 = -1 - \alpha - 2\beta$.

Consider now one of most important classes of linear equations: the homogeneous equations $A\mathbf{x} = \mathbf{0}$. If $\det(A) \neq 0$, then by Cramer's rule

$x_1 = x_2 = x_3 = \cdots = x_n = 0$. Thus, the only possibility for a nontrivial solution is $\det(A) = 0$. In this case, A is singular, no inverse exists, and nontrivial solutions exist but they are not unique.

• **Example 11.4.8**

Consider the two homogeneous equations:

$$x_1 + x_2 = 0 \tag{11.4.55}$$

$$x_1 - x_2 = 0. \tag{11.4.56}$$

Note that $\det(A) = -2$. Solving this system yields $x_1 = x_2 = 0$.

However, if we change the system to

$$x_1 + x_2 = 0 \tag{11.4.57}$$

$$x_1 + x_2 = 0 \tag{11.4.58}$$

which has the $\det(A) = 0$ so that A is singular. Both equations yield $x_1 = -x_2 = \alpha$, any constant. Thus, there is an infinite number of solutions for this set of homogeneous equations.

We close this section by outlining the proof of the theorem which we introduced at the beginning.

Consider the system $A\mathbf{x} = \mathbf{b}$. By elementary row operations, the first equation in this system can be reduced to

$$x_1 + \alpha_{12}x_2 + \cdots + \alpha_{1n}x_n = \beta_1. \tag{11.4.59}$$

The second equation has the form

$$x_p + \alpha_{2,p+1}x_{p+1} + \cdots + \alpha_{2n}x_n = \beta_2, \tag{11.4.60}$$

where $p > 1$. The third equation has the form

$$x_q + \alpha_{3,q+1}x_{q+1} + \cdots + \alpha_{3n}x_n = \beta_3, \tag{11.4.61}$$

where $q > p$, and so on. To simplify the notation, we introduce z_i where we choose the first k values so that $z_1 = x_1$, $z_2 = x_p$, $z_3 = x_q$, ... Thus, the question of the existence of solutions depends upon the three integers: m, n, and k. The resulting set of equations have the form:

$$\begin{pmatrix} 1 & \gamma_{12} & \cdots & \gamma_{1,k} & \gamma_{1,k+1} & \cdots & \gamma_{1n} \\ 0 & 1 & \cdots & \gamma_{2,k} & \gamma_{2,k+1} & \cdots & \gamma_{2n} \\ & & & \vdots & & & \\ 0 & 0 & \cdots & 1 & \gamma_{k,k+1} & \cdots & \gamma_{kn} \\ 0 & 0 & \cdots & 0 & 0 & \cdots & 0 \\ & & & \vdots & & & \\ 0 & 0 & \cdots & 0 & 0 & \cdots & 0 \end{pmatrix} \begin{pmatrix} z_1 \\ z_2 \\ \vdots \\ z_n \end{pmatrix} = \begin{pmatrix} \beta_1 \\ \beta_2 \\ \vdots \\ \beta_k \\ \beta_{k+1} \\ \vdots \\ \beta_m \end{pmatrix}. \tag{11.4.62}$$

Note that $\beta_{k+1}, \ldots, \beta_m$ need not be all zero.

There are three possibilities:

(a) $k < m$ and at least one of the elements $\beta_{k+1}, \ldots, \beta_m$ is nonzero. Suppose that an element β_p is nonzero $(p > k)$. Then the pth equation is

$$0z_1 + 0z_2 + \cdots + 0z_n = \beta_p \neq 0. \qquad \textbf{(11.4.63)}$$

However, this is a contradiction and the equations are inconsistent.

(b) $k = n$ and either (i) $k < m$ and all of the elements $\beta_{k+1}, \ldots, \beta_m$ are zero, or (ii) $k = m$. Then the equations have a unique solution which can be obtained by back-substitution.

(c) $k < n$ and either (i) $k < m$ and all of the elements $\beta_{k+1}, \ldots, \beta_m$ are zero, or (ii) $k = m$. Then, arbitrary values can be assigned to the $n-k$ variables $z_{k+1}, \ldots, z_n$. The equations can be solved for $z_1, z_2, \ldots, z_k$ and there is an infinity of solutions.

For homogeneous equations $\mathbf{b} = \mathbf{0}$, all of the β_i are zero. In this case, we have only two cases:

(b′) $k = n$, then (11.4.62) has the solution $\mathbf{z} = \mathbf{0}$ which leads to the trivial solution for the original system $A\mathbf{x} = \mathbf{0}$.

(c′) $k < n$, the equations possess an infinity of solutions given by assigning arbitrary values to $z_{k+1}, \ldots, z_n$. □

Problems

Solve the following systems of linear equations by Gaussian elimination:

1. $2x_1 + x_2 = 4, \qquad 5x_1 - 2x_2 = 1$
2. $x_1 + x_2 = 0, \qquad 3x_1 - 4x_2 = 1$
3. $-x_1 + x_2 + 2x_3 = 0, \quad 3x_1 + 4x_2 + x_3 = 0, \quad -x_1 + x_2 + 2x_3 = 0$
4. $4x_1 + 6x_2 + x_3 = 2, \quad 2x_1 + x_2 - 4x_3 = 3, \quad 3x_1 - 2x_2 + 5x_3 = 8$
5. $3x_1 + x_2 - 2x_3 = -3, \quad x_1 - x_2 + 2x_3 = -1, \quad -4x_1 + 3x_2 - 6x_3 = 4$
6. $x_1 - 3x_2 + 7x_3 = 2, \qquad 2x_1 + 4x_2 - 3x_3 = -1,$
   $-3x_1 + 7x_2 + 2x_3 = 3$
7. $x_1 - x_2 + 3x_3 = 5, \qquad 2x_1 - 4x_2 + 7x_3 = 7,$
   $4x_1 - 9x_2 + 2x_3 = -15$
8. $x_1 + x_2 + x_3 + x_4 = -1, \qquad 2x_1 - x_2 + 3x_3 = 1,$
   $2x_2 + 3x_4 = 15, \qquad -x_1 + 2x_2 + x_4 = -2$

Find the inverse of each of the following matrices by Gaussian elimination:

9. $\begin{pmatrix} -3 & 5 \\ 2 & 1 \end{pmatrix}$ 10. $\begin{pmatrix} 3 & -1 \\ -5 & 2 \end{pmatrix}$

11. $\begin{pmatrix} 19 & 2 & -9 \\ -4 & -1 & 2 \\ -2 & 0 & 1 \end{pmatrix}$ 12. $\begin{pmatrix} 1 & 2 & 5 \\ 0 & -1 & 2 \\ 2 & 4 & 11 \end{pmatrix}$

13. Does $(A^2)^{-1} = (A^{-1})^2$? Justify your answer.

11.5 EIGENVALUES AND EIGENVECTORS

One of the classic problems of linear algebra[4] is finding all of the λ's which satisfy the $n \times n$ system

$$A\mathbf{x} = \lambda \mathbf{x}. \tag{11.5.1}$$

The nonzero quantity λ is the *eigenvalue* or *characteristic value* of A. The vector $\mathbf{x}$ is the *eigenvector* or *characteristic vector* belonging to λ. The set of the eigenvalues of A is called the *spectrum* of A. The largest of the absolute values of the eigenvalues of A is called the *spectral radius* of A.

To find λ and $\mathbf{x}$, we first rewrite (11.5.1) as a set of homogeneous equations:

$$(A - \lambda I)\mathbf{x} = \mathbf{0}. \tag{11.5.2}$$

From the theory of linear equations, (11.5.2) has trivial solutions unless its determinant equals zero. On the other hand, if

$$\det(A - \lambda I) = 0, \tag{11.5.3}$$

there are an infinity of solutions.

The expansion of the determinant (11.5.3) yields an nth-degree polynomial in λ, the *characteristic polynomial*. The roots of the characteristic polynomial are the eigenvalues of A. Because the characteristic polynomial has exactly n roots, A will have n eigenvalues, some of which may be repeated (with multiplicity $k \leq n$) and some of which may be complex numbers. For each eigenvalue λ_i, there will be a corresponding eigenvector $\mathbf{x}_i$. This eigenvector is the solution of the homogeneous equations $(A - \lambda_i I)\mathbf{x}_i = \mathbf{0}$.

An important property of eigenvectors is their *linear independence* if there are n distinct eigenvalues. Vectors are linearly independent if the equation

$$\alpha_1 \mathbf{x}_1 + \alpha_2 \mathbf{x}_2 + \cdots + \alpha_n \mathbf{x}_n = \mathbf{0} \tag{11.5.4}$$

can be satisfied only by taking *all* of the α's equal to zero.

---

[4] The standard reference is Wilkinson, J. H., 1965: *The Algebraic Eigenvalue Problem*, Clarendon Press, Oxford.

To show that this is true in the case of n distinct eigenvalues $\lambda_1, \lambda_2, \ldots, \lambda_n$, each eigenvalue λ_i having a corresponding eigenvector $\mathbf{x}_i$, we first write down the linear dependence condition

$$\alpha_1\mathbf{x}_1 + \alpha_2\mathbf{x}_2 + \cdots + \alpha_n\mathbf{x}_n = \mathbf{0}. \tag{11.5.5}$$

Premultiplying (11.5.5) by A,

$$\alpha_1 A\mathbf{x}_1 + \alpha_2 A\mathbf{x}_2 + \cdots + \alpha_n A\mathbf{x}_n = \alpha_1\lambda_1\mathbf{x}_1 + \alpha_2\lambda_2\mathbf{x}_2 + \cdots + \alpha_n\lambda_n\mathbf{x}_n = \mathbf{0}. \tag{11.5.6}$$

Premultiplying (11.5.5) by A^2,

$$\alpha_1 A^2\mathbf{x}_1 + \alpha_2 A^2\mathbf{x}_2 + \cdots + \alpha_n A^2\mathbf{x}_n = \alpha_1\lambda_1^2\mathbf{x}_1 + \alpha_2\lambda_2^2\mathbf{x}_2 + \cdots + \alpha_n\lambda_n^2\mathbf{x}_n = \mathbf{0}. \tag{11.5.7}$$

In similar manner, we obtain the system of equations:

$$\begin{pmatrix} 1 & 1 & \cdots & 1 \\ \lambda_1 & \lambda_2 & \cdots & \lambda_n \\ \lambda_1^2 & \lambda_2^2 & \cdots & \lambda_n^2 \\ \vdots & \vdots & \vdots & \vdots \\ \lambda_1^{n-1} & \lambda_2^{n-1} & \cdots & \lambda_n^{n-1} \end{pmatrix} \begin{pmatrix} \alpha_1\mathbf{x}_1 \\ \alpha_2\mathbf{x}_2 \\ \alpha_3\mathbf{x}_3 \\ \vdots \\ \alpha_n\mathbf{x}_n \end{pmatrix} = \begin{pmatrix} 0 \\ 0 \\ 0 \\ \vdots \\ 0 \end{pmatrix}. \tag{11.5.8}$$

Because

$$\begin{vmatrix} 1 & 1 & \cdots & 1 \\ \lambda_1 & \lambda_2 & \cdots & \lambda_n \\ \lambda_1^2 & \lambda_2^2 & \cdots & \lambda_n^2 \\ \vdots & \vdots & \vdots & \vdots \\ \lambda_1^{n-1} & \lambda_2^{n-1} & \cdots & \lambda_n^{n-1} \end{vmatrix} = \begin{matrix} (\lambda_2 - \lambda_1)(\lambda_3 - \lambda_2)(\lambda_3 - \lambda_1)(\lambda_4 - \lambda_3) \\ (\lambda_4 - \lambda_2)\cdots(\lambda_n - \lambda_1) \neq 0, \end{matrix} \tag{11.5.9}$$

since it is a Vandermonde determinant, $\alpha_1\mathbf{x}_1 = \alpha_2\mathbf{x}_2 = \alpha_3\mathbf{x}_3 = \cdots = \alpha_n\mathbf{x}_n = 0$. Because the eigenvectors are nonzero, $\alpha_1 = \alpha_2 = \alpha_3 = \cdots = \alpha_n = 0$ and the eigenvectors are linearly independent. □

This property of eigenvectors allows us to express any arbitrary vector $\mathbf{x}$ as a linear sum of the eigenvectors $\mathbf{x}_i$ or

$$\mathbf{x} = c_1\mathbf{x}_1 + c_2\mathbf{x}_2 + \cdots + c_n\mathbf{x}_n. \tag{11.5.10}$$

We will make good use of this property in Example 11.5.3.

• Example 11.5.1

Let us find the eigenvalues and corresponding eigenvectors of the matrix

$$A = \begin{pmatrix} -4 & 2 \\ -1 & -1 \end{pmatrix}. \tag{11.5.11}$$

We begin by setting up the characteristic equation:

$$\det(A - \lambda I) = \begin{vmatrix} -4-\lambda & 2 \\ -1 & -1-\lambda \end{vmatrix} = 0. \qquad (11.5.12)$$

Expanding the determinant,

$$(-4-\lambda)(-1-\lambda) + 2 = \lambda^2 + 5\lambda + 6 = (\lambda+3)(\lambda+2) = 0. \quad (11.5.13)$$

Thus, the eigenvalues of the matrix A are $\lambda_1 = -3$ and $\lambda_2 = -2$.

To find the corresponding eigenvectors, we must solve the linear system:

$$\begin{pmatrix} -4-\lambda & 2 \\ -1 & -1-\lambda \end{pmatrix} \begin{pmatrix} x_1 \\ x_2 \end{pmatrix} = \begin{pmatrix} 0 \\ 0 \end{pmatrix}. \qquad (11.5.14)$$

For example, for $\lambda_1 = -3$,

$$\begin{pmatrix} -1 & 2 \\ -1 & 2 \end{pmatrix} \begin{pmatrix} x_1 \\ x_2 \end{pmatrix} = \begin{pmatrix} 0 \\ 0 \end{pmatrix} \qquad (11.5.15)$$

or

$$x_1 = 2x_2. \qquad (11.5.16)$$

Thus, any nonzero multiple of the vector $\begin{pmatrix} 2 \\ 1 \end{pmatrix}$ is an eigenvector belonging to $\lambda_1 = -3$. Similarly, for $\lambda_2 = -2$, the eigenvector is any nonzero multiple of the vector $\begin{pmatrix} 1 \\ 1 \end{pmatrix}$.

• **Example 11.5.2**

Let us now find the eigenvalues and corresponding eigenvectors of the matrix

$$A = \begin{pmatrix} -4 & 5 & 5 \\ -5 & 6 & 5 \\ -5 & 5 & 6 \end{pmatrix}. \qquad (11.5.17)$$

Setting up the characteristic equation:

$$\det(A - \lambda I) = \begin{vmatrix} -4-\lambda & 5 & 5 \\ -5 & 6-\lambda & 5 \\ -5 & 5 & 6-\lambda \end{vmatrix} = \begin{vmatrix} -4-\lambda & 5 & 5 \\ -5 & 6-\lambda & 5 \\ 0 & \lambda-1 & 1-\lambda \end{vmatrix} \qquad (11.5.18)$$

$$= (\lambda-1) \begin{vmatrix} -4-\lambda & 5 & 5 \\ -5 & 6-\lambda & 5 \\ 0 & 1 & -1 \end{vmatrix} = (\lambda-1)^2 \begin{vmatrix} -1 & 1 & 0 \\ -5 & 6-\lambda & 5 \\ 0 & 1 & -1 \end{vmatrix} \qquad (11.5.19)$$

$$= (\lambda-1)^2 \begin{vmatrix} -1 & 0 & 0 \\ -5 & 6-\lambda & 0 \\ 0 & 1 & -1 \end{vmatrix} = (\lambda-1)^2(6-\lambda) = 0. \quad (11.5.20)$$

Thus, the eigenvalues of the matrix A are $\lambda_{1,2} = 1$ (twice) and $\lambda_3 = 6$.

To find the corresponding eigenvectors, we must solve the linear system:

$$(-4 - \lambda)x_1 + 5x_2 + 5x_3 = 0, \tag{11.5.21}$$

$$-5x_1 + (6 - \lambda)x_2 + 5x_3 = 0 \tag{11.5.22}$$

and

$$-5x_1 + 5x_2 + (6 - \lambda)x_3 = 0. \tag{11.5.23}$$

For $\lambda_3 = 6$, (11.5.21)–(11.5.23) become

$$-10x_1 + 5x_2 + 5x_3 = 0, \tag{11.5.24}$$

$$-5x_1 + 5x_3 = 0 \tag{11.5.25}$$

and

$$-5x_1 + 5x_2 = 0. \tag{11.5.26}$$

Thus, $x_1 = x_2 = x_3$ and the eigenvector is any nonzero multiple of the vector $\begin{pmatrix} 1 \\ 1 \\ 1 \end{pmatrix}$.

The interesting aspect of this example involves finding the eigenvector for the eigenvalue $\lambda_{1,2} = 1$. If $\lambda_{1,2} = 1$, then (11.5.21)–(11.5.23) collapses into one equation

$$-x_1 + x_2 + x_3 = 0 \tag{11.5.27}$$

and we have *two* free parameters at our disposal. Let us take $x_2 = \alpha$ and $x_3 = \beta$. Then the eigenvector equals $\alpha \begin{pmatrix} 1 \\ 1 \\ 0 \end{pmatrix} + \beta \begin{pmatrix} 1 \\ 0 \\ 1 \end{pmatrix}$ for $\lambda_{1,2} = 1$.

In this example our 3×3 matrix has three *linearly independent* eigenvectors: $\begin{pmatrix} 1 \\ 1 \\ 0 \end{pmatrix}$ associated with $\lambda_1 = 1$, $\begin{pmatrix} 1 \\ 0 \\ 1 \end{pmatrix}$ associated with $\lambda_2 = 1$, and $\begin{pmatrix} 1 \\ 1 \\ 1 \end{pmatrix}$ associated with $\lambda_3 = 6$. However, with repeated eigenvalues this is not always true. For example,

$$A = \begin{pmatrix} 1 & -1 \\ 0 & 1 \end{pmatrix} \tag{11.5.28}$$

has the repeated eigenvalues $\lambda_{1,2} = 1$. However, there is only a single eigenvector $\begin{pmatrix} 1 \\ 0 \end{pmatrix}$ for *both* λ_1 and λ_2.

• **Example 11.5.3**

When we discussed the stability of numerical schemes for the wave equation in Section 7.6, we examined the behavior of a prototypical Fourier harmonic to variation in the parameter $c\Delta t/\Delta x$. In this example we shall show another approach to determining the stability of a numerical scheme via matrices.

Consider the explicit scheme for the numerical integration of the wave equation (7.6.11). We can rewrite that single equation as the coupled difference equations:

$$u_m^{n+1} = 2(1-r^2)u_m^n + r^2(u_{m+1}^n + u_{m-1}^n) - v_m^n \tag{11.5.29}$$

and

$$v_m^{n+1} = u_m^n, \tag{11.5.30}$$

where $r = c\Delta t/\Delta x$. Let $u_{m+1}^n = e^{i\beta\Delta x}u_m^n$ and $u_{m-1}^n = e^{-i\beta\Delta x}u_m^n$, where β is real. Then (11.5.29)–(11.5.30) becomes

$$u_m^{n+1} = 2\left[1 - 2r^2\sin^2\left(\frac{\beta\Delta x}{2}\right)\right]u_m^n - v_m^n \tag{11.5.31}$$

and

$$v_m^{n+1} = u_m^n \tag{11.5.32}$$

or in the matrix form

$$\mathbf{u}_m^{n+1} = \begin{pmatrix} 2\left[1 - 2r^2\sin^2\left(\frac{\beta\Delta x}{2}\right)\right] & -1 \\ 1 & 0 \end{pmatrix}\mathbf{u}_m^n, \tag{11.5.33}$$

where $\mathbf{u}_m^n = \begin{pmatrix} u_m^n \\ v_m^n \end{pmatrix}$. The eigenvalues λ of this *amplification matrix* are given by

$$\lambda^2 - 2\left[1 - 2r^2\sin^2\left(\frac{\beta\Delta x}{2}\right)\right]\lambda + 1 = 0 \tag{11.5.34}$$

or

$$\lambda_{1,2} = 1 - 2r^2\sin^2\left(\frac{\beta\Delta x}{2}\right) \pm 2r\sin\left(\frac{\beta\Delta x}{2}\right)\sqrt{r^2\sin^2\left(\frac{\beta\Delta x}{2}\right) - 1}. \tag{11.5.35}$$

Because each successive time step consists of multiplying the solution from the previous time step by the amplification matrix, the solution will be stable only if $\mathbf{u}_m^n$ remains bounded. This will occur only if all of the eigenvalues have a magnitude less or equal to one because

$$\mathbf{u}_m^n = \sum_k c_k A^n \mathbf{x}_k = \sum_k c_k \lambda_k^n \mathbf{x}_k, \tag{11.5.36}$$

where A denotes the amplification matrix and $\mathbf{x}_k$ denotes the eigenvectors corresponding to the eigenvalues λ_k. Equation (11.5.36) follows from our ability to express any initial condition in terms of an eigenvector expansion:

$$\mathbf{u}_m^0 = \sum_k c_k \mathbf{x}_k. \tag{11.5.37}$$

In our particular example, two cases arise. If $r^2 \sin^2(\beta\Delta x/2) \le 1$,

$$\lambda_{1,2} = 1 - 2r^2 \sin^2\left(\frac{\beta\Delta x}{2}\right) \pm 2ri \sin\left(\frac{\beta\Delta x}{2}\right) \sqrt{1 - r^2 \sin^2\left(\frac{\beta\Delta x}{2}\right)} \tag{11.5.38}$$

and $|\lambda_{1,2}| = 1$. On the other hand, if $r^2 \sin^2(\beta\Delta x/2) > 1$, $|\lambda_{1,2}| > 1$. Thus, we will have stability only if $c\Delta t/\Delta x \le 1$.

Problems

Find the eigenvalues and corresponding eigenvectors for the following matrices:

1. $A = \begin{pmatrix} 3 & 2 \\ 3 & -2 \end{pmatrix}$

2. $A = \begin{pmatrix} 3 & -1 \\ 1 & 1 \end{pmatrix}$

3. $A = \begin{pmatrix} 2 & -3 & 1 \\ 1 & -2 & 1 \\ 1 & -3 & 2 \end{pmatrix}$

4. $A = \begin{pmatrix} 0 & 1 & 0 \\ 0 & 0 & 1 \\ 0 & 0 & 0 \end{pmatrix}$

5. $A = \begin{pmatrix} 1 & 1 & 1 \\ 0 & 2 & 1 \\ 0 & 0 & 1 \end{pmatrix}$

6. $A = \begin{pmatrix} 1 & 2 & 1 \\ 0 & 3 & 1 \\ 0 & 5 & -1 \end{pmatrix}$

7. $A = \begin{pmatrix} 4 & -5 & 1 \\ 1 & 0 & -1 \\ 0 & 1 & -1 \end{pmatrix}$

8. $A = \begin{pmatrix} -2 & 0 & 1 \\ 3 & 0 & -1 \\ 0 & 1 & 1 \end{pmatrix}$

Project: Numerical Solution of the Sturm-Liouville Problem

You may have been struck by the similarity of the algebraic eigenvalue problem to the Sturm-Liouville problem. In both cases nontrivial solutions exist only for characteristic values of λ. The purpose of this project is to further deepen your insight into these similarities.

Consider the Sturm-Liouville problem:

$$y'' + \lambda y = 0, \qquad y(0) = y(\pi) = 0. \tag{11.5.39}$$

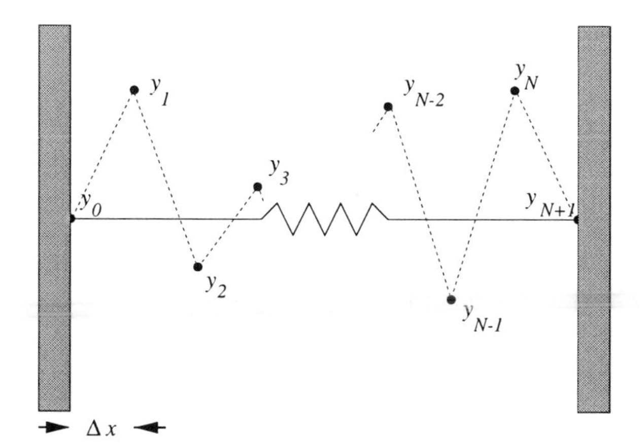

Figure 11.5.1: Schematic for finite-differencing a Sturm-Liouville problem into a set of difference equations.

We know that it has the nontrivial solutions $\lambda_m = m^2$, $y_m(x) = \sin(mx)$, where $m = 1, 2, 3, \ldots$

Step 1: Let us solve this problem numerically. Introducing centered finite differencing and the grid shown in Figure 11.5.1, show that

$$y'' \approx \frac{y_{n+1} - 2y_n + y_{n-1}}{\Delta x}, \qquad n = 1, 2, \ldots, N, \tag{11.5.40}$$

where $\Delta x = \pi/(N+1)$. Show that the finite-differenced form of (11.5.39) is

$$-h^2 y_{n+1} + 2h^2 y_n - h^2 y_{n-1} = \lambda y_n \tag{11.5.41}$$

with $y_0 = y_{N+1} = 0$ and $h = 1/(\Delta x)$.

Step 2: Solve (11.5.41) as an algebraic eigenvalue problem using $N = 1, 2, \ldots$. Show that (11.5.41) can be written in the matrix form of

$$\begin{pmatrix} 2h^2 & -h^2 & 0 & \cdots & 0 & 0 & 0 \\ -h^2 & 2h^2 & -h^2 & \cdots & 0 & 0 & 0 \\ 0 & -h^2 & 2h^2 & \cdots & 0 & 0 & 0 \\ \vdots & \vdots & \vdots & \vdots & \vdots & \vdots & \vdots \\ 0 & 0 & 0 & \cdots & -h^2 & 2h^2 & -h^2 \\ 0 & 0 & 0 & \cdots & 0 & -h^2 & 2h^2 \end{pmatrix} \begin{pmatrix} y_1 \\ y_2 \\ y_3 \\ \vdots \\ y_{N-1} \\ y_N \end{pmatrix} = \lambda \begin{pmatrix} y_1 \\ y_2 \\ y_3 \\ \vdots \\ y_{N-1} \\ y_N \end{pmatrix}. \tag{11.5.42}$$

Note that the coefficient matrix is symmetric. Except for very small N, computing the values of λ using determinants is very difficult. Consequently you must use one of the numerical schemes that have been

Table 11.5.1: Eigenvalues computed from (11.5.42) as a numerical approximation of the Sturm-Liouville problem (11.5.39).

N	λ_1	λ_2	λ_3	λ_4	λ_5	λ_6	λ_7
1	0.81057						
2	0.91189	2.73567					
3	0.94964	3.24228	5.53491				
4	0.96753	3.50056	6.63156	9.16459			
5	0.97736	3.64756	7.29513	10.94269	13.61289		
6	0.98333	3.73855	7.71996	12.13899	16.12040	18.87563	
7	0.98721	3.79857	8.00605	12.96911	17.93217	22.13966	24.95100
8	0.98989	3.84016	8.20702	13.56377	19.26430	24.62105	28.98791
20	0.99813	3.97023	8.84993	15.52822	23.85591	33.64694	44.68265
50	0.99972	3.99498	8.97438	15.91922	24.80297	35.59203	48.24538

developed for the efficient solution of the algebraic eigenvalue problem.[5] Packages for numerically solving the algebraic eigenvalue problem may already exist on your system or you may find code in a numerical methods book.

In Table 11.5.1 I have given the computed values of λ as a function of N using the IMSL routine EVLSF so that you may check your answers. How do your computed eigenvalues compare to the eigenvalues given by the Sturm-Liouville problem? What happens as you increase N? Which computed eigenvalues agree best with those given by the Sturm-Liouville problem? Which ones compare the worst?

Step 3: Let us examine the eigenfunctions now. First, reorder (if necessary) your eigenvectors so that each consecutive eigenvalue increases in magnitude. Starting with the smallest eigenvalue, construct an xy plot for each consecutive eigenvectors where $x_i = i\Delta x$, $i = 1, 2, \ldots, N$, and y_i are the corresponding element from the eigenvector. On the same plot, graph $y_m(x) = \sin(mx)$. Which eigenvectors and eigenfunctions agree the best? Which eigenvectors and eigenfunctions agree the worst? Why? Why are there N eigenvectors and an infinite number of eigenfunctions?

Step 4: The most important property of eigenfunctions is orthogonality. But what do we mean by orthogonality in the case of eigenvectors? Recall from three-dimensional vectors we had the scalar dot product:

$$\mathbf{a} \cdot \mathbf{b} = a_1 b_1 + a_2 b_2 + a_3 b_3. \tag{11.5.43}$$

[5] See Press, W. H., Flannery, B. F., Teukolsky, S. A., and Vetterling, W. T., 1986: *Numerical Recipes: The Art of Scientific Computing*, Cambridge University Press, Cambridge, chap. 11.

For n-dimensional vectors, this dot product is generalized to the inner product

$$\mathbf{x}\cdot\mathbf{y}=\sum_{k=1}^{n}x_k y_k. \tag{11.5.44}$$

Orthogonality implies that $\mathbf{x}\cdot\mathbf{y}=0$ if $\mathbf{x}\neq\mathbf{y}$. Are your eigenvectors orthogonal? How might you use this property with eigenvectors?

11.6 SYSTEMS OF LINEAR DIFFERENTIAL EQUATIONS

In this section we show how we may apply the classic algebraic eigenvalue problem to solve a system of ordinary differential equations.

Let us solve the following system:

$$x_1' = x_1 + 3x_2 \tag{11.6.1}$$

and

$$x_2' = 3x_1 + x_2, \tag{11.6.2}$$

where the primes denote the time derivative.

We begin by rewriting (11.6.1)–(11.6.2) in linear algebra notation:

$$\mathbf{x}' = A\mathbf{x}, \tag{11.6.3}$$

where

$$\mathbf{x}=\begin{pmatrix}x_1\\x_2\end{pmatrix} \quad \text{and} \quad A=\begin{pmatrix}1&3\\3&1\end{pmatrix}. \tag{11.6.4}$$

Note that

$$\begin{pmatrix}x_1'\\x_2'\end{pmatrix}=\frac{d}{dt}\begin{pmatrix}x_1\\x_2\end{pmatrix}=\mathbf{x}'. \tag{11.6.5}$$

Assuming a solution of the form

$$\mathbf{x}=\mathbf{x}_0 e^{\lambda t}, \qquad \text{where} \qquad \mathbf{x}_0=\begin{pmatrix}a\\b\end{pmatrix} \tag{11.6.6}$$

is a constant vector, we substitute (11.6.6) into (11.6.3) and find that

$$\lambda e^{\lambda t}\mathbf{x}_0 = A e^{\lambda t}\mathbf{x}_0. \tag{11.6.7}$$

Because $e^{\lambda t}$ does not generally equal zero, we have that

$$(A-\lambda I)\mathbf{x}_0=\mathbf{0}, \tag{11.6.8}$$

which we solved in the previous section. This set of homogeneous equations is the *classic eigenvalue problem*. In order for this set not to have trivial solutions,

$$\det(A-\lambda I)=\begin{vmatrix}1-\lambda & 3\\ 3 & 1-\lambda\end{vmatrix}=0. \tag{11.6.9}$$

Expanding the determinant,

$$(1-\lambda)^2 - 9 = 0 \quad \text{or} \quad \lambda = -2, 4. \tag{11.6.10}$$

Thus, we have two real and distinct eigenvalues: $\lambda = -2$ and 4.

We must now find the corresponding $\mathbf{x}_0$ or *eigenvector* for each eigenvalue. From (11.6.8),

$$(1-\lambda)a + 3b = 0 \tag{11.6.11}$$

and

$$3a + (1-\lambda)b = 0. \tag{11.6.12}$$

If $\lambda = 4$, these equations are consistent and yield $a = b = c_1$. If $\lambda = -2$, we have that $a = -b = c_2$. Therefore, the general solution in matrix notation is

$$\mathbf{x} = c_1 \begin{pmatrix} 1 \\ 1 \end{pmatrix} e^{4t} + c_2 \begin{pmatrix} 1 \\ -1 \end{pmatrix} e^{-2t}. \tag{11.6.13}$$

To evaluate c_1 and c_2, we must have initial conditions. For example, if $x_1(0) = x_2(0) = 1$, then

$$\begin{pmatrix} 1 \\ 1 \end{pmatrix} = c_1 \begin{pmatrix} 1 \\ 1 \end{pmatrix} + c_2 \begin{pmatrix} 1 \\ -1 \end{pmatrix}. \tag{11.6.14}$$

Solving for c_1 and c_2, $c_1 = 1$ and $c_2 = 0$ and the solution with this particular set of initial conditions is

$$\mathbf{x} = \begin{pmatrix} 1 \\ 1 \end{pmatrix} e^{4t}. \tag{11.6.15}$$

• Example 11.6.1

Let us solve the following set of linear ordinary differential equations:

$$x_1' = -x_2 + x_3, \tag{11.6.16}$$

$$x_2' = 4x_1 - x_2 - 4x_3 \tag{11.6.17}$$

and

$$x_3' = -3x_1 - x_2 + 4x_3; \tag{11.6.18}$$

or in matrix form,

$$\mathbf{x}' = \begin{pmatrix} 0 & -1 & 1 \\ 4 & -1 & -4 \\ -3 & -1 & 4 \end{pmatrix} \mathbf{x}, \qquad \mathbf{x} = \begin{pmatrix} x_1 \\ x_2 \\ x_3 \end{pmatrix}. \tag{11.6.19}$$

Assuming the solution $\mathbf{x} = \mathbf{x}_0 e^{\lambda t}$,

$$\begin{pmatrix} 0 & -1 & 1 \\ 4 & -1 & -4 \\ -3 & -1 & 4 \end{pmatrix} \mathbf{x}_0 = \lambda \mathbf{x}_0 \tag{11.6.20}$$

or

$$\begin{pmatrix} -\lambda & -1 & 1 \\ 4 & -1-\lambda & -4 \\ -3 & -1 & 4-\lambda \end{pmatrix} \mathbf{x}_0 = \mathbf{0}. \tag{11.6.21}$$

For nontrivial solutions,

$$\begin{vmatrix} -\lambda & -1 & 1 \\ 4 & -1-\lambda & -4 \\ -3 & -1 & 4-\lambda \end{vmatrix} = \begin{vmatrix} 0 & 0 & 1 \\ 4-4\lambda & -5-\lambda & -4 \\ -3+4\lambda-\lambda^2 & 3-\lambda & 4-\lambda \end{vmatrix} = 0 \tag{11.6.22}$$

and

$$(\lambda - 1)(\lambda - 3)(\lambda + 1) = 0 \quad \text{or} \quad \lambda = -1, 1, 3. \tag{11.6.23}$$

To determine the eigenvectors, we rewrite (11.6.21) as

$$-\lambda a - b + c = 0, \tag{11.6.24}$$

$$4a - (1 + \lambda)b - 4c = 0 \tag{11.6.25}$$

and

$$-3a - b + (4 - \lambda)c = 0. \tag{11.6.26}$$

For example, if $\lambda = 1$,

$$-a - b + c = 0, \tag{11.6.27}$$

$$4a - 2b - 4c = 0 \tag{11.6.28}$$

and

$$-3a - b + 3c = 0; \tag{11.6.29}$$

or $a = c$ and $b = 0$. Thus, the eigenfunction for $\lambda = 1$ is $\mathbf{x}_0 = \begin{pmatrix} 1 \\ 0 \\ 1 \end{pmatrix}$. Similarly, for $\lambda = -1$, $\mathbf{x}_0 = \begin{pmatrix} 1 \\ 2 \\ 1 \end{pmatrix}$ and for $\lambda = 3$, $\mathbf{x}_0 = \begin{pmatrix} 1 \\ -1 \\ 2 \end{pmatrix}$. Thus, the most general solution is

$$\mathbf{x} = c_1 \begin{pmatrix} 1 \\ 0 \\ 1 \end{pmatrix} e^{t} + c_2 \begin{pmatrix} 1 \\ 2 \\ 1 \end{pmatrix} e^{-t} + c_3 \begin{pmatrix} 1 \\ -1 \\ 2 \end{pmatrix} e^{3t}. \tag{11.6.30}$$

• **Example 11.6.2**

Let us solve the following set of linear ordinary differential equations:

$$x_1' = x_1 - 2x_2 \tag{11.6.31}$$

and

$$x_2' = 2x_1 - 3x_2; \tag{11.6.32}$$

or in matrix form,

$$\mathbf{x}' = \begin{pmatrix} 1 & -2 \\ 2 & -3 \end{pmatrix} \mathbf{x}, \qquad \mathbf{x} = \begin{pmatrix} x_1 \\ x_2 \end{pmatrix}. \tag{11.6.33}$$

Assuming the solution $\mathbf{x} = \mathbf{x}_0 e^{\lambda t}$,

$$\begin{pmatrix} 1-\lambda & -2 \\ 2 & -3-\lambda \end{pmatrix} \mathbf{x}_0 = \mathbf{0}. \tag{11.6.34}$$

For nontrivial solutions,

$$\begin{vmatrix} 1-\lambda & -2 \\ 2 & -3-\lambda \end{vmatrix} = (\lambda+1)^2 = 0. \tag{11.6.35}$$

Thus, we have the solution

$$\mathbf{x} = c_1 \begin{pmatrix} 1 \\ 1 \end{pmatrix} e^{-t}. \tag{11.6.36}$$

The interesting aspect of this example is the single solution that the traditional approach yields because we have repeated roots. To find the second solution, we try a solution of the form

$$\mathbf{x} = \begin{pmatrix} a+ct \\ b+dt \end{pmatrix} e^{-t}. \tag{11.6.37}$$

Equation (11.6.37) was guessed based upon our knowledge of solutions to differential equations when the characteristic polynomial has repeated roots. Substituting (11.6.37) into (11.6.33), we find that $c = d = 2c_2$ and $a - b = c_2$. Thus, we have one free parameter, which we will choose to be b, and set it equal to zero. This is permissible because (11.6.37) can be broken into two terms: $b \begin{pmatrix} 1 \\ 1 \end{pmatrix} e^{-t}$ and $c_2 \begin{pmatrix} 1+2t \\ 2t \end{pmatrix} e^{-t}$. The first term may be incorporated into the $c_1 \begin{pmatrix} 1 \\ 1 \end{pmatrix} e^{-t}$ term. Thus, the general solution is

$$\mathbf{x} = c_1 \begin{pmatrix} 1 \\ 1 \end{pmatrix} e^{-t} + c_2 \begin{pmatrix} 1 \\ 0 \end{pmatrix} e^{-t} + 2c_2 \begin{pmatrix} 1 \\ 1 \end{pmatrix} te^{-t}. \tag{11.6.38}$$

- **Example 11.6.3**

Let us solve the system of linear differential equations:

$$x_1' = 2x_1 - 3x_2 \tag{11.6.39}$$

and

$$x_2' = 3x_1 + 2x_2; \tag{11.6.40}$$

or in matrix form,

$$\mathbf{x}' = \begin{pmatrix} 2 & -3 \\ 3 & 2 \end{pmatrix} \mathbf{x}, \qquad \mathbf{x} = \begin{pmatrix} x_1 \\ x_2 \end{pmatrix}. \tag{11.6.41}$$

Assuming the solution $\mathbf{x} = \mathbf{x}_0 e^{\lambda t}$,

$$\begin{pmatrix} 2-\lambda & -3 \\ 3 & 2-\lambda \end{pmatrix} \mathbf{x}_0 = \mathbf{0}. \tag{11.6.42}$$

For nontrivial solutions,

$$\begin{vmatrix} 2-\lambda & -3 \\ 3 & 2-\lambda \end{vmatrix} = (2-\lambda)^2 + 9 = 0 \tag{11.6.43}$$

and $\lambda = 2 \pm 3i$. If $\mathbf{x}_0 = \begin{pmatrix} a \\ b \end{pmatrix}$, then $b = -ai$ if $\lambda = 2 + 3i$ and $b = ai$ if $\lambda = 2 - 3i$. Thus, the general solution is

$$\mathbf{x} = c_1 \begin{pmatrix} 1 \\ -i \end{pmatrix} e^{2t+3it} + c_2 \begin{pmatrix} 1 \\ i \end{pmatrix} e^{2t-3it}, \tag{11.6.44}$$

where c_1 and c_2 are arbitrary complex constants. Using Euler relationships, we can rewrite (11.6.44) as

$$\mathbf{x} = c_3 \begin{bmatrix} \cos(3t) \\ \sin(3t) \end{bmatrix} e^{2t} + c_4 \begin{bmatrix} \sin(3t) \\ -\cos(3t) \end{bmatrix} e^{2t}, \tag{11.6.45}$$

where $c_3 = c_1 + c_2$ and $c_4 = i(c_1 - c_2)$.

Problems

Find the general solution of the following sets of ordinary differential equations using matrix techniques:

1. $x_1' = x_1 + 2x_2$ $\qquad x_2' = 2x_1 + x_2$.
2. $x_1' = x_1 - 4x_2$ $\qquad x_2' = 3x_1 - 6x_2$.

3. $x_1' = x_1 + x_2 \qquad x_2' = 4x_1 + x_2.$

4. $x_1' = x_1 + 5x_2 \qquad x_2' = -2x_1 - 6x_2.$

5. $x_1' = -\frac{3}{2}x_1 - 2x_2 \qquad x_2' = 2x_1 + \frac{5}{2}x_2.$

6. $x_1' = -3x_1 - 2x_2 \qquad x_2' = 2x_1 + x_2.$

7. $x_1' = x_1 - x_2 \qquad x_2' = x_1 + 3x_2.$

8. $x_1' = 3x_1 + 2x_2 \qquad x_2' = -2x_1 - x_2.$

9. $x_1' = -2x_1 - 13x_2 \qquad x_2' = x_1 + 4x_2.$

10. $x_1' = 3x_1 - 2x_2 \qquad x_2' = 5x_1 - 3x_2.$

11. $x_1' = 4x_1 - 2x_2 \qquad x_2' = 25x_1 - 10x_2.$

12. $x_1' = -3x_1 - 4x_2 \qquad x_2' = 2x_1 + x_2.$

13. $x_1' = 3x_1 + 4x_2 \qquad x_2' = -2x_1 - x_2.$

14. $x_1' + 5x_1 + x_2' + 3x_2 = 0 \qquad 2x_1' + x_1 + x_2' + x_2 = 0.$

15. $x_1' - x_1 + x_2' - 2x_2 = 0 \qquad x_1' - 5x_1 + 2x_2' - 7x_2 = 0.$

16. $x_1' = x_1 - 2x_2 \qquad x_2' = 0 \qquad x_3' = -5x_1 + 7x_3.$

17. $x_1' = 2x_1 \qquad x_2' = x_1 + 2x_3 \qquad x_3' = x_3.$

18. $x_1' = 3x_1 - 2x_3 \qquad x_2' = -x_1 + 2x_2 + x_3 \qquad x_3' = 4x_1 - 3x_3.$

19. $x_1' = 3x_1 - x_3 \qquad x_2' = -2x_1 + 2x_2 + x_3 \qquad x_3' = 8x_1 - 3x_3.$

Answers
To the Odd-Numbered Problems

Section 1.1

1. $1+2i$

3. $-2/5$

5. $2+2i\sqrt{3}$

7. $4e^{\pi i}$

9. $5\sqrt{2}e^{3\pi i/4}$

11. $2e^{2\pi i/3}$

Section 1.2

1.
$$\pm\sqrt{2}, \qquad \pm\sqrt{2}\left[\frac{1}{2}+\frac{\sqrt{3}i}{2}\right], \qquad \pm\sqrt{2}\left[-\frac{1}{2}+\frac{\sqrt{3}i}{2}\right]$$

3.
$$i, \qquad -\frac{\sqrt{3}}{2}-\frac{i}{2}, \qquad z_2=\frac{\sqrt{3}}{2}-\frac{i}{2}$$

5.
$$w_1=\frac{1}{\sqrt{2}}\left[-\sqrt{\sqrt{a^2+b^2}+a}+i\sqrt{\sqrt{a^2+b^2}+a}\right], \qquad w_2=-w_1.$$

7. $z_{1,2} = \pm(1+i)$; $z_{3,4} = \pm 2(1-i)$

Section 1.3

1. $u = 2 - y,\ v = x$

3. $u = x^3 - 3xy^2,\ v = 3x^2y - y^3$

5. $f'(z) = 3z(1+z^2)^{1/2}$

7. $f'(z) = 2(1+4i)z - 3$

9. $f'(z) = -3i(iz-1)^{-4}$

11. $1/6$

13. $v(x,y) = 2xy + \text{constant}$

15. $v(x,y) = x\sin(x)e^{-y} + ye^{-y}\cos(x) + \text{constant}.$

Section 1.4

1. 0

3. $2i$

5. $14/15 - i/3$

Section 1.5

1. $(e^{-2} - e^{-4})/2$

3. $\pi/2$

Section 1.6

1. $\pi i/32$

3. $\pi i/2$

5. $-2\pi i$

7. $2\pi i$

9. -6π

Section 1.7

1.
$$\sum_{n=0}^{\infty}(n+1)z^n.$$

3.
$$f(z) = z^{10} - z^9 + \frac{z^8}{2} - \frac{z^7}{6} + \cdots - \frac{1}{11!z} + \cdots$$

We have an essential singularity and the residue equals $-1/11!$

5.
$$f(z) = \frac{1}{2!} + \frac{z^2}{4!} + \frac{z^4}{6!} + \cdots$$

We have a removable singularity where the value of the residue equals zero.

7.

$$f(z) = -\frac{2}{z} - 2 - \frac{7z}{6} - \frac{z^2}{2} - \cdots$$

We have a simple pole and the residue equals -2.

9.

$$f(z) = \frac{1}{2}\frac{1}{z-2} - \frac{1}{4} + \frac{z-2}{8} - \cdots$$

We have a simple pole and the residue equals $1/2$.

Section 1.8

1. $-3\pi i/4$ 3. $-2\pi i.$ 5. $2\pi i$ 7. $2\pi i$

Section 2.1

1.

$$f(t) = \frac{1}{2} - \frac{2}{\pi}\sum_{m=1}^{\infty}\frac{\sin[(2m-1)t]}{2m-1}$$

3.

$$f(t) = -\frac{\pi}{4} + \sum_{n=1}^{\infty}\frac{(-1)^n - 1}{n^2\pi}\cos(nt) + \frac{1-2(-1)^n}{n}\sin(nt)$$

5.

$$f(t) = \frac{4}{\pi^2}\sum_{m=1}^{\infty}\frac{\cos[(2m-1)\pi t]}{(2m-1)^2}$$

7.

$$f(t) = \frac{1}{\pi} + \frac{1}{2}\sin(t) - \frac{2}{\pi}\sum_{m=1}^{\infty}\frac{\cos(2mt)}{4m^2-1}$$

9.

$$f(t) = -\frac{4}{\pi^2}\sum_{m=1}^{\infty}\frac{(-1)^m}{(2m-1)^2}\sin[(2m-1)\pi t]$$

11.

$$f(t) = \frac{a}{2} - \frac{4a}{\pi^2}\sum_{m=1}^{\infty}\frac{1}{(2m-1)^2}\cos\left[\frac{(2m-1)\pi t}{a}\right] - \frac{2a}{\pi}\sum_{n=1}^{\infty}\frac{(-1)^n}{n}\sin\left(\frac{n\pi t}{a}\right)$$

13.

$$f(t) = \frac{L}{2\pi}\sin\left(\frac{\pi t}{L}\right) - \frac{2L}{\pi}\sum_{n=2}^{\infty}\frac{(-1)^n n}{n^2-1}\sin\left(\frac{n\pi t}{L}\right)$$

15.

$$f(t) = \frac{4a\cosh(a\pi/2)}{\pi}\sum_{m=1}^{\infty}\frac{\cos[(2m-1)t]}{a^2+(2m-1)^2}.$$

Section 2.3

1.

$$f(x) = \frac{\pi}{2} - \frac{4}{\pi}\sum_{m=1}^{\infty}\frac{\cos[(2m-1)x]}{(2m-1)^2}$$

$$f(x) = \frac{2}{\pi}\sum_{n=1}^{\infty}\frac{(-1)^{n+1}\sin(nx)}{n}$$

3.

$$f(x) = \frac{1}{4} - \frac{2}{\pi^2}\sum_{m=1}^{\infty}\frac{\cos[2(2m-1)\pi x]}{(2m-1)^2}$$

$$f(x) = \frac{4}{\pi^2}\sum_{m=1}^{\infty}\frac{(-1)^{m+1}\sin[(2m-1)\pi x]}{(2m-1)^2}$$

5.

$$f(x) = \frac{3}{4} + \frac{8}{\pi^2}\sum_{n=1}^{\infty}\frac{1}{n^2}\cos^2\left(\frac{n\pi}{4}\right)\cos\left(\frac{n\pi x}{2}\right)$$

$$f(x) = \frac{4}{\pi^2}\sum_{m=1}^{\infty}\frac{(-1)^{m+1}}{(2m-1)^2}\sin\left[\frac{(2m-1)\pi x}{2}\right] - \frac{2}{\pi}\sum_{n=1}^{\infty}\frac{(-1)^n}{n}\sin\left(\frac{n\pi x}{2}\right)$$

7.

$$f(x) = \frac{3}{4} + \frac{1}{\pi}\sum_{m=1}^{\infty}\frac{(-1)^m}{2m-1}\cos\left[\frac{(2m-1)\pi x}{a}\right]$$

$$f(x) = \frac{1}{\pi}\sum_{n=1}^{\infty}\frac{1+\cos(n\pi/2)-2(-1)^n}{n}\sin\left(\frac{n\pi x}{a}\right)$$

9.

$$f(x) = \frac{3a}{8} + \frac{2a}{\pi^2} \sum_{n=1}^{\infty} \frac{\cos(n\pi/2) - 1}{n^2} \cos\left(\frac{n\pi x}{a}\right)$$

$$f(x) = \frac{a}{\pi} \sum_{n=1}^{\infty} \left[\frac{2}{n^2\pi} \sin\left(\frac{n\pi}{2}\right) - \frac{(-1)^n}{n}\right] \sin\left(\frac{n\pi x}{a}\right)$$

11.

$$f(x) = \frac{1}{2} + \frac{2}{\pi} \sum_{m=1}^{\infty} \frac{(-1)^m}{m} \sin\left(\frac{m\pi}{2}\right) \cos\left(\frac{2m\pi x}{a}\right)$$

$$f(x) = \frac{4}{\pi} \sum_{m=1}^{\infty} \frac{(-1)^{m+1}}{2m-1} \sin\left[\frac{(2m-1)\pi}{4}\right] \sin\left[\frac{(2m-1)\pi x}{a}\right]$$

13.

$$f(x) = \frac{e^{ak} - 1}{ak} + 2ka \sum_{n=1}^{\infty} \frac{(-1)^n e^{ka} - 1}{k^2a^2 + n^2\pi^2} \cos\left(\frac{n\pi x}{a}\right)$$

$$f(x) = -2\pi \sum_{n=1}^{\infty} \frac{n[(-1)^n e^{ka} - 1]}{k^2a^2 + n^2\pi^2} \sin\left(\frac{n\pi x}{a}\right)$$

Section 2.4

1.

$$f(t) = \frac{1}{2} + \frac{2}{\pi} \sum_{n=1}^{\infty} \frac{\sin[(2n-1)t]}{2n-1}$$

$$f(t) = \frac{1}{2} + \frac{2}{\pi} \sum_{n=1}^{\infty} \frac{\cos[(2n-1)t - \pi/2]}{2n-1}$$

3.

$$f(t) = 2 \sum_{n=1}^{\infty} \frac{1}{n} \cos\left[nt + (-1)^n \frac{\pi}{2}\right]$$

$$f(t) = 2 \sum_{n=1}^{\infty} \frac{1}{n} \sin\left\{nt + [1 + (-1)^n]\frac{\pi}{2}\right\}$$

Section 2.5

1.

$$f(t) = \frac{\pi}{2} - \frac{2}{\pi} \sum_{m=-\infty}^{\infty} \frac{e^{i(2m-1)t}}{(2m-1)^2}$$

3.

$$f(t) = 1 + \frac{i}{\pi} \sum_{\substack{n=-\infty \\ n\neq 0}}^{\infty} \frac{e^{n\pi it}}{n}.$$

5.

$$f(t) = \frac{1}{2} - \frac{i}{\pi} \sum_{m=-\infty}^{\infty} \frac{e^{2(2m-1)it}}{2m-1}.$$

Section 2.6

1.

$$y(t) = A\cosh(t) + B\sinh(t) - \frac{1}{2} - \frac{2}{\pi} \sum_{n=1}^{\infty} \frac{\sin[(2n-1)t]}{(2n-1)+(2n-1)^3}$$

3.

$$y(t) = Ae^{2t} + Be^{t} + \frac{1}{4} + \frac{6}{\pi} \sum_{n=1}^{\infty} \frac{\cos[(2n-1)t]}{[2-(2n-1)^2]^2 + 9(2n-1)^2}$$
$$+ \frac{2}{\pi} \sum_{n=1}^{\infty} \frac{[2-(2n-1)^2]\sin[(2n-1)t]}{(2n-1)\{[2-(2n-1)^2]^2 + 9(2n-1)^2\}}$$

5.

$$y_p(t) = \frac{\pi}{8} - \frac{2}{\pi} \sum_{n=-\infty}^{\infty} \frac{e^{i(2n-1)t}}{(2n-1)^2[4-(2n-1)^2]}$$

7.

$$q(t) = \sum_{n=-\infty}^{\infty} \frac{\omega^2 \varphi_n}{(in\omega_0)^2 + 2i\alpha n\omega_0 + \omega^2} e^{in\omega_0 t}$$

Section 2.7

1. $x(t) = \frac{3}{2} - \cos(\pi x/2) - \sin(\pi x/2) - \frac{1}{2}\cos(\pi x)$

Section 3.3

1. $\pi e^{-|\omega/a|}/|a|$

Section 3.4

1. $-t/(1+t^2)^2$

3. $f(t) = \frac{1}{2}e^{-t}H(t) + \frac{1}{2}e^{t}H(-t)$

5. $f(t) = e^{-t}H(t) - e^{-t/2}H(t) + \frac{1}{2}te^{-t/2}H(t)$

7.
$$f(t) = \frac{i}{2}\text{sgn}(t)e^{-a|t|}, \quad \text{where} \quad \text{sgn(t)} = \begin{cases} 1, & t > 0 \\ -1, & t < 0. \end{cases}$$

9.
$$f(t) = \frac{1}{4a}(1 - a|t|)e^{-a|t|}$$

11.
$$f(t) = \frac{(-1)^{n+1}}{(2n+1)!}t^{2n+1}e^{-at}H(t)$$

13.
$$f(t) = \begin{cases} e^{2t}, & t > 0 \\ e^{-t}, & t < 0. \end{cases}$$

Section 3.6

1.
$$y(t) = [(t-1)e^{-t} + e^{-2t}]H(t)$$

3.
$$y(t) = \begin{cases} \frac{1}{9}e^{-t}, & t > 0 \\ \frac{1}{9}e^{2t} - \frac{1}{3}te^{2t}, & t < 0. \end{cases}$$

Section 4.1

1. $F(s) = s/(s^2 - a^2)$

3. $F(s) = 1/s + 2/s^2 + 2/s^3$

5. $F(s) = \left[1 - e^{-2(s-1)}\right]/(s-1)$

7. $F(s) = 2/(s^2+1) - s/(s^2+4) + \cos(3)/s - 1/s^2$

9. $f(t) = e^{-3t}$

11. $f(t) = \frac{1}{3}\sin(3t)$

13. $f(t) = 2\sin(t) - \frac{15}{2}t^2 + 2e^{-t} - 6\cos(2t)$

15. $sF(s) - f(0) = as/(s^2 + a^2) - 0 = \mathcal{L}[f'(t)]$

17. $F(s) = 1/(2s) - sT^2/[2(s^2T^2 + \pi^2)]$

Section 4.2

1. $f(t) = (t-2)H(t-2) - (t-2)H(t-3)$

3. $y'' + 3y' + 2y = H(t-1)$

5. $y'' + 4y' + 4y = tH(t-2)$

7. $y'' - 3y' + 2y = e^{-t}H(t-2)$

9. $y'' + y = \sin(t)[1 - H(t-\pi)]$

Section 4.3

1. $F(s) = 2/(s^2 + 2s + 5)$

3. $F(s) = 1/(s-1)^2 + 3/(s^2 - 2s + 10) + (s-2)/(s^2 - 4s + 29)$

5. $F(s) = 2/(s+1)^3 + 2/(s^2 - 2s + 5) + (s+3)/(s^2 + 6s + 18)$

7. $F(s) = e^6 e^{-3s}/(s-2)$

9. $F(s) = 2e^{-s}/s^3 + 2e^{-s}/s^2 + 3e^{-s}/s + e^{-2s}/s$

11. $F(s) = (1 + e^{-s\pi})/(s^2 + 1)$

13. $F(s) = 4(s+3)/(s^2 + 6s + 13)^2$

15. $f(t) = \frac{1}{2}t^2e^{-2t} - \frac{1}{3}t^3e^{-2t}$

17. $f(t) = e^{-t}\cos(t) + 2e^{-t}\sin(t)$

19. $f(t) = e^{-2t} - 2te^{-2t} + \cos(t)e^{-t} + \sin(t)e^{-t}$

21. $f(t) = e^{t-3}H(t-3)$

23. $f(t) = e^{-(t-1)}[\cos(t-1) - \sin(t-1)]H(t-1)$

25. $f(t) = \cos[2(t-1)]H(t-1) + \frac{1}{6}(t-3)^3e^{2(t-3)}H(t-3)$

27. $f(t) = \{\cos[2(t-1)] + \frac{1}{2}\sin[2(t-1)]\}H(t-1) + \frac{1}{6}(t-3)^3H(t-3)$

29. $f(t) = t[H(t) - H(t-a)]$; $F(s) = 1/s^2 - e^{-as}/s^2 - ae^{-as}/s$

31. $F(s) = 1/s^2 - e^{-s}/s^2 - e^{-2s}/s$

33. $F(s) = e^{-s}/s^2 - e^{-2s}/s^2 - e^{-3s}/s$

35. $Y(s) = s/(s^2+4) + 3e^{-4s}/[s(s^2+4)]$

37. $Y(s) = e^{-(s-1)}/[(s-1)(s+1)(s+2)]$

39. $Y(s) = 5s/[(s-1)(s-2)] + e^{-s}/[s^3(s-1)(s-2)]$
$+2e^{-s}/[s^2(s-1)(s-2)] + e^{-s}/[s(s-1)(s-2)]$

41. $Y(s) = 1/[s^2(s+2)(s+1)] + ae^{-as}/[(s+1)^2(s+2)]$
$-e^{-as}/[s^2(s+1)(s+2)] - e^{-as}/[s(s+1)(s+2)]$

43. $\lim_{s\to\infty} sF(s) = \lim_{s\to\infty} s^2/(s^2+a^2) = 1 = f(0)$.

45. $\lim_{s\to\infty} sF(s) = \lim_{s\to\infty} 3s/(s^2-2s+10) = 0 = f(0)$.

47. Yes

49. No

51. No

Section 4.4

1. $F(s) = \dfrac{1}{s^2+1}\coth\left(\dfrac{s\pi}{2}\right)$

3. $F(s) = \dfrac{1-(1+as)e^{-as}}{s^2(1-e^{-2as})}$

Section 4.5

1. $f(t) = e^{-t} - e^{-2t}$

3. $f(t) = \frac{5}{4}e^{-t} - \frac{6}{5}e^{-2t} - \frac{1}{20}e^{3t}$

5. $f(t) = e^{-2t}\cos\left(t + \frac{3\pi}{2}\right)$

7. $f(t) = 2.3584\cos(4t + 0.5586)$

9. $f(t) = \frac{1}{2} + \frac{\sqrt{2}}{2}\cos\left(2t + \frac{5\pi}{4}\right)$

Section 4.6

1.
$$\mathcal{L}(t) = \frac{1}{s^2} = \mathcal{L}(1)\mathcal{L}(1)$$

3.
$$\mathcal{L}\left(e^t - 1\right) = \frac{1}{s-1} - \frac{1}{s} = \frac{1}{s(s-1)} = \mathcal{L}(1)\mathcal{L}(e^t)$$

5.
$$\mathcal{L}[t - \sin(t)] = \frac{1}{s^2} - \frac{1}{s^2+1} = \frac{1}{s^2(s^2+1)} = \mathcal{L}(t)\mathcal{L}[\sin(t)]$$

7.

$$\mathcal{L}\left\{\frac{t^2}{a} - \frac{2}{a^3}[1 - \cos(at)]\right\} = \frac{2}{s^3}\left(\frac{a}{s^2 + a^2}\right) = \mathcal{L}(t^2)\mathcal{L}[\sin(at)]$$

9.

$$H(t-b) * H(t-a) = \int_a^t H(t-b-x)\,dx = -\int_{t-b-a}^{-b} H(\eta)\,d\eta,$$

if $t > a$ and $\eta = t - b - x$.

11.

$$f(t) = e^t - t - 1$$

13. Assuming that $a, b > 0$,

$$\int_0^t \delta(t-x-a)\delta(x-b)\,dx = \delta(t-b-a)$$

Section 4.7

1. $f(t) = 1 + 2t$

3. $f(t) = t + \frac{1}{2}t^2$

5. $f(t) = t^3 + \frac{1}{20}t^5$

7. $f(t) = t^2 - \frac{1}{3}t^4$

9. $f(t) = 5e^{2t} - 4e^t - 2te^t$

11. $f(t) = (1-t)^2 e^{-t}$

13. $f(t) = 4 + \frac{5}{2}t^2 + \frac{1}{24}t^4$

15. $x(t) = 2A\sqrt{t}/(\pi C) - Bt/(2C)$

17.

$$f(t) = \frac{\alpha}{\beta^2}\left(e^{\beta^2 t} - 1 + \frac{2\beta\sqrt{t}}{\sqrt{\pi}} - \frac{2e^{\beta^2 t}}{\sqrt{\pi}}\int_0^{\beta\sqrt{t}} e^{-u^2}\,du\right)$$

Section 4.8

1. $y(t) = \frac{5}{4}e^{2t} - \frac{1}{4} + \frac{1}{2}t$

3. $y(t) = e^{3t} - e^{2t}$

5. $y(t) = -\frac{3}{4}e^{-3t} + \frac{7}{4}e^{-t} + \frac{1}{2}te^{-t}$

7. $y(t) = \frac{3}{4}e^{-t} + \frac{1}{8}e^t - \frac{7}{8}e^{-3t}$

9. $y(t) = (t-1)H(t-1)$

11. $y(t) = e^{2t} - e^t + \left[\frac{1}{2} + \frac{1}{2}e^{2(t-1)} - e^{t-1}\right]H(t-1)$

13. $y(t) = \left[1 - e^{-2(t-2)} - 2(t-2)e^{-2(t-2)}\right] H(t-2)$

15. $y(t) = \left[\frac{1}{3}e^{2(t-2)} - \frac{1}{2}e^{t-2} + \frac{1}{6}e^{-(t-2)}\right] H(t-2)$

17. $y(t) = 1 - \cos(t) - [1 - \cos(t-T)]\, H(t-T)$

19.
$$\begin{aligned} y(t) &= e^{-t} - \tfrac{1}{4}e^{-2t} - \tfrac{3}{4} + \tfrac{1}{2}t \\ &\quad - \left[e^{-(t-a)} - \tfrac{1}{4}e^{-2(t-a)} - \tfrac{3}{4} + \tfrac{1}{2}(t-a)\right] H(t-a) \\ &\quad + a\left[\tfrac{1}{2}e^{-2(t-a)} + (t-a)e^{-(t-a)} - \tfrac{1}{2}\right] H(t-a). \end{aligned}$$

21. $y(t) = te^t + 3(t-2)e^{t-2}H(t-2)$

23. $y(t) = 3\left[e^{-2(t-2)} - e^{-3(t-2)}\right] H(t-2)$
$\quad + 4\left[e^{-3(t-5)} - e^{-2(t-5)}\right] H(t-5)$

25. $x(t) = 2e^{t/2} - 2 - t;\, y(t) = e^{t/2} - 1 - t$

27. $x(t) = \frac{1}{2}e^{-t} + \frac{3}{2}e^{-3t};\, y(t) = e^{-t} - 1$

Section 4.9

1. $G(s) = 1/(s+k)$ $\quad g(t) = e^{-kt}$

$a(t) = \left(1 - e^{-kt}\right)/k$

3. $G(s) = 1/(s^2 + 4s + 3)$ $\quad g(t) = \frac{1}{2}\left(e^{-t} - e^{-3t}\right)$

$a(t) = \frac{1}{6}e^{-3t} - \frac{1}{2}e^{-t} + \frac{1}{3}$

5. $G(s) = 1/[(s-2)(s-1)]$ $\quad g(t) = e^{2t} - e^t$

$a(t) = \frac{1}{2} + \frac{1}{2}e^{2t} - e^t$

7. $G(s) = 1/(s^2 - 9)$ $\quad g(t) = \frac{1}{6}\left(e^{3t} - e^{-3t}\right)$

$a(t) = \frac{1}{18}\left(e^{3t} + e^{-3t} - 2\right)$

9. $G(s) = 1/[s(s-1)]$ $\quad g(t) = e^t - 1$

$a(t) = e^t - t - 1$

Section 4.10

1. $f(t) = (2-t)e^{-2t} - 2e^{-3t}$

3. $f(t) = \left(\frac{1}{4}t^2 - \frac{1}{4}t + \frac{1}{8}\right)e^{2t} - \frac{1}{8}$

5. $f(t) = \left[\frac{1}{2}(t-1) - \frac{1}{4} + \frac{1}{4}e^{-2(t-1)}\right] H(t-1)$

7.

$$f(t) = \frac{e^{-bt}}{\cosh(ab)} - 8ab\sum_{n=1}^{\infty}(-1)^n \frac{\sin[(2n-1)\pi t/(2a)]}{4a^2b^2 + (2n-1)^2\pi^2} + 4\sum_{n=1}^{\infty}(-1)^n \frac{(2n-1)\pi\cos[(2n-1)\pi t/(2a)]}{4a^2b^2 + (2n-1)^2\pi^2}.$$

Section 5.1

1. $F(z) = 2z/(2z-1)$ if $|z| > 1/2$.

3. $F(z) = (z^6 - 1)/(z^6 - z^5)$ if $|z| > 0$.

5. $F(z) = (a^2 + a - z)/[z(z-a)]$ if $|z| > a$.

Section 5.2

1. $F(z) = zTe^{aT}/(ze^{aT} - 1)^2$ 3. $F(z) = z(z+a)/(z-a)^3$

5. $F(z) = [z - \cos(1)]/\{z[z^2 - 2z\cos(1) + 1]\}$

7. $F(z) = z[z\sin(\theta) + \sin(\omega_0 T - \theta)]/[z^2 - 2z\cos(\omega_0 T) + 1]$

9. $F(z) = z/(z+1)$ 11. $f_n * g_n = n + 1$ 13. $f_n * g_n = 2^n/n!$

Section 5.3

1. $f_0 = 0.007143$, $f_1 = 0.08503$, $f_2 = 0.1626$, $f_3 = 0.2328$

3. $f_0 = 0.09836$, $f_1 = 0.3345$, $f_2 = 0.6099$, $f_3 = 0.7935$

5. $f_n = 8 - 8\left(\frac{1}{2}\right)^n - 6n\left(\frac{1}{2}\right)^n$ 7. $f_n = (1 - \alpha^{n+1})/(1-\alpha)$

9. $f_n = \left(\frac{1}{2}\right)^{n-10} H_{n-10} + \left(\frac{1}{2}\right)^{n-11} H_{n-11}$

11. $f_n = \frac{1}{9}(6n-4)(-1)^n + \frac{4}{9}\left(\frac{1}{2}\right)^n$ 13. $f_n = a^n/(n!)$

Section 5.4

1. $y_n = 1 + \frac{1}{6}n(n-1)(2n-1)$
3. $y_n = \frac{1}{2}n(n-1)$
5. $y_n = \frac{1}{6}\left[5^n - (-1)^n\right]$
7. $y_n = (2n-1)\left(\frac{1}{2}\right)^n + \left(-\frac{1}{2}\right)^n$
9. $y_n = 2^n - n - 1.$
11. $x_n = 2 + (-1)^n; y_n = 1 + (-1)^n$
13. $x_n = 1 - 2(-6)^n; y_n = -7(-6)^n$

Section 5.5

1. marginally stable
3. unstable

Section 6.1

1. $\lambda_n = (2n-1)^2\pi^2/(4L^2)$ with $y_n(x) = \cos\left[(2n-1)\pi x/(2L)\right]$

3. $\lambda_0 = -1$, $y_0(x) = e^{-x}$ and $\lambda_n = n^2$, $y_n(x) = \sin(nx) - n\cos(nx)$

5. $\lambda_n = -n^4\pi^4/L^4$, $y_n(x) = \sin(n\pi x/L)$

7. $\lambda_n = k_n^2$, $y_n(x) = \sin(k_n x)$ with $k_n = -\tan(k_n)$

9. $\lambda_0 = -m_0^2$, $y_0(x) = \sinh(m_0 x) - m_0\cosh(m_0 x)$ with $\coth(m_0\pi) = m_0$ and $\lambda_n = k_n^2$, $y_n(x) = \sin(k_n x) - k_n\cos(k_n x)$ with $k_n = -\cot(k_n\pi)$

11.

(a) $\lambda_n = n^2\pi^2, \quad y_n(x) = \sin[n\pi\ln(x)]$

(b) $\lambda_n = (2n-1)^2\pi^2/4, \quad y_n(x) = \sin[(2n-1)\pi\ln(x)/2]$

(c) $\lambda_0 = 0, \quad y_0(x) = 1; \qquad \lambda_n = n^2\pi^2, \quad y_n(x) = \cos[n\pi\ln(x)]$

13. $\lambda_n = n^2 + 1$, $y_n(x) = x^{-1}\sin[n\ln(x)]$

Section 6.3

1.
$$f(x) = \frac{2}{\pi}\sum_{n=1}^{\infty}\frac{(-1)^{n+1}}{n}\sin\left(\frac{n\pi x}{L}\right)$$

3.
$$f(x) = \frac{8L}{\pi^2}\sum_{n=1}^{\infty}\frac{(-1)^{n+1}}{(2n-1)^2}\sin\left[\frac{(2n-1)\pi x}{2L}\right]$$

Section 6.4

1.
$$f(x) = \tfrac{1}{4}P_0(x) + \tfrac{1}{2}P_1(x) + \tfrac{5}{16}P_2(x) + \cdots$$

3.
$$f(x) = \tfrac{1}{2}P_0(x) + \tfrac{5}{8}P_2(x) - \tfrac{3}{16}P_4(x) + \cdots$$

5.
$$f(x) = \tfrac{3}{2}P_1(x) - \tfrac{7}{8}P_3(x) + \tfrac{11}{16}P_5(x) + \cdots$$

Section 7.3

1.
$$u(x,t) = \frac{4L}{c\pi^2}\sum_{m=1}^{\infty}\frac{1}{(2m-1)^2}\sin\left[\frac{(2m-1)\pi x}{L}\right]\sin\left[\frac{(2m-1)\pi ct}{L}\right]$$

3.
$$u(x,t) = \frac{9h}{\pi^2}\sum_{n=1}^{\infty}\frac{1}{n^2}\sin\left(\frac{2n\pi}{3}\right)\sin\left(\frac{n\pi x}{L}\right)\cos\left(\frac{n\pi ct}{L}\right)$$

5.
$$u(x,t) = \sin\left(\frac{\pi x}{L}\right)\sin\left(\frac{\pi ct}{L}\right) + \frac{4aL}{\pi^2 c}\sum_{n=1}^{\infty}\frac{(-1)^{n+1}}{(2n-1)^2}\sin\left[\frac{(2n-1)\pi}{4}\right]\sin\left[\frac{(2n-1)\pi x}{L}\right] \times \sin\left[\frac{(2n-1)\pi ct}{L}\right]$$

7.
$$u(x,t) = \frac{4L}{\pi^2}\sum_{n=1}^{\infty}\frac{(-1)^{n+1}}{(2n-1)^2}\sin\left[\frac{(2n-1)\pi x}{L}\right]\cos\left[\frac{(2n-1)\pi ct}{L}\right]$$

9.
$$u(x,t) = \frac{8L}{\pi^2}e^{-ht}\sum_{n=1}^{\infty}\frac{(-1)^{n+1}}{(2n-1)^2}\sin\left[\frac{(2n-1)\pi x}{2L}\right]\left\{\cos\left[t\sqrt{\lambda_n c^2 - h^2}\right] + h\sin\left[t\sqrt{\lambda_n c^2 - h^2}\right]\Big/\sqrt{\lambda_n c^2 - h^2}\right\},$$

where $\lambda_n = (2n-1)^2\pi^2/4L^2$.

Section 7.4

1.

$$u(x,t) = \sin(2x)\cos(2ct) + \cos(x)\sin(ct)/c$$

3.

$$u(x,t) = \frac{1+x^2+c^2t^2}{(1+x^2+c^2t^2)^2+4x^2c^2t^2} + \frac{e^x\sinh(ct)}{c}$$

5.

$$u(x,t) = \cos\left(\frac{\pi x}{2}\right)\cos\left(\frac{\pi ct}{2}\right) + \frac{\sinh(ax)\sinh(act)}{ac}$$

Section 7.5

1.

$$u(x,t) = \frac{4}{\pi^2}\sum_{m=1}^{\infty}\frac{\sin[(2m-1)\pi x]\sin[(2m-1)\pi t]}{(2m-1)^2}$$

3.

$$u(x,t) = \sin(\pi x)\cos(\pi t) - \sin(\pi x)\sin(\pi t)/\pi$$

5.

$$u(x,t) = xt - te^{-x} + \sinh(t)e^{-x} + \left[1 - e^{-(t-x)} + t - x - \sinh(t-x)\right]H(t-x)$$

7.

$$u(x,t) = \frac{gx}{\omega^2} - \frac{2g\omega^2}{L}\sum_{n=1}^{\infty}\frac{\sin(\lambda_n x)\cos(\lambda_n t)}{\lambda_n^2(\omega^4+\omega^2/L+\lambda_n^2)\sin(\lambda_n L)},$$

where λ_n is the nth root of $\lambda = \omega^2\cot(\lambda L)$.

9.

$$u(x,t) = E - \frac{4E}{\pi}\sum_{n=1}^{\infty}\frac{1}{2n-1}\sin\left[\frac{(2n-1)\pi x}{2\ell}\right]\cos\left[\frac{(2n-1)c\pi t}{2\ell}\right]$$

or

$$u(x,t) = E\sum_{n=0}^{\infty}(-1)^n H\left(t - \frac{x+2n\ell}{c}\right) + E\sum_{n=0}^{\infty}(-1)^n H\left\{t - \frac{[(2n+2)\ell - x]}{c}\right\}$$

11.

$$p(x,t) = p_0 - \frac{4\rho u_0 c}{\pi}\sum_{n=1}^{\infty}\frac{(-1)^n}{2n-1}\sin\left[\frac{(2n-1)\pi x}{2L}\right]\sin\left[\frac{(2n-1)c\pi t}{2L}\right]$$

13.

$$u(x,t) = \frac{gt^2}{2} - \frac{gL^2}{6c^2} - \frac{2gL^2}{c^2\pi^2}\sum_{n=1}^{\infty}\frac{(-1)^n}{n^2}\cos\left(\frac{n\pi x}{L}\right)\cos\left(\frac{n\pi ct}{L}\right)$$

Section 8.3

1.

$$u(x,t) = \frac{4A}{\pi}\sum_{m=1}^{\infty}\frac{\sin[(2m-1)x]}{2m-1}e^{-a^2(2m-1)^2t}$$

3.

$$u(x,t) = -2\sum_{n=1}^{\infty}\frac{(-1)^n}{n}\sin(nx)e^{-a^2n^2t}$$

5.

$$u(x,t) = \frac{4}{\pi}\sum_{m=1}^{\infty}\frac{(-1)^{m+1}}{(2m-1)^2}\sin[(2m-1)x]e^{-a^2(2m-1)^2t}$$

7.

$$u(x,t) = \frac{\pi}{2} - \frac{4}{\pi}\sum_{m=1}^{\infty}\frac{\cos[(2m-1)x]}{(2m-1)^2}e^{-a^2(2m-1)^2t}$$

9.

$$u(x,t) = \frac{\pi}{2} - \frac{4}{\pi}\sum_{n=1}^{\infty}\frac{\cos[(2n-1)x]}{(2n-1)^2}e^{-a^2(2n-1)^2t}$$

11.

$$u(x,t) = \frac{32}{\pi}\sum_{n=1}^{\infty}\frac{(-1)^n}{(2n-1)^3}\cos\left[\frac{(2n-1)x}{2}\right]e^{-a^2(2n-1)^2t/4}$$

13.

$$u(x,t) = \frac{4}{\pi}\sum_{n=1}^{\infty}\frac{\sin[(2n-1)x/2]}{2n-1}e^{-a^2(2n-1)^2t/4}$$

15.

$$u(x,t) = \sum_{n=1}^{\infty}\left[\frac{4}{2n-1} - \frac{8(-1)^{n+1}}{(2n-1)^2\pi}\right]\sin\left[\frac{(2n-1)x}{2}\right]e^{-a^2(2n-1)^2t/4}$$

17.

$$u(x,t) = \frac{T_0 x}{\pi} + \frac{2T_0}{\pi}\sum_{n=1}^{\infty}\frac{1}{n}\sin(nx)e^{-a^2n^2t}$$

19.

$$u(x,t) = h_1 + \frac{(h_2-h_1)x}{L} + \frac{2(h_2-h_1)}{\pi}\sum_{n=1}^{\infty}\frac{(-1)^n}{n}\sin\left(\frac{n\pi x}{L}\right)\exp\left(-\frac{a^2n^2\pi^2}{L^2}t\right)$$

21.

$$u(x,t) = h_0 - \frac{4h_0}{\pi}\sum_{n=1}^{\infty}\frac{1}{2n-1}\sin\left[\frac{(2n-1)\pi x}{L}\right]\exp\left[-\frac{(2n-1)^2\pi^2a^2t}{L^2}\right]$$

23.

$$u(x,t) = \frac{1}{3} - t - \frac{2}{\pi^2}\sum_{n=1}^{\infty}\frac{(-1)^n}{n^2}\cos(n\pi x)e^{-a^2n^2\pi^2t}$$

25.

$$u(x,t) = \frac{4}{\pi}\sum_{n=1}^{\infty}\frac{(-1)^{n+1}}{(2n-1)^4}\sin[(2n-1)x]\left[1 - e^{-(2n-1)^2t}\right]$$

27.

$$u(x,t) = \frac{A_0(L^2 - x^2)}{2\kappa} + \frac{A_0 L}{h} - \frac{2L^2 A_0}{\kappa} \sum_{n=1}^{\infty} \frac{\sin(\beta_n)}{\beta_n^4[1 + \kappa \sin^2(\kappa)/hL]} \cos\left(\frac{\beta_n x}{L}\right) \exp\left(-\frac{a^2 \beta_n^2 t}{L^2}\right),$$

where β_n is the nth root of $\beta \tan(\beta) = \kappa/hL$.

29.

$$u(r,t) = \frac{2}{\pi r} \sum_{n=1}^{\infty} \frac{(-1)^{n+1}}{n} \sin(n\pi r) e^{-a^2 n^2 \pi^2 t}$$

31.

$$u(r,t) = \frac{G}{4\rho\nu}(b^2 - r^2) - \frac{2Gb^2}{\rho\nu} \sum_{n=1}^{\infty} \frac{J_0(k_n r/b)}{k_n^3 J_1(k_n)} \exp\left(-\frac{\nu k_n^2}{b^2} t\right),$$

where k_n is the nth root of $J_0(k) = 0$.

Section 8.4

1.

$$u(x,t) = T_0\left(1 - e^{-a^2 t}\right)$$

3.

$$u(x,t) = x + \frac{2}{\pi} \sum_{n=1}^{\infty} \frac{(-1)^n}{n} \sin(n\pi x) e^{-n^2 \pi^2 t}$$

5.

$$u(x,t) = \frac{x(1-x)}{2} - \frac{4}{\pi^3} \sum_{m=1}^{\infty} \frac{\sin[(2m-1)\pi x]}{(2m-1)^3} e^{-(2m-1)^2 \pi^2 t}$$

7.

$$u(x,t) = x \operatorname{erfc}\left(\frac{x}{2\sqrt{t}}\right) - 2\sqrt{\frac{t}{\pi}} \exp\left(-\frac{x^2}{4t}\right)$$

9.

$$u(x,t) = \frac{u_0}{2} e^{-\delta x} \operatorname{erfc}\left(\frac{x}{2a\sqrt{t}} + \frac{a(1-\delta)\sqrt{t}}{2}\right) + \frac{u_0}{2} e^{-x} \operatorname{erfc}\left(\frac{x}{2a\sqrt{t}} - \frac{a(1-\delta)\sqrt{t}}{2}\right)$$

11.

$$u(x,t) = \frac{t(L-x)}{L} + \frac{Px(x-L)}{2a^2} - \frac{x(x-L)(x-2L)}{6a^2L}$$
$$- \frac{2PL^2}{a^2\pi^3} \sum_{n=1}^{\infty} \frac{(-1)^n}{n^3} \sin\left(\frac{n\pi x}{L}\right) \exp\left(-\frac{a^2n^2\pi^2 t}{L^2}\right)$$
$$+ \frac{2(P+1)L^2}{a^2\pi^3} \sum_{n=1}^{\infty} \frac{1}{n^3} \sin\left(\frac{n\pi x}{L}\right) \exp\left(-\frac{a^2n^2\pi^2 t}{L^2}\right).$$

13.

$$u(r,t) = \frac{r^2}{2} + 3t - \frac{3}{10} - \frac{2}{r} \sum_{n=1}^{\infty} \frac{\sin(\lambda_n r)}{\lambda_n^2 \sin(\lambda_n)} e^{-\lambda_n^2 t},$$

where $\tan(\lambda_n) = \lambda_n$.

15.

$$y(t) = \frac{4\mu A\omega^2}{mL} \sum_{n=1}^{\infty} \frac{\lambda_n e^{\lambda_n t}}{\lambda_n^4 - (\frac{2\mu}{mL})(1 + \frac{2\mu L}{m\nu})\lambda_n^3 + 2\omega^2\lambda_n^2 + \frac{6\omega^2\mu}{mL}\lambda_n + \omega^4},$$

where λ_n is the nth root of $\lambda^2 + 2\mu\lambda^{3/2}\coth(L\sqrt{\lambda/\nu})/(m\sqrt{\nu}) + \omega^2 = 0$.

17.

$$u(x,t) = 1 - 2e^{Vx/2 - V^2t/4}$$
$$\times \sum_{n=1}^{\infty} \frac{\lambda_n\{(V/2)\sin[\lambda_n(1-x)] + \lambda_n\cos[\lambda_n(1-x)]\}e^{-\lambda_n^2 t}}{(\lambda_n^2 + V^2/4)[\lambda_n\sin(\lambda_n) - (1+V/2)\cos(\lambda_n)]},$$

where λ_n is the nth root of $\lambda\cot(\lambda) = -V/2$.

19.

$$u(r,t) = \frac{a^2 - r^2}{4} - 2a^2 \sum_{n=1}^{\infty} \frac{J_0(k_n r/a)}{k_n^3 J_1(k_n)} e^{-k_n^2 t/a^2},$$

where k_n is the nth root of $J_0(k) = 0$.

Section 8.5

1.

$$u(x,t) = \tfrac{1}{2}\mathrm{erf}\left(\frac{b-x}{\sqrt{4a^2t}}\right) + \tfrac{1}{2}\mathrm{erf}\left(\frac{b+x}{\sqrt{4a^2t}}\right)$$

3.

$$u(x,t) = \tfrac{1}{2}T_0 \operatorname{erf}\left(\frac{b-x}{\sqrt{4a^2t}}\right) + \tfrac{1}{2}T_0 \operatorname{erf}\left(\frac{x}{\sqrt{4a^2t}}\right).$$

Section 8.6

1.

$$u(x,t) = \frac{4a^2\pi}{L^2} \sum_{m=1}^{\infty} (2m-1) \sin\left[\frac{(2m-1)\pi x}{L}\right] e^{-a^2(2m-1)^2\pi^2 t/L^2} \times \int_0^t f(\tau) e^{a^2(2m-1)^2\pi^2\tau/L^2}\, d\tau$$

3.

$$u(x,t) = \frac{2}{\sqrt{\pi}} \int_{x/\sqrt{4\nu t}}^{\infty} V\left(t - \frac{x^2}{4\nu\eta^2}\right) e^{-\eta^2}\, d\eta$$

Section 9.3

1.

$$u(x,y) = \frac{4}{\pi} \sum_{m=1}^{\infty} \frac{\sinh[(2m-1)\pi(a-x)/b] \sin[(2m-1)\pi y/b]}{(2m-1)\sinh[(2m-1)\pi a/b]}.$$

3.

$$u(x,y) = -\frac{2a}{\pi} \sum_{n=1}^{\infty} \frac{\sinh(n\pi y/a)\sin(n\pi x/a)}{n\ \sinh(n\pi b/a)}$$

5.

$$u(x,y) = \frac{4}{\pi} \sum_{n=1}^{\infty} (-1)^{n+1} \frac{\sinh[(2n-1)\pi y/2a] \cos[(2n-1)\pi x/2a]}{(2n-1)\sinh[(2n-1)\pi b/2a]}$$

7.

$$u(x,y) = 1$$

9.

$$u(x,y) = 1 - \frac{4}{\pi} \sum_{m=1}^{\infty} \frac{\cosh[(2m-1)\pi y/a] \sin[(2m-1)\pi x/a]}{(2m-1)\cosh[(2m-1)\pi b/a]}$$

11.
$$u(x, y) = 1$$

13.
$$u(x, y) = T_0 + \Delta T \cos(2\pi x/\lambda)e^{-2\pi y/\lambda}$$

15.
$$u(r, z) = 2a \sum_{n=1}^{\infty} \frac{\sinh(k_n z/a)\, J_0(k_n r/a)}{k_n^2 \cosh(k_n L/a) J_1(k_n)},$$

where k_n is the nth root of $J_0(k) = 0$.

17.
$$u(r, z) = \frac{2}{b^2 - a^2} \sum_{n=1}^{\infty} \frac{[bJ_1(k_n b) - aJ_1(k_n a)]J_0(k_n r)\cosh(k_n z)}{k_n \cosh(k_n d) J_0^2(k_n)},$$

where k_n is the nth root of $J_1(k) = 0$.

19.
$$u(r, z) = -\frac{2}{\pi} \sum_{n=1}^{\infty} \frac{(-1)^n I_1(n\pi r)\sin(n\pi z)}{n\, I_1(n\pi a)}$$

21.
$$u(r, z) = 2B \sum_{n=1}^{\infty} \frac{\exp[z(1 - \sqrt{1 + 4k_n^2}\,)/2] J_0\,(k_n r)}{(k_n^2 + B^2) J_0(k_n)},$$

where k_n is the nth root of $k\, J_1(k) = B\, J_0(k)$.

23.
$$u(r, \theta) = 50 \sum_{m=1}^{\infty} [P_{2m-2}(0) - P_{2m}(0)] \left(\frac{r}{a}\right)^{2m-1} P_{2m-1}[\cos(\theta)]$$

25. T_0

Section 9.4

1.
$$u(x, y) = \frac{1}{\pi}\left[\tan^{-1}\left(\frac{1 - x}{y}\right) + \tan^{-1}\left(\frac{x}{y}\right)\right]$$

3.

$$u(x,y) = \frac{T_0}{\pi}\left[\frac{\pi}{2} - \tan^{-1}\left(\frac{x}{y}\right)\right]$$

5.

$$\begin{aligned} u(x,y) = &\frac{T_0}{\pi}\left[\tan^{-1}\left(\frac{1-x}{y}\right) + \tan^{-1}\left(\frac{1+x}{y}\right)\right] \\ &+ \frac{T_1 - T_0}{2\pi}\, y \ln\left[\frac{(x-1)^2 + y^2}{x^2 + y^2}\right] \\ &+ \frac{T_1 - T_0}{\pi}\, x \left[\tan^{-1}\left(\frac{1-x}{y}\right) + \tan^{-1}\left(\frac{x}{y}\right)\right] \end{aligned}$$

Section 9.5

1.

$$\begin{aligned} u(x,y) = \frac{64\,R}{\pi^4\,T} &\sum_{n=1}^{\infty}\sum_{m=1}^{\infty} \frac{(-1)^{n+1}(-1)^{m+1}}{(2n-1)(2m-1)} \\ &\times \frac{\cos[(2n-1)\pi x/2a]\cos[(2m-1)\pi y/b]}{(2n-1)(2m-1)[(2n-1)^2/a^2 + (2m-1)^2/b^2]} \end{aligned}$$

Section 9.6

1.

$$u(x,y) = \frac{4}{\pi}\sum_{m=1}^{\infty}\frac{1}{2m-1}\exp\left[-\frac{(2m-1)\pi x}{a}\right]\sin\left[\frac{(2m-1)\pi y}{a}\right]$$

Section 10.1

1. $\mathbf{a} \times \mathbf{b} = -3\mathbf{i} + 19\mathbf{j} + 10\mathbf{k}$

3. $\mathbf{a} \times \mathbf{b} = \mathbf{i} - 8\mathbf{j} + 7\mathbf{k}$

5. $\mathbf{a} \times \mathbf{b} = -3\mathbf{i} - 2\mathbf{j} - 5\mathbf{k}$

7.

$$\begin{aligned} \mathbf{a} \times (\mathbf{b} \times \mathbf{c}) + \mathbf{b} \times (\mathbf{c} \times \mathbf{a}) + \mathbf{c} \times (\mathbf{a} \times \mathbf{b}) &= (\mathbf{a}\cdot\mathbf{c})\mathbf{b} - (\mathbf{a}\cdot\mathbf{b})\mathbf{c} + (\mathbf{b}\cdot\mathbf{a})\mathbf{c} \\ &\quad - (\mathbf{b}\cdot\mathbf{c})\mathbf{a} + (\mathbf{c}\cdot\mathbf{b})\mathbf{a} - (\mathbf{c}\cdot\mathbf{a})\mathbf{b} \\ &= 0 \end{aligned}$$

9.

$$\nabla f = y\cos(yz)\mathbf{i} + [x\cos(yz) - xyz\sin(yz)]\mathbf{j} - xy^2\sin(yz)\mathbf{k}$$

11.

$$\nabla f = 2xy^2(2z+1)^2\mathbf{i} + 2x^2y(2z+1)^2\mathbf{j} + 4x^2y^2(2z+1)\mathbf{k}$$

13. Plane parallel to the xy plane at height of $z = 3$, $\mathbf{n} = \mathbf{k}$

15. Paraboloid,

$$\mathbf{n} = -\frac{2x}{\sqrt{1+4x^2+4y^2}}\mathbf{i} - \frac{2y}{\sqrt{1+4x^2+4y^2}}\mathbf{j} + \frac{1}{\sqrt{1+4x^2+4y^2}}\mathbf{k}$$

17. A plane, $\mathbf{n} = \mathbf{j}/\sqrt{2} - \mathbf{k}/\sqrt{2}$

19. A parabola of infinite extent along the y-axis, $\mathbf{n} = -2x\mathbf{i}/\sqrt{1+4x^2} + \mathbf{k}/\sqrt{1+4x^2}$

21. $y = 2/(x+1); \quad z = \exp[(y-1)/y]$

23. $y = x; \quad z^2 = y/(3y-2)$

Section 10.2

1.

$$\nabla \cdot \mathbf{F} = 2xz + z^2$$

$$\nabla \times \mathbf{F} = (2xy - 2yz)\mathbf{i} + (x^2 - y^2)\mathbf{j}$$

$$\nabla(\nabla \cdot \mathbf{F}) = 2z\mathbf{i} + (2x + 2z)\mathbf{k}$$

3.

$$\nabla \cdot \mathbf{F} = 2(x-y) - xe^{-xy} + xe^{2y}$$

$$\nabla \times \mathbf{F} = 2xze^{2y}\mathbf{i} - ze^{2y}\mathbf{j} + \left[2(x-y) - ye^{-xy}\right]\mathbf{k}$$

$$\nabla(\nabla \cdot \mathbf{F}) = \left(2 - e^{-xy} + xye^{-xy} + e^{2y}\right)\mathbf{i} + \left(x^2e^{-xy} + 2xe^{2y} - 2\right)\mathbf{j}$$

5.

$$\nabla \cdot \mathbf{F} = 0$$

$$\nabla \times \mathbf{F} = -x^2\mathbf{i} + (5y - 9x^2)\mathbf{j} + (2xz - 5z)\mathbf{k}$$

$$\nabla(\nabla \cdot \mathbf{F}) = \mathbf{0}$$

7.

$$\nabla \cdot \mathbf{F} = e^{-y} + z^2 - 3e^{-z}$$

$$\nabla \times \mathbf{F} = -2yz\mathbf{i} + xe^{-y}\mathbf{k}$$

$$\nabla(\nabla \cdot \mathbf{F}) = -e^{-y}\mathbf{j} + (2z + 3e^{-z})\mathbf{k}$$

9.

$$\nabla \cdot \mathbf{F} = yz + x^3ze^z + xye^z$$

$$\nabla \times \mathbf{F} = (xe^z - x^3ye^z - x^3yze^z)\mathbf{i} + (xy - ye^z)\mathbf{j} + (3x^2yze^z - xz)\mathbf{k}$$

$$\nabla(\nabla \cdot \mathbf{F}) = \left(3x^2ze^z + ye^z\right)\mathbf{i} + \left(z + xe^z\right)\mathbf{j} + \left(y + x^3e^z + x^3ze^z + xye^z\right)\mathbf{k}$$

11.

$$\nabla \cdot \mathbf{F} = y^2 + xz^2 - xy\sin(z)$$

$$\nabla \times \mathbf{F} = [x\cos(z) - 2xyz]\mathbf{i} - y\cos(z)\mathbf{j} + (yz^2 - 2xy)\mathbf{k}$$

$$\nabla(\nabla \cdot \mathbf{F}) = [z^2 - y\sin(z)]\mathbf{i} + \left[2y - x\sin(z)\right]\mathbf{j} + \left[2xz - xy\cos(z)\right]\mathbf{k}$$

13.

$$\nabla \cdot \mathbf{F} = y^2 + xz - xy\sin(z)$$

$$\nabla \times \mathbf{F} = [x\cos(z) - xy]\mathbf{i} - y\cos(z)\mathbf{j} + (yz - 2xy)\mathbf{k}$$

$$\nabla(\nabla \cdot \mathbf{F}) = [z - y\sin(z)]\mathbf{i} + \left[2y - x\sin(z)\right]\mathbf{j} + \left[x - xy\cos(z)\right]\mathbf{k}$$

Section 10.3

1. $16/7 + 2/(3\pi)$ 3. $e^2 + 2e^8/3 + e^{64}/2 - 13/6$ 5. -4π

7. 0 9. 2π

Section 10.4

1. $\varphi(x, y, z) = x^2y + y^2z + 4z + \text{constant}$

3. $\varphi(x, y, z) = xyz + \text{constant}$

5. $\varphi(x, y, z) = x^2\sin(y) + xe^{3z} + 4z + \text{constant}$

7. $\varphi(x, y, z) = xe^{2z} + y^3 + \text{constant}$

9. $\varphi(x, y, z) = xy + xz + \text{constant}$

Section 10.5

1. $1/2$ 3. 0 5. $27/2$

7. 5 9. 0 11. $40/3$

13. $86/3$ 15. 96π

Section 10.6

1. -5 3. 1 5. 0

7. 0 9. -16π 11. -2

Section 10.7

1. -10 3. 2 5. π 7. $45/2$

Section 10.8

1. 3 3. -16 5. 4π 7. $5/12$

Section 11.1

1.
$$A + B = \begin{pmatrix} 4 & 5 \\ 3 & 4 \end{pmatrix} = B + A$$

3.
$$3A - 2B = \begin{pmatrix} 7 & 10 \\ -1 & 2 \end{pmatrix}, \qquad 3(2A - B) = \begin{pmatrix} 15 & 21 \\ 0 & 6 \end{pmatrix}$$

5.
$$(A + B)^T = \begin{pmatrix} 4 & 3 \\ 5 & 4 \end{pmatrix}, \qquad A^T + B^T = \begin{pmatrix} 4 & 3 \\ 5 & 4 \end{pmatrix}$$

7.
$$AB = \begin{pmatrix} 11 & 11 \\ 5 & 5 \end{pmatrix}, \qquad A^T B = \begin{pmatrix} 5 & 5 \\ 8 & 8 \end{pmatrix}$$
$$BA = \begin{pmatrix} 4 & 6 \\ 8 & 12 \end{pmatrix}, \qquad B^T A = \begin{pmatrix} 5 & 8 \\ 5 & 8 \end{pmatrix}$$

9.
$$BB^T = \begin{pmatrix} 2 & 4 \\ 4 & 8 \end{pmatrix}, \qquad B^T B = \begin{pmatrix} 5 & 5 \\ 5 & 5 \end{pmatrix}$$

11.

$$A^3 + 2A = \begin{pmatrix} 65 & 100 \\ 25 & 40 \end{pmatrix}$$

13. yes $\begin{pmatrix} 27 & 11 \\ 2 & 5 \end{pmatrix}$

15. yes $\begin{pmatrix} 11 & 8 \\ 8 & 4 \\ 5 & 3 \end{pmatrix}$

17. no

19.

$$5(2A) = \begin{pmatrix} 10 & 10 \\ 10 & 20 \\ 30 & 10 \end{pmatrix} = 10A$$

21.

$$(A + B) + C = \begin{pmatrix} 4 & 0 \\ 8 & 2 \end{pmatrix} = A + (B + C)$$

23.

$$A(B + C) = \begin{pmatrix} 9 & -1 \\ 11 & -2 \end{pmatrix} = AB + AC$$

25.

$$\begin{pmatrix} 3 & -1 \\ -5 & 2 \end{pmatrix} \begin{pmatrix} 2 & 1 \\ 5 & 3 \end{pmatrix} = \begin{pmatrix} 1 & 0 \\ 0 & 1 \end{pmatrix}$$

27.

$$\begin{pmatrix} 1 & -2 \\ 3 & 1 \end{pmatrix} \begin{pmatrix} x_1 \\ x_2 \end{pmatrix} = \begin{pmatrix} 5 \\ 1 \end{pmatrix}$$

29.

$$\begin{pmatrix} 0 & 1 & 2 & 3 \\ 3 & 0 & -4 & -4 \\ 1 & 1 & 1 & 1 \\ 2 & -3 & 1 & -3 \end{pmatrix} \begin{pmatrix} x_1 \\ x_2 \\ x_3 \\ x_4 \end{pmatrix} = \begin{pmatrix} 2 \\ 5 \\ -3 \\ 7 \end{pmatrix}$$

Section 11.2

1. 7

3. 1

5. −24

7. 3

Section 11.3

1. $x_1 = \frac{9}{5}$, $x_2 = \frac{3}{5}$

3. $x_1 = 0$, $x_2 = 0$, $x_3 = -2$

Section 11.4

1. $x_2 = 2,\ x_1 = 1$

3. $x_3 = \alpha,\ x_2 = -\alpha,\ x_1 = \alpha$

5. $x_3 = \alpha,\ x_2 = 2\alpha,\ x_1 = -1$

7. $x_3 = 2.2,\ x_2 = 2.6,\ x_1 = 1$

9. $A^{-1} = \begin{pmatrix} -1/13 & 5/13 \\ 2/13 & 3/13 \end{pmatrix}$

11. $A^{-1} = \begin{pmatrix} 1 & 2 & 5 \\ 0 & -1 & 2 \\ 2 & 4 & 11 \end{pmatrix}$

Section 11.5

1.
$$\lambda = 4, \quad \mathbf{x}_0 = \alpha \begin{pmatrix} 2 \\ 1 \end{pmatrix}; \qquad \lambda = -3 \quad \mathbf{x}_0 = \beta \begin{pmatrix} 1 \\ -3 \end{pmatrix}$$

3.
$$\lambda = 1 \quad \mathbf{x}_0 = \alpha \begin{pmatrix} -1 \\ 0 \\ 1 \end{pmatrix} + \beta \begin{pmatrix} 3 \\ 1 \\ 0 \end{pmatrix}; \qquad \lambda = 0, \quad \mathbf{x}_0 = \gamma \begin{pmatrix} 1 \\ 1 \\ 1 \end{pmatrix}$$

5.
$$\lambda = 1, \quad \mathbf{x}_0 = \alpha \begin{pmatrix} 1 \\ 0 \\ 0 \end{pmatrix} + \beta \begin{pmatrix} 0 \\ 1 \\ -1 \end{pmatrix}; \qquad \lambda = 2, \quad \mathbf{x}_0 = \gamma \begin{pmatrix} 1 \\ 1 \\ 0 \end{pmatrix}$$

7.
$$\lambda = 0, \quad \mathbf{x}_0 = \alpha \begin{pmatrix} 1 \\ 1 \\ 1 \end{pmatrix}; \ \lambda = 1, \quad \mathbf{x}_0 = \beta \begin{pmatrix} 3 \\ 2 \\ 1 \end{pmatrix}; \ \lambda = 2, \quad \mathbf{x}_0 = \gamma \begin{pmatrix} 7 \\ 3 \\ 1 \end{pmatrix}$$

Section 11.6

1.
$$\mathbf{x} = c_1 \begin{pmatrix} 1 \\ -1 \end{pmatrix} e^{-t} + c_2 \begin{pmatrix} 1 \\ 1 \end{pmatrix} e^{3t}.$$

3.
$$\mathbf{x} = c_1 \begin{pmatrix} 1 \\ 2 \end{pmatrix} e^{3t} + c_2 \begin{pmatrix} 1 \\ -2 \end{pmatrix} e^{-t}.$$

5.

$$\mathbf{x} = c_1 \begin{pmatrix} 1 \\ -1 \end{pmatrix} e^{t/2} + c_2 \begin{pmatrix} t \\ -1/2 - t \end{pmatrix} e^{t/2}.$$

7.

$$\mathbf{x} = c_1 \begin{pmatrix} 1 \\ -1 \end{pmatrix} e^{2t} + c_2 \begin{pmatrix} -1 + t \\ -t \end{pmatrix} e^{2t}.$$

9.

$$\mathbf{x} = c_3 \begin{pmatrix} -3\cos(2t) - 2\sin(2t) \\ \cos(2t) \end{pmatrix} e^t + c_4 \begin{pmatrix} 2\cos(2t) - 3\sin(2t) \\ \sin(2t) \end{pmatrix} e^t$$

11.

$$\mathbf{x} = c_3 \begin{pmatrix} 2\cos(t) \\ 7\cos(t) + \sin(t) \end{pmatrix} e^{-3t} + c_4 \begin{pmatrix} 2\sin(t) \\ 7\sin(t) - \cos(t) \end{pmatrix} e^{-3t}$$

13.

$$\mathbf{x} = c_3 \begin{pmatrix} -\cos(2t) + \sin(2t) \\ \cos(2t) \end{pmatrix} e^t + c_4 \begin{pmatrix} -\cos(2t) - \sin(2t) \\ \sin(2t) \end{pmatrix} e^t$$

15.

$$\mathbf{x} = c_1 \begin{pmatrix} -1 \\ 2 \end{pmatrix} e^{3t} + c_2 \begin{pmatrix} -3 \\ 2 \end{pmatrix} e^{-t}$$

17.

$$\mathbf{x} = c_1 \begin{pmatrix} 0 \\ 1 \\ 0 \end{pmatrix} + c_2 \begin{pmatrix} 0 \\ 2 \\ 1 \end{pmatrix} e^t + c_3 \begin{pmatrix} 2 \\ 1 \\ 0 \end{pmatrix} e^{2t}$$

19.

$$\mathbf{x} = c_1 \begin{pmatrix} 3 \\ -2 \\ 12 \end{pmatrix} e^{-t} + c_2 \begin{pmatrix} 1 \\ 0 \\ 2 \end{pmatrix} e^t + c_3 \begin{pmatrix} 0 \\ 1 \\ 0 \end{pmatrix} e^{2t}$$

Index

"In a Twilight, Feeling and Reasoning My Way"

Chapter 2

Abraham Lincoln
his hand and pen.
he will be good but
god knows When.[1]

These lines are the first selection in the eight-volume Rutgers edition of Lincoln's *Collected Works*. They were written in an arithmetic book Lincoln put together for himself. A typical piece of schoolboy doggerel, it probably was not composed by him. It is amusing to see what some writers of the pious school have done with this ditty. Dr. Chapman wrote: "It is profoundly significant that this child of destiny, at his life's early morning, in clumsy but impressive verse thus reverently coupled his name with that of his Creator."[2] If one searched long enough in the tangled jungle of the debate over Lincoln's religion one could probably find a representative of the skeptical school using these lines as proof positive of a dawning atheism, considering that "god" is not capitalized.

It is not in supposedly clairvoyant statements of his childhood that we are to anticipate the faith of his mature years, for with few exceptions Lincoln was reticent about discussing his religion directly. If we are to find shadows of coming events in his early years they must be in terms of his home environment and the frontier religion of the regions in which the Lincolns settled.

Research into Lincoln genealogy has established some facts that would have interested Lincoln. He admits in one of his autobiographical sketches that he could not learn much about his ancestors. His progenitor, Samuel Lincoln, came to the Massachusetts Bay Colony in 1637. He is reported to have helped in the building of the Old Ship Church in Hingham, claimed today to be the oldest church building in America in continual use. Mordecai Lincoln II, his great-great-grandfather, married a granddaughter of Abadiah Holmes, a Newport Baptist who was flogged on Boston Common for his dissenting opinions and practice. Lincolns who emigrated to the Shenandoah Valley of Virginia were also Baptists. Lincoln's grandfather Abraham donated a piece of his four-hundred-acre tract near Louisville, Kentucky, on which the Long Run Baptist Church was built. Here in its graveyard he lies, shot by an Indian while at work in his cornfield.

Although Lincoln's parents, Thomas and Nancy, were married by Jesse Head, a Methodist circuit rider, they appear to have become members of the Little Mount Separate Baptist

Church. The anti-slavery position of this Little Mount Church may throw some light on Lincoln's statement that his father left Kentucky "partly on account of slavery" and that he could not remember a time when he himself did not think slavery was wrong.

The Separate Baptists in Kentucky were distinguished from Regular Baptists in that they accepted no creed save the Bible itself while the latter group stood on the Philadelphia Confession of Faith. According to the custom of the times, young Abraham was probably carried horseback in his mother's arms when the family went to meeting.

In 1816 the Lincolns removed from the hilly limestone farms of Kentucky to the heavily wooded area of southern Indiana. Later Lincoln described their situation in verse:

> When first my father settled here,
> 'Twas then the frontier line;
> The panther's scream filled night with fear
> And bears preyed on the swine.[3]

Less poetically, his mother's cousin Dennis Hanks said, "We lived the same as the Indians, 'ceptin' we took an interest in politics and religion."

In their early Indiana years the family was cut off from much contact with itinerating parsons. The Bible was probably the only book this frontier family owned. Lincoln is said to have told a friend, "My mother was a ready reader and read the Bible to me habitually." Some Lincoln scholars maintain, however, that Nancy Hanks was illiterate. He learned by heart some of the biblical texts which his mother sang as she worked at chores in the cabin. One of the campaign biographies for which Lincoln furnished material describes mother, son, and daughter taking turns reading the Scriptures on the Sabbath. Their family Bible had been published in 1799 by the Society for the Propagation of Christian Knowledge. In addition to the text it had "arguments prefixed to the different books and moral and theological observations illustrating each chapter, composed by the Reverend Mr. Ostervald, Professor of Divinity." This was the battered old Bible from which Lincoln was seen reading in the White House.

When Abraham was nine years old death came to their Indiana cabin. Nancy Hanks died in an epidemic of the "milk sick." Her husband whipsawed rough boards into a respectable coffin and buried Nancy on a little knoll not far from the cabin as the autumn leaves were beginning to drop. The boy questioned why God had taken his mother away when he so desperately needed her. He searched but found no answer. After a time Parson David Elkin, whom the Lincolns had known at the Little Mount Church in Kentucky, visited them. In the forest clearing near Gentryville the minister conducted a memorial service for the pioneer mother, speaking words of Christian hope.

The traditional picture of the father Thomas as "shiftless" has been corrected by modern research. Within a year of Nancy's death he returned to Elizabethtown and induced Sarah Bush Johnston to become his second wife. In Indiana he helped establish the Pigeon Creek Baptist Church and became a leading member of the congregation. He served as moderator, trustee, and reconciler in matters of church discipline. He contributed to its upkeep.

It is difficult to establish the particular brand of Baptist beliefs that characterized this congregation. Lincoln's stepmother's statement to Herndon might be interpreted to mean that it was more liberal than the "hard-shell" variety. The evidence, however, in the Deerskin record book points toward "hard-shell" doctrine. Its dominant theological position was predestination, the belief that God had foreordained all events by divine decree from eternity. Its corollary was that only those predestinated would be saved. Sandburg pictures the boy Lincoln puzzling over that long word "predestination." Other characteristic features were foot washing, rejection of a paid and educated ministry, and opposition to musical instruments in worship. There was also among "hard-shells" resistance to missionary activity in foreign lands, and to such newfangled notions as Sunday schools for unregenerate children.

For some reason, possibly because of differences between "Separate" and "Regular" Baptists, Thomas Lincoln was not received as a member of the Pigeon Creek church by letter of transfer until June 7, 1823. Sister Lincoln was received the same day "by experience." Abraham's sister Sally was admitted to membership on April 8, 1826. It was of course an adult congregation with a membership that was mainly married. Sally's joining preceded her marriage to Aaron Grigsby by about four months. Abraham never joined, perhaps because he was not yet ready to marry and settle down or perhaps for reservations which he was keeping to himself. A visitor to the Pigeon Creek church in 1866 reported finding in the loft a record book with the entry "1 broom, 1/2 dozen tallow candles" and signed "Abe Lincoln, Sexton."

We know that he enjoyed mimicking these hell-fire and brimstone preachers who shouted and flailed the air as if "they were fighting bees." Ward Hill Lamon, whose account of the Hoosier years was based on reminiscences of old-time residents of Spencer County, described this spirited imitation. "On Monday mornings he would mount a stump, and deliver, with a wonderful approach to exactness, the sermon he had heard the day before. . . . His step sister, Matilda Johnston, says he was an indefatigable 'preacher.' When father and mother would go to church, Abe would take down the Bible, read a verse, give out a hymn, and we would sing. Abe was about fifteen years of age. He preached, and we would do the crying. Sometimes he would join in the chorus of tears."[4] This frontier preaching was a country lad's first introduction to public speaking and he lapped it up.

His simple, God-fearing parents were accustomed to say grace at meals. A visitor reported that Thomas' usual words were: "Fit and prepare us for Life's humblest service, for Christ's sake. Amen." Once the boy spoke up when the meal consisted of nothing but potatoes, "Dad, I call these mighty poor blessings."

Testimony from his relatives is conflicting about his own study of Scripture, but at some time and in some manner he acquired a profound knowledge of its contents. The proof of this is apparent from his later speeches and from the interest he displayed when President in tracking down special texts.

Lincoln's knowledge of the Bible far exceeded the content-grasp of most present-day clergymen.

Direct evidence for his acquaintance with the Bible at this period of his life can be found in his parody of biblical narration. *The Chronicles of Reuben* were written shortly before his first trip to New Orleans. This bit of buffoonery resulted from hard feeling between Lincoln and the Grigsby brothers. When two Grigsby brothers married sisters on the same day Lincoln took part in a plot which sent the young husbands into the bedrooms of the wrong brides. The trick was discovered in time, but it naturally created no little turmoil.

The raw lad commited a further indiscretion by putting the episode into writing as *The Chronicles of Reuben* in the heroic manner of the patriarchal narratives. Thus clumsily he showed a feeling for style in writing. The backwoods parson gave him his first picture of the public speaker and the imitation of the Bible provided his first faltering attempt at composition. Frontier religion was the vehicle of culture for this Hoosier youth as it was the chief civilizing influence in its region.

Lincoln helped his family pack up its belongings and move on into Illinois prairie country in 1830. He stayed with them until a cabin was constructed and a fair beginning made for planting. Then, having come of age, he left them to go off on his own. Neither farming nor carpentry appealed to him. His relations with his father may well have been strained. Lincoln's love of reading must have seemed to Thomas a useless luxury in the hard struggle to keep alive. One speculates as to whether the father may have justified his own ignorance on religious grounds and so proved a stumbling block to a son eager to learn more about his world. The frontier parsons he had met up to this time boasted that they had never attended college or seminary, which they looked upon as centers of devilish infidelity. For many of them a lack of education was next to Godliness.

The word "infidel" was freely and loosely used. A person who held that the earth traveled around the sun would be an "infidel." Anyone who had doubts about the Bible would be an "infidel." A Christian of another denomination was occasionally labeled an "infidel." The tactic of backwoods religion in meeting skeptical criticism was to shout it down as a work of the devil. There was no disposition to meet it on the level of debate, using the scientific understanding of the day to convince the skeptic. It would be a new experience for Lincoln to come to know in Springfield churches having settled pastors who were college-trained and who sought to reconcile the Bible with current historical and scientific knowledge.

"A piece of floating driftwood" was Lincoln's description of himself when he moved to New Salem in 1831 as the enterprising protégé of Denton Offutt, a small-town promoter. Offutt had been impressed with Lincoln's ingenuity in getting his flatboat over the dam at New Salem. He set Lincoln up to tend a store and a mill he had bought in this growing community of about twenty-five families.

There was no church building in New Salem, but the Rev. John M. Berry, one of the settlers, held services in his or neighboring houses. A leader of culture in the community, he removed to Iowa in 1849 and published there a book entitled *Lectures on the Covenants and Right to Church Membership.*

Berry was a Presbyterian as was Dr. Allen, a Yankee from Vermont and Dartmouth College, somewhat out of place in this Southern pioneer atmosphere. Allen gave his New England conscience scope, however, by founding a temperance society and by organizing the first Sunday school in New Salem. Mentor Graham, Lincoln's tutor, and a local schoolmaster, was a leading Baptist.

The Methodists turned out in yearly revivals. These camp meetings were often conducted by the roving, colorful Peter Cartwright. "Uncle Peter," as he was affectionately called, reports in his autobiography that at one of his services five hundred people started "jerking" at once. Women were particularly susceptible to this revivalistic enthusiasm. As their caps and combs came loose, Onstot, a local resident, reported, "so sudden would be the jerking of the head that their long loose hair would crack almost as loud as a waggoner's whip."

The crude emotionalism of these gatherings can hardly have commended itself to Lincoln with his strong sense of humor and of reserve in matters of the spirit. What must have disturbed him still more was the violent feuding between the jealous denominations. One form of Baptist predestinarian opinion held that its church members were created by God for heaven whereas the greater part of mankind had been destined for eternal flames. Methodist and Baptist denounced each other on whether the road to heaven passed over dry land or water. Local roughs tossed logs into the Sangamon River when baptisms were scheduled.

Some perspective on this loveless sectarianism was given by a Yale graduate who came to teach. "In Illinois," he wrote,

"I met for the first time a divided Christian community and was plunged without warning and preparation into a sea of sectarian rivalries, which was kept in constant agitation."

In his fascinating picture of the New Salem community Benjamin Thomas summarized Lincoln's reactions to his religious environment: "He entered with zest into the theological discussions of the community, and profited by the niceties of thought, the subtle distinctions and the fine spun argument that they necessitated. Yet, while he enjoyed them as a mental exercise, and while he eventually attained to a deep faith, emotionally the bitterness of sectarian prejudice must have been repellent to him, and was probably a cause of his lasting reluctance to affiliate with any sect."[5]

For most Christians today the Church is their chief avenue to strength in leading a Christian life and in offering God their worship. Bible reading and personal experience reinforce for them their basic orientation continuously being communicated through church life. This institutional element was lacking to a great degree in Lincoln. The Bible quite apart from the competing churches was his source of inspiration. Personal experience and reflection would give him ever deeper insights into the relevance of Scripture for personal decision and for understanding the meaning of history. The divisiveness of frontier denominationalism left a wound that never fully healed.

Any reservations Lincoln may have felt about the sectarian Christianity of New Salem would have been encouraged by association with another group of people in the neighborhood. These were called by their opponents "the infidels." They might have included Christians who wanted to reconcile the Bible with newer developments in science, rationalists who objected to the characteristic doctrines of the fighting sects, deists who advocated a religion of reason, humanists who argued that traditional religions endangered man's moral responsibility, and perhaps a few atheists of the village type. Any of these positions of dissent might also be labeled "atheist" by their opponents, who were not given to making precise differentiations in debate.

According to Herndon, Lincoln identified himself with this element in the community. There is undoubtedly some measure of truth here, but Herndon has got the picture badly out of focus. Later whispering campaigns against Lincoln as "an infidel," although untrue in their charge, must have been built on some positions taken by him at this time. Along with others in the community he apparently read Thomas Paine's *Age of Reason* and Volney's *Ruins,* discussing them cracker-barrel style around the fire at night. He became an enthusiastic member of a New Salem debating society in the winter of 1831–32. We are told that the club met twice a month with the chairman appointing the two debaters and assigning them their positions on the topic without regard to their own opinions on the matter. It is easy to see how positions on certain religious dogmas, taken simply for the purposes of debate by a brilliant youth in love with rhetoric and logic, might well be misunderstood later as expressing his own religious orientation. His handbill of 1846 describes a habit of arguing for some years in favor of "the doctrine of necessity."

Paine may well have spoken to Lincoln's doubts about the Bible. Many of these have ceased to be of concern for a modern Christian who accepts the methods of historical research in dealing with biblical narratives. The way, however, to a historical interpretation of the Bible, still being pioneered among Christian scholars in that day, could not have been a live option in New Salem. Many of the arguments against, for example, Mosaic authorship of the first five books of the Bible would in his community have characterized a person as "an infidel." By the next generation most educated Christians would have accepted multiple authorship of these books without feeling the loss of anything really central to faith.

Volney's discussion of the rise and fall of empires may have given the young Lincoln the intellectual excitement of the march of history. Accompanying this awakened interest in history may well have come for a time a certain relativism about biblical authority. Within a few years he would meet a scholarly defense of the Bible that would show how "unhistorical" was Volney's thesis that Jesus Christ had never existed. There was an equally absurd thesis fathered by an English "infidel" named Taylor that the Jews had never been a historic people but were an order of freemasons. It is doubtful that Lincoln ever fundamentally accepted Paine's and Volney's skeptical positions. He probably found that on some special points he agreed with them against what passed as orthodoxy among the sects.

Those who side with Herndon in his confusing picture of Lincoln as now "an atheist," now "a deist," and now "a freethinker" place great weight upon Lincoln's so-called "lost book on infidelity."[6] Always a very shaky thesis in the light of conflicting evidence, the tradition of the lost book was finally exploded by the discovery in 1941 of Lincoln's own statement denying in effect that he had ever held the "infidel" position.

The controversy about the "lost book" is worth, however, an examination. It drew forth from Lincoln's contemporaries fuller statements about his religious positions that can be found in any of his own documents for this period. Naturally they must be evaluated cautiously, for they were elicited by the bitter controversies following Holland's *Life*. The battle was renewed even more fiercely after Lamon's biography was published, largely from Herndon's collection of Lincoln material.[7]

Describing the impact of Paine and Volney upon the young storekeeper, Herndon wrote that "Lincoln read both these books, and assimilated them into his own being. He prepared an extended essay—called by many, a book—in which he made an argument against Christianity, striving to prove that the Bible was not inspired, and therefore not God's revelation, and that Jesus Christ was not the Son of God. The manuscript containing these audacious and comprehensive propositions he intended to have published or given a wide circulation in some other way. He carried it to the store, where it was read and freely discussed. His friend and employer, Samuel Hill, was among the listeners, and seriously questioning the propriety of a promising young man like Lincoln fathering such unpopular notions, he snatched the manuscript from his hands, and thrust it into the stove. The book went up in flames, and Mr. Lincoln's political future was secure. But his infidelity and his skeptical views were not diminished."[8]

No one ever offered direct evidence that he had himself read or heard read this supposed essay on "infidelity." There is another theory about the lost book reputedly burned by Hill. This version holds that youngsters in the village found a sizable letter written by Hill but somehow dropped by him. They returned it to Lincoln, the village postmaster, and as he was reading it aloud, possibly for purposes of identification, Hill, wanting to keep its contents a private matter, snatched the letter away and threw it into the flames. This version, however, has obvious difficulties.

Still another piece of testimony claims that Lincoln wrote an essay in defense of his own interpretation of Christianity. It would be easy in that contentious atmosphere to see how the later story of an infidel book might have grown up. In the light of Lincoln's own denial of infidelity in the handbill of 1846 this explanation has more to commend it than Herndon's dramatic tale or the "Hill letter" theory. It is further strengthened in coming directly from Mentor Graham, Lincoln's New Salem tutor and reputed helper in early speech writing.

". . . Abraham Lincoln was living at my house in New Salem, going to school, studying English grammar and surveying, in the year 1833. One morning he said to me, 'Graham, what do you think about the anger of the Lord?' I replied, 'I believe the Lord never was angry or mad and never would be; that His lovingkindness endures forever; that He never changes.' Said Lincoln, 'I have a little manuscript written, which I will show you'; and stated he thought of having it published. Offering it to me, he said he had never showed it to anyone, and still thought of having it published. The size of the manuscript was about one-half quire of foolscap, written in a very plain hand, on the subject of Christianity and a defense of universal salvation. The commencement of it was something respecting the God of the universe never being excited, mad, or angry. I had the manuscript in my possession some week or ten days. I have read many books on the subject of theology and I don't think in point of perspicuity and plainness of reasoning, I ever read one to surpass it. I remember well his argument. He took the passage, 'As in Adam all die, even so in Christ shall all be made alive,' and followed up with the proposition that whatever the breach or injury of Adam's transgressions to the human race was, which no doubt was very great, was made just and right by the atonement of Christ. . . ."[9]

This is hardly the statement of an "infidel" position. It reveals rather a mind dissatisfied with the sectarian theology of his community probing deeply into the Bible on its own. He wants to establish for himself a basic coherence between Christ's work as the Savior and God's ultimate plan for all mankind. Lincoln was actually in practice anticipating what would become the method of twentieth-century Christians. They would come to find God's word in the Bible although not everywhere identical with the words of the Bible.

Lincoln was questioning in this period the orthodox doctrine of scriptural authority by pressing on through "the logic of belief" to a deeper level. The unchanging affirmations for him in this process were man's need of salvation in terms of Adam's fall, God's loving purpose behind the infliction of punishment, and Christ's atoning work through his sacrificial death. Many of these themes would be repeated in later speeches or would provide the perspective for understanding his religious orientation.

These themes taken together in this New Salem period pointed him to the conviction, diametrically opposed to the popular Christian theology of his community, that God would through Christ ultimately save all men. Lincoln's doctrine of universal salvation through Christ would have seemed dangerous enough to the traditionally minded to warrant the label "infidel." To many this essay in defense of universalism in salvation would have been an "infidel book." These frontier preachers made the doctrine of endless punishment their chief whip to the leading of a good life on earth and to the acceptance in "faith" of the particular tenets of the sect.

Lincoln was probably led to his interpretation by two forces. First, the predestinarian cast of his youthful religious training convinced him, contrary to popular theology, that Christ's work had to become effectual to salvation for all men. If God had irresistibly decreed it, it would certainly come to pass. Secondly, Lincoln's rationalism required simple justice in the relations between God and man. He liked to repeat a jingle about the Indian Johnny Kongapod:

> "Here lies poor Johnny Kongapod.
> Have mercy on him, gracious God,
> As he would do if he was God
> And you were Johnny Kongapod."

The two forces of predestination and rationalism convinced him that salvation would be opened to all after proper reformatory punishment. It would not be restricted to a chosen few. The necessity in the first instance was that of logic, in the second that of ethics. Raised a predestinarian Baptist, Lincoln never became a Baptist, but he never ceased to be a predestinarian.

It is interesting to locate the section in Scripture which Lincoln used to support his interpretation of predestinated universal salvation. In the fifteenth chapter of I Corinthians, St. Paul describes a great cosmic restoration in the end-time. The first fruits of it are already present in the resurrection of Christ. Adam is responsible for the death that comes to all men, but Christ will deliver mankind from death into life everlasting in a cosmos purged of all opposition or resistance to the reign of God.

From the Graham statement and others made about Lincoln's belief in later years it would seem that Lincoln gave an absolute meaning to the phrase "in Christ shall all be made alive" quite apart from a number of qualifying themes in the Pauline passage.

"But now is Christ risen from the dead, and become the firstfruits of them that slept.

For since by man came death, by man came also the resurrection of the dead.

For as in Adam all die, even so in Christ shall all be made alive.

But every man in his own order: Christ the firstfruits; afterward they that are Christ's at his coming.

Then cometh the end, when he shall have delivered up the kingdom to God, even the Father; when he shall have put down all rule and all authority and power.

For he must reign, till he hath put all enemies under his feet. . . .

And when all things shall be subdued unto him, then shall the

Son also himself be subject unto him that put all things under him, that God may be all in all."[10]

The New Salem period, then, is hardly one of complete "infidelity," granting the conflicting meanings assigned to the word "infidel" at that time. This denial of the "infidel" position is in conflict with one held by many commentators. One group was determined to prove that Lincoln was once and always an "infidel." Another group was eager to stress a Prodigal Son period of grubbing in the husks of "infidelity" as a dramatic foil to a supposed later conversion to orthodox dogmatics. Both positions overstate the situation, although there are elements of truth badly out of focus in each. New Salem brought to Lincoln growing disillusionment with the fire and brimstone revivalists and their packaged theologies. Through doubts and questionings sharpened by discussions of Paine and Volney, Lincoln was struggling toward his own interpretation of God's purposes for man.

Lincoln occasionally stayed with the Rankin family in their home in Petersburg. Mrs. Rankin reported that he had had doubts at New Salem about his "former implicit faith in the Bible."

"Those days of trouble found me tossed amid a sea of questionings. They piled big upon me. . . . Through all I groped my way until I found a stronger and higher grasp of thought, one that reached beyond this life with a clearness and satisfaction I had never known before. The Scriptures unfolded before me with a deeper and more logical appeal, through these new experiences, than anything else I could find to turn to, or even before had found in them. I do not claim that all my doubts were removed then, or since that time have been swept away. They are not.

"Probably it is to be my lot to go on in a twilight, feeling and reasoning my way through life, as questioning, doubting Thomas did. But in my poor, maimed way, I bear with me as I go on a seeking spirit of desire for a faith that was with him of olden time, who, in his need, as I in mine, exclaimed, 'Help thou my unbelief.' "[11]

Then Lincoln added, "I doubt the possibility, or propriety, of settling the religion of Jesus Christ in the models of man-made creeds and dogmas. . . . I cannot without mental reservations assent to long and complicated creeds and catechisms."

Skepticism about the usefulness of dogma remained a settled conviction with him in adult life. It also partly explains why he never joined a church. These reservations about creeds, however, should not be understood as a denial of the biblical realities behind the dogmas. It was just because God and His purposes became so real to him that he felt no need for a dogmatic preservative or protective shell.

One of the ways in which Lincoln's faith was deepened was by making decisions about vocation and by committing himself to other people in responsible relationship. By this means he would come gradually to experience a sense of God's guidance in the events of personal life. The acceptance which he had won at New Salem whetted an appetite for politics, the fastest door at that time to success and recognition in a backwoods community. With this target in mind he set himself to the reading of law, borrowing legal books from many friends.

He decided to run for the state legislature and began a barnstorming campaign after his return from eight months' service in the Black Hawk War. His election as captain by his men pleased him greatly. He announced his platform in the *Sangamo Journal* on March 9, 1832. He favored internal improvements for the state, especially the deepening and straightening of the Sangamon River as less expensive than a railroad. He discussed the problems created by the loaning of money at exorbitant rates. Turning to education, he called it "the most important subject a people can be engaged in." It made it possible for men to study the history of their country. It gave them an appreciation of "the value of our free institutions . . . to say nothing of the advantages and satisfaction to be derived from all being able to read the Scriptures and other works, both of a religious and moral nature, for themselves."[12]

Then in a concluding paragraph he described his motivation in running for public office. "Every man is said to have his peculiar ambition. Whether it be true or not, I can say for one that I have no other so great as that of being truly esteemed of my fellow men, by rendering myself worthy of this esteem. How far I shall succeed in gratifying this ambition, is yet to be developed. I am young and unknown to many of you. I was born and have ever remained in the most humble walks of life. I have no wealthy or popular relations to recommend me. My case is thrown exclusively upon the independent voters of this county, and if elected they will have conferred a favor upon me, for which I shall be unremitting in my labors to compensate. But if the good people in their wisdom shall see fit to keep me in the background, I have been too familiar with disappointment to be very much chagrined."

Lincoln lost the election, running eighth of thirteen candidates, but his 277 of the 300 votes cast in the New Salem precinct were a sign of things to come. This defeat was his only one at the hands of the people. He turned again to the study of law.

His next attempt to be elected to the state legislature was successful and he served in all for four terms, becoming a member of the famous Long Nine in that body. He did the chores of a legislator faithfully and was instrumental in lining up support for the removal of the capital from Vandalia to Springfield.

When the legislature passed resolutions on the subject of slavery in terms of sacred property rights and with disapproval of abolitionist societies Lincoln and Dan Stone read a protest which was printed in the Journal of Proceedings. One of the tasks of this study will be to analyze Lincoln's religious opposition to slavery as it influenced his political positions on the issue. "They [Lincoln and Stone] believe that the institution of slavery is founded on both injustice and bad policy; but that the promulgation of abolition doctrines tends rather to increase than abate its evils. They believe that the Congress of the United States has no power, under the Constitution, to interfere with the institution of slavery in the different states."[13]

"Stand Still, *and See the Salvation of the Lord"*

Chapter 3

On April 15, 1837, on a borrowed horse and with his worldly possessions packed in his saddlebags, Lincoln rode into Springfield. Having passed his bar examinations, he was arriving to become the law partner of John T. Stuart, a polished Southerner who was already a powerful Whig leader. Lincoln's immediate future was secure and because of his influence in shifting the capital from Vandalia to Springfield he would be most acceptable to its progressive citizens.

Lincoln, however, was in a melancholy mood. Making his way to the store of Joshua Speed, a prosperous young merchant from an aristocratic Kentucky family, he asked how much the bedding and equipment for a single bedstead would cost. Dismayed at his resources, Lincoln said, "It is probably cheap enough; but I want to say that, cheap as it is, I have not the money to pay. But if you will credit me until Christmas, and my experiment here as a lawyer is a success, I will pay you then. If I fail in that I will probably never pay you at all."

Deeply moved by the melancholy face, Speed offered to share his quarters. "So small a debt seems to affect you so deeply, I think I can suggest a plan by which you will be able to attain your end without incurring any debt. I have a large room with a double bed upstairs, which you are welcome to share with me if you choose." Lincoln asked where the room was. Without a word he shouldered his saddlebags and went upstairs. Returning and smiling, he exclaimed, "Well, Speed, I'm moved."

With this incident began a deep friendship that lasted for life. The letters show a remarkable level of understanding and affectionate communication on such subjects as politics, women, and religion, still the hardy perennials in dormitory bull sessions. "When I knew him in early life," said Speed in a lecture after Lincoln's death, "he was a skeptic. He had tried hard to be a believer, but his reason could not grasp and solve the great problem of redemption as taught."

Lincoln's commitment to Speed passed beyond debate on things religious into a deepened experience of religious self-understanding. For four years these two roommates anxiously nursed each other over the humps in their courtship of the girls they left behind them and finally of the two they were to marry. Sandburg describes their friendship:

"Joshua Speed was a deep-chested man of large sockets, with broad measurement between the ears. A streak of lavender ran through him; he had spots soft as May violets. And he and Abraham Lincoln told each other their secrets about women.

Lincoln too had tough physical shanks and large sockets, also a streak of lavender, and spots soft as May violets.

" 'I do not feel my own sorrows more keenly than I do yours', Lincoln wrote to Speed in one letter. And again: 'You know my desire to befriend you is everlasting.' "[1]

Lincoln seems to have made some commitment close to an informal engagement with Mary Owens before settling in Springfield. He was miserable because he doubted whether he really loved her. Yet he felt bound to act according to honor. He described his melancholy to her. In one letter he revealed how a backwoods youth felt about the grand churches of Springfield with their settled pastors, their ordered worship, and their fashionable members. "I've never been to church yet, nor probably shall not soon. I stay away because I am conscious I should not know how to behave myself."[2]

The relationship with Mary Owens gradually petered out to the relief of both. Speed became the confidant of his friend in this matter as he would also share the ups and downs of a still more stormy courtship between Lincoln and Mary Todd.

There are two speeches of this period which are interesting from the point of view of future developments. The address of January 27, 1838, before the Young Men's Lyceum of Springfield suggests an understanding of the nation's history in comparison with religious institutions. Out of this small acorn would eventually grow a gnarled white oak resistant to wind and storm. The great Lincolnian theme of "this nation under God" would be anticipated formally, but as yet without dramatic content, in this essay, "The Perpetuation of Our Political Institutions."

"We find ourselves," he said, "under the government of a system of political institutions, conducing more essentially to the ends of civil and religious liberty, than any of which the history of former times tells us." He painted a picture of America as an impregnable fortress vulnerable only to violence from within. The lawlessness of the times was our greatest danger. He had doubtless chosen this subject because only three months before a howling mob had lynched the abolitionist editor Elijah Lovejoy at Alton, Illinois. His antidote for this was "reverence for the laws" which he wanted American mothers to breathe into "the lisping babe" and school and pulpit to inculcate. Then, using the metaphors of religion and its altars, he pleaded, "in short, let it become the *political religion* of the nation; and let the old and the young, the rich and the poor, the grave and the gay, of all sexes and tongues, and colors and conditions, sacrifice unceasingly upon its altars."

Americans had won their liberties by passionate feelings against injustice. They could no longer rely upon passion and the memory of the courageous actions of the Founding Fathers. Instead they must cultivate "cold, calculating reason." Here Lincoln manifested a strain of rationalism that was probably the by-product of a man of meager background struggling to become self-educated. It is interesting to find Lincoln dismissing those memories of a shared past that give continuity to a nation's history in favor of "calculating reason." In his more mature years he would evoke as no other speaker has ever done the "mystic chords of memory, stretching from every battlefield and patriot grave, to every living heart and hearthstone, all over this broad land."

Later on Lincoln would understand the nation's history in a continuity of purpose with the acts of God as interpreted by the biblical prophets. Here he offers something else. "Passion has helped us; but can do so no more. It will in future be our enemy. Reason, cold, calculating, unimpassioned reason, must furnish all the materials for our future support and defense. Let those (materials) be molded into *general intelligence,* (*sound*) *morality* and, in particular, *a reverence for the constitution and laws. . . .*" Then, reverting to the earlier parallelism, he closes: "Upon these let the proud fabric of freedom rest, as the rock of its basis; and as truly as has been said of the only greater institution, '*the gates of hell shall not prevail against it.*' "[3]

On Washington's Birthday in 1842, Lincoln delivered an address in the Second Presbyterian Church to the Washington Temperance Society. The speech shows such an intimate acquaintance with biblical language and incidents that, had its authorship been unknown, scholars would probably have assigned it to one professionally trained in the field of biblical study. While its perspective was a deep Christian compassion for one's fellow man, it rubbed many church members the wrong way. They just did not like having their virtue over against the drunkard explained away as "absence of appetite" rather than commended as "mental or moral superiority." Lincoln's political popularity suffered in a subsequent election partly, as he believed, because of this speech.

Lincoln began by accepting the principle of our present-day Alcoholics Anonymous, namely, that the reformed drunkard is himself the most persuasive helper for the alcoholic. Preachers and hired agents, Lincoln affirmed, had as a class a want of approachability fatal to their success. He criticized a temperance strategy that condemned the liquor manufacturer and merchant as the chief devils and showed no concern for dealing with present drunkards. Blazing condemnation thundered from pulpit and platform failed utterly to convince the wayward. It only alienated them because it was predicated on "a reversal of human nature, which is God's decree, and never can be reversed." He pointed out that the true approach to winning a man was first to "convince him that you are his sincere friend. Therein is a drop of honey that catches his heart, which, say what he will, is the great high road to his reason, and which, when once gained, you will find but little trouble in convincing his judgment of the justice of your cause, if indeed that cause really be a just one."

Not only was the widespread temperance program of denunciation impolitic. Lincoln argued that it was also unjust. This approach would have alienated those for whom moral questions were simple absolutes divorced from their total setting in life. Lincoln's point was that temperance reform was so new and the acceptance by mankind of drinking so universal in history and still the majority sentiment that the method of denunciation affronted men's opinions. In his development of this point he reveals an interesting attitude toward rational proof for God's existence. "The universal *sense* of mankind, on any subject, is an argument, or at least an *influence,* not easily overcome. The success of the argument in favor of the existence of an overruling Providence, mainly depends upon that sense; and men ought not, in justice, to be denounced for yielding to it in any case, or for giving it up slowly, *especially,* where they are backed by interest, fixed habits, or burning appetites."

Lincoln then turned to the highest ground in seeking to persuade his hearers that those who had never been afflicted with alcoholism should not refuse to join a society of reformed drunkards. His point was that the Washingtonian's compassion for the alcoholic was the same expression of concerned love that led a sovereign Creator to incarnate Himself in the form of man.

" 'But,' say some, 'we are no drunkards; and we shall not acknowledge ourselves such by joining a reformed drunkards' society, whatever our influence might be.' Surely no Christian will adhere to this objection. If they believe, as they profess, that Omnipotence condescended to take on himself the form of sinful man, and as such, to die an ignominious death for their sakes, surely they will not refuse submission to the infinitely lesser condescension, for the temporal, and perhaps eternal salvation, of a large, erring, and unfortunate class of their own fellow creatures. Nor is the condescension very great."

Some writers, intent on proving Lincoln a "freethinker," claim that Lincoln's use of "they" in this paragraph expresses his rejection of Christianity. It is a wrong conclusion for the simple reason that the word "they" is stylistic in use. A few paragraphs before Lincoln used "they" of the Washingtonians, the very group of which he was a member and to which he was speaking. Lincoln's whole setting of his essay is that the Washingtonian approach is the deeply Christian one of the incarnational principle. From this deeper Christian position he argues against the Christian pharisee of his day.

The essay concludes, however, on another note. There is a deification of "Reason," now somewhat more warmly commended than in the Lyceum address, in which the reason was of the "cold, calculating" type. There is, however, a hollow ring to his rhetorical peroration. Developments in his personal life would very soon convince him of the need for more than reason to solve personal dilemmas. From here the new openness would spread gradually to his interpretation of man's future. In this temperance address the analysis is biblically oriented, the program of action deeply Christian in context, but the motivation still remains rationalistic. "Happy day, when, all appetites controlled, all passions subdued, all matters subjected, *mind,* all-conquering *mind,* shall live and move the monarch of the world. Glorious consummation! Hail, fall of Fury! Reign of Reason, all hail!"[4]

Lincoln's bombastic apostrophe to "Reason" may have been a whistling in the dark, a self-prescription which he hoped would stabilize the deep unrest and melancholy within him. Having wriggled out of a commitment to Mary Owens, he soon found himself pledged to Mary Todd, a vivacious and cultured Kentuckian who had come to live with her sister, Mrs. Ninian W. Edwards, in Springfield. Coming from a genteel social background in Lexington, educated in the exclusive school of Madame Mentelle, Mary Todd enjoyed a wide popularity among the young bachelors who were admitted into the Edwards' social circle.

Abraham and Mary, perhaps because of the attraction of opposites, hit it off at once. They became engaged. Then something separated them on January 1, 1841. Herndon tells a vivid tale of the bride deserted at the altar, but the evidence points rather to Lincoln's breaking the engagement himself. He had become miserable in a siege of depression and doubted whether he could be happy with Mary or whether he could make her happy. Released apparently by Mary from his promise, instead of getting a hold on himself he burrowed ever more deeply in his gloom.

A glimpse into the blackness of his despair is afforded by letters to his friend John T. Stuart, then in Washington, imploring him to secure the postmastership at Springfield for his doctor. "I have, within the last few days, been making a most discreditable exhibition of myself in the way of hypochondriasm and thereby got an impression that Dr. Henry is necessary to my existence. Unless he gets that place he leaves Springfield. You therefore see how much I am interested in the matter."[5]

A second letter three days later further spelled out his agony. "I am now the most miserable man living. If what I feel were equally distributed to the whole human family, there would not be one cheerful face on the earth. Whether I shall ever be better, I cannot tell; I awfully forbode I shall not. To remain as I am is impossible; I must die or be better, it appears to me."[6]

The situation of the wretched suitor became even more lonely when Speed early in 1841 sold out his business and returned to Kentucky. He did, however, persuade Lincoln to visit him on the family plantation near Louisville from early August to mid-September. On his return Lincoln wrote to Joshua's half sister, referring to the Oxford Bible Speed's mother had given him: "I intend to read it regularly when I return home." Then he added, perhaps somewhat wistfully, "I doubt not that it is really, as she says, the best cure for the 'Blues' could one but take it according to the truth."[7]

Then occurred a change that brought a reversal of roles and made Lincoln counselor to Speed, now engaged to Fanny Henning. Drawn out of preoccupation with his own problems, in helping Speed, he began to help himself. Lincoln's reassuring letters comforted his friend's "nervous debility" and his questioning whether he was really in love with his fiancée.

When an illness of Fanny's frightened Speed into gloomy forebodings about her death Lincoln revealed his belief in God's immediate concern with and control over personal life. Lincoln argued that his friend's anguish over the possibility of his fiancée's death was clear evidence of his love for her. "I almost feel a presentiment that the Almighty has sent your present affliction expressly for that object. . . . Should she, as you fear, be destined to an early grave, it is indeed a great consolation to know that she is so well prepared to meet it. Her religion, which you once disliked so much, I will venture you now prize most highly."[8]

When Lincoln learned that the happiness of his newly married friend exceeded his expectation, he was overjoyed. It brought him more pleasure than had been his lot since "that fatal first Jany. '41" when he had asked to be released from his engagement to Mary Todd. "Since then, it seems to me, I should have been entirely happy, but for the never-absent idea, that there is *one* still unhappy whom I have contributed to make so. That still kills my soul. I cannot but reproach myself, for even wishing to be happy while she is otherwise. She accompanied a large party on the Rail Road cars, to Jacksonville last Monday; and on her return, spoke, so that I heard of it, of having enjoyed the trip exceedingly. God be praised for that."[9]

In another letter some months later Lincoln acknowledged the correctness of his friend's advice about his own unhappy situation with respect to Mary Todd: ". . . but before I resolve to do the one thing or the other, I must regain my confidence in my own ability to keep my resolves when they are made. In that ability, you know, I once prided myself . . ." Then he introduced the religious dimension. By less preoccupation with his own problem and by more reliance on God's guidance of his circumstances he expected to reach a solution. This religious awakening he clothed in predestinarian garb.

"The truth is, I am not sure there was any merit, with me, in the part I took in your difficulty; I was drawn to it as by fate; if I would, I could not have done less than I did." Then he gave a firm religious orientation to his fatalism. "I always was superstitious; and as part of my superstition, I believe God made me one of the instruments of bringing your Fanny and you together, which union, I have no doubt He had foreordained." Turning from the past and the belief that God foreordained his friend's marriage, he expressed for the immediate future a quiet acceptance of the deliverance that God's predestinating activity would bring him. "Whatever he designs, he will do for *me* yet. 'Stand *still,* and see the salvation of the Lord' is my text just now."[10]

Ruth Painter Randall in her charming *Courtship of Mr. Lincoln* finds evidence here of a sense of religious dependence often to reappear in the future. "It often happens that people who have come through a long, baffling siege of the spirit, grappling with problems that seem to have no solution, unable to decide which course of action to take, cease struggling to turn to their religious faith and rest their weary souls in waiting for a revelation of divine guidance. Such was Lincoln's state now. He was a man of deep religious feeling. All the rest of his life one finds incidents in which he placed his reliance on the will of God."[11]

The resolution of his love for Mary Todd was not to come, however, without a jolting experience that had all the features of a television re-creation of nineteenth-century romance. He nearly had to fight a duel, a situation which later on in life mortified him so deeply that he would never discuss the incident. The long and short of a very involved tale is that Lincoln wrote an anonymous letter to a local newspaper from a "Rebecca of the Lost Townships." In it he attacked a political rival, James Shields, very unfairly and with withering sarcasm. Others also wrote abusive letters under the same pseudonym, one coming from Mary Todd and a friend of hers.

Shields, a truculent Irishman, charged the whole authorship to Lincoln, who chivalrously would not reveal Mary's part in it. Although contrary to the law of Illinois, Lincoln felt he had to accept Shields's challenge to duel. The altercation was finally adjusted at the field proposed for combat. The episode, with its anonymous attacks on opponents comparable to *The Chronicles of Reuben,* reveals some of that raw frontier background. In later life Lincoln would learn the lesson of responsible opposition to his political adversaries without underhanded attacks and personal vilification. The immediate result of the code duello appears to have brought Abraham and Mary together again.

Once more he sought Speed's help in reassuring his own doubts. He wrote his friend on October 5, 1842, "to say something on that subject which you know to be of such infinite solicitude to me. The immense suffering you endured from the first days of September till the middle of February you never tried to conceal from me, and I well understood. You have now been the husband of a lovely woman nearly eight months. That you are happier now than you were the day you married her I well know; for without, you would not be living. But I have your word for it too; and the returning elasticity of spirits which is manifested in your letters. But I want to ask a closer question—'Are you now, in *feeling* as well as *judgment,* glad you are married as you are?' From anybody but me, this would be an impudent question not to be tolerated; but I know you will pardon it in me. Please answer it quickly as I feel impatient to know."[12]

Within a month, on November 4, 1842, Abraham and Mary were married in the Edwards' parlor with an Episcopal clergyman, the Rev. Charles Dresser, performing the ceremony.

There are at least two levels on which the marriage had significance for Lincoln's religious development. On the first level it brought Lincoln as a family man closer to conventional church relationships. Mary attended the Episcopal Church in Springfield and he accompanied her at times. After the death of their son Edward, Mary became a member of the Presbyterian Church. Her Episcopal rector had been out of town and the funeral service was performed for the sorrowing parents by the Rev. James Smith of the First Presbyterian Church. Lincoln became a regular attendant from that time on. Later, from the White House, Mrs. Lincoln wrote back to Springfield asking that "our particular pew" might be reserved for them until their return. Mary's religion clutched at tangible securities like the family pew.

The second level of religious influence exerted by the marriage was upon Lincoln's character. While the marriage was stable and blessed with mutual affection and comfort, there was also a darker side to it. Mary had a towering rage, was unduly concerned over little things, and seemed unable to achieve a satisfactory relationship to the children, now indulging them too much and now dealing far too harshly with them. There were already in Mary some of the tragic symptoms that after her cruel bereavements of three children and husband would require institutional care. Lincoln adored his children to the point of indulging them. At times he had to leave the house when Mary's anger and hysteria became too sharp. He learned forbearance and forgiveness, not as doctrines but in practice. Through it all Lincoln achieved a serenity of faith and a deepened understanding far beyond the level of his wife's. She had had perhaps greater initial commitment to the Christian faith than he had. Part of the tragedy of Mary Todd is that she did not have her husband's capacities for growth. Limited by the structure of an abnormal personality, her own faith seems brittle beside the flowering of his.

"Now Convinced of the Truth of the Christian Religion"

Chapter 4

Lincoln's race for Congress in 1846 was having its ups and downs. With only a few days left before the election, reports were reaching Lincoln that his political opponent on the Democratic ticket was whispering the charge of "infidelity" against him in the northern counties. Lincoln was accused of being a scoffer at religion.

Lincoln knew what this could mean for him. It had cost him a nomination once before in 1843. He had conducted a post-mortem of that Whig convention in a letter to Martin Morris: ". . . it was everywhere contended that no Christian ought to go for me, because I belonged to no church, was suspected of being a deist, and had talked about fighting a duel. With all these things Baker, of course, had nothing to do. Nor do I complain of them. As to his own church going for him, I think that was right enough, and as to the influences I have spoken of in the other, though they were very strong, it would be grossly untrue and unjust to charge that they acted upon them in a body or even very nearly so. I only mean that those influences levied a tax of considerable per cent upon my strength throughout the religious community."[1] He might have added that his temperance address had offended many.

The trouble now, however, was considerably more acute, for the charge was being made by no less a person than the popular, rugged Peter Cartwright, twenty-four years his elder and widely known all over Illinois. This excitable Methodist circuit rider was as strong in body as he was resourceful in spirit. Many times he had interrupted his revivalist message to collar some heckler or drunkard and toss him out of the meeting, returning to the stump with the words, "As I was saying . . ." Even the saints were not safe from chastisement. He rebuked a deacon who had offered a formal prayer: "Brother, three prayers like that would freeze hell over." Refuting a sermon on the indefectibility of grace, he gathered his hearers under a tree. Saying, "I promised I'd answer those who believe once in grace always in grace," he leaped up and hung for a minute to an overhanging limb. Then he let go, fell to the ground, and walked away, his sermon finished.

Sandburg paints a dramatic picture of an encounter that was supposed to have taken place between the two political rivals in the campaign.

"In spite of warnings he went anyhow to a religious meeting where Cartwright was to preach. In due time Cartwright said,

'All who desire to lead a new life, to give their hearts to God, and go to heaven, will stand,' and a sprinkling of men, women, and children stood up. Then the preacher exhorted, 'All who do not wish to go to hell will stand.' All stood up—except Lincoln. Then said Cartwright in his gravest voice, 'I observe that many responded to the first invitation to give their hearts to God and go to heaven. And I further observe that all of you save one indicated that you did not desire to go to hell. The sole exception is Mr. Lincoln, who did not respond to either invitation. May I inquire of you, Mr. Lincoln, where you are going?'

"And Lincoln slowly rose and slowly spoke, 'I came here as a respectful listener. I did not know that I was to be singled out by Brother Cartwright. I believe in treating religious matters with due solemnity. I admit that the questions propounded by Brother Cartwright are of great importance. I did not feel called upon to answer as the rest did. Brother Cartwright asks me directly where I am going. I desire to reply with equal directness: I am going to Congress.' The meeting broke up."[2]

The immediate threat from his antagonist's charge of "infidelity" lay in the fact that the time before the election was short. The charge, moreover, was not a public one. It was difficult to answer. Probably the damage was not yet widespread, but it might be dangerous not to do anything about it. Lincoln disliked intensely having to take public notice of it. He did not want to discuss his religion on the stump, but he also did not want to lose his chance to go to Washington.

His first strategy was to write friends in the northern counties a contradiction of Cartwright's charge, requesting that "they should publish it or not, as in their discretion they might think proper, having in view the extent of the circulation of the charge, as also the extent of credence it might be receiving." His friends decided not to publish Lincoln's statement. Then Lincoln learned that the whispering campaign had spread into other counties. With the election coming up on August 3, he had a handbill printed on July 31.

Within a few days the suspense was over. While Cartwright was a most effective evangelist he turned out to be a poor candidate for Congress. Lincoln received a sizable victory of 6340 votes over Cartwright's 4829. A reading of the returns, however, for Marshall and Woodford counties, which Cartwright had carried by decisive margins, led Lincoln to set the record straight for the future. He wrote to Allen Ford, the editor of the *Illinois Gazette,* about a week after the election asking that both letter and handbill be published in his paper. He strengthened the contents of his public statement by adding, "I here aver, that he, Cartwright, never heard me utter a word in any way indicating my opinions on religious matters, in his life."[3]

One of the most exciting discoveries in Lincoln scholarship was the finding of this letter and handbill by Harry E. Pratt in 1941. It demolished earlier interpretations which had tried to make Lincoln out an atheist or a deist. It is the only public document, moreover, in which Lincoln ever gave personal testimony about his religious views. The handbill is an authoritative refutation by Lincoln of the charge of "infidelity." It both casts light backward on the confusion of the New Salem period and provides motifs for understanding developments in the future. Its importance, however, must be limited by the qualification that here Lincoln speaks for 1846, not 1865. The fact that Lincoln was denying false charges made against him gives the document a necessarily negative character. He revealed just enough of his own religious orientation to make the required refutation. Beyond this minimum he offered very little. Hence it would not be correct to draw as a necessary conclusion, as a number of writers have, that Lincoln believed nothing beyond these statements. He was a politician refuting a local libel in this handbill, not a confessor delivering a testament of faith to all mankind.

"July 31, 1846

"TO THE VOTERS OF THE SEVENTH CONGRESSIONAL DISTRICT

"Fellow Citizens:

A charge having got into circulation in some of the neighborhoods of this District, in substance that I am an open scoffer at Christianity, I have by the advice of some friends concluded to notice the subject in this form. That I am not a member of any Christian Church, is true; but I have never denied the truth of the Scriptures; and I have never spoken with intentional disrespect of religion in general, or of any denomination of Christians in particular. It is true that in early life I was inclined to believe in what I understand is called the 'Doctrine of Necessity'—that is, that the human mind is impelled to action, or held in rest by some power, over which the mind itself has no control; and I have sometimes (with one, two or three, but never publicly) tried to maintain this opinion in argument. The habit of arguing thus however, I have, entirely left off for more than five years. And I add here, I have always understood this same opinion to be held by several of the Christian denominations. The foregoing, is the whole truth, briefly stated, in relation to myself, upon this subject.

"I do not think I could myself, be brought to support a man for office, whom I knew to be an open enemy of, and scoffer at, religion. Leaving the higher matter of eternal consequences, between him and his Maker, I still do not think any man has the right thus to insult the feelings, and injure the morals, of the community in which he may live. If, then, I was guilty of such conduct, I should blame no man who should condemn me for it; but I do blame those, whoever they may be, who falsely put such a charge in circulation against me."[4]

Here we find Lincoln publicly stating that he was not a church member. He never became one. Mrs. Rankin, who has already been quoted, recalled that Lincoln had stayed in her house at Petersburg about June of 1846. Lincoln spoke again of his reservations about "the possibility and propriety of settling the religion of Jesus Christ in the models of man-made creeds and dogmas." He then said something that he was accustomed to repeat throughout the rest of his life. The substance of it was that he would join a church that made the Savior's summary of the law of love to God and to neighbor its condition of membership. This apparently became a settled conviction with him as seen from his statement to Congressman Deming, who reported it in his address before the General Assembly of Connecticut in 1865:

"I am here reminded of an impressive remark he made to me upon another occasion, and which I shall never forget. He said, he had never united himself to any church, because he found difficulty in giving his assent, without mental reservations, to the long complicated statements of Christian doctrine which characterize their Articles of Belief and Confessions of Faith.

'When any church,' he continued, 'will inscribe over its altar as its sole qualification for membership the Savior's condensed statement of the substance of both the law and Gospel, Thou shalt love the Lord thy God with all thy heart, and with all thy soul, and with all thy mind, and thy neighbor as thyself,—that Church will I join with all my heart and soul.' "[5]

In his document of 1846, Lincoln immediately qualified his denial of church membership with the clause "but I have never denied the truth of the Scriptures." As we have seen before, the Bible rather than the Church remained his highroad to the knowledge of God. An interesting and important deepening of his trust in the Bible would shortly take place.

He did not like creedal tests for membership. This must not be interpreted to mean that he believed in a religion without content in the terms of a widespread contemporary approval of "creedless" religion, i.e., religiousness in general. Lincoln's beliefs were deeply biblical in their rootage. It was just because he took so seriously "the truth of the Scriptures" that he objected to man-made abstracts. His impatience with frontier squabbles over the minutiae of denominational differences doubtless was a strong conditioning element behind this conviction. Such contentious debates, he knew, obscured the essential emphasis of "the Savior" upon a faith that bore fruits.

Lincoln may not have sufficiently appreciated the necessity for a historical faith like Christianity to embody its central experience in creeds to help make it transmissible from one generation to another. He did, however, clearly see the danger of a creedal denominationalism that lacked love.

He further added in the handbill of 1846 that he had never

spoken "with intentional disrespect of religion in general, or of any denomination of Christians in particular." There is evidence from the time of presidency that he came later to have a more affectionate regard for the churches. His many responses to visiting delegations showed this. Fairly typical was his statement to the Methodists under Bishop Ames: "God bless the Methodist Church—bless all the churches—and blessed be God, Who, in this our great trial, giveth us the churches."[6]

Apparently aware of some of the tinder used by his opponents to kindle their charge of "infidelity," Lincoln explained his changing perspectives on what he understood was called the "Doctrine of Necessity." The issue here is not immediately clear. He defined his understanding of the phrase, about which he once debated in small groups, as the belief "that the human mind is impelled to action, or held in rest by some power, over which the mind itself has no control."

This definition by itself describes fatalism. Herndon said he was a fatalist. Henry C. Whitney, who rode with him on the circuit, wrote, "Mr. Lincoln was a fatalist: he believed, and often said, that 'There's a divinity that shapes our ends, Rough-hew them how we will.' "[7] Mrs. Lincoln interpreted her husband's lack of concern about threats of assassination as a conviction that he would die when it was foreordained. In a letter to Speed already quoted, Lincoln describes himself as a fatalist. To Congressman Arnold he said much later: "I have been all my life a fatalist," quoting Shakespeare's lines once again.

There are, however, many types of fatalism. They shade on the right from the view that God's will determines every event in man's life all the way over to the left with the view that in a universe of ironbound cause and effect all human actions are determined. The poles are theological predestination and philosophical determinism. Herndon catches the original religious rootage of this idea for Lincoln in the predestinarian debates among the Baptists in the churches of his childhood and youth. "His early Baptist training made him a fatalist to the day of his death. . . ."[8]

What probably happened is that Lincoln in the New Salem period revolted against the harsh predestinarian conclusions about the fate of unbelievers. He may have sought to restate the "Doctrine of Necessity" in more philosophical terms. Perhaps, as Roy Basler suggests, he was influenced by a climate of opinion stemming from William Godwin's *Political Justice*. Godwin, once a Calvinist, crossed his "Doctrine of Necessity" with ideas of human perfectibility from the French Enlightenment.

This interpretation would be consistent with Lincoln's appeal to "cold, calculating reason" and with his "deification" of "Reason" in the temperance address. Yet these appeals taken at face value would argue a view of the freedom of man's mind to choose and act by itself. This would, of course, conflict with Lincoln's statement in the handbill that he "was inclined to believe" in the "Doctrine of Necessity." The inconsistency here is a real one and Lincoln never clearly resolved the issue philosophically. The weight, however, was on the side of determination.

This can be proved by the very way he came to express his difference with the traditional theology of eternal damnation. Mentor Graham's statement shows that Lincoln's view of a universal restoration was predicated upon God's determining action: "In Christ shall all be made alive." This structural element of determination remained constant in Lincoln's thought. He would later describe the way God worked upon men's minds and upon their actions in terms nearly identical with those of the handbill. "The will of God prevails. . . . By His mere quiet power, on the minds of the now contestants, He could have either *saved* or *destroyed* the Union without a human contest."[9]

The direction of Lincoln's religious fatalism beyond the period of the handbill would be toward a deeper understanding of the biblical stress upon God's will and upon his own responsibility as "God's instrument." Although the Bible would be his primary inspiration for this insight he may have been helped somewhat in its articulation by the Calvinistic doctrine of predestination as expounded by two distinguished Old-School Presbyterian pastors under whom he would sit in Springfield and in Washington.

This is not to state that Lincoln accepted the system of Calvinism. It is merely to point out that Presbyterian preaching on this theme may have helped him to give greater logical precision to a fundamental insight acquired in youth from his Baptist environment, modified in a philosophical direction in New Salem and in the early years at Springfield, and again reestablished in his later life as a primary biblical belief of his own.

In the handbill of 1846, as though to lessen any possible exploitation by his political opponents of his "Doctrine of Necessity" as a sign of Infidelity, Lincoln adds: "I have always understood this same opinion to be held by several of the

Christian denominations." He goes on to point out that he himself could not support for public office a man he "knew to be an open enemy of, and scoffer at, religion."

This statement alone demolishes the numerous attempts to paint Lincoln as a "freethinker," unless of course the theory is held that everything Lincoln said about God and his faith was politically motivated deception. Such a bizarre theory contradicts the one fundamental trait that all find in Lincoln—a bedrock of integrity. Even if the theory be entertained for the sake of argument, it is completely self-defeating because a man of Lincoln's intelligence could have done a much better job of deceiving religious people by greater conformity to conventional religion than he ever showed.

From the point of view of advancing Lincoln's political career, his term in the House of Representatives was a definite failure. The chief reason for this lay in the resolutions he introduced into the House demanding to know the "spot" where the Mexican War had started. He held Democratic President Polk responsible for starting the war unconstitutionally. Lincoln's attitude was decidedly unpopular at home where expansionist sentiment was rampant. On his return to Illinois he also discovered that he had lost any power of patronage from his own Whig party. He therefore turned away from politics and concentrated on his practice of the law.

"The decision," writes Benjamin Thomas, "made the years of his political retirement, from 1849 to 1854, among the most fruitful of his life. For as he put aside all thought of political advancement and devoted himself to personal improvement, he grew tremendously in mind and character."[10]

The Lincoln family, with its breadwinner increasingly successful as a lawyer of wide reputation, was suddenly brought up short against the cruel accidents of life. Their second son Edward Baker died on February 1, 1850, just under the age of four, after an illness of fifty-two days. The grief-stricken parents, unable to locate the Episcopal clergyman who had married them and whose church they occasionally attended, turned to the Rev. James Smith of the First Presbyterian Church to conduct the funeral. A rugged Scotsman with a fine mind and an interesting fund of anecdotes, he was a helpful pastor to the bereaved parents. The Lincolns soon took a pew with an annual rental of fifty dollars and became regular attendants. Mrs. Lincoln became a member.

The loss of Eddie turned Lincoln's thoughts to the great question of life after death. He had known the dreadful emptiness that came to a sobbing nine-year-old boy when his mother had died in that Indiana cabin. Again he felt the chill hand of death when his sister died shortly after her marriage. There is evidence that he felt keenly the death of Ann Rutledge at New Salem. These experiences, added to the tendency toward melancholy in his nature, explain his fondness for poetry that expressed the transience of life, its subjection to sudden changes in fortune, and its vulnerability before death. He said many times that his favorite poem was William Knox's *Oh, Why Should the Spirit of Mortal Be Proud?* although he did not know the author. He quoted Grey's *Elegy,* liked Holmes's *Last Leaf,* and earlier had tried to express his own feelings of sad fatalism in verse:

> "I range the fields with pensive tread,
> And pace the hollow rooms;
> And feel (companions of the dead)
> I'm living in the tombs."[11]

A few days after the funeral there appeared in the *Illinois Journal* these unsigned lines entitled *Little Eddie.*

> The angel death was hovering nigh,
> And the lovely boy was called to die.
> Bright is the home to him now given,
> For "of such is the kingdom of heaven."

They were probably written by Abraham or Mary, especially since they had the phrase "Of such is the Kingdom of Heaven" engraved on Edward Baker's white marble stone.

The clearest expression we have of Lincoln's view of personal immortality comes less than a year after Eddie's death in a letter to his stepbrother John Johnston when he heard of the serious illness of his father. The letter suggests strained relations between father and son, but closes on a deeply biblical note. "I sincerely hope Father may yet recover his health; but at all events tell him to remember to call upon, and confide in, our great, and good, and merciful Maker; who will not turn away from him in any extremity. He notes the fall of a sparrow, and numbers the hairs of our heads; and He will not forget the dying man, who puts his trust in Him. Say to him that if we could meet now, it is doubtful whether it would not be more painful than pleasant; but that if it is to be his lot to go now, he will soon have a joyous [meeting] with many loved ones gone before; and where [the rest] of us, through the help of God, hope ere-long [to join] them."[12]

A strong friendship grew up between Lincoln and his pastor Dr. Smith that lasted until Lincoln's death. Describing Smith to Seward as "an intimate personal friend of mine," he appointed him American consul at Dundee when the distinguished preacher retired to his native Scotland. After Lincoln's death the family sent Dr. Smith one of the President's gold-headed canes as a token of their esteem and affection.

More significant, however, for Lincoln's religious development was the intellectual challenge of his friend's mind. Shortly after Eddie's death the Lincolns went to Kentucky to stay with the Todd relatives. Here Lincoln found and partly read a learned work entitled *The Christian's Defense* by the Rev. James Smith, published in Cincinnati in 1843.

The book was the product of three weeks of formal debates at Columbus, Mississippi, between the author and a popular "freethinker" named Olmstead. The author was, of course, the man who had comforted them in bereavement. This personal tie and the drama of the religious debate undoubtedly drew Lincoln's interest. Upon his return to Springfield Lincoln secured a copy of the book and studied it carefully.

Dr. Smith described the result to Herndon, who had written to him in Scotland. "It was my honor to place before Mr. Lincoln arguments designed to prove the divine authority and inspiration of the Scriptures, accompanied by the arguments of infidel objectors in their own language. To the arguments on both sides Mr. Lincoln gave a most patient, impartial, and searching investigation. To use his own language, he examined the arguments as a lawyer who is anxious to investigate truth investigates testimony. The result was the announcement made by himself that the argument in favor of the divine authority and inspiration of the Scriptures was unanswerable."[13]

This 600-page tome is worthy of careful study, for it is one of the very few technical books on theology read by Lincoln. Despite the discovery of new archaeological evidence and the acceptance of quite different historical perspectives than those available then, the book is still an interesting study. Most of the defenses of Christianity in that day were addressed to the already converted. From his frontier background Lincoln knew the abuse heaped upon the "infidel" in standard dogmatics. Rather than be given an answer, he would be consigned to Satan.

But here was a welcome change. The author had himself been a skeptic. He quoted extensively from Paine, Volney, Robert Taylor, and other objectors to the authority of the Bible. Lincoln was himself beginning to quote more from his opponents as a preliminary to fairer and more effective rebuttal. Smith then endeavored to give a logical and even legal reply according to rules of testimony. This two-way conversation apparently interested Lincoln a great deal. In an appendix Dr. Smith drew from a standard legal work, Starkie's *Practical Treatise on the Law of Evidence,* to refute Hume's arguments about testimony.

Here Lincoln found convincing explanations of some of the points made by Paine and Volney that may have troubled him through the years. While the author's answers contain much that could not be accepted today they were advanced then in good faith. There was an attempt to show that modern geological discoveries and classical historians both constituted external testimony to the truth of Scripture. Since the language of the Bible was at times metaphorical it was unreasonable, urged Dr. Smith, to demand that it speak like Newtonian science.

It was argued that God accommodated Himself in revelation to the cultural level of his people, suggesting thereby a view of development that the author never articulated himself. He did, however, have a feeling for the sweep of human history from its earliest beginnings. The history of the Jews and of the early Christian community were the keys to its understanding. Illustrations of the significance of holidays like Independence Day for the American people were creatively used to show how Sunday was the celebration of the Christian act of redemption. There is no wall between an embalmed "sacred" history and current "secular" history. History is all of one piece, with the Bible its key.

There are some interesting parallels between positions advanced by Dr. Smith and those later stated by Lincoln. Whether or not there is direct influence is incapable of proof at this date. The author's argument that the universal practice of sacrifice is a testimony to the Atonement can serve both as a sample of his writing and as an illustration of a possible area of influence. The Honorable Orlando Kellogg came to the

White House in the interests of a wounded soldier who had previously deserted. Lincoln, granting a pardon, asked a question: "Kellogg . . . isn't there something in Scripture about the 'shedding of blood' being 'the remission of sins'?"

"Is there," argued Dr. Smith, "any rational mode of explaining the universality of this practice save that it originated in a divine appointment . . . and that it continued to be regarded as a divine institution wherever the true God was known, until its whole design was consummated in the voluntary offering up of Christ himself at Calvary! Admit this view of the case, and the whole history of animal sacrifice is plain and satisfactory. The providence of God, watching over the perpetuity of his own institutions, secures the universal observance of this practice; and in the practice itself we have the great leading principle of his government, as exercised towards a fallen and guilty race, 'that without the shedding of blood there is no remission.' "[14]

More significant, however, than parallels in content is the possibility of a direct influence in method. Smith's procedure was to establish everything that he could in Scripture by the use of reason before appealing to faith. Lincoln apparently felt in his New Salem years and for some time after that reason and faith were opposed to each other. Increasingly he came to regard them as complementary. His own clearest statement about the interrelations of reason and faith was made to Speed in the summer of 1864. Finding Lincoln reading the Bible in his room at the Soldiers' Home, Speed said, "If you have recovered from your skepticism, I am sorry to say that I have not." Then, reported Speed, "Looking me earnestly in the face, and placing his hand on my shoulder, he said, 'You are wrong, Speed; take all of this Book upon reason that you can, and the balance on faith, and you will live and die a happier and better man.' "[15]

Thomas Lewis, a Springfield lawyer and deacon of the church Lincoln attended, and John T. Stuart, his early law partner, both spoke of the impact of Dr. Smith's book on Lincoln. Ninian W. Edwards, his brother-in-law, reported Lincoln told him, "I have been reading a work of Dr. Smith on the evidences of Christianity, and have heard him preach and converse on the subject and am now convinced of the truth of the Christian religion."[16]

A few months after Mrs. Lincoln became a member of the First Presbyterian Church in 1852, Lincoln upon invitation of the session lectured on the Bible. Perhaps we have the outline of that address embedded in a letter Dr. Smith wrote to Herndon. His pastor reported Lincoln to have said about the Bible:

"It seems to me that nothing short of infinite wisdom could by any possibility have devised and given to man this excellent and perfect moral code. It is suited to men in all the conditions of life, and inculcates all the duties they owe to their Creator, to themselves, and to their fellow men."[17]

Sometime during this period Lincoln read Robert Chambers's *Vestiges of the Natural History of Creation,* first published in Edinburgh in 1844. Its aim was to reconcile the newly discovered materials in geology, paleontology, and biology with the Christian doctrine of creation. The book passed through many revisions, each time incorporating new scientific data. Herndon described its influence upon Lincoln. "The treatise interested him greatly, and he was deeply impressed with the notion of the so-called 'universal law'—evolution; he did not greatly extend his researches, but by continual thinking in a single channel seemed to grow into a warm advocate of the new doctrine."[18]

In a letter, Herndon wrote that Lincoln subsequently read the sixth edition of *Vestiges* and as a result "adopted the progressive and development theory as taught more or less directly in that work. He despised speculation, especially in the metaphysical world. He was purely a practical man."[19] The significance of this is that Lincoln thought his way through to the acceptance of an evolutionary framework from a writing that was not anti-Christian but instead showed the living God at work in the processes described by science.

More of this liberal religious approach he could discover in the books of Theodore Parker, pressed upon him by Herndon, or in the collected works of William Ellery Channing which Jesse Fell presented to him "some eight or ten years prior to his death."[20]

"Stamped with the Divine Image and Likeness"

Chapter 5

Lincoln had to shout to make himself heard above the shuffling crowd as it surged away from the House of Representatives in Springfield. From the vantage point of the stairway he announced that he would answer Judge Douglas's speech the next day. Douglas had just addressed a vast crowd gathered on October 3, 1854, for the Springfield Fair. He had boldly defended his role in the passage of the Kansas-Nebraska Act. His justification was the principle of popular sovereignty whereby the settlers would themselves determine whether or not to become slave states. He argued against his opponents that his stand was not a repeal of the Missouri Compromise and its subsequent reaffirmation in 1850, but a truly democratic extension of its very principles.

Douglas's Nebraska bill, Lincoln later said, "roused him as

nothing had before." It drew him away from the law at a time when "he had been losing interest in politics." A marked change, however, characterized Lincoln on his return to politics in 1854 from his virtual withdrawal from them in 1849. Then he had been the local boy who made good by hard work and pleasing personality, but above all by working with the Whig machine. When Lincoln returned to politics he still had the drive of ambition and the hunger for recognition, but these were not now paramount: they were held in check and ennobled by dedication to a cause. The authority of moral leadership is increasingly to be recognized in his speeches. In finding a cause that was bigger than himself Lincoln actually found himself.

His biographer, Benjamin Thomas, evaluates the new Lincoln: "The impact of a moral challenge, purging Lincoln of narrow partisanship and unsure purpose, is about to transform an honest, capable, but essentially self-centered small-town politican of self-developed but largely unsuspected talents into a statesman who will grow to world dimensions."[1]

Here also is the key to the religious development of Lincoln up to the time of the presidency. While the documents of this period offer little in the way of a personal statement of faith, they are vibrant with indignation over Douglas's indifference to the moral question of slavery. He was doubly disturbed when this moral indifference was offered as a calculated policy for political action. To state that Lincoln's opposition to slavery was rooted in his conception of morality is only, however, to give a part of the picture. His conception of morality itself was derived from religion. He would later say that without the revelation of the Bible man could not distinguish between right and wrong. Lincoln did not believe in an independent system of moral values. The good for Lincoln was ultimately anchored in the will of God, not subject to human likes or dislikes.

The motivating forces in his religious view of the world that shaped his moral decision on slavery are distinguishable in at least two directions. One was a theological understanding of the significance of work that was quarried from the Bible. "In the sweat of thy face shalt thou eat bread." This came increasingly to the fore in the later years of the presidency. The second was a passion for the fundamental equality of all men "in the sight of God" that Lincoln believed had been covenanted in the Declaration of Independence. That document had for him the force of revelation itself. It was this second point that he would affirm again and again in the classical debates with Douglas.

Far too many studies of Lincoln's religion, in their concentration upon personal piety, have neglected the deep, spiritual dynamic that underlay his attitude toward slavery. Lincoln's religion cannot be hermetically sealed off from his social, economic, and political attitudes. His political action, as revealed by his own words, was ultimately the social expression of an understanding of God and of man that demanded responsible activity. This is contrary to a widespread modern opinion that religion should be a separate interest or even hobby in life and should not be allowed to influence fields like politics.

No person as deeply immersed as Lincoln in the biblical

faith could possibly take such a view. His criticism of the churches of his day was that they neglected this fundamental love of God and of neighbor by too much introverted attention upon correctness in theological opinion. Prophetically Lincoln saw that concern for orthodoxy substituted for practical obedience.

This perspective in Lincoln must not be confused, however, with another view that the churches themselves should mix in politics. He took a dim view of preachers who used the pulpit for politics and said he preferred those who preached "the gospel." By this, however, he meant that the layman's task was to put this gospel to work not merely in individual piety, although certainly there, but also in responsible political activity. As a biblical believer Lincoln saw God dealing with men not merely as isolated individuals with capacities for piety but as men in social orders that are answerable to the Almighty.

Lincoln was very far from the modern heresy of believing that religion itself is a very good thing irrespective of its relation to everyday life. A Tennessee woman once sought a pardon for her husband, who was held prisoner in the Johnson's Island Camp, on the ground that her husband was a religious man. Lincoln granted the pardon but replied, "You say your husband is a religious man; tell him when you meet him, that I say I am not much of a judge of religion, but that, in my opinion, the religion that sets men to rebel and fight against their government, because, as they think, that government does not sufficiently help *some* men to eat their bread in the sweat of *other* men's faces, is not the sort of religion upon which people can get to heaven!"[2]

The religious substratum of Lincoln's attitude toward slavery needs to be highlighted today for still another reason. It is in danger of being wholly left out of the picture in the concentration of a number of recent studies on his political maneuvering and on the problems about the constitutionality of the issue. What is needed is to see the way in which the religious ground is expressed in the shifting political attitudes and strategies. The full picture requires all the components and not just selected ones.

On the fourth of October in a suffocating heat wave Lincoln, without coat and tie, stood up before a packed audience to answer Douglas. By invitation the judge sat before him in the front row. The report of this Springfield address appears in much-abbreviated form in the *Illinois Journal,* but the three-hour speech was fortunately repeated on October 15 in Peoria and recorded.

Lincoln argued that the founders of the Republic had had to accept the slavery that existed among them. In prohibiting slavery in the Northwest Territories, however, by the Ordinance of 1787 they adopted a principle of putting it in course of ultimate extinction. This principle of the non-extension of slavery to territories had, Lincoln argued, been consistently held. The Missouri Compromise of 1820 and its reaffirmation in 1850 resulted from a partial giving and taking on both sides.

Now, however, the Douglas Nebraska bill in effect destroyed the previous agreements by providing for squatter sovereignty under the high-sounding principle of democracy. "This *declared* indifference, but as I must think, covert *real zeal* for the spread of slavery, I can not but hate. I hate it because of the monstrous injustice of slavery itself. I hate it because it deprives our republican example of its just influence in the world . . ."[3] Because of the slavery issue, Americans were driven into war upon civil liberty and into contempt for the Declaration of Independence.

He wanted to assure the South that he was not prejudiced against them. The problem was national, not sectional. "They are just what we would be in their situation. If slavery did not now exist amongst them, they would not introduce it. If it did now exist amongst us, we should not instantly give it up. This I believe of the masses north and south."

If Lincoln had the power, he would not know how to deal with the existing situation. Probably he would free all the slaves and send them to Liberia. His own feelings would not allow him to make them politically and socially his equals. He wanted the South to know that he stood by their constitutional right to own slaves in the slave states and to have a fugitive slave law enforced. He was concerned only to prevent its spread. He wanted to keep it just where the Founding Fathers had placed it—in course of ultimate extinction "in God's good time."

His quarrel was with Douglas's interpretation of a great principle. "The doctrine of self-government is right—absolutely and eternally right—but it has no just application, as here attempted. Or perhaps I should rather say that whether it has such just application depends upon whether a negro is *not* or *is* a man. If he is *not* a man, why in that case, he who *is* a man may, as a matter of self-government, do just as he pleases with him. But if the negro *is* a man, is it not to that extent a total destruction of self-government, to say that he too shall not govern *him-*

self? When the white man governs himself that is self-government; but when he governs himself, and also governs *another* man, that is *more* than self-government—that is despotism. If the negro is a *man*, why then my ancient faith teaches me that 'all men are created equal'; and that there can be no moral right in connection with one man's making a slave of another.

"Judge Douglas frequently, with bitter irony and sarcasm, paraphrases our argument by saying, 'The white people of Nebraska are good enough to govern themselves, *but they are not good enough to govern a few miserable negroes!*'

"Well I doubt not that the people of Nebraska are, and will continue to be as good as the average of people elsewhere. I do not say the contrary. What I do say is, that no man is good enough to govern another man, *without that other's consent.* I say this is the leading principle—the sheet anchor of American republicanism. Our Declaration of Independence says:

" 'We hold these truths to be self evident: That all men are created equal; that they are endowed by their Creator with certain inalienable rights; that among these are life, liberty and the pursuit of happiness. That to secure these rights, governments are instituted among men, DERIVING THEIR JUST POWERS FROM THE CONSENT OF THE GOVERNED.' "

Later in the speech he said, "To deny these things is to deny our national axioms, or dogmas." Here is the key that unlocks Lincoln's fundamental attitude toward slavery. The axiom that all men are created equal had for him the force of religious dogma. Jefferson was Lincoln's mentor in political philosophy, but Lincoln's religious perspective was more concrete than Jefferson's. Lincoln went deeper than "self-evidence" for these truths, even further than the somewhat deistic phrase "endowed by their Creator." For Lincoln the Creator was the living God of history, revealed in the Bible, Whose judgments were continuously written on the pages of history and recorded in the human conscience.

He drove the problem back to a view of human nature. "Slavery is founded in the selfishness of man's nature—opposition to it, in his love of justice. . . . Repeal the Missouri Compromise . . . repeal all past history, you still can not repeal human nature." Then in the phraseology of the Bible he took a stand on bedrock: "It still will be the abundance of man's heart, that slavery extension is wrong; and out of the abundance of his heart, his mouth will continue to speak."

The proof that Lincoln understood the Declaration not merely as a national axiom but as a universal dogma is found in his Lewistown, Illinois, speech of August 17, 1858. It is a shame that this speech is so little known. Here the Declaration was specifically identified with the Genesis doctrine of man as created in the divine image. After quoting the same section of the Declaration that he had used in the Peoria speech, he went on to establish its theological foundation.

"This was their majestic interpretation of the economy of the Universe. This was their lofty, and wise, and noble understanding of the justice of the Creator to His creatures. Yes, gentlemen, to *all* His creatures, to the whole great family of man. In their enlightened belief, nothing stamped with the Divine image and likeness was sent into the world to be trodden on, and degraded, and imbruted by its fellows. They grasped not only the whole race of man then living, but they reached forward and seized upon the farthest posterity. They erected a beacon to guide their children and their children's children, and the countless myriads who should inhabit the earth in other ages. Wise statesmen as they were, they knew the tendency of prosperity to breed tyrants, and so they established these great self-evident truths, that when in the distant future some man, some faction, some interest, should set up the doctrine that none but rich men, or none but white men, were entitled to life, liberty and the pursuit of happiness, their posterity might look up again to the Declaration of Independence and take courage to renew the battle which the fathers began—so that truth, and justice, and mercy, and all the humane and Christian virtues might not be extinguished from the land; so that no man would hereafter dare to limit and circumscribe the great principles on which the temple of liberty was being built."[4]

Lincoln's astute rival in the great debates of 1858 understood Lincoln's theological perspective on the slavery issue. Three times Douglas attacked Lincoln on just this point. At Ottawa, Douglas accused him of having "learned by heart Parson Lovejoy's catechism." He said that Lincoln "holds that the negro was born his equal and yours, and that he was endowed with equality by the Almighty, and that no human law can deprive him of these rights, which were guaranteed to him by the Supreme Ruler of the Universe."

In the seventh and last debate at Alton, Douglas summarized his stand with merciless clarity. "I care more for the great principle of self-government, the right of the people to rule, than I do for all the negroes in Christendom. [Cheers.] I would not endanger the perpetuity of this Union. I would not blot out the great inalienable rights of white men for all the negroes that ever existed. [Renewed applause.]"

Judge Douglas had pointed out that because the Founding Fathers did not abolish slavery they understood "all men" in the Declaration to mean "white men." Lincoln repudiated this narrowed interpretation in his Chicago speech, strengthening the religious dimension in his view with Scripture.

"We had slavery among us, we could not get our Constitution unless we permitted them to remain in slavery, we could not secure the good we did secure if we grasped for more, and having by necessity submitted to that much, it does not destroy the principle that is the charter of our liberties. Let that charter stand as our standard.

"My friend has said to me that I am a poor hand to quote Scripture. I will try it again, however. It is said in one of the admonitions of the Lord, 'As your Father in Heaven is perfect, be ye also perfect.' The Savior, I suppose, did not expect that any human creature could be perfect as the Father in Heaven; but He said, 'As your Father in Heaven is perfect, be ye also perfect.' He set that up as a standard, and he who did most towards reaching that standard, attained the highest degree of moral perfection. So I say in relation to the principle that all men are created equal, let it be as nearly reached as we can."[5]

In this quotation Lincoln demonstrated his conviction that God's basic law of freedom in the creation of man was meant to be dynamically and progressively embodied in the positive legislation of states. T. Harry Williams, an editor of a selection of Lincoln's writings, establishes the religious orientation to Lincoln's theory of jurisprudence. "The doctrine of the existence of a supernatural fundamental law and the corollary that fundamental human law approximated the higher code was an important part of Lincoln's philosophy."[6]

In a speech at Cincinnati he warned those who justified slavery as a biblical ordinance that the institution mentioned in the Bible was *white* slavery, not *black* slavery. He doubted whether white slaveholders really wanted to defend their peculiar institution by a point that might make them the slaves of the next more powerful white man.

The Rutgers editors have followed Nicolay and Hay in assigning to the period of the great debates a remarkable fragment on pro-slavery theology. With the scorn of an Old Testament prophet Lincoln lays bare the taint of self-interest that corrupts man's judgments. His reference to Dr. Ross is probably to the pamphlet *Slavery as Ordained of God,* by the Rev. Frederick A. Ross of Alabama, published in 1857.

"Suppose it is true, that the negro is inferior to the white, in the gifts of nature; is it not the exact reverse justice that the white should, for that reason, take from the negro, any part of the little which has been given him? '*Give* to him that is needy' is the Christian rule of charity; but 'Take from him that is needy' is the rule of slavery.

"The sum of pro-slavery theology seems to be this: 'Slavery is not universally *right,* nor yet universally *wrong;* it is better for *some* people to be slaves; and, in such cases, it is the Will of God that they be such.'

"Certainly there is no contending against the Will of God; but still there is some difficulty in ascertaining, and applying it, to particular cases. For instance we will suppose the Rev.

Dr. Ross has a slave named Sambo, and the question is 'Is it the Will of God that Sambo shall remain a slave, or be set free?' The Almighty gives no audible answer to the question, and his revelation—the Bible—gives none, or, at most, none but such as admits of a squabble, as to its meaning. No one thinks of asking Sambo's opinion on it. So, at last, it comes to this, that *Dr. Ross* is to decide the question. And while he considers it, he sits in the shade, with gloves on his hands, and subsists on the bread that Sambo is earning in the burning sun. If he decides that God wills Sambo to continue a slave, he thereby retains his own comfortable position; but if he decides that God wills Sambo to be free, he thereby has to walk out of the shade, throw off his gloves, and delve for his own bread. Will Dr. Ross be actuated by that perfect impartiality, which has ever been considered most favorable to correct decisions?

"But, slavery is good for some people!!! As a *good* thing, slavery is strikingly peculiar, in this, that it is the only good thing which no man ever seeks the good of, for *himself*.

"Nonsense! Wolves devouring lambs, not because it is good for their own greedy maws, but because it is good for the lambs!!!"[7]

To his best friend Joshua Speed, a slaveholder, he had written straight from the shoulder: "It is hardly fair for you to assume, that I have no interest in a thing which has, and continually exercises, the power of making me miserable. You ought rather to appreciate how much the great body of the Northern people do crucify their feelings, in order to maintain their loyalty to the Constitution and the Union." Later in the same letter he related his feeling on slavery to the question of Speed about his political affiliation.

"I am not a Know-Nothing. That is certain. How could I be? How can any one who abhors the oppression of negroes, be in favor of degrading classes of white people? Our progress in degeneracy appears to me to be pretty rapid. As a nation, we began by declaring that *'all men are created equal.'* We now practically read it 'all men are created equal, *except negroes.'* When the Know-Nothings get control it will read 'all men are created equal, except negroes, *and foreigners, and Catholics.'* When it comes to this I should prefer emigrating to some country where they make no pretence of loving liberty—to Russia, for instance, where despotism can be taken pure, and without the base alloy of hypocrisy."[8]

About the same time he wrote to George Robertson, professor of law at Transylvania College, confessing that he could not see how the problem of slavery, the basic self-contradiction of American freedom, could be solved. He appealed beyond human measures to the God Who directs history. "Our political problem now is 'Can we, as a nation, continue together *permanently—forever*—half slave, and half free?' The problem is too mighty for me. May God, in his mercy, superintend the solution."[9]

In the Second Inaugural he would show how God had dealt with this cancer in the nation. In the presidential years he could move forward on the slavery issue only when he had first resolved the conflict between his oath to uphold the Constitution with its entrenched institution of slavery and the practical war measures which he felt obliged to take in order to preserve and then restore the Union. Behind this dramatic struggle, however, this theological orientation on slavery continually exerted pressure on him.

He believed slavery was a moral evil; he believed it was against the constitution of the universe; he believed it was a mocking of the God Who had created the Negro in His own image as He had all men. He believed that the injustice of slavery invited the wrath of God.

He expressed this last conviction strongly at Columbus, Ohio, on September 16, 1859, in a contrast he drew between Jefferson and Douglas: ". . . but that man did not take exactly this view of the insignificance of the element of slavery which our friend Judge Douglas does. In contemplation of this thing, we all know he was led to exclaim, 'I tremble for my country when I remember that God is just!' We know how he looked upon it when he thus expressed himself. There was danger to this country—danger of the avenging justice of God in that little unimportant popular sovereignty question of Judge Douglas. He supposed there was a question of God's eternal justice wrapped up in the enslaving of any race of men, or any man, and that those who did so braved the arm of Jehovah—that when a nation thus dared the Almighty every friend of that nation had cause to dread His wrath. Choose ye between Jefferson and Douglas as to what is the true view of this element among us."[10]

The biblical coloration of this motivating conviction would become even stronger in his many expositions of the text: "In the sweat of thy face, thou shalt eat bread." To a degree he anticipated this later position in a fragment on free labor, probably from his Cincinnati speech of September 1859. "As labor is the common *burthen* of our race, so the effort of *some* to shift their share of the burthen on to the shoulders of *others*, is the great, durable, curse of the race. Originally a curse for transgression upon the whole race, when, as by slavery, it is concentrated on a part only, it becomes the double-refined curse of God upon his creatures."[11]

Lincoln's use of the titles "Lord" and "Savior" in the passages quoted and in other references during this late Springfield period is about as far as the direct evidence goes for any distinctly Christian orientation at this time. It is important, however, not to underestimate the significance of these ascriptions. In his many quotations of Christ's words Lincoln does not use them as the sayings of "the great teacher" in a humanist sense but as the teaching of one who is "the Lord" and "the Savior" in an implicitly Christological sense.

There are, moreover, some important comments by friends of his that, if evaluated with caution, help to give more background to his understanding of Christ as "the Savior" and "the Lord." There is a high probability in this evidence because it builds with some consistency on the statement of Mentor Graham, previously quoted, that Lincoln wrote an essay about 1833 on predestinated universal salvation in criticism of the orthodox doctrine of endless punishment.

It is also consistent with the evidence that in 1850 Lincoln, through the reading of his pastor's *The Christian's Defense* and his own wrestling with the problem, became convinced intellectually of the validity of the biblical revelation. This would not have meant that Lincoln accepted the Calvinistic theology of his clergyman friend. Lincoln's conviction that God would restore the whole of creation as the outcome of Christ's Atonement would have been in itself a bar to membership in the Springfield church he attended. His often expressed impatience with creedal formularies as a condition for church membership surely reflects the struggle between his independent analysis of biblical teaching and the pressures upon him to conform to conventional denominationalism. The weight of this evidence is, moreover, considerably strengthened when the Jesse Fell statement, often thought to be impossible of reconciliation with those of Isaac Cogdal and Jonathan Harnett, is seen to be much less of a problem than supposed.

Isaac Cogdal, who had known Lincoln from the time of the New Salem period, recalled a discussion on religion in Lincoln's office in 1859. ". . . Herndon was in the office at the time. Lincoln expressed himself in about these words: He did not nor could not believe in the endless punishment of any one of the human race. He understood punishment for sin to be a Bible doctrine; that the punishment was parental in its object, aim, and design, and intended for the good of the offender; hence it must cease when justice is satisfied. He added that all that was lost by the transgression of Adam was made good by the atonement: all that was lost by the fall was made good by the sacrifice, and he added this remark, that punishment being a 'provision of the gospel system, he was not sure but the world would be better off if a little more punishment was preached by our ministers, and not so much pardon of sin.' "[12]

This last comment has all the earmarks of an authentic

Lincoln utterance. It shows the independence with which he handled biblical evidence in reaching a conclusion at odds with current preaching practice. His basic acceptance of "the provisions of the gospel system" gave him this leverage against popular religion. The prophetic note of a God of mercy Who punishes the sins of men in the judgments of history with a view to reformation would become a dominant theme in his later religious utterances, especially in his presidential proclamations.

Here Lincoln saw much more clearly than most parsons of his day that there is an unbiblical preaching of pardon for sin that, by extricating the individual man from his historical and social setting, gives him illusions about punishment in this world and the next. Lincoln understood the gospel to mean the salvation of all men in both a this-worldly and a next-worldly framework. Many ministers had reduced Christianity to a message of escape for individuals from punishment in the next world. Lincoln's emphasis upon sacrifice as a key to "the economy of the universe" would be immeasurably deepened in the crucible of the war years.

The second statement was one dictated by Jonathan Harnett of Pleasant Plains, describing a theological discussion in 1858 in Lincoln's office. "Lincoln covered more ground in a few words than he could in a week, and closed up with the restitution of all things to God, as the doctrine taught in the scriptures, and if anyone was left in doubt in regard to his belief in the atonement of Christ and the final salvation of all men, he removed those doubts in a few questions he answered and propounded to others. After expressing himself, some one or two took exceptions to his position, and he asked a few questions that cornered his interrogators and left no room to doubt or question his soundness on the atonement of Christ, and salvation finally of all men. He did not pretend to know just when that event would be consummated, but that it would be the ultimate result, that Christ must reign supreme, high over all, The Saviour of all; and the supreme Ruler, he could not be with one out of the fold; all must come in, with his understanding of the doctrine taught in the scriptures."[13]

The Harnett dictation strengthens the testimony to Lincoln's belief in Christ as ultimately the Savior of all and points to that fifteenth chapter of I Corinthians previously discussed in connection with Mentor Graham's statement. This was clearly the biblical locus for Lincoln's arguments on this theme.

The Jesse Fell statement is better known than the two just quoted chiefly because it was written for Colonel Ward Lamon's *Life of Abraham Lincoln* and was reprinted by Herndon in his biography. The problem with this statement is that, taken in isolation from other evidence, it tends to suggest to the reader more than it actually says. If this were the sole statement on Lincoln's religion it might be concluded that Lincoln wholly rejected Christianity for a practical belief "summed up . . . in these two propositions: the Fatherhood of God, and the brotherhood of man."

Fell was a pioneer in introducing the liberal Christianity of William Ellery Channing into Illinois, being responsible for establishing a Unitarian Church in Bloomington shortly after 1859. Lincoln's habit of saying that he would join a church that inscribed over its altars as a condition of membership "the

Savior's summary of the Gospel" obviously opened an area of discussion between Lincoln and Fell, but there was still a major difference between the two. Lincoln adopted a liberal attitude that brought him into disagreement with the "orthodox" religion on such points as limited atonement and eternal punishment. But it was on the very anvil of a belief in Christ as atoner and God as punisher for purposes of reformation that he hammered out his criticism. In other words, Lincoln took a liberal attitude toward biblical orthodoxy as he understood it.

Fell, on the other hand, went beyond the liberal attitude of Lincoln and began to substitute for biblical orthodoxy a set of distinctly liberal dogmas such as the fatherhood of God and the brotherhood of man. Fell said later that had the circumstances of his life been different he would have wanted to be a Unitarian minister. Channing's Unitarianism, moreover, was of the "high" variety. It was theistic and not humanist. It had a Christology of its own.

Fell says of Lincoln's religion in the Springfield period: "whilst he held many opinions in common with the great mass of Christian believers, he did not believe in what are regarded as the orthodox or evangelical views of Christianity.

"On the innate depravity of man, the character and office of the great head of the Church, the Atonement, the infallibility of the written revelation, the performance of miracles, the nature and design of present and future rewards and punishments (as they are popularly called) and many other subjects, he held opinions utterly at variance with what are usually taught in the Church."[14]

Here is where the unwary reader may read more out of the

passage than Fell meant to put in. Fell is not saying that Lincoln did not believe *at all* in Christ as Atoner nor that Lincoln did not *at all* accept the belief that God punishes. He says that *on* these subjects his views were "utterly at variance" with current orthodoxy. The word "utterly" is perhaps too strong in its suggestions, but it is still true in perspective. The Calvinist theology of that day stressed an atonement by Christ limited to those predestined to be saved. The fate of unbelievers in eternal flames was a commonplace of the preacher of that period.

Fell completed the substance of his description as follows: "He fully believed in a superintending and overruling Providence that guides and controls the operations of the world, but maintained that law and order, not their violation or suspension, are the appointed means by which this Providence is exercised." The theme of God's providential guidance would be Lincoln's clue to interpreting religiously the meaning of the nation's history in the presidential years.

The trouble with so many studies of Lincoln's religion is that they soon bog down in the clash of other people's testimony about his belief. It is therefore a relief to turn from the Cogdal, Harnett, and Fell statements to an exciting new source revealed in 1957 with the discovery of a devotional book first published in 1852 and inscribed with Lincoln's name. In an introduction to the reprinting of *The Believer's Daily Treasure; or, Texts of Scripture Arranged for Every Day in the Year,* Sandburg suggests that either the book was given to Lincoln by a person for whom he cared or that he held the book itself in high regard, for he seldom wrote his name, even for purposes of identification, in his own books.

It was of course in 1852 that his wife joined the First Presbyterian Church and that he attended some of the inquirers' sessions in addition to their regular attendance at Sunday worship. Possibly his wife gave him this tiny book to carry with him on the judicial circuit. Possibly his pastor gave him this biblical devotional. It is just the book that could have fed Lincoln's soul. Completely free of denominational dogmas, it presents the evangelical core of the Bible as a practical guide to everyday life.

Its many references from the Psalms may throw light on a remark Lincoln later made to a nurse in the White House: "They are the best, for I find in them something for every day in the week."[15] Although we do not have any statement of Lincoln's as to how much or little he used this devotional manual, there may well have been considerable influence. The following quotations have been selected to give the flavor of the book and to show some parallelism with the great religious themes of his later state papers.[16]

"Believer's Evidences

FEBRUARY 27 — LOVE TO ENEMIES

Love ye your enemies, and do good, and lend, hoping for nothing again; and your reward shall be great, and ye shall be the children of the Highest: for he is kind unto the unthankful and to the evil. Luke 6:3

Lord, shall thy bright example shine
In vain before my eyes?
Give me a soul akin to thine,
To love my enemies.

Duties of the Believer — in the Church

MAY 13 — MUTUAL CANDOUR

Judge not, that ye be not judged. Why beholdest thou the mote that is in thy brother's eye, but considerest not the beam that is in thine own eye? Matt. 7:1,3

Make us by thy transforming grace,
Great Saviour, daily more like thee
Thy fair example may we trace,
To teach us what we ought to be.

Joys of the Believer

JULY 20 — TRIBULATION A SOURCE OF JOY

We glory in tribulation: knowing that tribulation worketh patience; and patience, experience; and experience, hope. Rom. 5:3,4."

When the balloting came for the Illinois legislature which would itself elect the United States senator, the Republican candidates received four thousand more votes than their opponents. An unrepresentative system of apportionment, however, made possible the election of Douglas. Lincoln had lost the senatorship. Again he turned to his much-neglected law practice. He did not of course realize that, while he could not be senator in 1859, he would, chiefly by the fame of these debates, become the President in 1861.

To his personal friend and physician he wrote, "I am glad I made the late race. It gave me a hearing on the great and durable question of the age, which I could have had in no other way; and though I now sink out of view, and shall be forgotten, I believe I have made some marks which will tell for the cause of civil liberty long after I am gone."[17]

Contrary to his prophecy, he did not sink out of view. In the excitement leading up to Lincoln's nomination at Chicago and even in the presidential campaign itself, since by the conventions of that day candidates did not campaign for themselves, there is little in Lincoln's own words that throws light on his continuing religious development. Since the slavery theme has been chosen to illustrate the theological rootage of Lincoln's attitude on the great question of his age it will be well to note a few developments here. The debates with Douglas convinced Lincoln that he needed to do basic research on the attitude of the Founding Fathers to "the peculiar institution." Long hours were spent poring over materials in the Illinois State Library and in digesting Elliot's *Debates on the Federal Constitution.* The fruit of this scholarly research strengthened the Cooper Union address in New York in February 1860. This speech made him better known in the East.

A new incisiveness and a rhetoric strengthened by homespun but graphic illustrations show his intellectual growth in this period. At Hartford, Connecticut, on March 5, 1860, he stated with a succinctness never reached in his debates with Douglas the Republican opposition to the spread of slavery.

"For instance, out in the street, or in the field, or on the prairie I find a rattlesnake. I take a stake and kill him. Everybody would applaud the act and say I did right. But suppose the snake was in a bed when children were sleeping. Would I do right to strike him there? I might hurt the children; or I might not kill, but only arouse and exasperate the snake, and he might bite the children. Thus, by meddling with him here, I would do more hurt than good. Slavery is like this. We dare not strike at it where it is. The manner in which our Constitution is framed constrains us from making war upon it where it already exists. The question that we now have to deal with is, 'Shall we be acting right to take this snake and carry it to a bed where there are children?' The Republican party insists upon keeping it out of the bed."[18]

At New Haven he gave a humorous twist to his point that the defense of slavery was vitiated by ideological taint. His story parallels his earlier one about Dr. Ross, Sambo, and pro-slavery theology.

"The slaveholder does not like to be considered a mean fellow for holding that species of property, and hence he has to struggle within himself and sets about arguing himself into the belief that slavery is right. The property influences his mind. The dissenting minister, who argued some theological point with one of the established church, was always met with the reply, 'I can't see it so.' He opened the Bible, and pointed him to a passage, but the orthodox minister replied, 'I can't see it so.' Then he showed him a single word—'Can you see that?' 'Yes, I see it,' was the reply. The dissenter laid a guinea over the word and asked, 'Do you see it now?' [Great laughter.]

So here. Whether the owners of this species of property do really see it as it is, it is not for me to say, but if they do, they see it as it is through 2,000,000,000 of dollars, and that is a pretty thick coating. [Laughter.]"[19]

He felt keenly the inertia and resistance of the clergy on the moral issue of slavery. Shortly before the election in October 1860 he and Newton Bateman thumbed through a list of how Springfield residents were proposing to vote. When he concluded that twenty out of twenty-three ministers would vote against him, he asked sadly how it could be possible for men with the Bible in their hands to support candidates who were in effect pro-slavery.

Most of the incidents and themes in Lincoln's religious development in the Springfield period are summed up in his moving Farewell Address. Here was his deep sense of belonging to the people and the sadness of leaving friends. He recalled the birth of his children and the death of one of them. Seven states were already in secession and many federal forts and arsenals had been seized in the South. Lincoln spoke in sober tones of the responsibility placed upon him, the great and deepening theme of the years to come. Here was the foreboding that he might not return, that premonition that would later be expressed as a doubt that he himself would outlive the national conflict.

But if man by himself is weak God is all-powerful. If the "Doctrine of Necessity" once may have had the impersonal cast of determinism upon it, it had now been replaced by a faith in the God Who is personally and intimately concerned for every man and Who providentially brings to pass His designs for good in man's history. He commended his friends to God's care and asked for their prayers for him. "Doctor," said Lincoln to his friend and pastor at their last meeting before the departure, "I wish to be remembered in the prayers of yourself and our church members."

In the cold drizzle of that February day, with an expression of tragic sadness on his face, Lincoln spoke from the platform of the special train. The stubby locomotive tested its steam and the people peered upward from under their sheltering umbrellas.

"My friends—No one, not in my situation, can appreciate my feeling of sadness at this parting. To this place, and the kindness of these people, I owe everything. Here I have lived a quarter of a century, and have passed from a young to an old man. Here my children have been born, and one is buried. I now leave, not knowing when, or whether ever, I may return, with a task before me greater than that which rested upon Washington. Without the assistance of that Divine Being who ever attended him, I cannot succeed. With that assistance, I cannot fail. Trusting in Him who can go with me, and remain with you, and be everywhere for good, let us confidently hope that all will yet be well. To His care commending you, as I hope in your prayers you will commend me, I bid you an affectionate farewell."[20]

"An Humble Instrument in the Hands of the Almighty"

Chapter 6

The acceleration in Lincoln's religious development that came with his assuming the burdens of the presidency in a time of civil conflict may be traced to two sources. The first was the personal anguish of the death of friends and the tragic loss of his beloved Willie. The second was the suffering and pain that tore at the nation's life in crisis after crisis and cried aloud for some interpretation. The erosion of these forces may be traced in the deepening facial lines of almost every subsequent photograph of Lincoln. The first pressure turned Lincoln toward a deeper piety than he had known before. The second inspired him to probe beneath the seeming irrationality of events for a prophetic understanding of the nation's history. The first was clearly reflected in the Farewell Address at Springfield and the second began to find expression in the First Inaugural.

In this conciliating speech designed to allay Southern fears Lincoln argued for the perpetuity of the Union. The Union preceded even the adoption of the Constitution, which was itself established "to form a more perfect Union." Government must proceed by majority rule. The unwillingness of a minority to accept majority rule was the very principle of anarchy and despotism since a secession from the Union would itself be disrupted within a few years by the momentum of dissent. The facts of geography, moreover, made any separation unviable. "They cannot but remain face to face; and intercourse, either amicable or hostile, must continue between them. . . . Can aliens make treaties easier than friends can make laws? . . . Suppose you go to war, you cannot fight always; and when, after much loss on both sides, and no gain on either, you cease fighting, the identical old questions, as to terms of intercourse are again upon you."

Lincoln then expressed his conviction of the essential rightness and wisdom of the people if allowed time to register their verdict. His confidence in the people had its roots in religious reality and its presupposition in God. It was not the secular theory that the will of the people constituted right. That is the principle of mobocracy. *Vox populi, vox dei* meant for Lincoln that, if not thwarted by man's rebellion, God so guided the consciences of men in history that the people's verdict was properly their response to His guidance. Even the qualification in this last sentence, "if not thwarted by man's rebellion," needs further modification, for Lincoln held that God was constantly overruling those designs which were at cross-purposes with His own.

This theme was no isolated one for Lincoln. He had expressed it again and again on that circuitous train journey from Springfield to Washington. In Buffalo he gave it this form: "For the ability to perform it, I must trust in that Supreme Being who has never forsaken this favored land, through the instrumentality of this great and intelligent people."[1] In Trenton he told the New Jersey senators what Weems's *Life of Washington* had meant to him as a boy. The Revolutionary fathers had struggled for more than just national independence. It was "something that held out a great promise to all the people of the world to all time to come" and Lincoln would be most happy if he might "be an humble instrument in the hands of the Almighty, and of this, his almost chosen people, for perpetuating the object of that great struggle."[2] The inspired phrase, "his almost chosen people," will be discussed more fully in Chapter 8.

For Lincoln, confidence in the rightness of democratic process had as its presupposition faith in the history-molding God of justice and mercy. Not a pious addendum or a rhetorical tailpiece, Lincoln's faith in the Almighty Ruler of nations provided the dynamic and justification for a high confidence in the essential integrity of the people. He would spell out this theme in many ways in the years ahead.

"Why should there not be a patient confidence in the ultimate justice of the people? Is there any better, or equal hope, in the world? In our present differences, is either party without faith of being in the right? If the Almighty Ruler of nations, with his eternal truth and justice, be on your side of the North, or on yours of the South, that truth, and that justice, will surely prevail, by the judgment of this great tribunal, the American people."

Lincoln had earlier in the speech defined "the substantial dispute" as between the section that "believes slavery is *right,* and ought to be extended" and the other that "believes it is *wrong,* and ought not to be extended." He then took the seeming moral impasse between the contestants and placed it beyond the merely moral. Since each party believed it was right and shared a common faith in the God Who ruled the nations he appealed to each to accept God's verdict of rightness which, if allowed expression in the maintenance of the Union, would sooner or later be made evident by "the judgment of this great tribunal, the American people."

Then, spelling out the same conviction in a slightly different way, he counseled the South not to take precipitate action. If they who were dissatisfied were in the right, it was clear that God Himself would so guide the nation that adjustments would be made. This awareness sprang not from any conviction of the inherent virtue of one side in the controversy offering relief to dissatisfied opponents, a self-righteous condescension likely to exacerbate controversy, but from the perspective of the God of truth and justice Who would overrule both sides. From this religious point of vantage the bitter strife between two sides equally convinced of their own integrity and virtue was moderated by an appeal to a higher court without, however, loss of responsibility.

"Intelligence, patriotism, Christianity, and a firm reliance on Him, who has never yet forsaken this favored land, are still competent to adjust, in the best way, all our present difficulty."

Then, reworking Seward's somewhat prosaic suggestions for a final paragraph with the alchemy of his touch, Lincoln appealed to the shared memories of the past. In the coming years he would as no other President has ever done deepen for the American people the "mystic chords of memory" that make them a community.

"I am loth to close. We are not enemies, but friends. We must not be enemies. Though passion may have strained, it must not break our bonds of affection. The mystic chords of memory, stretching from every battlefield, and patriot grave, to every living heart and hearthstone, all over this broad land, will yet swell the chorus of the Union, when again touched, as surely they will be, by the better angels of our nature."[3]

The hope expressed in the First Inaugual that there would be no violence was shattered by the hostilities at Fort Sumter within a little more than a month. Virginia seceded and the war was on in earnest. Then toward the end of July came the humiliating disaster at Bull Run. In response to a resolution of both Houses of Congress, Lincoln appointed a national fast day in a proclamation signed by Seward as Secretary of State and by himself.

The themes of this remarkable document are a further development of his prophetic understanding of the nation's history. They crowd so fast one upon the other that the reader often fails to appreciate their amazing scope. Governments must acknowledge the lordship of God over their life. They must contritely confess their faults "as a nation and as individuals." The present conflict was the chastisement of God. Men would do well "to recognize the hand of God in this terrible visitation."

Just how men should interpret this "visitation" was not spelled out. Much of Lincoln's subsequent meditation would center around this idea until a full and clear answer would be given in the Second Inaugural. God guided the fathers into the pathway of civil and religious liberty. Prayer was offered for the restoration of the Union in "the . . . conviction that the fear of the Lord is the beginning of wisdom."

The style of Lincoln's proclamations differs, naturally, from that of his speeches. Daniel Dodge, in his *Abraham Lincoln: Master of Words,* suggested the influence of the Book of Common Prayer through Seward, who was an Episcopalian. The frequent repetition of the same idea by the use of words in pairs ("fit and becoming") is a characteristic of the liturgical rhythm of the Prayer Book, but it is also typical legal style in which the conjunction of synonyms gives greater precision in meaning. While Seward's influence may well be present, the ideas expressed are rooted in the biblical substratum of Lincoln's faith. The proclamations provide much of the background for understanding the Second Inaugural and are links in a chain leading up to it. Passing far beyond the bare language of the congressional resolution, which was a somewhat impenitent request for protection, Lincoln set the whole matter in a biblical frame of reference.

"And whereas it is fit and becoming in all people, at all times, to acknowledge and revere the Supreme Government of God; to bow in humble submission to His chastisements; to confess and deplore their sins and transgressions in the full conviction that the fear of the Lord is the beginning of wisdom; and to pray, with all fervency and contrition, for the pardon of their past offenses, and for a blessing upon their present and prospective action:

"And whereas, when our own beloved Country, once, by the blessing of God, united, prosperous and happy, is now afflicted with faction and civil war, it is peculiarly fit for us to recognize the hand of God in this terrible visitation, and in sorrowful remembrance of our own faults and crimes as a nation and as individuals, to humble ourselves before Him, and to pray for His mercy,—to pray that we may be spared further punishment, though most justly deserved; that our arms may be blessed and made effectual for the re-establishment of law, order and peace, throughout the wide extent of our country; and that the inestimable boon of civil and religious liberty, earned under His guidance and blessing, by the labors and sufferings of our fathers, may be restored in all its original excellence . . ."[4]

The harsh reality of death had always troubled Lincoln. He had known its coldness when as a lad of nine his mother breathed no more in the rude Indiana cabin. Then his sister had died. Still later their beloved Eddie was buried. Now with the outbreak of war Lincoln would feel the presence of death on an unprecedented scale. Elmer Ellsworth, his friend and former law student, was shot in the occupation of Alexandria and became one of the first casualties of the war. Lincoln wept at his funeral, held in the White House, and wrote a note of consolation to the boy's parents. "In the untimely loss of your noble son, our affliction here, is scarcely less than your own. . . . May God give you that consolation which is beyond all earthly power."[5]

This faith in God's healing power in bereavement would be critically tested when their son Willie fell dangerously ill early in 1862. After four or five final days of suffering and delirium, with the father sharing the night watch, the lad died on February 20. Although Tad was also gravely ill and not for a time expected to recover, Mrs. Lincoln removed herself from the sick child and took to her bed in grief. Friends worried for the President, so utterly crushed did he appear, but within two days of the funeral he called a cabinet meeting and carried on with the business of state.

Mrs. Rebecca Pomeroy, from Chelsea, Massachusetts, acted as special nurse for Willie and Tad. She spoke to Lincoln about the prayers of Christians all over the land for him in his bitter loss. "I am glad to hear that," she reported his words. "I want them to pray for me. I need their prayers. I will try to go to God with my sorrows. . . . I wish I had that childlike faith you speak of, and I trust He will give it to me. I had a good Christian mother, and her prayers have followed me thus far through life."[6]

Ida Tarbell describes the impact of this blow upon Lincoln. "There is ample evidence that in this crushing grief the President sought earnestly to find what consolation the Christian religion might have for him. It was the first experience of his life, so far as we know, which drove him to look outside of his own mind and heart for help to endure a personal grief. It was the first time in his life when he had not been sufficient for his own experience. Religion up to this time had been an intellectual interest. . . . From this time on he was seen often with the Bible in his hand, and he is known to have prayed frequently. His personal relation to God occupied his mind much."[7]

In his *Six Months at the White House* the painter Carpenter tells of a visit by the Rev. Francis Vinton of Trinity Church, New York, to the grief-stricken father. "Your son is *alive,* in Paradise," said Dr. Vinton, whom Mrs. Lincoln had asked to call. "Do you remember that passage in the Gospels: 'God is not the God of the *dead* but of the living, for *all* live unto him'?" Lincoln had a discussion with the rector and asked that his sermon on the subject of eternal life be sent to him. Carpenter relates that Lincoln read the sermon many times, having a copy made for his own use. "Through a member of the family, I have been informed that Mr. Lincoln's views in relation to spiritual things seemed changed from that hour."[8]

The flowering of this richer faith and Lincoln's deep human sympathy were expressed in his famous letter to the widow Bixby. Lincoln had been told that five of her sons had died in battle. The correct number was actually two. Its dominating thought is the Gettysburg speech in microcosm, personal loss sacrificially related to the preservation of freedom. Sandburg has called it "a piece of the American Bible . . . blood-color syllables of a sacred music."

"I feel how weak and fruitless must be any words of mine which should attempt to beguile you from the grief of a loss so overwhelming. But I cannot refrain from tendering to you the consolation that may be found in the thanks of the Republic they died to save. I pray that our Heavenly Father may assuage the anguish of your bereavement, and leave you only the cherished memory of the loved and lost, and the solemn pride that must be yours, to have laid so costly a sacrifice upon the altar of Freedom."[9]

Accompanying his deepened appreciation of the reality of eternal life came increased attention to prayer. Mrs. Lincoln is reported to have said that the morning of the ceremony of the inauguration Lincoln read the conclusion of the address to the family and then, when they had withdrawn from the room, prayed audibly for strength and guidance.[10]

In a letter, Noah Brooks wrote of Lincoln: "He said that after he went to the White House, he kept up the habit of daily prayer. Sometimes it was only ten words, but those ten words he had."[11]

John Nicolay, one of his secretaries, reported: "Mr. Lincoln was a praying man. I know that to be a fact and I have heard him request people to pray for him, which he would never have done had he not believed that prayer is answered. . . . I have heard him say that he prayed."[12]

For Lincoln the purpose of prayer was not to get God to do man's bidding but to place man where he might come to see God's purposes and to experience the strength of relying on the everlasting arms. A graphic picture of this resort to prayer is given by General James Rusling, who stood with the President in a Washington hospital at the bedside of General Sickles. The incident is doubly attested and has, moreover, the same authentic ring in the idiom about "a solemn vow to Almighty God" that had earlier marked his approach to emancipating the slaves. General Sickles had been wounded at Gettysburg and was recovering from a leg amputation.

"In reply to a question from General Sickles whether or not

the President was anxious about the battle at Gettysburg, Lincoln gravely said, 'No, I was not; some of my Cabinet and many others in Washington were, but I had no fears.' General Sickles inquired how this was, and seemed curious about it. Mr. Lincoln hesitated, but finally replied: 'Well, I will tell you how it was. In the pinch of the campaign up there, when everybody seemed panic-stricken, and nobody could tell what was going to happen, oppressed by the gravity of our affairs, I went to my room one day, and I locked the door, and got down on my knees before Almighty God, and prayed to Him mightily for victory at Gettysburg. I told Him that this was His war, and our cause His cause, but we couldn't stand another Fredericksburg or Chancellorsville. And I then and there made a solemn vow to Almighty God, that if He would stand by our boys at Gettysburg, I would stand by Him. And He *did* stand by your boys, and I *will* stand by Him. And after that (I don't know how it was, and I can't explain it), soon a sweet comfort crept into my soul that God Almighty had taken the whole business into his own hands and that things would go all right at Gettysburg. And that is why I had no fears about you.'"[13]

Prayer became increasingly a resource of strength to him. He began to speak more openly and in an unembarrassed way about it. This was in distinction to an earlier stage when he had maintained a wall of reserve about his religion and had chosen very carefully the men with whom he discussed it. To his newspaper friend Noah Brooks he once remarked: "I have been driven many times upon my knees by the overwhelming conviction that I had nowhere else to go."[14]

In the light of this evidence and of much more like it the popular image about Lincoln in the White House as a man of prayer is shown to have a solid basis in fact. This facet of Lincoln's religion has been graphically portrayed by Herbert S. Houck in his statue in the Washington Cathedral of Lincoln kneeling at prayer.

The relations of Lincoln as President with the clergy and churches of the country are fully detailed by Edgar DeWitt Jones in *Lincoln and the Preachers.* Here only those incidents will be described that throw light upon Lincoln's own faith or perhaps explain why he never became a church member. The pastor of the New York Avenue Presbyterian Church, Dr. Phineas D. Gurley, was, like his Springfield pastor, a Presbyterian of the Old School. Lincoln once told him that he could not perhaps accept all the doctrines in the Confession of Faith, but "if all that I am asked to respond to is what our Lord said were the two great commandments, to love the Lord thy God with all thy heart and mind and soul and strength, and my neighbor as myself, why, I aim to do that."

The Lincolns rented the eighth pew from the pulpit in Dr. Gurley's church and attended with regularity, their children sometimes accompanying them. Occasionally Lincoln went to the midweek prayer meeting. He must have gone often enough to make necessary the expedient of having him sit alone in the pastor's room, from which he could hear but not be seen by the congregation. This arrangement made it difficult for the curious to ruin the effect of the service and saved the President from their importunities afterward.

Delegations of churchmen brought him resolutions on the

war and on slavery. Some complained of the suspected heresies of men appointed as military chaplains. He listened with attention to them all, not hesitating at times to lecture those who lectured him.

To a committee of the Reformed Presbyterian Synod that petitioned him for emancipation he replied that their difference with him was not over the moral issue of slavery but over the best means to get rid of it. "Were an individual asked whether he would wish to have a wen on his neck, he could not hesitate as to the reply; but were it asked whether a man who has such a wen should at once be relieved of it by the application of the surgeon's knife, there might be diversity of opinion, perhaps the man might bleed to death, as the result of such an operation.

"Feeling deeply my responsibility to my country and to that God to whom we all owe allegiance, I assure you I will try to do my best, and so may God help me."[15]

Asked by the general superintendent of the Christian Commission to preside at a Washington's Birthday mass meeting in the House of Representatives in 1863, Lincoln declined, but approved its "worthy objects." His response, however, carried far beyond the conventional terms of the invitation, which had suggested in effect that the public meetings in other cities had served "to check distrust and disloyalty and to restore confidence and support to the Government."

The President replied in the spirit of the New Testament, "Whatever shall tend to turn our thoughts from the unreasoning, and uncharitable passions, prejudices, and jealousies incident to a great national trouble, such as ours, and to fix them upon the vast and long-enduring consequences, for weal, or for woe, which are to result from the struggle; and especially, to strengthen our reliance on the Supreme Being, for the final triumph of the right, can not but be well for us all."[16]

A Southern newspaper gave wide publicity to one of Lincoln's encounters with a clergyman in a delegation. The minister said he hoped "the Lord was on our side." When Lincoln rejoined, "I don't agree with you," the mouths of all were stopped. The President made it clearer. "I am not at all concerned about that, for I know that the Lord is *always* on the side of the *right*. But it is my constant anxiety and prayer that *I* and *this nation* should be on the Lord's side."[17]

One of the delegations drew from him the statement that he wished he were a more devout man. It is all of a piece with what he had told Generals Rusling and Sickles. It came shortly before his Gettysburg Address. This whole period was a time of greater consecration for Lincoln. His pastor Dr. Gurley had introduced the Baltimore Presbyterian Synod to the President and their moderator had offered their "respects" to him. "I was early brought to a living reflection," he replied, "that nothing in my power whatever, in others to rely upon, would succeed without the direct assistance of the Almighty, but all must fail. I have often wished that I was a more devout man than I am. Nevertheless, amid the greatest difficulties of my Administration, when I could not see any other resort, I would place my whole reliance in God, knowing that all would go well, and that He would decide for the right."[18]

Some of the delegations must have sorely tried his patience. Carpenter tells of an anti-slavery group from New York headed by a somewhat pompous and cocksure clergyman. "Well, gentlemen," said Lincoln in a deflating tone, "it is not often one is favored with a delegation *direct* from the Almighty." To another clergyman who on gaining admittance to his office said that he had merely called to pay his respects, Lincoln relaxed and said: "I am very glad to see you indeed. I thought you had come to preach to me!"

To the delegation that protested the choice of "unorthodox" chaplains by the regiments, Lincoln told the story of the little colored boy who was asked what he was making out of mud. The boy showed his questioner the little church building, the pews, and the pulpit—all made of mud. Asked then why he hadn't made a preacher, the boy smiled and said, "I hain't got *mud* enough."

Nicolay told his fiancée, "Going into his room this morning to announce the Secretary of War I found a little party of Quakers holding a prayer meeting around him, and he was compelled to bear the affliction until the 'spirit' moved them to stop. Isn't it strange that so many and such intelligent people often have so little common sense?"[19]

"The Best Gift God Has Given to Man"

Chapter 7

There undoubtedly was an influence of the churches and of the clergy upon Lincoln as President, but it cannot have been a dominant one in the development of his faith. He was strengthened in realizing that most Northern church opinion supported his steps toward emancipation. The churches were valued as interpreters of God's plan for the nation. He was grateful for their humanitarian work and their pastoral concern for the wounded and the bereaved. Yet the rock on which he stood was the Bible. Here he found help in interpreting the nation's history in the light of its great motifs of judgment, punishment, justice, mercy, and reconciliation. No President has ever had the detailed knowledge of the Bible that Lincoln had. No President has ever woven its thought and its rhythms into the warp and woof of his state papers as he did.

The "First Lecture on Discoveries and Inventions" which he delivered in 1858 before the Young Men's Association of Bloomington is chiefly an exegesis of the Bible, particularly of Genesis, to show the origin of clothing, weaving, toolmaking, and transportation. He must have written this with Cruden's Concordance open on his desk, for there are at least thirty-four references to the Bible in this manuscript in his own hand. The type of argument employed by his Springfield pastor, Dr. James Smith, in *The Christian's Defense* is used by Lincoln to reconcile newer scientific findings with biblical chronology. "The . . . mention of '*thread*' by Abraham is the oldest recorded allusion to spinning and weaving; and *it* was made about two thousand years after the creation of man, and now, near four thousand years ago. Profane authors think these arts originated in Egypt; and this is not contradicted, or made improbable, by anything in the Bible; for the allusion of Abraham, mentioned, was not made until after he had sojourned in Egypt."[1]

Lincoln's knowledge of the Bible was so thorough that his political opponents generally found themselves on dangerous ground when they quoted it against him. When Judge Douglas somewhat fantastically cited Adam and Eve as the first beneficiaries of his doctrine of "popular sovereignty" Lincoln corrected him. "God did not place good and evil before man, telling him to make his choice. On the contrary, he did tell him there was one tree, of the fruit of which he should not eat, upon pain of certain death." Then added Lincoln pointedly, "I should scarcely wish so strong a prohibition against slavery in Nebraska."[2]

News was brought to Lincoln that the dissident Republicans

had nominated Fremont at their Cleveland convention in May 1864. When told that instead of the thousands that had been advertised as participants only four hundred had actually come, he immediately found the number significant. Reaching for his Bible, he turned quickly to I Samuel 22:2 and read about David in the Cave of Adullam: "And every one that was in distress, and every one that was in debt, and every one that was discontented, gathered themselves unto him; and he became a captain over them: and there were with him about four hundred men."[3]

So much a part of him was the world of the Bible with its stories and characters that much of his humor flowed between its familiar banks. A typical story is the incident related by Senator Henderson of Missouri, who called on Lincoln some months before the emancipation of the slaves. Lincoln complained of being hounded by the abolitionists spearheaded by Charles Sumner, Henry Wilson, and Thaddeus Stevens. Just then as Lincoln looked out of the window he saw the three determined congressmen crossing the White House lawn to his door. His sad face lit up and a sparkle came into his eyes.

"Henderson, did you ever attend an old blab school? Yes? Well, so did I, and what little schooling I got in early life was in that way. I attended such a school in a log schoolhouse in Indiana where we had no reading books or grammars, and all our reading was done from the Bible. We stood in a long line and read in turn from it. One day our lesson was the story of the three Hebrew children and their escape from the fiery furnace. It fell to a little towheaded fellow who stood next to me to read for the first time the verse with the unpronounceable names.

He made a sorry mess of Shadrach and Meshach, and went all to pieces on Abednego. Whereupon the master boxed his ears until he sobbed aloud. Then the lesson went on, each boy in the class reading a verse in turn. Finally the towheaded boy stopped crying, but only to fix his gaze upon the verses ahead, and set up a yell of surprise and alarm. The master demanded the reason for this unexpected outbreak. 'Look there, master,' said the boy, pointing his finger at the verse which in a few moments he would be expected to read, and at the three proper names which it contained, 'there comes them same damn three fellows again!' "[4]

Lincoln enjoyed quoting a text as his immediate response to something said to him. He deflated the somewhat pompous Lord Lyons, the British ambassador, who made an official call to announce formally to the President in the name of his gracious sovereign Queen Victoria the betrothal of the Prince of Wales to the Princess Alexandra of Denmark. Said Lincoln to the bachelor ambassador when he had finished his communication, "Go thou and do likewise."

Next to this type of repartee, he liked to quote Scripture in answer to Scripture. Hugh McCulloch, an official of the Treasury Department, once introduced a delegation of New York bankers with much deference. Speaking of their patriotism and loyalty in holding the securities of the nation, he clinched his commendation of them with the text: "Where the treasure is there will the heart be also." Lincoln, like a crack of the whip, rejoined, "There is another text, Mr. McCulloch, which might apply, 'Where the carcass is, there will the eagles be gathered together.' "[5]

It would be possible to multiply illustrations of his sharpness in tracking down biblical texts. More significant than these items for his maturing faith is his view of the authority of the Bible as revelation for him. His words to Speed—to the effect that Speed should take all of the Bible that he could on reason and the rest on faith and he would live and die a happier man—have already been recorded. Lincoln spelled out this conviction with even greater clarity in his response on September 7, 1864, to the loyal colored people of Baltimore, who presented him with a magnificently bound Bible as a token of their appreciation of his work for the Negro. This is one of the great documents of Lincoln's religious confession. His reference to the Savior's communication of good to the world, his conviction that the knowledge of right and wrong springs from revelation, and his belief that human destiny is illuminated by its teachings need more emphasis than they have generally had in studies of his religion.

"In regard to this Great Book, I have but to say, it is the best gift God has given to man. All the good the Savior gave to the world was communicated through this book. But for it we could not know right from wrong. All things most desirable for man's welfare, here and hereafter, are to be found portrayed in it."[6]

This public statement of belief in the Bible was carried further in a conversation which L. E. Chittenden, the register of the Treasury, recorded in his *Recollections*.

"The character of the Bible," said Lincoln, "is easily established, at least to my satisfaction. We have to believe many things that we do not comprehend. The Bible is the only one that claims to be God's Book—to comprise his law—his history. It contains an immense amount of evidence of its own authenticity. It describes a Governor omnipotent enough to operate this great machine, and declares that He made it. It states other facts which we do not fully comprehend, but which we cannot account for. What shall we do with them?

"Now," continued Lincoln, "let us treat the Bible fairly. If we had a witness on the stand whose general story we knew was true, we would believe him when he asserted facts of which we had no other evidence. We ought to treat the Bible with equal fairness. I decided a long time ago that it was less difficult to believe that the Bible was what it claimed to be than to disbelieve it. It is a good book for us to obey—it contains the Ten Commandments, the Golden Rule, and many other rules which ought to be followed. No man was ever the worse for living according to the directions of the Bible."[7]

The Second Inaugural Address was to be the climactic expression of this biblical faith. It reads like a supplement to the Bible. In it there are fourteen references to God, four direct quotations from Genesis, Psalms, and Matthew, and other allusions to scriptural teaching. The London *Spectator* commented prophetically on this Scripture-steeped masterpiece: "We cannot read it without a renewed conviction that it is the noblest political document known to history, and should have for the nation and the statesmen he left behind him something of a sacred and almost prophetic character. Surely, none was ever written under a stronger sense of the reality of God's government. And certainly none written in a period of passionate conflict ever so completely excluded the partiality of victorious faction, and breathed so pure a strain of mingled justice and mercy."[8]

Sandburg shows an artist's appreciation of unity when he brackets Lincoln's laughter and his religion in a one-chapter discussion in Volume III of *The War Years.* They belong together. Humor is the threshold to the religious despite a widespread belief derived from a dour Puritanism that they are poles apart. Philosophically they are close, but they diverge at a critical point. In his *Concluding Unscientific Postscript* the Danish thinker Soren Kierkegaard describes humor as a boundary zone between an ethical evaluation of life and a religious one.

In humor a man is beginning to sit loose to his own anxious hold on life. There are, of course, stages in this development. Some people can laugh at others but not at the foibles and pretensions of the self. Such laughter is usually ironic and isolating. Others can laugh at themselves. This is real humor, free of bitter irony, that binds men in a community of appreciation. While Lincoln on occasion did employ irony and sarcasm, his basic humor was one that sat so loose to the claims of the self that it deflated those at odds with him. This is exactly why, besides just enjoying laughter for its own sake, Lincoln told stories. They saved time, enabling him to make his point unmistakably clear without subjecting his listeners to any self-righteousness of his own in the position he wished to maintain. Like parables, they shift responsibility from the narrator to his hearers.

In 1863, according to Colonel Silas W. Burt, a military delegation waited on Lincoln. When they prepared to leave, one who had had too much to drink slapped the President on the leg and asked for one of his "good" stories. Turning to the embarrassed group, Lincoln said, "I believe I have the popular reputation of being a storyteller, but I do not deserve the name in its general sense, for it is not the story itself, but its purpose or effect that interests me. I often avoid a long and useless discussion by others, or a laborious explanation on my own part, by a short story that illustrates my point of view. So too, the sharpness of a refusal or the edge of a rebuke may be blunted by an appropriate story so as to save wounded feelings and yet serve the purpose. No, I am not simply a storyteller, but storytelling as an emollient saves me friction and distress."[9]

Herndon, who was constitutionally incapable of appreciating Lincoln's religion, was just as obtuse on his humor. After a long-winded disquisition the partner concluded in a psychosomatic diagnosis that Lincoln told jokes because they served as a "stimulant, sending more blood to the brain, [which] aroused the whole man to an active consciousness—sense of his surroundings."[10]

An excellent illustration of how Lincoln employed humor to cool down opposition comes from an encounter with Senator Fessenden of Maine. The story also shows his ability as a humorist in exploiting the incongruities of life. He deflates his angry friend by linking types of profanity with types of denominational confession. Fessenden, enraged over an issue of patronage, exploded to Lincoln, losing control of himself in abusive profanity. Lincoln waited until his rage was spent and then asked softly, "You are an Episcopalian, aren't you,

Senator?" "Yes, sir," said his opponent stiffly, "I belong to that church." "I thought so," said Lincoln. "You Episcopalians all swear alike. Seward is an Episcopalian. But Stanton is a Presbyterian. You ought to hear him swear." He then analyzed blasphemy into its major theological variants. A somewhat chastened and sheepish Fessenden sat down then to talk quietly about patronage.[11]

The story illustrates the non-pharisaic perspective of Lincoln and provides the point of transition from humor to religion. Humor partially frees us from our anxious clutching at existence by providing an outlook independent, to a degree, of the self. Religion deepens the perspective by making the self see that God is in control of the world, thereby reducing the self-assertion of the ego. Humor deals with the proximate incongruities of existence, religion with the ultimate incongruities. Reinhold Niebuhr puts the relationship in a graphic image: ". . . there is laughter in the vestibule of the temple, the echo of laughter in the temple itself, but only faith and prayer, and no laughter, in the holy of holies."[12]

One of the greatnesses of Lincoln was the way he held to strong moral positions without the usual accompaniment of self-righteousness or smugness. He expressed this rare achievement provisionally in his humor and in an ultimate dimension in his religious evaluations. To the Pennsylvania delegation that congratulated him after the inauguration he said, urging forbearance and respect for differences of opinion between the states, "I would inculcate this idea, so that we may not, like Pharisees, set ourselves up to be better than other people."[13]

The Puritan in religion has found it difficult to submit himself in his moral rectitude to higher judgment when he judges his fellows. This accounts for much of the pride and lovelessness of organized Christianity. Lincoln expressed the antipharisaic side of the Christian gospel more poignantly and winsomely than most ecclesiastics. For him humor was closely related to his religious perspective. He is said to have enjoyed and even encouraged the spreading of a newspaper story about him. Two Quaker women were overheard on a train in the following dialogue:

"I think Jefferson will succeed."

"Why does thee think so?"

"Because Jefferson is a praying man."

"And so is Abraham a praying man."

"Yes, but the Lord will think Abraham is joking."

This story perfectly illustrates the relationship of humor and faith in Lincoln. The fact that Lincoln also appreciated it tends to show that he too was aware of their closeness in his make-up. Sandburg after recording this incident speculates as to whether there was here in germinal form that dilemma wherein "both read the same Bible and pray to the same God; and each invokes his aid against the other."

The picture then of Lincoln in the White House as the man of prayer needs the supplement of Lincoln the man of laughter. This was the Lincoln who served as a nucleus for many joke books, the Lincoln with a fund of anecdotes and especially a plentiful supply of preacher stories, the Lincoln who roared at the writings of Artemus Ward and kept the comic narratives about the Rev. Petroleum V. Nasby in his desk with his state papers. Once when he had been reading a selection from Ward to the Cabinet he looked up to see a circle of unsmiling and uncomprehending faces. "Gentlemen," he said, "why don't you laugh? If I didn't laugh under the strain that is upon me day and night, I should go mad. And you need that medicine as well as I."

There was development in his humor as in his faith. His first stories were told chiefly to amuse and divert his audience. Gradually his stories took on more purposiveness and served as vehicles to communicate his point of view quickly and effectively. Two examples will be given, the first from his New Salem period and the second from the presidency.

In New Salem Lincoln described an old-line Baptist preacher who chose as his text: "I am the Christ whom I shall represent today." His shirt and pants of coarse linen were anchored with one button at the waist and another at the neck. As he waxed eloquent, a blue lizard worked its way up his leg. Slapping did no good. Finally a desperate wrench and a kick resulted in nothing less than the loss of his pants. Repeating his text, the preacher valiantly struggled on, but now the lizard was tickling his back. A second convulsion fixed the lizard, but also resulted in the loss of the tow shirt. The congregation sat in silent horror. Then an old lady arose and addressed herself to the unclad preacher: If you represent Christ, then I'm done with the Bible."[14]

A prominent Presbyterian preacher of Southern sympathies, the Rev. Dr. McPheeters, found that he had been locked out of his St. Louis church by the commanding general of the area. Both factions in the congregation sent representatives to the President, who listened attentively to their conflicting statements and positions. It reminded Lincoln of an Illinois farmer and his son John who were hunting down a hog that had repeatedly ravaged their melon patch. The trail led up a muddy creek with the hog continually crossing the stream to the exhaustion and impatience of the hunters. "John," said the father, "you cross over and go up on that side, and I'll keep on this side, for I believe the old fellow is on both sides." "Gentlemen," said Lincoln to his quarrelsome delegation, "that is just where I stand in regard to your controversies in St. Louis. I am on both sides. I can't allow my generals to run the churches, and I can't allow you ministers to preach rebellion. Go home, preach the gospel, stand by the Union, and don't disturb the Government any more."[15]

Both the humor and the faith endeared Lincoln to the people. Both were complementary aspects of his nature. "Two strains," said Benjamin Thomas in a skillful study of his humor, "—pioneer exaggeration and Yankee laconicism—met in him. In his humor, as in his rise from obscurity to fame and in his simple, democratic faith and thought, he epitomized the American ideal."[16]

1862—"The Will of God Prevails"

Chapter 8

In Lincoln's wrestling with the responsibilities of the presidency his religious evaluation of his own role and of the nation's history became more clearly defined and then conjoined. A deepened personal faith led to a surer grasp of the rootage of American democracy in the will of God. Lincoln was aware of a process of development going on within himself, for he told his friend Noah Brooks that "he did not remember any precise time when he passed through any special change of purpose, or of heart; but he would say that his own election to office, and the crisis immediately following, influentially determined him in what he called 'a process of crystallization,' then going on in his mind."[1]

Here is the explanation for the many speeches and comments on that long train trip from Springfield to Washington which some historians of politics have found so barren. They criticize him for a lack of policy during this period. His own Secretary of State within a month after the beginning of his administration complained of the same problem. The truth of the matter is that Lincoln was never a person with a doctrinaire program ready for all possible events. There was a strong element of pragmatism in his make-up that was actually one of his greatest strengths. This pragmatism was rooted not primarily in political opportunism, or in a weak vacillation of decision, but in an inner resolution to obey the will of God. Because he believed deeply that God guided history he sought humbly and patiently to learn his assignment in the drama. His faith freed him from the illusion of titanism and from the frantic stance of an Atlas bearing the burden of the world alone upon his bruised shoulders.

What he said again and again on that trip to the nation's capital was that reliance on the divine will was the nation's real strength and that the American people and the President were instruments in the hands of the God of history. This was the background for his emphasis upon national renewal and national regeneration. These themes have been analyzed in the First Inaugural and in the Fast Day Proclamation after the disaster at Bull Run. They were pithily condensed into the concluding line of his message to the special session of Congress in July 1861: ". . . let us renew our trust in God and go forward without fear, and with manly hearts."[2] They would rise again to sublime expression two years later at Gettysburg.

There is testimony to the effect that Lincoln found the book of Job a source of strength in the White House. That ancient patriarch, stripped of possessions, family, and health, dared in dialogue to challenge the Almighty. Out of the whirlwind God replied that the mystery of suffering was too great for one as finite as Job to grasp, but if he could not understand the divine answer he was at least blessed in his resolute faith, which had retained its fidelity and integrity when all outward supports had been smashed.

Undoubtedly Lincoln resembled Job in being a wrestler with God. The match, however, was of a different order. Lincoln never questioned the ultimate justice of God. That was a settled conviction that gave his struggle a different twist than Job's. Lincoln's battle was to read "the signs of the times," to learn what the will of God actually demanded in the conflicting events of his day. Lincoln did not question the supremacy of the God before Whom "the nations are as the small dust of the balance." He questioned how to know that divine will in the day-to-day responsibilities of a nation at war with itself. He sought to avoid both the futility and the rebellion of opposing God's purposes in history. He would not use his ax vainly to split across the grain. The problem was to know the direction of the grain when so many honest men disagreed.

He told his callers many times that his concern was not to get God on his side, but to be quite sure that he and the nation were on God's side. An interview in June 1862 with a delegation from Iowa led by Congressman James Wilson threw more light on this point. It revealed again Lincoln's strong predestinarian conviction about God's will. A member was pressing Lincoln for more resolute action on emancipation, saying, "Slavery must be stricken down wherever it exists. If we do not do right I believe God will let us go our own way to our ruin. But if we do right, I believe He will lead us safely out of this wilderness, crown our arms with victory, and restore our now dissevered Union."

With a sparkle in his eyes and with his right arm outstretched toward the speaker, Lincoln agreed about the judgments of the God of history but added with great emphasis: "My faith is greater than yours. . . . I also believe He will compel us to do right in order that He may do these things, not so much because we desire them as that they accord with His plans of dealing with this nation, in the midst of which He means to establish justice. I think He means that we shall do more than we have yet done in furtherance of His plans, and He will open the way for our doing it. I have felt His hand upon me in great trials and submitted to His guidance, and I trust that as He shall further open the way I will be ready to walk therein, relying on His help and trusting in His goodness and wisdom." Wilson recorded that his manner of delivery was most impressive. Lincoln continued, saying that military reverses were to be expected. "Sometimes it seems necessary that we should be confronted with perils which threaten us with disaster in order that we may not get puffed up and forget Him who has much work for us yet to do."[3]

In the documents of 1861 Lincoln had urged the nation to see the hand of God in the outbreak of the war and in the Union reverses. His concern beyond that period was to understand more concretely the will of God in respect to the causation and continuance of the bloody conflict. The mounting lists of the dead and the wounded drove him in anguish to ultimate answers about the purpose of the war. He began to find these in terms of the providence of God in respect to slavery and the future of democratic government in all the earth. Among his papers after his death was discovered an undated document which he had not intended, said Nicolay and Hay, "to be seen of men." It was a meditation on the prevailing of God's will. It brought the themes already discussed—predestination, human instrumentalities, and God's effectual direction of men's actions—into burning focus around God's purpose for the nation. His secretaries gave a late September 1862 date to the autograph, but the Rutgers editors have chosen an early one in September. This would associate the document with Lincoln's despair following the second battle of Bull Run when Attorney General Bates reported that he "seemed wrung by the bitterest anguish." This date would also provide a point of transition to his decision in late September that he must free the slaves to save the Union. The meditation reveals theological profundity and legal precision of definition.

"The will of God prevails. In great contests each party claims to act in accordance with the will of God. Both *may* be, and one *must* be wrong. God cannot be *for*, and *against* the same thing at the same time. In the present civil war it is quite possible that God's purpose is something different from the purpose of either party—and yet the human instrumentalities, working just as they do, are of the best adaptation to effect His purpose. I am almost ready to say this is probably true—that God wills this contest, and wills that it shall not end yet. By His mere quiet power, on the minds of the now contestants, He could have either *saved* or *destroyed* the Union without a human contest. Yet the contest began. And, having begun, He could give the final victory to either side any day. Yet the contest proceeds."[4]

By September 22, 1862, Lincoln had determined to issue a preliminary proclamation of emancipation. He had, as was discussed in Chapter I, "made a solemn vow before God" to act for the slaves as soon as a Union victory should give stature to his action, following on this point the earlier advice of Seward. Antietam had been won in the previous week and Lee's army had withdrawn across the Potomac into Virginia. The making of this vow was by no means the irresponsible resort to superstition that it has seemed to some. The method obviously had dangers if it were lightly undertaken as a flipping of a coin for a divine decision, but such an irresponsible procedure was not Lincoln's. His action was preceded by the blood, sweat, and tears of a struggle to know the will of God on the Union and on slavery amid the problems of his constitutional powers and oath and the effect of such contemplated action on the North, on the border states, on the South, and finally on the world. The historian Randall comments perceptively on Lincoln's decision: "If these deliberations had given him humility, and a sense of association with Divine Purpose (which was more than once indicated), they had also given executive confidence. In reaching his important decision there is ample reason to believe that Lincoln had not only endured anxious hours, but had undergone a significant inner experience from which he emerged with quiet serenity."[5]

The solemn vow and covenant may have been more in

conformance with Old Testament than with New Testament religion, but the practice was imbedded in Lincoln's biblical piety and came to him as part of the religious heritage of the nation.

Behind Lincoln's act in establishing covenant with the God of nations on the issue of emancipation lay the conviction that America was a chosen nation destined to further God's plan for mankind. This faith, rooted in the development of New England theology, must be studied in some detail if we are rightly to understand Lincoln's convictions. In an essay entitled "The Union as Religious Mysticism" Edmund Wilson claims that "Lincoln's conception of the progress and meaning of the Civil War was indeed an interpretation that he partly took over from others but that he partly made others accept. . . . Like most of the important products of the American mind at that time, it grew out of the religious tradition of the New England theology of Puritanism."[6]

The American dream was the later flowering of the Puritan conception of New England as God's new land of promise and of the colonization of these rocky shores as the new exodus over the Red Sea and Jordan, bringing religious liberty to God's elect. The kernel was already present in the text from which John Cotton preached to the emigration under Governor Winthrop. "Moreover I will appoint a place for my people Israel, and will plant them, that they may dwell in a place of their own, and move no more; neither shall the children of wickedness afflict them any more . . ."[7]

Combining these special interpretations with the rigor of Calvin's doctrine of predestination, the New England divines believed fervently in the religious destiny of their commonwealth. It became important to keep open the channels through which God could make known His will to His people in the day-to-day events that were occurring. King Philip's savage Indian war was minutely studied to understand why God was punishing His people in its disasters while its deliverances were gratefully recorded as "particular providences."

The outlook on life furthered by this sense of mission and destiny, while it might have doctrinaire concepts, was actually the very opposite of a closed or static system. Each new configuration of revelatory events required a fresh openness for man to understand God's current visitation. Hence an attitude of pragmatism sprang up, with the result that developing forms of democratic government became increasingly responsive to this fact. In the complicated currents of the eighteenth century a humanitarian rationalism began to erode the explicitly religious upland of the previous century. It did not, however, change the forward-looking orientation or the pragmatic spirit. Indeed it reinforced it with perfectionist impulses of its own. This interesting evolution can be charted in two dimensions: (1) the function of the people in a democracy and (2) the democratic means to further this function.

In the Puritan interpretation the people became aware that they were instruments of Providence. This was slowly transmuted into a reliance upon the people as the corporate bearer of God's wisdom. The people's wisdom would be expressed in the long run by means of majority rule. Here is the original root for the adage that the will of the people is the will of God. We have had numerous examples of Lincoln's confidence in the essential rightness and wisdom of the people. This explains his belief, expressed at Buffalo, that God's will is ultimately to be known *through* the people. "I must trust in that Supreme Being who has never forsaken this favored land, through the instrumentality of this great and intelligent people."

He had expressed this belief as early as 1850 in his eulogy at the death of President Taylor. "Yet, under all circumstances, trusting to our Maker, and through His wisdom and beneficence, to the great body of our people, we will not despair, nor despond."[8]

This is also the philosophy behind the well-known remark attributed to him about fooling the people. Governor Fifer of Illinois claimed he had heard Milton Hay quote Lincoln: "You can fool some of the people all of the time and all of the people some of the time, but you can't fool all of the people all of the time." It is behind his statement that "we cannot escape history." In the long run there is no appeal from history to any higher court for the simple reason that history has been woven on God's loom.

The Puritan conception of the people's task required in addition that the will of God could only be known provided all channels of communication were kept open. The resolution not to allow the Spirit to be fenced in was gradually expressed in governmental structure by granting to religious groups freedom from state control. This freedom was in turn transmuted into the obligation so eloquently stated by Jefferson that a majority must not suppress its minorities. Only in such a free atmosphere could truth itself be furthered by the clash of opinion. The eighteenth century expressed the older concept of openness to the Spirit's direction in terms of the essential equality of all men in possessing certain inalienable rights. These were, of course, spelled out in the Bill of Rights as amendments to the Constitution.

Lincoln's passages in the First Inaugural about majority rule illustrated perfectly the point that the wisdom of the people could be expressed only through majority rule that preserved openness for minority groups. "A majority held in restraint by constitutional checks, and limitations, and always changing easily with deliberate changes of popular opinions and sentiments, is the only true sovereign of a free people. Whoever rejects it does, of necessity, fly to anarchy or to despotism. . . . Why should there not be a patient confidence in the ultimate justice of the people? Is there any better or equal hope in the world?"[9]

One of the means for holding a majority to responsibility was the system of checks and balances in the Constitution to keep power dispersed as much as possible. The doctrine of man behind this is, of course, not the view of his essential goodness which has sometimes been offered as the rationale for democracy, but a realistic view of the selfishness in human nature. Lincoln argued that human nature as such could not be changed. "The Bible says somewhere that we are desperately selfish. I think we would have discovered that fact without the Bible."[10] In other words, man is good enough to make democracy possible and bad enough to make it necessary.

The Puritan heritage distilled through the eighteenth-century patriots without, however, loss of its original religious strength explains many features in Lincoln's thought. It is the background for the predestinating will of God, for corporate and individual responsibility, for the direction of democracy as a way, for America as "God's almost chosen people," for belief in the wisdom of the people, for the possibility of making a solemn vow and covenant with God and observing its historical results, for the importance of "discerning the signs of the times," and even for the Gettysburg note, not of victory, but of "testing" the nation's vocation. In subsequent speeches and letters he would articulate, deepen, and reattest this heritage until he became in Sidney Mead's perceptive phrase "the spiritual center of American history."[11]

About a month after issuing the preliminary proclamation of emancipation Lincoln had an interesting interview with a distinguished Quaker lady. Mrs. Eliza Gurney was the widow of Joseph Gurney, the English Friend, writer, and philanthropist. She came to the White House and talked with the President about seeking divine guidance. Lincoln had great sympathy for the dilemma of the Friends. Pacifist on the one hand, they hated war. Strongly anti-slavery on the other, they saw that only on the anvil of war could emancipation be actually forged. Lincoln had a way of transcending the clash of religious absolutes while not withdrawing from responsible and resolute action. He would explain this dilemma in his Second Inaugural.

After Mrs. Gurney knelt and "uttered a short but most beautiful, eloquent and comprehensive prayer that light and wisdom might be shed down from on high, to guide our President," Lincoln gave the answer that has been preserved in the Lincoln papers:

"I am glad of this interview, and glad to know that I have your sympathy and prayers. We are indeed going through a great trial—a fiery trial. In the very responsible position in which I happen to be placed, being a humble instrument in the hands of our Heavenly Father, as I am, and as we all are, to work out His great purposes, I have desired that all my works and acts may be according to His will, and that it might be so, I have sought His aid—but if after endeavoring to do my best in the light which He affords me, I find my efforts fail, I must believe that for some purpose unknown to me, He wills it otherwise. If I had had my way, this war would never have been commenced; if I had been allowed my way this war would have been ended before this, but we find it still continues; and we must believe that He permits it for some wise purpose of His own, mysterious and unknown to us; and though with our limited understandings we may not be able to comprehend it, yet we cannot but believe, that He who made the world still governs it."[12]

There is an interesting sequel to this dialogue. One year later Mrs. Gurney sent a letter to the President by the hand of a mutual friend, Isaac Newton. She wrote "to give thee the assurance of my continued hearty sympathy in all thy heavy burthens and responsibilities and to express, not only my own earnest prayer, but I believe the prayer of many thousands whose hearts thou hast gladdened by the praiseworthy and *successful* effort to 'burst the bands of wickedness, and let the oppressed go free' that the Almighty . . . may strengthen thee to accomplish *all* the blessed purposes, which, in the unerring counsel of his will and wisdom, I do assuredly believe he did design to make thee instrumental in accomplishing, when he appointed thee thy present post of vast responsibility as the Chief Magistrate. . . ."

Lincoln wrote an answer that shows his deep respect for his correspondent and breathes an air of religious confidence at the very time that Union hopes had been dashed by Early's raid to the outskirts of Washington, the failure of Burnside at the Petersburg crater, and widespread Copperhead disturbances in the North.

"My esteemed friend. I have not forgotten—probably never shall forget—the very impressive occasion when yourself and friends visited me on a Sabbath forenoon two years ago. Nor has your kind letter, written, nearly a year later, ever been forgotten. In all, it has been your purpose to strengthen my reliance on God. I am much indebted to the good Christian people of the country for their constant prayers and consolations; and to no one of them, more than to yourself. The purposes of the Almighty are perfect, and must prevail, though we erring mortals may fail to accurately perceive them in advance. We hoped for a happy termination of this terrible war long before this; but God knows best, and has ruled otherwise. We shall yet acknowledge His wisdom and our own error therein. Meanwhile we must work earnestly in the best light He gives us, trusting that so working still conduces to the great ends He ordains. Surely He intends some great good to follow this mighty convulsion, which no mortal could make, and no mortal could stay.

"Your people—the Friends—have had, and are having, a very great trial. On principle, and faith, opposed to both war and oppression, they can only practically oppose oppression by war. In this hard dilemma, some have chosen one horn and some the other. For those appealing to me on conscientious grounds, I have done, and shall do, the best I could and can, in my own conscience, under my oath to the law. That you believe this I doubt not; and believing it, I shall still receive, for our country and myself, your earnest prayers to our Father in Heaven. Your sincere friend—A. Lincoln."[13]

In a talk once with his register of the Treasury, L. E. Chittenden, Lincoln spelled out further his sense of being directed by God's will. His reported words have a concreteness about them that marks a considerable advance beyond the view of his 1846 handbill, but the structure of analysis is continuous with it.

"That the Almighty does make use of human agencies, and directly intervenes in human affairs, is one of the plainest statements in the Bible. I have had so many evidences of His direction, so many instances when I have been controlled by some other power than my own will, that I cannot doubt that this power comes from above. I frequently see my way clear to a decision when I am conscious that I have no sufficient facts upon which to found it. But I cannot recall one instance in which I have followed my own judgment, founded upon such a decision, when the results were unsatisfactory; whereas, in almost every instance when I have yielded to the views of others, I have had occasion to regret it. I am satisfied that, when the Almighty wants me to do or not to do, a particular thing, he finds a way of letting me know it."[14]

Lincoln's concern at this period to be open to the divine will found expression about a month after the preliminary proclamation of emancipation was issued. It took the form of an order for Sabbath observance in the armed forces and listed among the supporting reasons "a due regard for the Divine will." Lincoln turned to General Washington for precedent. "The first General Order issued by the Father of his country after the Declaration of Independence, indicates the spirit in which our institutions were founded and should ever be defended. 'The General hopes and trusts that every officer and man will endeavor to live and act as becomes a Christian soldier defending the dearest rights and liberties of his country.' "[15]

The high peak, however, in Lincoln's desire to obey the divine will was revealed in his annual message to Congress on December 1, 1862. Here his imagination soared as he pictured the special destiny of America freed of slavery as a means to the advance of freedom and democracy over all the earth. He suggested three constitutional amendments for the gradual and compensated emancipation of slaves owned by loyal masters and for voluntary colonization abroad for freedmen. The theme of slavery began to yield to the larger theme of democracy "as the last, best hope of earth." This association had been made before by Lincoln, but henceforth it appears with more persuasiveness and conviction as one of God's purposes being forged in the fires of civil conflict. The religious impulse was the same for the drive to eliminate slavery as it was for the realization of a brave new world. It was the conviction of the declaration that all men are created equal regarded progressively as the expression of God's will for "the great family of man." It meant every human son of man under whatever type of government. This vision was based not on humanistic utopianism but on an apocalyptic revelation of God's purpose for mankind.

This speech is supported by such artistry of words that it remains the masterpiece of Lincoln's longer speeches. Its opening paragraph provides the theological framework and points toward the great peroration at the end. God's guidance is sought for responsible action.

"Since your last annual assembling another year of health and bountiful harvests has passed. And while it has not pleased the Almighty to bless us with a return of peace, we can but press on, guided by the best light He gives us, trusting that in His own good time, and wise way, all will yet be well."

The sureness of his touch can be seen as he uses a biblical quotation to underscore his point on the physical inseparability of the United States.

"A nation may be said to consist of its territory, its people, and its laws. The territory is the only part which is of certain durability. 'One generation passeth away, and another generation cometh, but the earth abideth forever.' It is of the first importance to duly consider, and estimate, this ever-enduring part."

After pleading for a fresh approach to action on the slavery issue he begins the great concluding paragraph with its oft-quoted phrase, "we cannot escape history." Lincoln might have said, "we cannot escape God," for what he means is that history is controlled and determined by God and that it is futile for man to oppose its plan. Lincoln speaks as a prophet disclosing the oracle of God for the Union and for the world.

"Fellow citizens, *we* cannot escape history. . . . The fiery trial through which we pass, will light us down, in honor or dishonor, to the latest generation. We *say* we are for the Union. The world will not forget that we say this. We know how to save the Union. The world knows we do know how to save it. We—even *we here*—hold the power, and bear the responsibility. In *giving* freedom to the *slave,* we *assure* freedom to the *free*—honorable alike in what we give, and what we preserve. We shall nobly save, or meanly lose, the last, best hope of earth. Other means may succeed; this could not fail. The way is plain, peaceful, generous, just—a way which, if followed, the world will forever applaud, and God must forever bless."[16]

On January 1, 1863, Lincoln issued the Final Proclamation of Emancipation, freeing forever the slaves within the areas then in rebellion against the United States. He gratefully accepted Secretary Chase's suggestion of the words "a solemn recognition of responsibility before men and before God" in the document. He incorporated Chase's wording verbatim except for clarifying his view that the action was constitutional as a war measure. "And upon this act, sincerely believed to be an act of justice, warranted by the Constitution, upon military necessity, I invoke the considerate judgment of mankind and the gracious favor of Almighty God."[17]

1863—"This Nation, under God"

Chapter 9

The year 1863 began with great military anxieties. At Murfreesboro, Tennessee, Confederate General Bragg smashed the right wing of Rosecrans's Army of the Cumberland. After days of savage fighting and heavy casualties General George Thomas stabilized the situation for the North as the Confederates withdrew toward Chattanooga. In the East with the Army of the Potomac the immediate but enduring crisis was one of command. Burnside's earlier defeat at Fredericksburg had cost him the confidence of his subordinates. Finally Lincoln had to dismiss him, taking in his stead Joseph Hooker. The President reprimanded Hooker in the famous letter for his disloyalty to Burnside and for his foolish clamoring for a dictator.

The immediate impact of emancipation was disappointing. Since it was unaccompanied by Union victories it was ineffectual in the only areas in which it was authoritative. Democrats accused the President of tricking them into supporting a war for the Union in order to convert it into a crusade against slavery. When Lincoln came to see that plans for sending the Negroes to other areas as colonists could not be implemented, he began to use them as soldiers in order to help the Union and to give them increased stature among their white neighbors.

Slowly expressions of support for emancipation began to trickle in from England, France, and other countries. The workingmen of Manchester, England, had been thrown into dire want and widespread unemployment by the blockade on Southern cotton. Yet they supported Lincoln's stand. The President expressed his gratitude for their resolutions "as an instance of sublime Christian heroism which has not been surpassed in any age or in any country."[1]

At the suggestion of Senator Harlan of Iowa, Congress passed a resolution calling upon the President to set apart a day for national prayer and humiliation. The text of this Senate document was somewhat remarkable for its explicitly Christian reference to seeking "Him for succour according to His appointed way, through Jesus Christ." "Fully concurring in the views of the Senate," Lincoln named April 30, 1863, as a National Fast Day. He counseled personal and national repentance. The Bible and the course of history, he argued, showed the necessity for a nation to acknowledge God.

". . . it is the duty of nations as well as of men, to own their dependence upon the overruling power of God, to confess their sins and transgressions, in humble sorrow, yet with assured hope that genuine repentance will lead to mercy and pardon;

and to recognize the sublime truth, announced in the Holy Scriptures and proven by all history, that those nations only are blessed whose God is the Lord."[2]

He then passed beyond the terms of his proclamation after Bull Run by assigning definite reasons for this punishment of war. The entire nation and not just the South was guilty of presumptuous sins which had to be corrected. The document resembled a page from Amos, or Isaiah, or Jeremiah.

"And, insomuch as we know that, by His divine law, nations like individuals are subjected to punishments and chastisements in this world, may we not justly fear that the awful calamity of civil war, which now desolates the land, may be but a punishment, inflicted upon us, for our presumptuous sins, to the needful end of our national reformation as a whole People?"

Unwilling to allow even this statement of sin to remain in the abstract, Lincoln confronted the nation with forgetfulness of God, pride, foolish imagination, and the illusion of Lordship. It was as though he had paraphrased Luther's statement that the essence of sin is the sinner's unwillingness to admit that he is a sinner, when he charged the nation with forgetting its need for redemption.

"We have been the recipients of the choicest bounties of Heaven. We have been preserved, these many years, in peace and prosperity. We have grown in numbers, wealth and power, as no other nation has ever grown. But we have forgotten God. We have forgotten the gracious hand which preserved us in peace, and multiplied and enriched and strengthened us; and we have vainly imagined, in the deceitfulness of our hearts, that all these blessings were produced by some superior wisdom and virtue of our own. Intoxicated with unbroken success, we have become too self-sufficient to feel the necessity of redeeming and preserving grace, too proud to pray to the God that made us!"

Lincoln then named the date for the national fast and urged the people to keep "the day holy to the Lord" by appropriate exercises in church and in home. He closed with the assurance that God would accept and bless a repentant people.

"All this being done, in sincerity and truth, let us then rest humbly in the hope authorized by the Divine teachings, that the united cry of the Nation will be heard on high, and answered with blessings, no less than the pardon of our national sins, and the restoration of our now divided and suffering Country, to its former happy condition of unity and peace."

Within a month of the issuance of this Proclamation, Hooker crossed the Rappahannock and the Rapidan. Because he had divided his forces and could not bring them into battle at the right place he was badly mauled by Lee at Chancellorsville. On May 6 came a dispatch that Hooker had retreated across the Rappahannock. With head bent and hands clasped behind him, Lincoln paced the floor, saying, "My God! My God! What will the country say? What *will* the country say?"

Lee soon crossed upriver and lunged across the Potomac into Maryland toward Pennsylvania. Panic broke out in the North. In a huff Hooker resigned almost on the eve of battle. Lincoln, ignoring cries for the return of McClellan, placed General George Gordon Meade in charge, the fifth general to command in the East within a year. The climactic battle of Gettysburg was joined on July 1 and continued with staggering losses on both sides for the next two days. On July 4 at 10 A.M. Lincoln announced to the country a bloody victory for the North with the wish "that on this day, He whose will, not ours, should be forever done, be everywhere remembered and reverenced with profoundest gratitude."[3]

Within a few days he learned that Vicksburg had fallen to Grant on Independence Day. On July 15, 1863, Lincoln issued a Proclamation of Thanksgiving for these Union victories, acknowledging also the mourning of the nation for the lost. "It is meet and right to recognize and confess the presence of the Almighty Father and the power of His Hand equally in these triumphs and in these sorrows."

The text contains a moving invocation of the Holy Spirit as he called upon the people to "render the homage due to the Divine Majesty, for the wonderful things He has done in the Nation's behalf, and invoke the influence of His Holy Spirit to subdue the anger, which has produced, and so long sustained a needless and cruel rebellion, to change the hearts of the insurgents, to guide the counsels of the Government with wisdom adequate to so great a national emergency, and to visit with tender care and consolation throughout the length and breadth of our land all those who, through the vicissitudes of marches, voyages, battles, and sieges, have been brought to suffer in mind, body, or estate, and finally to lead the whole nation, through the paths of repentance and submission to the Divine Will, back to the perfect enjoyment of Union and fraternal peace."[4]

During the next two months Copperhead activity rose to a peak in the North. Ohio Democrats nominated for governor Clement Vallandigham, who had been arrested by the military and deported to the South, but who now from a base in Canada urged that peace be concluded with the Confederates. A noisy "peace" meeting denouncing Lincoln's "tyranny" was staged in Springfield, Illinois.

To counter this demonstration the Republicans of Illinois campaigned as the National Union party, inviting all loyal Democrats to join them. This fusion group planned a meeting for September 3 in Springfield and invited the President to address them. Unable to be present because of the press of affairs, Lincoln wrote an important letter to his friend James C. Conkling to be read at the Union rally.

Lincoln expressed gratitude for the support of all Union-loving men and justified his military pressure on the South as the only practical means of restoring the Union. He argued that in the eyes of his military commanders emancipation constituted the heaviest blow yet dealt to the rebellion.

"The signs look better. The Father of Waters again goes unvexed to the sea. . . . It is hard to say that anything has been more bravely, and well done, than at Antietam, Murfreesboro, Gettysburg, and on many fields of lesser note. Nor must Uncle Sam's web-feet be forgotten. At all the watery margins they have been present. Not only on the deep sea, the broad bay, and the rapid river, but also up the narrow muddy bayou, and wherever the ground was a little damp, they have been, and made their tracks. Thanks to all. For the great republic—for the principle it lives by, and keeps alive—for man's vast future,—thanks to all."

Here again was the theme of "this nation, under God" with its democratic principles as "the last, best hope of earth." One can almost sense the coming of the Gettysburg Address. Lincoln went on to say that peace "does not appear so distant as it did.

"Still let us not be over-sanguine of a speedy final triumph. Let us be quite sober. Let us diligently apply the means, never doubting that a just God, in His own good time, will give us the rightful result."[5]

One of the notes that rings through this letter is thanksgiving for the hard work and sacrifice of the people. It was said of Lincoln that whenever he was congratulated on a Union victory he always spoke at once of the men who by fighting and dying had actually won it. Once an officer reported to him about conditions among the freedmen on the North Carolina coast, telling of a patriarch who told his people: "Massa Linkum, he eberywhar. He know eberyting. He walk de earf like de Lord!" Lincoln did not smile, but after walking around a bit in silence he remarked: "It is a momentous thing to be the instrument, under Providence, of the liberation of a race."[6]

The note of gratitude was a marked characteristic of Lincoln. He knew how to say thank you to men and to offer praise to God. It was therefore entirely natural that he should accept the suggestion of Sarah Hale and make Thanksgiving a national holiday. Succeeding Presidents have followed the precedent established by him in 1863 and fixed the November date by proclamation. The mood of thankfulness revealed in the Conkling letter found a more solemn and liturgical expression in the official proclamation.

"The year that is drawing towards its close, has been filled with the blessings of fruitful fields and healthful skies. To these bounties, which are so constantly enjoyed that we are prone to forget the source from which they come, others have been added, which are of so extraordinary a nature, that they cannot fail to penetrate and soften even the heart which is habitually insensible to the ever watchful providence of Almighty God."

Lincoln spoke of the maintenance of peace with other countries and the advance of the armed forces. Then in vivid, concrete imagery he summed up the growth of the nation and its expectation.

"Needful diversions of wealth and of strength from the fields of peaceful industry to the national defense, have not arrested the plough, the shuttle or the ship; the axe has enlarged the borders of our settlements, and the mines, as well of iron and coal as of the precious metals, have yielded even more abundantly than heretofore. Population has steadily increased, notwithstanding the waste that has been made in the camp, the siege and the battlefield; and the country, rejoicing in the consciousness of augumented strength and vigor, is permitted to expect continuance of years with large increase of freedom."

Then Lincoln provided the theological interpretation of the nation's situation.

"No human counsel hath devised nor hath any mortal hand worked out these great things. They are the gracious gifts of the Most High God, who, while dealing with us in anger for our sins, hath nevertheless remembered mercy. It has seemed to me fit and proper that they should be solemnly, reverently and gratefully acknowledged as with one heart and one voice by the whole American People."

Lincoln then named the last Thursday in November as "a day

of Thanksgiving and Praise to our beneficent Father who dwelleth in the Heavens." He asked the people to commend "to His tender care all those who have become widows, orphans, mourners or sufferers in the lamentable civil strife in which we are unavoidably engaged, and fervently implore the interposition of the Almighty Hand to heal the wounds of the nation and to restore it as soon as may be consistent with the Divine purposes to the full enjoyment of peace, harmony, tranquillity and Union."[7]

The next significant document of 1863 was of course the Gettysburg Address. It followed naturally upon the Thanksgiving Proclamation. Lord Curzon ranked it and the Second Inaugural as two of the three greatest examples of eloquence in the English language. Edward Everett, whose task it had been to deliver a two-hour oration preceding the President's remarks, recognized it as a jewel. He wrote Lincoln shortly after the dedication of the cemetery, wishing that he might have been able to come as near to the central idea of the occasion in two hours as the President had in two minutes.

Some commentators have dismissed the religious implications of the address by pointing out that there are no biblical quotations in it and that the phrase "under God" was an afterthought. It would be difficult to make a more superficial comment. While it is technically true that there are no texts from Scripture in it, the language is at times biblical and the solemn style has the cadences of the King James Version.

"Four score and seven years ago" is an inspired adaptation of Old Testament counting. The image of the birth of the nation is expressed in the verb "brought forth" borrowed from the common biblical phrase: "she brought forth a son." "The new birth," or the image of biblical regeneration, is the controlling concept behind the picture of purposeful sacrifice—"that this nation, under God, shall have a new birth of freedom." The first and second drafts of his address and the transcriptions of the reporters present show that he must have added the phrase "under God" as he spoke. In the three subsequent copies Lincoln included this inspired addition of his.

But this discussion of its affinities with the language and style of Scripture only begins to scratch the surface of its true religious depth. The central image behind the whole speech is the rite of baptism or the solemn dedication of children to God. The New Testament describes baptism as a dying to sin with Christ in his death and as a rising to newness of life in the power of his resurrection. Lincoln conflates the themes of the life of man in birth, baptismal dedication, and spiritual rebirth with the experience of the nation in its eighty-seven years of history.

Although founded on the proposition that all men are created equal, the nation had denied its universal dedication by narrowing "all" to "all white men." At Gettysburg the old man of sin died that the nation might be reborn in the truth of a growing democracy for all men everywhere.

Basler comments on the Address: "To it he brought the fervor of devoutly religious belief. Democracy was to Lincoln a religion, and he wanted it to be in a real sense the religion of his audience. Thus he combined an elegiac theme with a patriotic theme, skillfully blending the hope of eternal life with the hope of eternal democracy."[8]

His use of the word "proposition" has sometimes been questioned. It has been said that Matthew Arnold read no further. Here of course the image is a proposition in Euclidean geometry that requires proof. Lincoln is supposed to have carried a copy of Euclid with him on the circuit and to have laughingly approved the suggestion of a clergyman to have the Tract Society print the Greek geometrician. The verb "testing" further depicts this process of demonstration although it takes on the overtones of a test of faith such as Abraham's or Job's in the Old Testament. For Lincoln the truth that all men are created equal was not a static concept predicated upon the base of self-evidence.

Here was a basic difference in emphasis between Jefferson and Lincoln. For Jefferson the phrase was a geometrical *axiom* needing no proof, but resting on the basis of "a self-evident truth." For Lincoln in his maturer years the phrase was a *proposition* that was in continual process of being demonstrated. Lincoln's way of rooting democracy in the will of God made it a dynamic faith to live by. The Civil War was "a test" or "trial" of that faith. The theme of being justified by faith, an echo of the Puritan religious inheritance, is also in the background of this language about the testing of a proposition.

The religious images behind the Gettysburg Address are all of one piece with Lincoln's deepened sense of consecration at this time. Some of this evidence has already been pointed out. He told General Sickles that in the crisis of the battle he made a solemn vow that if God would stand by the soldiers he would stand by Him. He told the Baltimore Presbyterians that he wished he was a "more devout" man. The Gettysburg Address partly articulates this "increased devotion" in Lincoln himself as he summons the nation to a higher dedication of its life.

"Four score and seven years ago our fathers brought forth on this continent, a new nation, conceived in Liberty, and dedicated to the proposition that all men are created equal.

"Now we are engaged in a great civil war, testing whether that nation, or any nation so conceived and so dedicated, can long endure. We are met on a great battle-field of that war. We have come to dedicate a portion of that field, as a final resting place for those who here gave their lives that that nation might live. It is altogether fitting and proper that we should do this.

"But, in a larger sense, we can not dedicate—we can not consecrate—we can not hallow—this ground. The brave men, living and dead, who struggled here, have consecrated it, far above our poor power to add or detract. The world will little note, nor long remember what we say here, but it can never forget what they did here. It is for us the living, rather, to be dedicated here to the unfinished work which they who fought here have thus far so nobly advanced. It is for us the living, rather, to be here dedicated to the great task remaining before us—that from these honored dead we take increased devotion to that cause for which they gave the last full measure of devotion—that we here highly resolve that these dead shall not have died in vain—that this nation, under God, shall have a new birth of freedom—and that government of the people, by the people, for the people, shall not perish from the earth."[9]

1864–65—"With Malice toward None,
With Charity for All,
With Firmness in the Right"

Chapter 10

Lincoln continued to feel a personal responsibility for the problem of slavery. The institution affronted his deepest religious and moral convictions and contradicted his understanding of American destiny. Yet his own power to deal with the evil was limited by his responsibilities as a President under oath to uphold the Constitution and to public opinion as a representative leader of the nation. By an "act of military necessity" he had forever freed the slaves in those areas of the Confederacy at war with the United States on January 1, 1863. It left unsolved, of course, the problem of slavery in the border states and in certain areas of the Confederacy already pacified by military government at the date of the Emancipation Proclamation.

He had urged the Congress in his Annual Message of December 1862 to pass constitutional amendments looking toward federal compensation for those states that undertook gradual emancipation. He had assembled the congressional delegations from the border states and pleaded with them to work for state emancipation. He had written many letters to influential men in those states, urging them to mold public opinion in a way favorable to this work. The response to Lincoln's initiative, however, had been meager.

Lincoln finally decided that the only way out of the impasse was congressional initiative for a constitutional amendment, to be ratified by the states, that would forever abolish slavery from the Union. On April 8, 1864, such an amendment passed the Senate, but then failed in the House to secure the necessary two-thirds vote. He suggested that this amendment be incorporated in the Republican platform at the Baltimore convention. This was accordingly done.

In his Annual Message of December 6, 1864, he recommended another vote on the measure, arguing that the results of the fall election showed that in the next Congress the amendment would pass. Since this was inevitable he urged the present Congress to pass the amendment so that the states might the earlier receive the proposed change and begin the process of ratification or rejection.

"It is the voice of the people now, for the first time, heard upon the question. In a great national crisis, like ours, unanimity of action among those seeking a common end is very desirable—almost indispensable. And yet no approach to such unanimity is attainable, unless some deference shall be paid to the will of the majority, simply because it is the will of the majority. In this case the common end is the maintenance of the Union; and, among the means to secure that end, such will, through the election, is most clearly declared in favor of such constitutional amendment."[1]

Lincoln was not, however, content to rest his case on the persuasion of statesmanship alone, where it would have been precarious in the extreme. He showed himself once again the astute politician who had won himself a reputation for log-rolling in the Illinois legislature. He determined to win the needful Democratic votes by promising the patronage that was his to give. One Democratic congressman was to receive a federal appointment for his brother. Another whose election was contested was promised support. In a dramatic roll-call vote before packed galleries the amendment passed by four votes. It was then submitted to the states to become finally the Thirteenth Amendment prohibiting slavery.

Charles A. Dana, who had once been the President's agent in lining up votes for the admission of Nevada to the Union, described Lincoln's prowess: "[He] was a supreme politician. He understood politics because he understood human nature . . . . there was no flabby philanthropy about Abraham Lincoln. He was all solid, hard, keen intelligence combined with goodness."[2]

There are two interesting documents that bring out the religious overtones of Lincoln's action during this period just surveyed. One is a letter to Albert Hodges, editor of the Frankfort (Kentucky) *Commonwealth.* It summarized Lincoln's conversation with a Kentucky delegation that had protested the enlisting of Negroes as soldiers. Many of its points anticipated the Second Inaugural and re-emphasized his faith in God's immediate control of the events of history and in His purpose to accomplish justice. Lincoln often repeated the adage about man as proposing, but God as disposing.

"I am naturally anti-slavery. If slavery is not wrong, nothing is wrong. I cannot remember when I did not so think, and feel. And yet I have never understood that the Presidency conferred upon me an unrestricted right to act officially upon this judgment and feeling. . . . I have done no official act in mere deference to my abstract judgment and feeling on slavery. . . .

"I add a word which was not in the verbal conversation. In telling this tale I attempt no compliment to my own sagacity. I claim not to have controlled events, but confess plainly that events have controlled me. Now, at the end of three years' struggle the nation's condition is not what either party, or any man devised, or expected. God alone can claim it. Whither it is tending seems plain. If God now wills the removal of a great wrong, and wills also that we of the North, as well as you of the South, shall pay fairly for our complicity in that wrong, impartial history will find therein new cause to attest and revere the justice and goodness of God."[3]

The second document was a response to a delegation of the American Baptist Home Missionary Society. It had called upon him to present a series of resolutions supporting Lincoln's stand and apparently favoring the end of all slavery. His written reply two days later is about as sharp a criticism of Southern Christians as he ever made, tempered at the end by the Savior's warning on judgment.

Evidence has already been presented that he preferred in his later years to express his objection to slavery in terms of a biblical understanding of work rather than in his earlier derivation of it from the "self-evident truths" of creation. Lincoln felt strongly about the essential importance of labor to society and liked to make it concrete by referring to the injunction on work in Genesis. He had known in early life what it meant to earn bread in the sweat of his brow. He was offended by the arrogant complacency of the planter interests and especially by their mouthpieces in the clergy. To him slavery was a form of stealing. He exposed its pretense by invoking the Golden Rule of the Savior. The passage breathed the passion of an Amos against hypocrisy and injustice, but it was redeemed from sheer denunciation by a Christian perspective on the self and judgment.

". . . I can only thank you for thus adding to the effective and almost unanimous support which the Christian communities are so zealously giving to the country, and to liberty. Indeed it is difficult to conceive how it could be otherwise with anyone professing Christianity, or even having ordinary perceptions of right and wrong. To read in the Bible, as the word of God himself, that 'In the sweat of *thy* face thou shalt eat bread,' and to preach therefrom that, 'In the sweat of *other men's* faces shalt thou eat bread,' to my mind can scarcely be reconciled with honest sincerity. When brought to my final reckoning, may I have to answer for robbing no man of his goods; yet more tolerable even this, than for robbing one of himself, and all that was his. When, a year or two ago, those professedly holy men of the South, met in semblance of prayer and devotion, and, in the Name of Him who said, 'As ye would all men should do unto you, do ye even so unto them,' appealed to the Christian world to aid them in doing to a whole race of men, as they would have no man do unto themselves, to my thinking, they contemned and insulted God and His church, far more than did Satan when he tempted the Savior with the Kingdoms of the earth. The devil's attempt was no more false, and far less hypocritical. But let me forbear, remembering it is also written 'Judge not, lest ye be judged.' "[4]

In the last of his Thanksgiving Day proclamations Lincoln repeated the motifs which are now thoroughly familiar to the reader—the nation devoted to a cause larger than its own life, gratefulness in concrete detail for special providences, penitence as the proper approach of man to God, and the vocation of the American people in the new promised land. There is, however, no dull repetition about the document, for he succeeded in saying it with freshness and in a poetic style that drew both imagery and rhythm from the Bible.

"It has pleased Almighty God to prolong our national life another year, defending us with his guardian care against unfriendly designs from abroad, and vouchsafing to us in His mercy many and signal victories over the enemy, who is of our own household. It has also pleased our Heavenly Father to favor as well our citizens in their homes as our soldiers in their camps and our sailors on the rivers and seas with unusual health. He has largely augmented our free population by emancipation and by immigration, while He has opened to us new sources of wealth, and has crowned the labor of our workingmen in every department of industry with abundant rewards. Moreover, He has been pleased to animate and inspire our minds and hearts with fortitude, courage and resolution sufficient for the great trial of civil war into which we have been brought by our adherence as a nation to the cause of Freedom and Humanity, and to afford us reasonable hopes of an ultimate and happy deliverance from all dangers and afflictions."

He then designated the last Thursday in November as "a day of Thanksgiving and Praise to Almighty God, the beneficent Creator and Ruler of the Universe. And I do farther recommend to my fellow citizens aforesaid that on that occasion they do reverently humble themselves in the dust and from thence offer up penitent and fervent prayers and supplications to the Great Disposer of events for a return of the inestimable blessings of Peace, Union and Harmony throughout the land, which it has pleased Him to assign as a dwelling place for ourselves and for our posterity throughout all generations."[5]

One of the elements of perennial newness in Lincoln's statements about God is the abundant wealth of his titles and attributes in describing the Creator. They can all be summarized under a phrase in his Second Inaugural, "believers in a Living God." For Lincoln the "givenness" of God and God's nearness to him in immediate relationship called forth a tribute of poetic praise. Mrs. Lincoln spoke of his religion as poetry. The devout St. Francis, who sang "The Canticle to the Sun" and imaginatively invested all creation with the breath of personal life, was paralleled by the President, who praised God in a wealth of concrete images. The following list has been selected from the Rutgers edition of his works:

"Almighty, Almighty Architect, Almighty Arm, Almighty Being, Almighty Father, Almighty God, Almighty Hand, Almighty Power, Almighty and Merciful Ruler of the Universe, Beneficent Creator and Ruler of the Universe, Great Disposer of Events, Divine Being, Divine Guidance, Divine Power, Divine Providence, Divine Will, Father, Beneficent Father Who Dwelleth in the Heavens, Common God and Father of All Men, Father in Heaven, Father of Mercies, Great Father of Us All, God of Hosts, God of Right, God of Nations, Most High God, Holy Spirit, Living God, Great and Merciful Maker, Maker of the Universe, Most High, Supreme Being, Supreme Ruler of the Universe."

Any study of Lincoln's religion should examine the practical expression of religious thought and devotion in his everyday life. This is the measuring rod that Lincoln himself used in digging down through the sands of denominationalism to the bedrock of the Savior's life and teaching. Most Christians believe that they should express their love for God by loving their neighbor, but often their religion remains on the level of officially defined concept and piously articulated intention. Lincoln expressed the fruit of Christian living with rare integrity and charm.

Tolstoy was so overwhelmed by this dimension in Lincoln that he could call him "A Christ in miniature." Jesse Fell, in a statement already surveyed, put the matter in a way that compels agreement: ". . . his principles and practices and the spirit of his whole life were of the kind we universally agree to call Christian."

His qualities of forbearance, patience, simple honesty, forgivingness, humility, and kindliness are documented in the stories known to every schoolboy and simply multiplied by his biographers. What Lincoln says about humility and charity rings with a more authentic sound when it is remembered that he could accept calmly the snub of General McClellan in retiring to bed without allowing the waiting President to see him. Its echo rings clearly again in his patient dealing with the vain Chase, who intrigued to get the Republican nomination for himself in 1864.

Lincoln could write in a letter: "I am a patient man—always willing to forgive on the Christian terms of repentance; and also to give ample *time* for repentance."[6] To another correspondent he wrote, "I shall do nothing in malice. What I deal with is too vast for malicious dealing."[7]

From a window of the White House to a group who serenaded him on his re-election, he confessed: "So long as I have been here I have not willingly planted a thorn in any man's bosom. While I am deeply sensible to the high compliment of a re-election; and duly grateful, as I trust, to Almighty God for having directed my countrymen to a right conclusion, as I think, for their own good, it adds nothing to my satisfaction that any other man may be disappointed or pained by the result. May I ask those who have not differed with me, to join with me, in this same spirit towards those who have?"[8]

Lincoln's own spirit of a basic charity toward all men that excluded vindictiveness or malice was expressed in terms of God's will for the whole nation in the Second Inaugural Address. That document with its authentic rootage in a biblical understanding of God, man, and history has become for the whole world a charter of Christian statesmanship. It perfectly expressed the religion of its author, so soon to be struck down, who had said at the Baltimore Sanitary Fair, ". . . I am responsible . . . to the American people, to the Christian world, to history, and on my final account to God."[9]

The address began on a somewhat pedestrian note with Lincoln disclaiming the need for any new statement of policy and summarizing the military situation. Then each new sentence began to mount higher as on eagles' wings.

"Both parties deprecated war; but one of them would *make* war rather than let the nation survive; and the other would *accept* war rather than let it perish. And the war came.

"One eighth of the whole population were colored slaves, not distributed generally over the Union, but localized in the Southern part of it. These slaves constituted a peculiar and powerful interest. All knew that this interest was, somehow, the cause of the war. To strengthen, perpetuate, and extend this interest was the object for which the insurgents would rend the Union, even by war; while the government claimed no right to do more than to restrict the territorial enlargement of it. Neither party expected for the war, the magnitude, or the duration, which it has already attained. Neither anticipated that the *cause* of the conflict might cease with, or even before, the conflict itself should cease. Each looked for an easier triumph, and a result less fundamental and astounding. Both read the same Bible, and pray to the same God; and each invokes His aid against the other. It may seem strange that any men should dare to ask a just God's assistance in wringing their bread from the sweat of other men's faces; but let us judge not that we be not judged. The prayers of both could not be answered; that of neither has been answered fully. The Almighty has His own purposes. 'Woe unto the world because of offenses! For it must needs be that offenses come; but woe to that man by whom the offense cometh!' If we shall suppose that American Slavery is one of those offenses which, in the providence of God, must needs come, but which, having continued through His appointed time, He now wills to remove, and that He gives to both North and South, this terrible war, as the woe due to those by whom the offense came, shall we discern therein any departure from those divine attributes which the believers in a Living God always ascribe to Him? Fondly do we hope—fervently do we pray—that this mighty scourge of war may speedily pass away. Yet, if God wills that it continue, until all the wealth piled by the bond-man's two hundred and fifty years of unrequited toil shall be sunk, and until every drop of blood drawn with the lash, shall be paid with another drawn with the sword, as was said three thousand years ago, so still it must be said 'the judgments of the Lord, are true and righteous altogether.'

"With malice toward none; with charity for all; with firmness in the right, as God gives us to see the right, let us strive on to finish the work we are in; to bind up the nation's wounds; to care for him who shall have borne the battle, and for his widow, and his orphan—to do all which may achieve and cherish a just, and a lasting peace, among ourselves, and with all nations."[10]

Comment on this address seems almost blasphemous, but it may be helpful to see it as the climax of Lincoln's religious development. Earlier, in the proclamation after Bull Run, he had urged the nation to see the hand of God in the visitation of war. Increasingly he defined the theological issue in the conflict.

His papers and letters show his belief that God willed "His almost chosen people" like Israel of old to be the bearer of new freedom to all men everywhere. The Puritan background of Lincoln's confidence in American destiny under God had become rationalized in the eighteenth and nineteenth centuries into the dream of world democracy with the original religious perspective rapidly disappearing into the distance.

With his incisive logic Lincoln gave definition to America's hope for democracy in terms compelling to his contemporaries, but he also sustained that vision in its original religious rootage and reference to God's will. The religious interpretation was organic and fundamental. It never savored of a pious but irrelevant afterthought as it does in so much contemporary political and pulpit oratory. The depth of Lincoln's religious interpretation of the nation's history, however, was not merely an inheritance from the past but a living power of rekindled insight.

Lincoln's gift of mystical intuition led him into a specific explanation of the slavery issue in terms of the Old Testament prophets. His Puritan forebears would have called it a "discerning of the signs of the times," a feeling for "particular providences." As Lincoln analyzed God's intention to lead men into larger freedom and appropriated to his use the language of the Declaration of Independence and the sayings of the Founding Fathers he came in the context of events to regard slavery as a contradiction of God's will. This defiance of God's justice had been built into the life of the nation and was therefore subject to God's judgment.

The long, unhappy debates over the slavery issue had weakened the nation and made it seem hypocritical to aspiring men in other countries. For somewhat more than the last year of his life Lincoln understood the tragedy and suffering of the Civil War as God's judgment upon this evil and as punishment to bring about its removal. The judgment fell upon both sides, for slavery was a national and not merely a sectional evil. The North had also prospered from the cheap raw materials that slave labor fed into its factories. It was conceivable to Lincoln that a just God might allow the war to continue "until all the wealth piled by the bond-man's two hundred and fifty years of unrequited toil shall be sunk." In the severe language of Scripture Lincoln held the nation under judgment: "the judgments of the Lord are true and righteous altogether."

But the judgments of God have as their purpose the reformation of His people. The renewal of an America newly dedicated to the increase of freedom had been his theme at Gettysburg and must be understood as the implied correlative here of the emphasis upon judgment.

The leaven of Christianity that is at work in this address and carries the analysis of judgment beyond that of an Amos or a Jeremiah is expressed in the Savior's warning about the peril of passing judgment. It is also present in the almost scriptural paraphrase of the Savior's summary of the law and of St. Paul's famous chapter on love that Lincoln achieves in his phrases "With malice toward none; with charity for all." The disclaimer of human judgments on the opponent does not lead, however, to irresolution in action. The very opposite is the case. Understanding the perspective of two antagonists before the judgment seat of God and thereby freed from the tyranny of self-righteous fanaticism, there comes to Lincoln the resource of "firmness in the right, as God gives us to see the right."

This document is one of the most astute pieces of Christian theology ever written. It is also a charter of Christian ethics. It may seem strange to call Lincoln a theologian, for he was obviously not one in any technical sense. There are many profundities in the Christian religion which he never did illuminate, but in the area of his vision he saw more keenly than anyone since the inspired writers of the Bible. He knew he stood under the living God of history.

He understood, for example, the finiteness of man's religious perspective without thereby becoming either a relativist or a skeptic. He achieved a religious perspective above partisan strife that was not shared by most of the Christian theologians of his day or any day, who often express Christian insights in obsolete terms in a way that ultimately deifies the position of man. Lincoln could detach himself from his own interested participation in events and submit all, including himself as judging, to higher judgment. The highly tentative nature of his own judgments do not argue uncertainty or irresolution but simply confess that God is God and man remains on every level man. The utter priority and "givenness" of God had for Lincoln as its corollary the utter dependence of man upon God.

The Second Inaugural illuminates the finiteness of man when in sincerity men embrace opposite courses of action under the conviction that each is responsive to God's will. "Both read the same Bible, and pray to the same God; and each invokes His aid against the other." It has often been pointed out that religion has exacerbated human strife by clothing diametrically opposite lines of activity with the sanctions of absolute holiness.

There is a fanaticism in many people that blinds them to the truth of their position, increases their self-righteousness, and isolates them from their fellows. It is the fallacy of supposing that a sincere intention to do God's will guarantees that what I sincerely do is the will of God. Lincoln could detach himself from the element of pretense in the idealistic claims made by both sides. In his meditation on the divine will Lincoln opened up horizons beyond the simple but all too common analysis—that one side is right and the other is wrong. Both could not be right. Each might be partly right and partly wrong and God might use both sides as His instruments to effect a result not foreseen by either side. Lincoln argued that this was the case with the Civil War. "Only God can claim it," he once said, thereby stating that the complexities of historical events are so involved that finite man cannot claim infallible insight for his own interpretations. Lincoln expressed this again in his own comments on the Second Inaugural in a letter written to Thurlow Weed:

"I expect the latter to wear as well as—perhaps better than—anything I have produced; but I believe it is not immediately popular. Men are not flattered by being shown that there has been a difference of purpose between the Almighty and them. To deny it, however, in this case, is to deny that there is a God governing the world. It is a truth which I thought needed to be told; and as whatever of humiliation there is in it, falls most directly on myself, I thought others might afford for me to tell it."[11]

The fact, however, that the divine will remains in the

realm of ultimate mystery need not lead to a despairing or resigned agnosticism. Because Lincoln was aware that Providence would overrule the element of pretension in the highly idealistic claims made by each side he was not therefore led to the conclusion that responsible action undertaken by either side would necessarily be irrelevant to the moral issue. The mystery is illuminated by meaning like lightning in the night sky. Knowing that God and not Lincoln would have the final word in the dilemma of the two contestants at prayer, he could yet venture provisional judgments and act resolutely in their light. "It may seem strange that any men should dare to ask a just God's assistance in wringing their bread from the sweat of other men's faces . . . let us strive on to finish the work we are in." Lincoln's sympathy for the dilemma of the Quakers, caught between pacificism and emancipation as an act in war, is another illustration of the depth of his biblical spirituality on this point.

He could appreciate the sincerity of his foe although he believed him wrong, but because of his religious perspective he could deal magnanimously and forgivingly without the self-righteousness of the victor. This point of vantage beyond the strife of factions did not, however, kill the nerve of resolution. Since God was supreme a man might act without fanaticism and hatred on the one hand and without torturing doubt or irresolution on the other.

This is the charter of responsible action for the Christian citizen. St. Francis, the loving, gentle imitator of Christ, will always be dear to the hearts of men. The appeal, however, of the ascetic who leaves the responsibilities of society for the cultivation of individual piety and ecstatic person-to-person relationships has an element of romanticism in it. Lincoln represents the responsible Christian citizen of this world struggling to be responsive to the guidance of God amid the challenges of the full historic setting of man's life. Both the Christian saint and the Christian statesman are needed, but there can be little doubt as to which one more fully represents the demands of a God revealed through the Bible as the "Great Disposer of Events."

Lincoln's religious analysis of the nation's history is as relevant today as it was when he painfully developed it. One has only to substitute the phrase "segregation in public education" for Lincoln's word "slavery" and his theological analysis becomes luminous. America stands today among the nations of the world as the self-proclaimed defender of democracy in areas that are being progressively inundated by totalitarianism. Our professed defense of the right of all men everywhere to be free sounds increasingly hollow to multitudes in Asia and Africa because of our racial discrimination at home. If the rising colored peoples of the world turn from democracy to communism partly because of our racialism the security of America will become a very frail thing. And yet "the judgments of the Lord are true and righteous altogether."

In our competition with the Soviet Union we proclaim ourselves the defenders of religion against atheistic materialism, but if infatuation with our own materialist standard of living makes us unwilling to use our resources to lift the economy of less developed nations we may find ourselves without friends in a world increasingly dominated by communism.

Yet "the judgments of the Lord are true and righteous altogether."

To a religious understanding of the nation's destiny that is as meaningful today as for the crisis of the Civil War Lincoln adds a dynamic for responsible action that is equally helpful. Courageous action prayerfully undertaken in openness and in unself-righteous concern for all men is the leaven that can alter the whole lump. Commenting on Lincoln's religious insights into the moral dilemmas of his day, Reinhold Niebuhr in his *Irony of American History* writes as follows:

"This combination of moral resoluteness about the immediate issues with a religious awareness of another dimension of meaning and judgment must be regarded as almost a perfect model of the difficult but not impossible task of remaining loyal and responsible toward the moral treasures of a free civilization on the one hand while yet having some religious vantage point over the struggle."[12]

It was such a perspective that enabled Lincoln not only to serve the Union without self-righteous stances but also, therefore, to serve it more effectively. Those who wield great power are often snared in traps of their own making when they become blinded by considerations of prestige.

Lincoln could concentrate on practical actions needed to restore the Union because he was free of the Radical Republican determination to punish the South and rationalize its own ideology of the war. Christian statesmanship marks the address he made on the evening of April 11, 1865, his last public utterance. He appealed to the people to support his principles and action for reconstruction.

"We all agree that the seceded States, so called, are out of their proper practical relation with the Union; and that the sole object of the government, civil and military, in regard to those States is to again get them into that proper practical relation. I believe it is not only possible, but in fact, easier, to do this, without deciding, or even considering, whether these States have even been out of the Union, than with it. Finding themselves safely at home, it would be utterly immaterial whether they had ever been abroad. Let us all join in doing the acts necessary to restoring the proper practical relations between these States and the Union; and *each forever after, innocently indulge his own opinion* whether, in doing the acts, he brought the States from without, into the Union, or only gave them proper assistance, they never having been out of it."[13]

The problems of national reconciliation in a spirit of forgiveness were uppermost in his mind when the bullet struck him on Good Friday. He carried to his death another item of unfinished business mentioned four days before in this last speech. "He, from whom all blessings flow, must not be forgotten. A call for a national thanksgiving is being prepared, and will be duly promulgated."[14]

Any study of Lincoln's religion should lead to some brief characterization of his position. Since his religious development was a continuing process there is always danger of freezing prematurely upon some formula which then becomes a caricature. Dodge has pointed to the slow maturing of his literary style. Others have traced the gradual evolution of his views on slavery. Judging by the acceleration noticeable in his last years it would be folly to assume that he had reached his religious maturity when the assassin's bullet cut short his earthly span.

Francis Grierson in his deeply moving *The Valley of the Shadows* characterized Lincoln as a "practical mystic." Many have been content to stop here, although the word "mystic" demands further definition.

Nathaniel Stephenson concluded that "his religion continues to resist intellectual formulation. He never accepted any definite creed. To the problems of theology, he applied the same sort of reasoning that he applied to the problems of the law. He made a distinction, satisfactory to himself at least, between the essential and incidental, and rejected everything that did not seem to him altogether essential."[15] Granting the deeply personal and individual character of Lincoln's religion, is it possible to give a more precise description than this?

Lincoln was unquestionably our most religious President. Professor Randall concluded the fourth volume of his great study with a chapter entitled "God's Man." "Lincoln was a man of more intense religiosity than any other President the United States ever had. . . . Surely, among successful American politicians, Lincoln is unique in the way he breathed the spirit of Christ while disregarding the letter of Christian doctrine. And the letter killeth, but the spirit giveth life."[16]

Randall's judgment was anticipated by a number of Lincoln's contemporaries. Henry Whitney, his circuit-riding friend, wrote: "The conclusion is inevitable, that Mr. Lincoln was practically and essentially, though not ritualistically a Christian."[17] Isaac Arnold, congressman from Illinois and a close friend, shared the same estimate: "No more reverent Christian than he ever sat in the executive chair. . . . It is not claimed that he was orthodox. For creeds and dogmas he cared little. But in the great fundamental principles of religion, of the Christian religion, he was a firm believer."[18]

There is some evidence that Lincoln may have been considering membership in the church. His friend Noah Brooks, who was to have become his secretary, made this point. "In many conversations with him, I absorbed the firm conviction that Mr. Lincoln was at heart a Christian man, believed in the Savior, and was seriously considering the step which would formally connect him with the visible Church on earth."[19] In the light of Lincoln's unorthodox attitude on universal salvation and his distaste for creedal definitions it is not likely that he would have sought membership in the Washington church which he was attending. Considering the vow which he made before the battle of Gettysburg, the weight of evidence points more toward Lincoln's feeling a concern to make a public confession of his faith rather than to seek membership in any one church.[20]

By the creedal standards of the churches of his day he was not "an orthodox Christian." His wife said that he was not "a technical Christian." Our definition today is a more liberal one. David Mearns has aptly called him a "Christian without a Creed."[21] This may be the least inadequate phrase. It would, however, be a serious mistake to interpret it to mean that his faith was without specific content.

A better phrase might be "a biblical Christian." It would at once shift the emphasis from the institutional side of Christianity, in which his religion was defective, to its bedrock foundation in Scripture. From the Bible in a quite independent way he quarried granite to support a religious interpretation of American history and of man's vast future.

To use a still more precise definition, Lincoln was "a biblical prophet" who saw himself as "an instrument of God" and his country as God's "almost chosen people" called to world responsibility.

In *The Soul of Abraham Lincoln* William Barton reported his ownership of a half page of note paper in Lincoln's handwriting. On it Lincoln had copied out, presumably for his own meditation, a passage from the Puritan theologian, Richard Baxter. The passage was about man's assurance of salvation. That assurance was grounded not in man's subjective feelings but in the sense of direction to his life. These wise words may have spoken to Lincoln's understanding of his own religious situation:

"It is more pleasing to God to see his people study Him and His will directly, than to spend the first and chief of their effort about attaining comfort [i.e., assurance of belief] for themselves. We have faith given us, principally that we might believe and live by it in daily applications of Christ. You may believe immediately (by God's help), but getting assurance of it may be the work of a great part of your life."[22]

Appendix I

In 1920, William Barton concluded *The Soul of Abraham Lincoln* with a creed expressed entirely in Lincoln's own words with only slight changes for grammatical reasons. It is printed here as a classic tour de force in studies of Lincoln's religion. Its danger is that, separated from its context in his life and writings, it overemphasizes the doctrinal element. He did, after all, protest against complicated creeds as requirements for church membership. Its strength, however, is to underscore the biblically rooted faith of the man in opposition to recent interpretations of his religion as "creedless" in the sense of being "wholly free from any doctrinal commitment."

The Creed of Abraham Lincoln in His Own Words

I believe in God, the Almighty Ruler of Nations, our great and good and merciful Maker, our Father in Heaven, who notes the fall of a sparrow, and numbers the hairs of our heads.

I believe in His eternal truth and justice.

I recognize the sublime truth announced in the Holy Scriptures and proven by all history that those nations only are blest whose God is the Lord.

I believe that it is the duty of nations as well as of men to own their dependence upon the overruling power of God, and to invoke the influence of His Holy Spirit; to confess their sins and transgressions in humble sorrow, yet with assured hope that genuine repentance will lead to mercy and pardon.

I believe that it is meet and right to recognize and confess the presence of the Almighty Father equally in our triumphs and in those sorrows which we may justly fear are a punishment inflicted upon us for our presumptuous sins to the needful end of our reformation.

I believe that the Bible is the best gift which God has ever given to men. All the good from the Saviour of the world is communicated to us through this book.

I believe the will of God prevails. Without Him all human reliance is vain. Without the assistance of that Divine Being, I cannot succeed. With that assistance I cannot fail.

Being a humble instrument in the hands of our Heavenly Father, I desire that all my works and acts may be according to His will; and that it may be so, I give thanks to the Almighty, and seek His aid.

I have a solemn oath registered in heaven to finish the work I am in, in full view of my responsibility to my God, with malice toward none; with charity for all; with firmness in the right as God gives me to see the right. Commending those who love me to His care, as I hope in their prayers they will commend me, I look through the help of God to a joyous meeting with many loved ones gone before.

Appendix II

Jay Monaghan's *A Lincoln Bibliography, 1839–1939* listed nearly four thousand books and pamphlets on Lincoln. The output has not slackened in the last two decades. The Library of Congress catalogues about fifty items on Lincoln's religion. The aim of the following list is to winnow the chaff from the grain and present some suggestions to beginners in this vast field.

Lincoln's religion should be studied as part of his life. The best one-volume study representing current Lincoln scholarship is Benjamin Thomas's *Abraham Lincoln* (1952). While it predates critical Lincoln studies, Lord Charnwood's *Abraham Lincoln* (1917) is most valuable for its analysis of his character and of his significance as a world figure.

Beyond the one-volume scope the reader has interesting choices. He may select the ten-volume *Abraham Lincoln: A History* by Nicolay and Hay (1890). Because it was written by Lincoln's private secretaries and was submitted for approval to Robert Lincoln, the President's son, it has often been called the "official" biography. It has by no means been replaced by later studies. It is not, however, particularly perceptive on the subject of Lincoln's faith.

Carl Sandburg's two-volume *Prairie Years* (1926) and the four-volume *War Years* (1939) constitute an artistic masterpiece and should become an American classic. Filled with local color and the flavor of the times, Sandburg's books show a lively interest in Lincoln's religion. His good chapter of Lincoln's laughter and religion has been mentioned previously. In 1954, Sandburg published a one-volume condensation of his work.

James Randall's four-volume study, *Lincoln the President* (1955) is the work of a gifted writer and a professional historian. It is a needed corrective to the mythical picture of Lincoln that underlay so many older studies of his religion. Professor Randall's chapter on "God's Man" shows an appreciative interest in Lincoln's faith, but it is not integrated with the larger study itself in a way that would bring out the developmental aspect of Lincoln's religion. Professor Richard Current has edited Randall's writings to produce *Mr. Lincoln* (1957), a one-volume concentration on Lincoln the man.

Professor Allan Nevins's four-volume *Ordeal of the Union* (1947) and *Emergence of Lincoln* (1950) are helpful for understanding the history of the United States from 1846 to 1861.

Paul Angle's *A Shelf of Lincoln Books: A Critical, Selective Bibliography of Lincolniana* (1946) is our best introduction to eighty-one books dealing with special Lincoln topics. Benjamin Thomas's *Portrait for Posterity* (1947) studies very interestingly the biographers of Lincoln and outlines briefly their battles over Lincoln's religion.

The basic source for Lincoln's religion must remain the eight-volume edition of his collected works published by Rutgers in 1953 with Roy Basler as editor. Nearly all of the significant material has been reprinted in the present, which the writer has designed as an anthology of Lincoln's writings on religion as well as an interpretation of his religion. Most of the early books failed to understand that Lincoln is a better interpreter of his own faith than either Herndon or Holland ever could be. David Donald's *Lincoln's Herndon* (1948) throws a critical spotlight over secondary and tertiary levels of testimony. Albert House has studied "The

Genesis of the Lincoln Religious Controversy" in *Proceedings of the Middle States Association of History and Social Science Teachers* (1938). A full-length study of all the sources for Lincoln's religion is much needed.

Early works that are still useful would include William Herndon and Jesse Weik, *Herndon's Life of Lincoln,* particularly the edition of 1930 with a critical introduction by Paul Angle. F. B. Carpenter's *Six Months at the White House* (1866) is an extremely valuable source, especially when the artist records his firsthand experiences. Douglas C. McMurtrie in 1936 edited *Lincoln's Religion,* which contains an address by the Rev. James Reed that was first published in *Scribner's* in 1873. It also contains Herndon's answering lecture to Reed, which was first printed in the *State Register of Springfield. Lincoln's Religious Belief,* by B. F. Irwin, was first published in the *Illinois State Journal* May 16, 1874, and may be consulted in Appendix VI to Barton's *Soul of Abraham Lincoln.*

Two twentieth-century studies are worth mentioning. William J. Johnstone, *Abraham Lincoln: The Christian* (1913), is a carefully documented study, but it uses materials in a very naïve way. Its interpretations far outrun the evidence. John Wesley Hill, *Abraham Lincoln, Man of God* (1920), fortunately does not mold his materials as excessively as Johnstone, but the author has uncritically accepted items such as the Lincoln letter of March 2, 1837, to the Rev. James Lemen, now regarded as a forgery.

Two examples of extreme positions are interesting. Ervin Chapman's *Latest Light of Lincoln* (1917) has already been quoted as a representative of the pious school, bordering more on hagiography than history. John E. Remsburg, on the other side, devoted over three hundred pages of his *Six Historic Americans* (1906), which he wrote for the Freethinker press, to a ponderous and shrill argument that Lincoln was once and always an infidel. Remsburg and Herndon corresponded sympathetically on this theme.

The great classic in the field has been William E. Barton's

The Soul of Abraham Lincoln (1920), which has not yet been superseded in its encyclopedic criticism of sources. New materials have, however, come to light and Barton's book has a cluttered aspect that makes it tedious to read. The element of source criticism overwhelms Barton's own interpretation of Lincoln's religion. Its appendices make available a number of useful documents to which reference has already been made.

Edgar DeWitt Jones's *Lincoln and the Preachers* (1948) presents a good survey of just what the title indicates. There is a useful appendix, "Who's Who of the Preachers in the Lincoln Story." H. H. Horner's *The Growth of Lincoln's Faith* (1939) and Ralph Lindstrom's *Lincoln Finds God* (1958) are interesting short interpretations. Clarence Macartney's *Lincoln and the Bible* (1949) is a valuable and interestingly written study.

Other significant books and manuscript sources used in the preparation of this book are mentioned in the footnotes. *Lincoln Lore,* published by the Lincoln National Life Foundation and now edited by Gerald McMurtrie, contains interesting material from time to time on the subject of Lincoln's religion.

Footnotes and References

Chapter 1

1. Francis B. Carpenter, *Six Months at the White House with Abraham Lincoln* (1866), pp. 89–90.

2. Robert B. Warden, *Life of S. P. Chase,* pp. 481–82, quoted in Nicolay and Hay, *Abraham Lincoln: A History* (1890), Vol. VI, pp. 159–60.

3. "The Diary of Gideon Welles," *Atlantic Monthly,* 1909, p. 369.

4. *The Collected Works of Abraham Lincoln* (CWAL, henceforth) (Rutgers University Press, 1953), Vol. V, pp. 388–89. Lincoln's mistakes in spelling, which were properly retained by the Rutgers editors, have been corrected in the quotations to allow the reader to concentrate on the thought. Slight changes have at times been made in punctuation.

5. CWAL, Vol. V, pp. 419–25.

6. Narrated by William Johnstone in *Abraham Lincoln: The Christian* (1913), pp. 91–93. Johnstone interviewed one of the Beecher grandsons in Philadelphia.

7. Ward H. Lamon, *The Life of Abraham Lincoln* (1872), p. 489. Also quoted in Carl Sandburg, *The War Years,* Vol. IV, pp. 244–45. Lincoln's statement that dreams are nowadays seldom told has been corrected with the advent of Freud!

8. Stephen Vincent Benét, *John Brown's Body* (1928), p. 213.

Chapter 2

1. CWAL, Vol. I, p. 1.
2. Ervin Chapman, *Latest Light on Lincoln,* (1917), p. 315.
3. CWAL, Vol. I, p. 386.
4. Lamon, *The Life of Abraham Lincoln,* p. 39.
5. Benjamin Thomas, *Lincoln's New Salem* (1934), p. 89.
6. Benjamin Thomas, in *Portrait for Posterity* (1947), has sketched the battle among Lincoln biographers over his religion. A full-length study of this area is much needed, although the primary data for Lincoln's religion must always remain his own speeches and writings. David Donald's careful study, *Lincoln's Herndon* (1948), has made important contributions to the question of Herndon's reliability on the religious issue. For example, Donald points out that Holland's use of a famous Bateman interview with Lincoln was based on an eight-page letter from Bateman. It did not depend upon Holland's imaginative recollection of a conference with Bateman as Herndon and most Lincoln scholars have supposed. Thomas's evaluation is as follows:

"The controversy over Lincoln's religion seems now to have been a bandying of words. Neither side defined its terms. Had they done so, they might have found they were not too far apart. But, in its later stages, what was originally an honest-if-violent difference of opinion, degenerated into a promotional intrigue with Black [ghost writer of Lamon's *Life of Abraham Lincoln*] the venal prompter and Herndon his gullible stooge" (p. 89).

7. David Donald has a perceptive comment on this debate: "It is a mistake to consider these two main streams of tradition as representing respectively the 'ideal' and the 'real' Lincoln. Each was legendary in character. The conflict in Lincoln biography between the Holland-Hay-Tarbell school and the Herndon-Lamon contingent was not essentially a battle over factual differences; it was more like a religious war. One school portrayed a mythological patron saint; the other, an equally mythological frontier hero. Gradually the two conceptions began to blend" (*Lincoln's Herndon,* p. 372).

8. William Herndon and Jesse Weik, *Herndon's Life of Lincoln* (1949), p. 355. There is a helpful introduction by Paul Angle.
9. Letter from Mentor Graham to B. F. Irwin, March 17, 1874.
10. I Corinthians 15:20–25, 28.
11. Henry Rankin, *Personal Recollections of A. Lincoln* (1916), pp. 324–26. Rankin claimed that his mother's memory was of a high order and that she gave great care to her report of Lincoln's words.
12. CWAL, Vol. I, p. 8.
13. Ibid., p. 75.

Chapter 3

1. Carl Sandburg, *The Prairie Years* (1926), Vol. I, p. 264.
2. CWAL, Vol. I, p. 78.
3. Ibid., pp. 108–15.
4. Ibid., pp. 271–79.
5. Ibid., p. 228.
6. Ibid., p. 229.
7. Ibid., p. 261.
8. Ibid., pp. 267–68.
9. Ibid., p. 282.
10. Ibid., p. 289.
11. Ruth Painter Randall, *The Courtship of Mr. Lincoln* (1957), p. 168. Mrs. Randall published an excellent short article in the New York *Times*, February 7, 1954, entitled "Lincoln's Faith Was Born of Anguish."
12. CWAL, Vol. I, p. 303.

Chapter 4

1. CWAL, Vol. I, p. 320.
2. Sandburg, *The Prairie Years,* Vol. I, pp. 336–37.
3. CWAL, Vol. I, p. 384.

4. Ibid., p. 382.
5. Henry C. Deming, *Eulogy upon Abraham Lincoln before the General Assembly of Connecticut* (1865), p. 42.
6. CWAL, Vol. VII, p. 351.
7. Henry C. Whitney, *Life on the Circuit with Lincoln* (1892), p. 267.
8. *Herndon's Life of Lincoln,* p. 56.
9. CWAL, Vol. V, pp. 403–4.
10. Benjamin Thomas, *Abraham Lincoln,* p. 130.
11. CWAL, Vol. I, p. 368.
12. Ibid., Vol. II, p. 97.
13. *Scribner's Magazine,* July 1873, p. 333.
14. James Smith, *The Christian's Defense* (1843), p. 96.
15. Joshua Speed, *Lecture on Abraham Lincoln,* pp. 32–33.
16. Statement of December 24, 1872, quoted in William Barton, *Soul of Abraham Lincoln* (1920), p. 164.
17. James Smith to Herndon, January 24, 1867, first published in Springfield *Daily Illinois Journal,* March 12, 1867.
18. *Herndon's Life of Lincoln,* p. 354.
19. J. E. Remsburg, *Six Historic Americans* (1906), pp. 114–15.
20. Lamon, *Life of Abraham Lincoln,* pp. 490–92.

Chapter 5

1. Thomas, *Abraham Lincoln,* p. 143.
2. CWAL, Vol. VIII, p. 155.
3. Ibid., Vol. II, pp. 247–83. The Peoria Speech.
4. Ibid., pp. 546–47.
5. Ibid., p. 501.
6. T. Harry Williams, *Selected Writings and Speeches of Abraham Lincoln* (1943), p. xviii, Introduction.
7. CWAL, Vol. III, pp. 204–5.
8. Ibid., Vol. II, pp. 320–23.

9. Ibid., p. 318.

10. Ibid., Vol. III, p. 410.

11. Ibid., p. 462.

12. Letter of Isaac Cogdal to B. F. Irwin, April 10, 1874.

13. Dictated and signed by Jonathan Harnett to B. F. Irwin, and sent by latter to *State Journal,* April 20, 1874.

14. Robert D. Richardson, *Abraham Lincoln's Autobiography* (1947), Pt. II, pp. 33–34.

15. Letter of Mrs. Rebecca R. Pomeroy in *Lincoln Scrapbook,* Library of Congress, p. 54.

16. *Lincoln's Devotional* (1957) was originally entitled *The Believer's Daily Treasure* and published by the Religious Tract Society (London) (1852).

17. CWAL, Vol. III, p. 339.

18. Ibid., Vol. IV, p. 5.

19. Ibid., p. 16.

20. Ibid., pp. 190–91. This version follows the one in Lincoln's and Nicolay's handwriting set down after the event and is the one preferred by the Rutgers's editors. The two alternate versions included by the editors use such terms as "Divine Providence," "Almighty Being," "the great God," "the God of our fathers," and "the same omniscient mind and Almighty arm."

Chapter 6

1. CWAL, Vol. IV, p. 220.

2. Ibid., p. 236.

3. Ibid., pp. 262–71. The First Inaugural.

4. Ibid., p. 482.

5. Ibid., p. 386.

6. Letter of Mrs. Rebecca R. Pomeroy, *Lincoln Scrapbook,* Library of Congress, p. 54.

7. Ida Tarbell, *The Life of Abraham Lincoln* (1902), Vol. II, pp. 89–92. She fails to give enough emphasis to Lincoln's earlier religious awaken-

ing, recorded in his letter to Speed during his broken engagement to Mary Todd.

8. Carpenter, *Six Months at the White House,* pp. 117–19.

9. CWAL, Vol. VIII, p. 117.

10. *Scribner's Monthly,* 1873, p. 343.

11. Letter of Noah Brooks to J. A. Reed, December 31, 1878.

12. William E. Curtis, *The True Abraham Lincoln* (1903), pp. 385–86.

13. James F. Rusling, *Men and Things I Saw in Civil War Days* (1899), p. 15. The description of this incident in Johnstone's *Abraham Lincoln: The Christian,* p. 113, has a facsimile endorsement by General Sickles (February 11, 1911) and by General Rusling (February 17, 1910).

14. Noah Brooks, *Harper's Monthly.* July 1865.

15. CWAL, Vol. V, p. 327.

16. Ibid., Vol. VI, p. 114.

17. Carpenter, op. cit., p. 282.

18. CWAL, Vol. VI, pp. 535–36. Version printed in the Washington *National Republican.* October 24, 1863.

19. Thomas, *Abraham Lincoln,* p. 459.

Chapter 7

1. CWAL, Vol. II, p. 438.

2. Ibid., p. 278.

3. Nicolay and Hay, *Abraham Lincoln,* Vol. IX, p. 40.

4. Adlai Ewing Stevenson, *Something of Men I Have Known,* p. 352. Told by Senator Henderson to Vice-President Stevenson.

5. John Wesley Hill, *Abraham Lincoln, Man of God* (1920), pp. 269–70.

6. CWAL, Vol. VII, p. 542. The Ostervald Bible of his parents which Lincoln used in the White House printed the following statement: "The Scriptures therefore are the most valuable blessing God ever bestowed upon us, except the sending his Son into the world; they are a treasure containing everything that can make us truly rich and truly happy."

7. L. E. Chittenden, *Recollections of President Lincoln and His Administration* (1891), pp. 448–50.
8. Carpenter, *Six Months at the White House,* p. 31.
9. Colonel Silas Burt, *Century Magazine,* February 1907. Reprinted in Rufus Wilson, *Lincoln among His Friends* (1942), p. 333.
10. Letter of Herndon to Weik, December 22, 1888.
11. Carl Sandburg, *The War Years* (1939), Vol. III, p. 369.
12. Reinhold Niebuhr, *Discerning the Signs of the Times* (1946), p. 131.
13. CWAL, Vol. IV, p. 274.
14. *Herndon's Life of Lincoln,* p. 67. Told to Herndon by his cousin.
15. Edgar DeWitt Jones, *Lincoln and the Preachers* (1948), p. 148.
16. Benjamin Thomas in Abraham Lincoln Association Papers, 1935.

Chapter 8

1. Carpenter, *Six Months at the White House,* p. 189.
2. CWAL, Vol. IV, p. 441.
3. James Wilson, *North American Review,* December 1896, p. 667.
4. CWAL, Vol. V, pp. 403–4.
5. James G. Randall, *Lincoln the President* (1945), Vol. II, p. 161.
6. Edmund Wilson, *Eight Essays* (1954), p. 189. In 1922, Nathaniel Stephenson wrote in his *Lincoln* (p. 267): "It was a lofty but grave religion that matured in his final stage. Was it due to far-away Puritan ancestors?"
7. II Samuel 7:10.
8. CWAL, Vol. IV, p. 220, and Vol. II, p. 89.
9. Ibid., Vol. IV, pp. 268, 270.
10. Ibid., Vol. III, p. 310.
11. Sidney Mead, "Abraham Lincoln's 'Last Best Hope of Earth': The American Dream of Destiny and Democracy," *Church History,* March 1954, p. 3. This is an excellent study.
12. CWAL, Vol. V, p. 478. There is an interesting account of this inter-

view in the Lincoln Papers, Library of Congress, under the date 1862, September (28).

13. CWAL, Vol. VII, p. 535. Mrs. Gurney's letters are in the Robert Todd Lincoln Collection in the Library of Congress.

14. Chittenden, *Recollections,* pp. 448ff.

15. CWAL, Vol. V, pp. 497–98.

16. Ibid., pp. 518–37. Annual Message to Congress, December 1862.

17. Ibid., Vol. VI, p. 30. The Final Proclamation of Emancipation.

Chapter 9

1. CWAL, Vol. VI, p. 64.

2. Ibid., pp. 155–56.

3. Ibid., p. 314.

4. Ibid., pp. 332.

5. Ibid., pp. 406–10. Letter to James C. Conkling.

6. Carpenter, *Six Months at the White House,* pp. 208–9.

7. CWAL, Vol. VI, pp. 496–97. First National Proclamation of Thanksgiving, 1863.

8. Roy P. Basler, *Abraham Lincoln, His Speeches and Writings* (1946), p. 42.

9. CWAL, Vol. VII, p. 23. This was the final text of the address that Lincoln sent to Alexander Bliss sometime after March 4, 1864, for the Baltimore Sanitary Fair.

Chapter 10

1. CWAL, Vol. VIII, p. 149. Annual Message, December 1864.

2. Thomas, *Abraham Lincoln,* p. 494.

3. CWAL, Vol. VII, pp. 281–83. Letter to Albert Hodges.

4. Ibid., Vol. VII, p. 368.

5. Ibid., Vol. VIII, pp. 55–56. Thanksgiving Proclamation of 1864.

6. Ibid., Vol. V, p. 343.
7. Ibid., p. 346.
8. Ibid., Vol. VIII, p. 101.
9. Ibid., Vol. VII, p. 302.
10. Ibid., Vol. VIII, pp. 332–33. Second Inaugural Address.
11. Ibid., p. 356. Letter to Thurlow Weed.
12. Reinhold Niebuhr, *The Irony of American History,* p. 172.
13. CWAL, Vol. VIII, p. 403. Italics added.
14. Ibid., p. 399.
15. Nathaniel Stephenson, *Lincoln* (1922), p. 264.
16. James Randall and Richard Current, *Lincoln the President* (1955), Vol. IV, pp. 375–77.
17. Whitney, *Life on the Circuit with Lincoln,* p. 246.
18. Isaac N. Arnold, *The Life of Abraham Lincoln* (1885), p. 446. Arnold's statement continued: "Belief in the existence of God, in the immortality of the soul, in the Bible as the revelation of God to man, in the efficacy and duty of prayer, in reverence toward the Almighty, and in love and charity to man, was the basis of his religion."
19. Noah Brooks's letter of December 31, 1872, to J. A. Reed.
20. This interpretation also finds some support in the statement of Dr. Gurley, Lincoln's Washington pastor: ". . . after the death of his son Willie, and his visit to the battlefield of Gettysburg, he said, with tears in his eyes, that he had lost confidence in everything but God, and that he now believed his heart was changed, and that he loved the Savior, and if he was not deceived in himself, it was his intention soon to make a profession of religion." (Quoted from Reed's lecture, p. 326, in Barton's Appendix.)
21. David C. Mearns, "Christian without a Creed," a printed address in the Library of Congress to the Young People of St. John's Church, February 13, 1955. Mearns's phrase is taken from a penciled note of Nicolay's: ". . . the world can utter no other verdict than this—He was a Christian without a creed" (p. 14).
22. Barton, *The Soul of Abraham Lincoln,* p. 289. The section in brackets is the writer's explanation of Baxter's meaning for the word "comfort."

Index

A 32